HVACR

Level 3

SIXTH EDITION

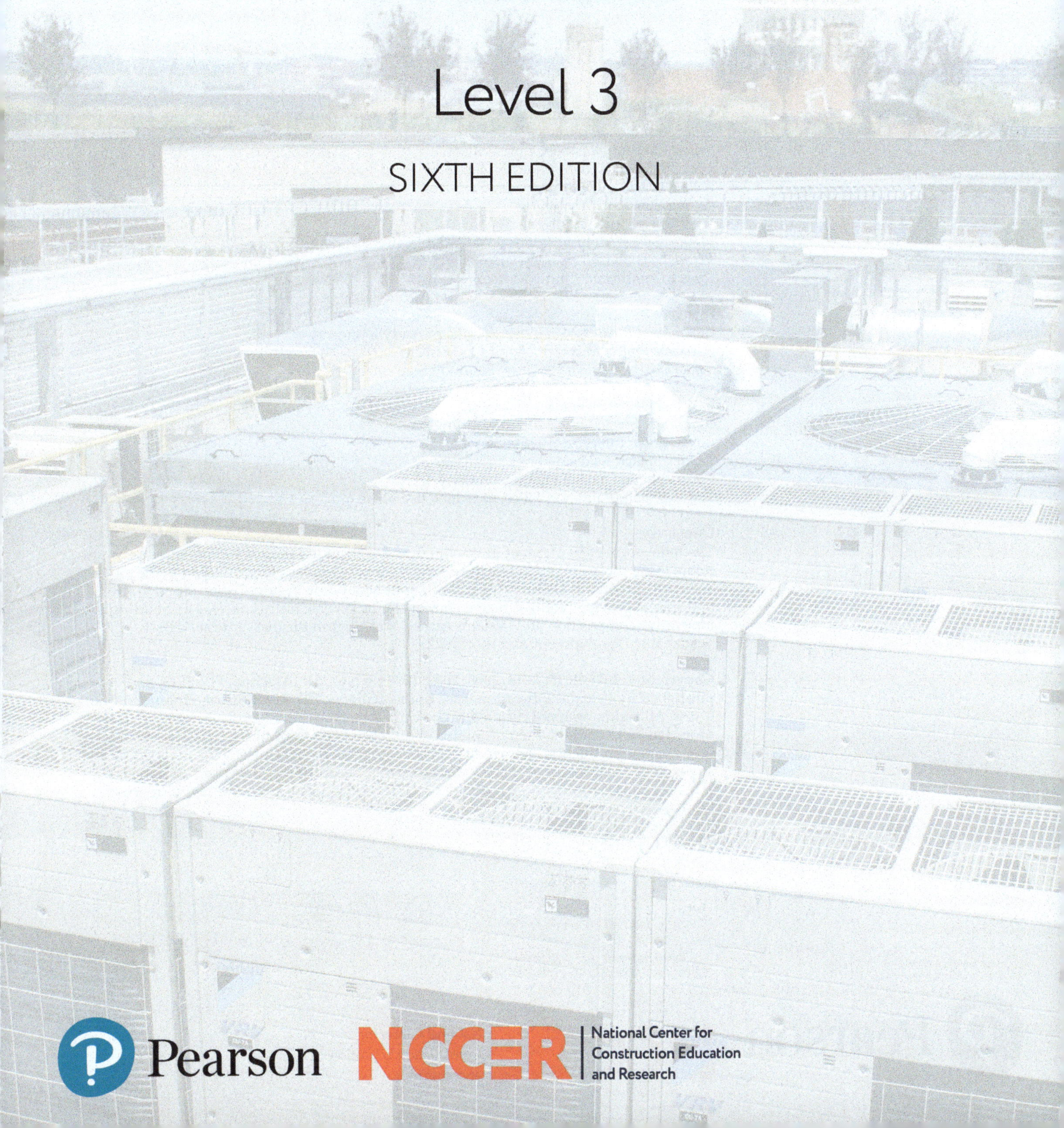

Pearson
NCCER National Center for Construction Education and Research

NCCER
President and Chief Executive Officer: Boyd Worsham
Vice President of Innovation and Advancement: Jennifer Wilkerson
Chief Learning Officer: Lisa Strite
Senior Curriculum Leader: Chris Wilson
HVACR Project Manager: Patrick Bruce
Technical Writing Manager: Gary Ferguson
Technical Writer: Troy Staton
Art Manager: Bree Rodriguez
Production Artist: Judd Ivines, Lisa Wailes, Chris Kersten
Permissions Specialists: Sherry Davis
Project Coordinator: Emma Morse

Pearson
Commercial Product Manager-Associations: Andy Dunaway
Associate Project Manager: Monica Perez-Kim
Senior Digital Content Producer: Shannon Stanton
Content Producer: Alexandrina Wolf
Designer: Mary Siener
Rights and Permissions: Kaitlyn Coddington
Composition: Integra Software Services
Printer/Binder: Lakeside Book Company
Cover Printer: Lakeside Book Company
Text Typefaces: Palatino LT Pro and Helvetica Neue

Cover Image
The renovation of 80 M Street in Washington, DC included the addition of three new floors using mass timber architecture—the first mass timber office building in the city. Air-cooled, modular Daikin VRV™ multi-split heat pump systems were added to the building to provide highly-efficient comfort to the diverse office spaces. The building's innovative composition and systems earned Washington Business Journal's Best Sustainable Development Award in 2022.

10 9 8 7 6 5 4 3 2 1

6 2026

Pearson

NCCER National Center for Construction Education and Research

ISBN-10: 0-13-820974-X
ISBN-13: 978-0-13-820974-2

PREFACE

To the Trainee

The heating, ventilating, air-conditioning, and refrigeration (HVACR) industry doesn't just impact the comfort of our indoor spaces. It also preserves the quality of our food supply and supports manufacturing at many levels. The spread of COVID-19 in recent years elevated the importance of indoor air quality and filtration. HVACR technicians install and maintain these systems, ensuring our medical institutions, restaurants, and food processing facilities are maintained to the required standards. HVACR technicians serve a vital role in keeping many industries operational and contributing to the business sectors they serve.

Competent technicians are required to install, maintain, and troubleshoot HVACR systems. They must be able to read and interpret drawings or other specifications to maintain these systems. Technicians understand air movement, temperature and pressure relationships, and heat transfer applications. They often need basic knowledge in other crafts such as sheet metal, welding, pipefitting, and electrical practices to succeed in their chosen field.

The rapid advancement of technology and consistent changes in regulations prompt employers to recognize the importance of continuous education and keeping up with the latest equipment and techniques. Technical training and apprenticeship programs provide employers with skilled craftworkers who are well prepared to lead the way and confront new challenges.

NCCER's HVACR program has been designed by highly qualified subject matter experts. The four levels present an apprentice approach to the HVACR field, including both theoretical and practical skills essential to your success as an HVACR technician.

We wish you the best as you begin an exciting and promising career. This newly revised curriculum will help you enter the workforce with the knowledge and skills needed to perform productively in any facet of the HVACR craft.

New with *HVACR Level 3*

NCCER is proud to present *HVACR Level 3*. This new edition includes updates to the curriculum that will help engage you and provide the best training possible. These updates include video resources that depict relevant tasks that HVACR technicians must routinely complete.

All modules have received updated art and tables. Module 03311 received additional insight on the development of heat pumps and new environmental considerations. Heat pump technology and efficiency has advanced in recent years, allowing them to serve in new environments and applications with the full support of the US Department of Energy.

We wish you success as you progress through this training program. NCCER appreciates and welcomes its customers' feeback. If you have any comments on how NCCER might improve upon this textbook, please complete the Online Curriculum Feedback Form using the QR code on this page. You may submit yours by emailing **support@nccer.org**. When doing so, please identify feedback on this title by listing *#HVACRL3* in the subject line.

Our website, **www.nccer.org**, has information on the latest product releases and training.

SCAN ME

NCCER Standardized Curricula

NCCER is a not-for-profit 501(c)(3) education foundation established in 1996 by the world's largest and most progressive construction companies and national construction associations. It was founded to address the severe workforce shortage facing the industry and to develop a standardized training process and curricula. Today, NCCER is supported by hundreds of leading construction and maintenance companies, manufacturers, and national associations.

The NCCER Standardized Curricula was developed by NCCER in partnership with Pearson, the world's largest educational publisher.

Some features of the NCCER Standardized Curricula are as follows:

- An industry-proven record of success
- Curricula developed by the industry, for the industry
- National standardization providing portability of learned job skills and educational credits
- Compliance with the Office of Apprenticeship requirements for related classroom training (*OSHA Standard CFR 29:29*)
- Well-illustrated, up-to-date, and practical information

NCCER maintains a secure online database that provides certificates, digital credentials, transcripts, and wallet cards to individuals who successfully complete programs under an NCCER-accredited organization or through one of NCCER's self-paced, online programs. This system also allows individuals and employers to track and verify industry-recognized credentials and certifications in real time.

For information on NCCER's credentials, contact NCCER Customer Service at 1.888.622.3720 or visit **www.nccer.org**.

Digital Credentials

Show off your industry-recognized credentials online with NCCER's digital credentials!

NCCER is now providing online credentials. Transform your knowledge, skills, and achievements into digital credentials that you can share across social media platforms, send to your network, and add to your resume. For more information, visit **www.nccer.org**.

Cover Image Provider

DAIKIN

As the only company in the world to manufacture both air conditioners and refrigerants, Daikin provides products and services utilizing environmental technologies to people around the world. Daikin's core air conditioning business represents a critical component of social infrastructure. Together with transforming the indoor environment in hot regions, Daikin has contributed to people's health through heatstroke prevention and improvement of air quality, which has also helped to increase labor efficiency.

DESIGN FEATURES

Content is organized and presented in a functional structure that allows trainees to access the information where they need it.

Source: Darkdiamond67/iStock

Objectives

Successful completion of this module prepares you to do the following:

1. Identify the characteristics of heat pumps and describe common heat pump operating modes.
   a. Describe heat pump characteristics and identify different heat pump types.
   b. Describe the common heat pump cooling, heating, and defrost modes of operation.
2. Recount the sequence of operation for the common heat pump operating modes.
   a. Recount the sequence of operation for the cooling mode.
   b. Recount the sequence of operation for the heating mode.
   c. Recount the sequence of operation for the defrost mode.
   d. Recount the sequence of operation for dual-fuel systems.
   e. Describe how to evaluate electronic controls in heat pumps.
3. Explain how to test and troubleshoot various functions and components of heat pump systems.
   a. Explain how to test field and factory wiring.
   b. Explain how to test and troubleshoot heat pump thermostats.
   c. Explain how to test various valves.
   d. Explain how to test defrost controls.

Performance Tasks

Under supervision, you should be able to do the following:

1. Demonstrate the ability to diagnose and identify the proper solution for an instructor-selected mechanical problem in a heat pump.
2. Demonstrate the ability to diagnose and identify the proper solution for an instructor-selected electrical problem in a heat pump.
3. Initiate and evaluate the defrost mode of a heat pump.

Overview

This module reviews the operation of heat pumps and presents methods and procedures for their troubleshooting and repair. Heat pumps contain more controls and have more complex circuits than a cooling-only system. Understanding how they differ from cooling-only systems and what to expect during the heating mode is essential to successful troubleshooting and repair.

Digital Resources for HVACR

Scan this code using the camera on your phone or mobile device to view the digital resources related to this craft.

SCAN ME

Trainees can navigate *HVACR Level 3* using color coded tabs on the upper right hand corner.

The Objectives list the skills and knowledge trainees need in order to complete the module successfully.

The Performance Tasks give you an opportunity to apply your knowledge to real-world tasks.

Section Openers provide a visual organizational structure for the information. Objectives and Performance Tasks are broken out for each section.

3.0.0 Testing and Troubleshooting Heat Pumps

Objective

Explain how to test and troubleshoot various functions and components of heat pump systems.

a. Explain how to test field and factory wiring.
b. Explain how to test and troubleshoot heat pump thermostats.
c. Explain how to test various valves.
d. Explain how to test defrost controls.

Performance Tasks

1. Demonstrate the ability to diagnose and identify the proper solution for an instructor-selected mechanical problem in a heat pump.
2. Demonstrate the ability to diagnose and identify the proper solution for an instructor-selected electrical problem in a heat pump.
3. Initiate and evaluate the defrost mode of a heat pump.

To provide sufficient heat, air-source heat pumps typically require a supplemental source of heat. The supplemental heat provides additional heating capacity below the heat pump's **balance point** as well as an alternative form of heat if the heat pump fails. When a system thermostat is switched to the Emergency Heat mode, the heat pump is locked out and the alternate heat source is used.

Dual-fuel systems, which typically combine a heat pump with a fossil-fuel heat source, typically have a changeover outdoor temperature that determines which heat source is used. In most cases, the furnace and heat pump can't operate at the same time. Below the balance point, the system usually relies on the secondary source. Above the balance point, the heat pump has sufficient capacity to handle the heating load alone.

Balance point: The outdoor temperature at which the heating capacity of a heat pump is equal to the heat loss of the structure it serves.

Trade Terms appear on the page adjacent to the text where they are first presented.

Step-by-step presentations help make the concepts clear and easy to grasp.

LOTO Procedure

To prepare for LOTO, locate and identify all the relevant isolating devices. Be certain which switch(es), valve(s), or other energy-isolating devices apply to the equipment to be serviced. Remember that more than one energy source may be involved.

The following procedure outlines the general steps for LOTO:

Step 1 Notify all affected workers of the LOTO and the purpose. The authorized worker must know the types of energy that are relevant and how to disable them.

Step 2 If the machine or equipment is running, shut it down using the normal procedure. For example, use the thermostat to shut off an air conditioning system before de-energizing the system.

Step 3 Operate the switch, valve, or other energy-isolating device(s) so that the equipment is isolated from its energy source(s). Any stored energy, such as pressure in a hydraulic reservoir, must be dissipated or restrained by repositioning, blocking, or bleeding down. Examples of stored energy are springs, elevated machine members, rotating flywheels, hydraulic pressure, and air, gas, steam, or water pressure.

Step 4 Lock out and tag the energy isolating devices with assigned lock(s) and tag(s).

Step 5 After making sure that no workers are exposed, operate the start button or other normal operating controls to ensure the equipment doesn't operate. Then return the operating controls to the neutral or Off position.

SCAN ME

QR codes link trainees directly to digital resources that highlight current content.

Important information is highlighted, illustrated, and presented to facilitate learning.

Placement of images near the text description and details such as callouts and labels help trainees absorb information.

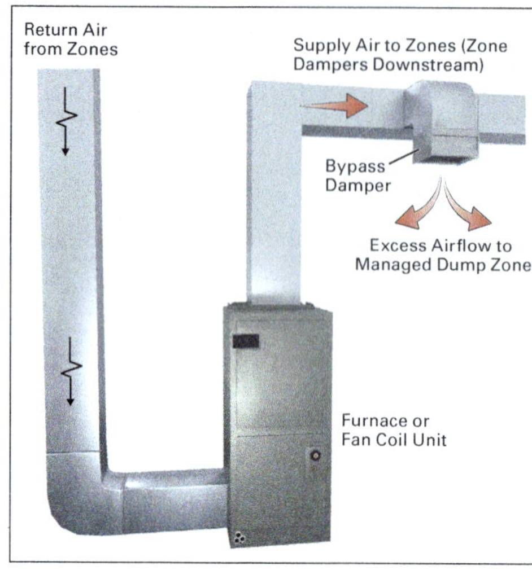

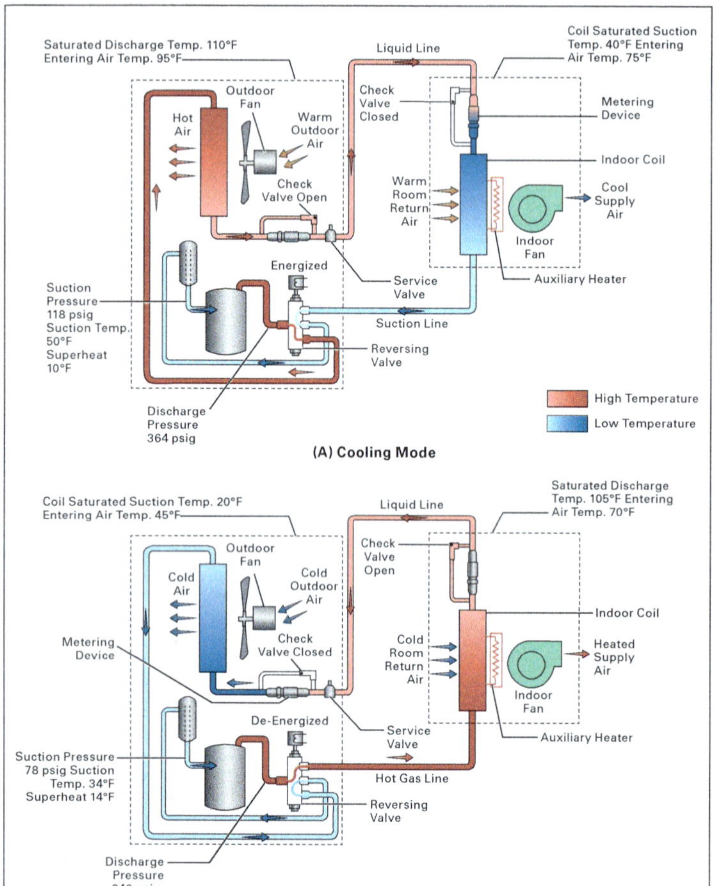

(A) Cooling Mode

(B) Heating Mode

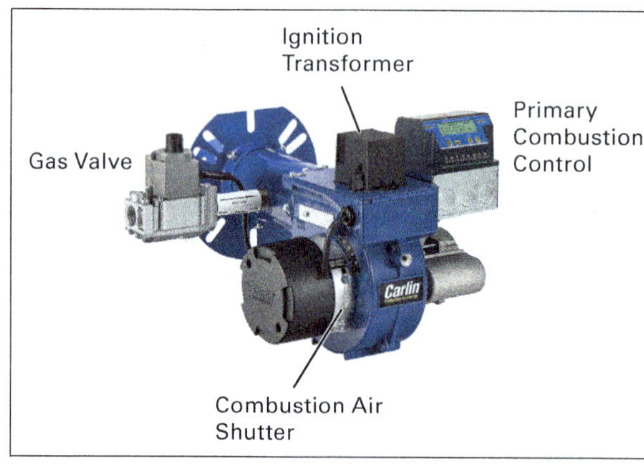

New boxes highlight safety and other important information for trainees. Warning boxes stress potentially dangerous situations, while Caution boxes alert trainees to dangers that may cause damage to equipment. Notes boxes provide additional information on a topic.

WARNING!

An energized circuit can deliver a severe and possibly fatal electric shock. Always use extreme caution when working with energized circuits. Refer to the arc-flash protection requirements associated with electrical systems found in National Fire Protection Agency (NFPA) 70E, *Standard for Electrical Safety in the Workplace*®.

CAUTION

Before installing any field wiring, consult the manufacturer's literature to ensure the wire is the correct type and appropriate size for the application. Most field-installed control wiring circuits require 18 AWG wire, but the length of the run is always a factor in electrical wiring. Also pay attention to the rolls of thermostat wire you purchase, as many distributors routinely sell thermostat cable smaller than 18 AWG.

NOTE

Familiarize yourself with different troubleshooting guides offered by manufacturers. Manufacturers also release bulletins periodically regarding field problems that are also helpful.

Going Green

Heat Pump Pool Heaters

Heat pumps also make efficient pool heaters. Heat is extracted from the air and transferred to the swimming pool by a refrigerant-to-water heat exchanger. Heat pumps offer an energy-efficient alternative to traditional fossil-fuel pool heaters. Hybrid units that combine gas heating with a heat pump are also available.

Going Green looks at ways to preserve the environment, save energy, and make good choices regarding the health of the planet.

On Site

High-Point Air Vents

In large systems, there may be many high points—points at which the piping system rises to its highest level. Air in the system naturally attempts to reach these high points.

When a manual air vent is opened to purge a system or terminal, it is left open until a steady stream of water is obtained. This happens very quickly after a system is full and only a little air remains. However, when filling a new system, it can take some time and multiple visits to the same vent.

Catch the water in a suitable container if it is not piped to a drain. Move from vent to vent in order, opening and closing each vent and collecting the vented water. Repeat this process several times until all the air is purged from the system.

Automatic air vents may also be installed. The outlet of an automatic air vent should be piped to a drain to capture the bits of water it will release.

Know the Code

Low-Water Cutoff Controls

The 2021 *International Mechanical Code*® requires both steam and hot-water boilers to have low-water cutoff protection that disables the burner or heating elements when activated. However, the code makes an exception for boilers with coil and finned-tube heat exchangers that require water flow through the appliance to activate the burner. This exempts most wall-mounted models that operate on demand and modulate the burner based on flow rate. Flow switches protect this type of boiler or water heater.

The *IMC*® also requires hot-water boilers to have a temperature gauge and a pressure gauge, or a combination thereof.

These boxed features provide additional information that enhances the text.

Did You Know?

Fiberglass

Fiberglass has been commercially manufactured and marketed for more than 60 years. Fiberglass is used to make many products. The material absorbs sound, impedes heat transfer, removes impurities from liquids and gases, and reinforces other materials such as boat hulls. Uncontaminated fiberglass alone does not support the growth of mold or bacteria.

These qualities make fiberglass an important material in the HVACR industry.

Source: PragasitLalao/iStock/Getty Images

Review questions at the end of each section and module allow trainees to measure their progress

3.0.0 Section Review

1. If a heat pump outdoor unit is located 150' from the air handling unit, the control circuit wire size will likely be _____.
   a. 24 AWG
   b. 22 AWG
   c. 18 AWG
   d. 16 AWG

2. To what does *power stealing* refer?
   a. A method used to power a defrost control board
   b. Sharing a single power source for a heat pump and a furnace
   c. A method used to power a thermostat without a Common wire
   d. A reduction in compressor current when the electric heaters are energized

3. When testing a reversing valve for internal leakage, above what temperature differential would you consider the valve to be leaking?
   a. 1°F
   b. 3°F
   c. 5°F
   d. 10°F

4. A demand defrost approach incorporates which of the following?
   a. Two thermistors
   b. A pressure differential switch
   c. A variable-speed outdoor fan motor
   d. A timer to establish the defrost frequency

Module 03311 Review Questions

1. If a heat pump works normally in the heating mode but will not provide cooling or a defrost mode, which of the following components or their related controls is *most likely* defective?
   a. Compressor
   b. Reversing valve
   c. Room thermostat
   d. Compressor contactor

2. Beyond the reversing valve, what is the other major difference between a heat pump and a cooling-only system?
   a. Heat pumps require two metering devices.
   b. Only cooling systems require a filter-drier.
   c. A heat pump requires a head-pressure control.
   d. Heat pumps only use fixed-orifice metering devices.

3. True or False? Heat pumps do not need to defrost above the balance point.
   a. True
   b. False

4. Generally, in the cooling mode, the reversing valve solenoid is _____.
   a. bypassed
   b. energized
   c. back-seated
   d. de-energized

NCCERconnect

This interactive online course is a unique web-based supplement in the form of an electronic book that provides a range of visual, auditory, and interactive elements to enhance training. Visit **www.nccerconnect.com** for more information!

Forced-air heating.

Instructions: Drag each label to its correct location. Then click Submit.

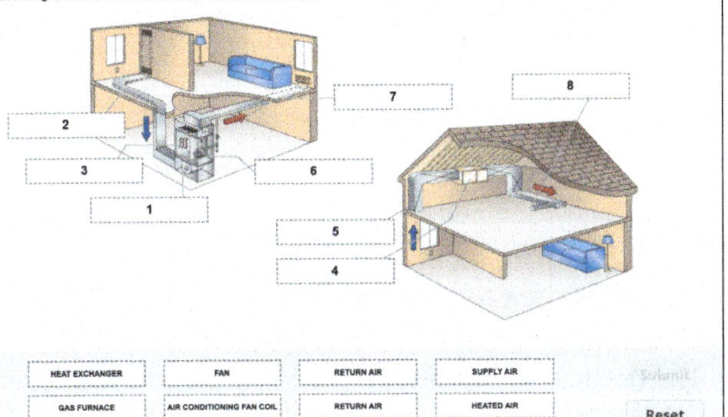

1.0.0 Basic HVACR Principles

can be expensive, electric-resistance heating alone is no longer common in cold climates. It is more likely to be used in warmer climates where heat is seldom required.

Figure 3 Electric furnace.

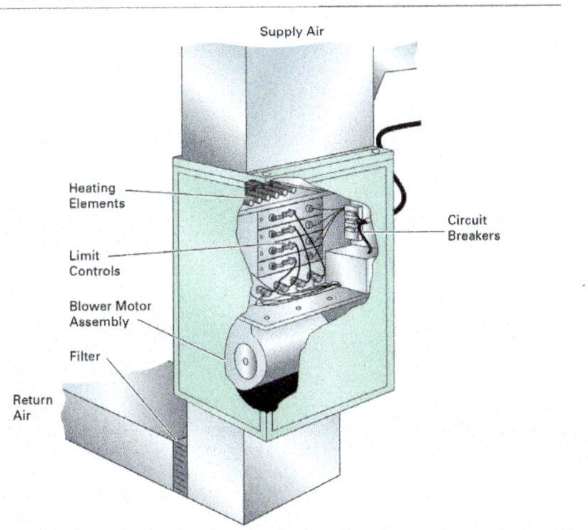

5: Introduction to Cooling (03107)

Refrigeration is a term that describes the transfer of heat from one place, where it is not wanted, to another place where the added heat is insignificant or unimportant. Air conditioning units and refrigerators simply transfer heat out of a space or object and move it elsewhere (*Figure 1*). In most cases, the heat is transferred outdoors. Domestic refrigerators and small commercial reach-in cases reject heat to the space. This heat must be moved outdoors by the building cooling system.

Figure 1 Mechanical refrigeration systems transfer heat from one location to another.

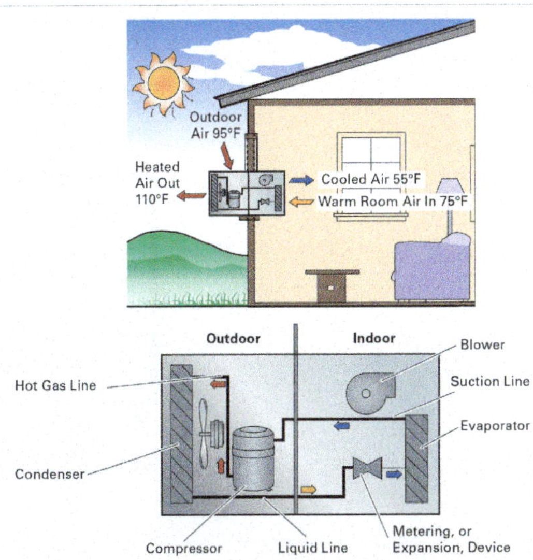

A refrigerant circulates through a refrigerant circuit to enable the process. The refrigerant absorbs

ACKNOWLEDGMENTS

This curriculum was revised as a result of the farsightedness and leadership of the following sponsors:
AGC of Oklahoma Education Foundation
Brown and Root Industrial Services
College of Southern Maryland
Construction Industry Training Council of Washington
Epling Distributing
HVAC Technical Institute
International Code Council
LaGuardia Community College
Lennox Industries
Mielke Holdings
Pacific Northwest National Laboratory
Service Experts
Standard Heating and Air Conditioning
Southern Careers Institute
Vista College

This curriculum would not exist were it not for the dedication and unselfish energy of those volunteers who served on the Authoring Team. A sincere thanks is extended to the following:
Allen Heath
David Norman
Frank Calderone
James Koehn
Jamie Kono
Jim Cika
Jim Hughes
Joseph Moravek
Kasey Wade
Paul Ringley
Terry Mielke

A final note: This book is the result of a collaborative effort involving the production, editorial, and development staff at Pearson Education, Inc., and NCCER. Thanks to all of the dedicated people involved in the many stages of this project.

NCCER PARTNERS

To see a full list of NCCER Partners, please visit **www.nccer.org/about-us/partners**.

NCCER

SCAN ME

CONTENTS

Module 03317 Control Circuit Troubleshooting

Module 03318 Motor Testing and Troubleshooting

Module 03210 Troubleshooting Cooling

Module 03311 Troubleshooting Heat Pumps

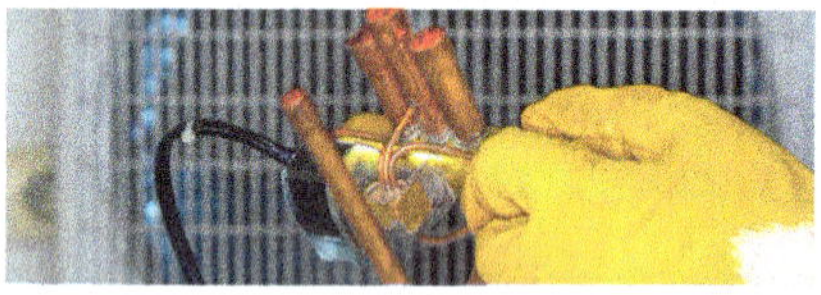

Module 03209 Troubleshooting Gas Heating

Module 03310 Troubleshooting Oil Heating

Module 03312 Troubleshooting Accessories

Module 03213 Sheet Metal Duct Systems

Module 03203 Introduction to Hydronic Systems

Module 03306 Steam Systems

Module 03214 Fiberglass and Fabric Duct Systems

Module 03304 Retail Refrigeration Systems

Module 03315 Zoning, Ductless, and Variable Refrigerant Flow Systems

MODULE 03317

Control Circuit Troubleshooting

Source: justocker/iStock

Objectives

Successful completion of this module prepares you to do the following:

1. Identify and describe the operation of common HVACR control circuit devices.
   a. Describe relays, contactors, and motor starters and explain how to test them.
   b. Identify, describe, and test common pressure and temperature control devices.
2. Describe the operation, installation, and testing of thermostats.
   a. Describe basic thermostats and operating characteristics.
   b. Explain how to install and wire room thermostats.
   c. Explain how to troubleshoot room thermostats.
3. Explain how to troubleshoot common electrical circuits.
   a. Identify safety practices related to troubleshooting HVACR power and control circuits.
   b. State the sequence of operation for basic heating and cooling systems.
   c. Explain how to approach HVACR problems and troubleshoot in an organized manner.
   d. Explain how to measure voltage and current imbalance.
   e. Explain how to troubleshoot control circuits and low-voltage transformers.
   f. Explain how to test resistive and inductive loads.
   g. Identify and describe the function of various hydronic control system components.

Performance Tasks

Under supervision, you should be able to do the following:

1. Wire, test, and complete the installer setup of a thermostat.
2. Successfully diagnose assigned control circuit problems and failures.
3. Perform electrical tests and/or troubleshooting procedures on the following:
   - Relays and contactors
   - Thermistors
   - Room thermostats
   - Resistive loads

Overview

Many problems that occur in HVACR systems are caused by a failure in the control circuits or in the electrical loads they serve. Technicians must understand how various control components function and interpret equipment schematics, wiring diagrams, and other service literature. Competent service technicians combine this knowledge and their troubleshooting skills every day to consistently solve simple and complex problems in the field.

Digital Resources for HVACR

Scan this code using the camera on your phone or mobile device to view the digital resources related to this craft.

SCAN ME

NCCER Industry-Recognized Credentials

If you are training through an NCCER-accredited sponsor, you may be eligible for credentials from NCCER. The ID number for this module is 03317. Note that this module may have been used in other NCCER curricula and may apply to other level completions. Contact NCCER at 1.888.622.3720 or go to **www.nccer.org** for more information.

You can also show off your industry-recognized credentials online with NCCER's digital credentials. Transform your knowledge, skills, and achievements into credentials that you can share across social media platforms, send to your network, and add to your resume. For more information, visit **www.nccer.org**.

1.0.0 Common HVACR Control Devices

Performance Tasks

2. Successfully diagnose assigned control circuit problems and failures.
3. Perform electrical tests and/or troubleshooting procedures on the following:
   - Relays and contactors
   - Thermistors

Objective

Identify and describe the operation of common HVACR control circuit devices.

a. Describe relays, contactors, and motor starters and explain how to test them.
b. Identify, describe, and test common pressure and temperature control devices.

HVACR control systems are designed to automatically control heating, cooling, and ventilation systems. The controls for a small system, such as a window air conditioner, are very simple—a thermostat, a few switches, and safety devices are all that is needed. However, as systems become larger and more complicated, the complexity of the control system increases.

The use of electronic controls is now widespread across all HVACR product lines. To find a modern unit that does not contain some solid-state controls is rare. Manufacturers use them to increase flexibility and reliability. Electronic controls are often designed to perform more than one task and function in a variety of products.

To understand control systems and circuits, you must first recognize and understand the common components. All automatic control systems have the following basic characteristics in common:

- One or more sensing elements, such as thermistors, pressure sensors, and humidistats, monitor changes in temperature, pressure, and humidity.
- One or more control mechanisms, such as wall thermostats, translate these changes into actions that switch loads such as contactors, motor starters, and other relays.
- The connecting wiring, pneumatic piping, and/or mechanical linkages transport the energy required to energize loads as well as other control devices. Load devices are energized and de-energized as necessary to accomplish the work needed to heat or cool the space.

Although this relationship may sound complicated, it is actually quite simple. Essentially, the need to change current environmental conditions is transmitted through a control system to load devices that can make those changes happen.

1.1.0 Relays, Contactors, and Motor Starters

HVACR control circuits rely on various relays, contactors, and motor starters to transfer both low- and line-voltage power. These devices are physically different in size and configuration, but their principles of operation are essentially the same. Because of the critical role they play in system operation, technicians must understand how these devices work. Once you understand the individual devices, **wiring diagrams** begin to make more sense.

Wiring diagram: A troubleshooting aid that provides pictorial representation of the physical wiring and connections in a unit.

1.1.1 Relays

A relay is designed to enable or disable the flow of electricity to both loads and other controls. A relay is sometimes used to reroute the flow of electricity in a different direction. Because most relays are designed to conduct a limited electrical current, some electrical loads must be transferred to devices with heavier contacts, such as contactors.

Relays come in many shapes and sizes. *Figure 1* shows a few of the most common relays found in HVACR systems. You may also encounter plug-in relays (*Figure 2*). Plug-in relays have a mating socket or base into which they fit, which makes troubleshooting and replacement easier. All wiring terminations are made at the socket or base.

In HVACR equipment today, relays that were once stand-alone components are often soldered to a printed circuit (PC) board (*Figure 3*). If a relay soldered directly to a board fails, it is generally best to replace the board itself. Although field replacement of the relay may be possible, it is not typically practical.

NOTE

Refer to NCCER Module 03317, *Appendix 03317A, Common Schematic Symbols*, for a diagram of common wiring diagram symbols.

Figure 1 Common relay styles.
Source: HVACGuru

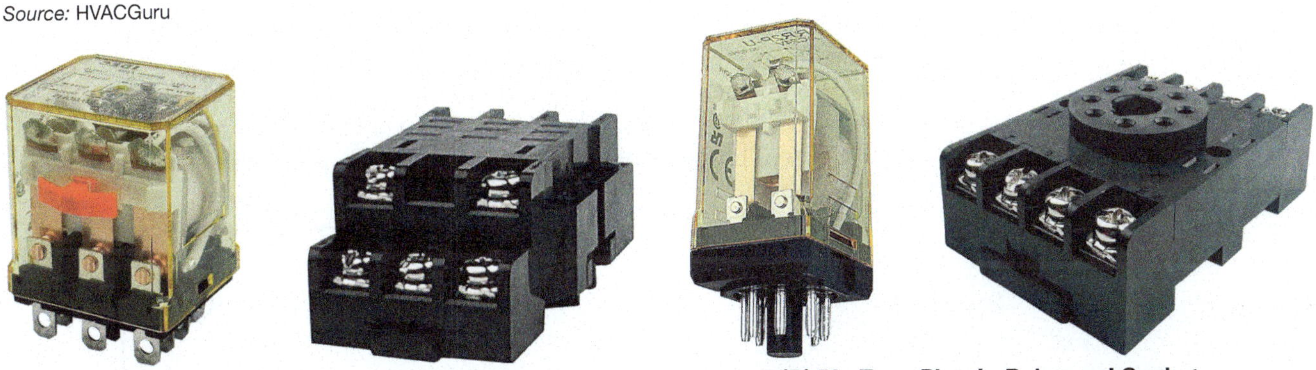

(A) Blade-Type Plug-In Relay and Socket

(B) Pin-Type Plug-In Relay and Socket

Figure 2 Plug-in relays.
Source: Courtesy of Kele, Inc.

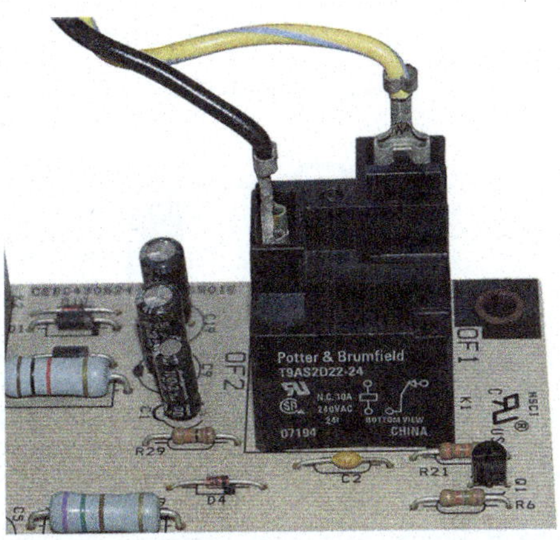

Figure 3 Relay affixed to a PC board.
Source: HVACGuru

On Site

The Origin of Relays

The use of an electric relay to transmit coded messages, one of the first applications, is largely credited to Samuel Morse, one of the inventors of the telegraph. With the help of colleague Leonard Gale and based on preliminary work done by Princeton University's Joseph Henry, Morse applied the relay to communicate by code. An incoming telegraph signal activated an electromagnet that then actuated a set of contacts to close another circuit, thereby transmitting the signal to the next relay.

Source: Studio-Annika/iStock

CAUTION

Many control relays are rated to switch voltages up to 250VAC. If a voltage of 277VAC or higher must be switched, ensure that the relay used is appropriately rated.

A basic relay consists of two parts: an electromagnetic coil and one or more sets of contacts. When the coil is energized with the proper voltage, it causes the position of the relay's contacts to change. Most relays you encounter in HVACR equipment operate on 24VAC power, while the contacts can switch either line voltage or low voltage loads. However, in larger commercial equipment and systems, you will encounter relays with line-voltage coils.

Relay contact positions are identified as being either normally open (NO) or normally closed (NC). The position designation always refers to the position when the relay coil is de-energized. All contacts shown on schematic diagrams are shown with the contacts in their de-energized position.

Figure 4 shows the open and closed contacts of a typical relay. It also shows the schematic symbol for NO and NC relay contacts.

Figure 5 shows a schematic diagram for a simple relay with an operating coil and one set of NO contacts. It is classified as a single-pole, single-throw (SPST) relay because it only has one set of contacts (single pole) with a single current path (single throw).

Figure 4 Normally open (NO) and normally closed (NC) relay contacts.

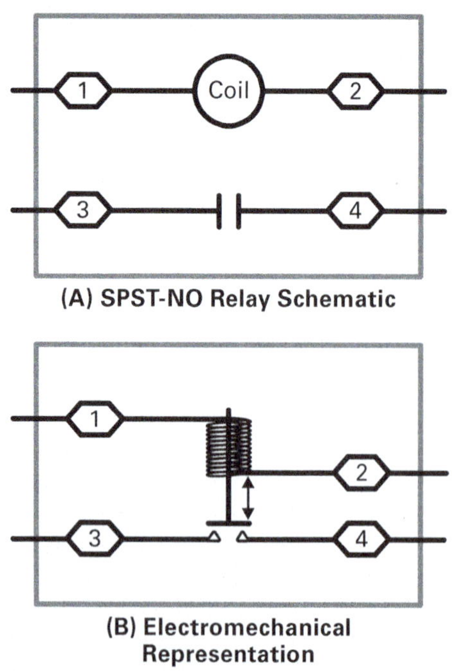

Figure 5 Single-pole, single-throw (SPST) relay.

An electromechanical representation of the SPST-NO relay is also shown in *Figure 5*. Although it is never drawn this way on a diagram, it shows what is really happening inside the relay. The electromagnet pulls the linkage toward it or allows it to move away through the action of a spring.

Many other relay forms exist. The direct counterpart of the first relay introduced is the SPST-NC relay. The only difference is that the contacts are normally closed. Another common relay form is the single-pole, double-throw (SPDT) relay (*Figure 6*). In a SPDT relay, power entering the single pole can be directed to one of two paths. In this case, one set of contacts is NO while the other is NC.

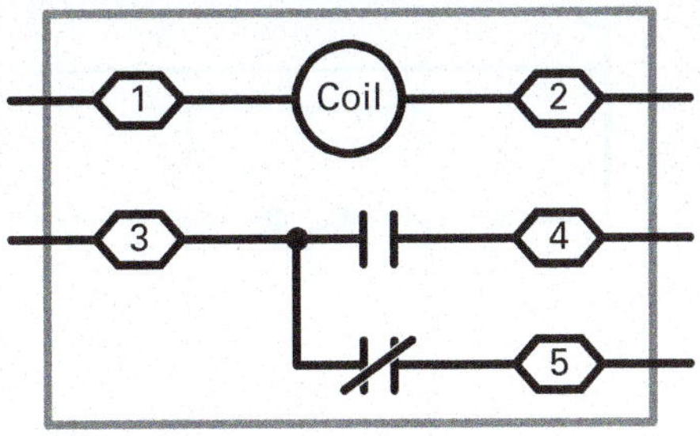

Figure 6 Single-pole, double-throw (SPDT) relay schematic.

Another relay form is the double-pole, double-throw (DPDT) relay. As shown in *Figure 7*, a second pole with double-throw contacts is added to a basic SPDT relay to form a DPDT relay. The two poles can be energized from the same power source or be completely independent and possible managing two different voltages.

You can see how more poles can be added to create the needed configuration. Additional relay configurations include a three-pole, double-throw (3PDT) relay and even a four-pole, double-throw (4PDT) relay. Any of these configurations can also be found on printed circuit boards in both electromechanical and solid-state forms.

Because each pole in a relay is electrically isolated, poles from a single relay can be wired into different branches of the equipment's control circuits. Even though the coil and poles are physically housed together, the two are usually shown in different areas of a **ladder diagram**, because they are related to different voltages. Ladder diagrams are presented later in this module.

Ladder diagram: An electrical diagram that provides a schematic of power and control circuits with the loads arranged like the rungs of a ladder between vertical lines representing the power source.

Because the relay coil and the contact sets are often shown in different areas of a schematic or ladder diagram, a method of associating related coils and contacts across the diagram is needed. An example of how this is done is shown in *Figure 7*, where the coil is designated R1 (relay 1) and all its related contact sets are identified with numbers that begin with R1. As shown, the first pole is designated as R1-1 and R1-2, which refers to relay 1, contact set 1 and relay 1, contact set 2. The second pole is designated as R1-3 and R1-4. Any set of contacts identified with an R1 prefix is understood to be part of relay R1 and controlled by the R1 operating coil.

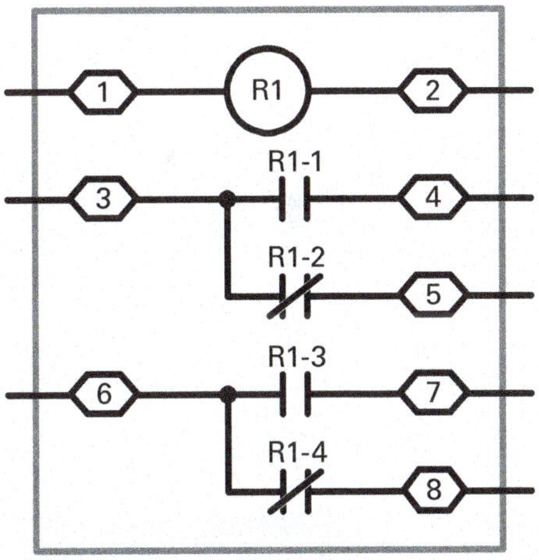

Figure 7 Double-pole, double-throw (DPDT) relay schematic.

Many possible designations for relay and contactor coils are available. For example, CR (control relay), CC (compressor contactor), and IFR (indoor fan relay) are common acronyms found on diagrams. Always consult the legend provided with a wiring or ladder diagram to properly identify a component. It is best to avoid making assumptions about the meaning of an acronym, as manufacturers do vary in their approach to wiring diagrams.

Time-Delay Relays

The purpose of a time-delay relay is to delay contact movement for some length of time. The range of the delay timing depends entirely on the timing device. A delay can vary from a fraction of a second to hours. Many time-delay relays have an adjustable delay period, but some designed for specific applications have a fixed delay setting.

While time-delay relays can be electromechanical in nature, most of today's time-delay relays are electronic. Indeed, the logic found in many PC boards often performs timing functions, with no external timing device required.

One common use of a time-delay relay or logic function is to delay the start of an indoor blower to avoid blowing cold air on occupants. The blower is also allowed to run several minutes longer at the end of a heating cycle, to extract all the remaining heat from the heat exchanger.

Electrical systems serving several large motors may use time delays to start the motors in a sequence, limiting the inrush current on the power system. For example, the schematic drawing in *Figure 8* shows four compressors using time-delay relays to start sequentially. Note that the first stage of the thermostat controls two compressors. Thus, only three thermostat stages are needed. Although this isn't a common approach with today's electronics and processors, it demonstrates how time-delay relays can be applied.

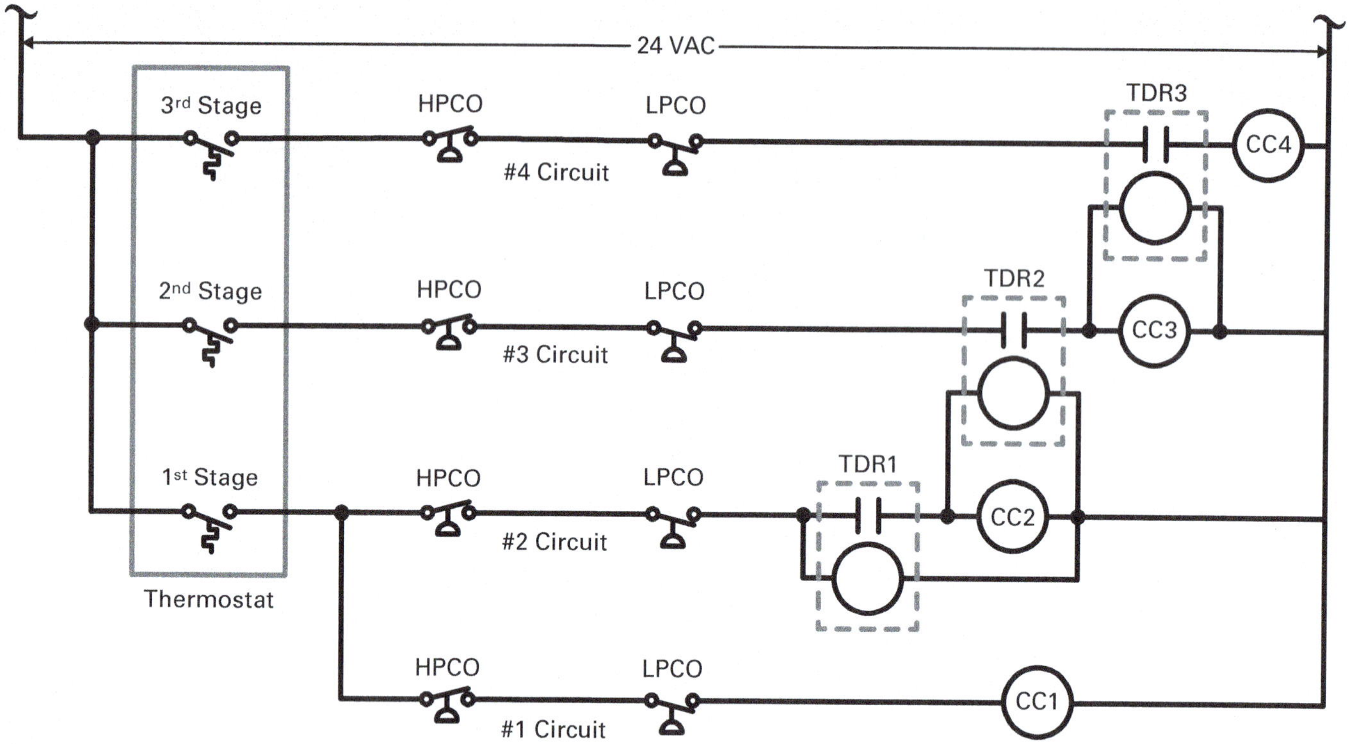

Figure 8 Four-compressor control circuit with time-delay relays.

If all three stages of the thermostat contacts close at once, power is supplied to both CC1 and CC2. CC1 is energized immediately, starting the related compressor. But the open contacts of TDR1 prevent power from being applied to the CC2 coil. Once the preset delay elapses, the TDR1 contacts close to energize CC2. This sequence continues until all four compressors are running. This represents a *delay-on-make* timing action.

A compressor short-cycle protection circuit contains a timing function that prevents the compressor contactor from re-energizing for a specified period after the compressor shuts off. This allows time for the pressures to equalize across the compressor, as compressor motors struggle to get started against a significant pressure differential. This is referred to as a *delay-on-break* application since the delay period begins when the contactor control circuit opens (de-energized). A delay-on-make relay begins its timing period when the circuit is first energized. If a delay-on-make relay is used for short-cycle protection, the timing period will not start until the system calls for the compressor to start. This results in an undesirable delay; the delay period before the compressor can restart should begin when it shuts down.

A compressor short-cycle delay period is typically 3 to 5 minutes. Short-cycle time-delay relays (*Figure 9*) are typically wired into the control circuit adjacent to the contactor. These relays are sold as optional and after-market accessories. The type pictured can accept control voltages between 18VAC and 240VAC. For voltages of 120VAC and above, the red jumper is cut to enable a resistive circuit internally.

Short-cycle protection is also a common feature in today's digital thermostats. The feature can usually be enabled or disabled through the thermostat's set-up menu. However, the protection provided by a wall thermostat neither protects the compressor from short-cycling due to a low- or high-pressure condition, nor from the opening of any other external control.

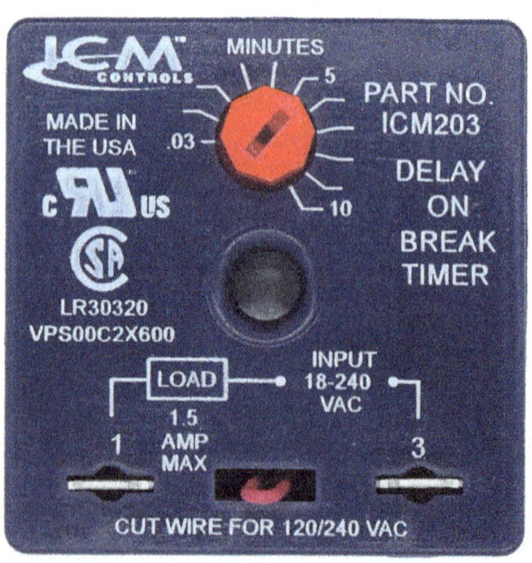

Figure 9 Compressor short-cycle protection relay (delay on break).

Lockout Relays and Circuits

The purpose of the lockout relay in a condensing unit control circuit is to prevent the automatic restart of HVACR equipment after an open safety control has reset itself. The need to reset a system manually is generally associated with the word *lockout*. Once the lockout relay has been activated, the system can only be reset by interrupting the power supply to the control circuit. For example, turning the room thermostat function switch Off and then back On, or cycling the circuit breaker serving the control transformer, will reset the lockout relay and allow the system to start.

Lockout relays have a coil with a significantly higher resistance, or *impedance*, than a common control relay. Because of the increased resistance, this type of relay is called an *impedance relay*.

In the circuit shown in *Figure 10*, the lockout relay coil, due to its high resistance, is not energized during normal operation. The easier path through the circuit is through the closed contacts of each safety device. However, when any one of the safety controls opens, the only remaining current path is through the lockout relay coil. The lockout relay is then energized, and its NC contacts open.

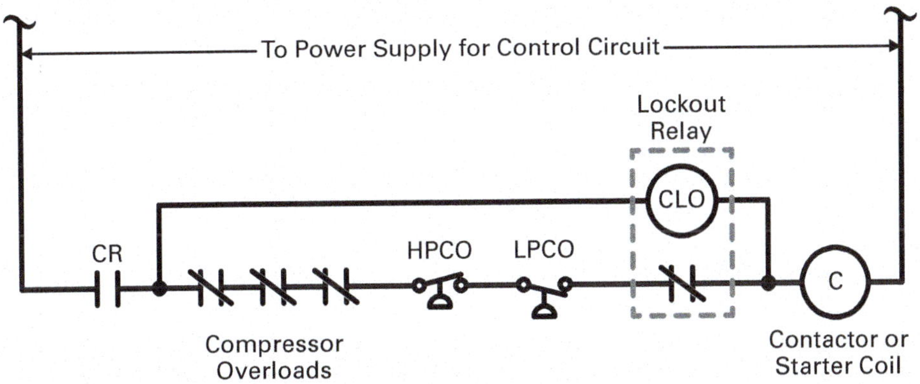

Figure 10 Lockout relay in an HVACR control circuit.

Figure 11 Solid-state relay (SPST-NO).
Source: Courtesy of Sensata Technologies

These contacts remain open as long as the control circuit is energized. Even if the safety control that opened originally closes again, the open contacts of the lockout relay prevent the system from restarting. De-energizing the control circuit allows the lockout relay contacts to return to their NC position.

The proper operation of the lockout relay relies on the resistance of the lockout relay coil being significantly higher than the resistance of the compressor contactor coil. If a lockout relay fails, it should be replaced with an exact duplicate to ensure the coil resistance is correct.

Solid-State Relays

Unlike electromechanical relays, solid-state relays (*Figure 11*) have no moving parts. Solid-state relays use semiconductors called **triacs** to control current flow. Light-emitting diodes (LEDs) typically indicate the status. Solid-state relays are used in general control applications as well as in motor starting circuits.

Triacs: Three-terminal electronic semiconductors that support current flow in either direction when enabled by an input to the gate electrode.

Some advantages of solid-state relays are longer life, improved reliability, high-speed switching, and excellent resistance to shock and vibration. The absence of mechanical contacts eliminates contact chattering, arcing as contacts open and close, and the hazards of explosives and flammable gases near traditional contacts. Many can handle significantly more current than the average electromechanical relay.

Solid-state relays can accommodate a wide range of control voltages, including both AC and DC. Remember, though, that they do not have magnetic solenoid coils inside like a traditional relay, because there are no mechanical parts inside to be repositioned.

On Site

Flammable Refrigerants

Given that new HVACR systems developed for use with flammable refrigerants must be designed to eliminate and control electrical arcs, you will likely see an increase in the use of solid-state relays in HVACR systems.

CAUTION

Never place additional lockout relays, lights, or any other loads in parallel with the lockout relay coil. Doing so will interfere with lockout relay operation and prevent it from protecting the equipment.

National Electrical Manufacturers Association (NEMA): An electrical standards-developing organization made up of business leaders, electrical experts, engineers, scientists, and technicians from the electrical industry.

Figure 12 Typical definite-purpose contactor.
Source: Schneider Electric

1.1.2 Contactors and Motor Starters

A contactor (*Figure 12*) can be described as a heavy-duty relay used to control high-current loads. Contactors are used to start larger motors, compressors, and electric heaters. Like relays, contactors do not provide any level of overload protection. Thus, overload protection must be provided separately for motors that require it.

Most HVACR contactors are classified as *definite-purpose contactors*. Years ago, contactors that conformed to **National Electrical Manufacturers Association (NEMA)** standards were too expensive for most HVACR equipment manufacturers and applications. The NEMA contactors easily outlasted the rest of the equipment. Thus, the industry created a new class of contactors to satisfy the unique needs of HVACR manufacturers, leading to the term *definite purpose*. Definite-purpose contactors are not as durable and long-lasting as NEMA-rated contactors, yet they serve the HVACR industry very well.

Many contactors can be fitted with accessory contact sets (*Figure 13*), commonly referred to as *auxiliary contacts*. This allows the contactor to control other things besides the primary load. For example, when the contactor is energized and the main contacts close, a set of auxiliary contacts can open to de-energize a crankcase heater circuit.

Auxiliary contacts generally mount on one or both sides of a contactor. Rather than being electrically energized like a relay, a tab on the movable portion of the contactor (referred to as the *armature*) physically engages the contact linkage and pushes the contacts closed. Both NO and NC contact sets are available, and more than one set can often be added to a contactor. Note that they are not generic; auxiliary contacts must be compatible with the contactor in use and come from the same manufacturer.

Figure 13 Accessory contact set.
Source: Courtesy of Eaton

On Site

Contactor Upgrades

Contactors can suffer from a variety of problems. Some species of ants, for example, are attracted to them. Ants often swarm a contactor and foul the contacts. Normal arcing as the contacts open and close under load also causes the contacts to burn.

Versatile upgrade contactors are now available. The White-Rodgers by Copeland SureSwitch® is a sealed universal contactor that can replace many definite-purpose contactors. These contactors offer longer life with circuitry that inhibits arcing across the contacts and a multivoltage operating coil. Other desirable features include fully sealed contacts, line-voltage brownout protection, integrated compressor short-cycle protection, and a tricolor LED to indicate operating status.

Source: Courtesy of White-Rodgers by Copeland

A motor starter (*Figure 14*) is basically a contactor mated with an overload relay to provide motor overload protection. Thermal overloads or current sensors in the overload relay sense excess current to a motor and open the contactor control circuit, de-energizing the motor. Once the overload relay has tripped, a

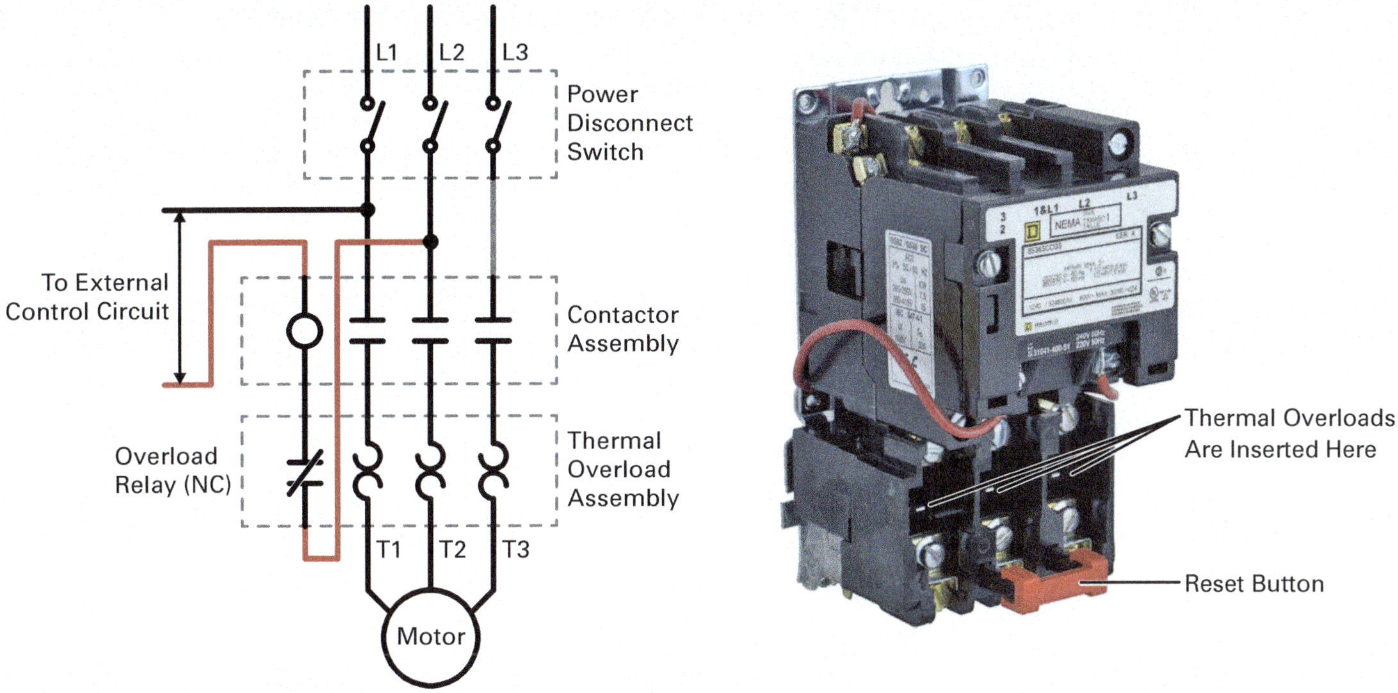

Figure 14 NEMA motor starter with attached overload relay.
Source: Schneider Electric (right)

manual reset is generally required to restore operation. This procedure ensures that a technician is on hand to monitor motor operation and troubleshoot the problem.

The motor starter shown in *Figure 14* requires the installation of a thermal element to monitor each phase of power serving the motor. These elements must be very carefully sized to accommodate the motor being protected. Motor starters are often purchased with a matching enclosure.

Thermal element selection is primarily based on two things: the full load current rating of the motor, and the temperature of the area where the starter is in relation to the motor environment. Since the overload device is thermal in nature, the temperature of the motor starter environment must be considered. If the motor starter location is significantly warmer than the motor location, the overload relay will trip prematurely if the environment is not considered when selecting the thermal elements.

Thermal elements are designed to warm to their critical temperature as the current passing through them exceeds a specific level. Once they reach the critical temperature, a soft metal that holds the trip linkage in position melts. The trip linkage is released, causing the overload relay to open the control circuit contacts. Once current flow stops, the metal begins to cool and harden, allowing the overload relay to be reset shortly thereafter. Manufacturers provide charts for the selection of thermal elements.

Electronic overload relays offer a range dial to set the desired overload current value; no thermal elements are required. Electronic overload relays (*Figure 15*) offer greater reliability and faster response to phase loss and phase imbalances. Many also offer remote reset capability, which is not available in electromechanical overload relays with thermal elements.

Motor starters often have auxiliary contact sets built in. The auxiliary contacts may be used to report a problem through a light or siren or to enable another control action. *Combination motor starters* offer a complete assembly with a contactor and overload relay mounted in an enclosure that may also house circuit protection (fuses or a circuit breaker) as well as a means of disconnect (*Figure 16*).

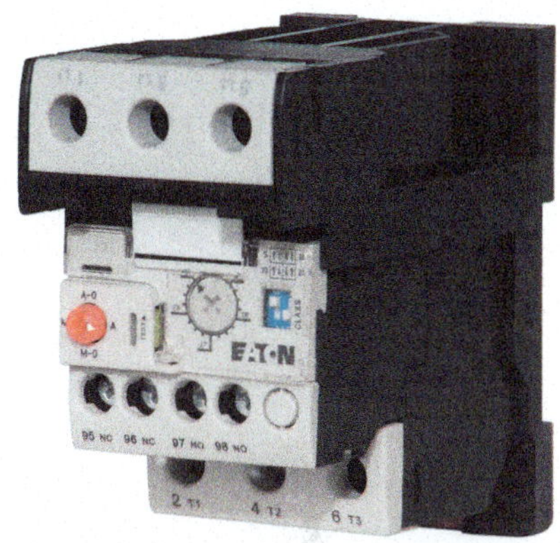

Figure 15 Electronic overload relay.
Source: Courtesy of Eaton

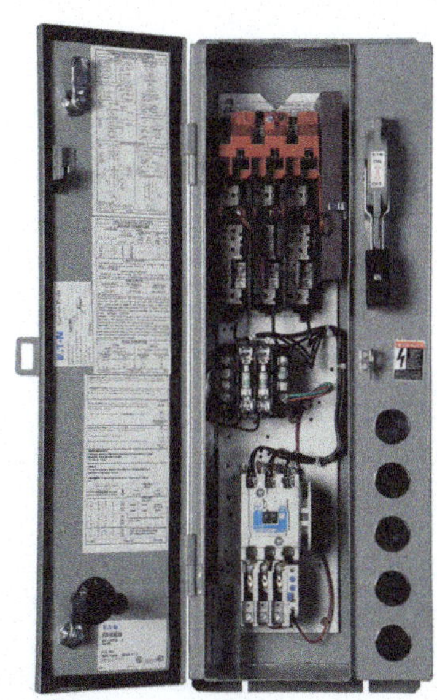

Figure 16 Combination motor starter with disconnect switch and fuses.
Source: Courtesy of Eaton

1.1.3 Checking Switch, Relay, and Contactor Contacts

Once a switch or set of contacts has been identified as a possible cause of an open circuit, the contacts are tested to confirm their position and/or presence of power. Determining the position of a set of contacts is relatively simple.

Switches and contacts can be tested to determine their position by placing voltmeter leads across energized contacts (*Figure 17*). If the switch contacts are open, the voltmeter indicates the applied voltage (240VAC in the example shown in *Figure 17A*). If the switch contacts are closed, the voltmeter should indicate 0 volts.

WARNING!

An energized circuit can deliver a severe and possibly fatal electric shock. Always use extreme caution when working with energized circuits. Refer to the arc-flash protection requirements associated with electrical systems found in National Fire Protection Agency (NFPA) 70E, *Standard for Electrical Safety in the Workplace*®.

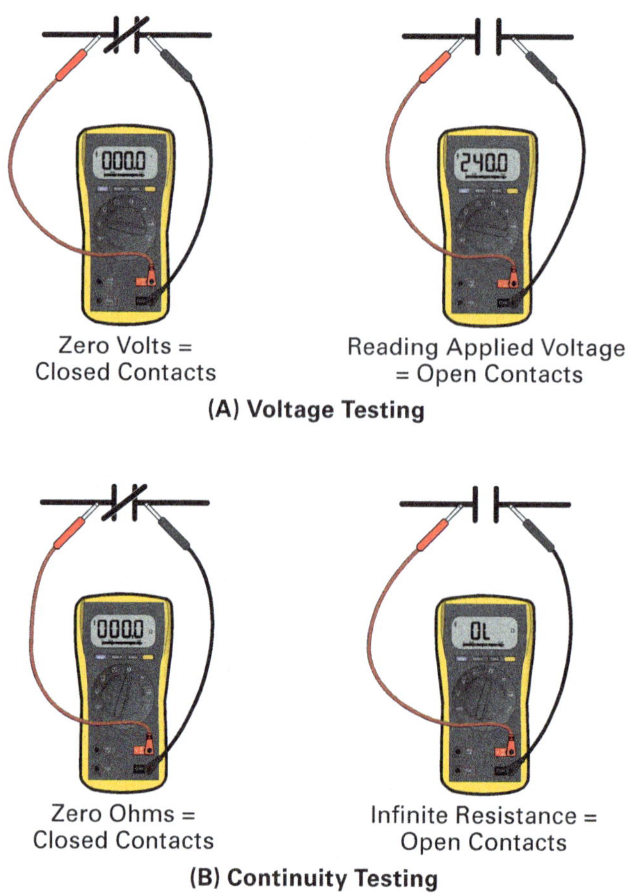

Figure 17 Voltage and continuity tests for contacts.

CAUTION

Remember that only de-energized circuits can be checked with an ohmmeter. Applying an ohmmeter to an energized electrical circuit can permanently damage the meter.

Contact position can also be tested by checking for continuity through them with power off, as shown in *Figure 17B*. If the contacts are open, the ohmmeter indicates an infinite resistance reading. If the contacts are closed, the ohmmeter indicates continuity.

Checking Contacts for Voltage Drop

Contactors carry the heaviest and most important loads in HVACR systems. The electrical load of the compressor is generally the most significant. It's important to ensure that the total voltage drop in the power supply is minimized to 3% of the voltage delivered by the utility. Excessive voltage drop in the power supply under load is often caused by poor terminations or undersized wiring, and it increases the operating current of inductive loads.

However, this voltage drop is not related to the voltage drop that can occur across the contactor contacts. At the contactor itself, there must be little or no voltage drop across the contacts. Just a 1V drop across the contacts is excessive.

To check the total voltage drop in the power supply under load, begin by measuring the voltage at the line side of the contactor without a load applied. If you are working with equipment that has multiple stages of operation, such as multiple compressors or multiple electric heating elements, ensure that all stages are enabled to maximize the load. Start the equipment and recheck the voltage on the line side of the contactor. The voltage drop in the power supply is the difference between the two measurements.

For example, if the voltage is 240VAC with no load, the maximum allowable voltage drop would be 3% of 240VAC, or 7.2 V, when the maximum electrical load is applied. Any amount of voltage drop less than 7.2V is considered acceptable. Less voltage drop is always better, as a reduced voltage causes an increase in the current of inductive loads like motors. A small voltage drop, however, is common.

To check the voltage drop across the contactor contacts, measure the voltage across each individual contact. Place one lead of the voltmeter on the line side of the contact, and the other on the load side of the same contact while the contactor is energized and current is flowing. Any voltage drop detected is most likely to be measured in millivolts, so set and read the meter accordingly. Measure each contact present in the same manner.

A measurement across the contacts that even approaches 1V is cause for concern, meaning the contactor should be replaced. A great deal of heat can be generated at the contacts by a tiny voltage drop because the contacts have become a load device that generates heat. Many craft professionals choose to replace a contactor that shows anything more than 0.5V of voltage drop.

1.2.0 Other Control Circuit Devices

HVACR control circuits normally include several different types of control and safety switches. Pressure and temperature switches are very common examples.

Many systems use one or more pressure switches in the control circuit as a safety device. Others are used to control multiple condenser fans or prove that a furnace induced-draft fan is operating. Small commercial refrigeration systems may use a pressure switch as the primary compressor control, instead of a thermostat.

A variety of temperatures must be monitored or maintained within a specified range. Both electromechanical thermostats and thermistors are used. Thermistors monitor temperatures and provide a signal (resistance) that varies with the temperature. This resistance provides an **analog** input that electronic controls use along with logic to make decisions and react accordingly.

Electromechanical thermostats only open or close a set of contacts at predetermined temperatures. Devices that change their contact positions from open to closed, or closed to open, are used as **binary** inputs to other devices. *Binary* means there are only two possible conditions.

Pressure measurement devices also provide analog or binary inputs. Devices that measure pressure and provide a variable analog signal are called **pressure transducers**.

Analog: Refers to continuously variable signals that represent measured quantities such as voltage, temperature, and pressure.

Binary: Refers to only two possible conditions, such as open/closed or off/on.

Pressure transducers: Devices that convert a monitored pressure to a variable analog signal.

1.2.1 Pressure Switches and Transducers

Electromechanical pressure switches use a bellows that expands or contracts in response to pressure change. The bellows presses against linkage to open or close the contacts. Refer to *Figure 18* for examples of electromechanical pressure switches.

High- and low-pressure safety switches are common examples of fixed pressure switches. While many pressure switches have a fixed pressure setting, adjustable pressure switches are generally needed for applications other than safety controls. *Differential pressure switches*, like those used as oil-pressure safety switches, require two pressure inputs. Differential pressure switches respond to the difference between the two pressures.

(A) Fixed Pressure Switches

(B) Adjustable Low-Pressure Cycling Control

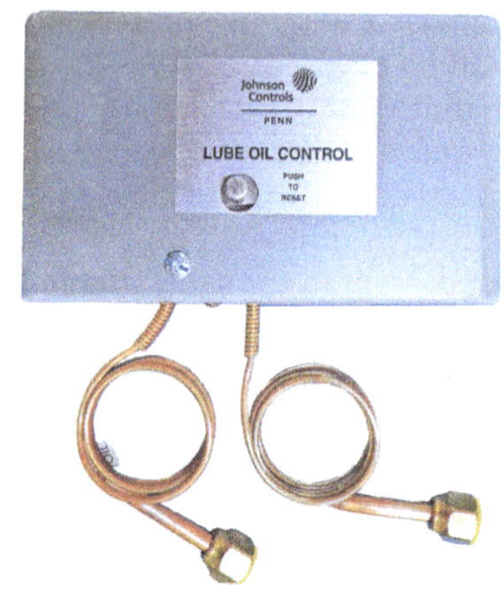

(C) Oil Pressure Differential Safety Switch

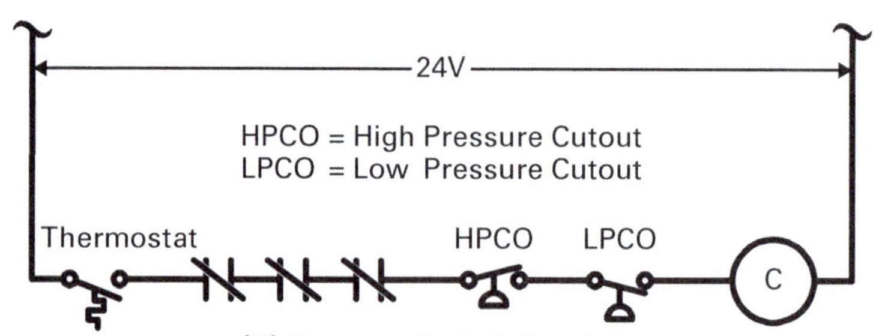

(D) Pressure Switch Symbols

Figure 18 Electromechanical pressure switches.

Another electromechanical pressure switch is commonly used as a draft-inducer proving switch in gas furnaces (*Figure 19*). In this case, the pressure in the heat exchanger is being compared to the atmospheric pressure at the switch. One simple way to check the operation of an inducer pressure switch is to disconnect the pressure input tube from the switch and set the furnace thermostat to call for heat. The furnace burner should not ignite.

Note that pressure switches rarely fail. Generally, when a system shuts down due to an open pressure switch, there is a legitimate reason. Always verify the failure of a pressure switch with accurate gauges and test instruments before condemning the switch. This is true of all safety and limit switches. Basic troubleshooting should always begin with the assumption that safety devices are working properly until otherwise proven through testing.

A pressure transducer (*Figure 20*) may be hard to distinguish from a standard pressure switch. Pressure transducers do not have internal contacts, although many do have a diaphragm. While some transducers provide a variable resistance signal like a thermistor, most pressure transducers used in HVACR applications provide a variable 4 to 20 mA or 0 to 10V signal.

Figure 19 Draft-inducer pressure switch.
Source: HVACGuru

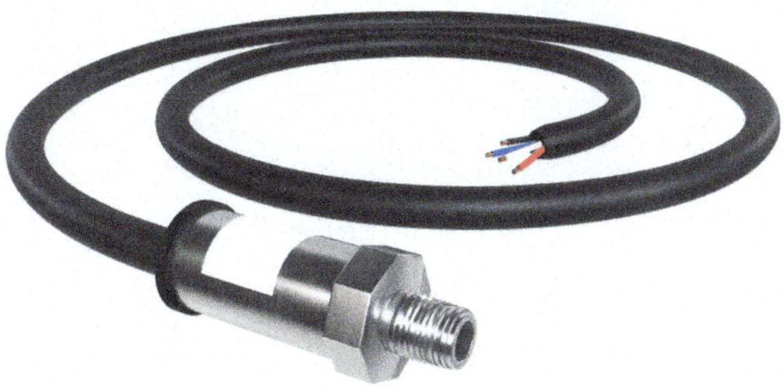

Figure 20 Pressure transducer.

The contacts of pressure switches are tested just like the contacts of relays. Testing the accuracy of a transducer requires a chart from the component or equipment manufacturer that shows the expected signal at various pressures. The pressure is measured with an accurate gauge and the output signal is compared to the value on the chart.

1.2.2 Temperature Switches

Temperature switches and sensors play a significant role in many control schemes. As you know, temperature switches are often referred to as *thermostats*. However, temperature switches and sensors discussed here refer to other devices beyond wall-mounted thermostats that have primary control over heating and cooling equipment operation.

Freezestats

A *freezestat* is a temperature safety switch designed to provide protection against freezing. For example, in a chiller, the water temperature must not be allowed to fall to the freezing point as this will severely damage the chiller. A freezestat is usually a normally closed, bimetal switch. When the temperature falls to the setpoint, the contacts open.

On a chiller, the freezestat is typically installed on the leaving chilled water line (*Figure 21*). The freezestat is typically set to no lower than 35°F (2°C), but often is set slightly higher. In today's chillers that incorporate a great deal of solid-state control, thermistors are used to monitor the chilled water temperature and control chiller operation. However, the protective freezestat may still be an electromechanical switch. The freezestat sensing element (bulb or thermistor) is inserted into a piping *thermowell* (*Figure 22*). The thermowell isolates the sensing element from the fluid, but transfers heat readily.

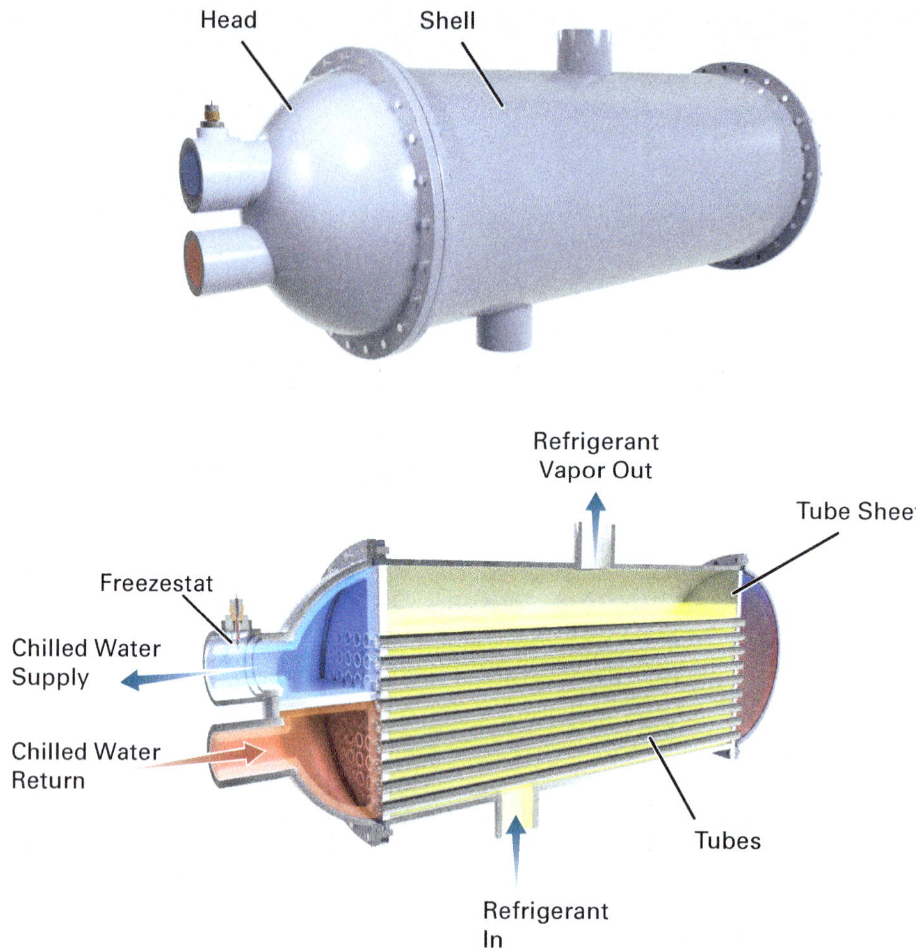

Figure 21 Chiller freezestat location.

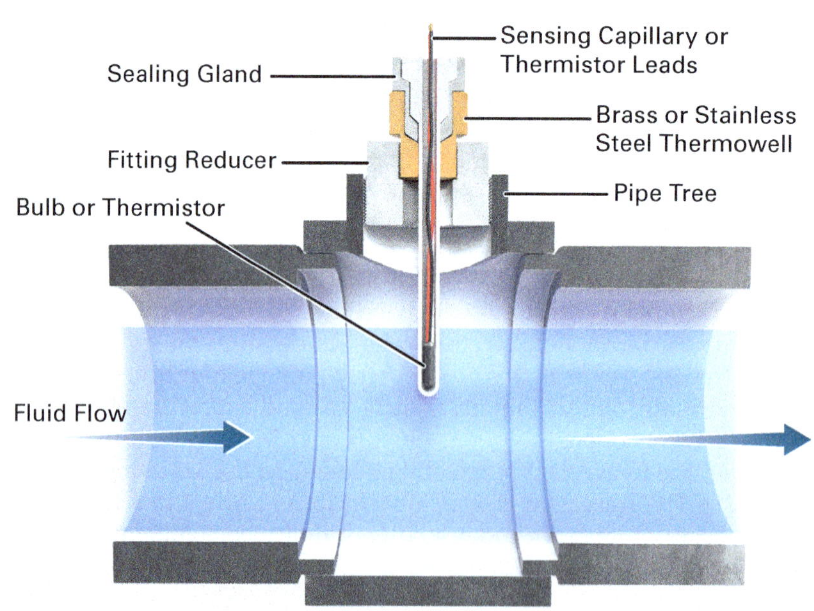

Figure 22 Thermowell installation.

Freezestats are also commonly found on water coils that are exposed to outside air, such as an outside air reheat coil (*Figure 23*). This coil is used to preheat cold outside air before it mixes with return air and passes over the primary heating coils. A freezestat opening forces the outside air dampers closed, isolating the coil from the outside air mass if the hot water source fails and/or the coil is in danger of freezing.

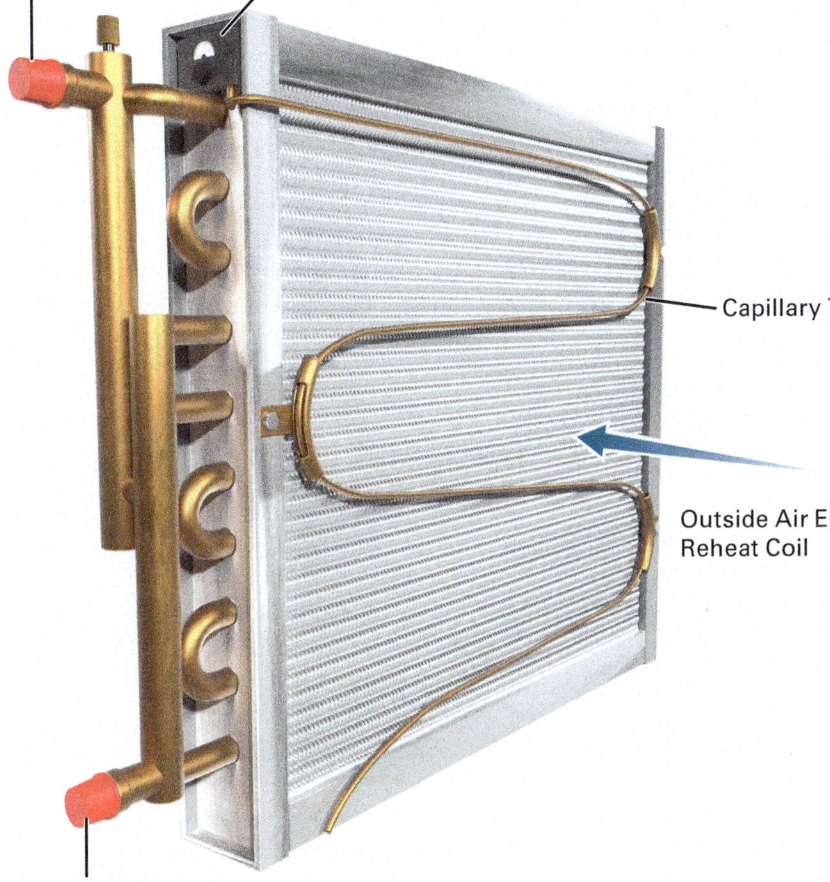

Figure 23 Outside air reheat coil with freezestat.

The type of freezestat shown in *Figure 23* is electromechanical with a very long sensing capillary that is laced across the face of the coil. These long capillary sensing tubes are called *averaging elements*. However, the term *averaging* may be misleading because the freezestat will activate if any small segment of the capillary tube is exposed to the setpoint temperature. Temperatures sensed along its length are not averaged by logic or any other electronic means.

Defrost Thermostats

Defrost thermostats (*Figure 24*) may be used instead of thermistors in some heat pumps to control the initiation and termination of the defrost cycle. The body of this bimetal switch is designed to be strapped tightly to a small refrigerant line or coil U-bend. The opening and closing temperatures are preset and cannot be changed.

A set of contacts like those in a temperature switch provide a *binary input* (either open or closed) to a defrost control board, while a thermistor provides an *analog input* (a varying value). Most microprocessors in HVACR systems are highly dependent on inputs—binary and/or analog—to determine an appropriate output.

When a defrost thermostat is replaced, the replacement must have precisely the same temperature settings. Problems with the defrost cycle are very likely otherwise.

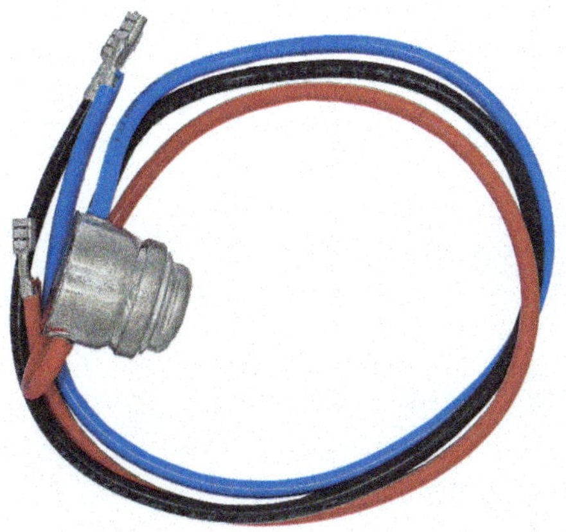

Figure 24 Defrost thermostat.
Source: HVACGuru

Furnace Limit Switches

Furnace limit controls are another type of temperature switch with a bimetal sensing element. Limit switches are positioned near furnace heat exchangers and electric heating elements. These switches are normally wired in series with the low-voltage control circuit of the furnace. If the temperature inside the heat exchanger, control vestibule, or plenum reaches the setpoint, the control circuit is interrupted (switch opens) to prevent overheating. Two limit switches are shown in *Figure 25*. Limit switches like these are typically wired in series with each other and have only a single set of contacts inside.

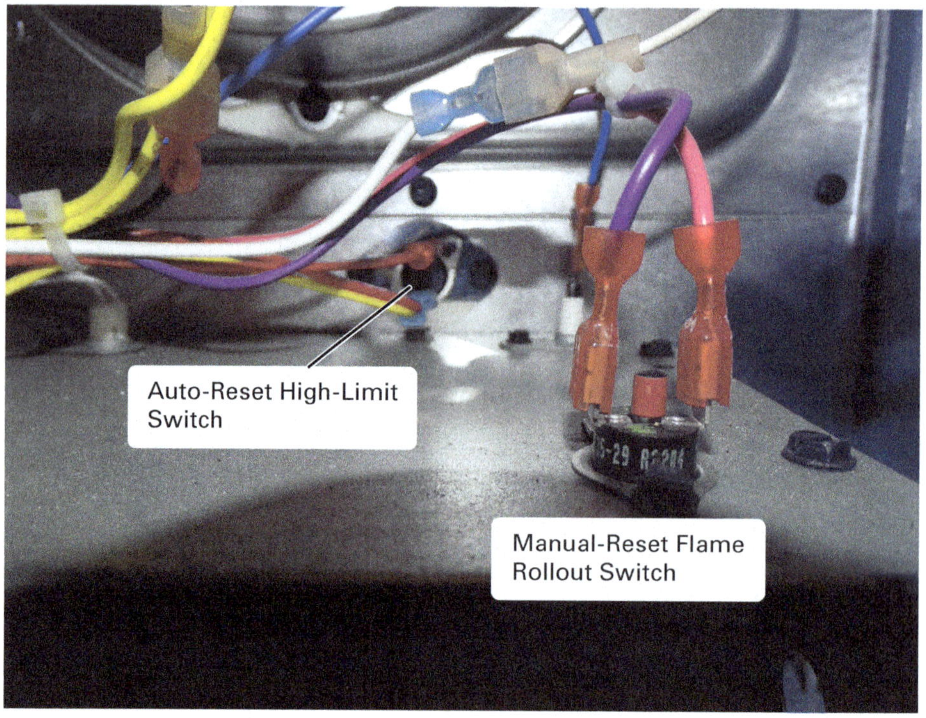

Figure 25 Gas furnace limit switches.

As shown in *Figure 25*, limit switches reset either automatically or manually. Limit switches offer protection from transient events as well as more serious developments. For example, a dirty air filter may cause a furnace to cycle on and off on the high-limit switch. This type of limit switch usually resets automatically. When the dirty filter is replaced, the problem goes away. However, a soot-blocked heat exchanger may cause the burner flame to roll out into the furnace vestibule. This condition cannot be allowed to repeat itself for obvious reasons. To ensure a technician gets involved before the furnace can operate again, a manual-reset rollout limit switch is used.

Limit switches typically have a fixed setting. Thus, like defrost thermostats, limit switches must be replaced with a switch that has precisely the same settings. Similar switches are often used in electric heat assemblies.

1.2.3 Thermistors and Thermocouples

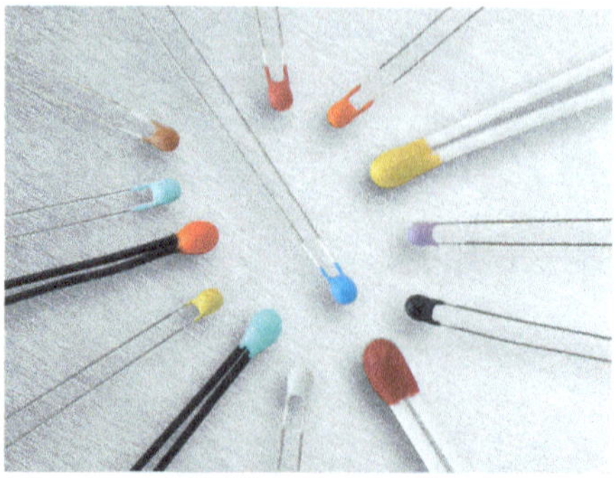

Figure 26 Thermistors.
Source: U.S. Sensor Corp

Thermistors (*Figure 26*) are temperature-sensitive semiconductors. Thermistor resistance varies with variations in temperature in a predictable way. Thermistors are used in a variety of HVACR control applications, including heat pump defrost control, and are classified as *resistive temperature devices* (RTDs).

Thermocouples (*Figure 27*) are a junction of dissimilar metals that generate a predictable voltage from thermal energy (heat). This design means that thermocouples are classified as transducers.

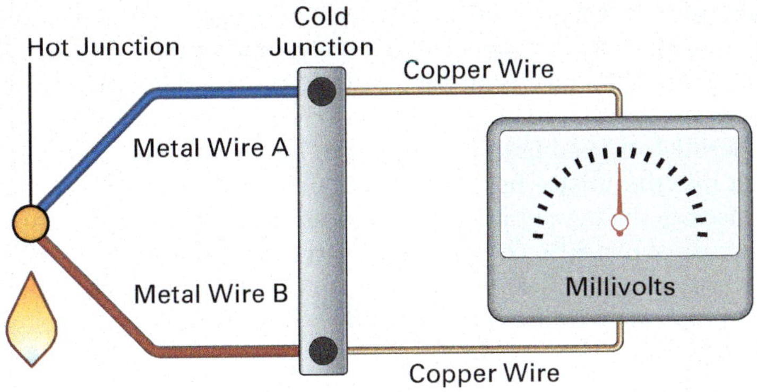

Figure 27 Thermocouple operation.

On Site

Thermocouples in Gas Furnaces

In the HVACR craft today, thermocouples often serve as temperature probes connected to electronic thermometers. But in the past, thermocouples were also used as a power source in gas furnaces.

Standing-pilot, natural-draft furnaces used a thermocouple to prove that a pilot flame was established. These thermocouples generated about 30mV of power, with the power fed to the main gas valve and used to hold a safety valve open. If the pilot flame went out, the voltage was lost, and the gas valve closed.

Wall furnaces and floor furnaces without blowers used what was known as a *thermopile*. The thermopile generated about 750mV of power. Thermopiles powered the entire furnace, with one leg of power routed through a wall thermostat. The only significant load was the gas valve itself.

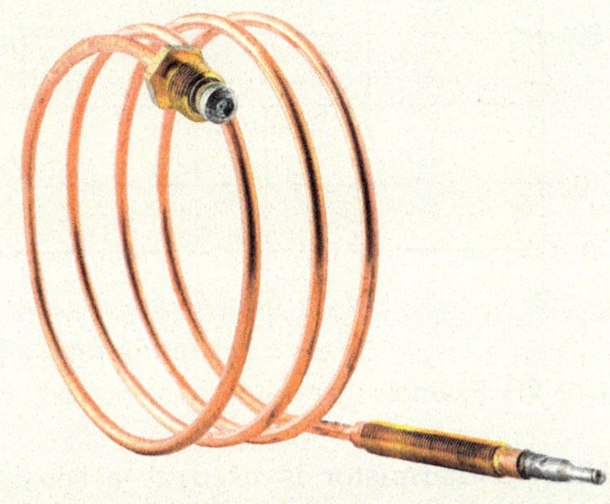

Source: Kinek00/Shutterstock

Thermocouples respond to change slightly faster than RTDs and can measure much higher temperatures. Most RTDs are limited to temperatures up to about 750°F (~400°C) while some thermocouples can be used at temperatures exceeding 3,600°F (~2,000°C). RTDs generally offer a higher degree of accuracy and precision than thermocouples but are also more expensive. In the HVACR craft, you are more likely to work with thermistors than thermocouples.

Thermistors

Thermistors have either a positive or a negative coefficient of resistance. If the resistance increases as temperature rises, it has a positive coefficient of resistance. If the resistance increases as the temperature drops, it has a negative coefficient of resistance. Typical applications of thermistors include the following:

- Electronic thermometer probes
- Integral and remote room thermostat sensors
- Duct temperature sensors
- Thermal-electric expansion valve sensors
- Heat pump temperature sensing for defrost control
- Large motor temperature sensors, embedded into the motor windings

In a heat-pump defrost control circuit, a minimum of two thermistors are generally used. One thermistor senses outdoor coil temperature and the other senses air temperature. Additional thermistors are used in some defrost control schemes.

Manufacturers often provide resistance charts based on temperature for checking thermistors used in their products (*Figure 28*). Based on the chart, you can see that this thermistor has a negative coefficient of resistance—the resistance increases as the temperature falls. Also note that this particular thermistor has little use above 80°F (27°C), as the curve becomes very flat. However, it would function well in a heat-pump defrost control scheme. Note that a table instead of a graph is often provided instead.

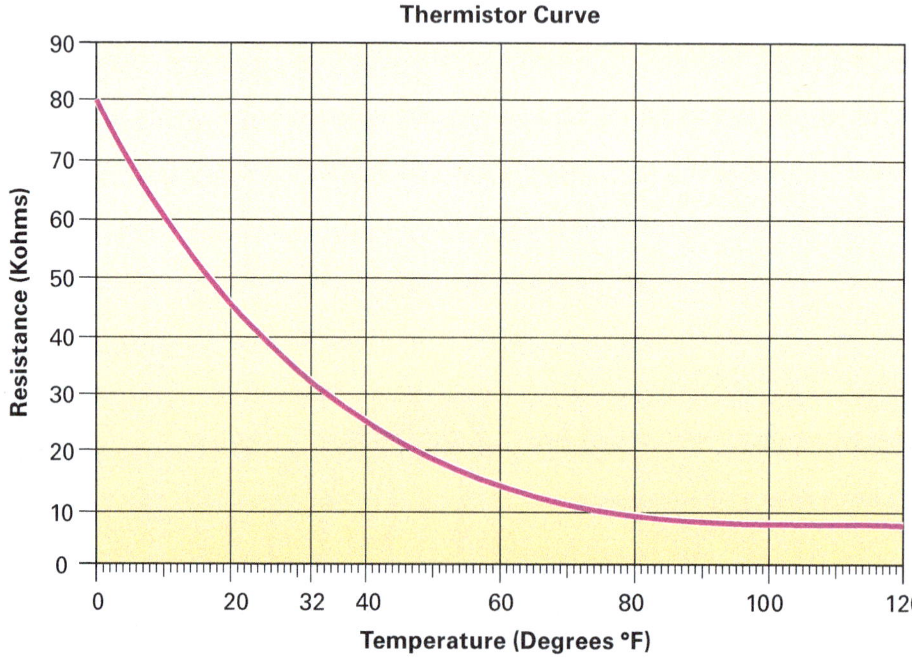

Figure 28 Example thermistor resistance graph.

When a thermistor is used as a sensing device for the operation of a thermal-electric expansion valve (TEEV), the thermistor can be checked as follows. Refer to *Figure 29*:

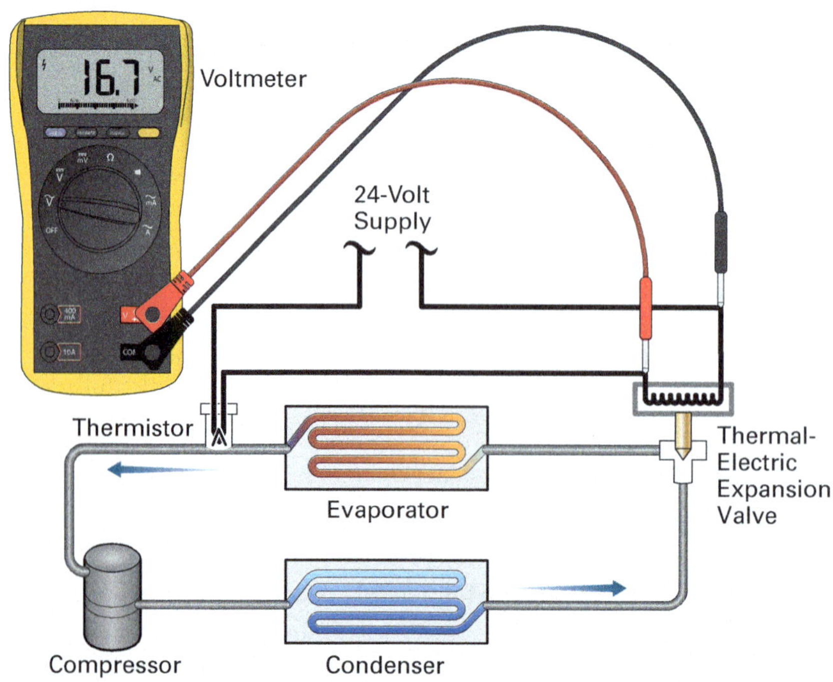

Figure 29 Testing a TEEV.

- With the leads of the voltmeter placed across the terminals of the TEEV and the system operating at or near peak load (maximum refrigerant flow), the voltmeter reading should fall within the range of 15V to 20V.
- With the system operating at a reduced load (reduced refrigerant flow), the voltage reading should remain between 8V and 14V.

These changes in voltage result from the changing resistance of the thermistor. Note that the voltage specifications can differ from valve to valve. Therefore, the manufacturer's specifications should always be consulted. The readings should also be observed for several minutes to ensure consistency. This procedure tests the valve as well as the heater in the TEEV, but the thermistor and heater can be disconnected for detailed testing with the power off.

Thermistors can be embedded in the windings of large motors for temperature protection. An electronic module monitors the thermistor inputs and shuts the motor down if it overheats. In many cases, these thermistors are wired in series. Thermistor groups can be checked by taking a resistance reading across the entire series (*Figure 30*). An open circuit is indicated by a lack of continuity. Proceed as follows:

Step 1 De-energize the system and allow the equipment to cool to the ambient temperature.

Step 2 Connect the ohmmeter leads as indicated in *Figure 30* and record the resistance in ohms.

Step 3 Compare the resistance to the appropriate chart.

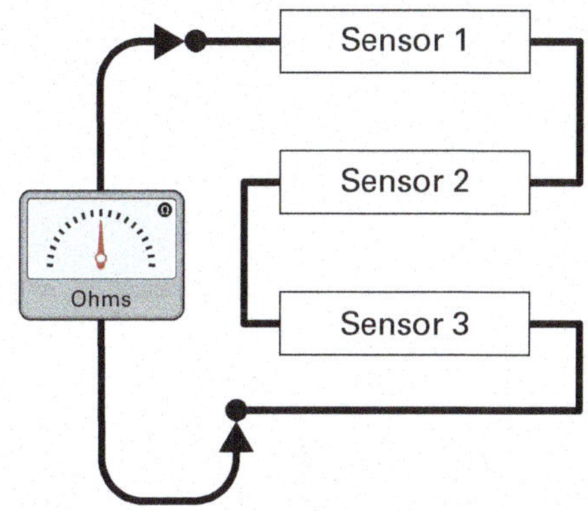

Figure 30 Motor protection thermistor test.

The resistance will change as the temperature changes. The resistance of the thermistor package will vary by manufacturer and application, so the correct specifications for the motor must be consulted. These thermistors are installed at the factory, and are not field replaceable.

Thermistors are very simple devices and do not fail often. Before condemning a thermistor, make sure that the leads are securely connected to the control device, and that the thermistor is making good thermal contact if it is monitoring a surface. You must also be certain of the temperature at its location during testing.

1.0.0 Section Review

1. You measure 230VAC across a set of NO relay contacts while the relay coil is energized. What does this likely indicate?
   a. A short exists in the control circuit.
   b. The relay contacts are grounded.
   c. The contacts have closed, meaning the relay is operating as expected.
   d. The contacts are open, meaning the relay is likely defective.

2. A set of contacts that can only be in one of two positions (open or closed) and is part of a microprocessor-based control circuit is referred to as a(n) ______.
   a. analog input
   b. binary input
   c. dependent output
   d. analog output

2.0.0 Room Thermostats

Performance Tasks

1. Wire, test, and complete the installer setup of a thermostat.
2. Successfully diagnose assigned control circuit problems and failures.
3. Perform electrical tests and/or troubleshooting procedures on the following:
   - Room thermostats

Objective

Describe the operation, installation, and testing of thermostats.

a. Describe basic thermostats and operating characteristics.
b. Explain how to install and wire room thermostats.
c. Explain how to troubleshoot room thermostats.

Room thermostats provide the interface between the user and the HVACR equipment. While most room thermostats today are electronic, it is helpful to first understand how simple electromechanical thermostats work. This section will describe the operating characteristics of simple thermostats before providing detailed information about electronic thermostats.

2.1.0 Room Thermostat Types

Invar®: A steel alloy containing 36% nickel. Invar® is one of the two metals in a bimetal thermostatic element.

The simplest room thermostats, as well as other temperature controls, rely on a bimetal element to sense changes in temperature. A bimetal element (*Figure 31*) is composed of two different metals bonded together. One of the metals is usually copper or brass (nonferrous). The other, an alloy called **Invar®**, contains 36% nickel. When heated, the copper or brass expands more rapidly than the Invar®, causing distortion in the shape of the element. This change in shape moves switch contacts in relation to each other.

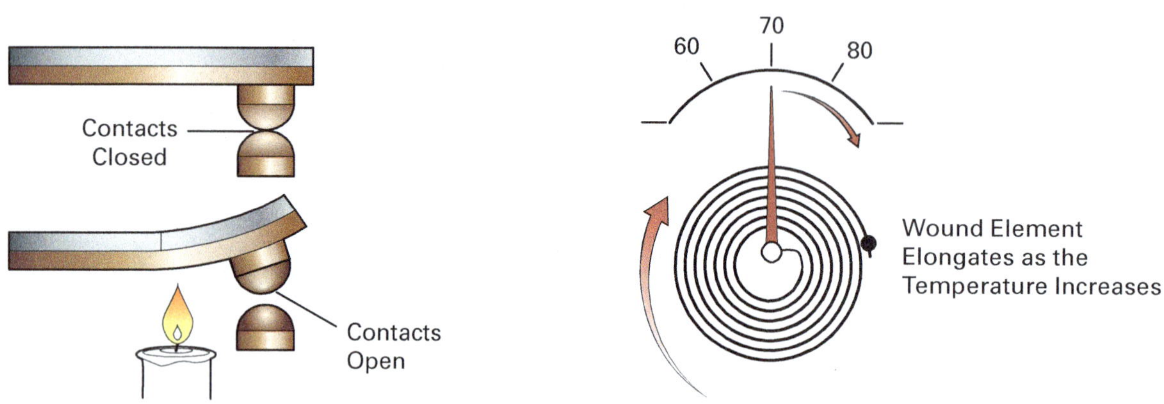

(A) Bimetal Contact Movement

(B) Bimetal Temperature Indicator

Figure 31 Examples of bimetal elements.

Droop: A condition that causes a thermostat's control point to drift away from its setpoint, caused by heat developed from current flow through the thermostat.

Bimetal sensing elements suffer from a mechanical condition known as *thermal offset*, commonly called **droop**. Droop causes the thermostat control point to drift away from the selected setpoint. Droop is caused by excessive heat developed from the current flow through the thermostat. However, droop was not a significant factor unless the thermostat was calling for heat for long periods.

Even though thermostat contacts switch a small current at a low voltage, arcing can still damage the small contacts. A magnet is used in most bimetal thermostats to help snap the contacts together to reduce arcing. These bimetal thermostats are referred to as *snap-action thermostats*.

For many years, room thermostats used mercury encased in glass to complete a circuit and contain the arc, eliminating traditional contacts (*Figure 32*). As the bimetal element changes shape, the mercury rolls from one end of the glass bulb to the other, opening or closing the circuit. Due to the hazardous nature of mercury, these thermostats have been eliminated from the market. Failed mercury-bulb thermostats must be replaced with an electronic model, and the old thermostat must be disposed of properly.

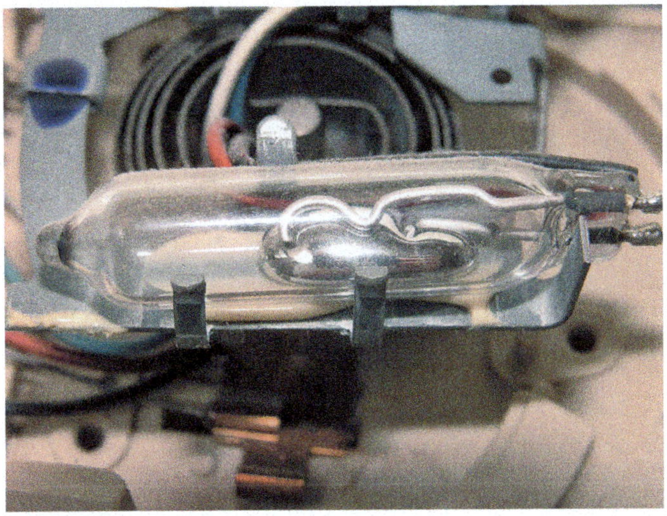

Figure 32 Mercury in a thermostat bulb.
Source: PhotoEuphoria/iStock

WARNING!

Mercury is a toxic heavy metal. Even short-term exposure can result in damage to the lungs and central nervous system. Do not dispose of thermostats containing mercury bulbs along with common trash or garbage. Contact your local waste management authority, or the Thermostat Recycling Corporation (TRC) at *https://www.thermostat-recycle.org/* for mercury disposal and recycling guidance.

Modern room thermostats are electronic, although bimetal thermostats are still available for simple applications. Although most room thermostats operate on a 24VAC power source, *line voltage thermostats* are still used occasionally. These thermostats are commonly used for controlling electric baseboard heaters and wall heaters. Traditional line-voltage room thermostats relied on a bimetal sensing element. However, electronic and programmable line-voltage thermostats are also available.

2.1.1 Heating Thermostats

A wall-mounted heating-only thermostat closes on a drop in temperature (*Figure 33*). When the temperature in the conditioned space reaches the thermostat setpoint (rises), the thermostatic switch will open. Note how the symbol for the heating thermostat is drawn—the switch bar is above the two terminals, mimicking the switch action. As the temperature drops, so does the bar, which closes the contacts.

Because of residual heat in the heat exchangers, the continued operation of the fan, and the time it takes for the thermostat to respond to the changing room temperature, the room temperature can overshoot the setpoint. To avoid discomfort, the thermostat contains an adjustable *heat anticipator.*

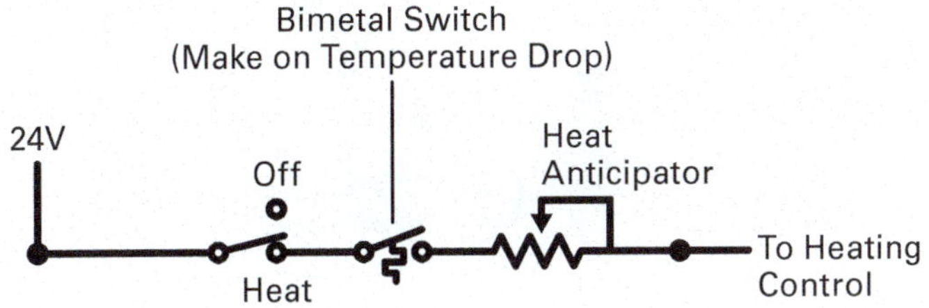

Figure 33 Heating thermostat schematic.

Did You Know?

Mercury

The Thermostat Recycling Corporation (TRC) determined that roughly 15 tons of mercury were installed in thermostats in 2001 alone. Since each thermostat required roughly 3–4 grams of mercury, that's a lot of thermostats.

The TRC, a joint effort supported by HVACR industry titans, collects mercury-bulb thermostats. The mercury is recovered and recycled before disposing of the remaining parts. Many HVACR vendors and parts distributors participate in the TRC program, so it is not difficult to find a drop-off location. Do not casually dispose of any component that contains mercury.

The heat anticipator is a small adjustable resistor in series with the switch contacts. The resistor warms when the heating circuit is energized, and it warms the bimetal strip. This helps keep the thermostat more accurate and responsive. Heat anticipators are typically adjustable and are set based on the current draw of the heating control circuit flowing through the thermostat.

2.1.2 Cooling Thermostats

The cooling-only thermostat (*Figure 34*) is the opposite of a heating thermostat. When the bimetal coil warms and unwinds, the switch closes its contacts and starts the compressor.

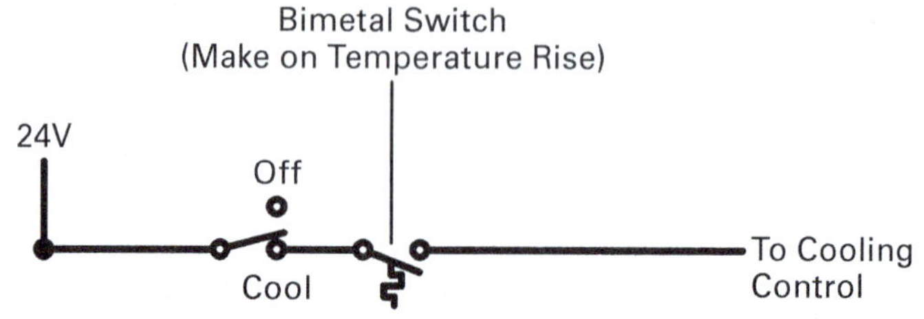

Figure 34 Cooling thermostat schematic.

Cooling compensator: A fixed resistor installed in a thermostat to act as a cooling anticipator.

Instead of a heat anticipator, the thermostat has a **cooling compensator** to help improve accuracy and response. The cooling compensator is a fixed resistor wired in parallel with the thermostatic switch and is not adjustable. No current flows through the compensator when cooling is on because it has a much higher resistance than the switch contacts. When the thermostat circuit is open, however, a small current flows through the compensator to warm the thermostat during the off cycle. The cooling compensator helps account for the lag between the call for cooling and the time when the system begins to effectively cool the space.

Heat anticipators and cooling compensators indirectly control the cycle rate of the equipment. The cycle rate refers to the number of times per hour that the system starts to maintain the setpoint. Longer cycles are generally healthier for the equipment and less irritating than having the system cycle often, but longer cycles may also lead to the system overshooting the temperature setting.

NOTE

It is important to understand that only electromechanical thermostats are fitted with a heating anticipator and a cooling compensator. They are unnecessary in electronic thermostats.

2.1.3 Heating and Cooling Thermostats

When heating and cooling systems are combined for year-round comfort, it isn't practical to use a separate thermostat for each system. Therefore, the two are combined into a single thermostat (*Figure 35*). The thermostat shown is electronic, meaning that it does not contain mercury and there are no moving parts or linkage related to temperature sensing. However, some do have microrelays installed on the thermostat PC board.

A manual switch or button is used to change the operating mode. The same switch (the Mode button in *Figure 35*) can also be used to set the mode to Off. The button in the center controls the function of the fan. In the Auto position, the indoor fan cycles on and off with the heating or cooling equipment. In the On position, the indoor fan runs continuously, regardless of whether the unit is in heating or cooling.

The thermostat shown in *Figure 35* is also equipped with a Filter Reset button. A timer in the thermostat can be set for 30 to 180 days. It is important to

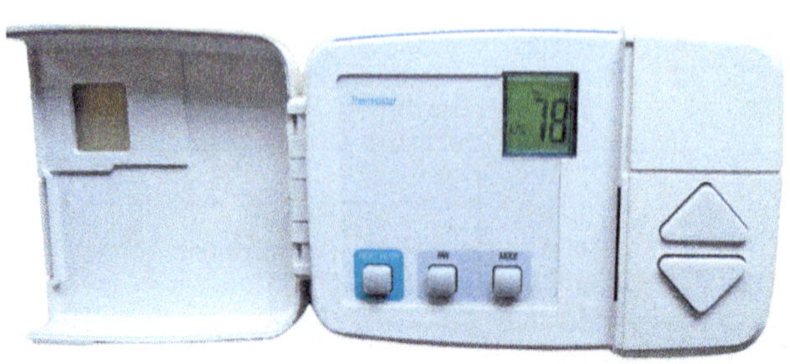

Figure 35 Standard heating and cooling thermostat.

understand the thermostat has no knowledge of the filter condition; the filter monitor merely tracks time. At the end of the set period, the thermostat will remind the user to change the air filter. After it is changed, the user resets the reminder, and the timer starts over.

Automatic-Changeover Thermostats

A disadvantage of a standard heating and cooling thermostat is that the user must determine which operating mode is needed and change it manually. An *automatic-changeover thermostat* selects the operating mode automatically, based on the heating and cooling setpoints. These thermostats have an extra Mode switch or menu item. In addition to Heat, Cool, and Off, an Auto position is available. However, the user can still select heating or cooling independently.

Older electromechanical thermostats contain a built-in mechanical **differential**, which is the difference between the cut-in and cut-out points of a thermostat. The differential is normally 2° to 3°F (1.1° to 1.6°C). For example, if the heating setpoint is 70°F (21.1°C), the furnace may cycle on at 69°F (20.5°C) and run until the temperature is 71° to 72°F (21.6° to 22.2°C). In electronic thermostats, the same feature is applied in the thermostat firmware.

Differential: The difference between the cut-in and cut-out temperature of a temperature control. Differentials also apply to pressure controls.

Automatic-changeover thermostats also have a minimum interlock setting, commonly known as the **deadband**. The deadband is a feature that prevents the heating and cooling setpoints from being too close together. If the setpoints are not sufficiently separated, the thermostat may bounce back and forth between heating and cooling modes. Although a 3°F (1.6°C) deadband is a common default value and serves most applications well, the value can sometimes be adjusted, depending on the thermostat's features. Defeating the minimum deadband setting altogether is not usually allowed by the thermostat logic.

Deadband: A temperature range in wall thermostats that separates heating and cooling modes in an automatic changeover thermostat. The deadband ensures that the system does not consistently alternate between modes due to competing setpoints.

2.1.4 Multistage Thermostats

A two-stage room thermostat is used to control equipment that has more than one stage of heating or cooling. Examples of such equipment include heat pumps, gas furnaces with staged gas burners, and cooling equipment using multiple compressors or compressors with capacity control.

For example, when a heat pump calls for heat, the first stage of the room thermostat controls the compressor, providing compression heat. If compression heat cannot satisfy the room thermostat, the temperature continues to drop and the second stage of the room thermostat, usually preset 1°–2°F (0.6°–1.1°C) below the first stage, energizes supplementary heat to assist compression heat. The same approach is applied to systems with multiple stages of cooling.

2.1.5 Electronic Thermostats

Electronic thermostats (*Figure 36*) are now the standard of the industry. Early electromechanical thermostats that were programmable contained a motor-driven time clock that raised and lowered temperatures at desired intervals. That technology is now obsolete.

Modern electronic programmable thermostats use microprocessors and integrated circuits to provide the needed control and energy-saving features. The interface typically relies on touch-screen technology and soft buttons. Generally, the more sophisticated and expensive the thermostat, the more features it offers users. The following are some of the features electronic thermostats typically offer:

- *Improved temperature control* — Temperature setpoints are generally maintained within a tighter range and more accurately than was possible with older thermostats. Some versions also offer humidity monitoring and can control a humidifier. Many also accommodate remote sensors that can monitor several indoor areas and average the values and/or monitor the outdoor temperature.
- *Installation versatility* — Wireless versions of electronic thermostats are available for applications where it is difficult or impossible to pull thermostat cable. A built-in transmitter sends information to a remote receiver connected to the equipment being controlled.

Figure 36 Wi-Fi-enabled, electronic programmable thermostat.

- *Override control* — This feature allows the occupant to override the current program as needed. For example, the night setback on Monday night could be overridden so the thermostat is not lowered before the football game is over.
- *Multiple setback programs* — This feature allows the occupant to design and select different schedules for different days. For example, a special schedule for a vacation away from home could be programmed. In addition, the user can set a different temperature than what is programmed, then use a hold feature to hold that temperature indefinitely or until the next scheduled event occurs.
- *Battery backup* — This feature prevents program loss in the event of a power failure. Wireless models must operate solely on batteries.
- *Maintenance tracking* — This feature indicates when maintenance is to be performed, for example, when to replace the air filters. Maintenance timers can also be used to remind users of the need to replace UV lamps or similar tasks.
- *Wi-Fi connectivity* — Many of today's thermostats connect to the local network and can be accessed through both the network and its internet connection. This allows the thermostat's software to update as needed in the background. It also provides thermostat control to the user wherever an internet connection is available.

The potential savings available from programmable thermostats are significant. For example, it is estimated that a setback of 10°F (5.5°C) for both daytime and nighttime temperatures can result in a 20% energy savings. A 5°F (2.8°C) setback should yield a 10% energy savings.

2.2.0 Thermostat Installation

Even the best thermostat won't perform well if it is improperly installed. Selecting the proper location for the thermostat is the first step.

2.2.1 Mounting a Thermostat

The thermostat must be installed in the conditioned space unless remote sensors are used. For proper operation, electromechanical thermostats must be mounted level on a wall that is reasonably plumb. Electronic thermostats function properly regardless of level or plumb, but are mounted level for appearance.

Use care when unpacking the new thermostat. Rough handling and static electricity can damage it. Save all the literature and hardware for future use.

Locate the thermostat 5' above the finished floor (the common mounting height) in a central location with good air circulation, and on an inside wall. Locating the thermostat near a return air grille is also helpful. Avoid the following locations and conditions:

- Consistently drafty or stagnant spots
- Where air from supply registers and grilles will blow on the thermostat
- Areas affected by radiant heat from the sun or from appliances
- Walls that enclose a chimney or flue vent
- Outside walls and walls that are adjacent to unconditioned spaces

To replace or install a new thermostat, first de-energize the equipment that provides the low-voltage power source. Lock out and tag the equipment.

Before removing an existing thermostat, record the thermostat wire color and corresponding thermostat terminal designation. This helps prevent miswiring of the new thermostat. Ideally, the color matches the terminal letters, but it very often does not match perfectly.

Install the new subbase on the wall. If the subbase is mounted on a junction box, you may need an accessory adapter to mount the thermostat. If there is no junction box provided, use wall anchors that suit the wall type. The subbase provides terminals for the wiring connections.

Ensure the thermostat subbase is mounted level (*Figure 37*). It is best to hold the subbase on the wall and level it while marking the screw locations with a pencil or scribe. Once the screws are installed, level the thermostat before tightening them.

Subbases have a specific spot for placement of the level, as shown in *Figure 37*. In this case, there are two points across which the level is placed. Some have a flat area on the top of the subbase to accommodate the level. As mentioned earlier, the operation of an electronic thermostat is not affected by being out of level. Visually, however, a thermostat that is not level looks tacky and unprofessional.

CAUTION

Thermostats containing solid-state devices are sensitive to static electricity. Before handling the thermostat, discharge static electricity from your body through grounding. Touch only the plastic housing and similar nonmetallic areas when holding the thermostat whenever possible.

NOTE

The National Electric Code (*NEC*®) does not require a mandatory mounting height for thermostats, but local codes may address the topic. Installations subject to the provisions of the Americans with Disabilities Act (ADA), thermostats cannot be mounted more than 48" above the finished floor.

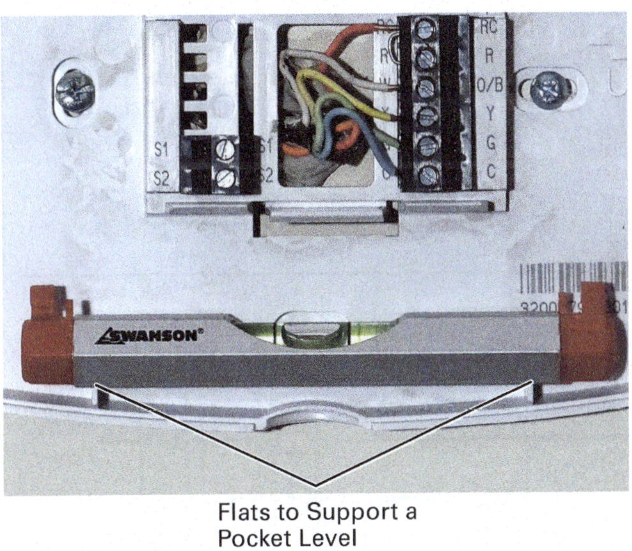

Figure 37 Leveling a thermostat.

2.2.2 Thermostat Wiring

The thermostat is connected through multi-conductor thermostat wire to a terminal strip or junction box in the air conditioning unit. A standard coding method is most often used to designate popular wiring terminals and wire colors. Note that some of today's systems, including ductless units, communicate and control themselves differently. In this section, traditional control wiring practices that apply to most residential and light commercial systems are presented.

The color code and corresponding terminals of a standard heat/cool thermostat are shown in *Figure 38*. These terminal designations are standardized among most equipment manufacturers. However, it is not safe to assume that an existing system is wired according to that color code. Additional wire colors and terminal designations are used with more complex thermostats, and sometimes installers use a different color for a valid reason.

Thermostat Wiring Codes

Terminal Designation	Wire Color	Function
C	Black or Blue	The Common Lead of the Control Transformer
R and R_C	Red	Power
G	Green	Fan Control
Y	Yellow	Cooling Control Y1 = Stage 1 Y2 = Stage 2
W	White	Heating Control W1 = Stage 1 W2 = Stage 2
O (B)	Orange (Brown)	Heat Pump Reversing Valve or Other Device Controlled as the Operating Mode Changes
E	N/A	Related to the Emergency Heat Function of a Heat Pump Thermostat
L	N/A	Usually Related to the Illumination of a Light
S	N/A	Typically Related to the Connection of Thermistors for Remote Temperature Sensing

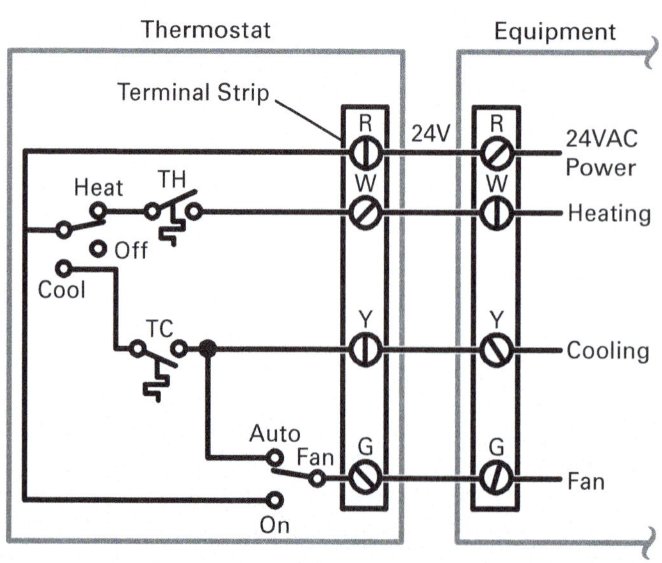

Figure 38 Common thermostat terminals and colors.

For example, when there are multiple stages of heating or cooling, those terminals are designated with the appropriate letter plus a number, such as W1 and W2. However, there will only be one white wire in the bundle. Wiring terminals for external temperature sensors are often present as well.

Many electronic thermostats also contain a C (Common) terminal and require 24VAC power for the internal electronics. Traditional thermostats rarely need a C terminal to function. As a result, when replacing a thermostat where there is no wire left in the bundle to use for C, a different model of thermostat may be needed. In some cases, the G wire (fan control) can be swapped to serve as a C wire, while the G wire from the equipment is connected at the same terminal as Y. However, this can only be done if the thermostat is not required to start the fan in the heating mode, as is the case with most gas furnaces. The user also loses the ability to operate and control the fan independently.

Memorizing the color code for thermostat wiring is not nearly as important as understanding what each wire does. It is important to maintain the color code when possible, and you will find that easy to do in many cases. But it is far more important to understand the function of the terminals. Color simply provides a way of distinguishing one wire from another.

Thermostat wiring must be done in accordance with national and local electrical codes. Standard thermostat wire does not have the insulating characteristics as conductors used for higher voltages. Thus, the *NEC®* does not allow Class 2 control wiring (the class into which thermostat wiring falls) to be placed inside conduit, junction boxes, or raceways where a line voltage is present.

Thermostat wiring cable with multiple conductors is readily available from all HVACR distributors. A No. 18 wire size is adequate for most installations. However, long runs of thermostat wire can cause excessive voltage drop that interferes with equipment operation. If the thermostat wire run is over 100′ (30 m), use No. 16 or even heavier wire to reduce the voltage drop.

To avoid damaging the small conductors, use a stripping tool designed to strip No. 18 wire. Splicing should be avoided whenever possible. If splicing is necessary, soldered splices are best, and all splices must be made in an accessible location. After terminating the wires at the thermostat, remember to seal the opening in the wall behind the thermostat so it is not affected by drafts from the wall cavity (*Figure 39*). You can use a small ball of fiberglass insulation or putty to seal the opening.

NOTE

Some manufacturers offer a "C-wire adapter" that can be installed to resolve the problem of insufficient wires at the thermostat. It is always best to know how a thermostat is currently wired before planning for its replacement. Consult the manufacturer's documentation for the new thermostat to determine acceptable solutions.

Figure 39 Sealing the wall opening.

Wireless thermostats need a base for communication located at or near the equipment. The thermostat wiring is connected to the equipment interface module (*Figure 40*). The interface module accepts inputs from the thermostat and responds accordingly. A fully wireless setup makes thermostat relocation possible without running new wires.

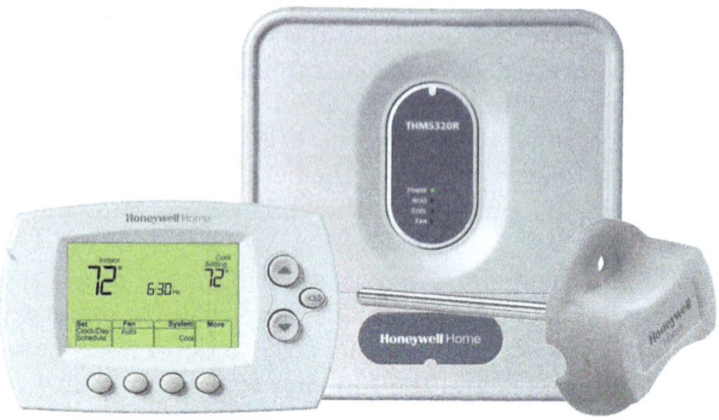

Figure 40 Honeywell wireless thermostat, equipment interface module, and remote temperature sensor.
Source: Jackson Systems, LLC.

2.2.3 Electronic Thermostat Installer Setup

Digital thermostats offer a significantly higher level of flexibility than their predecessors. A single thermostat model works with a variety of systems. However, it also places a greater level of responsibility on the installer to ensure the thermostat settings are appropriate for the system and meets the user's preferences in operation.

The installer setup menu for a thermostat is not evident to users. Accessing the installer setup generally requires a unique series of key presses or entries, and sometimes specific keys must be held down for much longer than a user would normally hold them. For each thermostat model, the installation manual should be consulted for information to access the menu. The point is to make the menu challenging to access and impossible to access by accident, as users who may not understand the entries can make changes that will cause significant functional problems.

It is also likely that you will need the installation instructions in hand as you set up the thermostat. Manufacturers identify the menu items in a variety of ways. Many use a numeric code or a single number to identify each menu item, with no text on screen to indicate the feature being adjusted. Arrow keys are commonly used to change the setting of a menu item. Some thermostats do offer a clear language display that is a bit easier to understand, but it is rare, and a full description of the setting is often needed to make an informed decision.

The following provides some insight into the settings that installers must address before starting the system:

- *System type* — The choices here often include gas or oil warm-air furnaces, heat pumps, or electric resistance heat. The choice made here typically affects the system switching menu available to the user.
- *System stages* — Today's systems may have multiple stages of both heating and cooling. You must know how many stages are available to the thermostat.
- *Fan/Blower control* — This generally refers to control of the indoor blower during the heating mode. Thermostats always energize the indoor fan (G) on a call for cooling, but control of the fan in heating can differ. A gas furnace, for example, typically starts the blower through its onboard controls after a short delay. In this case, you do not want the thermostat to start the indoor blower when it calls for heat.

- *Reversing valve operation* — When a heat pump system is selected, this option allows you to determine whether the valve is energized by the thermostat in the heating or cooling mode.
- *Programming* — This refers to the ability to schedule temperature setback and setup. It can be set to non-programmable if the user wishes. When programming is allowed, you may need to choose between two or four periods per day and whether all seven days can be programmed individually. Alternatively, you can usually allow Monday through Friday to carry the same schedule while Saturday and Sunday can be programmed individually. This may be referred to as *5+1+1 programming.*
- *Smart/Intelligent recovery* — These terms refer to the behavior of the system as it recovers from a temperature setback or setup. For example, assume a system is set to 68°F (20°C) for sleeping in the heat mode, then returns to a 73°F (23°C) setting in the morning. With smart or intelligent recovery selected, the thermostat will learn over time how long it must run to return to the 73°F setpoint, then start at an appropriate time ahead of schedule to reach 73°F.
- *Energy-managed recovery* — This generally applies to heat pumps equipped with supplemental heat sources. In the heating mode, selecting this feature disables the supplemental heat source and forces the system to try and recover using the heat pump only.
- *Compressor short-cycle protection* — Most thermostats offer short-cycle protection that prohibits the operation of the system compressor (cooling or heat pump) for several minutes after it shuts down. Unless short-cycling protection is provided some other way, such as with a delay-on-break timer in the outdoor unit, it is best to use this feature. In some cases, you can select the duration of the delay.
- *Auto-changeover* — This feature enables or disables the thermostat's ability to change modes (heating or cooling) automatically.
- *Cycle rate* — Electronic thermostats do not use traditional heating or cooling anticipation to eliminate short-cycling and improve accuracy. Instead, they control the system *cycle rate,* which is defined as the number of cycles per hour (cph) allowed. When this feature is adjustable, it typically ranges from 2 cph to 6 cph. A setting of 3 cph is usually a good place to start and is usually the default setting in a new thermostat. In some thermostats, the cycle rate is fixed.
- *Keypad lockout* — Some thermostats offer various levels of keypad lockout to prevent unauthorized tampering. This feature is more likely to be used in a commercial setting.

Always go through the installer setup carefully and consider how each choice will impact both the system and the user. It is also important to involve the user in some of the decisions, such as those related to programming. Remember that every thermostat brand and model may have unique features and options—review the manufacturer's installation manual carefully.

Wi-Fi Setup

Connecting to a home or business network has become second nature to many of us. Wi-Fi thermostats must be connected to the home or business network. To do so, it is likely you will need the assistance of the user to provide or input the password for a private network.

Like most smart devices, the thermostat should display a list of networks that are within range. Make sure you choose the correct network, as some have similar names. Enter the network password or allow the user to step up and make the entry to ensure the network remains secure.

An internet connection is also necessary. With an internet connection provided through the network, the user will be able to use a smartphone app or the manufacturer's website software to access and control the thermostat from a remote location. Once the network connection is made and the user has downloaded any apps needed, it is usually best to allow them to complete the thermostat registration process. Provide assistance as needed. In some cases, a media access control (MAC) address and a cyclic redundancy check (MAC CRC) address are needed. Although this information can usually be found on the back of the thermostat or in the About section of a thermostat menu, it is usually provided on the thermostat packaging as well. These addresses provide the manufacturer with a path for communication for firmware updates.

2.2.4 Final Testing

The final step is to confirm that the thermostat functions properly in all operating modes. With system power restored, place the thermostat in the Heat mode and the Fan switch in the Auto position. Adjust the setpoint to at least 5°F (3°C) above the room temperature. The heating equipment should start quickly (unless short-cycle protection is active). The blower will often start after a short delay.

Next, adjust the setpoint to 5°F below the room temperature. The primary heat source should shut off immediately, but the blower may continue to run for several minutes to maximize the transfer of heat from the heat exchanger or indoor coil.

Now test cooling operation in the same basic manner. Ensure the equipment starts as expected. Once cooling operation has been tested, turn the mode switch to Off and set the Fan to On. Ensure that the blower starts and runs to circulate air.

CAUTION

To avoid compressor damage, do not check compressor operation when the outdoor temperature is below 60°F (16°C) unless you are confident that refrigerant has not condensed in the compressor suction line. This is to prevent compressor damage due to a flooded start.

2.3.0 Thermostat Troubleshooting

Troubleshooting thermostats involves both visual and electrical testing. Remember to handle electronic thermostats carefully when removing from the subbase.

As a first step, visually check the thermostat for problems that might affect the operation. Ensure it is in a good location and is not being influenced by system airflow or heat-producing appliances. Also look for physical damage.

Next, check the system control voltage. If the control circuit voltage (usually 24VAC) is not adequate, it can prevent relays and contactors from operating as they should. Measure the control circuit voltage at the R and C terminals on the furnace or air handler. If no voltage is measured, troubleshoot the control transformer and check its fuse (if equipped). Later, as you test the various functions, you may also need to check the control voltage directly at key relays and contactors. The voltage may be dropping significantly under load. This can cause relays and contactors to "chatter" or fail to activate.

When working with electronic thermostats, it is a good idea to review the installer setup. Ask the user if they have made any changes in the setup recently. An error in the setup can cause unusual operation. For example, consider what happens if the reversing valve is set up to energize in the wrong mode. Review the setup before troubleshooting begins, assuming the thermostat is responsive to inputs. Also ensure the batteries, if installed, are live.

Some electronic thermostats are self-diagnosing to some degree. Error codes might be displayed. For instance, a blinking battery icon indicates a weak or dead battery. Other error codes identify faulty components such as an open or shorted temperature sensor. Manufacturers of electronic thermostats often provide troubleshooting aids. Review the guidance before troubleshooting begins.

Various functions of the thermostat can be checked by removing the thermostat from the wall and using jumper wires to simulate the different thermostat functions. Use an insulated jumper wire(s) with alligator clips connected across the thermostat's R terminal and fan (G), heat (W), or cool (Y) terminal individually.

WARNING!

Use jumper wires in this manner only on low-voltage (<50VAC) control circuits. Jumper wires should be used only for troubleshooting purposes. Never use a jumper wire to bypass a safety control or to power a line-voltage load.

If the thermostat is suspected to be the cause of a problem, remove the thermostat from its subbase and use the jumper wire across the appropriate terminals. This testing can also be conducted at the indoor unit low-voltage terminal board. However, the thermostat should be removed from the subbase first to avoid any possibility of damage. If the use of jumpers confirms that all the wiring and equipment functions normally except when the thermostat should be energizing the control circuits, replace the thermostat.

CAUTION

Never jump or connect terminals R and C to each other, as this represents a direct short of the transformer secondary windings. If the transformer is not fused, it will be immediately damaged.

2.3.1 Fan Switching

If the thermostat Fan switch will not start the indoor fan, access the furnace or fan coil unit low-voltage terminal board. With the control circuit energized and the fan switch in the On position, use a multimeter to test for voltage between Common (C) and G at the terminal board. If the thermostat is working, 24VAC will be present (*Figure 41*).

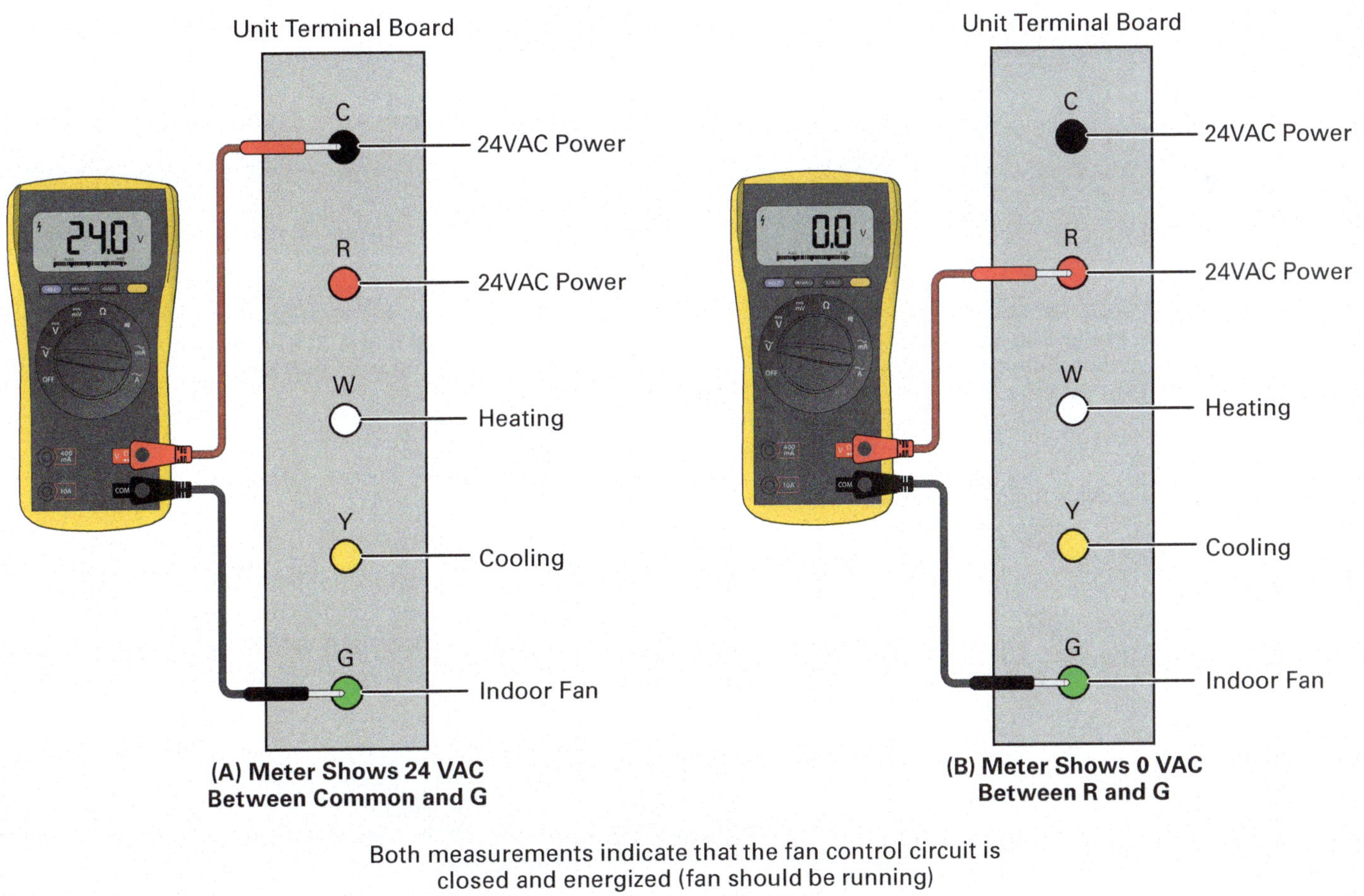

Figure 41 Checking the function of the Fan switch (control circuit energized).

Alternatively, or as confirmation, you can test for voltage between R and G—if the thermostat circuit is closed and intact, there will be no voltage reading here.

If the R-G circuit is not energized as it should be, a defective thermostat is a possibility. However, you must also ensure the circuit wiring is intact before condemning it.

2.3.2 Cooling Operation

If the thermostat won't start the cooling equipment, move to the unit terminal board for testing. With the control circuit energized, the thermostat set to the cooling mode, and the setpoint well below the room temperature, use a multimeter to test for voltage between C and Y (*Figure 42*). A completed circuit through the thermostat should result in a reading of 24VAC. You can also test for voltage between R and Y for confirmation—no voltage should be present.

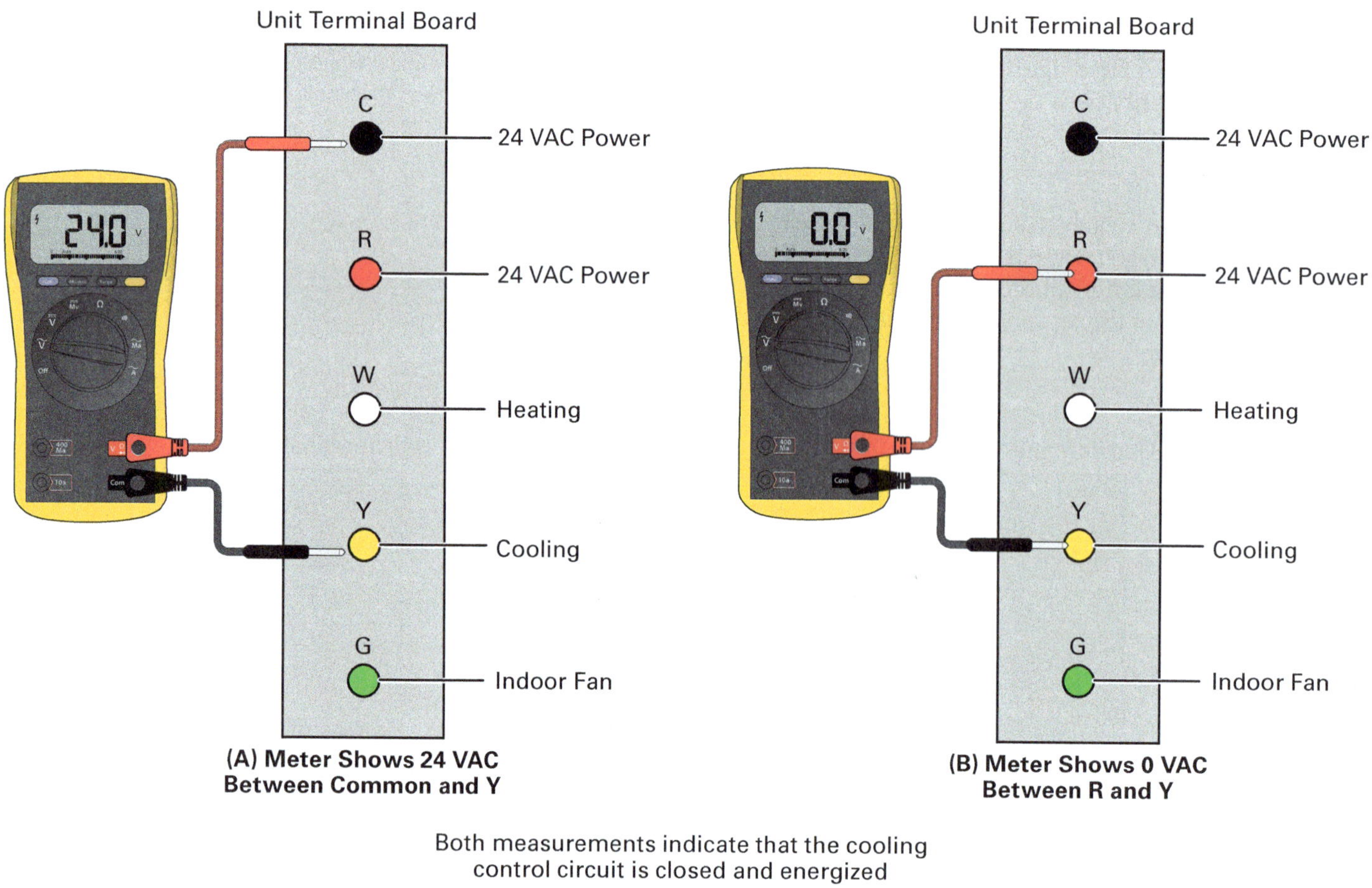

Figure 42 Thermostat cooling control circuit test.

Remember, though, that the thermostat could be delaying the compressor operation if the compressor short-cycle protection feature is activated. Many thermostats begin a protective delay as soon as they are energized. Before testing, make sure any such time delay has elapsed or has been temporarily disabled.

2.3.3 Heating Operation

If the thermostat won't start the heating equipment, use the same approach. With the control circuit energized, the thermostat set to the heating mode, and the setpoint well above the room temperature, use a multimeter to test for voltage between C and W (*Figure 43*). A completed circuit through the thermostat should result in a reading of 24VAC. You can also test for voltage between R and W for confirmation. There should be no voltage present.

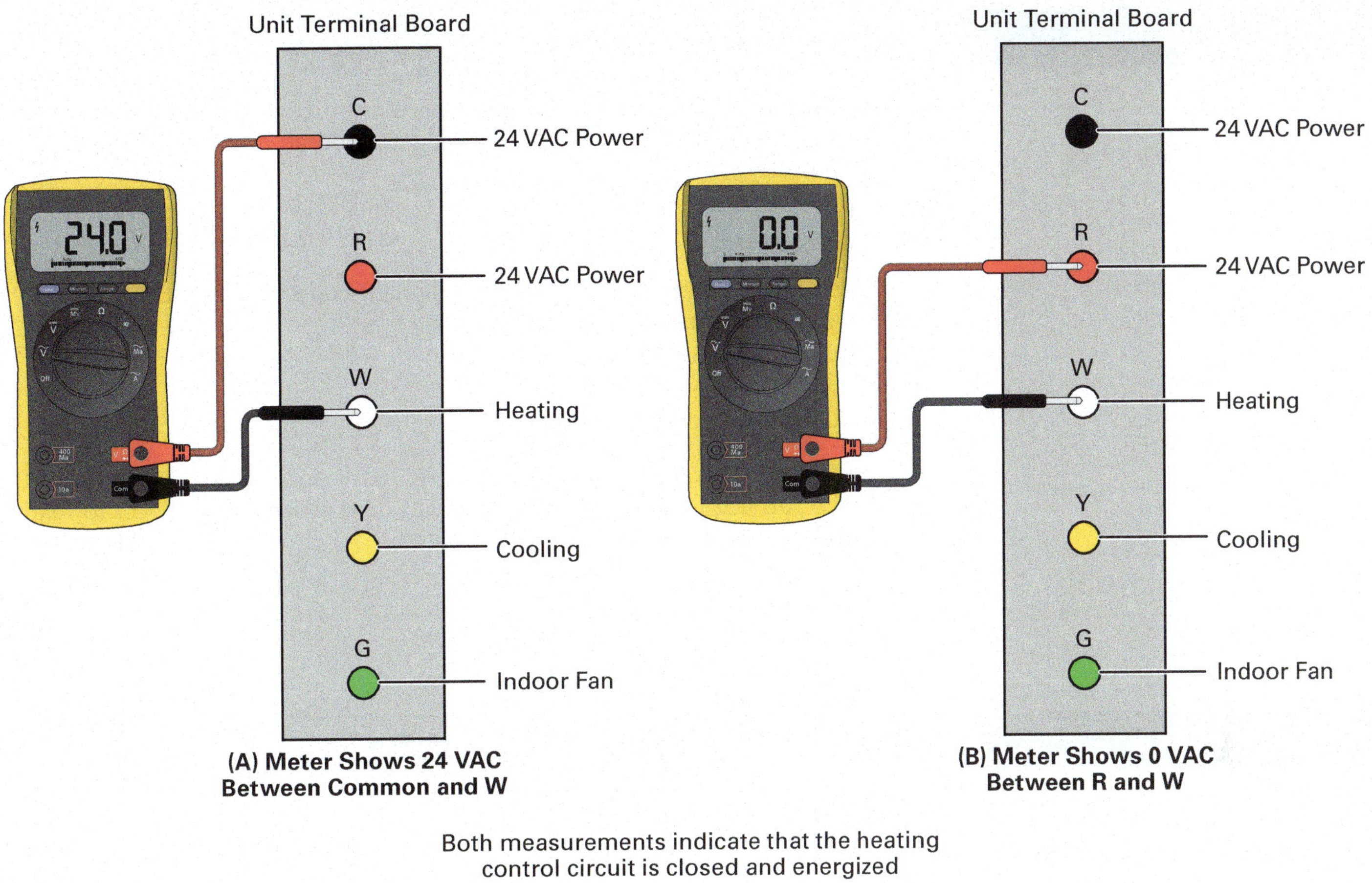

Figure 43 Thermostat heating control circuit test.

On Site

Sophisticated Front Ends

With the rise of variable refrigerant volume (VRV) systems (also referred to as variable refrigerant flow, or VRF systems), temperature control is becoming increasingly sophisticated. But control systems are also more flexible and offer comprehensive system management.

The tablet-style panel shown can be wall-mounted and can control up to 650 points, including the VRV indoor and outdoor units. It is classified as an *integrated building management system* (Intelligent Touch Manager™) that can control not only the VRV and ventilation equipment, but other equipment, sensors, alarms, and pumps.

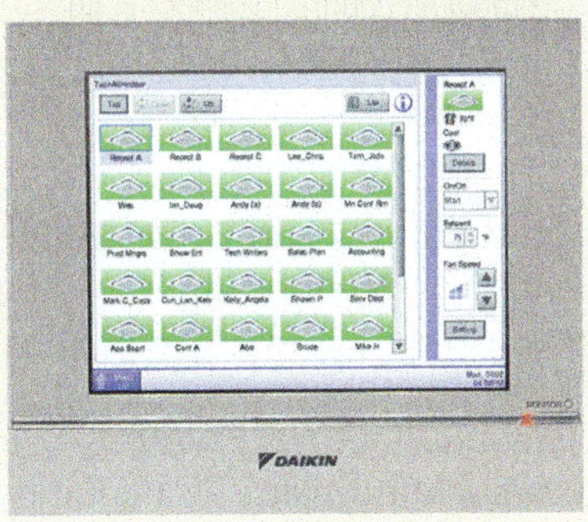

Source: Daikin Comfort Technologies North America, Inc.

2.0.0 Section Review

1. What is the default deadband setting that prevents heating and cooling setpoints from being set too close together on an auto-changeover thermostat?
   a. 1°F (0.6°C)
   b. 2°F (1.1°C)
   c. 3°F (1.6°C)
   d. 5°F (2.8°C)

2. What should you do before physically handling an electronic thermostat?
   a. Read the instructions
   b. Ground the thermostat
   c. Install the batteries
   d. Ground yourself

3. If a thermostat is calling for cooling, between what two terminals of the furnace or fan coil unit terminal board would you expect to read 24VAC?
   a. C and Y
   b. C and W
   c. R and Y
   d. R and G

3.0.0 Troubleshooting Circuits

Performance Tasks

2. Successfully diagnose assigned control circuit problems and failures.
3. Perform electrical tests and/or troubleshooting procedures on the following:
   - Relays and contactors
   - Room thermostats
   - Resistive loads

Objective

Explain how to troubleshoot common electrical circuits.

a. Identify safety practices related to troubleshooting HVACR power and control circuits.
b. State the sequence of operation for basic heating and cooling systems.
c. Explain how to approach HVACR problems and troubleshoot in an organized manner.
d. Explain how to measure voltage and current imbalance.
e. Explain how to troubleshoot control circuits and low-voltage transformers.
f. Explain how to test resistive and inductive loads.
g. Identify and describe the function of various hydronic control system components.

Control circuits can be relatively simple, yet they often challenge new technicians. Understanding a basic control circuit and when devices should be energized or de-energized is an essential skill for an HVACR technician. Troubleshooting begins with understanding a system's sequence of operation.

Troubleshooting HVACR systems covers a wide range of electrical and mechanical problems. Of course, not all problems fit neatly into one of these two categories. For example, a loose or corroded wire may cause a compressor in a cooling system to cycle on and off intermittently. Although the problem is electrical, it creates a mechanical refrigeration problem. Determining why a failure occurred is as important to your success as determining what failed.

The methods used to troubleshoot electrical problems and components are very similar regardless of the type of HVACR equipment being serviced. This section describes concepts and procedures that are common to troubleshooting electrical problems in most systems.

3.1.0 Safety Practices

Practically all electric shocks are due to human error, rather than equipment failure. Nearly all electrocutions are due to someone's failure to observe safety precautions or failure to repair equipment with electrical defects. You must embrace and develop safe electrical work practices to protect yourself and others.

3.1.1 Electrical Safety Precautions

The following is a list of recommended safety practices. Always follow the guidance provided in NFPA 70E, *Standard for Electrical Safety in the Workplace®*, as well as all OSHA guidelines:

- Always turn off power at the main disconnect before working on an electrical circuit or device. Follow the appropriate lockout/tagout (LOTO) procedure for the equipment and jobsite. When working with larger and more complex systems, ensure that all forms of energy are disabled and rendered harmless. This may include sources of hydraulic, pneumatic, steam, and mechanical energy.
- Use a voltmeter to verify that a circuit controlled by a disconnect switch is de-energized. Remember that, even though the power may be switched off, electrical potential still exists at the input side of the disconnect switch.
- Never switch equipment on or off while standing in water or touching a wet surface.
- If it is necessary to work on live electrical wiring, try to only use one hand inside the panel. If a shock is received while using only one hand, current will probably flow through the hand and arm, then down the side and through the feet to the ground. If a shock is conducted through both hands and arms, the electrical path passes through the heart and lungs.
- Use appropriate electrical PPE for the voltage present, such as rubber gloves and insulated boots.
- Use tools with the appropriate electrical insulation in sound condition.
- Remove metal jewelry such as rings and watches when working with electrical circuits.

The *National Electrical Code®* (*NEC®*) provides the minimum requirements for the installation of electrical systems. Use the edition of the *NEC®* currently adopted in your state as your on-the-job reference, in conjunction with local codes. The code specifies the minimum requirements for protecting people and property from electrical hazards. In some areas, different editions of the code may be in use, so be sure to use the edition specified by your employer.

CAUTION

Solid-state devices can be damaged by electrostatic charges. These devices should be handled in the manner specified by the manufacturer. Always ground yourself to discharge any stored static charge before handling solid-state controls.

3.1.2 Precautions When Working with Solid-State Controls

Solid-state equipment and/or components can be damaged if improperly handled or installed. The following guidelines should be followed when handling, installing, or otherwise working with solid-state equipment or components.

WARNING!

Solid-state equipment may contain components that produce potentially hazardous leakage current, even in the Off state. Allow solid-state components to rest for several minutes after they are de-energized, providing time for stored energy in capacitors and similar devices to dissipate.

Static Electricity

Microprocessors and other integrated circuit components are very sensitive to static electricity. When handling circuit boards, avoid touching the components, printed circuits, and connector pins whenever possible. Always ground yourself before touching a PC board.

Disposable grounding wrist straps are sometimes supplied with boards containing electrostatic-sensitive components. The wrist straps consistently bleed away any static electricity your body naturally accumulates. Store unused boards inside special metallized static-shielding bags and/or in a nonconductive tote box (*Figure 44*).

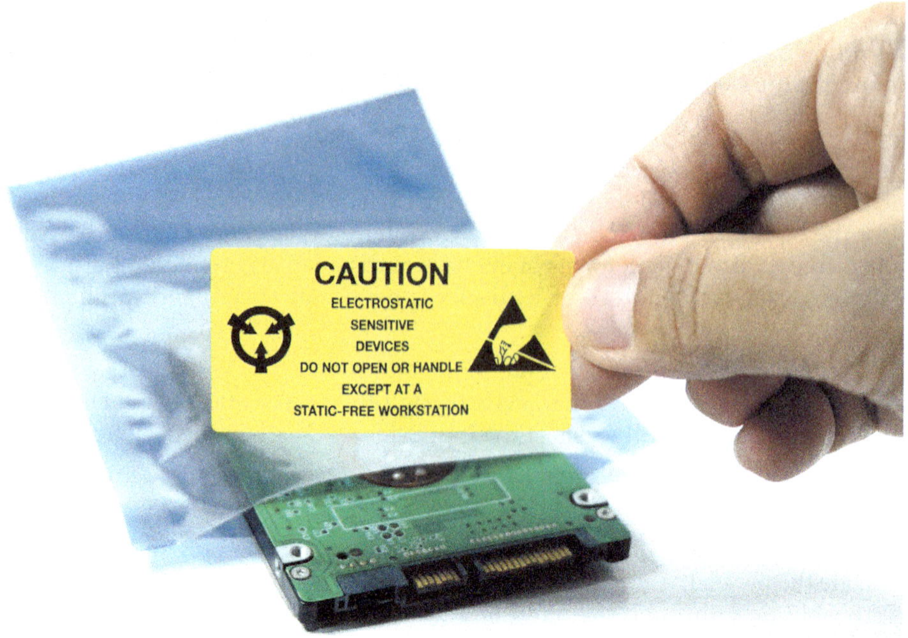

Figure 44 Electrostatic discharge (ESD) shielding bag.
Source: PJ_joe/iStock

Polarity

Solid-state devices can be damaged by the application of power with the incorrect polarity or phase sequence. Input power and control signals must be applied with the polarity and phase sequence as specified by the manufacturer. Be certain that cable plugs are aligned correctly before bringing them together.

Environmental Conditions

Ensure that the solid-state devices are operated within the minimum and maximum ambient operating temperatures as specified by the manufacturer. The relative humidity of the environment may also be a factor.

Follow the manufacturer's recommendations pertaining to the selection of enclosures and ventilation requirements. On existing equipment, make sure that ventilation passages remain clean and open. Also remember that excessive shock or vibration may damage solid-state devices, but the devices often prove more durable than their electromechanical counterparts.

Moisture, corrosive gases, dust, and other contaminants can have adverse effects on components that are not adequately protected against such contaminants. These contaminants can support electrical *bridging* between the circuits on a board. A thick layer of dust interferes with natural cooling.

Electrical Noise

Electrical noise, often caused by other devices, can affect the operation of solid-state controls. Sources of electrical noise include machines with large, fast-changing voltages or currents when energized or de-energized, such as motor starters and welding equipment. The following are basic steps to help minimize electrical noise in control circuits:

- Maintain sufficient physical separation between electrical noise sources and sensitive equipment to ensure that the noise will not cause malfunctions or unintended actuation of the control.
- Maintain a physical separation between sensitive signaling wires and electrical power conductors. For example, low-voltage control and signaling wires should not be placed inside of, or strapped to, conduit carrying line voltage power.
- Use shielded wire (*Figure 45*) in signaling circuits to minimize magnetic interference as directed by component manufacturers.

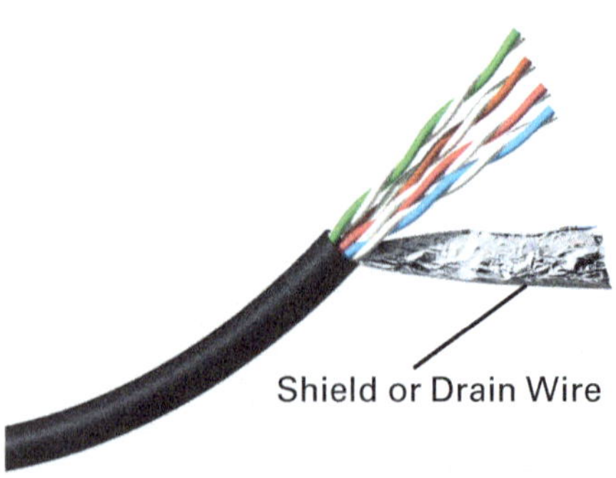

Figure 45 Shielded communication wire.

3.1.3 OSHA Lockout/Tagout Rule

OSHA's LOTO rule covers the procedures to be followed for the "servicing and maintenance of machines and equipment in which the unexpected energization or startup of the machines or equipment, or releases of stored energy, could cause injury to employees." The standard establishes minimum performance requirements for the control of hazardous energy.

The purpose of the OSHA procedure is to make sure that machinery is isolated from all hazardous energy sources, and those sources are locked out and tagged before employees perform service or maintenance activities. Remember that OSHA rules exist for your protection and the protection of your co-workers.

NOTE

The OSHA procedure provides only the minimum requirements for LOTO procedures. Workers should always consult and follow the LOTO procedures provided by their employer and jobsite. Remember that lives depend on these procedures.

LOTO Procedure

To prepare for LOTO, locate and identify all the relevant isolating devices. Be certain which switch(es), valve(s), or other energy-isolating devices apply to the equipment to be serviced. Remember that more than one energy source may be involved.

The following procedure outlines the general steps for LOTO:

NOTE

This procedure is provided only as an example. Each employer can designate who is qualified to conduct the procedure on behalf of the company. Not every employee has the authorization to do so in some companies and jobsites.

Step 1 Notify all affected workers of the LOTO and the purpose. The authorized worker must know the types of energy that are relevant and how to disable them.

Step 2 If the machine or equipment is running, shut it down using the normal procedure. For example, use the thermostat to shut off an air conditioning system before de-energizing the system.

Step 3 Operate the switch, valve, or other energy-isolating device(s) so that the equipment is isolated from its energy source(s). Any stored energy, such as pressure in a hydraulic reservoir, must be dissipated or restrained by repositioning, blocking, or bleeding down. Examples of stored energy are springs, elevated machine members, rotating flywheels, hydraulic pressure, and air, gas, steam, or water pressure.

Step 4 Lock out and tag the energy isolating devices with assigned lock(s) and tag(s).

Step 5 After making sure that no workers are exposed, operate the start button or other normal operating controls to ensure the equipment doesn't operate. Then return the operating controls to the neutral or Off position.

WARNING!

Never try to remove a lock or tag from any switch, valve, or energy-isolating device without the specific consent of the party installing the devices.

After the service is complete, the equipment is ready to be released. Use the following procedure to restore the machines or equipment to their normal operating condition:

Step 1 Check the area around the equipment to make sure that everyone is at a safe distance.

Step 2 Remove all tools from the machine or equipment and reinstall any guards. Make a final visual inspection to make sure all tools and parts are clear.

Step 3 Remove the LOTO devices.

Step 4 Operate the energy isolating devices to restore energy to the machine or equipment as needed.

3.2.0 System Operating Sequences

The system operating sequence determines the layout of a control circuit, as each step must be executed in the proper order at the proper time. When sitting down to design a control circuit, the first thing an engineer needs to know is the needed sequence of operation. HVACR technicians also need to understand and envision the sequence of operation to successfully troubleshoot a system.

An automatic cooling control circuit can be quite simple, and thus it is a good place to begin an analysis of control circuits. *Figure 46* shows a very basic

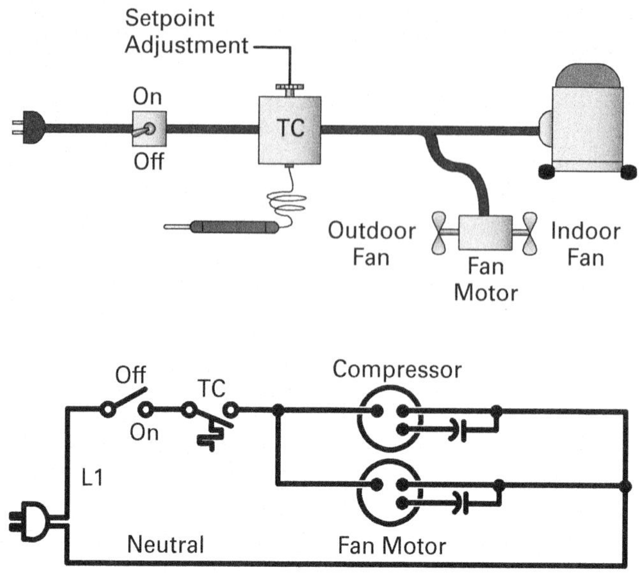

Figure 46 Basic window-unit control circuit.

cooling control circuit that might apply to an old window unit. It operates with the following sequence of operation:

1. Power through the power cord is applied to the On-Off switch.
2. When the switch is in the On position, power is passed to the cooling thermostat (TC).
3. When the thermostat closes on a rise in temperature, both the compressor and fan motor are energized.
4. When the thermostat opens on a fall in temperature, both motors are de-energized.

Analyze the control diagram in *Figure 46* and visualize that operating sequence taking place. Note that the neutral conductor in this 120VAC power source always remains unbroken; only the energized conductor is switched to make a particular action occur.

Figure 47 shows a control circuit that is more typical of a simple residential system. Some additional features have been added to this circuit. A transformer has been added because 24VAC is required for the control circuit. This makes the control circuit components and wiring less expensive, and easier to install and service. An indoor fan motor (IFM) has also been added to the circuit. Because the control circuit is now operating at 24VAC, the control relays (C and IFR) are added to provide control for the motors. The following is the sequence of operation of the circuit shown in *Figure 47*:

1. Line voltage is applied at terminals L1 and L2. Voltage from L1 is present at one set of contacts of compressor contactor (C), the contacts of indoor fan relay (IFR), and the primary winding of the control transformer. Voltage from L2 is present at the other set of contacts of compressor contactor (C), the outdoor fan motor (OFM), and the indoor fan motor (IFM), and the primary winding of the system transformer. The transformer is energized and producing 24VAC as long as line voltage is applied.
2. With the transformer energized, power from the 24VAC Common lead of the secondary winding is applied to the coils of the indoor fan relay (IFR) and compressor contactor (C). The other secondary winding lead, identified as R, applies power to the Fan terminal of the thermostat Fan switch and to the thermostat On-Off switch.
3. When the Fan switch is in the On position, power is applied through the first control circuit path to the coil of the indoor fan relay (IFR), energizing the relay. The IFR contacts to close, applying line voltage to the indoor fan motor (IFM);

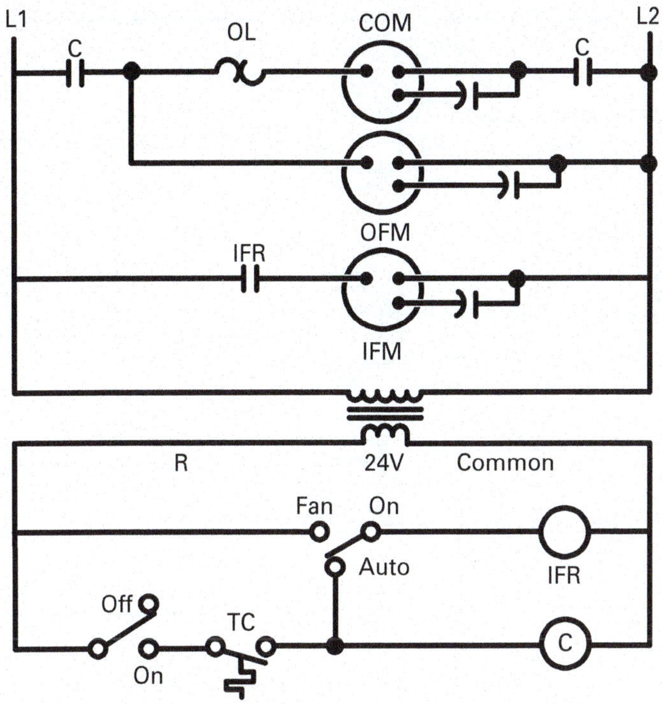

Figure 47 Basic cooling system control circuit.

it starts in response and runs continuously. When the Fan switch is in the Auto position, the IFR coil is energized only when the thermostat calls for cooling.

4. With the thermostat On-Off switch closed, the setpoint below room temperature, and the Fan switch in the Auto position, the contactor coil (C) is energized along with the indoor fan relay coil (IFR). With the contactor energized, the Compressor (COM) and the outdoor fan motor (OFM) are energized. The indoor fan relay (IFR) contacts close, and the indoor fan motor (IFM) starts in response.

On Site

Running the Indoor Blower Consistently

Some users like to operate the indoor fan constantly, and at certain times this helps create a more even temperature in the building. However, there is a negative effect to operating the indoor fan constantly during the cooling mode.

During cooling operation, moisture condenses on the evaporator coil and collects in the condensate drain pan beneath it. Water is still present on the coil face and in the pan when the compressor shuts down. With the blower continuing to run, the moisture is quickly re-evaporated by the air stream and returned to the conditioned space. This can significantly increase the relative humidity in the conditioned space.

This sequence of operation may be a little more detailed than those you'll see in the manufacturer's literature, but examining the diagram as you read it will help you visualize all these actions taking place.

3.2.1 Packaged Heating and Cooling Unit Sequence

A wiring diagram for a packaged heating/cooling unit is shown in *Figure 48*. This circuit is more complex and closer to what you will encounter, but at its heart is the simple circuit previously described. Heating controls have been added near the bottom of the diagram.

The features of this heating/cooling unit include the following:

- A single room thermostat provides temperature control for both heating and cooling.
- The cooling control circuit contains compressor short-cycle protection and a current-sensitive overload device.

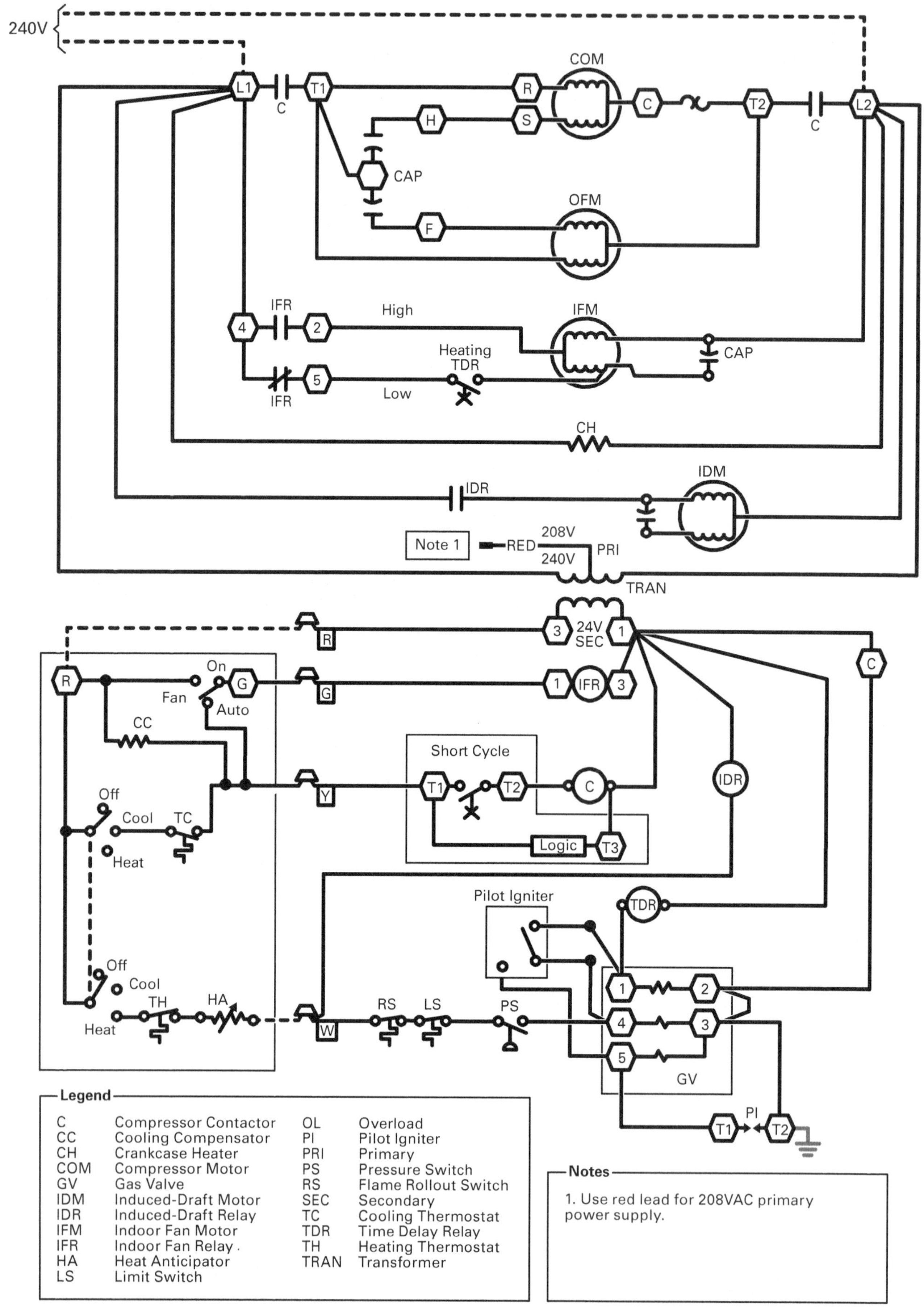

Figure 48 Wiring diagram for a packaged heating/cooling unit.

- A two-speed indoor fan that runs at its high speed for cooling and at its low speed for heating. In heating, the time-delay relay controls the indoor fan.
- Operation of the inducer fan must be proven before the gas valve is energized. The inducer pressure switch (PS) closes when the induced-draft motor is running at the required speed. If the induced-draft fan stops running, the pressure switch will open, de-energizing the gas valve. The induced-draft motor (IDM) is energized by the induced-draft relay (IDR) as soon as the thermostat closes.
- The heating section has two additional safety devices in series with the gas valve. A flame rollout switch (RS) will open if burner flames escape from the combustion chamber. The limit switch (LS) is a heat-sensitive switch that disables burner operation if the heat exchanger becomes too hot, which usually indicates an airflow problem.

Refer to the circuit diagram in *Figure 48* to follow the cooling and heating operating sequences that follow.

Cooling Sequence of Operation

Assume that the line voltage is adequate and available to the unit. This also means that the 24VAC control transformer and the crankcase heater (CH) are energized. The room thermostat mode selector must be switched to Cool. The Fan switch is set to Auto so that the indoor fan cycles on when the thermostat calls for cooling. The thermostat starts the indoor fan by transferring power from the R terminal to both Y and G on a call for cooling.

1. As the room temperature rises, thermostat contacts (TC) close, supplying 24VAC power to the coil of the indoor fan relay (IFR) and to the logic circuit of the compressor short-cycle timer. The indoor fan relay energizes, closing the NO IFR contacts at terminals 2 and 4. This provides line-voltage power to the indoor fan motor (IFM), which then starts and runs at its high speed. At the same time, the NC IFR contacts at terminals 4 and 5 open. This isolates the low-speed windings of the indoor fan motor that are used for the heating mode, preventing any possibility of the motor having both windings energized.
2. Once the logic in the short-cycle timer times out (2 to 5 minutes), the contacts across terminals T1 and T2 close to energize the contactor (C). On the line-voltage side of the circuit, contactor contacts L1-T1 and L2-T2 close to supply power to the compressor (COM) and outdoor fan motor (OFM). The outdoor fan motor is energized whenever the compressor is energized, except when there are multiple condenser fans with one or more motors staged.
3. The compressor, outdoor fan motor, and indoor fan motor will continue to run until the room thermostat is satisfied. When the thermostat contacts (TC) open, the contactor and indoor fan relay are both de-energized and all motors shut down.

Heating Sequence of Operation

To start heating operation, set the room thermostat mode selector switch to Heat. Set the Fan function switch to Auto. However, in a gas heating system, the furnace enables the indoor blower motor. In other words, the G terminal at the thermostat will not be energized along with W when the thermostat calls for heating.

1. As the room temperature falls, thermostat contacts (TH) close, supplying 24VAC power to the coil of the induced-draft relay (IDR). In the line-voltage side of the circuit, the NO contacts of the induced-draft relay (IDR) close, applying power to the induced-draft motor (IDM).
2. When the induced-draft motor reaches sufficient speed and is moving enough air, the pressure switch (PS) closes to energize the gas pilot valve (GV) and pilot igniter (PI). The gas pilot valve delivers gas to the pilot assembly where a high-voltage arc from the pilot igniter ignites it. Once the pilot igniter module verifies that the pilot is sufficient through its flame rectification sensor

Think About It

Analyzing a Circuit Diagram

Use the diagram in *Figure 48* to determine the answers to these questions.

1. Is the induced-draft motor de-energized if the flame rollout switch (RS) opens? Why or why not?
2. If you connect a voltmeter between L2 and terminal 4 of the IFR while the compressor contactor is de-energized, what voltage, if any, would the meter read?
3. Is the indoor fan relay (IFR) energized or de-energized in the heating mode?

to light the main burners, a switch closes to energize the main gas valve. It opens to provide gas to the burners and the main burners light.

3. When the main gas valve is energized by the pilot igniter, the time-delay relay (TDR) is energized and begins timing. On the line-voltage side of the circuit, the TDR contacts are in series with the NC contacts of the indoor fan relay (IFR) and the indoor fan motor (IFM). The purpose of the time delay is to allow the heat exchanger time to warm up so that cold air is not blown into the structure. Once the delay elapses, the TDR contacts close, and the indoor fan motor starts and runs at low speed.
4. The unit will continue to run in this manner until the room thermostat is satisfied. When thermostat contacts TH open, the gas valve and induced-draft relay are de-energized, shutting off the burners. The time-delay relay is also de-energized, and an off time-delay starts; the contacts remain closed until the off delay has elapsed. This allows the indoor fan motor to continue to run after the burner shuts off to make use of the residual heat left in the heat exchanger. Once the off delay has elapsed, the indoor fan motor is de-energized, ending the cycle.

On Site

Pneumatic Controls

Pneumatic control systems use compressed air to supply energy for the operation of valves, motors, relays, and the related control components. They were used primarily in large commercial and industrial systems. Pneumatic control circuits consist of air piping, valves, orifices, and similar mechanical devices. Thermostats control the pressure in the air lines to downstream controls. Using pneumatic-electric (PE) and electric-pneumatic (EP) switches, air can be used to control electrical devices, and vice versa.

Pneumatic control systems were once popular in commercial office buildings, but the systems are expensive to install and more challenging to service than electronics, which require far less attention. However, pneumatic control systems are still used in a few unique applications.

Shown here are a pneumatic damper actuator and a pneumatic valve actuator. Most devices of this type have an internal spring that positions the valve or actuator to a default position. Air pressure is then used to overcome the spring pressure and reposition the device.

3.3.0 An Organized Approach to Electrical Troubleshooting

Troubleshooting can be defined as a procedure through which a technician locates the source of a problem. Once the problem is identified, the technician then makes the repairs and/or adjustments to correct the cause of the problem.

Most problems can be identified quickly when a systematic approach is used. Troubleshooting can be divided into the five basic parts:

- Customer interviews
- Physical examination of the system

- Basic system analysis
- Use of manufacturer's troubleshooting aids
- Fault isolation in equipment problem area

3.3.1 Customer Interviews

The troubleshooting procedure should begin with the technician learning as much as possible from the customer. This preliminary discussion can provide valuable information that identifies where to begin troubleshooting.

The first evidence of trouble is often a complaint from one or more people who are too cold, too hot, or bothered by drafts. In many cases, the problem is not related to a breakdown or component failure. If the system appears to be operating correctly but individuals are complaining about comfort, check for one or more of the following conditions:

- *Air distribution and circulation problems* — People near the perimeter of an area controlled by a thermostat may feel discomfort. The thermostat senses only the temperature at its location (unless it has optional remote sensors). Temperatures farther from the thermostat are subject to greater variation. Running the blower continuously when the area is occupied can help, although the draft it might create could also work against you. When possible, redirect the airflow from supply air grilles and diffusers when the draft is objectionable.
- *False heat loads* — Direct sunlight or heat from lamps, appliances, and pipes can make the thermostat warmer than its environment. In the heating mode, this results in a colder space. In the cooling mode, the area is likely to be overcooled.
- *Covered grilles and diffusers* — Occupants frequently cover the face of a supply air grille or diffuser, possibly creating a problem for others.
- *Overcrowding* — Overheating will result if more people or mechanical equipment occupy a conditioned area than the space was designed to hold.
- *Stale air* — A stuffy or smoky atmosphere usually results from an insufficient fresh air introduction, air that is too humid, or air that has inadequate exhaust. Stale air can also be caused by overcrowding or low air circulation. Check the area outdoors around the outside air intake when unusual odors persist.
- *System size* — Extreme weather conditions may exceed the capacity of the heating or cooling equipment. The equipment may also be oversized for load, which results in short run cycles. Oversized cooling equipment often results in poor humidity control. Running the indoor fan constantly in the cooling mode can also cause excessive humidity, as moisture collected at the evaporator coil is returned to the space when the compressor is not running.

3.3.2 Physical Examination of the System

Many problems can be identified by simply using sight, touch, and smell to check a system. As you look the system over, inspect for the following conditions:

- Signs of physical damage or tampering
- Oil stains on coils or tubing that often indicate a refrigerant leak
- Dirt accumulation on filters and coils that can affect system operation
- Unusual sounds and vibrations that could indicate a problem
- Unusual odors, especially for the odor of the heating fuel (gas or oil) and burned electrical components that typically produce strong, acrid odors

3.3.3 Basic System Analysis

The second part of the diagnosis is to find out what the system is actually doing and what symptoms it exhibits. If the system is running, make electrical, temperature, pressure, and/or airflow measurements at key points in the system. The measured values can then be compared with typical readings for a properly operating system as previously recorded on system operating logs or in the manufacturer's service literature. This simple process can often help identify where troubleshooting should begin.

3.3.4 Using Manufacturers' Troubleshooting Aids

To aid in the isolation of faults, many manufacturers provide troubleshooting information marked on the equipment or contained in the service instructions for a particular product. This information typically includes the following:

- **Label diagrams**
- Troubleshooting tables
- **Fault isolation diagrams**
- Diagnostic equipment and tests

Label diagrams: Troubleshooting aids placed in a convenient location inside the equipment. These diagrams typically consist of a wiring diagram, a component arrangement diagram, a legend, and notes pertaining to the equipment and optional components.

Fault isolation diagrams: Troubleshooting aids often contained in the manufacturer's installation, operation, and maintenance (IOM) instructions for equipment. Fault isolation diagrams typically outline failure symptoms and guide the technician through a logical decision and action process to isolate the cause of failure.

Label diagrams (*Figure 49*) are usually placed in a convenient location inside the equipment, such as on the inside of a service access panel. Label diagrams normally show a component arrangement diagram, wiring diagram, legend, and notes pertaining to the equipment.

The component arrangement diagram shows where the components are physically located in the unit. It is useful because it helps locate and identify the components shown on the wiring diagram. The legend identifies the meanings of the symbols and abbreviations used on the label diagram.

The wiring diagram, also called a *schematic*, provides a picture of how the unit functions electrically and shows the actual external and internal wiring of the unit. Another schematic is shown in *Figure 50*. Schematic wiring diagrams and ladder diagrams are primary troubleshooting aids for isolating electrical problems.

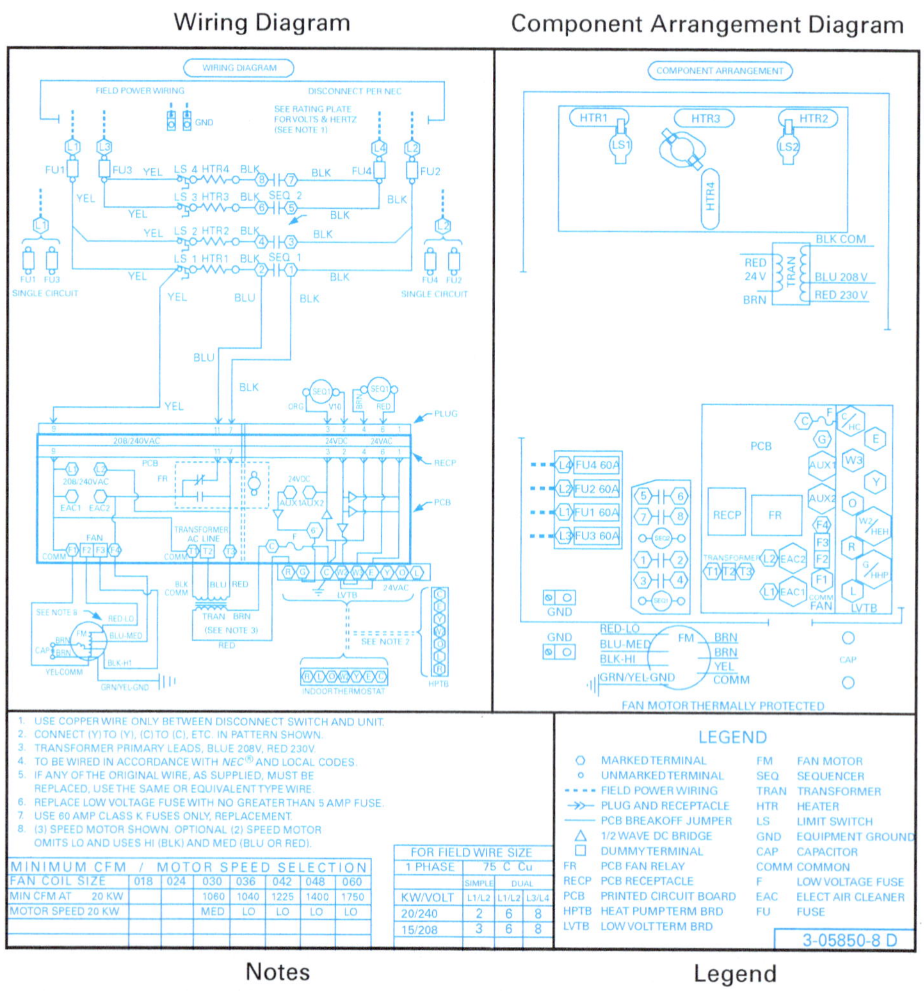

Figure 49 Typical label diagram.
Source: Carrier Corporation

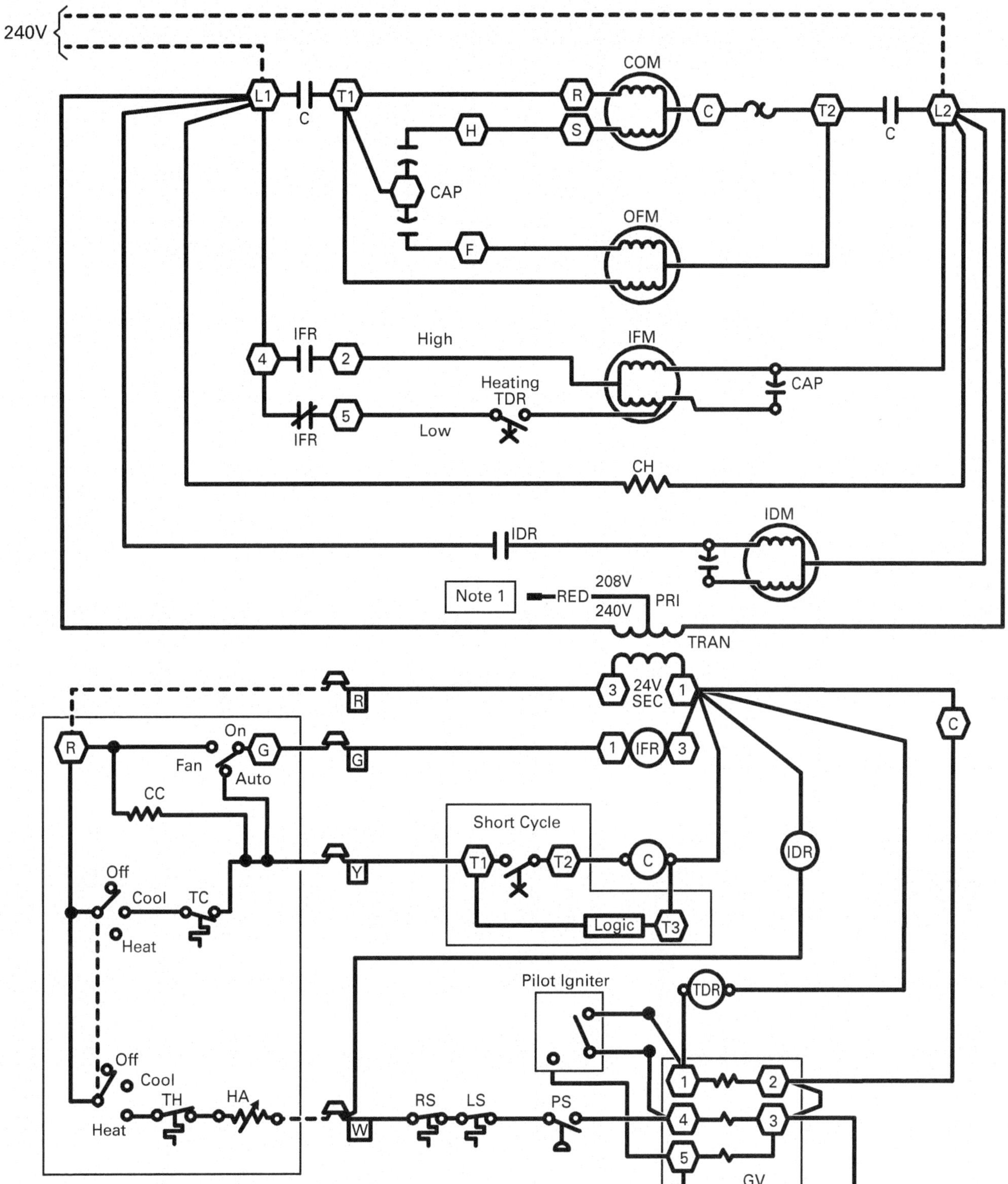

Legend

C	Compressor Contactor	OL	Overload
CC	Cooling Compensator	PI	Pilot Igniter
CH	Crankcase Heater	PRI	Primary
COM	Compressor Motor	PS	Pressure Switch
GV	Gas Valve	RS	Flame Rollout Switch
IDM	Induced-Draft Motor	SEC	Secondary
IDR	Induced-Draft Relay	TC	Cooling Thermostat
IFM	Indoor Fan Motor	TDR	Time Delay Relay
IFR	Indoor Fan Relay	TH	Heating Thermostat
HA	Heat Anticipator	TRAN	Transformer
LS	Limit Switch		

Notes

1. Use red lead for 208VAC primary power supply.

Figure 50 HVAC unit schematic electrical diagram.

Troubleshooting tables and/or fault isolation diagrams are usually contained in the manufacturer's installation, startup, and service instructions for a particular product. Like the example shown in *Figure 51*, troubleshooting tables are intended to guide technicians to a corrective action based on observations of system operation. Fault isolation diagrams, as shown in *Figure 52*, also called *troubleshooting trees*, normally start with a symptom and guide technicians through a logical decision-action process to help isolate the failure.

Many manufacturers incorporate automatic or semi-automatic testing features into their equipment to help isolate malfunctions. Depending on the equipment, these built-in diagnostic devices can be simple or complex. Some units contain microprocessor controllers that can run a check of primary system functions, and then report the results through a numeric or flashing LED or LCD display. These built-in test functions normally isolate the fault to a functional problem area rather than isolating the problem to a specific component.

For example, if a test indicates a compressor failure, this means only that the failure has been isolated to the compressor and its related electrical components. If it isn't running when it should, there are many possible causes beyond a defective compressor. The technician must perform additional troubleshooting to identify the real fault.

In addition to built-in diagnostics, manufacturers have developed modular electronic testers that are available to troubleshoot specific components in the manufacturer's product line. ECM motor testers are one example. Use modular testers according to the instructions provided.

Malfunction	Probable Cause	Corrective Action
Compressor motor and condenser motor will not start, but fan/coil unit (blower motor) operates normally.	Check the thermostat system switch to ascertain that it is set to COOL.	Make necessary adjustments to settings.
	Check the thermostat to make sure that it is set below room temperature.	Make necessary adjustments.
	Check the thermostat to see if it is level. Most thermostats must be mounted level; any deviation will ruin their calibration.	Remove cover plate, place a spirit level on top of the thermostat base, loosen the mounting screws, and adjust the base until it is level; then tighten the mounting screws.
	Check all low-voltage connections for tightness.	Tighten.
	Make a low-voltage check with a voltmeter on the condensate float switch; the condensate may not be draining.	The float switch is normally found in the fan/coil unit. Repair or replace.
	Low air flow could be causing the trouble, so check the air filters.	Clean or replace.
	Make a low-voltage check of the antifrost control.	Replace if defective.
	Check all duct connections to the fan-coil unit.	Repair if necessary.
Compressor, condenser, and fan/coil unit motors will not start.		Adjust as necessary.

Figure 51 Example troubleshooting table.

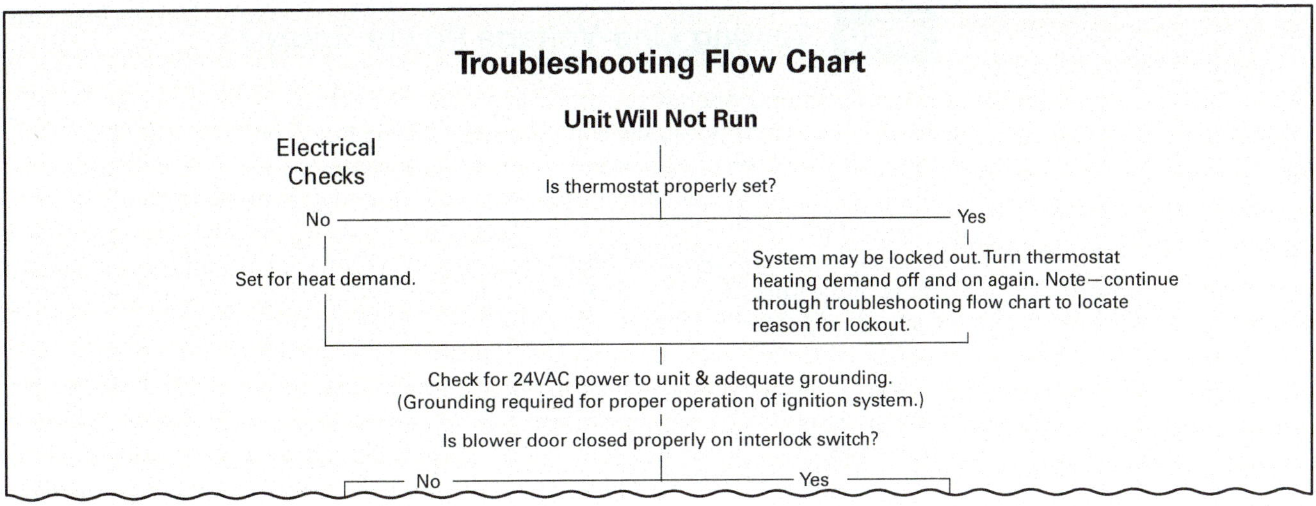

Figure 52 Example fault isolation diagram.

Think About It

Wiring Diagram Exercise

Using the schematic diagram of the packaged heat pump included here, answer the following questions:

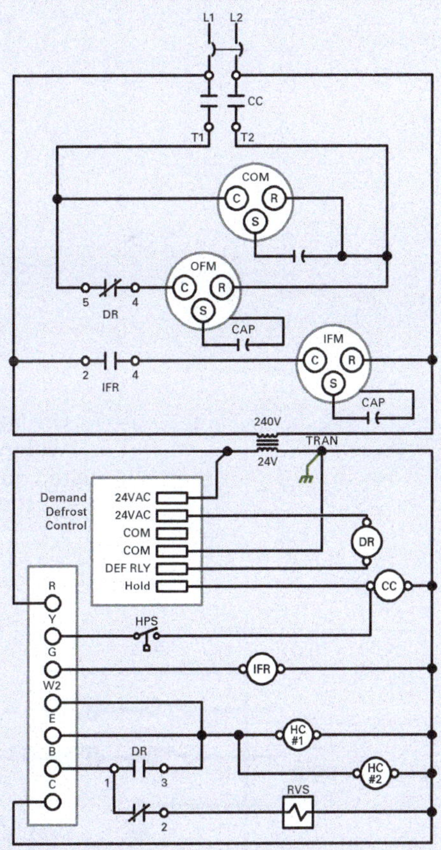

1. Where would voltmeter leads be placed to check for the correct incoming voltage to the unit?
2. Placing voltmeter leads across terminals R and C of the low-voltage terminal strip yields a reading of 24VAC. What does this indicate?
3. During defrost, the defrost relay (DR) is energized. During a normal defrost, voltmeter leads are placed across terminal 2 of the defrost relay and terminal X on the low-voltage terminal strip. What should the voltmeter read?
4. Assume a room thermostat is connected to the low-voltage terminal strip. Across which terminals on the low-voltage terminal strip would voltmeter leads have to be placed to confirm that a call for second-stage heat was coming from the room thermostat? What voltage should be measured?
5. During defrost, the defrost relay (DR) is energized. Knowing this, what happens to the outdoor fan motor during defrost?
6. It must be known if the demand defrost control is receiving the correct input power. Across which terminals of the demand defrost control should power be checked? What is the expected voltage reading?
7. The compressor in this unit occasionally won't start and trips the power supply circuit breaker. Compressor current draw during startup must be checked. What test instrument is required and where in the circuit would the instrument be attached or applied?
8. HC #1 and HC #2 relays control the operation of electric resistance heating elements. How do those relay coils receive power during a defrost cycle? Are both energized?
9. If the compressor fails in this heat pump, it is still possible to heat the home. How does the occupant make this happen? At what terminals on the low-voltage terminal strip would you expect to find 24VAC that proves the thermostat is calling for the electric heat to operate?
10. Since this unit energizes the defrost relay during heat pump defrost operation, what does this fact tell about the state of the reversing valve solenoid (RVS) during cooling operation? Is the RVS energized or de-energized during the cooling mode?

3.4.0 Testing Line-Voltage Power Sources

HVACR systems operate on either single-phase or three-phase line-voltage circuits. Power circuits usually consist of the field-installed power wiring from the main electrical service (breaker panel) to a disconnect switch located near the unit, and from the disconnect switch to the equipment. The distribution circuits include protective devices, such as fuses and/or circuit breakers.

All HVACR equipment is designed to operate within a specific range of voltages including a safety factor, which is typically ±10% of the design voltage. This safety factor is added to compensate for temporary voltage fluctuations that might occur.

For example, when the contactor closes in a condensing unit and applies power to inductive loads such as the compressor and outdoor fan motor, the voltage may drop a few volts from the measured open-circuit voltage (circuit unloaded). This is due to the increased current flow to start the motors. This is expected as long as the voltage does not drop below the manufacturer's stated minimum voltage. Once the motors are up to the operating speed, the current demand drops sharply, and the voltage rises closer to the open-circuit voltage.

An operating voltage that is too high or too low can cause overheating and the failure of motors and other inductive loads. Resistive loads are not affected as much. The power supply voltage applied to the equipment should be measured and checked against the supply voltage indicated on the unit nameplate (*Figure 53*).

If a low voltage condition exists at the equipment, the voltage should be measured at the electrical service entrance to make sure that the problem doesn't exist in the branch circuit or feeder that supplies the equipment. Excessive voltage drop in a branch circuit under load often indicates that the conductors are too small for the load. A licensed electrician should evaluate the wire size in this case. Loose connections at the circuit breaker panel can also cause an excessive voltage drop under load.

Operating voltages must be within the range of values shown on the motor nameplate. The voltage tolerances used for most HVACR motors are as follows:

- *Single-voltage rated motors* — The input supply voltage should be within ±10% of the nameplate voltage. For example, a motor with a nameplate single voltage rating of 230V should have an input voltage that ranges between 207V (–10% of 230V) and 253V (+10% of 230V).

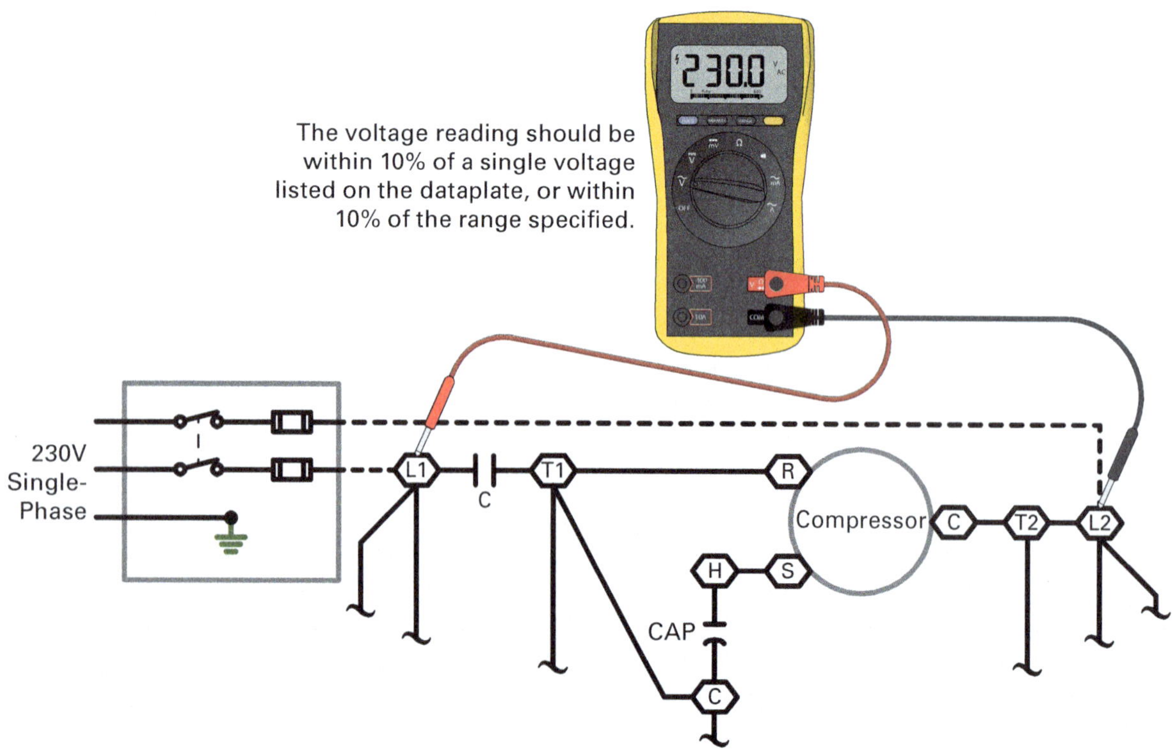

Figure 53 Single-phase voltage test.

- *Dual-voltage rated motors* — The input supply voltage should be within ±10% of the nameplate voltage. For example, a motor with a nameplate dual voltage rating of 208/230V should have an input voltage that ranges between 187V (–10% less than 208V) and 253V (+10% more than 230V). It is important to note, however, that a reduced voltage results in an increase in operating current for inductive loads.

3.4.1 Voltage and Current Imbalance

Voltage imbalance is important when working with three-phase equipment. A small imbalance in the phase-to-phase voltage provided results in a much greater current imbalance. The resulting current imbalance increases the heat generated in the motor windings. Both excessive current and heat causes overload relays to trip and reduce motor life.

For this reason, the voltage imbalance between any two legs of a three-phase power source should not exceed 2%. If a voltage imbalance of more than 2% exists, the problem must be corrected by licensed electricians or the local utility before operating the equipment.

Figure 54 shows how the percentage of voltage imbalance is determined in a three-phase system using the following formula:

$$\text{Voltage Imbalance}(\%) = \frac{\text{Maximum Deviation from Average}}{\text{Average Voltage}} \times 100$$

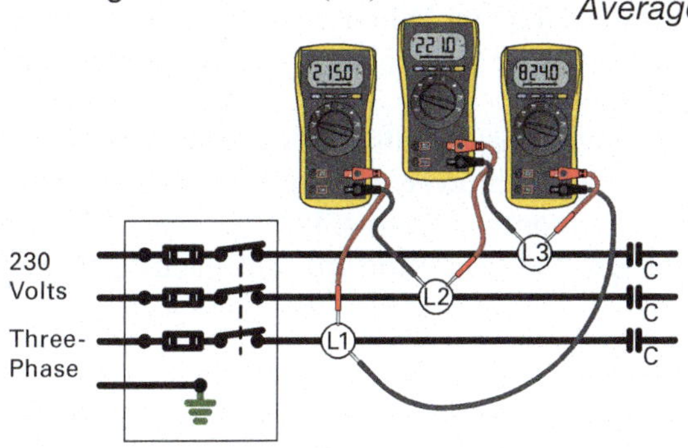

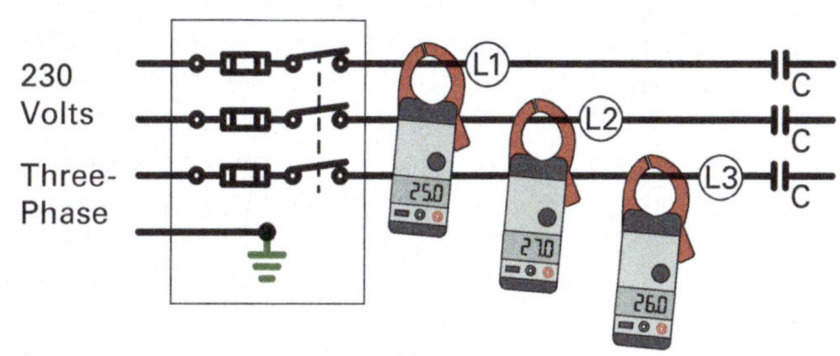

1)

Phase	**Measured Reading**
L1-L2	215V
L2-L3	221V
L3-L1	224V

2) Average Voltage $= \frac{215 + 221 + 224}{3} = 220V$

3) Individual Phase Imbalance from Average:
L1 to L2 = 220 – 215 = 5V
L2 to L3 = 221 – 220 = 1V
L3 to L1 = 224 – 220 = 4V

4) 5V = Maximum Imbalance

5) % Imbalance $= \frac{\text{Maximum Imbalance}}{\text{Average Voltage}} \times 100$

% Imbalance $= \frac{5V}{220V} \times 100 = 2.27\%$ **(Out of Balance)**

Maximum voltage imbalance between any two legs must not exceed 2%.

(A) Calculating Voltage Imbalance

1)

Phase	**Measured Reading**
L1	25A
L2	27A
L3	26A

2) Average Current $= \frac{25 + 27 + 26}{3} = 26A$

3) Individual Phase Imbalance from Average:
L1 = 26 – 25 = 1A
L2 = 27 – 26 = 1A
L3 = 26 – 26 = 0A

4) 1A = Maximum Imbalance

5) % Imbalance $= \frac{\text{Maximum Imbalance}}{\text{Average Current}} \times 100$

% Imbalance $= \frac{1A}{26A} \times 100 = 3.87\%$ **(Acceptable)**

Maximum current imbalance between any two legs must not exceed 10%.

(B) Calculating Current Imbalance

Figure 54 Three-phase voltage and current tests.

The percentage of current imbalance is calculated using the same equation but with current replacing voltage:

$$\text{Current Imbalance}(\%) = \frac{\text{Maximum Deviation from Average}}{\text{Average Current}} \times 100$$

The current imbalance between legs of a three-phase system should never exceed 10%. Note that a current imbalance can occur without a voltage imbalance present. This can occur when an electrical terminal or contact becomes loose or corroded, causing a high resistance in that leg of power. Since current follows the path of least resistance, the current in the other two legs will increase to compensate, causing more heat to be generated in the load.

NOTE

The system must be operating under a normal load when voltage and current imbalance tests are performed.

3.5.0 Control Circuits and Low-Voltage Power Sources

Control circuits provide a link between loads and the input power. Control circuits start, stop, or otherwise control the operation of a load. Control circuits contain one or more control devices, such as relays, contactors, and thermostats that energize or de-energize loads and other control devices. The more complex the system, the more control devices it will have. When a load such as a motor is not working, it must be determined whether the problem is with the load itself, or in the circuits controlling the load.

Control circuits are generally low-voltage circuits that operate at 24VAC. In some larger systems using three-phase line power, a control voltage of 120VAC or higher may be used for some of the control circuits. For this reason, always be certain of the control circuit voltage. Never assume the control circuit voltage is a low voltage.

3.5.1 Isolating to a Faulty Circuit

Isolating a faulty control circuit can be relatively easy based on an analysis of the symptoms and the process of elimination.

For example, on a service call related to the packaged heat pump system with its ladder diagram shown in *Figure 55*, the customer complains that the unit is switched to the cooling mode and run, but it is blowing warm air instead of cool air. A preliminary check of system operation reveals that the compressor is not running when the thermostat is calling for cooling.

Begin fault isolation through the process of elimination. Because the indoor fan motor runs, you know the unit is powered. Next, find out if the compressor will run in the heating mode by setting the thermostat to call for heating.

If it does start and run, then the compressor load circuit and everything in the compressor contactor control circuit, including the control relay (CR), can be eliminated. It must be related to the cooling control circuit alone, and the thermostat or the interconnecting wiring may be the problem. There aren't many other possibilities. Of course, if the compressor does not run in either the heating or cooling mode, many possible causes exist.

3.5.2 Isolating to a Faulty Circuit Component

Once the source of an electrical problem has been isolated to a particular area or circuit, the next step is to make a series of voltage measurements across the components in the circuit to find the fault.

Figure 56 shows several different testing scenarios on an energized control circuit. Measurements are made until either no voltage is observed, or until the voltage has been measured across all circuit components. When there are many devices in the circuit being tested, the measurements should begin at the midpoint of the circuit, then proceed in either direction depending upon the response (voltage or no voltage). This effectively eliminates half of the possibilities each time the technique is applied. Note that this method of using the voltmeter, with one probe remaining in a fixed position while the other probe is moved from point to point, is sometimes referred to as *hopscotch testing*.

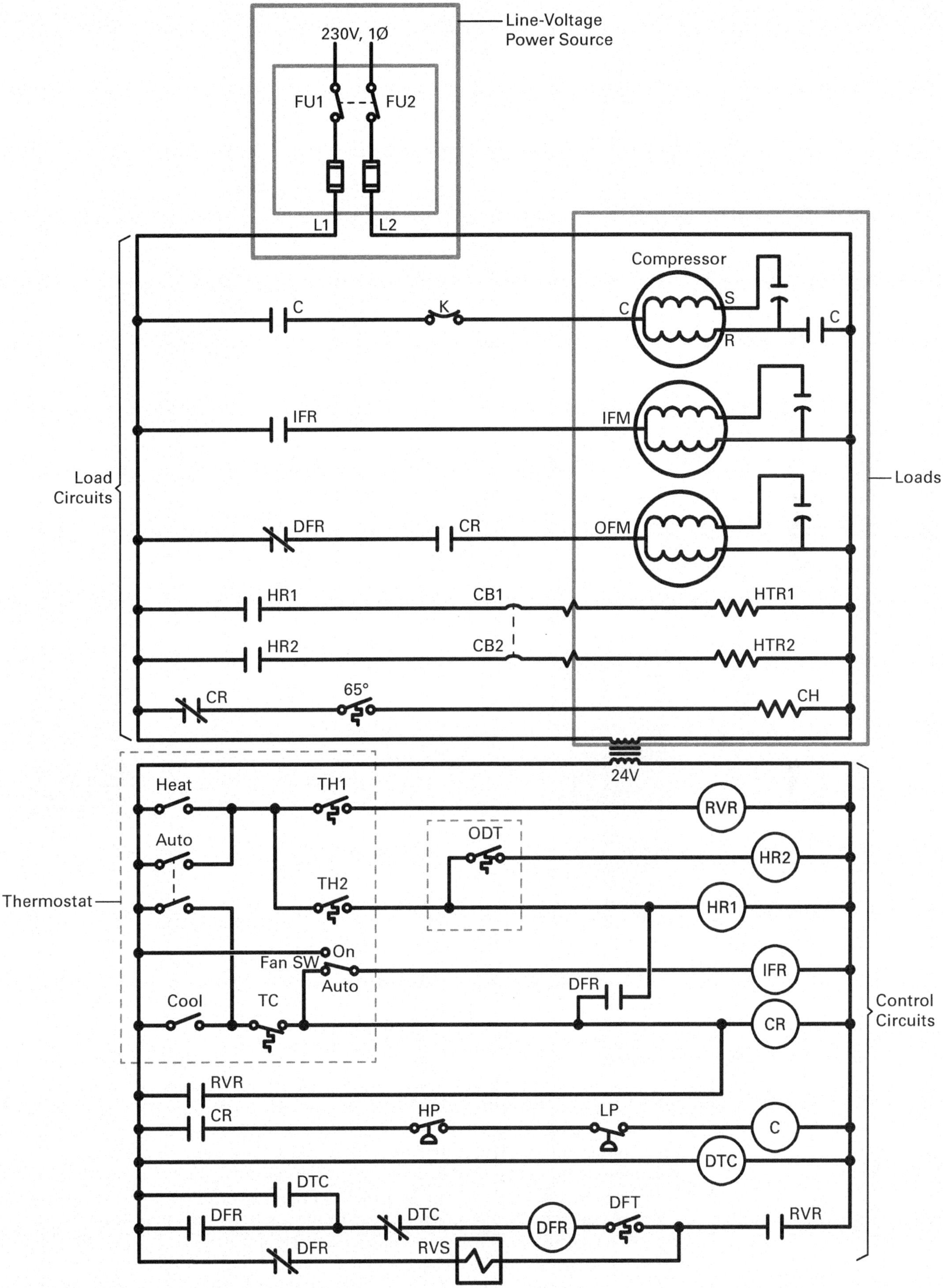

Figure 55 Heat pump ladder diagram.
Source: Carrier Corporation

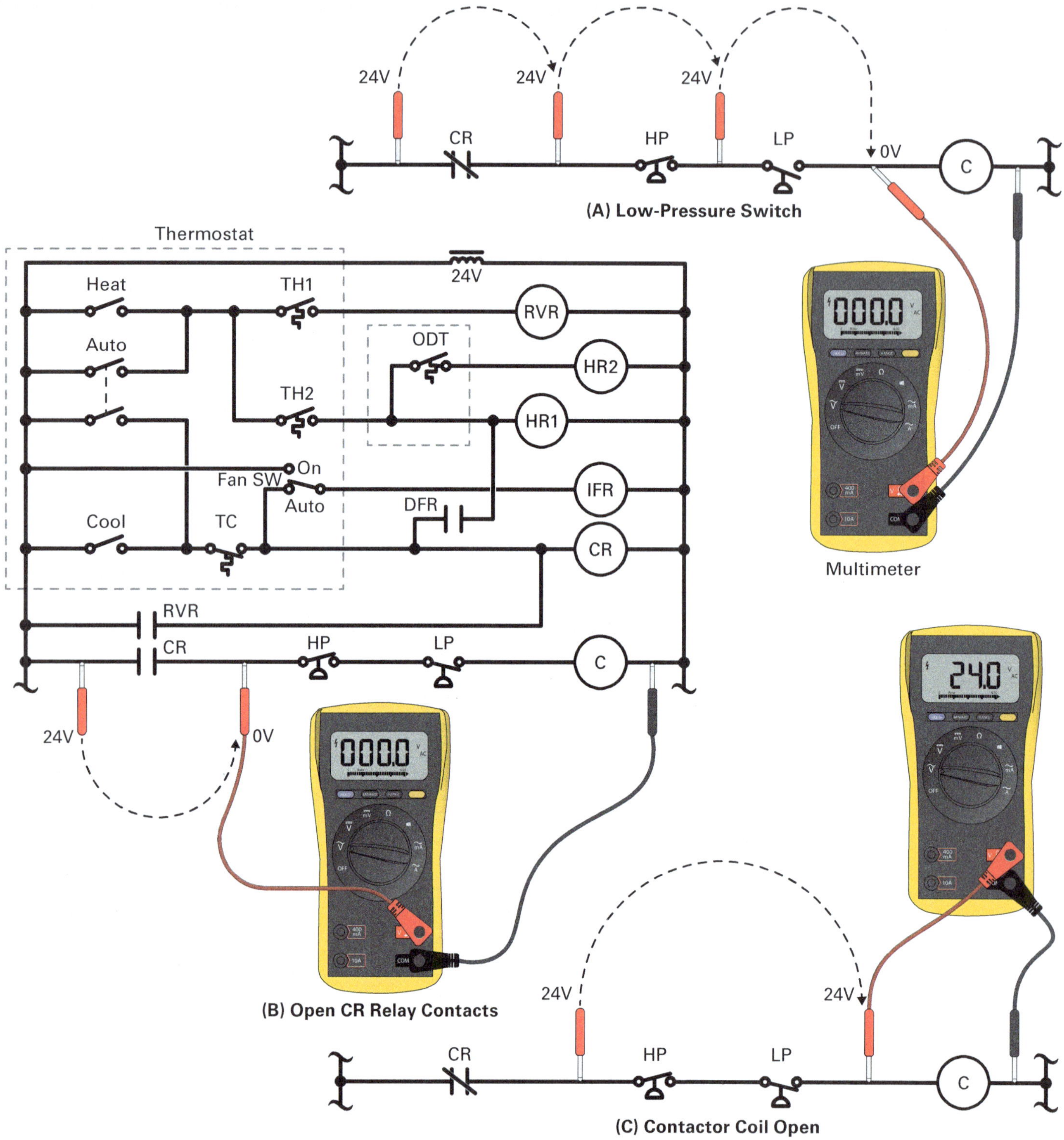

Figure 56 Isolating to a faulty circuit component.

At some point during this test, no voltage will be indicated on the meter. This pinpoints an open circuit between the last measurement point and the previous measurement point. *Figure 56A* is an example of this situation. In this case, the contacts of the low-pressure switch (LP) are open, preventing the compressor contactor coil (C) from energizing. In this case, the next step would be to connect a gauge manifold and check the refrigerant charge.

In *Figure 56B*, the contacts of the control relay (CR) are open, preventing the compressor contactor (C) from being energized. The NO contacts close when the control relay coil (CR) is energized. The next step would be to see if voltage is being applied to the CR coil. If not, then troubleshooting continues with the thermostat and the field-installed control wiring. If voltage is present at the CR coil, it would suggest the relay coil is open or the relay has failed in some other way. Confirm the diagnosis by checking continuity through the coil with the circuit de-energized. Although it is rare, the coil may be intact, while the contacts or tiny relay linkage inside have failed.

Figure 56C shows the readings seen when a relay or contactor coil is open. Voltage is evident all the way through the circuit and at the coil itself. Again, confirm the condition of the coil by checking its continuity with the system de-energized. If it is intact but the contacts do not change position when it is energized, the relay has suffered internal damage to the contacts or contact linkage.

3.5.3 Transformer Testing

Control transformers are usually checked by measuring the voltage at the primary and secondary windings (*Figure 57*). The secondary winding is typically measured first. The multimeter should be set to measure AC voltage on a range that is higher than the expected voltage. If the voltage measured across the secondary winding is within ±10% of the required voltage (21.6 to 26.4V), the transformer is good. However, a reduced voltage, even within the required range, could cause control problems over a significant run of control wiring where the voltage drop may be significant.

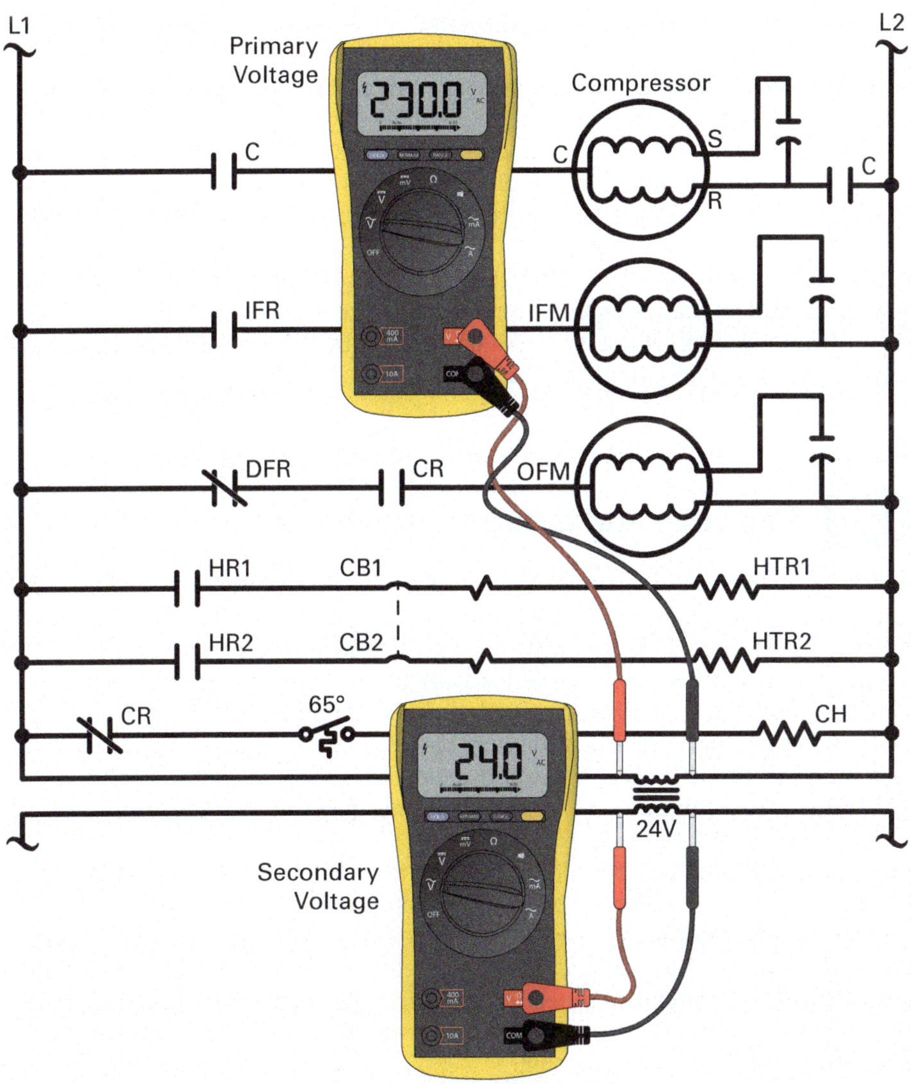

Figure 57 Transformer voltage testing.

If there is no voltage from the secondary, check to ensure the transformer is energized. If it is energized, then the transformer windings must be tested with an ohmmeter while it is de-energized.

Many transformers are equipped with multiple primary winding connections to accommodate different voltages. For example, a multitap transformer may have connections for a 120VAC, 208VAC, and 240V power supply. Four primary leads would be provided on this transformer, one of which is the common lead and is used with any incoming voltage.

The start-up technician must ensure that this connection is correct. For example, if the transformer is factory-connected for 240VAC and the unit's supply voltage is 208VAC, the secondary voltage will be too low. When the secondary voltage is lower than expected, check the wiring connections to the primary winding first and ensure it is wired for the correct line voltage.

If the voltage measured at the primary winding is within ±10% of the required voltage, but the secondary voltage is still low or zero, check the following:

- Check the transformer for overheating or physical damage.
- Ensure that all the electrical connections are sound.
- Check the fuse or circuit breaker in the transformer's secondary (if present and accessible). If the fuse is external, replace it. Reset the circuit breaker if it has opened. If the fuse is buried in the transformer windings, replace the transformer. However, the reason for the overcurrent condition that caused the problem must still be determined.

When there is no voltage produced on the secondary winding, continuity checks will confirm if the primary or secondary windings are open. To perform a continuity check, de-energize the system and disconnect one lead on each winding of the transformer to isolate it from the rest of the circuit. Use an ohmmeter set on a low scale to measure continuity.

Replace any transformer that has open windings. If the transformer is fused internally, do not attempt to replace the fuse. It is generally soldered into the winding. It is best to simply replace the transformer, saving field labor that may not prove productive in the end.

A circuit breaker may also be built-in to the transformer. There will be a small reset button that pops up to indicate the breaker has tripped. However, this type of protection is not common on small transformers and are generally found on transformers rated at 75VA and higher.

3.6.0 Resistive and Inductive Loads

Loads are devices that consume power to do work. Compressor motors, fan motors, heater elements, and the primary winding of transformers are all loads normally found in cooling and heating units. Technically, even the coil of a relay or contactor represents an electrical load, although the load is small when compared to a motor. If power is consumed, the device represents a load.

Crankcase heaters and electric heating elements are two examples of resistive loads found in HVACR equipment. Inductive loads include contactor and relay coils, transformers, and motors. HVACR systems generally contain more inductive loads than resistive loads.

Once a resistive or inductive load has been identified as the probable cause of an electrical problem, it should be tested to confirm its electrical condition. The best way to test these loads is by checking for continuity through their windings at the electrical terminals. Be sure to electrically isolate the component being measured by first disconnecting the wiring leads from the circuit. This is important to achieve an accurate resistance reading, and to ensure power is not accidently applied to the ohmmeter.

As shown in *Figure 58*, a reading of zero ohms usually indicates the windings are shorted, while a reading of infinite resistance indicates an open winding.

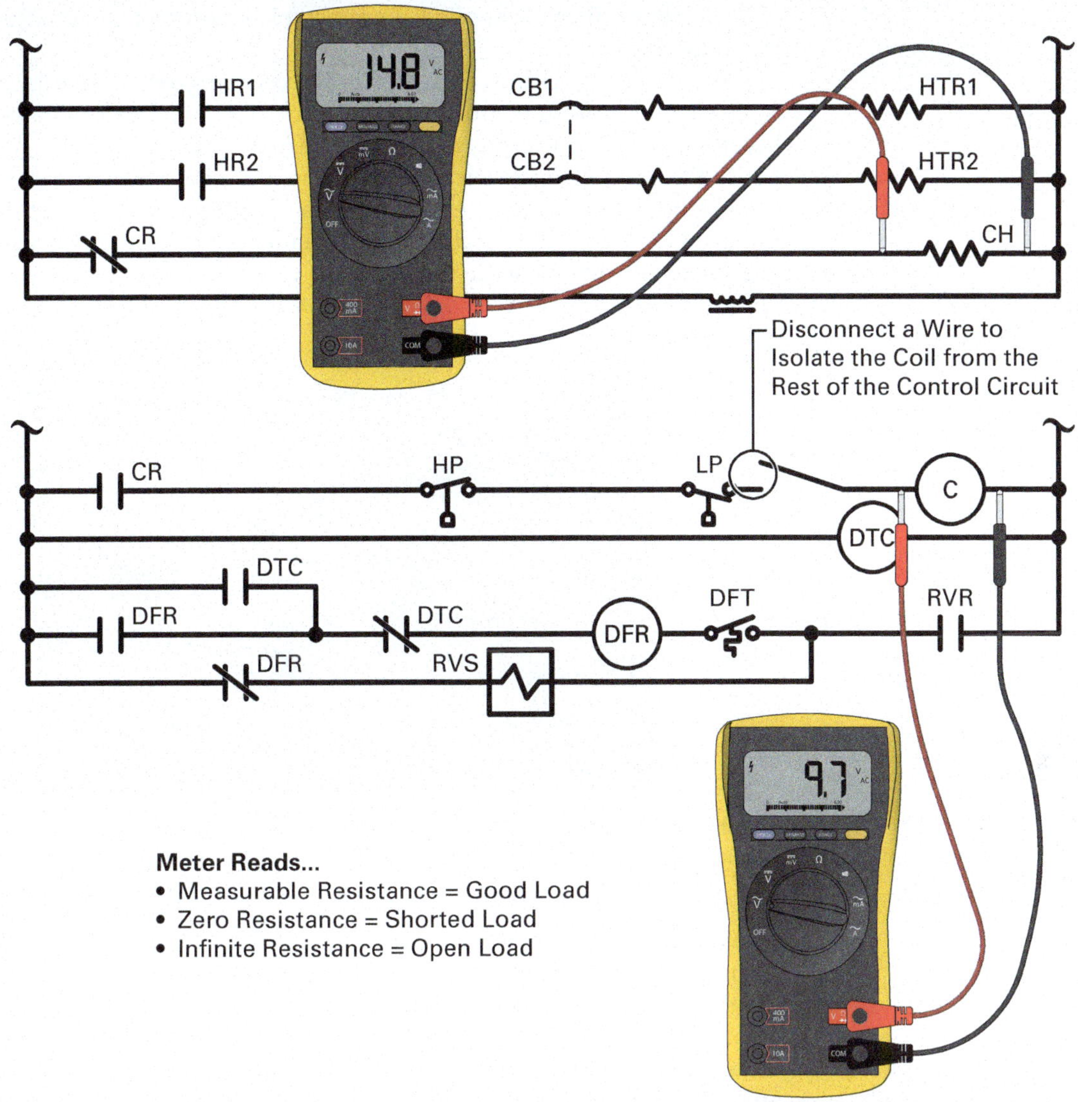

Figure 58 Resistive and inductive load resistance checks.

In either case, the component must be replaced. When the multimeter indicates a measurable resistance, although it may be small, it usually indicates that the device is operable.

You can also test for grounded windings on various solenoid coils by placing one meter probe on an unpainted part of the unit cabinet (ground). Then touch the other probe to the coil terminals. If any resistance is measured from the terminals to ground, the component is grounded and must be replaced. The ohmmeter should always read infinity (OL) when testing for grounds. Grounded devices in a control circuit usually cause a fuse to blow when the circuit is energized, due to the excessive current. A fuse protecting the secondary winding of the control transformer (if equipped) would likely open very quickly.

The actual resistance value measured for functional resistive and inductive loads can vary widely depending on the type of device. Ideally, the exact resistance value for the device can be found in the manufacturer's service literature. Another way to judge the resistance reading is by comparing the resistance of the device being tested with that of a similar device that is known to be good. In most troubleshooting cases though, knowing the precise resistance is not as important as whether there is continuity through the windings.

3.6.1 Resistive Load Testing

Heating elements are common resistive loads in HVACR. Resistive loads do not experience an inrush of high current when energized, which is common for inductive loads. An electric heating element has a certain resistance value that when a given voltage is applied, a predictable amount of current flows and heat is produced.

Resistive loads tend to fail by simply becoming open circuits. An example is an electric heating element that melts in one spot and separates (*Figure 59*). Insufficient heat during a heat pump defrost cycle may be the first symptom noticed from a failed electric heating element. Because heat pump systems often have more than one heating element, the user may not even notice. Thus, it is important to check the condition and operation of all electric heating elements in a system during periodic maintenance visits.

Figure 59 Failed electric heating element.
Source: SimplyCreativePhotography/iStock

With an electric heating element energized, a clamp-on ammeter placed around one of the power leads can determine if it is drawing current. If a heating element has voltage available to it but is not drawing current, the circuit through it is open at some point. That can be confirmed by de-energizing the heater, disconnecting at least one lead from the circuit, and checking for continuity through the element with an ohmmeter.

Manufacturers rarely provide the exact resistance value of resistive devices. The voltage and power consumption, in watts, of a heating element is often provided, but not the resistance value. On the rare occasion that the exact resistance of a heating element is needed, measure the resistance of a similar heater that is known to be functional. Ohm's law can also be used to calculate the resistance. Most of the time though, you are simply looking for continuity to determine the component's condition (intact or open circuit).

3.7.0 Hydronic Controls

In the northern portion of the United States, hydronic heating is very popular. Hot water is circulated through pipes to terminals in different parts of the building. Boilers are pressurized vessels with unique safety and installation requirements.

A simple hydronic system, such as the one shown in *Figure 60*, contains the following major components:

- A boiler where the water is heated to 140°F–180°F (60°C–82°C).
- One or more circulating pumps to move the heated water through the system.
- An expansion tank to accommodate additional water volume in the closed system to accommodate the expansion of heated water.
- An aquastat to control the leaving water temperature to meet the heating demand.

NOTE

Hydronic systems and components are presented in detail in NCCER Module 03203, *Introduction to Hydronic Systems* and NCCER Module 03305, *Commercial Hydronic Systems*.

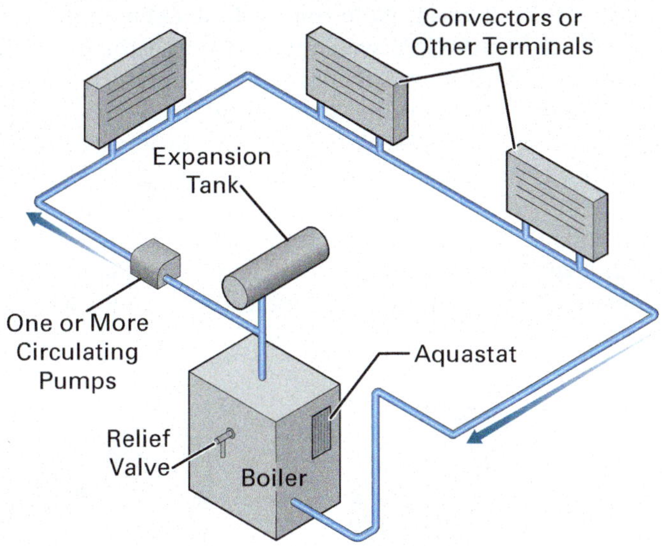

Figure 60 Simple hydronic system.

- Terminals to exchange heat between the water and the air. Some systems use an underfloor network of tubing in place of terminals.
- A relief valve to relieve excess pressure if the system pressure reaches a hazardous level.

The water may be heated electrically or by a fossil-fuel burner. The fossil-fuel burners and the many associated controls are similar to the burner controls on gas and oil-fired warm air furnaces.

3.7.1 Aquastat

The aquastat (*Figure 61*) is usually mounted on the boiler and has a temperature-sensing element that is inserted in a well or secured to the leaving water line.

Traditional boilers maintain a reservoir of hot water that can be circulated on demand. Many of today's boilers, however, are *instantaneous boilers*, meaning that the boiler fires only when there is a heating demand. The burner modulates its capacity to match the water flow rate and maintain the water temperature set at the aquastat. Today's boilers are very similar to today's instantaneous domestic water heaters.

The aquastat shown in *Figure 61* is also capable of providing low-water protection for the boiler. All boilers require safety controls that prevent the burner from firing when there isn't enough water in the boiler (traditional boilers) or when there is no water flow (instantaneous boilers).

Another feature of the aquastat shown in *Figure 61* is the ability to reset the water temperature based on the heating demand and outdoor conditions. The concept of water temperature reset is presented in the next section.

Figure 61 Digital aquastat.
Source: Hydrolevel Company

3.7.2 Water Temperature Reset

To achieve maximum fuel efficiency, some boilers incorporate a *water reset control*. This device monitors the outdoor temperature and adjusts the boiler water temperature accordingly to maximize energy savings. The control maintains a higher water temperature on colder days to supply more heat but reduces the water temperature on milder days. The reset controller uses an outside temperature sensor that is typically mounted on the north side of the structure, away from direct sunlight.

Some reset controls also incorporate the heating demand into the calculations to determine the optimum heating water temperature.

3.7.3 Low-Water Cutoff

Insufficient water in a traditional boiler can have catastrophic results. The boiler may overheat, produce steam, and develop explosive pressures. To prevent this, a low-water cutoff device is used. Note that instantaneous boilers use a flow switch to ensure there is sufficient water flow through the boiler to prevent overheating, as well as a high-temperature safety switch.

The simplest devices use a float that opens a safety switch when the water level drops below a preset level. The switch is in the burner circuit and shuts off the burner to prevent overheating. An alarm or warning light is often activated as well.

Electronic versions (*Figure 62*) have a probe that is inserted in the boiler to monitor the water level. Mineral contamination in areas with poor water quality will affect the operation of the control. Periodic testing, maintenance, and cleaning of the probe is required.

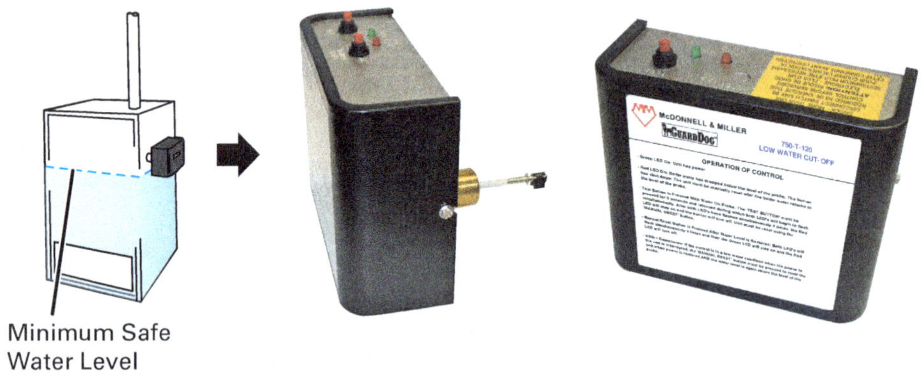

Figure 62 Electronic low-water cutoff.
Source: Courtesy of Xylem Inc.

3.7.4 Circulator Pumps

The circulator pumps in a hydronic system perform a job similar to the blower in a forced air furnace. Larger circulating pumps are mounted on the floor, while inline pumps are mounted in a run of piping (*Figure 63*). The preferred location for a single circulating pump is in the supply line leaving the boiler.

Zoned systems with multiple thermostats and circulators are common today, often paired with an instantaneous boiler. Each zone has its own small circulating pump that is typically controlled by a zone thermostat (*Figure 64*). As more zones call for heat and more circulators start, the boiler increases the firing rate to continue providing the needed water volume at the correct temperature.

Figure 63 Inline circulating pump.
Source: Courtesy of Xylem Inc.

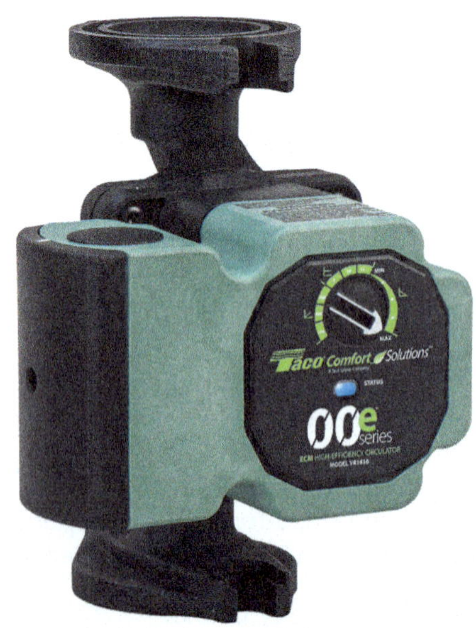

Figure 64 Zone circulating pump.
Source: Courtesy of Taco, Inc.

3.7.5 Zone Valves

Zoned control often requires a zone valve in addition to a zone pump. It is energized along with the pump on a call for heat. Without a zone valve closing when the zone thermostat is satisfied, hot water may still enter the piping and overheat the space. *Figure 65* shows a typical zone valve.

Zone valves are typically motorized valves that turn a ball 90 degrees to allow or block flow. They open and close rather slowly to avoid water hammer in the piping system.

In residential applications, zone valves are typically powered by the 24VAC control circuit. Most valves open on a call for heating from the zone thermostat, but zone valves can be NO or NC. Some valves are equipped with an end-switch that closes as the valve opens. The switch closure may be used to energize the zone circulator pump, or it may be used to start the boiler burner.

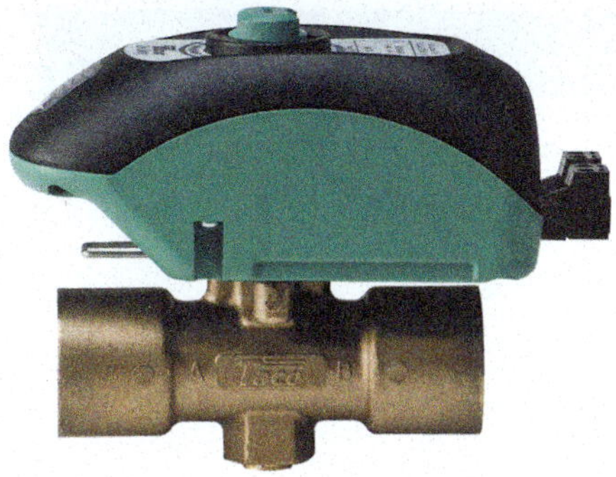

Figure 65 Hydronic zone valve.
Source: Courtesy of Taco, Inc.

3.7.6 Zoned Hydronic System Operating Sequence

In many ways, the operating sequence of a hydronic system is like the operating sequence of a forced-air heating system. The fossil-fuel burners used in both systems operate similarly and may even use identical components.

As is the case with any HVACR system, understanding the sequence of operation before troubleshooting is important. The following sequence of operation assumes a traditional residential packaged boiler (gas-fired) equipped with an aquastat with low-and high-temperature limits. Assume the structure is divided into three heating zones, each controlled by its own 24VAC zone thermostat.

Assume there is a call for heat from one of the three zones. The zone thermostat closes its contacts between R and W, energizing the zone valve. As the motorized zone valve opens, it closes a set of end-switch contacts. The contact closure sends 24VAC power back to the aquastat, which then energizes a relay and starts the zone circulating pump. The pump begins to move water through the terminal associated with the zone.

On a call for heat, the aquastat must also energize the boiler burner. Assuming all the safety switches are closed, such as the flame rollout and high-temperature limit safety switches, the ignition sequence begins and the burner lights to heat the water.

If another zone thermostat calls for heat, its zone valve will open and the aquastat will start its circulating pump. Additional cool water will begin to flow, and the burner will fire at a higher rate as necessary to maintain the desired water temperature supplied to the zones.

As the various zone thermostats are satisfied and open the circuit between R and W, the zone valves close. The valves are generally powered open but close when de-energized by a spring that pulls the valve back to its normal position. As the last zone is satisfied, the burner is also de-energized.

Troubleshooting the components in a hydronic control system is similar to troubleshooting components in any other HVACR control system. You simply apply the same principles—listen carefully to the complaint and use a process of elimination to determine where to begin targeted testing as you search for the source of problem.

NOTE

This sequence of operation is just one example of many possibilities. It is essential that you familiarize yourself with the specific system being serviced by reviewing the wiring diagrams and physical wiring connections. The aquastat in use has a significant effect on how a hydronic control system is developed and wired, and numerous proprietary zone-control systems are on the market as well.

3.0.0 Section Review

1. To maximize electrical safety when working on live electrical circuits, ______.
   a. try to work with one hand only inside the panel
   b. keep one foot on an insulated surface
   c. lay insulation material under your knees
   d. wear metal jewelry to dissipate the current

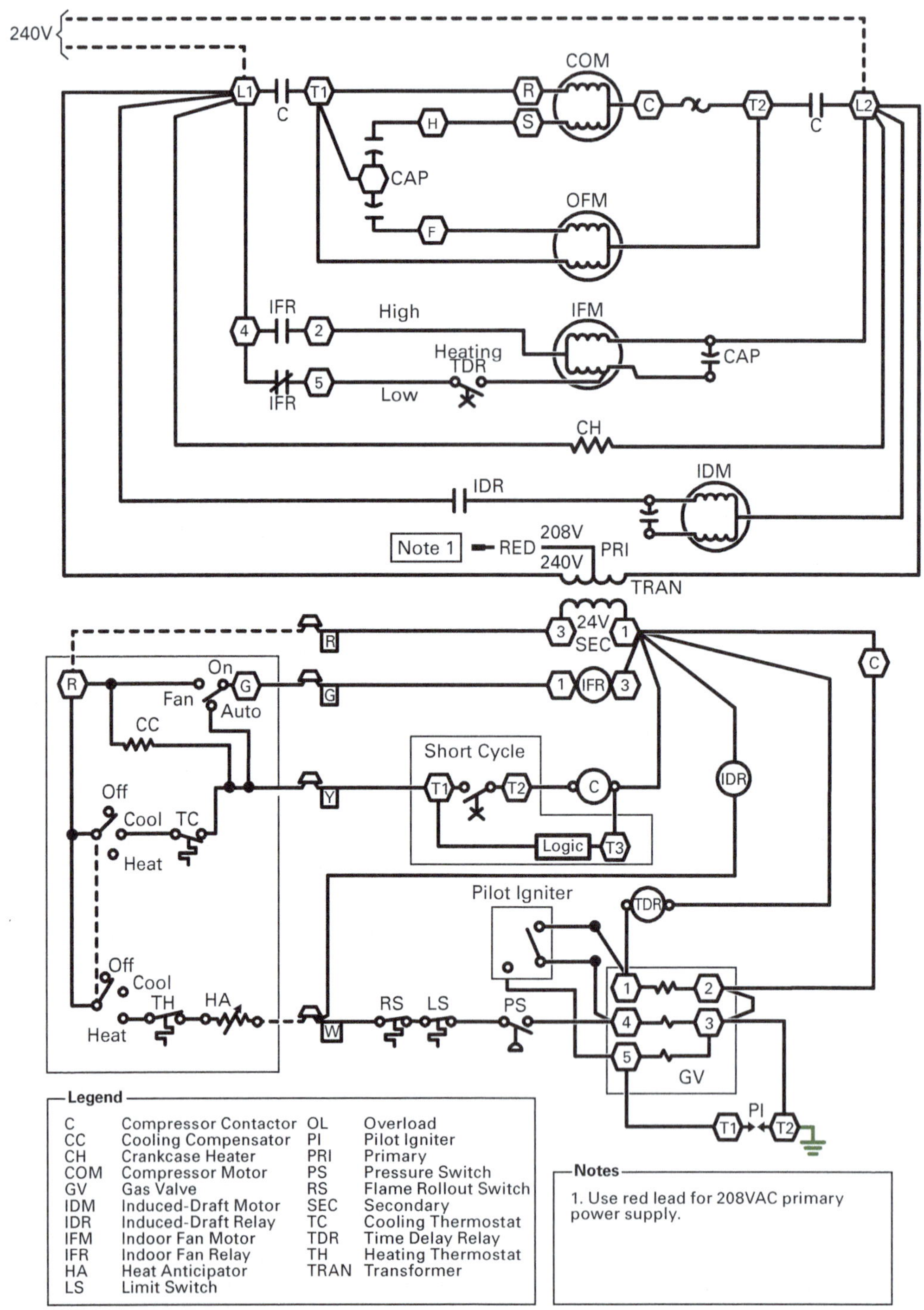

Figure SR01

2. Refer to *Figure SR01*. What effect does the limit switch (LS) have on the operation of the induced draft motor (IDM), if the switch opens?
   a. It breaks the line voltage circuit serving the IDM.
   b. It prevents the indoor draft relay (IDR) from energizing to start the motor.
   c. It has no effect on IDM operation when it opens.
   d. It de-energizes the time-delay relay, which would prevent the IDM from starting.

3. What is the purpose of the legend on a label diagram?
   a. It provides potential solutions to common control problems.
   b. It identifies the meaning of symbols and abbreviations.
   c. It is used to locate components in the equipment cabinet.
   d. It isolates a specific section of a ladder diagram.

4. For equipment that operates on three-phase power, what is the maximum current imbalance allowed?
   a. 2%
   b. 5%
   c. 8%
   d. 10%

5. If a 24VAC control transformer reads only 21V on the secondary winding while the unit is idle, which of the following is the most likely problem and should be checked first?
   a. A device in the control circuit is drawing more current than it should.
   b. The secondary winding is wired for the wrong control voltage.
   c. The primary winding is wired for the wrong line voltage.
   d. The transformer secondary winding is grounded.

6. True or False? The coils of relays and contactors are examples of electrical loads.
   a. True
   b. False

7. Safety controls will prevent an instantaneous boiler burner from operating when ______.
   a. there is no water flow
   b. the incoming water temperature is below 55°F
   c. the demand for hot water is excessive
   d. only one zone valve is open

Module 03317 Review Questions

1. What is the purpose of a compressor short-cycle timer?
   a. To prevent compressor restart before the system pressures equalize.
   b. To ensure the compressor runs for at least five minutes before stopping.
   c. To track how often the compressor is energized each hour.
   d. To run the compressor periodically when it has been idle for an extended period.

2. What is the purpose of a lockout relay in a condensing unit control circuit?
   a. To prevent unauthorized equipment use.
   b. To prevent compressor operation until the indoor blower starts.
   c. To protect the compressor motor from a current surge.
   d. To prevent automatic equipment restart when a tripped safety control resets.

3. The purpose of a motor starter is to _____.
   a. apply additional current to a motor as it starts
   b. vary the speed of a motor as conditions change
   c. monitor motor voltage as it runs
   d. energize a motor and provide overload protection

4. When checking the voltage drop across the contacts of a compressor contactor, you should replace the contactor at what level of voltage drop?
   a. 0.5V to 1V
   b. Any reading over 3V
   c. 2% of the line voltage
   d. 3% of the line voltage

5. To what does the term *differential* in a thermostat refer?
   a. The difference between a thermostat's cut-in and cut-out points.
   b. The difference enforced between the cooling and heating setpoints.
   c. The difference between the setpoint and the actual space temperature.
   d. The difference between the heating cycle rate setting and the actual number of starts per hour.

6. Which of the following is *not* a common feature of an electronic thermostat?
   a. Battery backup to prevent loss of programs
   b. A heat anticipator
   c. Programmed period override
   d. Maintenance tracking

7. Which of the following is a reasonable location for a room thermostat?
   a. Anywhere that it will receive direct sunlight.
   b. About 5' above the finished floor on an outside wall.
   c. About 5' above the finished floor on an inside wall.
   d. Near a corner where there is minimum air circulation.

8. In a standard thermostat wiring scheme, the R wire represents _____.
   a. the indoor fan control wire
   b. one leg of the 24VAC power source
   c. the heat pump reversing valve control wire
   d. the compressor control wire

9. In the electrical control circuit of a packaged heating/cooling system, the outdoor fan motor typically ______.
   a. has a continuous-on mode for ventilation
   b. runs at low speed in the heating mode
   c. runs whenever the compressor is energized
   d. runs whenever the indoor blower is energized

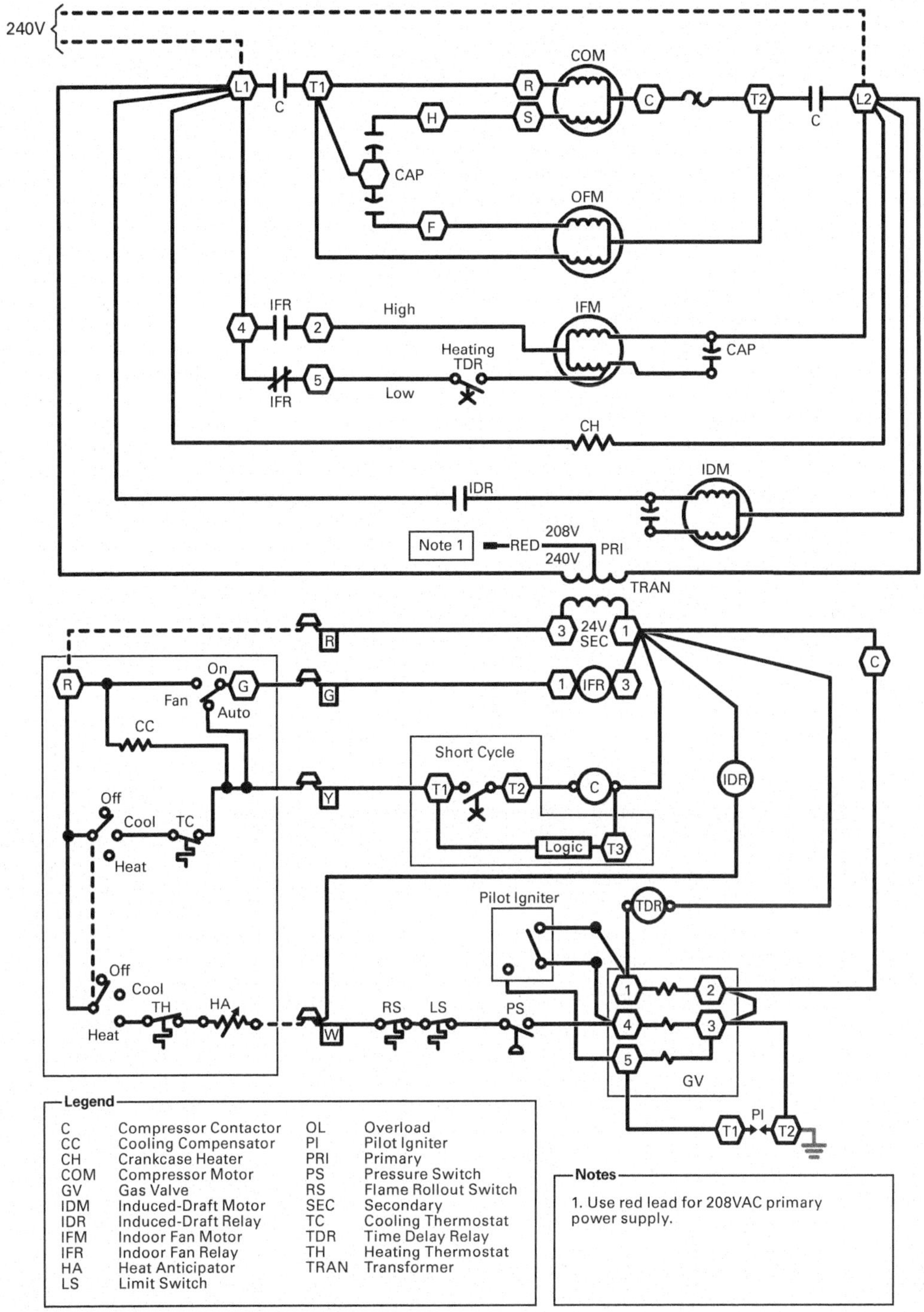

Figure RQ01

10. Refer to *Figure RQ01*. Which component can prevent the compressor from starting immediately on a call for cooling from the thermostat?
   a. Time-delay relay
   b. Indoor fan relay
   c. Contactor
   d. Short-cycle timer

11. Refer to *Figure RQ01*. Which component causes the indoor fan to cycle on and off with the compressor in the cooling mode?
    a. Time-delay relay
    b. Thermostat Fan switch
    c. Contactor
    d. Short-cycle timer

12. Refer to *Figure RQ01*. Which component prevents the low- and high-speed windings of the indoor fan motor from being energized at the same time?
    a. TDR
    b. IFR
    c. PS
    d. IDR

13. Refer to *Figure RQ01*. Which component prevents the main gas valve (GV at terminals 1 and 2) from being energized on a call for heat from the thermostat immediately after the pressure switch (PS) closes?
    a. Time-delay relay
    b. Flame rollout switch
    c. Pilot igniter
    d. Room thermostat

14. Refer to *Figure RQ01*. When is the time-delay relay energized?
    a. Immediately on a call-for-cooling.
    b. Immediately on a call-for-heating.
    c. When the pilot igniter proves a pilot flame.
    d. When the compressor contactor is energized.

15. When a contactor coil somehow becomes grounded, what would you expect to discover as you begin troubleshooting?
    a. A blown control circuit fuse
    b. A grounded compressor motor winding
    c. An open compressor winding
    d. A compressor running continuously

Answers to odd-numbered review questions are found in the Module Review Answer Key at the back of this book.

Answers to Section Review Questions

Answer	Section	Objective
Section One		
1. d	1.1.3	1a
2. b	1.2.0	1b
Section Two		
1. c	2.1.3	2a
2. d	2.2.1	2b
3. a	2.3.2	2c
Section Three		
1. a	3.1.1	3a
2. c	3.2.1	3b
3. b	3.3.4	3c
4. d	3.4.1	3d
5. c	3.5.3	3e
6. a	3.6.0	3f
7. a	3.7.1	3g

Motor Testing and Troubleshooting

Source: Smoczyslaw/iStock

Objectives

Successful completion of this module prepares you to do the following:

1. Describe the construction of various motors and explain how to electrically test them.
   a. Describe the construction of common single-phase motors.
   b. Explain how to electrically test motors for open, shorted, and grounded windings.
   c. Explain how to test start capacitors, run capacitors, start relays, and start thermistors.
   d. Explain how to test three-phase motor windings and inverter power modules.
2. Identify and explain how to service electronically commutated motors (ECMs).
   a. Identify and relate basic ECM operating concepts.
   b. Explain how to install and set up an ECM.
   c. Explain how to test and troubleshoot an ECM.
3. Describe the operation of variable-frequency drives (VFD) and their selection considerations.
   a. Describe the operation of a VFD.
   b. Identify and describe common programmable VFD parameters.
   c. Describe the important considerations for the selection of a VFD.
   d. Define and describe dynamic motor-braking processes.

Performance Tasks

Under supervision, you should be able to do the following:

1. Test single-phase induction motor windings for opens, shorts, and grounds.
2. Perform a megohm test on a motor.
3. Test motor start and run capacitors.
4. Test an ECM motor.

Overview

Motors are the workhorses in HVACR systems. Motors also represent the most significant electrical loads in most systems. As a technician, your ability to test and troubleshoot motors will play a significant role in your success. New technologies such as variable-frequency speed control and electronically commutated motors have improved system performance and efficiency, but these products also require a different approach to testing and troubleshooting. This module will build on what you have learned previously and ensure that you are well prepared to service commonly encountered motors.

Digital Resources for HVACR

Scan this code using the camera on your phone or mobile device to view the digital resources related to this craft.

SCAN ME

1.0.0 Basic Motor Testing

Performance Tasks

1. Test single-phase induction motor windings for opens, shorts, and grounds.
2. Perform a megohm test on a motor.
3. Test motor start and run capacitors.

Objective

Describe the construction of various motors and explain how to electrically test them.

a. Describe the construction of common single-phase motors.
b. Explain how to electrically test motors for open, shorted, and grounded windings.
c. Explain how to test start capacitors, run capacitors, start relays, and start thermistors.
d. Explain how to test three-phase motor windings and inverter power modules.

Motors represent the most significant electrical loads in HVACR systems. Motors are also responsible for the hard work required for a system to operate and successfully transfer heat. As a technician, your ability to test and troubleshoot motors will play a significant role in your success.

Basic motor construction has been presented in previous modules, including NCCER Module 03206, *Alternating Current*. Before presenting basic motor testing skills, motor construction will be reviewed briefly.

1.1.0 Single-Phase Motor Construction

Six types of single-phase motors are commonly used in HVACR equipment:

- Shaded pole
- Split phase
- Permanent split capacitor (PSC)
- Capacitor start
- Capacitor start, capacitor run (CSCR)
- Electronically commutated motors (ECMs).

Shaded-pole and split-phase motors are very simple in construction and limited in the work they can perform. Shaded-pole motors, when found, are typically used in low-torque applications such as small direct-drive fans. On the other hand, PSC and CSCR motors are routinely used for fans and blowers as well as in single-phase hermetic compressors because of their operating characteristics and reasonable efficiency.

Indoor and outdoor fan and blower motors may be single- or multispeed PSC motors or electronically commutated motors (ECMs). ECMs are now common in new equipment, especially for use with indoor blowers. ECMs can be used to retrofit equipment with PSC motors, but a conversion to ECMs in existing equipment is not very common, primarily due to the cost.

Both PSC (*Figure 1*) and CSCR motors (*Figure 2*) have at least three external leads connected to two internal windings. The run winding (R) contains relatively few turns of heavy gauge wire. The start winding (S) contains a greater number of turns of lighter gauge wire. The point where the two windings meet internally is referred to as common (C).

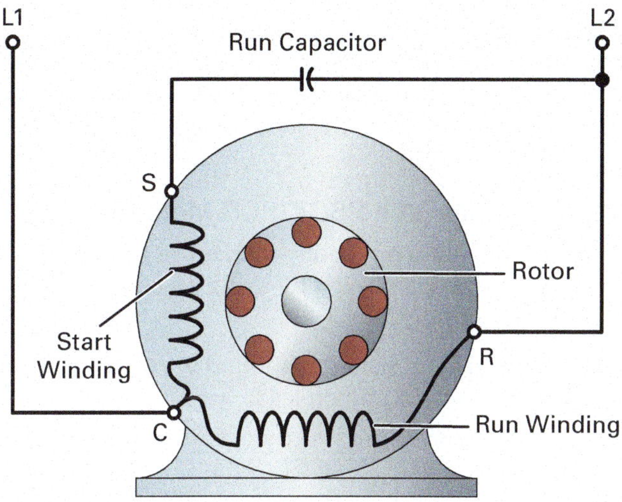

Figure 1 PSC motor configuration.

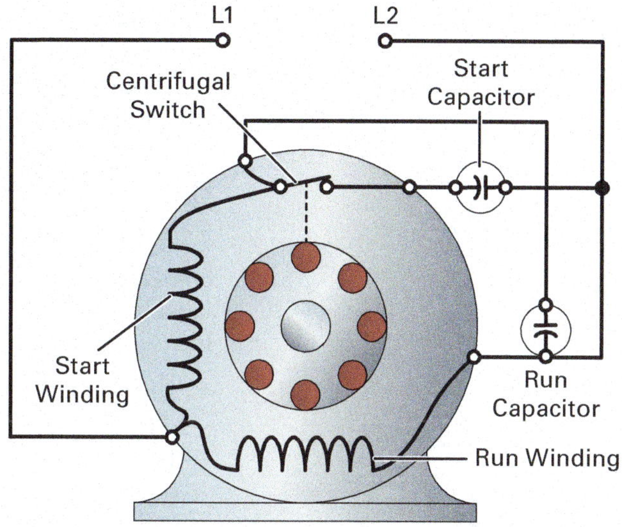

Figure 2 CSCR motor configuration with a centrifugal switch.

The physical arrangement of motor windings used in PSC and CSCR motors is the same. The configuration of the motor as a PSC or CSCR motor is evidenced by the run and/or start circuit components used with the motor. The PSC motor has a run capacitor connected in series with the start winding. The CSCR uses a start capacitor to aid in starting, in addition to the run capacitor.

As shown in *Figure 2*, a centrifugal switch is used to disconnect the start capacitor from the circuit as the CSCR motor comes up to speed. As the name implies, the switch opens when the rotational speed is high enough to develop enough centrifugal force to physically separate the contacts.

A start relay can also be used in the start capacitor circuit instead of a centrifugal switch. The start capacitor and the contacts of the start relay are then connected in parallel with the run capacitor circuit. In either case, the start capacitor is removed from the circuit when the motor reaches about 75% of full speed. A centrifugal switch is most often used when the motor drives a fan or blower, while a start relay is used with CSCR-driven compressors.

Multispeed PSC motors (*Figure 3*) used to drive fans and/or blowers in HVACR equipment can operate at two or more speeds. The motor's speed changes as different *taps* are energized. These motors have the same number of windings as their single-speed counterparts (two), but the run winding has leads connected in multiple locations.

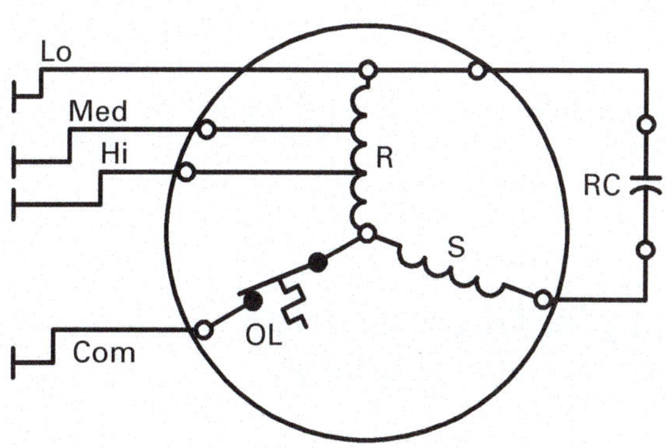

Figure 3 PSC multispeed motor configuration.

The motor speed is determined by the operating mode. When working with a fossil-fuel furnace, lower fan speeds are generally used for heating and higher speeds are used for cooling, but that may not be the case for every system. Fan speeds are usually reduced in the dehumidification mode if the system is equipped for that mode. A control relay with contacts located in the motor load circuit is normally used to prevent more than one motor-winding speed tap from being energized at the same time—a condition that would destroy the motor. Multiple speed taps cannot be energized at the same time.

Although multispeed motors use tapped windings, these motors can fail to run at any speed. When troubleshooting multispeed motors, it is important to eliminate the speed selection circuits external to the motor as the cause of the problem before condemning the motor itself.

1.1.1 Identifying Unmarked Terminals of a PSC/CSCR Motor

Sometimes the leads on a single-phase motor are poorly marked, or perhaps the label has been damaged. The leads can be identified by using an ohmmeter to measure the resistance of the motor windings:

- The two leads across which the highest resistance is found are the run and start windings. The remaining lead is the common lead.
- The lead with the highest resistance measured to the common lead is the start winding.

Start by using the ohmmeter to find the two leads across which the highest resistance is measured (*Figure 4*). These two represent the run (R) and start (S) windings. The remaining lead is the common (C) lead.

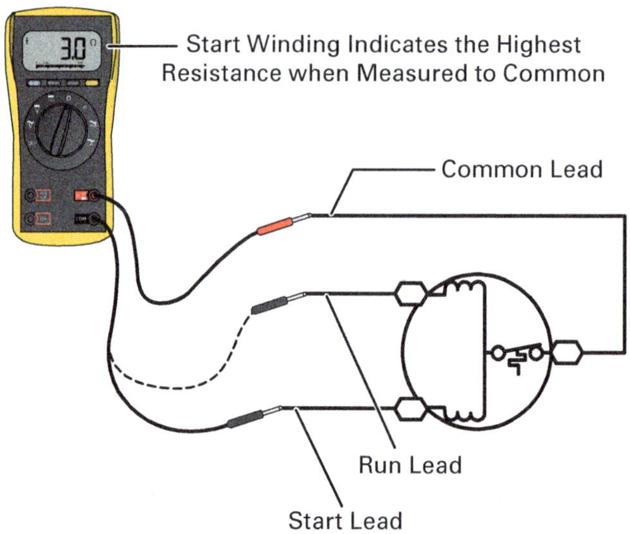

Figure 4 Identifying unmarked terminals of a PSC/CSCR motor.

Next, put one probe of the multimeter on the common (C) lead, and determine which of the remaining leads has the highest resistance reading to common. This identifies the start winding. The remaining lead belongs to the run winding. The resistance between the common and run winding is the lowest of the possible combinations.

1.2.0 Testing for Open, Shorted, and Grounded Windings

Motors are tested for open, shorted, and/or grounded windings with the equipment de-energized, the capacitors discharged, and all the motor leads disconnected.

1.2.1 Testing for Open or Shorted Windings

Testing for shorted or open windings is done by measuring the resistance of the windings with an ohmmeter. Be sure to use a multimeter that has an accurate low-range (R × 1Ω) resistance scale. This is necessary because the resistance of some undamaged windings can be as low as 0.5Ω.

One probe of the multimeter is connected to one of the motor leads, as shown in *Figure 5*. Touch the other probe to the remaining motor leads, one lead at a time, and observe the meter indication. If the multimeter reads a measurable resistance, the windings are at least intact. If the ohmmeter reads zero resistance through two or more leads, the motor has a shorted winding. If it reads an infinite resistance (OL), the winding is open.

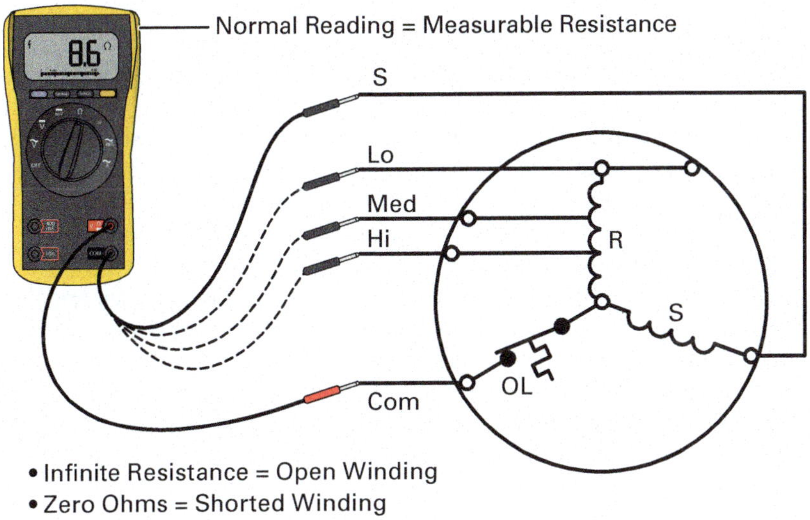

Figure 5 Testing for open or shorted windings.

WARNING!

The terminals of hermetic and semi-hermetic compressor motors in a pressurized system have been known to blow out when disturbed. To avoid injury, do not disturb wiring at the compressor terminals unless it is necessary. When electrically testing compressors, test from the ends of the connected leads. It is safer to connect/disconnect wiring at the compressor terminals when the system is not under pressure.

Here are some general rules commonly used to judge the condition of windings:

- Remember that an open internal thermal-overload device will necessarily open the common winding, resulting in an OL reading to any other winding. If the motor is at ambient temperature, then the device should have reset automatically. If the motor is hot, you must wait for the motor to cool before testing, which can take several hours. A thermal overload that will not close is also a reason to condemn small motors. Large motors may be worth the repair investment. Such a repair is typically done by a competent motor shop.

WARNING!

The surface of an overheated motor can be dangerously hot. Use a gloved hand or a thermometer to check the surface temperature before handling an installed motor.

- When testing a single-phase PSC/CSCR compressor motor, the resistance of the start winding is typically three to five times that of the run winding. However, the winding resistance can vary widely depending on the motor design.
- When testing a single-phase, multispeed motor run winding, the highest resistance is normally measured between the common lead and the low (LO) speed lead, and the lowest resistance between the common lead and the high (HI) speed lead. The resistance measured between the common lead and the medium speed lead (MED) should be somewhere between that measured for the LO and HI leads.
- When testing a three-phase motor, the resistance measured across each pair of windings should be nearly identical.

1.2.2 Testing for Grounded Windings

Testing a motor for grounded windings is done by measuring the resistance of each winding to a reliable ground. Set the meter on the R × 10,000Ω scale if it does not have an auto-ranging feature. Touch one probe of the multimeter to a reliable ground point, such as the motor or compressor frame, or to the compressor discharge or suction line. Poor electrical contact because of a coat of paint, layer of dirt, or corrosion can cause an inaccurate measurement and potentially hide a grounded winding.

The other meter probe is then touched to each of the motor leads, one lead at a time (*Figure 6*). An infinite or very high resistance should be measured from each lead to ground. If a high resistance reading is indicated, it should not be less than 1,000Ω per volt of the motor's voltage reading. For example, on a 230VAC motor, the resistance should not be less than 230,000Ω (230V × 1,000Ω/V = 230,000Ω). Although this indicates that the motor winding is not currently grounded, it may indicate that the motor's electrical insulation is deteriorating. Additional tests over time using a **megohmmeter (megger)** is recommended, especially if the motor supports a critical application.

Megohmmeter (megger): A test instrument used to test high-resistance circuits using a high-voltage test signal.

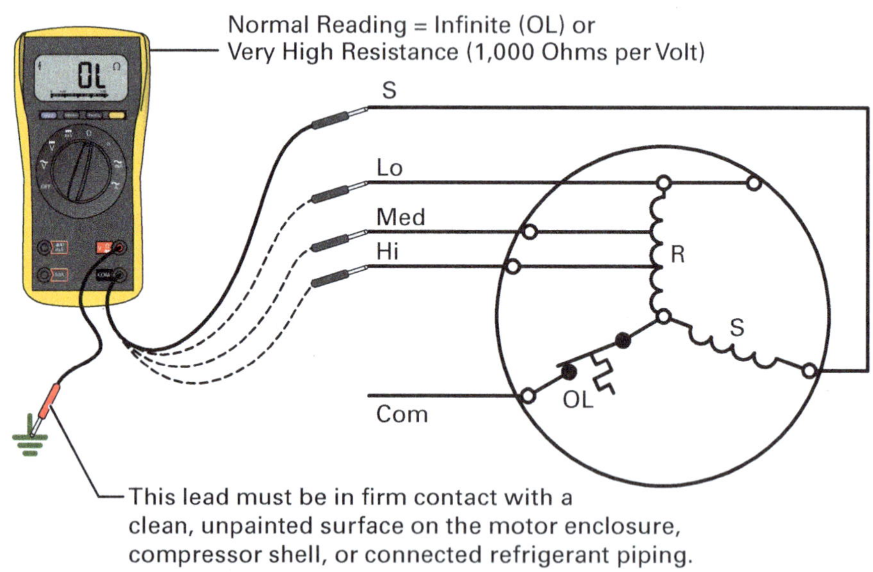

Figure 6 Testing for a grounded winding.

If zero resistance (OL) or a resistance that is less than 1,000Ω per volt is measured, this indicates that the motor winding has a path to ground. Note, however, that erroneous resistance readings to ground may be seen in hermetic and semi-hermetic compressors if liquid refrigerant is present inside the compressor shell. For example, a compressor installed outdoors without a crankcase heater in winter conditions, and that hasn't been operated recently, may contain condensed liquid refrigerant. In this case, it is recommended that the refrigerant be recovered from the outdoor unit or system. Then retest the compressor before condemning it.

1.2.3 Testing with a Megohmmeter

Normally, a standard ohmmeter is not used to look for extremely high resistance values. In the previous example, it is likely that you were unable to read the actual resistance between any motor winding and ground. The ohmmeter simply displayed OL. You are detecting an open circuit between each winding and ground.

To test motor winding resistance to ground more effectively, a much higher electrical potential than that furnished by a standard ohmmeter is needed. Much higher test voltages, typically up to 5,000VDC (5kVDC) can be supplied by a megohmmeter. With a standard ohmmeter, you are relying on the voltage of the instrument's installed batteries alone. Note that the test voltage a megger produces is DC, not AC.

Meggers measure resistance in the megohm (MΩ) range. One megohm equals 1,000,000Ω. There are small, handheld models that can provide around 1kVDC for testing (*Figure 7*). These handheld models are sufficient for most HVACR work. Larger battery-operated meggers and those that require an AC power source (also shown in *Figure 7*) can develop much higher test voltages.

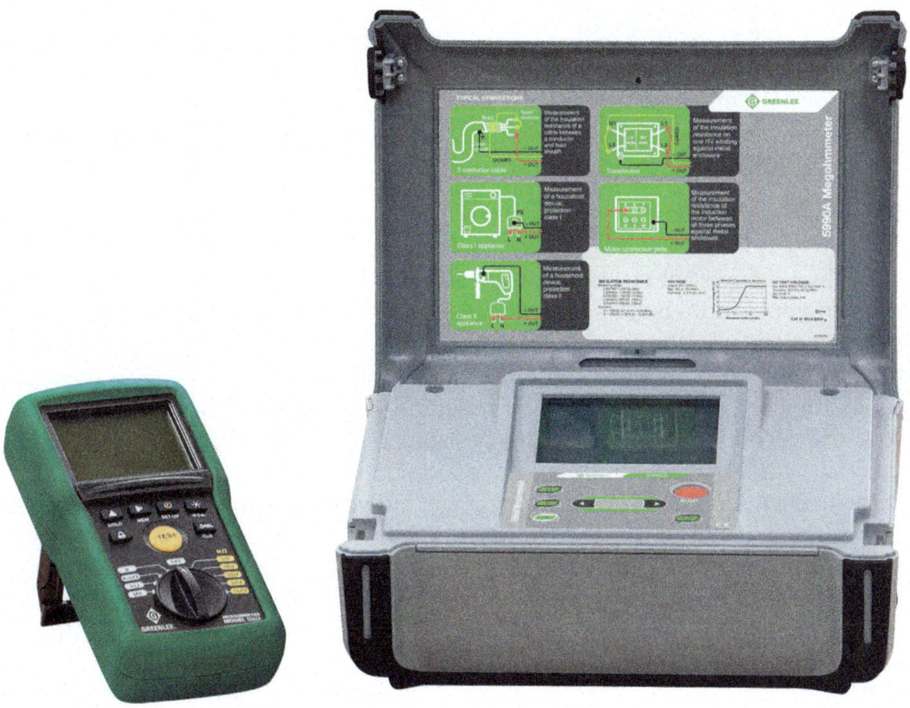

Figure 7 Examples of meggers.
Source: Ridge Tool Company

WARNING!

When using a megger, you can be injured or damage the equipment due to the extreme voltage. Follow all the manufacturer's safety guidelines.

Here are some basic safety guidelines to remember when using a megger:

- Use meggers for high-resistance measurements only, such as motor insulation measurements or to test two separate conductors in a cable. Never use a megger for general resistance measurements and continuity testing.
- Never touch the conductive portion of the test leads or alligator clips.

- De-energize the circuit and fully disconnect the tested component from other electric circuits before using the meter.
- Never test a motor, such as one contained in a compressor, while it is exposed to a vacuum. This can cause internal high-voltage arcs that seriously damage the motor windings.

Meggers perform motor insulation tests that can reveal degradation of the motor winding insulation. The probes are used just as you would when using a standard ohmmeter (*Figure 8*). However, the meter leads generally terminate in alligator clips to ensure positive and secure electrical contact. When a motor winding is grounded and causes a circuit breaker or fuse to open, any ohmmeter should detect the grounded condition as the resistance from the motor winding to ground will be relatively low.

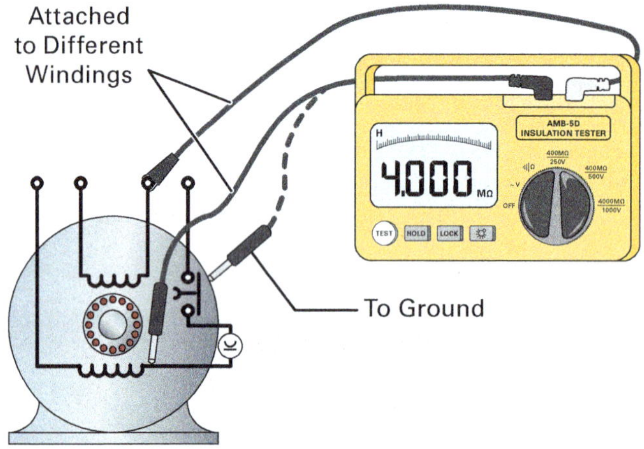

Figure 8 Using a megger.

Acid in a refrigerant circuit, though, can slowly damage the insulation coating on motor windings. A series of megger tests can detect the reduction in insulation quality, while a common ohmmeter cannot. After a few periodic tests with a megger, you can see any trend of deterioration.

A motor needs to be replaced or rewound if the resistance reading drops below the recommended minimum resistance for the motor. As mentioned previously, undamaged windings have a resistance to ground that measures at least 1,000Ω per volt of operating voltage. However, you can also contact the motor manufacturer for a specific minimum value. A higher resistance than the minimum reading typically identifies reliable winding insulation. Any resistance below that threshold is cause for concern. Note that a motor with sound winding insulation often has a resistance reading many times the minimum acceptable resistance.

1.3.0 Testing Capacitors, Start Relays, and Start Thermistors

Capacitors, start relays, and start thermistors are common accessories used with single-phase motors to assist them in starting and running more efficiently. Many motors are condemned each year when one of these accessories is truly the source of the problem. Thus, it is important that HVACR technicians know how to properly test and troubleshoot these devices.

1.3.1 Start and Run Capacitor Testing

Start and run capacitors affect the wattage, current, torque, speed, and efficiency of a motor. Run capacitors remain connected in the motor circuit as the motor runs and therefore are referred to as *continuous-duty capacitors*. For that reason, capacitors are typically filled with a **dielectric** fluid that dissipates heat.

Dielectric: A material that strongly resists the passage of current.

A failed capacitor may provide a visual indication of its failure. For example, the "pop-out" hole on the top of a start capacitor may appear bulged or blown out. A run capacitor may bulge and/or be visibly leaking. Any capacitor found to be defective should be replaced with one that has the proper rating as shown on the motor nameplate or specified by the manufacturer. Assuming that the capacitor you are replacing has the correct rating can lead to additional failures.

When testing capacitors, it is best to use a capacitor tester (*Figure 9*). Although there are other ways to test the basic condition, a capacitor tester provides all the information needed. Follow the meter manufacturer's instructions to perform the test. In many cases, the probes must be held in place for several seconds while the meter battery charges the capacitor. For motor capacitors, the reading will be in microfarads (µF).

The measured value for a start capacitor should be ±10% of the value shown on the capacitor label. This is the default tolerance unless a different tolerance is specified on the label. Start capacitors, though, are rated for a capacitance range, such as 88µF–108µF. In this case, the allowed tolerance is –10% of 88µF and +10% of 108µF.

For a run capacitor, the measured value should be ±10% of the nameplate value. If the measured value is outside this range, replace the capacitor.

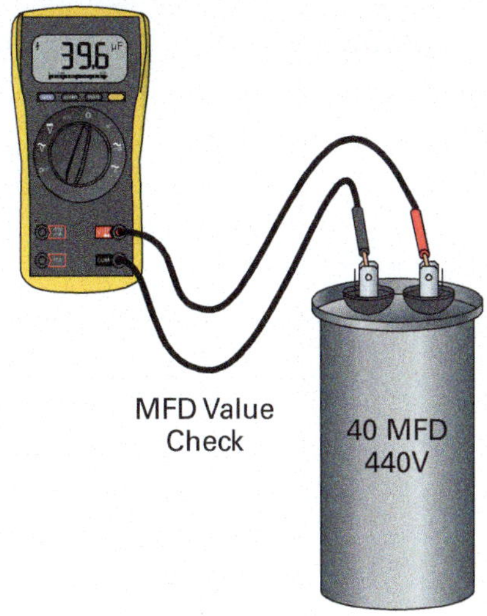

Figure 9 Capacitor testing.

1.3.2 Start Relay Testing

A start relay is used to disconnect the start capacitor from the motor windings when the motor reaches about 75–80% of its operating speed (*Figure 10*). Some motors use a centrifugal switch inside the motor to serve this purpose. Start relays are external to the motor enclosure.

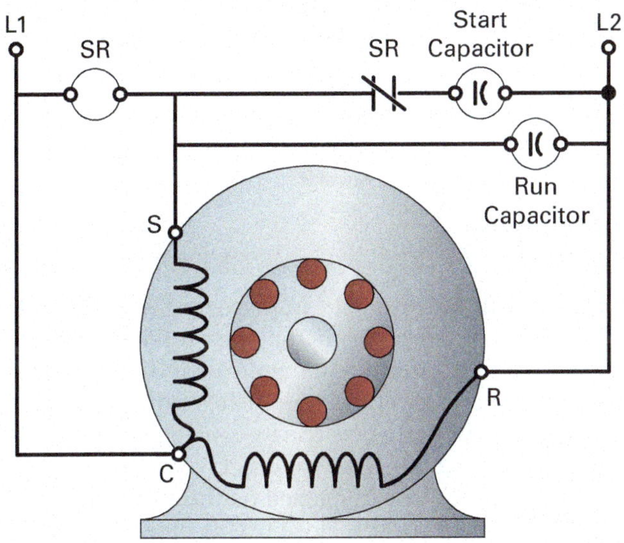

Figure 10 Schematic of a capacitor-start, capacitor-run motor using a start relay.

Start relays can be actuated by either current or voltage. The start relays used with HVACR-related motors are normally voltage-actuated, meaning that the relays respond to a unique voltage value. This type, the most common type of start relays, is referred to as a **potential relay**.

Potential relays have a specific **pickup voltage**. This is the voltage at which a potential relay coil receives enough energy to reposition the relay contacts. Note that this voltage is higher than the voltage of the power supply. As the motor increases speed, the voltage sent to the coil increases, and the contacts eventually open. This removes the start capacitor from the circuit. At that point, the momentum of the rotor allows it to continue rotating through the magnetic fields and reach full speed.

Start relays tend to fail with their contacts closed, as they are normally closed (NC) relays. This results in the start capacitor remaining in the circuit, causing

Potential relay: A relay energized by back EMF voltage generated during startup by the start winding in a capacitor-start motor.

Pickup voltage: The specific voltage at which a potential relay coil is sufficiently energized to reposition the relay contacts.

the motor windings to overheat and eventually fail. It may also cause the start capacitor to fail.

Always make sure to use a start relay that fits the motor specifications. The substitution of a relay with a different pickup voltage can damage both the start capacitor and the motor. Also ensure the replacement relay is wired into the circuit exactly like the original.

Start relay operation can be tested by monitoring the motor start-winding current with a clamp-on ammeter. When using a digital clamp-on ammeter, use the MIN/MAX current feature. Testing begins by first finding the full load amperage (FLA) rating for the motor as marked on the compressor or motor nameplate. If the meter does not have an auto-ranging feature, set it to a range that is high enough for the motor FLA.

With the power turned off, the ammeter jaws are placed around the wire that connects the motor start capacitor to the start relay contacts, as shown in *Figure 11*. Power is turned on to the unit while watching the ammeter display to observe the current flow in the start capacitor circuit as the motor starts.

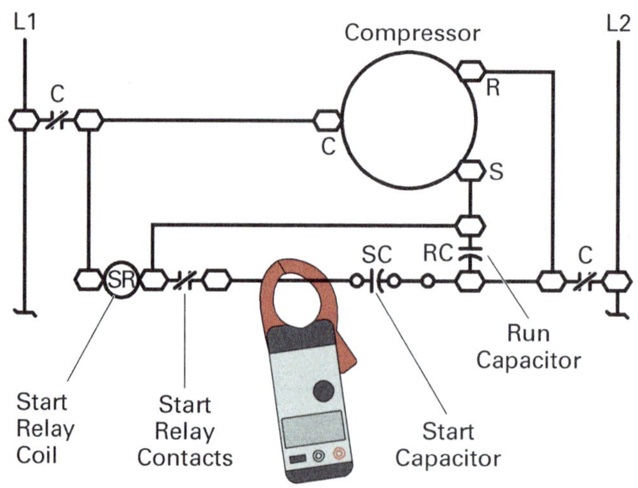

Figure 11 Start relay testing.

Energize the motor with the ammeter in place on the lead between the start relay contacts and the start capacitor. If the start relay is operating properly, the clamp-on ammeter display should momentarily indicate the inrush current, and then fall back to zero as the motor approaches full speed. This shows that the start relay is operating properly because its contacts were closed but opened as they should. Note that this happens very quickly.

If a current value continues to be displayed on the ammeter after the motor is up to speed, the relay contacts are still closed. This means the relay is defective in some way and should be replaced. It may be the coil, or it may be damage to the contacts. The specific mode of failure isn't that important; an unreliable start relay is unacceptable.

Of course, this response to a failed test assumes that the wiring is correct, and the connections are sound. A poor wiring connection at the relay coil, for example, can cause the same apparent failure.

If during startup, the motor struggles to rotate and no current is shown on the clamp-on ammeter, several possible causes exist. The start relay contacts may be stuck open, or the start capacitor may have failed open. As mentioned previously, incorrect wiring and poor connections can also create problems. In this case, check the relay contacts with the system de-energized and the contact leads removed. If the contacts are open, there will be no continuity.

1.3.3 Start Thermistor Testing

A start thermistor (*Figure 12*) is another device used to provide starting torque for single-phase compressors, without using a start capacitor or relay. A start

Figure 12 PTC start thermistor.

thermistor is a temperature-sensitive device that changes its electrical resistance in response to temperature change. Positive temperature coefficient (PTC) thermistors increase their resistance to current flow with an increase in temperature.

Figure 13 shows a PSC compressor motor with a PTC start thermistor. As shown, the PTC thermistor is wired in parallel to the run capacitor.

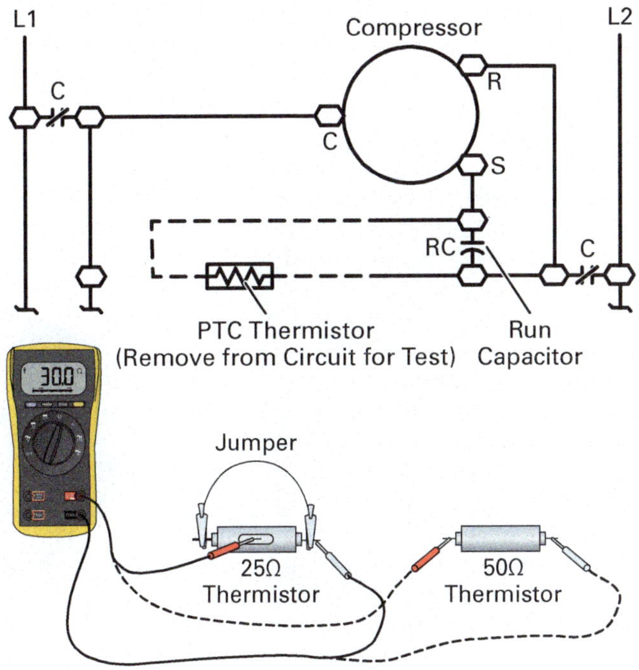

Figure 13 Start thermistor test.

At room temperature, the start thermistor resistance is very low. Those used for HVACR compressors generally have a resistance of about 25Ω or 50Ω. When the compressor is energized, a surge of current is applied to the start winding, because the low resistance of the PTC thermistor is effectively bypassing the run capacitor. This surge increases motor starting torque. The high current causes the PTC thermistor resistance to increase very rapidly to several thousand ohms.

As the resistance rises, current flow through the thermistor ceases and takes the path of least resistance—through the run capacitor. The motor then runs as a normal PSC motor. A small leakage current through the thermistor keeps it warm and the resistance high while the motor is running. Circuit operation remains this way until the motor is de-energized. After a short cool-down period, the thermistor's resistance will once again be low enough and ready for another operating cycle.

WARNING!

A PTC start thermistor quickly becomes hot when energized, reaching temperatures exceeding 200°F (93°C). Allow the thermistor to cool before handling.

The thermistor is tested with the equipment de-energized, any capacitors discharged, and the thermistor isolated from the rest of the circuit. However, allow the thermistor to cool before handling, as it becomes very hot during an operating cycle.

With the thermistor at or very near the ambient temperature, measure the resistance. If the thermistor has three terminals, measure the resistance between the center terminal and the end, and add a jumper wire across the ends, as shown in *Figure 13*. The cold resistance of any PTC thermistor should be about 100% to 180% of the resistance rating. For example, a thermistor rated at 25Ω should have a cold resistance of 25Ω to 45Ω. If the PTC thermistor resistance is lower, or more than 200% higher, than its rating, it should be replaced.

Note that, if the resistance of a start thermistor is acceptable and yet the compressor still does not start consistently, a start capacitor and start relay may have to be installed to ensure sufficient compressor starting energy is consistently provided. PTC start thermistors are not as effective at moving the motor as start capacitors. While hard-start kits are available from manufacturers, they are also available as aftermarket accessories through HVACR wholesalers (*Figure 14*).

Figure 14 Combination relay/start capacitor hard-start kit.
Source: RectorSeal, LLC

1.4.0 Three-Phase Motors

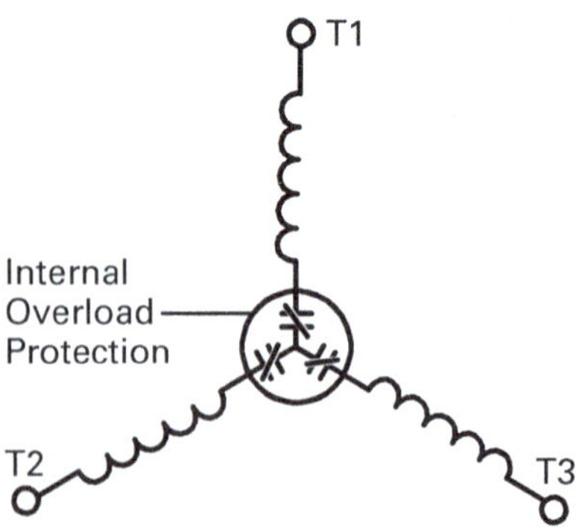

Figure 15 Three-lead, wye-connected, single-voltage three-phase motor.

Part-winding start: An approach to starting a three-phase motor that uses only a portion of the motor windings to initiate rotation, reducing starting current and heat production. Motors must be wound for part-winding start and have more leads than a standard three-phase motor.

Three-phase motors (*Figure 15*) are used when greater horsepower (hp) is needed. Three-phase motors range in size from 1 hp to 400 hp. Three-phase motors offer superior starting torque, without the need for start accessories, such as capacitors and thermistors. When compared to a single-phase motor of the same horsepower, a three-phase motor uses significantly less current on each leg, which reduces the required wire size.

All three-phase motors have at least three windings, with each winding having an equal resistance and the same size and number of wire turns. Six- and nine-lead three-phase motors are also found in large applications where **part-winding start** is necessary to reduce the initial inrush current at startup. This method uses one-half or two-thirds of the motor winding during startup to initiate rotation, before energizing the full windings.

The advantage is a reduction in starting current as well as a reduction in the heat produced. Motors that divide each winding in half will see a 50% reduction in starting current. Motors that use two-thirds of each winding experience a current reduction of 33%.

A typical part-winding start, single-voltage motor has six leads instead of three, as shown in *Figure 16*. Two contactors with a slight time delay between their activation are often used to start the motor. The start contactor cannot remain connected longer than one or two seconds. In some cases, rotation does not occur immediately, but the magnetic fields are partially established to assist starting once the full windings are energized.

Three-phase motors designed to operate at more than one voltage require more motor leads. *Figure 17* shows a dual-voltage, nine-lead, wye-connected

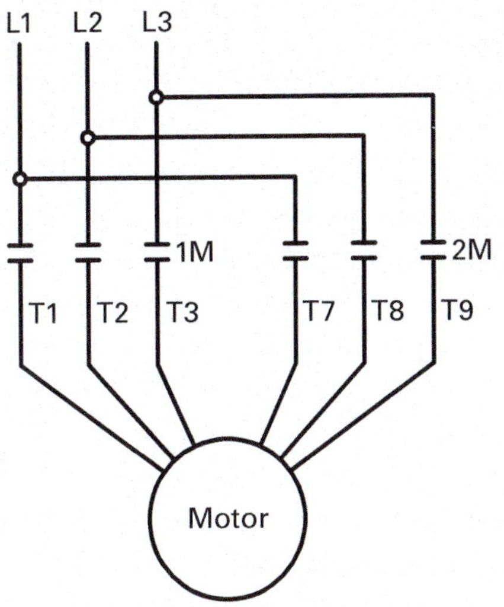

Figure 16 Six-lead, three-phase motor connected for part-winding start.

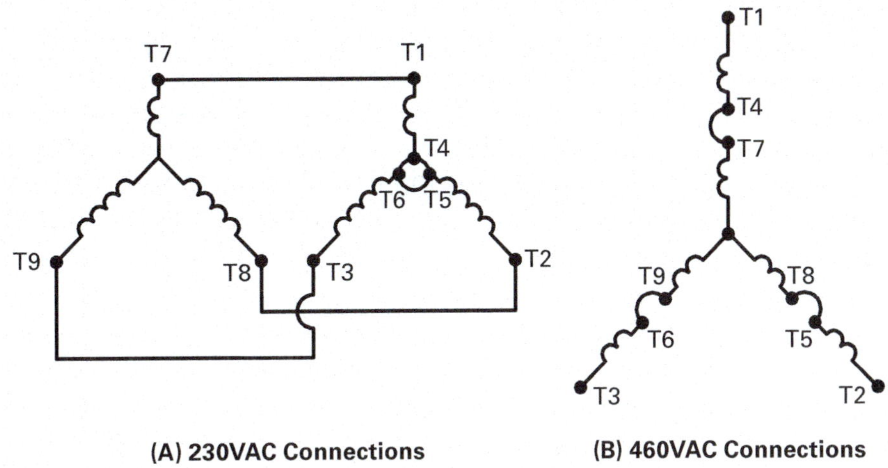

Figure 17 Nine-lead, dual-voltage, wye-connected three-phase motor.

three-phase motor designed to operate at either 230VAC or 460VAC. *Figure 18* shows a dual-voltage, nine-lead delta-connected three-phase motor. For 230VAC operation, the windings are connected in parallel. For 460VAC operation, the windings are connected in series.

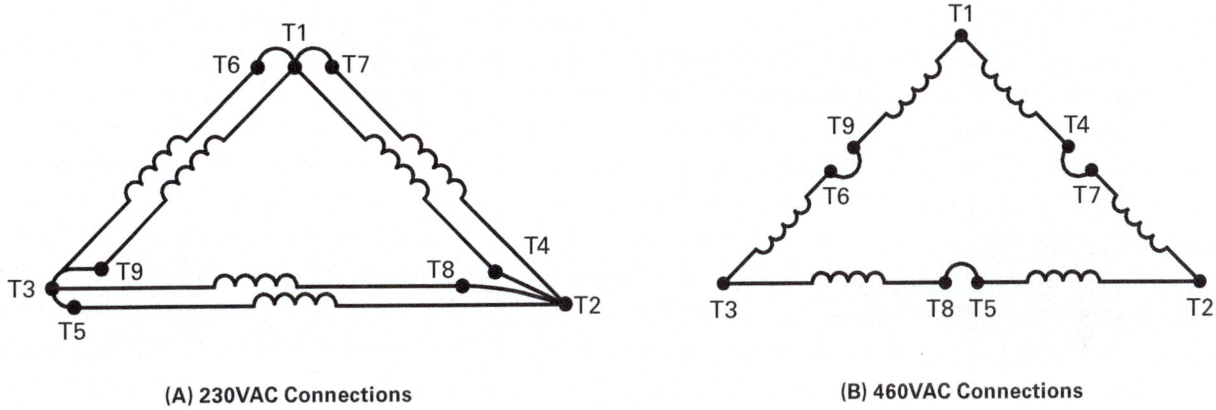

Figure 18 Nine-lead, dual-voltage, delta-connected three-phase motor.

Three-phase motors, including those used in compressors, have another advantage when compared to single-phase motors. The rotation of a three-phase motor is easily changed by swapping any two of the three leads that power the motor (*Figure 19*). This can be done at the contactor or motor starter connections, on either the line side or load side of the contactor.

CAUTION

Part-winding and dual-voltage motors must be wired carefully, and the diagrams provided with them can be confusing. Wire these motors carefully, in accordance with both the motor and equipment manufacturer's guidance, to ensure proper operation. Improper wiring connections can quickly damage a motor when it is energized.

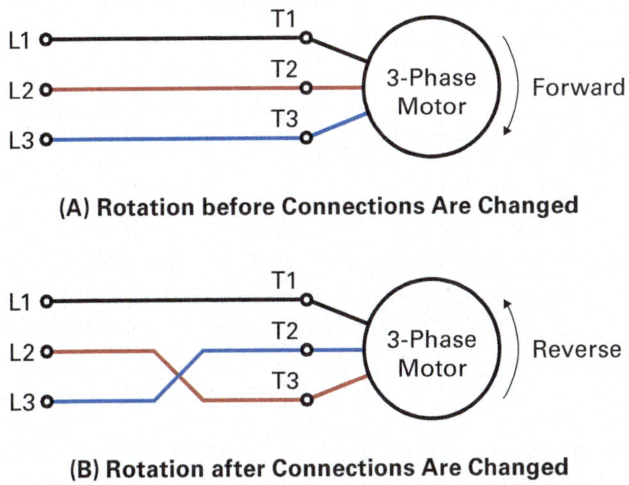

Figure 19 Changing the direction of rotation in a three-phase motor.

1.4.1 Testing Three-Phase Motor Windings

Testing the windings of a three-lead, three-phase motor is very easy. This is also a task done more often today, due to the increased use of variable-speed, inverter-driven compressors.

Inverter-driven compressors (*Figure 20*) are found in most ductless systems, as well as in deluxe cooling and heat pump systems that operate well above the minimum level of efficiency. The compressors used are driven by what is essentially a simple three-phase AC motor.

To test the integrity of the windings in a three-phase motor, simply check the resistance between each pair of windings—T1 to T2, T1 to T3, and T2 to T3 (*Figure 21*). The resistance should be the same winding-to-winding. Smaller motors will have a higher resistance across the windings than larger motors. The resistance should never be 0 ohms (Ω), nor should any winding be open unless an internal overload is open. Test for grounded windings using the same method used for single-phase motors.

Figure 20 Variable-speed, inverter-driven scroll compressor and inverter power module (IPM).
Source: Courtesy of Emerson

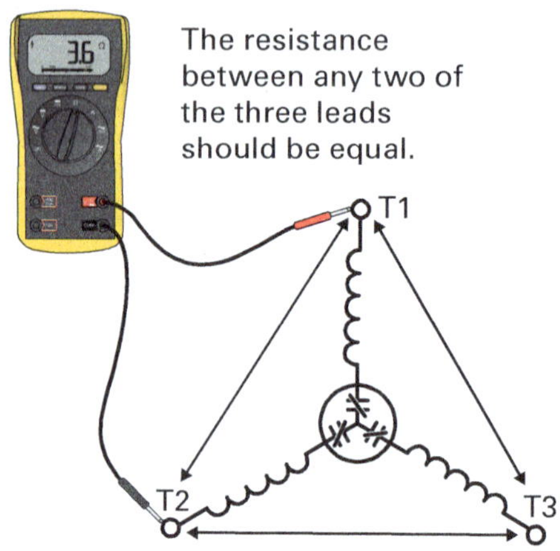

Figure 21 Testing three-phase motor windings.

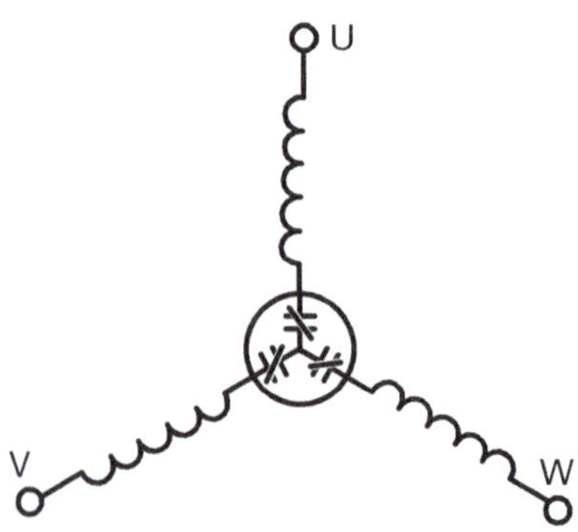

Figure 22 Inverter-driven compressor motor windings.

The windings of an inverter-driven compressor are tested the same way. The resistance between the windings should be the same for each pair. The one difference you will encounter is the labeling of the windings. The windings for an inverter-driven compressor are labeled as U, V, and W (*Figure 22*).

1.4.2 Testing Inverter Power Modules

The development of variable-speed compressors has brought huge gains in energy efficiency to the industry. Using a three-phase AC compressor, an inverter power module (IPM) rectifies AC to DC, processes and filters the DC output, then inverts the DC output back to a simulated AC for consumption. By doing so, the compressor can be driven at an infinite number of speeds within its capacity. An example of an IPM is shown in *Figure 23*.

In a traditional system, the compressor is simply cycled on and off according to demand. An inverter-driven compressor varies the capacity to precisely track the load. Most can operate down to as low as 20% of the rated full speed. IPMs offer *soft start*, meaning that the current to the compressor is initiated slowly. Most also accept inputs from safety devices, such as the high-pressure switch, and protect themselves from an excessively high or low AC input voltage.

These compressors are not typically equipped with an external crankcase heater. IPMs routinely use *stator heating* to keep the compressor warm. This comes from a very low current passed through the stator that is not enough to cause the compressor to rotate but does generate a small amount of heat.

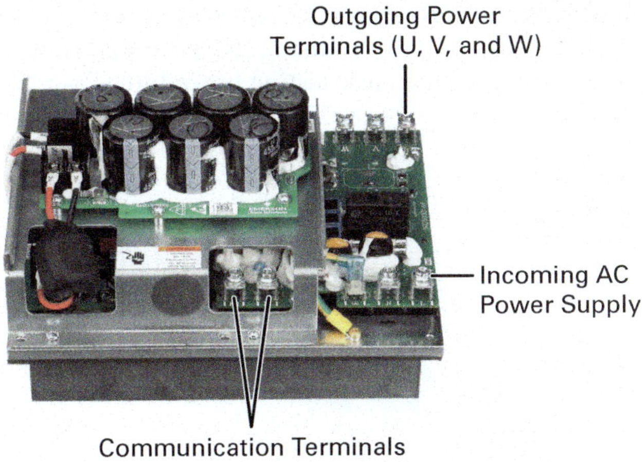

Figure 23 Inverter power module.
Source: Courtesy of White-Rodgers by Copeland

IPMs are similar from unit to unit, although not precisely the same. Specific IPM test procedures can be conducted using a standard multimeter. However, the tests are somewhat complex and are best conducted in accordance with a manufacturer's detailed instructions for their product. Some IPMs do have built-in diagnostics and report specific problems using codes on a small display or through blinking LEDs.

However, test devices are made specifically for testing the IPM output to the compressor. The test kit shown in *Figure 24* comes with the necessary cables and probes to test the output, as well as the instrument itself. Note that this device can also test the serial communications between the indoor and outdoor units of a system. Using this type of tester is also a safer option than making a number of electrical measurements at various points on the IPM with a multimeter while the module is energized.

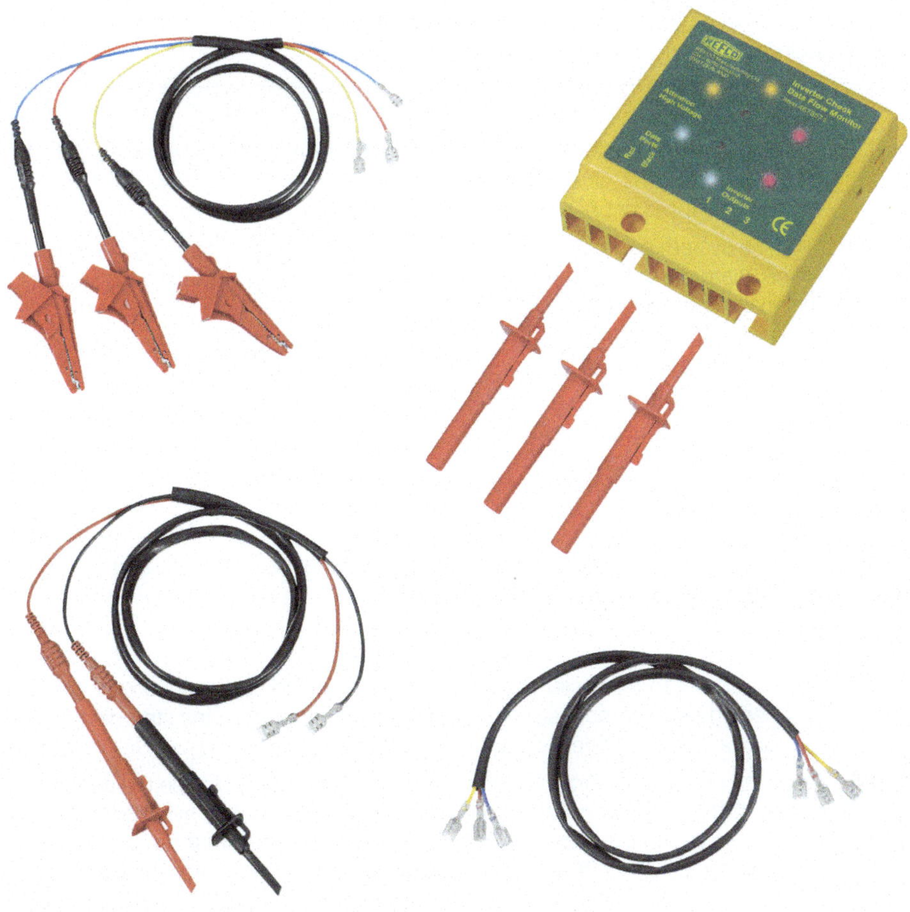

Figure 24 Inverter phase-testing module and probe kit.
Source: REFCO Manufacturing Ltd.

NOTE

Guidance in using this device to test serial communications is found in NCCER Module 03315, *Zoning, Ductless, and VRF Systems*.

CAUTION

Before using the tester, de-energize the system and test the compressor windings for shorts and grounds. An electrically damaged compressor may cause damage to the IPM.

The unit shown uses a series of LEDs to indicate the state of the three-phase power outputs (*Figure 25*). Before using the tester, carefully read and follow the operating instructions. The basic testing procedure is as follows:

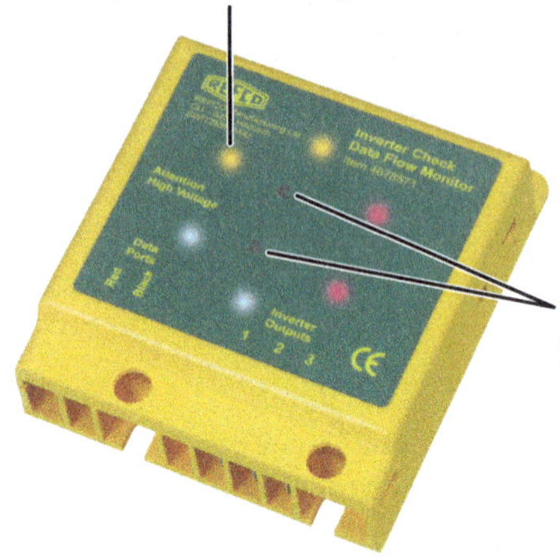

Figure 25 Phase-testing LED array.
Source: REFCO Manufacturing Ltd.

WARNING!

IPMs are equipped with large capacitors that can deliver a powerful shock. The module is typically protected by an outer cover. Before removing the cover, de-energize the system, then lock out and tag all forms of energy. Use a voltmeter to ensure the power is off. Allow a minimum of three minutes to elapse before touching or removing the module cover, providing time for the capacitors to discharge.

Step 1 With the cover off, look for signs of damage or failure, such as burn marks or broken circuit paths. Check the fuses mounted on the board, if any, although many of the fuses onboard are not designed for field replacement.

Step 2 With the system de-energized, label the U, V, and W power leads connected to the compressor from the IPM. The leads must be reconnected to the same terminals. Disconnect the leads from the compressor.

Step 3 Connect the three probes from the inverter tester kit to the U, V, and W power leads from the IPM. The probes should be connected to the tester inverter outputs labeled 1, 2, and 3. Place the tester in a safe location, preferably in a shaded area so the LEDs are more clearly visible.

WARNING!

Ensure the connections between the probes and the power leads are secure and protected from touching metal or grounded surfaces.

Step 4 Restore power to the system and set the thermostat to call for heating or cooling. Remember that most systems will not start immediately; the start delay typically ranges from three to five minutes.

Step 5 Once the power leads from the IPM are energized, you should see the six red, yellow, and blue LEDs in the circular pattern begin to flash in a symmetrical pattern. As the output to the compressor increases thus increasing its speed, the lights typically stop flashing and appear steady. This indicates the IPM is performing as expected. Asymmetrical flashing patterns or the failure of one or more LEDs to illuminate indicate a fault in the IPM.

Some IPMs will not allow for a valid test to be conducted without the compressor connected. When no power is being consumed, some boards will disable the power output. However, this same test can be conducted with the compressor connected. As noted previously, test the compressor windings for shorts and grounds before reconnecting it to the IPM for further testing.

The IPM outgoing power terminals U, V, and W often have a place for a probe to be securely inserted, or terminals to which alligator clips can be connected while the compressor leads are also in place. Make sure the system is de-energized and the capacitors have discharged before connecting the tester leads to the IPM outgoing power terminals.

The tester should display symmetrically flashing LEDs when the compressor is energized. The condenser fan usually starts at or near the same time. As the compressor speed increases (assuming the load indicates the need for additional capacity), you will see the LEDs transition from flashing to steady illumination.

While a field tester like this one can't identify a specific fault in the IPM, it can determine if the output to the compressor is correct. Since field repair of the IPM is not practical, the module must be replaced if it is not functioning as it should. Keep in mind that vendors may offer credit for a faulty module when a new module is purchased. The old one is then tested and repaired at the factory. Parts like these are expensive, so it is important to be certain that your diagnosis is correct. Test and retest as necessary.

NOTE

The IPM may only energize the power leads for a few seconds without the compressor connected, or it may fail to energize the leads at all. When the IPM energizes the compressor but does not sense the expected current flow, it assumes there is a problem and may respond by de-energizing the power leads.

1.0.0 Section Review

1. Which of the following describes the construction of the windings in a PSC motor?
   a. The start winding contains the fewest number of turns, using heavy gauge wire.
   b. The common winding has more turns than either of the other two windings.
   c. The run winding is constructed using conductive metal bars banded in a circular pattern.
   d. The start winding contains the most turns and is constructed of lighter gauge wire.

2. Which of the following results would you expect when testing the windings of a healthy three-phase motor with three leads?
   a. Each of the three windings should have a path to ground.
   b. Only one of the three windings should have a clear path to ground.
   c. The resistance measured across each pair of windings should be nearly identical.
   d. The resistance measured across one pair of windings should double and triple as you test the remaining pairs.

3. At what range of operating speed should a start relay for a single-phase motor be energized to open its contacts?
   a. 95% to 100%
   b. 75% to 80%
   c. 35% to 40%
   d. 15% to 20%

4. True or False? The direction of rotation for a three-phase motor is determined at the factory and cannot be changed in the field without removing the motor end bell.
   a. True
   b. False

2.0.0 Electronically Commutated Motors

Performance Task

4. Test an ECM motor.

Objective

Identify and explain how to service electronically commutated motors (ECMs).

a. Identify and relate basic ECM operating concepts.
b. Explain how to install and set up an ECM.
c. Explain how to test and troubleshoot an ECM.

ECMs are now widely used in residential and commercial HVACR equipment. These motors offer several benefits to the end user including the following:

- *Increased energy savings* — An ECM consumes a fraction of the power consumed by a conventional PSC motor.
- *Improved comfort* — When coupled with modulating or multistage heating and cooling products, an ECM can adjust the fan speed to match the heating or cooling capacity. Fan speed can also be reduced to allow for better dehumidification in the cooling mode.
- *Quieter operation* — ECMs start at a low speed by design and ramp up to the required speed. ECM construction also provides quieter operation than PSC motors.

Although it is possible to replace PSC motors with ECMs in the field, conversions are rare due to the cost of both parts and labor. Despite the added efficiency, there is little return on the investment.

2.1.0 ECM Characteristics and Operation

ECMs can be described as three-phase, direct-current (DC) motors with a permanent magnet rotor. Most ECM motors consist of two sections—the motor section and the control module (*Figure 26*).

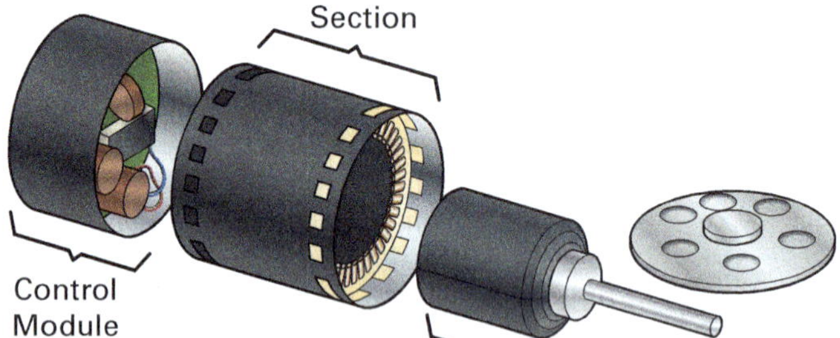

Figure 26 ECM major components.

The control module serves two purposes. First, it converts the single-phase AC power provided to DC. The control module can work with more than one AC input voltage by simply adding or removing a jumper on the five-pin connector in the motor control section. The control module converts the AC to three-phase DC, which is then fed to the motor. Second, the control module accepts and translates external inputs that determine its output to the motor.

The permanent magnet rotor is rotated by sequentially energizing the three sets of windings in the stator to move the rotor in a clockwise direction (*Figure 27*). The signal applied to each stator is a square-wave voltage produced from the module's rectified and conditioned DC power output.

To advance the rotor, the square-wave voltage is first applied to the A-B windings that move the rotor clockwise. Power is then applied to the B-C windings,

Did You Know?

ICM Versus ECM

Early HVACR system literature and manufacturers referred to ECMs as integrated control module (ICM) motors. The words themselves indicate that a module was incorporated into a motor for its control. Today, these motors are referred to as ECMs (electronically commutated motors).

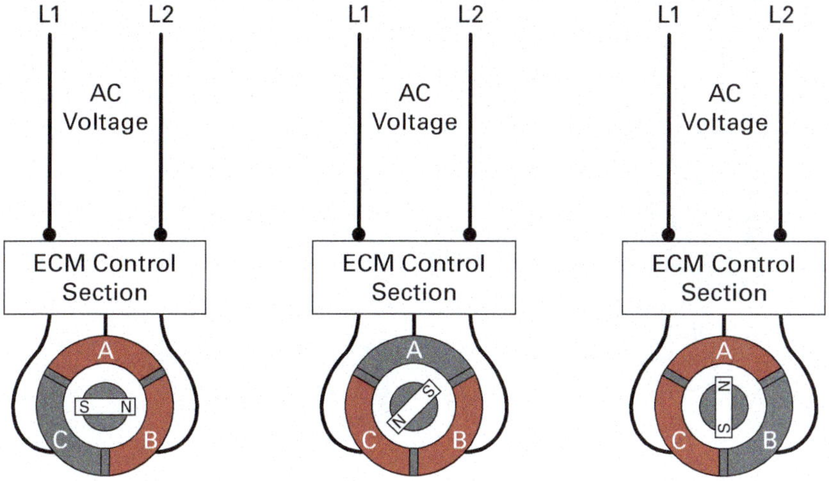

Figure 27 ECM motor operation.

and then to the A-C windings to complete a full revolution of the rotor. The speed at which the windings are energized determines the rotational speed of the motor.

ECM motors look much like conventional PSC motors, but have an additional section bolted to the motor on the end opposite the shaft. The added section is the control module. The size of the control module has been reduced over the years as technology advances.

The three primary versions contain a five-pin connector where power is applied (*Figure 28*). The addition or deletion of a jumper on the five-pin connector sets the motor's AC operating voltage. The connector used to carry the various signals between the motor and the equipment is either a sixteen-pin connector (Versions 2.3 and 2.5) or a four-pin connector (Version 3.0). Version 3.0 communicates in a different way, thus the need for fewer wires.

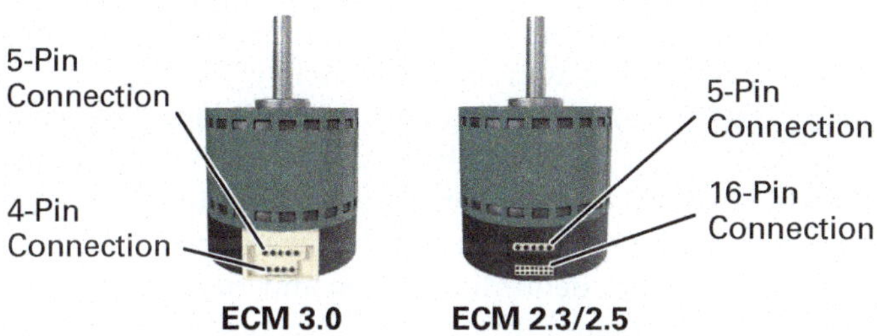

Figure 28 ECM identification.

ECMs have another characteristic that can help identify them. When turning the motor shaft by hand, points of high and low resistance can be felt. That is caused by the constant magnetic field of the permanent magnet rotor. The shaft of a typical AC induction motor rotates smoothly and easily.

Unlike conventional PSC motors, power is always available to the control module through the five-pin connector as long as the equipment is energized. After additional speed control inputs are received through the sixteen-pin or four-pin connectors, the motor will start and run. Although power is applied to the module, it cannot function without external inputs that provide instructions.

CAUTION

Version 3.0 motors may not be used to replace Version 2.3 or Version 2.5, due to the difference in communication formats. A Version 3.0 motor can only be used as a replacement for earlier versions if the equipment control board with which the motor module communicates is replaced as well.

2.1.1 ECM Speed Control

The speed of an ECM is controlled by the firmware in the control module and the input signals from the electronic interface in the equipment. For example, if a gas furnace has multiple heating stages (high- and low-fire modes), the furnace control board may request a lower fan speed during low-fire operation. When the equipment is installed, the furnace board is configured by the installer to relate the desired fan speeds with the furnace's operating modes.

Figure 29 shows how an ECM installed in a two-stage furnace responds to a call for heat. As the room temperature falls, the first stage of the room thermostat sends power to the furnace control board via W/W_1 (or via a DC signal in some systems).

On Site

Rock and Roll!

When an ECM motor starts, it may rock back and forth a few times before beginning to spin in the right direction. This is considered normal, as it is verifying which direction the motor should be turning before committing to it.

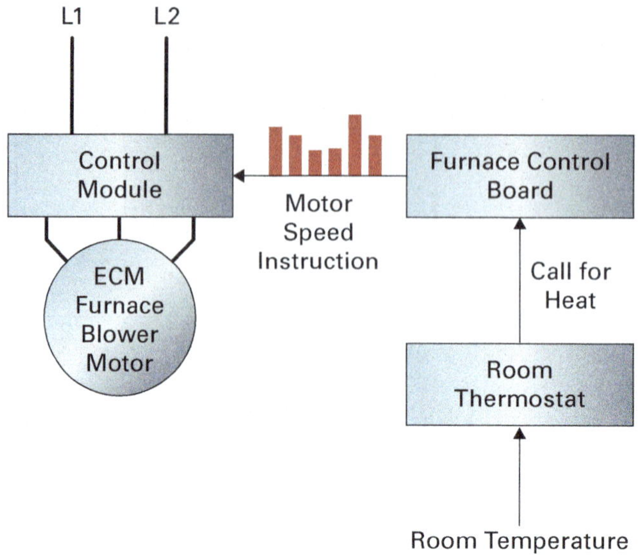

Figure 29 ECM speed control.

Because the call is for first-stage heat only, the furnace control board responds by sending a signal to the ECM requesting low-speed operation. These signals are typically in the form of pulses. The frequency of the pulses tells the motor what rotational speed is needed. If the room thermostat calls for second-stage heat, the furnace control board would send new speed information to the control module.

2.2.0 ECM Installation and Setup

Other than replacement scenarios, ECMs are factory-installed. No setup is required for the motor itself when the equipment is installed. Any setup is done on the equipment interface for the ECM. This setup is essential, as ECMs are installed in a variety of systems with varying airflow needs.

For example, a 100,000 Btuh gas furnace with an ECM may be installed with a 2-, $2^1/_2$-, or 3-ton cooling unit. Each size has its own airflow requirements based on a nominal airflow of 350 to 450 cubic feet per minute (cfm) per ton. If that same gas furnace has multistage heating, each stage of heating has its own airflow needs. If the system has a dehumidification and/or a humidification mode, more factors must be considered. Continuous fan operation may also have a dedicated fan speed. The control that sends signals to the ECM must be set up to ensure that all the possible operating modes are considered in the dataset.

The furnace control board on a premium furnace or air handling unit can accommodate all those diverse airflow requirements. How that is accomplished varies by manufacturer. Airflow setup can take the form of rotary selection switches, dual inline package (DIP) switches (*Figure 30*), microswitches, or small jumper pins. Some boards today are set up by advancing through an LED display and selecting a code that corresponds to the range of air volume needed. More sophisticated systems allow setup to be done with a wired or wireless user interface such as a laptop computer, smartphone, or tablet. The furnace board then relays the needed information to the motor control module based on the mode of operation.

Figure 31 shows an interface board used on a packaged heating/cooling unit equipped with an ECM. On the board shown, equipment capacity and auxiliary electric heater size settings, determined using microswitches, are factors that help determine blower speed. A portion of the twelve-pin connector shown is dedicated to ECM communication, using the four-pin configuration of a Version 3.0 ECM.

If an ECM must be replaced, use the correct replacement based on the version number. This is essential due to the differences in communication protocol. No universal ECM replacements currently exist. Even when an exact replacement is provided, it is important to read and apply the instructions provided with the motor. Changes in design or wiring may have taken place since the original equipment and motor were placed in service. A replacement motor can always

Figure 30 DIP switches.
Source: Mihailgrey/iStock

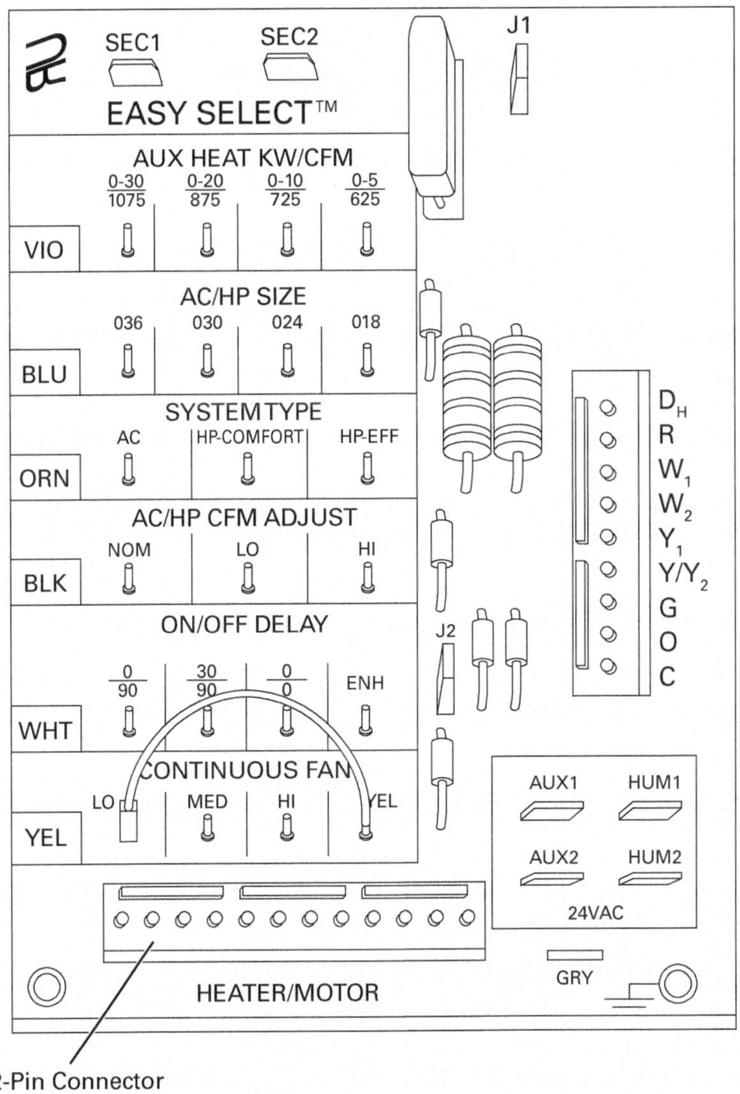

Figure 31 ECM equipment interface board.
Source: Carrier Corporation

be obtained from the equipment manufacturer. HVACR wholesalers also offer replacement motors, with cross-reference charts available for their stocked brand.

The only motor setup required with the proper motor in hand is to ensure the five-pin connector is configured for the correct input voltage. The jumper is typically installed for 120VAC operation and removed for 240VAC operation, but other configurations may exist. Note that the two connectors—the power connector and the module communications connector—must be compatible with the connectors provided on the equipment. The connectors are keyed so that each connector can only be mated with a compatible connector of the same design, and can only be connected one way. This helps eliminate any error in connections.

When a new motor is installed, be sure to position the motor to allow the wiring connections to hang down between the 4 o'clock and 8 o'clock positions (*Figure 32*). This helps prevent any moisture from entering the connectors. As the wires exit the motor, form a drip loop that encourages water to roll down the wiring rather than into the motor.

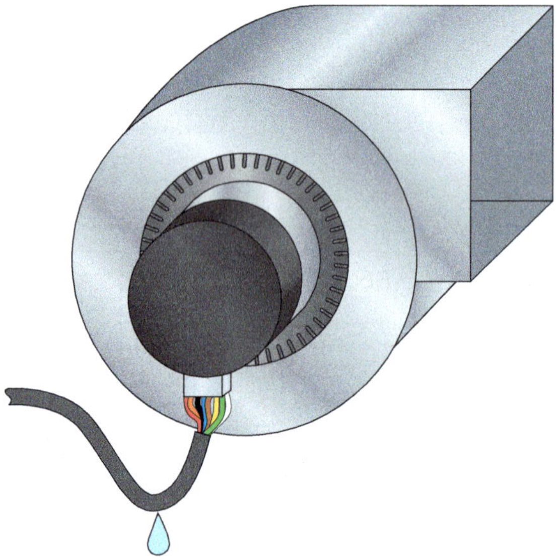

Figure 32 Forming a drip loop.

Once the new motor is installed, check the operation thoroughly in all operating modes to ensure the motor's response to the operating mode is correct. With fossil-fuel furnaces, ensure that the temperature rise in the heating mode also remains within the allowable range. The best practice is to measure the static pressure on the system and also to calculate the system air volume.

2.3.0 ECM Troubleshooting

Testing or troubleshooting an ECM often happens during a routine service call, especially one where airflow is an issue. Where comfort and airflow are involved (as opposed to complete failure), always check for the obvious problems first, such as the following:

- Dirty air filters, closed dampers, or blocked supply registers or return grilles. It may be necessary to measure system air volume to confirm that the proper volume of air is being delivered.
- Low refrigerant charge.
- Incorrect thermostat setpoint or calibration issues.

If the equipment control board has a self-diagnostic feature or displays fault codes, use those features to help identify a problem.

If no obvious problems are found, check the equipment to confirm the control board has been properly configured for the airflow requirements. It is common for some manufacturers to have default settings on the controller from the factory. Those default settings are not appropriate for every application. An incorrect setup in an ECM-equipped system can result in poor system performance.

2.3.1 Motor Testing

The following troubleshooting procedures are general in nature and are not meant to replace specific troubleshooting procedures provided by an equipment or motor manufacturer. However, they do apply to most ECMs serving residential and light commercial systems.

If the ECM is not running when it should be running, de-energize the furnace or air handling unit. Then disconnect the five-pin connector and restore power. Check for the correct voltage at the connector pins (*Figure 33*). With a 120VAC power supply, the polarity of the voltage must also be correct. The applied voltage should be ±15% of the nameplate voltage. Note that the motor also requires a reliable ground connection for proper operation.

CAUTION

Always remove power from the equipment before connecting or disconnecting ECM electrical harnesses. Never connect or disconnect ECM wiring harnesses with power applied to the equipment.

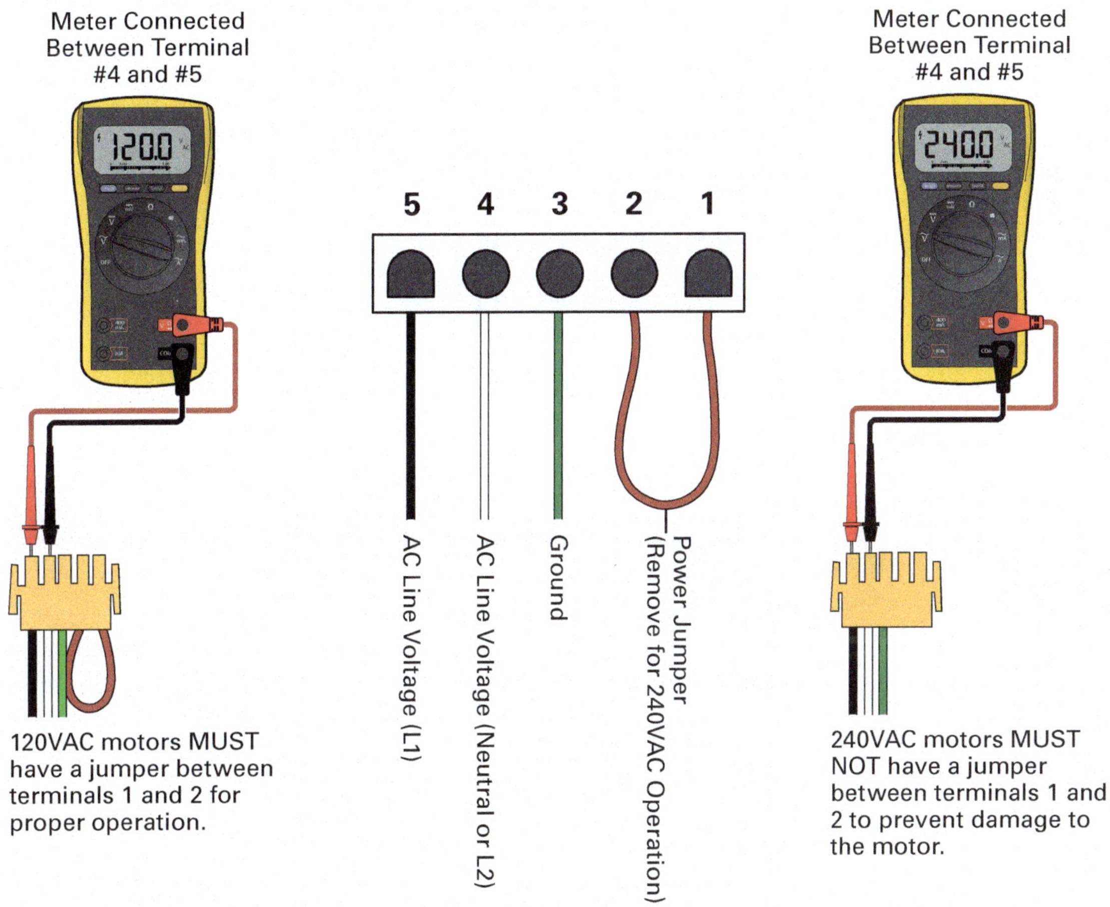

Figure 33 Input voltage check.

Equipment manufacturers often supply charts listing the voltages that should be read across the sixteen-pin or four-pin communication connectors of the motor. Voltages can be AC or DC. The chart also may supply the conditions under which the various voltages should be read. These signals and voltages can be tested when specific information is available. This process determines whether the equipment control module is functional and sending the expected signals.

Devices that simulate communication input signals are commonly used to test the motor (*Figure 34*). The tester shown can be used with Version 2.3/2.5 motors with sixteen-pin connectors, and Version 3.0 motors that have four-pin connectors. The tester simulates signals from the control board in the equipment. Testers similar to the one shown in *Figure 34* are made by several manufacturers and are available through all HVACR parts and equipment wholesalers.

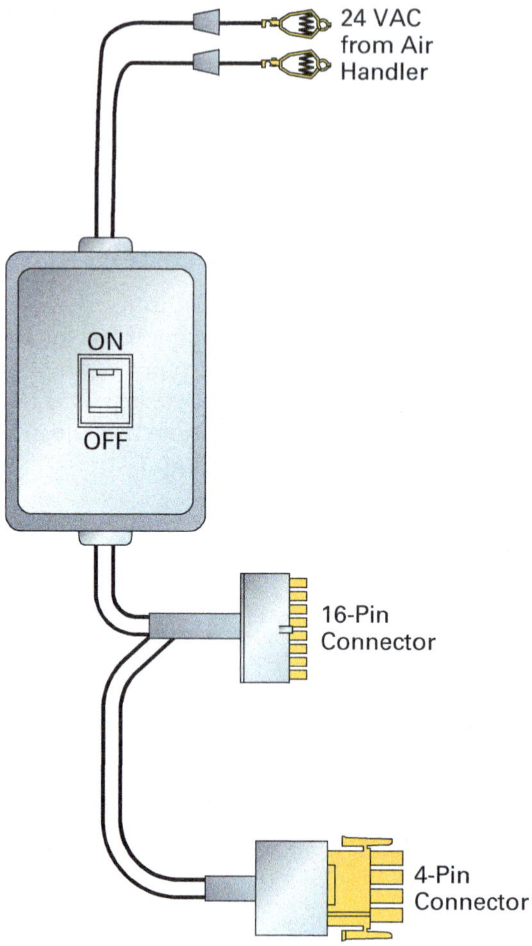

Figure 34 ECM tester.

Using a similar tester as the one shown in *Figure 34*, follow these steps to test an ECM:

Step 1 De-energize the equipment. Disconnect the sixteen-pin or four-pin connector from the motor but leave the five-pin power supply connector attached to it.

Step 2 Ensure the tester switch is off. Connect the alligator leads on the tester to the R and C terminals on the control transformer or terminal board to obtain 24VAC. This powers the tester.

Step 3 Connect the sixteen-pin or the four-pin connector on the tester to the matching connector on the motor control module.

Step 4 Switch the tester on.

- If the motor runs when the tester is turned on, the problem lies elsewhere. Potential causes include a failed air handler control board, or a loose connection. If the motor does not run, the problem may be within the motor control section or the motor itself. Remove the tester and restore all normal connections when done.
- If the motor did not run, testing must proceed to the motor and its control module.

Step 5 De-energize the equipment. Disconnect the sixteen-pin or four-pin communications connector from the motor as well as the five-pin power supply connector. Then wait at least 5 minutes before proceeding to the next step.

WARNING!

After disconnecting the power connector from an ECM motor, wait at least five minutes before accessing the control module. This allows time for the internal capacitors to discharge. If the capacitors are not allowed sufficient time to discharge first, an electrical shock may result.

Step 6 Remove the bolts securing the control module to the back of the motor. The motor section can usually remain in place in the blower assembly during this procedure.

Step 7 Disconnect the three-pin connector between the motor and module. Place the control module aside.

Step 8 The three-pin connector on the motor represents the three-phase motor windings. Use an ohmmeter set on its lowest scale to check the resistance between each pair of windings (*Figure 35*). Each winding should measure less than 20Ω. There should be no more than a ±10% difference between the three resistance readings.

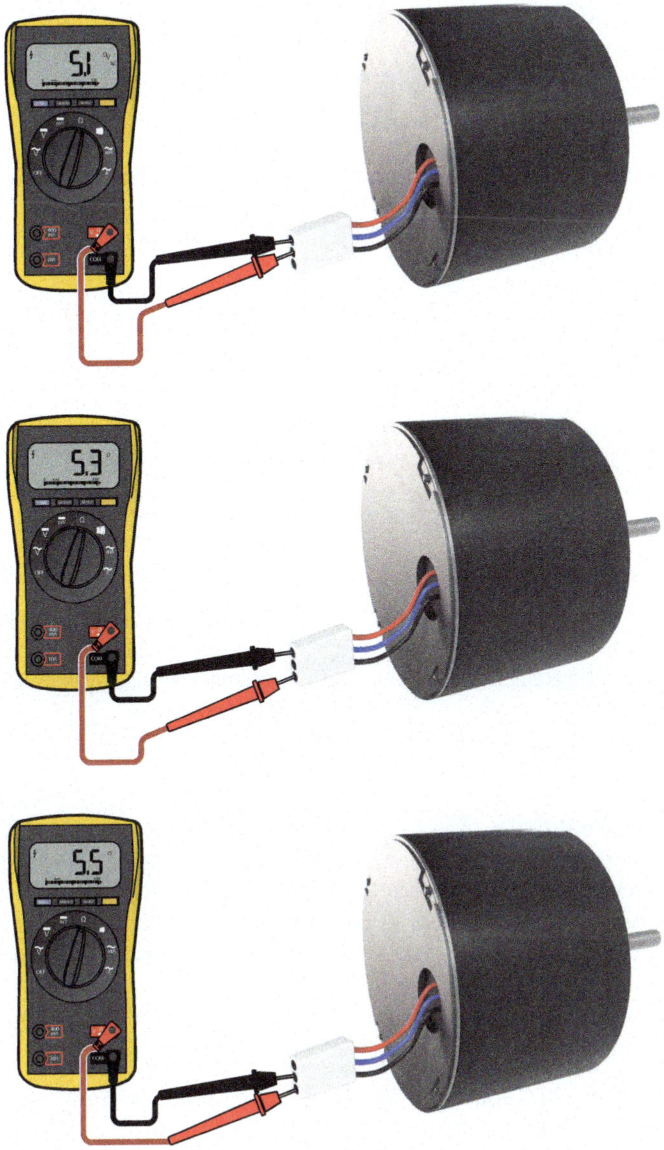

Figure 35 ECM winding resistance test.

Step 9 Set the ohmmeter on its highest scale to test for grounded motor windings. Place one lead on the frame of the motor (ground) and the other lead on each of the pins in the three-pin motor connector. Ideally, all readings will show no continuity (OL). Replace any motor with a winding that reads less than 100,000Ω to ground.

If the motor windings test as required but the motor didn't run when commanded using the tester, the control module is likely defective. In this case, only the control module needs to be replaced. The motor itself can remain. If the motor section is defective (grounded or open windings), replace the entire ECM assembly.

2.0.0 Section Review

1. What type of signal is applied to the stator of an ECM?
   a. Pure sine wave
   b. Modulated AC signal
   c. Current-limited AC signal
   d. Square-wave DC voltage

2. When installing an ECM, in which direction should you position the motor wiring connections?
   a. At the bottom of the motor
   b. At the top of the motor
   c. Only to the left of the motor
   d. Parallel with the motor shaft

3. What is the maximum variance allowed in the resistance measured between ECM motor windings?
   a. ±1%
   b. ±10%
   c. ±15%
   d. ±20%

3.0.0 Variable-Frequency Drives

Performance Tasks

There are no Performance Tasks in this section.

Objective

Describe the operation of variable-frequency drives (VFD) and their selection considerations.

a. Describe the operation of a VFD.
b. Identify and describe common programmable VFD parameters.
c. Describe the important considerations for the selection of a VFD.
d. Define and describe dynamic motor-braking processes.

A variable frequency drive (VFD), sometimes referred to as an *adjustable frequency drive* (AFD), first rectifies three-phase AC, 60Hz input power to DC. The DC output is then filtered and inverted back to AC at an adjustable frequency to control the speed of an AC motor. VFD-controlled motors provide energy savings by eliminating most of the energy losses associated with older methods of modulating motor speed.

Many benefits can be derived from using a VFD (*Figure 36*). These benefits include the following:

- *Controlled starting* — Capable of limiting the starting current and reducing the resulting power line disturbance.
- *Controlled acceleration* — Provides soft start, adjustable acceleration based on time or load imposed, and a reduced motor size for pure inertial load acceleration.
- *Adjustable operating speed* — Modulated speed optimizes pump and blower operation to match the load. Lower speeds provide energy savings.
- *Adjustable torque-limiting* — Current limiting provides accurate torque control, protecting the machinery or process from damage resulting from excessive torque.
- *Controlled stopping* — Provides soft and timed stopping, as well as fast reversal of rotation with much less stress on an AC motor than **plugging**.

Plugging: An electric motor braking method in which reverse torque is applied by reversing the voltage or the phase sequence serving the motor. Although it provides rapid braking, the process places a great deal of stress on both the motor and the controlling device.

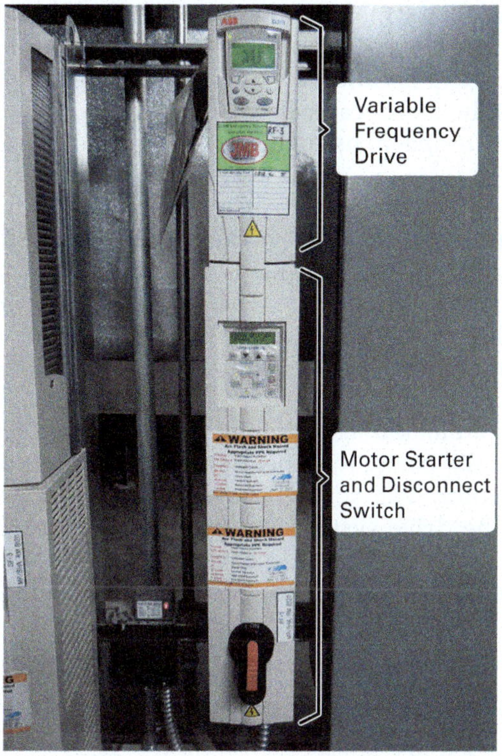

Figure 36 VFD with a matched combination disconnect/ motor starter serving a large fan.

3.1.0 Basic VFD Operation

The following formula determines the speed of an AC induction motor:

Motor speed (rpm) = 120 × frequency (Hz) ÷ number of motor poles

For any motor, the number of poles is a constant because the poles are integral to its construction. The formula, then, states that the speed of the motor is proportional to the applied frequency. As the frequency increases, the motor speed increases. Conversely, as the frequency decreases, the motor speed decreases.

On Site

Variable Frequency Drives

VFDs come in many shapes and sizes to serve a wide variety of applications.

Source: Schneider Electric

Pulse-width modulated (PWM): A method of controlling the average power or amplitude of an electrical signal, providing control of an electrical load. The speed at which the modulated signal changes (frequency of change) depends on the type of load, but for motor drives, the frequency of change is typically 4 kilohertz (kHz) to 16 kHz.

CAUTION

While VFDs can control the speed of a motor beyond the nameplate rated load current (RLA), the motor must have a service factor higher than 1.0 to safely make use of the additional current. In addition, motor-driven loads also have a maximum rotational speed that should never be exceeded. Before entering settings that operate a motor at a current exceeding the nameplate rating, ensure both the motor and load can accommodate the change.

To maintain a constant motor torque, VFDs maintain a consistent relationship between the voltage and frequency of the output applied to the motor. This relationship is called the *volts per hertz ratio (V/Hz)*. Done effectively, the speed of the motor can be controlled over a very wide range, even beyond the nameplate speed, using the proper torque characteristics for the application.

VFDs come in a wide variety of sizes and features. Today, programmable, microprocessor-based, **pulse-width-modulated (PWM)** VFDs are the most common. Pulse-width modulation helps eliminate dirty power problems, such as electromagnetic interference (EMI). Powerline problems were often associated with older VFDs that were state-of-the-art some years ago. For this reason, the remainder of this section will focus on the newer PWM-type of VFD.

Figure 37 shows a block diagram of a typical PWM VFD system. It consists of three primary parts: the operator control panel, the VFD controller, and the AC motor. The control panel provides an interface between the operator and the drive system. At the control panel, the operator can input commands to start, stop, and change the speed and/or direction of the motor. The operator can also monitor the VFD and motor status through the control panel.

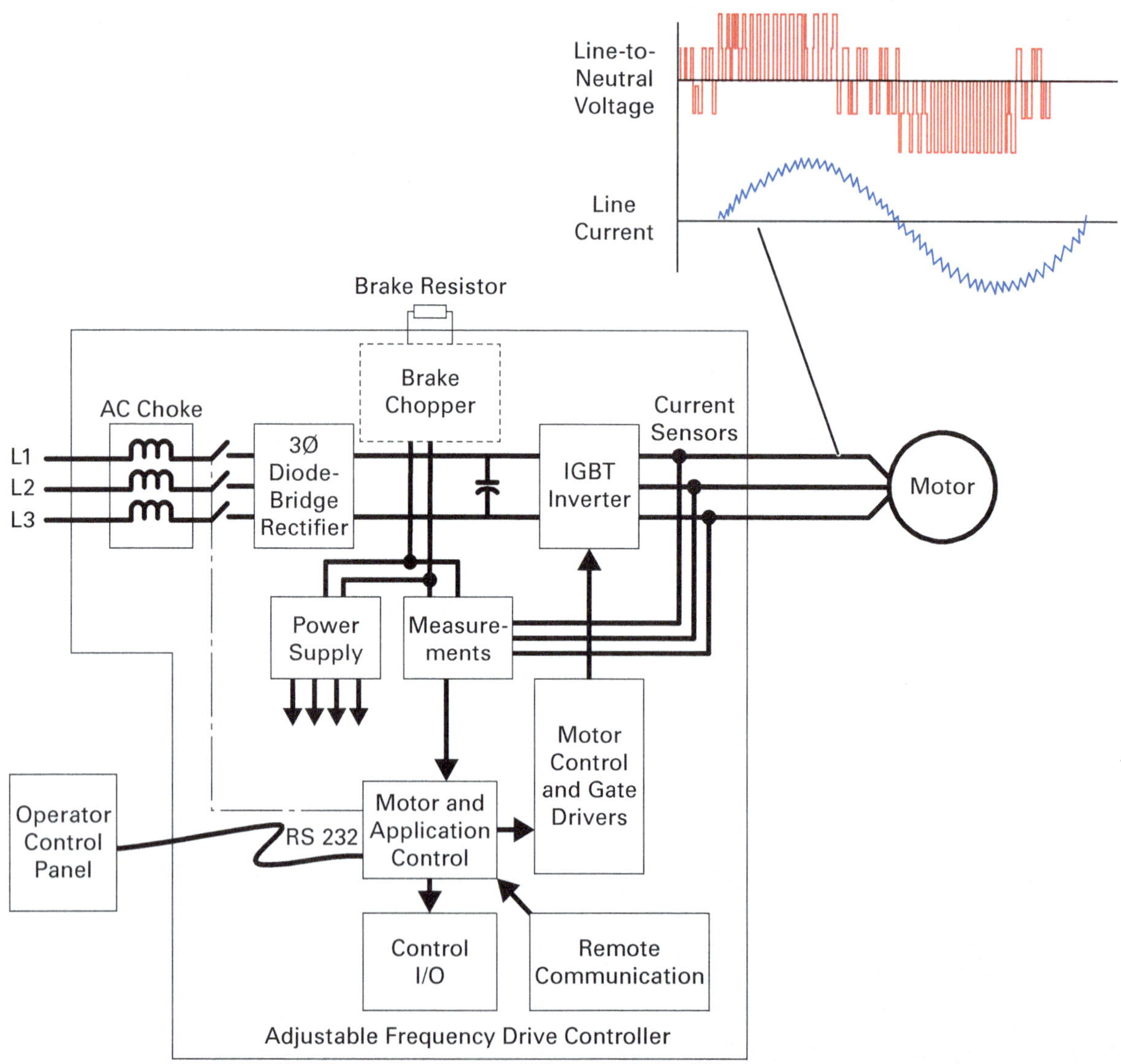

Figure 37 Pulse-width modulated (PWM) VFD schematic.

The control panel can often be removed and mounted away from the VFD itself. An input/output (I/O) interface board accepts external inputs and allows communication with personal computers (PCs) or external programmable logic controllers (PLCs).

The typical motor used with a VFD in the US is a standard NEMA Design A or Design B, squirrel-cage induction motor rated for 240V or 480V, three-phase, 60Hz operation.

The actual range of frequencies used by the VFD is programmable and largely depends on the motor and its application. The VFD also regulates the output voltage in proportion to the output frequency to provide a constant V/Hz ratio as required by the motor type.

The three-phase AC choke shown on the left side of *Figure 37* levels out any high-frequency (HF) disturbances from the utility, as well as any HF disturbances generated by the drive that could feed back to the utility. Think of the AC choke as a trap for these HF disturbances.

Since the inverter section is powered by a fixed DC voltage, the maximum amplitude of its output waveform is fixed. The **insulated gate bipolar transistor (IGBT)** inverter changes the DC input from the rectifier into a three-phase, PWM, AC voltage output to the motor. The IGBT is designed to provide the high-speed switching necessary for PWM VFD operation. The IGBT can switch on and off several thousand times a second.

Insulated gate bipolar transistor (IGBT): A three-terminal power semiconductor that is highly efficient (low power losses) and designed to operate at high switching frequencies.

Going Green

VFD Pumping Applications

VFDs have become very common in HVACR hydronic systems. To maintain steady flow rates and pressures in a large hot water heating or chilled water system, three-way valves are often used to control the flow of water through the various coils. When a zone no longer needs heating or cooling water, the water was diverted around the coil without affecting the flow rate. The pump operates continuously at a steady speed.

With a VFD, two-way valves may be used instead of three-way valves. The VFD modulates the pump motor speed to maintain a constant pressure while the flow rate varies. As valves serving the various loads close, the pump simply slows down to keep the pressure consistent. This saves a great deal of energy. However, some systems are still designed around three-way valves, and other systems may have one or more three-way valves strategically positioned in the system to allow a minimum amount of flow to continue when all the loads are satisfied. This keeps hot or chilled water readily available in the loop.

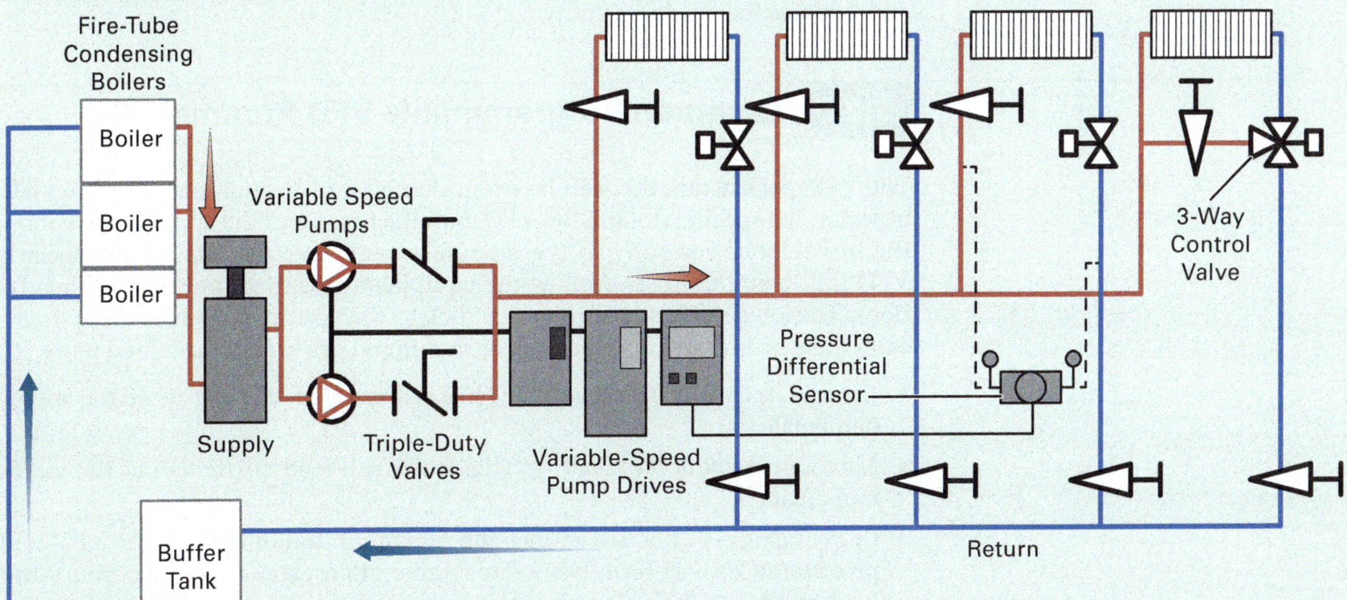

Primary Variable Flow System

VFDs typically have a range of output frequencies over which the V/Hz ratio remains constant, and a second range of output frequencies over which the voltage remains constant while the frequency varies (*Figure 38*). The frequency at which the transition from one range to the other occurs is called the *base frequency*. Between the base frequency and the maximum frequency points, the rated output voltage remains constant. Below the base frequency point, the output voltage is less than the rated voltage and is based on the V/Hz ratio. Depending on the motor characteristics, these two ranges correspond to *constant torque* (constant V/Hz) and *constant power* (constant voltage with an adjustable frequency).

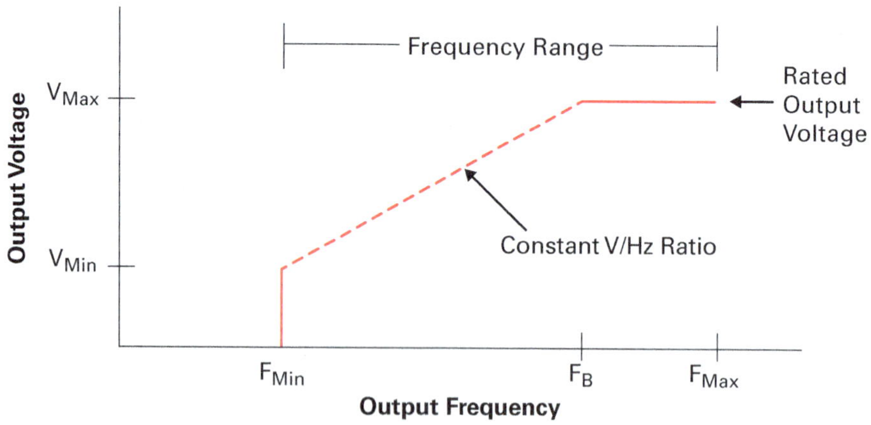

V_{Min} = Minimum Output Voltage
V_{Max} = Maximum Output Voltage
F_{Min} = Minimum Frequency
F_{Max} = Maximum Frequency
F_B = Base Frequency

Figure 38 VFD operating frequency range.

The VFD microprocessor controls the motor according to inputs from various sensors and according to its programmed settings. New commands can be entered at the control panel or through a remote device (if connected). The microprocessor calculates the changes needed to carry out the new command and initiates the needed activity.

3.2.0 Common Programmable VFD Parameters

The VFD parameters that can be programmed and/or monitored vary widely based on the specific design, the VFD manufacturer, and the application. Follow the instructions carefully in the manufacturer's user manual for the specific VFD unit being used, as well as the equipment manufacturer's recommendations. The job specifications may also dictate VFD settings. Some of the parameters that can be programmed and/or monitored are briefly described here.

- *Nominal voltage of the motor* — This is the rated voltage from the motor nameplate.
- *Nominal current of the motor* — This is the full load current from the motor nameplate.
- *Current limit* — This determines the maximum motor current the VFD will provide on a short-term basis. This value often exceeds the maximum full load current on the nameplate. However, the service factor of the motor must be greater than 1.0, and the maximum rotational speed of the load must never be exceeded.

- *Nominal speed of the motor* — This is the rotational speed from the motor nameplate.
- *Nominal frequency of the motor* — This is the frequency shown on the motor, which is typically either 60Hz, or 50Hz for motors designed for use in many countries outside the US.
- *Supply voltage* — This is usually a range that encompasses the nominal supply voltage to the VFD.
- *Minimum and maximum frequency* — This defines the frequency limits for VFD operation.
- *Acceleration and deceleration time* — Acceleration time defines the time required for the output frequency to accelerate from the set minimum frequency to the set maximum frequency. Deceleration time defines the time required to decelerate from maximum to minimum frequency.
- *V/Hz ratio* — Typically there are two V/Hz settings: *linear* and *squared*. The linear setting is used with applications requiring constant torque. In this mode, the voltage supplied to the motor changes in a linear manner with the frequency, from 0Hz to the nominal frequency of the motor. With the squared setting, the voltage to the motor changes following a squared curve from 0Hz to the nominal frequency of the motor. In this mode, the motor runs below its normal frequency and thus produces less torque. The squared mode is commonly used in applications where the torque demand from the load is proportional to the square of the speed, such as with centrifugal fans and pumps.
- *V/Hz optimization* — This mode provides an automatic torque boost. The voltage to the motor can be adjusted automatically, allowing the motor to produce sufficient torque to start and run at low frequencies when necessary. The allowable voltage increase depends on the motor type and horsepower. Automatic torque boost can be used in applications where the starting torque due to friction or pressure differential is high, such as with a conveyor.

On Site

PWM VFD Outputs Over Long Cable Runs

When a PWM VFD output is applied to a motor over a long cable run, the difference in the motor and cable impedance can cause voltage overshoots at the motor terminals. These overshoots can even exceed 1,000VAC—enough to easily damage the motor's winding insulation.

For small motors, this typically occurs when cable lengths exceed 33'. For larger motors, cable lengths exceeding 100' often cause a problem. This is one of the reasons you'll generally find the VFD close to the motor. It is always best to confer with the manufacturer and/or the system engineer in these conditions.

If you experience recurring motor failures, this is one of the conditions to look for as the cause. There are steps that can be taken to address long cable runs between the two components, and VFDs that are designed to accommodate long runs.

3.3.0 VFD Selection Considerations

Selecting a VFD for a particular application requires a deep understanding of VFD capabilities and the characteristics of the application. For this reason, evaluating the needs of a drive system and selecting the components is a task done by an electrical engineer, a drive manufacturer application engineer, or other trained and qualified persons. While this module introduces technicians to VFDs, far more training and experience is needed to make a practical selection. Most VFD manufacturers offer training that is specific to their product line to help both technicians and engineers understand and service their product.

NCCER Module 03318, *Appendix 03318A, Sample VFD Application Checklist* shows a typical checklist one might use to gather the data needed to evaluate a VFD selection. The remainder of this section briefly presents some of the factors that can affect VFD selection.

3.3.1 VFD Selection Basics

The basic requirement of VFD sizing is to match the torque-versus-speed capability of the VFD to the torque-versus-speed requirement of the driven load.

If the load torque exceeds 150% for constant-torque drives or 110% for variable-torque drives during startup, or intermittently while the motor is running, oversizing of the drive may be required. The minimum and maximum motor speeds required in the application will determine the drive's base speed. Each HVACR pump and blower has a range of rpm in which it is designed to operate in each system. For example, to deliver the minimum and maximum amount of chilled water, a given pump may need to operate within a speed range of 950 to 1,750 rpm.

The full-load current ratings of typical VFDs are matched to the full-load, three-phase motor current ratings in current *National Electrical Code*® (*NEC*®) tables. Generally, a VFD with a given horsepower rating will be adequate for a motor with an equal rating, but the actual motor current required under operating conditions is the final determining factor for VFD sizing. If the motor will operate at full load, the VFD output current rating must be equal to or greater than the motor nameplate current. If the motor is oversized to provide a wide speed range, the VFD should be sized to provide the current required by the motor at the maximum operating torque.

Motor oversizing should be limited to an increase of 1 hp. Gross oversizing is not recommended and results in wasted energy.

3.3.2 Compatible Motors

The motors used with VFDs are often standard National Electrical Manufacturers Association (NEMA) Design A or Design B motors. The use of other motor designs may require special drive characteristics.

Some VFD manufacturers recommend motors that are *inverter-duty grade*. However, these motors are expensive and are rarely stocked. For most HVACR applications, it's best to select a VFD that works with standard motors. Note that nameplates rarely indicate that a motor is designed for inverter duty. In this case, the manufacturer must be consulted.

The motor design letter assigned by NEMA identifies certain characteristics, such as its slip factor and locked rotor torque. Many motors are self-ventilated, meaning that the motor's rotor is designed to induce airflow through the motor. Operating self-ventilated motors at reduced speeds can cause a rise in the operating temperature. As the speed of such a motor is reduced, motor cooling is reduced due to the reduced speed of the rotor. Thus, this limitation determines the minimum allowable motor speed.

Derating of the motor (reducing its assigned horsepower) or forced ventilation may be necessary to achieve the needed torque output at reduced speeds without damaging the motor. For retrofit VFD applications, especially those involving older motors), some VFD manufacturers require that existing motors be replaced to ensure that the VFD is properly sized and not affected by previous motor abuse or deficient rewinding.

3.3.3 Motor and VFD Environment

The environment in which both the motor and VFD operate must be considered to properly select the equipment and VFD enclosure. The ambient temperature, provisions for ventilation, humidity, and the presence of gaseous and particulate contaminants must all be considered. If the VFD is located outdoors, consideration must be given to the proper enclosure and whether the VFD will be exposed to direct sunlight.

Heat rejection is a major concern with VFDs. The heat produced in the VFD enclosure can be substantial and lead to a short life. In some cases, a VFD manufacturer will allow a product to be installed in a warmer environment than what it was designed for, but the product will likely be derated in terms of capacity. For example, a manufacturer may derate a 10-hp VFD to 7.5 hp for use in a hot environment. Note that the capacity and ratings of the motor are also affected by the local environment.

3.3.4 Torque Requirements

The starting, peak, and running torque requirements must be considered when selecting a drive. Starting torque requirements can vary from a small percentage of the motor's full-load torque to a level several times that of the full-load torque. The peak torque also varies with changes in the load conditions. For example, a hot water pump running with a maximum heating load, with every connected terminal calling for heat, will require far more motor torque than when it is running to satisfy only a few terminals. The selected motor torque rating must exceed the torque needed by the machine from start to full speed. Motors that develop significantly more torque than the load requires can accelerate more rapidly.

3.3.5 Duty Cycle

The *duty cycle*, which is defined as a repetitive load pattern over a given period of time, is expressed as the ratio of on-time to the complete cycle period. When the operating cycle is such that the drive operates at idle or at reduced load for more than 25% of the time, the duty cycle becomes an even more important factor in selecting the proper drive.

Selecting the proper drive also depends on whether the load is steady, varies, follows a repetitive cycle of variation, or is completely random. Most HVACR-related loads, though, follow some pattern. Certain applications may require continuous reversals, long acceleration times at high torque, frequent rapid acceleration, or even cyclic overloads. These are all factors in the selection process, but rarely need to be considered in HVACR applications.

3.4.0 VFD Motor Braking

Braking refers to stopping an AC motor when it is necessary for it to stop faster than by removing the power and letting the motor wind down on its own. Braking can be accomplished in several ways. Some common methods of motor braking are:

- Electromechanical braking
- DC injection braking
- Dynamic braking

3.4.1 Electromechanical Braking

Electromechanical braking is accomplished using friction brakes, similar to the brakes on a car. Friction brakes are used for applications such as printing presses, small mobile cranes, overhead doors and hoisting equipment, and machine tools. Because electromechanical braking is not used in HVACR applications, only dynamic braking is discussed in this section. Note that braking is not required in many VFD applications.

Plugging, which was defined earlier in this module, is another means of braking. But it is not a method preferred for most applications, and plugging is not used in HVACR applications. It is considered the most aggressive means of stopping a motor electrically and results in a great deal of stress.

Note that dynamic braking cannot occur unless the power source and control circuit are operational. For this reason, a mechanical brake, used in case of a power or equipment failure, may be required in addition to a dynamic braking system. This would only be necessary if allowing the motor or load to coast to a stop would be hazardous, to either the machinery or personnel.

3.4.2 DC Injection Braking

One common method of braking an AC motor is by using a DC electric braking circuit. This is done by momentarily connecting a DC voltage to two of the three leads of the three-phase motor stator winding. This is referred to as *DC injection braking*.

When DC is applied to the stator winding of an AC motor, the stator poles become electromagnets. The magnetic field of the rotor is attracted to the magnetic field produced in the stator. The attraction of these two magnetic fields produces a braking action in the motor.

An advantage of using this braking method is that motors can be stopped quickly without using friction brakes. The primary factor affecting the speed of the braking action is the DC voltage applied. More voltage creates a stronger magnetic field, and thus the motor stops faster. As long as the DC is applied to the stator, the rotor will resist rotation. Thus, DC injection can also be used to hold the rotor in position.

DC braking cannot be used to hold a suspended load, however. Mechanical brakes must be employed when a load must be suspended, as with a crane or hoist, due to the hazard it represents.

3.4.3 Dynamic Braking

Dynamic braking takes a very different approach. Dynamic braking can be described as a means of converting the rotational energy of the motor to heat, which is then dissipated.

Dynamic braking is accomplished by continuing to power the motor after it is essentially de-energized but continues to rotate. The power sent to the motor though, is not enough to keep it spinning. If an AC motor is turning faster than its rotating magnetic field in the stator, it then becomes a generator. The generated power is sent back to the VFD. This process transfers the kinetic energy of the rotor into electrical power. Since the turning rotor is now under a load, without an input of energy to keep it turning, it comes to a stop. This is like driving a car downhill in a lower gear so that the car's engine acts as a brake.

The electrical energy produced by the motor during the braking process must go somewhere. Braking resistors are switched into the circuit by the *brake chopper* of the VFD. The resistor(s) used for dynamic braking must be capable of absorbing at least six times the stored rotational energy of the motor at its **base speed**. The generated power is released in the form of heat as it passes through the resistor.

Base speed: The rotational speed listed on a motor's nameplate at which the motor will develop its rated horsepower at its rated load and voltage.

3.0.0 Section Review

1. To maintain a constant torque, between which two parameters are VFDs designed to maintain a consistent ratio?
   a. Voltage and current
   b. Speed and current
   c. Frequency and voltage
   d. Frequency and current

2. True or False? The current limit setting on a VFD must never exceed the maximum full load current shown on the motor nameplate.
   a. True
   b. False

3. Given that motors are occasionally oversized to provide a wider speed range in a VFD application, the increase should be limited to _____.
   a. ½ hp
   b. 1 hp
   c. 2 hp
   d. 5 hp

4. How is DC injection braking applied to slow an AC motor?
   a. DC is applied to electric brake calipers.
   b. DC is applied to the rotor of the motor.
   c. DC is applied to two leads of the motor stator.
   d. DC has no practical use in slowing or stopping an AC motor.

Module 03318 Review Questions

1. How many windings are present in a three-speed multispeed motor?
   a. Two
   b. Three
   c. Four
   d. Five

2. On a single-phase motor, if the internal overload device is open, what resistance reading would you expect if you apply your ohmmeter probes to common and any other winding?
   a. Continuity with the motor's specified resistance for the winding
   b. Continuity with an extreme but measurable resistance
   c. Continuity with no measurable resistance
   d. Infinity (OL)

3. When checking a 440V motor for grounded windings, what is the minimum measured resistance that would indicate the motor is *not* grounded?
   a. 1,000Ω
   b. 230,000Ω
   c. 440,000Ω
   d. Infinity (OL)

4. Which of the following is a correct statement regarding meggers?
   a. Never test a motor while it is exposed to a vacuum.
   b. Portable meggers can generate as much as 100kV of AC voltage.
   c. Motors should remain wired into the equipment circuits during testing.
   d. A megger is better than an ohmmeter for routine resistance and continuity testing.

5. When troubleshooting the operation of a compressor start relay circuit by measuring the start winding current with a clamp-on ammeter, no current is measured on the ammeter when the compressor is energized. Which of the following is a possible cause?
   a. The start relay contacts are stuck open.
   b. The start relay contacts are stuck closed.
   c. The run capacitor has failed.
   d. The start relay coil is open.

6. What type of motor is used in inverter-driven compressors?
   a. Shaded pole AC
   b. Three-phase DC
   c. Three-phase AC
   d. Permanent split capacitor (AC)

7. Which of the following statements about ECM motors is correct?
   a. ECMs start at a low speed by design.
   b. Each ECM has three wiring harnesses connected to it.
   c. ECMs are routinely used as field replacements for PSC motors.
   d. The ECM motor and control module cannot be replaced separately.

8. Which of the following describes an ECM?
   a. A single-phase AC motor
   b. A single-phase DC motor
   c. A three-phase DC motor
   d. A split-phase AC motor

9. How do you set up an ECM to operate at a different AC voltage?
   a. By reversing any two of three motor leads.
   b. By reversing the four- or sixteen-pin connector halves.
   c. By repositioning a DIP switch on the equipment interface.
   d. By adding or removing a jumper at the five-pin connector.

10. True or False? A Version 3.0 ECM can replace any previous version as well as failed Version 3.0 models.
    a. True
    b. False

11. The voltage applied to an ECM should be within what percentage of the nameplate voltage?
    a. 5%
    b. 10%
    c. 15%
    d. 25%

12. When testing the windings of an ECM motor to ground, the motor should be replaced if the resistance is less than ______.
    a. 400,000Ω
    b. 300,000Ω
    c. 250,000Ω
    d. 100,000Ω

13. To control speed, today's VFDs primarily rely on what approach?
    a. Current source inversion (CSI)
    b. Pulse-width modulation (PWM)
    c. Variable voltage inversion (VVI)
    d. Silicon-controlled rectification (SCR)

14. VFDs typically have a range of output frequencies over which the V/Hz ratio remains constant, and a second range of output frequencies over which the voltage remains constant while ______.
    a. the frequency varies
    b. the resistance varies
    c. the polarity varies
    d. the current varies

15. Which of the following describes the dynamic braking feature of a VFD?
    a. The direction of rotation of the motor is reversed.
    b. DC is momentarily applied to the AC motor windings.
    c. Friction brakes on the motor shaft are actuated to slow the motor.
    d. The motor is turned into a generator, creating a load that causes it to slow down.

Answers to odd-numbered review questions are found in the Module Review Answer Key at the back of this book.

Answers to Section Review Questions

Answer	Section	Objective
Section One		
1. d	1.1.0	1a
2. c	1.2.1	1b
3. b	1.3.2	1c
4. b	1.4.0	1d
Section Two		
1. d	2.1.0	2a
2. a	2.2.0; *Figure 26*	2b
3. b	2.3.1	2c
Section Three		
1. c	3.1.0	3a
2. b	3.1.0	3b
3. b	3.3.1	3c
4. c	3.4.2	3d

MODULE 03210

Troubleshooting Cooling

Source: Ritchie Engineering Co, Inc./ YELLOW JACKET

Objectives

Successful completion of this module prepares you to do the following:

1. Describe the operation of the refrigeration cycle and identify common problems.
   a. Identify and describe the function of the primary components found in a refrigerant circuit.
   b. Recount key events in the typical refrigeration cycle.
   c. Explain how to analyze refrigerant circuit operation.
   d. Identify possible causes for abnormal pressure and temperature conditions.
   e. Explain how condenser and evaporator air volume affect the refrigeration cycle.
   f. Identify and describe problems related to fixed metering devices.
   g. Identify and describe problems related to expansion valves and distributors.
   h. Identify and describe other problems related to the refrigerant circuit.
2. Explain how to troubleshoot and replace a compressor.
   a. Identify common problems that can lead to compressor failure.
   b. Explain how to identify compressor mechanical problems.
   c. Explain how to replace a hermetic compressor.
   d. Identify additional steps often required to replace a compressor following a motor failure.

Performance Task

Under supervision, you should be able to do the following:

1. Demonstrate the ability to diagnose and resolve at least four of the following malfunctions:
   - Compressor failures
   - System-related compressor problems
   - Refrigerant undercharge and overcharge
   - Evaporator and condenser fouling or blockage
   - Metering device problems
   - Refrigerant line and filter-drier blockage
   - Refrigerant circuit contamination, such as acid and non-condensables

Overview

The refrigerant circuit is a closed-loop system, so identifying the source of a problem is not always easy. Because the operation of a refrigerant circuit is affected by external conditions, it can also be challenging to determine if a problem is inside the system or the result of an external influence. Being able to see through the tubing would certainly help. An experienced technician, though, can visualize the activity inside.

A thorough understanding of the refrigerant circuit provides the basis for matching specific symptoms to probable causes. With a little practice, your gauges and thermometer can help you see through the tubing and know precisely what is happening on the other side.

Digital Resources for HVACR

Scan this code using the camera on your phone or mobile device to view the digital resources related to this craft.

SCAN ME

NCCER **Industry-Recognized Credentials**

If you are training through an NCCER-accredited sponsor, you may be eligible for credentials from NCCER. The ID number for this module is 03210. Note that this module may have been used in other NCCER curricula and may apply to other level completions. Contact NCCER at 1.888.622.3720 or go to **www.nccer.org** for more information.

You can also show off your industry-recognized credentials online with NCCER's digital credentials. Transform your knowledge, skills, and achievements into credentials that you can share across social media platforms, send to your network, and add to your resume. For more information, visit **www.nccer.org**.

1.0.0 Refrigeration Cycle Operation and Troubleshooting

Performance Task

1. Demonstrate the ability to diagnose and resolve at least four of the following malfunctions:
   - Compressor failures
   - System-related compressor problems
   - Refrigerant undercharge and overcharge
   - Evaporator and condenser fouling or blockage
   - Metering device problems
   - Refrigerant line and filter-drier blockage
   - Refrigerant circuit contamination, such as acid and non-condensables

Objective

Describe the operation of the refrigeration cycle and identify common problems.

a. Identify and describe the function of the primary components found in a refrigerant circuit.
b. Recount key events in the typical refrigeration cycle.
c. Explain how to analyze refrigerant circuit operation.
d. Identify possible causes for abnormal pressure and temperature conditions.
e. Explain how condenser and evaporator air volume affect the refrigeration cycle.
f. Identify and describe problems related to fixed metering devices.
g. Identify and describe problems related to expansion valves and distributors.
h. Identify and describe other problems related to the refrigerant circuit.

An understanding of the basic refrigeration cycle is critical for technicians to effectively troubleshoot cooling systems. Many different systems are used to provide cooling for personal comfort, food preservation, and industrial processes. However, each of these systems relies on the basic vapor-compression refrigeration cycle to produce the cooling.

The refrigeration cycle is basically the same for these systems. Factors that vary from system to system are the type of refrigerant used, the size and style of the components, pressures and temperatures, and the installed locations of the components.

The basic refrigeration cycle and its components are covered in NCCER Module 03107, *Introduction to Cooling*. A review is presented here as a foundation for troubleshooting cooling systems. Note that the information in this module is based primarily on air-cooled systems.

1.1.0 Refrigeration Cycle Components

A basic refrigerant circuit (*Figure 1*) includes these four major components:

- *Evaporator* — A heat exchanger in which the heat from the area or substance being cooled is absorbed and transferred to the refrigerant.
- *Compressor* — Creates the pressure differential needed to make the refrigeration cycle work.

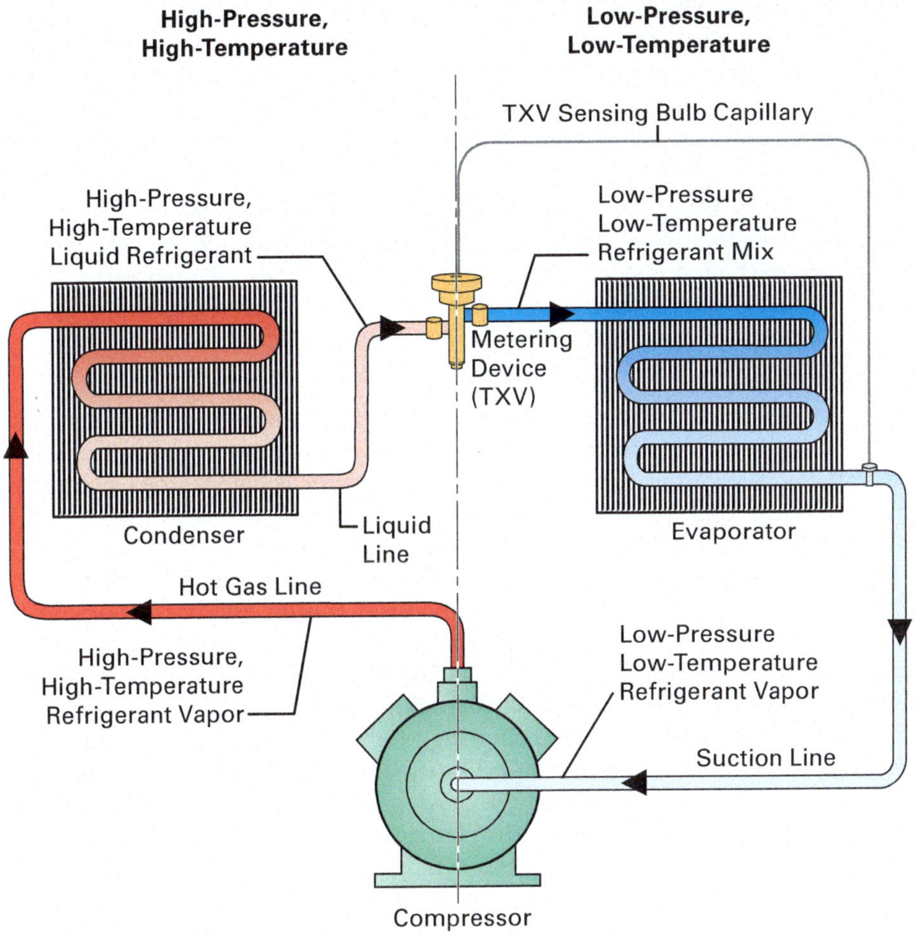

Figure 1 Basic refrigeration cycle.

- *Condenser* — A heat exchanger in which the heat absorbed by the refrigerant at the evaporator is transferred to outdoor air, or to another cooler substance.
- *Metering device* — Also referred to as an *expansion device.* Controls the flow of refrigerant to the evaporator. The metering device provides a pressure drop that lowers the boiling point of the refrigerant before it enters the evaporator.

The refrigerant piping used to connect the basic components is also shown in *Figure 1.* The components and lines form a closed circuit. The three primary refrigerant lines are identified as follows:

- *Suction line* — Carries refrigerant vapor from the evaporator to the compressor.
- *Hot gas line* — Carries refrigerant vapor from the compressor to the condenser; also called the *discharge line.*
- *Liquid line* — Carries liquid refrigerant to the metering device.

Throughout this module, the terms *high temperature, low temperature, high pressure,* and *low pressure* are used often to differentiate between the two sides of the system. These terms are always used relative to the topic. For clarity, these terms mean the following in the context of the refrigerant cycle:

- *High temperature* — Describes temperatures that are comparable to the saturation temperature at the condenser, although the temperatures can be significantly higher. A high temperature in the refrigerant circuit is generally higher than 80°F (27°C).
- *Low temperature* — Describes temperatures that are comparable to the saturation temperature at the evaporator. A low temperature in the refrigerant circuit is generally below 60°F (16°C).

- *High pressure* — Describes pressures that are comparable to the discharge pressure from the compressor. Due to the characteristics of refrigerants, it must be considered in the context of the system and the refrigerant in use. For example, the high-pressure side of a centrifugal chiller using HCFC-123 refrigerant may operate at 3 psig, while the high-pressure side of a comfort cooling system using HFC-410A may operate at 400 psig.
- *Low pressure* — Describes pressures that are comparable to the suction pressure of the compressor. Like the term *high pressure*, it must be considered in the context of the system and the refrigerant in use. For example, the low-pressure side of a centrifugal chiller using HCFC-123 refrigerant will be below atmospheric pressure, while the low-pressure side of a comfort cooling system using HFC-410A may operate at 120 psig.

On a call for cooling, the control circuit starts the compressor. The pressure difference between the compressor inlet and outlet begins to build immediately, causing refrigerant to flow. The incoming vapor is compressed, creating a high-pressure, high-temperature vapor stream at the compressor discharge port. The vapor travels to the condenser via the hot gas line.

The condenser is a series of tubing coils through which the refrigerant flows. As cooler air moves across the condenser tubing, the hot vapor cools off. As it gives up heat to the air outdoors, it cools to its saturation temperature and begins to condense to its liquid state. Eventually, all the refrigerant vapor condenses. Any decrease in temperature below the saturation temperature is referred to as *subcooling*. The resulting high-temperature, high-pressure subcooled liquid travels through the liquid line to the inlet of the metering device.

The metering device regulates the flow of refrigerant into the evaporator. The metering device also significantly decreases the refrigerant pressure because it restricts flow. Some of the refrigerant flashes to its vapor state, absorbing heat from the remaining liquid. This results in a dramatic reduction in refrigerant temperature. The outlet of the metering device represents the beginning of the low-temperature, low-pressure side of the system. As the refrigerant exits the metering device, it is saturated, meaning that both liquid and vapor are present in some ratio. A ratio of 80% liquid and 20% vapor is a reasonable expectation.

The evaporator receives the saturated low-temperature, low-pressure saturated refrigerant from the metering device. The evaporator is a series of tubing coils that expose the cooler refrigerant to the warmer air passing through. Heat is transferred through the tubing to the cooler refrigerant, causing it to boil and continue vaporizing. A low-temperature, low-pressure refrigerant vapor finally exits the evaporator. Any increase in the vapor temperature above the saturation temperature is referred to as *superheat*.

The vapor then returns to the inlet of the compressor, where the cycle repeats.

1.2.0 Refrigeration Cycle for a Typical HFC-410A Cooling System

A basic HFC-410A refrigerant circuit is shown in *Figure 2*. The pressure on the high-pressure side can be referred to as the *head pressure, discharge pressure*, or *high-side pressure*. The pressure on the low-pressure side is generally referred to as the *suction pressure* or *low-side pressure*. The vertical dashed line represents the division between the two sides of the system. Major pressure changes occur at the compressor and metering device.

This section describes a typical air conditioner operating on HFC-410A. This example demonstrates the temperature/pressure relationships typical of those at key points in the circuit. For this example, assume an indoor air temperature of 75°F (24°C) and an outdoor air temperature of 95°F (35°C). The temperatures and pressures used in the example are representative values only and will vary due to differences in the equipment and load conditions.

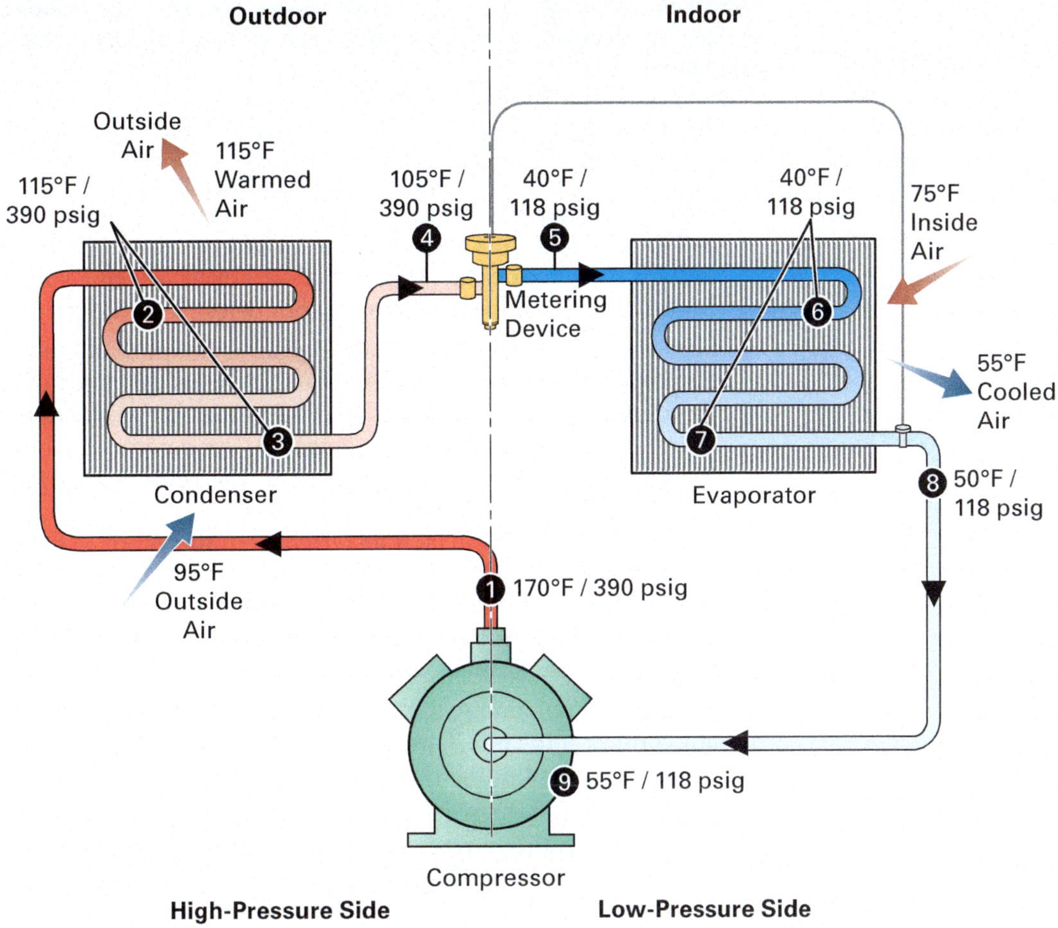

Temperature °F	Pressure HFC-410A
27	91.0
28	92.9
29	94.9
30	96.8
31	98.8
32	100.9
33	102.9
34	105.0
35	107.1
36	109.2
37	111.4
38	113.6
39	115.8
40	118.1
41	120.3
42	122.7
43	125.0
44	127.4
45	129.8
46	132.2
47	134.7
48	137.2
49	139.7

Temperature °F	Pressure HFC-410A
50	142.2
55	155.5
60	169.6
65	184.5
70	200.4
75	217.1
80	234.9
85	253.7
90	273.5
95	294.4
100	316.4
105	339.6
110	364.1
115	389.9
120	416.9
125	445.4
130	475.4
135	506.9
140	540.1
145	575.1
150	611.9

Temperature / Pressure Chart

Figure 2 Typical air conditioning cycle for HFC-410A.

Load conditions refer to the climatic conditions, both indoors and outdoors. The humidity level of the indoor air plays a significant role in the load. Since a great deal of energy is required to condense airborne moisture, moist indoor air increases the refrigeration load. Water vapor condensing on the evaporator coil surface represents a latent load, which is a substantial part of the total refrigeration load.

The numbered descriptions below correspond to the numbers shown at key points in *Figure 2*. Follow along on the figure as system operation is described.

1. After compression, the highly superheated gas from the compressor flows through the hot gas line to the condenser. The hot gas may be around 170°F at 390 psig, and the superheat here would be around 55°F because the saturation temperature is 115°F (170°F – 115°F). All this superheat must be transferred out of the refrigerant vapor before it can begin to condense.
2. Because the refrigerant in the condenser is hotter than the warmer outside air passing over the condenser, it quickly gives up its superheat as it flows through. The temperature falls to 115°F in the early condenser passes. In this example, 115°F represents the saturation temperature of the refrigerant. At the saturation temperature, and only at this temperature for the pressures shown in *Figure 2*, liquid and vapor coexist.

   As heat continues to be transferred from the vapor to the outside air, the vapor begins to condense. By the time the refrigerant has traveled about three-quarters of the distance through the condenser, all the refrigerant has condensed into a liquid. In reality, the point where the last bit of liquid vaporizes depends on a variety of factors.
3. During the remaining travel through the condenser, the liquid refrigerant continues to cool. This lowers its temperature to about 105°F. In other words, the liquid refrigerant is subcooled 10°F since the saturation temperature at the given pressure is 115°F (115°F – 105°F = 10°F subcooling).
4. Subcooled liquid refrigerant from the condenser flows through the liquid line to the metering device. Note that the subcooling may increase or decrease due to line length, the environment, and elevation change. For this example, no change in temperature has occurred, but it is an important concept to remember.
5. The metering device controls the flow of refrigerant to the evaporator. Subcooled liquid from the condenser enters at a temperature of 105°F and a pressure of 390 psig. A very significant pressure drop occurs at the metering device. The pressure drop from flowing through the small orifice reduces the pressure of the liquid refrigerant from 390 psig to 118 psig at the outlet.

   At this new pressure condition, the refrigerant temperature is 40°F and in its saturated state. For this to happen, some of the liquid refrigerant must flash to vapor instantly, absorbing a great deal of heat in the process. This accounts for the cooling effect on the remaining liquid. The resulting mixture is about 80% liquid and 20% vapor, and the refrigerant is now at saturation. The low-temperature, low-pressure refrigerant mixture from the metering device then enters the evaporator coil.
6. The saturated refrigerant mixture enters the evaporator. The pressure of 118 psig corresponds to the 40°F saturation temperature for HFC-410A. This is typical for evaporators used for comfort cooling, but the exact temperature and pressure will vary.
7. Because the refrigerant flowing through the evaporator is colder than the room air passing through, it absorbs heat, causing the liquid refrigerant to boil and vaporize. After traveling roughly 90% of the way through the evaporator, all the refrigerant has ideally vaporized. It can then begin to increase in temperature.
8. During the remaining 10% of travel through the evaporator, the vapor continues to absorb heat from the warmer air, raising its temperature to 50°F. In other words, the vapor is now superheated 10°F, since the saturation temperature here is 40°F (50°F – 40°F = 10°F superheat). The superheated vapor flows through the suction line and to the inlet of the compressor.

9. The superheated vapor entering the compressor may lose or gain superheat before it reaches the compressor. In cooling applications, an increase in temperature is far more likely. However, in a large food processing application, the suction line may pass through cold rooms that can cool the refrigerant further. This can be a significant variable from one system to another. Like the liquid line, it depends largely on the line length and its environment. Never assume that the refrigerant temperatures at the ends of the liquid line and suction line are the same.

The pressures and temperatures in this walk-through illustrate just one example of system operating characteristics. Many possible combinations of temperatures and pressures exist, each of which represents proper operation for the refrigerant, application, and environmental conditions.

1.3.0 Analyzing Operating Conditions

When problems arise in the refrigeration cycle, careful analysis and evaluation is needed to determine the cause. Often what seems to be the problem is only a symptom. Different problems can produce the same or similar symptoms. Isolating the problem requires a logical, systematic approach, which generally begins with a preliminary inspection.

1.3.1 Preliminary Inspection

A reasonable preliminary inspection of the refrigerant circuit includes the following steps:

WARNING!

Be sure all electrical power to the equipment is turned off. Open, lock, and tag disconnects, then use a meter to confirm the power is off. Use extreme caution around pressurized or hot components. Follow all safety instructions labeled on the equipment and provided in the manufacturer's service manual.

Step 1 Make a quick visual inspection of the system, looking for any obvious signs of refrigerant leakage or external damage. Look for any fault codes that may be displayed on indoor and outdoor unit control boards.

Step 2 Turn off the equipment. Lockout and tag disconnect switches as needed. Remember there may be multiple sources of energy.

Step 3 Look for the following piping problems and issues:

- Oversized, undersized, and deformed refrigerant lines
- Very long runs of piping and the environment through which the piping is routed
- Excessive number of elbows or fittings
- Uninsulated suction line
- Buried refrigerant lines

Step 4 At the evaporator and condenser, check for the following:

- Elevation of the evaporator and condenser coils relative to the compressor
- Corroded fin collars
- Dirty or damaged coils
- Dirty or missing air filters
- Dirty blower wheel
- Incorrect blower speed selection
- Loose drive belt(s)
- Signs of frost formation on the evaporator

Step 5 As applicable, check the following at the compressor:

- Service valves fully open
- Mounting bolts are positioned per manufacturer's instructions (loosened to allow limited movement)
- Crankcase heater is functional
- Scorched or blistered cylinder heads
- Low or high oil level on open and semi-hermetic compressors

The preliminary inspection often reveals one or more issues. A puddle of oil under a fitting, for example, shows that at least one leak is present. Beyond possible problems, you may also discover factors that could affect refrigerant pressures and temperatures, providing you with information to make an informed operating analysis later. For example, refrigerant lines traveling 50' across a hot attic can significantly reduce subcooling.

Some items may need to be resolved before testing begins. It's best not to assume that a system has only one problem. The preliminary inspection may not reveal the most significant problem, but it may reveal one or more issues of concern.

1.3.2 Analyzing System Operating Conditions

To determine if the system or compressor has a functional problem, you must know the operating conditions. If the compressor is operable, essential temperatures and pressures can be measured and recorded. The measurements can then be compared to your expectations of system operation at the current operating conditions. For example, if the indoor air is 85°F, you would not expect to see only 3°F of superheat, regardless of the metering device in use. An air conditioning evaporator with this much load imposed on it should never be operating with such a low superheat.

The following steps describe one approach to this process:

WARNING!

Use extreme caution around rotating, pressurized, and hot components. Follow all safety instructions labeled on the equipment and provided in the manufacturer's service manual.

Step 1 Connect a gauge manifold to the system ports. Be prepared to go to several different system locations to take ambient air, supply and return air, and refrigerant line temperature measurements.

Step 2 Check and record the line voltage at the compressor contactor.

Step 3 Energize the system and set the thermostat to call for cooling.

Step 4 Start the system and listen for abnormal sounds.

- Check for excessive vibration in the compressor, piping, motors, and fans.
- Note any knocks or rattles from the compressor that may indicate **floodback**. The noise may also indicate that mechanical damage to the compressor has already occurred.

Floodback: A condition in which liquid refrigerant is returned to an operating compressor.

Step 5 Check the compressor oil sight glass (if equipped). Heavy foaming at the sight glass should clear within a few minutes after startup. If it does not, this may also indicate refrigerant floodback, or that an abnormal amount of refrigerant has been condensing in the oil during the off cycle.

WARNING!

The compressor terminals are under pressure inside the shell. The seal around the terminal assembly can be damaged. Disturbing the terminals to take measurements could cause them to blow out, causing injury. When making electrical measurements on a compressor in a pressurized system, take the measurements at the opposite end of the leads. Visually inspect the compressor terminals carefully when necessary to ensure the connections are sound and undamaged.

Step 6 Measure and record the compressor voltage and current at the contactor. The voltage should be compared to the voltage recorded before startup to see if there is a significant voltage drop under load. Note that this does not test for voltage drop through the contactor itself; the measured voltage drop in this case is the total voltage for the branch circuit.

- The measured voltage should be within ±10% of the motor nameplate value.
- In a three-phase motor, the voltage imbalance between any two phases should not exceed 2%. The current imbalance between phases must not exceed 10%.

Step 7 Ensure all fans and motors are operational and moving the expected amounts of air. Calculate the air volume moving through the system.

Step 8 The following should be measured or calculated for your analysis of system operation:

- Indoor dry-bulb and wet-bulb temperatures
- Outdoor dry-bulb temperature
- Suction line pressure
- Saturated suction temperature
- Calculate superheat at both the outlet of the evaporator and the inlet of the compressor or condensing unit (split systems)
- Discharge pressure
- Saturation temperature of the discharge vapor
- Discharge line temperature
- Liquid line pressure
- Calculate subcooling at both the condenser outlet and the metering device inlet (split systems)
- Compression ratio
- Net oil pressure (open and semi-hermetic compressors; subtract the suction pressure from the measured oil pressure to determine net oil pressure)
- Temperature at the bottom of the compressor cylinder heads (open and semi-hermetic compressors)
- Temperature at the top and bottom of the compressor motor housing (semi-hermetic compressors)

Once the actual conditions are known, they can be compared against normal operating parameters and your expectations for the conditions. *Table 1* shows an example of a form you can use to document system parameters and compare them to the example system in *Figure 2*. Remember, though, that readings vary widely among systems due to differences in application, environmental conditions, and refrigerant type.

If possible, refer to any existing operating logs or the manufacturer's service information to find typical readings for the specific equipment you are servicing. It's always a good idea to document the operating information for a system when it is originally started and during maintenance inspections. The information is a valuable reference later when troubleshooting is required.

1.4.0 Causes of Abnormal Pressures and Temperatures

Now that you have gathered the operating information, you can analyze the information to determine what is abnormal. Is the head pressure unusually high? Is the superheat low or nonexistent? Ask yourself questions about the operating conditions, using the knowledge you have gained previously about the refrigerant circuit.

Next, consult the manufacturer's troubleshooting aids to identify the likely causes of the problem you see. Most refrigeration cycle problems result in

TABLE 1 Operating Data Form

System Parameter	Typical HFC-410A System*	Your System
Suction Pressure (psig)	118 psig	_____
Suction Line Temperature (°F)	55°F	_____
Saturated Suction Temperature (°F)	40°F	_____
Superheat (°F)	15°F	_____
Discharge Pressure (psig)	390 psig	_____
Discharge Line Temperature (°F)	170°F	_____
Saturated Discharge Temperature (°F)	115°F	_____
Liquid Temperature Leaving Condenser (°F)	105°F	_____
Liquid Line Temperature Entering Expansion Device (°F)	105°F	_____
Subcooling (°F)	10°F	_____
Air Temperature Entering Evaporator (DB °F)	75°F	_____
Air Temperature Entering Evaporator (WB °F)	62°F	_____
Air Temperature Leaving Evaporator (DB °F)	55°F	_____
Air Temperature Drop Across Evaporator (DB °F)	20°F	_____
Outdoor Temperature (DB °F)	95°F	_____
Outdoor Temperature Rise Across Condenser (DB °F)	20°F	_____
Compressor Compression Ratio (Absolute Discharge Pressure ÷ Absolute Suction Pressure)	3.05:1	_____
Compressor Net Oil Pressure (Open and Semi-Hermetic Compressors; Measured Oil Pressure — Suction Pressure)	34 psig	_____
Temperature at the Bottom of the Compressor Cylinder Heads (Open and Semi-Hermetic Compressors)		_____
Temperature at the Top and Bottom of the Compressor Motor Housing (Semi-Hermetic Compressors)		_____

*HFC-410A pressures and temperatures shown are typical for many systems but should be considered examples only. The pressures and temperatures shown here do not fit every application.

abnormal system pressures and/or temperatures. *Figure 3* summarizes the common system problems related to high and low pressures. However, it is very important to understand that abnormal pressure on one side often affects the other side. For example, in a system with a fixed restrictor for a metering device, a high head pressure typically results in a higher suction pressure, because the restrictor can't respond in any way to changing conditions.

The information provided in *Figure 3* and in manufacturers' troubleshooting aids often helps isolate the problem to a specific component or area. In most cases, troubleshooting aids only provide some clues that require further investigation. Investigate the most likely causes first, based on your knowledge of the system and its service history. Use the process of elimination to isolate the problem.

The sections that follow explore some of the common causes of refrigerant circuit problems and the related symptoms.

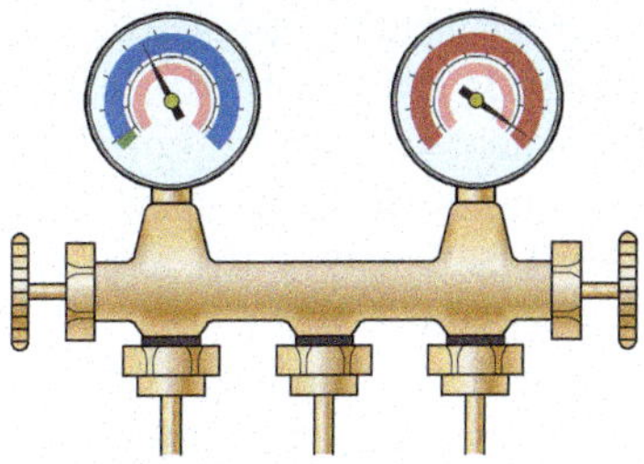

High Head Pressure

- Air in the System
- Dirty Condensor Coil
- Refrigerant Overcharge
- Condenser Fan Motor Failure
- Condenser Fan Blade Damaged or Direction of Rotation Incorrect
- Condenser Air Being Recirculated
- High Ambient Temperature

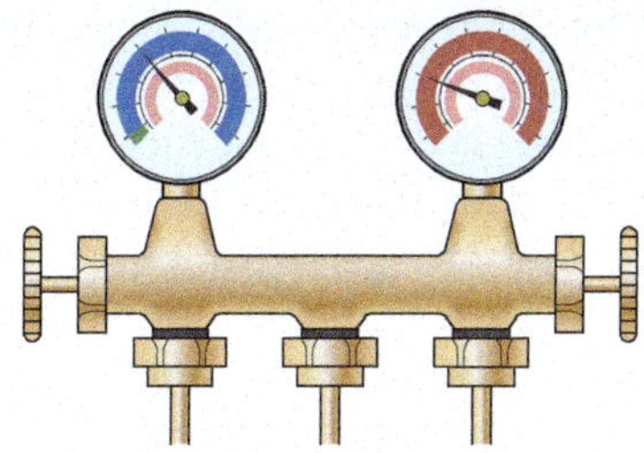

Low Head Pressure

- Refrigerant Undercharge
- Damaged Compressor Valves or Scroll Plates
- Low Ambient Temperature

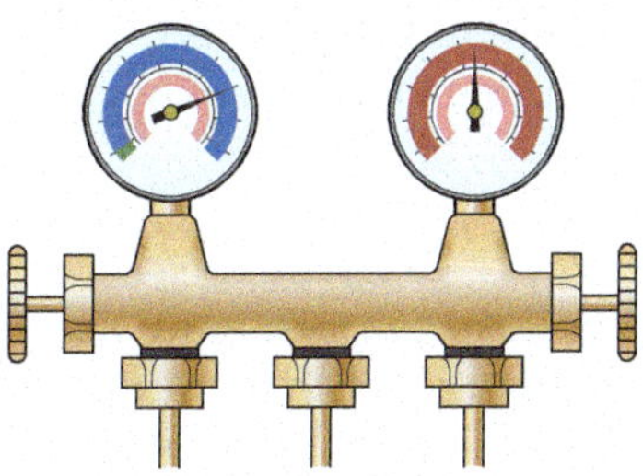

High Suction Pressure

- Refrigerant Overcharge
- Abnormal Load on the Evaporator
- Damaged or Incorrect Metering Device
- Damaged Compressor Valves or Scroll Plates
- Worn Piston Rings (Reciprocating Compressors)

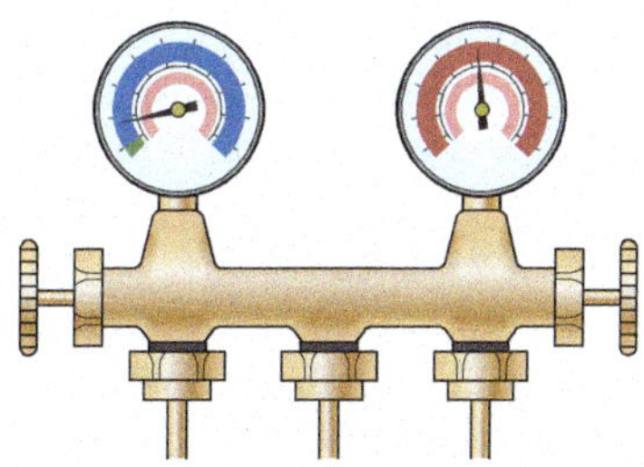

Low Suction Pressure

- Indoor Blower Failure
- Indoor Blower Speed Insufficient
- Dirty Air Filters
- Dirty or Frozen Evaporator Coil
- Refrigerant Undercharge
- Damaged, Fouled, or Incorrect Metering Device
- Kinked or Restricted Liquid Line
- Clogged Filter-Drier

Figure 3 Common causes of abnormal refrigerant pressures.

1.4.1 Low or Excessive Refrigerant Charge

An incorrect refrigerant charge is a very common problem. It is important for technicians to understand how an incorrect charge affects system operation so that the problem is more easily recognized. However, it is equally important not to jump to conclusions, simply because a low refrigerant is frequently encountered.

Refrigerant Undercharge

A loss of refrigerant is the most frequent charge-related problem. The customer may complain of deteriorating cooling performance over a period of days or weeks. Low refrigerant usually results from leaks that occur due to poor installation, physical damage, or a factory defect. Some leaks are small enough that the loss of charge only becomes significant over the course of months.

A low charge can also result from incorrect charging techniques at startup or during a previous service call. For example, if a system with a thermostatic expansion valve (TXV) is charged to only a 5°F subcooling value on a very cool day with a low evaporator load, it will not have enough refrigerant to operate properly when the outdoor temperature and evaporator load rises. As the refrigerant flow rate increases to accommodate the load, the subcooling will diminish until the TXV is receiving a liquid/vapor mixture instead of the solid column of liquid it needs to operate properly.

A low charge typically causes both the suction and discharge pressures to be lower than normal. How far the pressures fall depends on how much refrigerant is lost. With a fixed metering device, the superheat will be higher than it should be for the conditions. A fixed metering device is more sensitive to the refrigerant charge. With a TXV, the subcooling will be low, but the superheat may still be reasonable if only a small amount of refrigerant is lost.

Systems with an automatic expansion valve (such as some window and through-wall units) may not exhibit a low suction pressure because the valve is operating wide open, trying to raise the evaporator pressure.

Once you determine that the refrigerant charge is low, consider why that is the case. Although leaks occur often, perhaps it was not properly charged during a previous service visit. If that is a possibility, add refrigerant as necessary, but only after you check for leaks in the most likely places, such as tubing joints and service valves. Repair any leaks prior to recharging the system, because the charge may otherwise need to be recovered to repair a leak.

Do not simply add refrigerant to a system without considering why the charge is low. It is important to be certain—repeated service calls to recharge a leaking system will always result in customer dissatisfaction and discomfort. Explain the approach you are taking to your client, so they understand and know what you plan to do if the problem recurs.

Refrigerant Overcharge

The symptoms of a refrigerant overcharge are similar from system to system. Both the suction and discharge pressures are typically higher than normal. The superheat may be low or nonexistent on a system with a fixed metering device, with floodback a possibility.

A system with an expansion valve may have an elevated suction pressure as well. The subcooling will be high, as refrigerant tends to back up in the condenser and continues to cool. However, unless the system is grossly overcharged, the valve will likely maintain its superheat setting.

If the system is overcharged, recover a significant portion of the charge. You can then restart the system and add back what is needed to reach the appropriate superheat or subcooling value. Remember that systems with a fixed restrictor are charged by superheat, while those with an expansion valve are charged by subcooling.

Remember that many blends tend to fractionate. As a result, even when subcooling is present, bubbles might appear in a liquid-line sight glass. This is especially true when the system is lightly loaded, and the refrigerant is moving slowly through the system. Sight glasses are of little value in charging blends, but they do still provide a useful moisture indicator.

NOTE

When charging a system with HFC-410A refrigerant as well as other refrigerants that fractionate, always charge with liquid from the cylinder. When adding a charge to an operating system, charge liquid into the suction side of the system using your gauge manifold valve as a metering device.

1.5.0 Evaporator and Condenser Problems

Both the evaporator and the condenser can experience problems with reduced airflow. Both coils must be clean, and the fan or blower must be delivering the appropriate amount of air for a system to operate as it should.

1.5.1 Evaporator Airflow Problems

A cooling system indoor blower must move about 400 cubic feet of air per minute (cfm) per ton of cooling. Multispeed and variable-speed motors should be checked and set up properly during the initial startup, but the settings often need to be checked again during a service call. Even when the set speeds are correct, the coil may become fouled over time, reducing the air volume through the system.

Both too much and too little airflow indoors cause comfort problems. Too much air flowing across the evaporator coil typically results in poor humidity control. As the air volume increases, the temperature difference across the coil (ΔT) decreases. Beyond that, excess airflow actually increases cooling capacity slightly, but the supply air temperature rises.

NOTE

Systems operating in special modes, such as dehumidification, will move more or less air per ton. Ductless systems typically operate at roughly 250 cfm per ton by design, but these systems are also capable of operating in a variety of modes that affect indoor air volume.

Too little airflow is a more common problem. A dirty evaporator coil (*Figure 4*) impedes airflow as well as heat transfer. A reduced suction pressure and low superheat results when a fixed metering device is used. The low superheat is one way to differentiate this problem from a low refrigerant charge. An expansion valve can often compensate and maintain some superheat but still at a reduced suction pressure.

Aside from an improper blower speed selection, the following conditions are often the cause of low airflow:

- Air filter and/or evaporator coil dirty
- Blower rotating in the wrong direction
- Blower drive belt(s) loose
- Dislodged duct liner or similar obstruction
- Closed diffusers and grilles
- Poor air distribution system design

Figure 4 Dirty evaporator coil.
Source: Tanut Nitkumhan/iStock

Low airflow may appear to have the same symptoms as a low refrigerant charge. Technicians can easily misdiagnose the problem, as they often expect to find a system suffering from a low refrigerant charge. It is important to fully investigate a problem rather than embrace the first cause that comes to mind.

Common symptoms of a low refrigerant charge are a low suction pressure coupled with a low head pressure. However, when low airflow is the real problem, the discharge pressure typically remains close to normal.

You will eventually encounter a frozen evaporator coil in a system. A coil covered in ice, similar to the one in *Figure 5*, will obviously cause an airflow problem. But what does a frozen coil tell you?

Figure 5 Ice on an evaporator coil.
Source: bombermoon/iStock

A frozen evaporator coil alone isn't a diagnosis. It shouldn't be interpreted immediately as a low refrigerant charge or as a problem with the airflow. Consider it only as an indication that the coil temperature dropped below freezing and water vapor collected and froze as a result. The source of the problem must still be discovered.

To determine the source of the problem, all the ice must be removed first. Even a small amount remaining can affect your readings. Once the coil is clear, restart the system and check both the airflow and the refrigerant circuit operating characteristics. Only then will you know what caused the problem. Indeed, the system may be suffering from both a low charge and low airflow.

On Site

ECM Motor Speed

When analyzing the airflow in a system, remember that electronically commutated motors (ECMs) are commonly used with indoor blowers. When checking airflow in a system using an ECM, ensure that the blower is operating in the appropriate mode before checking the airflow. Advanced systems offer multiple operating modes, such as dehumidification. To maximize the condensation of water vapor, the ECM slows down in the dehumidification mode to enhance moisture removal.

Going Green

Microchannel Coils

Microchannel coils are a new way to get more from less. Microchannel coils eliminate the standard round copper or aluminum tubing. In place of round tubing are flattened tubes with multiple small passages inside. This increases the refrigerant surface area in contact with the heat transfer surface while also minimizing the internal volume, reducing the size of the refrigerant charge.

Microchannel coils generally increase efficiency while maintaining or reducing the size of the coil. Condenser coils were first fitted with microchannel coils, but microchannel coils are also now being used for evaporators.

Note that many manufacturers insist that microchannel coils be cleaned carefully, without the use of harsh alkaline- or acid-based cleaners. Plain water, or a mild soap mixed with water, is best. Many microchannel coils have a special coating that is damaged by chemical cleaning. As is true of all coils, thorough rinsing is necessary when any kind of cleaner is used to protect the coil from chemical attack.

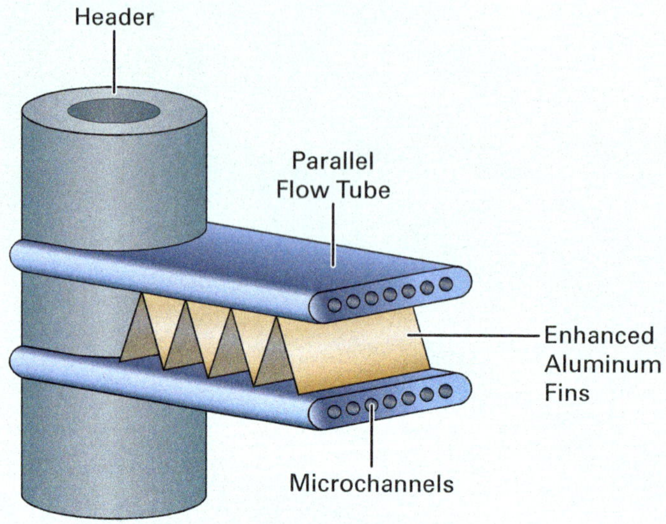

1.5.2 Condenser Airflow Problems

For a system to operate properly, the condenser must be able to reject the heat absorbed in the evaporator, as well as the heat of compression. Thus, the condenser must reject more heat than the evaporator absorbs. Roughly 15% of the condenser's capacity to reject heat is related to the heat of compression.

When heat rejection is inadequate, the system pressures and temperatures typically rise. If the problem is severe enough, the high-pressure switch or compressor discharge temperature safety switch will shut the compressor off. Here are some of the possible causes for poor condenser coil performance and elevated high-side pressures:

- Condenser coil clogged with debris or blocked
- Failure of the condenser fan motor or its related wiring and controls

- Incorrect direction of condenser fan rotation
- Condenser fan blade not properly positioned in the fan orifice
- Incorrect condenser fan motor or blade
- Recirculation of condenser discharge air, such as a deck or overhang
- High condenser entering-air temperature due to external sources, such as dryer vents or adjacent condensing units
- Unusually high temperatures in hot climates that exceed the manufacturer's operating range

Many commercial condenser coil assemblies are constructed of two separate coils in layers (*Figure 6*). The fins of condenser coils are set close together, but dust and dirt can easily make its way between the coil fins, where they either pass through or become trapped.

With layered coil construction, a great deal of the debris becomes trapped against the face of the inner coil. This happens easily because the fins of the two coils are never aligned. Cleaning the face of the coil often results in little or no improvement in performance. To clean layered condenser coils, the two coils must be carefully separated. Then the face of the inner coil is accessible for cleaning.

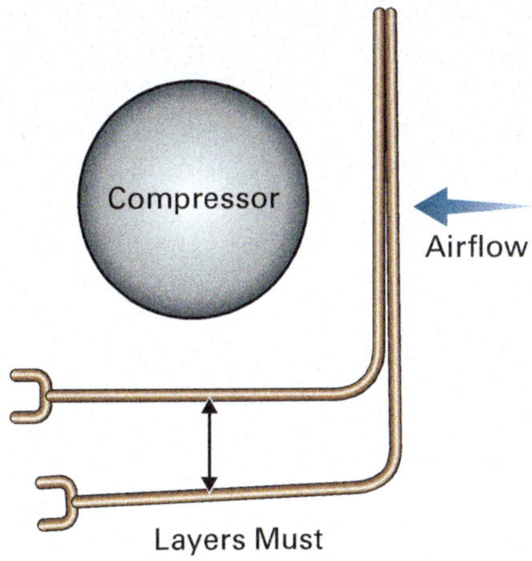

Figure 6 Cleaning a layered condenser coil.

1.6.0 Fixed-Orifice Metering Device Problems

Fixed-orifice metering devices are simple with little that can go wrong. As long as the proper size is used to accommodate the outdoor unit/indoor coil combination, these devices provide reliable service. However, the orifice can become fouled or clogged if the circuit is dirty.

On systems with a single fixed metering device, clogging is rare due to the size of the opening. This is especially true when the proper steps are taken to keep the system clean and install a filter-drier upstream. However, several fixed metering devices are sometimes used on commercial cooling-system indoor coils. The orifice in each one is typically smaller in diameter and more easily fouled when several are working together. The orifices are also very difficult to access, as they are generally found inside brazed joints.

The size of the fixed metering device is factory-selected for use with a particular combination of condensing units and indoor coils. If the orifice is too small, the evaporator will be starved for refrigerant. Excess liquid will then build up in the condenser. The initial effects will be inadequate cooling, high superheat, and low suction pressure. After the unit runs for a while, the liquid trapped in the condenser cools and the head pressure will be close to normal.

If the metering device orifice is too large, excess liquid will be fed to the evaporator. This condition causes a reduced head pressure with an elevated suction pressure and low superheat. Refrigerant floodback to the compressor is possible.

Although the condition is rare, excess oil in circulation displaces some of the liquid refrigerant and reduces the volume of refrigerant flowing through the orifice. This can starve the evaporator. Excess oil in a system can occur when the compressor is replaced multiple times, or when too much oil has been added to accommodate exceptionally long refrigerant lines.

Did You Know?

Metering Devices on New Equipment

To meet higher energy efficiency standards, many manufacturers are moving away from fixed metering devices on the indoor coils of some product lines. TXVs enable a system to reach higher energy efficiency ratios (EERs) and offer much better performance over a broad set of operating conditions.

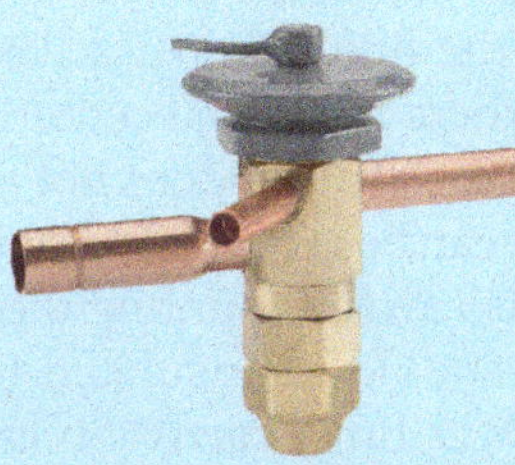

Source: Courtesy of Sporlan Division – Parker Hannifin Corporation

On Site

Frost on Suction Lines and Compressors

For technicians working with comfort-cooling systems, the appearance of ice on suction lines and compressor inlets indicates trouble. Typical comfort systems are designed to operate at evaporator temperatures above freezing. However, frost on the suction line and around the compressor inlet is not unusual in refrigeration applications.

For a walk-in freezer to maintain a box temperature of 10°F (–12°C), a common evaporator temperature—the saturated suction temperature—would be around 0°F (–18°C). Even if the system is operating with a superheat value of 10°F (5.5°C), the suction line temperature will be only 10°F (–12°C)—well below the freezing mark. Thus, ice is going to naturally form. In fact, the *absence* of frost on the suction line often suggests a problem.

This is a perfect example of why technicians must understand their craft and the thermodynamic processes at work. To evaluate a system, one must first see the big picture and consider the application.

Tip: When working on a low-temperature system, scrape the frost from the suction line in the immediate area before measuring the suction temperature. Otherwise, the reading will be inaccurate.

1.7.0 Thermostatic Expansion Valve Problems

TXVs are often blamed for problems caused by other sources. This is because TXVs are often misunderstood.

Troubleshooting TXV problems should always begin with measuring the superheat and subcooling, and then considering whether the saturated suction temperature is appropriate for the conditions. If too little refrigerant is fed to the evaporator, the superheat will be high. If too much refrigerant is fed, the superheat will be low. Although these symptoms can be caused by a faulty TXV, the problem is often elsewhere.

It is important to remember that TXVs, by design, will try to maintain the superheat setting under all operating conditions. When the system charge is slightly low, the TXV can often compensate and maintain the superheat, although at a reduced suction pressure. The TXV can simply open wider and feed enough refrigerant. Thus, a small loss of refrigerant may not be obvious in the superheat reading. However, when the charge is low enough to result in little or no subcooling, the TXV can't compensate, and the superheat rises.

The TXV is even better at masking a refrigerant overcharge. It will only open as far as necessary to maintain the superheat; any excess refrigerant simply backs up in the condenser (or the receiver, if equipped). It may require a considerable overcharge to cause the high-side pressure to rise to the point of being obvious. In this case, an excessive amount of subcooling indicates that there is more refrigerant in the system than needed. No other symptoms may be noted unless the overcharge is massive.

Some system problems caused by TXVs are the result of poor installation. Make sure the sensing bulb is securely fastened to a clean, straight, horizontal section of the suction line, close to the evaporator outlet. On suction lines that are $\frac{7}{8}$" OD and larger, the bulb should be installed at the 4 o'clock or 8 o'clock position on the line (*Figure 7*). For smaller suction lines, the bulb can be positioned anywhere above the centerline of the tubing.

When the installation does not allow for mounting the sensing bulb on a horizontal line, it can be mounted on a descending section of the suction line, but never on a riser. Also avoid mounting the bulb on or after a trap in the suction line.

The bulb should be insulated to minimize the effect of the air temperature around it. Over time, corrosion can take place between the bulb and the line, impeding heat transfer. The contact surfaces of the bulb and tubing should be clean. When a TXV is acting erratically, check the insulation and ensure the bulb and tube are clean and making good contact.

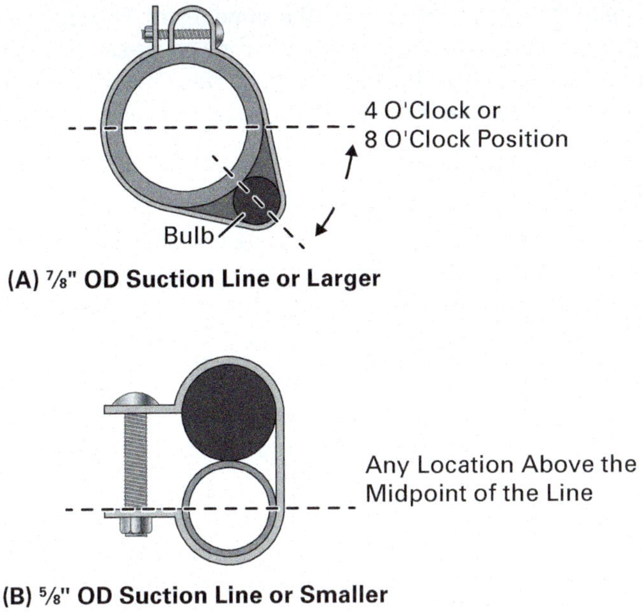

Figure 7 TXV sensing bulb mounting position.

Problems like these are often blamed on TXVs, but there are other possible causes to consider:

- Overfeeding (low or no superheat)
- Underfeeding (high superheat)
- Erratic operation (**hunting**)

Hunting: Erratic changes in refrigerant flow rate through a TXV as it attempts to find the proper valve position to maintain the superheat setting. Hunting is characterized by consistent swings in superheat of several degrees, up and down.

1.7.1 Refrigerant Overfeed

Refrigerant overfeed can be caused by the following conditions. Note that none of the conditions on this list are related to a faulty TXV:

- *Improper superheat setting* — This is generally an error that occurs when adjustments are made without a proper understanding of the valve and the refrigerant circuit. Many TXVs, however, especially those used by equipment manufacturers, are not adjustable.
- *Poor thermal contact between the sensing bulb and the suction line* — This is one of the more common problems. When the sensing bulb is unable to respond to the temperature in the suction line, the TXV typically feeds too much refrigerant. This is due to the air temperature around the bulb being higher than the temperature of the suction line.
- *Improper location of the sensing bulb* — When the sensing bulb is mounted in the wrong place, it may not be sensing the presence of liquid leaving the evaporator.
- *Oversized TXV* — An oversized TXV tends to hunt, as it can't stabilize the flow rate. This is because the valve is forced to operate very close to its seat.
- *Insufficient refrigerant load* — TXVs are excellent metering devices for systems with varying loads. But even a TXV will struggle to maintain the superheat when the load is very low. A severely reduced load has the same effect as an oversized TXV.
- *Wrong power element for the system* — The power element, which houses the diaphragm and includes the capillary and sensing bulb, is selected for one or more refrigerants and applications. TXVs with replaceable elements can be matched to power elements with different characteristics. The result of an incorrect power element may be refrigerant overfeed, but it can also result in a starved evaporator.

A faulty TXV can be the cause of refrigerant overfeed, but it is very rare. The only failures that would lead to overfeeding are fouling that might cause the valve to stick open or damage to the moving parts with the same result. Fouling can occur from repeated compressor electrical failures and the debris generated from them, or from copper oxides that develop from brazing and soldering. However, both sources are avoided through the protection of a filter-drier and proper service practices.

1.7.2 Refrigerant Underfeed

Refrigerant underfeed, or *starving*, can also be caused by a fouled or damaged TXV. The valve can stick in any position, but again, the condition is rare. Some TXVs have strainers at their inlet, which can become clogged with debris resulting from poor service practices. This would impede the flow of liquid into the valve, thus starving the evaporator.

Refrigerant underfeed generally occurs for one the following reasons:

- *Shortage of refrigerant* — A shortage of refrigerant that results in little or no subcooling leaves the TXV without enough refrigerant to keep up with the load. The superheat rises as a result.
- *Improper superheat setting* — This is generally an error that occurs when adjustments are made without a proper understanding of the valve and the refrigerant circuit. Many TXVs, however, especially those used by equipment manufacturers, are not adjustable.
- *Undersized TXV* — An undersized TXV may maintain its superheat setting under some conditions, but it is unable to feed enough refrigerant when the system is under a significant load. The superheat will rise when the system is under a significant load.
- *Loss of charge from the TXV power element* — When the power element loses its charge and there is no pressure on the diaphragm to help open the valve, the TXV simply closes. There is very little refrigerant in the assembly. Thus, if it is damaged or leaking, the charge is gone quickly and completely. Test for a loss of charge in the power element by removing the bulb from the suction line and holding it tightly in your hand. If the valve fails to feed refrigerant, the bulb has probably lost its charge. Leaking power elements must be replaced.
- *Obstructed liquid line or filter-drier* — An obstruction anywhere along the liquid line, depending on its severity, acts like a second metering device. There can be a significant pressure drop, and some of the liquid may flash. This results in a loss of subcooling and a reduced pressure at the inlet of the expansion valve. When the system is under load, a lack of subcooled refrigerant affects cooling performance, and the superheat rises. When the obstruction is significant, you can often feel the reduction in temperature from one side of the obstruction to the other. Frost may even form.
- *Undersized or exceptionally long liquid line* — The symptoms for an undersized or exceptionally long liquid line are similar to those resulting from an obstruction. Essentially, an undersized liquid line becomes a very long capillary tube.
- *Exceptionally high liquid-line riser* — The pressure drop per foot of liquid line riser (assuming it is properly sized for the length of the run) depends on the refrigerant and its density. HFC-410A experiences a pressure drop of about 0.43 psi per foot of riser. Thus, an HFC-410A riser 25' high experiences a pressure drop of about 11 psi. When there is insufficient subcooling, the liquid refrigerant may begin to flash before it reaches the expansion valve. Because risers are often unavoidable, increasing the subcooling to accommodate the loss is generally required. For a residential split system, this means an increase in the refrigerant charge.

1.7.3 Hunting

Hunting describes consistent changes in refrigerant flow rates through the TXV as it attempts to find a stable valve position. All TXVs may hunt to some degree under varied operating conditions, but not significantly. Hunting is characterized by relatively rapid swings in superheat of several degrees, up and down.

Hunting can only be verified by making several suction line temperature or pressure measurements over time. If there is a repetitive pattern, such as that shown in *Figure 8*, the valve is hunting.

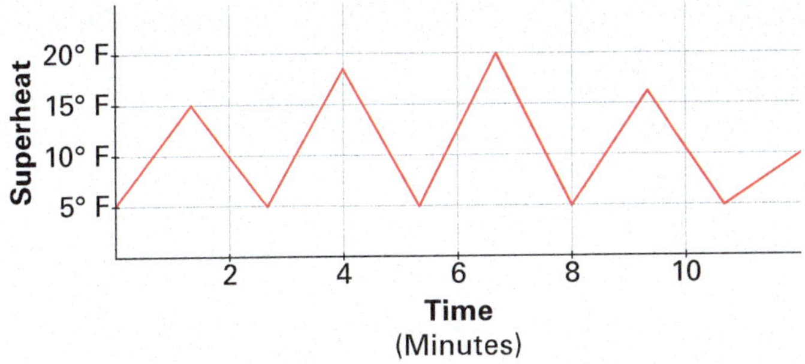

Figure 8 Superheat variations from hunting.

Hunting can be caused by an oversized expansion valve or sticking and binding of the TXV internal parts. Beyond the expansion valve itself, the following conditions can cause excessive hunting:

- Very light load at the evaporator, especially when the outdoor temperature and the head pressure are elevated
- Rapid or erratic changes in the load
- Rapid changes in the condensing pressure and/or temperature
- Excessively long liquid line
- Obstructions in the liquid line

1.8.0 Restrictions and Contaminants

Restrictions and contaminants in the refrigerant circuit are relatively common issues. Although a discussion of restrictions was included in the discussions of metering device problems, some additional information is provided in this section.

Split HVAC systems are more susceptible to refrigerant line restrictions and damage because the field-installed interconnecting tubing may be routed through areas where the systems are subject to accidents, abuse, or even vandalism. The refrigerant lines of packaged HVACR products are all contained within the cabinet, so the lines are less susceptible to damage. Restrictions in packaged systems are far more likely to be found at an accessory such as the filter-drier.

1.8.1 Refrigerant Line Restrictions

A restriction in a refrigerant liquid line or related accessory causes a pressure drop and reduces the refrigerant flow rate. When the restriction is significant, it is easily located.

A system with a liquid line restriction is sometimes diagnosed as a low refrigerant charge. This is because the symptoms of low suction pressure and high superheat are similar. Subcooling can be used to help determine if the problem is a restriction rather than a low charge.

Systems with a significant restriction in the liquid line usually show normal or elevated subcooling at the condenser, with little or no subcooling at the metering device inlet. Because of the restriction, refrigerant backs up in the condenser and/or receiver, where it is further subcooled. However, because the restriction acts like a metering device, some refrigerant may flash as it passes through. As a result, the subcooling is either severely reduced or nonexistent by the time the refrigerant reaches the metering device.

A restriction can often be detected by measuring the pressure drop across the accessory or tubing suspected to have the restriction; however, this is not always practical because gauge ports are not usually available along the length of the line.

An easier way to locate the restriction is to measure the temperature drop at various points along the line. Depending on the severity of the restriction, a technician can sometimes feel a temperature change in the line, or even observe frost or sweat just downstream of the problem area. If the restriction is very slight, measuring the temperature at various points is usually the only way to find the restriction.

Restrictions are often found at a clogged or undersized filter-drier. On a newly installed system, the filter-drier may be fouled shortly after startup with contaminants that entered the system during installation. If a system has a sight glass downstream of the filter-drier, flash gas (bubbles) usually appears in the sight glass. To confirm that a filter-drier is restricted, start by measuring the temperature drop across it. A temperature drop of 0.5°F is normal.

Measure the pressure drop if ports are available on each side of the filter-drier. Filter-drier selection charts generally reflect the correct size at 1 psi of pressure drop. A higher pressure drop indicates the filter-drier is either restricted or too small for the system.

The service valves represent another potential obstruction in split systems. During normal operation, ensure that both the liquid line and suction line service valves are fully open. During installation or service processes, technicians sometimes fail to fully open the valves, causing a restriction.

1.8.2 Air and Moisture Contamination

Refrigeration systems should contain only refrigerant and oil. Anything else is considered a contaminant. Air and moisture are the most common contaminants in a refrigeration or air conditioning system.

Air is a non-condensable gas, which means that it does not change states at common operating temperatures and pressures. Air entering the system can also contain water vapor. When air is in a system, it accumulates at the top of the condenser, taking up valuable condensing surface area. This increases both the head pressure and the condensing temperature.

Under the heat of compression, moisture can react with the refrigerant to form acids. These acids cause corrosion and breakdown of the motor winding insulation. Moisture in the refrigerant can also result in *sludge*, which reduces the lubricating properties of the oil and plugs oil passages in the compressor. Moisture can also freeze at the metering device, especially in refrigeration systems where the evaporator temperature must be colder.

Air and moisture enter a system in the following ways:

- Improper or no evacuation/dehydration process at installation or following refrigerant circuit work
- Leaks in the refrigerant circuit during an evacuation
- Failure to purge hoses prior to charging refrigerant

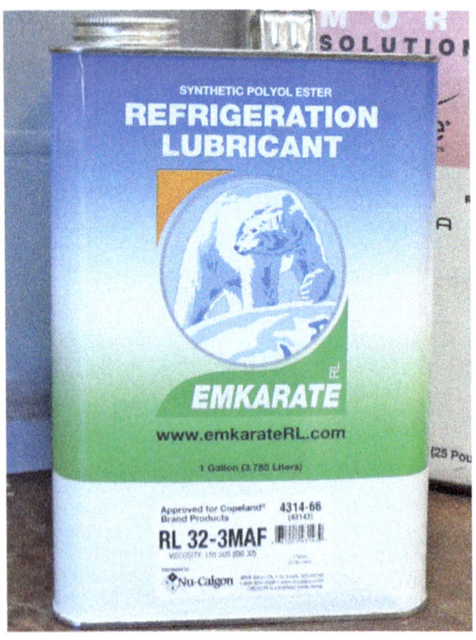

Figure 9 Polyol ester (POE) lubricating oil.

Systems using extremely hygroscopic lubricants (*Figure 9*), such as polyol ester (POE) and polyalkylene glycol (PAG) types, must be protected from moisture contamination. If the refrigerant circuit is left open, the oil will quickly absorb water vapor from the air. The moisture cannot be removed with a vacuum pump due to the strong molecular bond. Although filter-driers can usually separate the water, it is best to avoid the problem altogether. A circuit left open for an extended period on a rainy day will be very difficult to dehydrate.

On Site

Preventing Non-Condensables in a System

Do you want to prevent non-condensables in an HVACR system? The answer is simple. If the sealed refrigeration system must be opened for any reason, close it as quickly as possible. Evacuation removes non-condensables most of the time. Always use a well-maintained vacuum pump and evacuate the system to 500 microns before recharging. Allow the pump to run for a significant length of time at this vacuum level; water doesn't vaporize instantly. The triple-evacuation method, using dry nitrogen between each evacuation, is recommended.

One piece of advice from an industry veteran: Treat open cans of oil, as well as the refrigerant circuit, as if it is full of angry bees! Open and close it as quickly as possible. If the pump doesn't seal securely at the neck, use tape or rags around the neck to help restrict the movement of air and moisture into the container.

1.8.3 Acid Contamination

Acid is rarely introduced into a system, although it can be inadvertently added from contaminated recovery cylinders. It generally forms inside an operating system by the reaction of air or moisture with the refrigerant and oil. It can also be formed during a compressor motor failure, as the insulation on the motor windings breaks down in the extreme heat. An acidic sludge can result, which blocks oil passages, metering devices, and other components. The acid also attacks the winding insulation on the replacement compressor, breaking it down prematurely.

Acid is usually created as a compressor fails, resulting from the destruction of the motor windings and the insulating material. This is often referred to as a **burnout**. Burnouts are classified as either mild or severe based on the amount of acid present in the system. A mild burnout tends to yield little acid because it often occurs during a start attempt before the compressor is effectively pumping.

Burnout: A compressor failure resulting from the breakdown of the motor winding insulation, grounding the motor windings, and releasing contaminants into the refrigerant circuit. The breakdown of the insulation typically results in the formation of strong acids.

A severe burnout typically occurs over a longer period, while the compressor is running, allowing a significant amount of acid and other contaminants to be pumped through the system. This may also be referred to as a *running burn*.

WARNING!

When working on any pressurized system containing refrigerant, eye and hand protection must always be worn to prevent personal injury. This is especially true when working with acid-contaminated systems. The acids can cause burns or eye damage. Strong odors from the refrigerant circuit may also cause nausea.

Before replacing a compressor that has suffered a burnout, the system must be tested for acid. A sealed-tube acid/moisture test kit (*Figure 10*) is connected to a system service port to obtain a sample of the refrigerant vapor. Oil test kits (*Figure 11*) require that the test be performed on a sample of the compressor oil. In either case, follow the test kit manufacturer's instructions for using the kit and for determining the level of contamination. Oil test kits must be compatible with oil type in use.

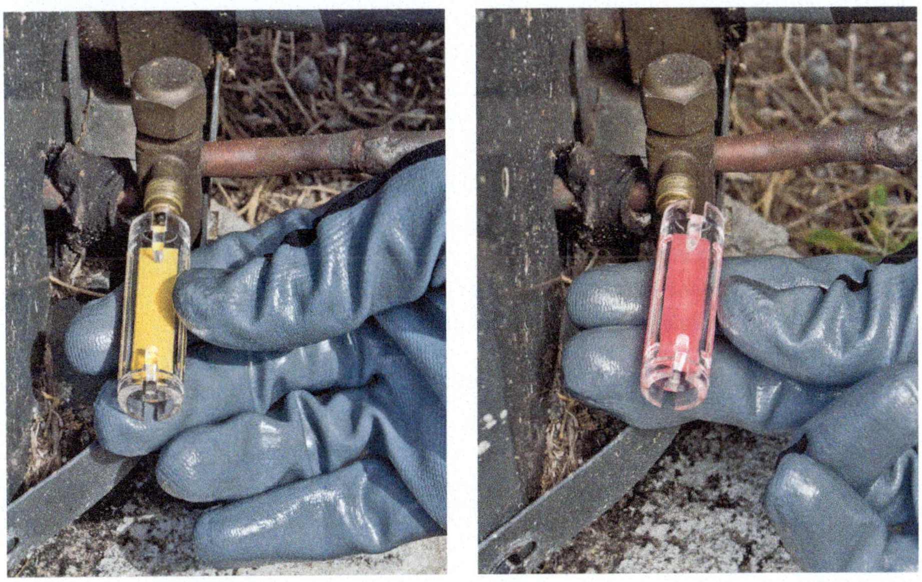

Before

After
(Positive Result)

Sealed-Tube Test Kit

Figure 10 Refrigerant vapor test kit.
Source: Courtesy of QuikProducts by Mainstream Engineering Corp.

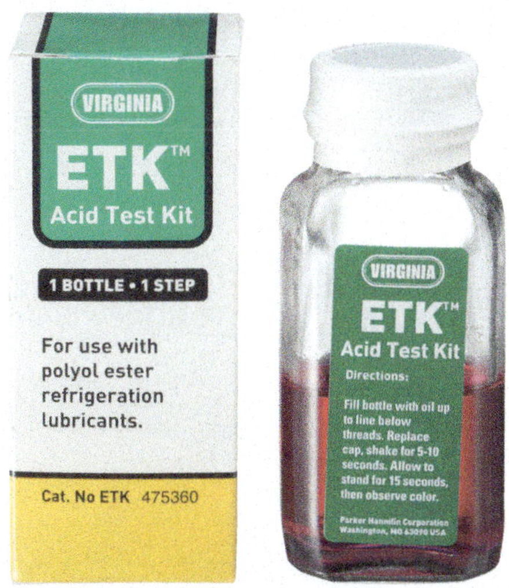

Figure 11 Oil test kit.
Source: Courtesy of Sporlan Division – Parker Hannifin Corporation

On Site

Refrigerant Smell Can Indicate a Burnout

Your nose can often provide the first indication of a burnout. Refrigerant from a system that has had a moderate to severe compressor motor burnout may have a very strong, pungent odor. Once you've smelled it, you won't likely forget it.

Burning some refrigerants can create a substance known as *phosgene gas*, once used in warfare to sicken enemy troops. It can form inside the circuit during a severe burnout. It can also be formed when using a torch in the presence of escaping refrigerant vapors.

1.0.0 Section Review

1. Which of the following components takes high-pressure, high-temperature refrigerant vapor and cools it to form liquid refrigerant?
   a. Compressor
   b. Evaporator
   c. Metering device
   d. Condenser

2. What temperature on the hot gas line would you consider normal in a properly operating HFC-410A cooling system?
   a. 80°F
   b. 120°F
   c. 170°F
   d. 230°F

3. Knocks and rattles from a compressor likely indicate which of the following conditions?
   a. Floodback
   b. High head pressure
   c. Undersized metering device
   d. Dirty condenser

4. One of the possible causes of a low suction pressure would be ______.
   a. indoor blower failure
   b. defective compressor valves
   c. an excessive refrigerant charge
   d. the recirculation of condenser air

5. On a unit with a fixed metering device, low airflow is generally indicated by a ______.
   a. high superheat condition
   b. low superheat condition
   c. high subcooling condition
   d. low subcooling condition

6. A fixed metering device that is too large will result in ______.
   a. low suction pressure
   b. high head pressure
   c. high superheat
   d. low superheat

7. On a unit equipped with a TXV, which of the following can be the cause of refrigerant overfeeding refrigerant?
   a. A plugged filter-drier
   b. A shortage of refrigerant
   c. Poor thermal contact between the bulb and suction line
   d. An excessively long liquid-line riser

8. True or False? A major restriction in the liquid line will cause a loss of subcooling downstream.
   a. True
   b. False

2.0.0 Compressor Troubleshooting and Replacement

Objective

Explain how to troubleshoot and replace a compressor.

a. Identify common problems that can lead to compressor failure.
b. Explain how to identify compressor mechanical problems.
c. Explain how to replace a hermetic compressor.
d. Identify additional steps often required to replace a compressor following a motor failure.

Performance Task

2. Demonstrate the ability to diagnose and resolve at least four of the following malfunctions:
   - Compressor failures
   - System-related compressor problems
   - Refrigerant undercharge and overcharge
   - Evaporator and condenser fouling or blockage
   - Metering device problems
   - Refrigerant line and filter-drier blockage
   - Refrigerant circuit contamination, such as acid and non-condensables

Compressors, being electromechanical devices, do occasionally fail due to flaws in manufacturing or design. However, in most cases, other problems in the cooling system directly or indirectly lead to a failed compressor. If the cause of a compressor failure is not found and eliminated, the replacement compressor will likely fail as well. *Figure 12* summarizes some of the symptoms exhibited by the compressor in response to common system problems.

Electrically troubleshooting the compressor motor is not significantly different than troubleshooting other motors. Compressor testing and electrical measurement has been covered in other modules within this curricula. At this point in your training, you should be able to measure compressor voltage, current, and winding resistance.

Slugging: The entrance of excess liquid refrigerant and/or oil into the compression area of a compressor Slugging can occur at startup due to condensed refrigerant and/or oil collected near the compressor inlet.

2.1.0 Problems for Compressors

Problems discussed in detail in this section include the following:

- **Slugging**
- Floodback
- **Flooded starts**
- Overheating

Although overheating can be a symptom of an internal problem, it can also result from an external problem. Slugging, floodback, and flooded starts are problems that are system related.

Flooded starts: Compressor starts during which slugging, oil foaming, and/or inadequate lubrication occur due to refrigerant migrating to the crankcase and condensing during the off cycle. When the compressor starts and the pressure in the crankcase drops rapidly, the condensed refrigerant flashes to the vapor state and causes the oil to foam, forcing oil into the compression area and reducing lubricity.

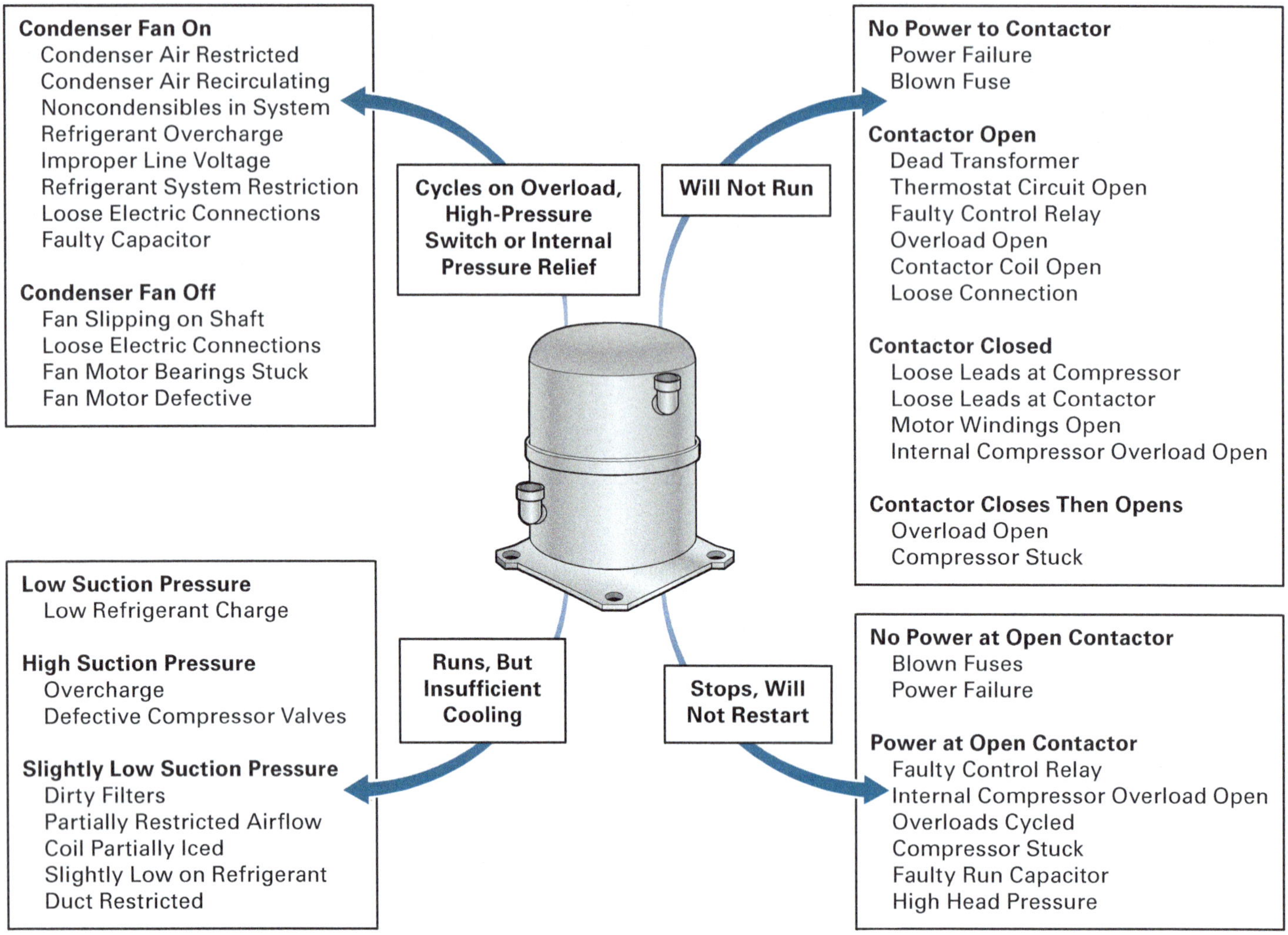

Figure 12 Compressor symptoms and causes.

2.1.1 Slugging

Slugging occurs when a compressor tries to compress liquid instead of vapor. Slugging typically involves a large dose of liquid entering the compression chamber or cylinder. *Wet vapor*, or vapor carrying microdroplets of refrigerant with little superheat, returning to the compressor doesn't result in slugging nor does it cause the same level of damage. Think of wet vapor as a dense fog. Refrigerant vapor is more likely to be in this condition when it is very close to its saturation temperature.

Slugging usually occurs at startup, but a rapid and dramatic change in system operating conditions can cause it as well.

Slugging is usually accompanied by a loud knocking noise from the compressor. Due to the extreme stress, slugging often results in blown gaskets or broken valves, piston rods, and crankshafts. This type of damage can happen during a single event. Reciprocating compressors are more susceptible to damage than scroll compressors, as the unique design of the scroll plates allows them to accommodate some liquid passing through. However, scroll compressors are not immune to the potential for damage.

Common causes of slugging include the following or a combination thereof:

- Buried refrigerant lines
- Excessive or trapped oil
- Overcharge of refrigerant
- Oversized, damaged, or failed metering device
- Failed crankcase heater
- Low evaporator load

Condensed refrigerant many be present in any cold area of a system during the off cycle. Buried refrigerant lines often collect condensed refrigerant that can return to the compressor as floodback on startup, leading to slugging. When buried lines are unavoidable, installing a suction accumulator is an option that offers valuable protection.

Slugs of oil can be trapped in the suction line when the suction gas does not have enough velocity to return the oil to the compressor. This often happens with compressors that have unloaders or similar systems for capacity control. When operating at part load, the velocity in the suction line is significantly reduced. At startup or when the compressor is fully loaded, slugs of oil can return to the compressor in sufficient volume to cause damage.

A refrigerant overcharge can overwhelm a fixed metering device to cause floodback, which may result in slugging. This is less likely to happen with expansion valves, however, unless the valve is stuck open. A failed crankcase heater can allow refrigerant to condense in the oil. When the oil foams at startup, slugs can be carried into the compression area.

2.1.2 Floodback

Floodback has been discussed previously in this module. Floodback may result in slugging, but not necessarily.

Refrigerant floodback dilutes and washes oil out of the crankcase, resulting in the oil foaming and failing to lubricate bearing surfaces and other moving parts. Floodback can lead to damaged piston rings and bearings as liquid refrigerant washes the oil from the mating surfaces.

Flooding is most often caused by one or more of the following conditions:

- Overcharge of refrigerant, especially on systems using a fixed metering device
- Oversized metering device
- Superheat setting too low for all operating conditions
- Broken, improperly located, improperly mounted, or poorly insulated expansion valve power element or temperature sensor
- Low evaporator load

2.1.3 Flooded Starts

Flooded starts are often caused by refrigerant condensing in the crankcase where it mixes with the oil. This is more likely during long off cycles at low ambient temperatures. Refrigerant will naturally migrate to, and condense in, a cold compressor shell.

A crankcase heater is the most important defense against refrigerant migrating to the compressor. However, a flooded start can also occur due to refrigerant condensing in a suction line during the off cycle; a crankcase heater can't prevent condensation from occurring outside of the compressor. Ensure the crankcase heater is working or add one, if necessary. Crankcase heaters warm the oil during the off cycle and prevent refrigerant condensation.

Many equipment manufacturers specify that power must be applied to the equipment for up to 24 hours before the initial startup or after a prolonged winter shutdown. This is done to allow the crankcase heater enough time to drive any accumulated refrigerant out of the compressor oil to prevent a flooded start.

Another way to prevent the migration of refrigerant through the system during the off cycle is to use a liquid-line solenoid valve (*Figure 13*). The valve can be wired to operate on a call for cooling along with the compressor. Once the need for cooling is satisfied, the valve closes as the compressor shuts down. Liquid-line solenoid valves are most often used in refrigeration or commercial applied systems, but these valves can be applied to smaller systems as well.

On some systems, the solenoid may close before the compressor is shut down. This allows the compressor to pump the refrigerant remaining in the

Figure 13 Liquid line solenoid.
Source: Courtesy of Sporlan Division – Parker Hannifin Corporation

evaporator and suction line out and into the condenser and/or receiver before it shuts down. This prevents any significant amount of refrigerant from condensing in these areas. A pressure switch is then used to turn off the compressor. When a solenoid valve is used in this way, it is referred to as a *pump-down solenoid*.

2.1.4 Overheating

Compressors naturally generate heat and are designed to handle it. However, when discharge temperatures from the compressor exceed 250°F (121°C), moving parts may seize due to thermal expansion. The oil also begins to break down and form a residue known as *coke*. It is well known that overheating, regardless of the reason, is the primary cause of grounded compressors and burnouts.

A sample of oil can provide some visual clues. Dark, black oil is a sign that the compressor has been running too hot. The oil will also show this dark, black color following a severe burnout, due to contaminants from the winding insulation and as well as coke residue.

It is important to remember that most compressors depend on the cool refrigerant gas returning to absorb a great deal of the heat produced. Thus, a system operating at a high superheat can lead to compressor overheating.

Case History

Overheated Compressors

A dozen new 20-ton packaged rooftop units were experiencing rapid-fire compressor failures. In this industrial plant, a great deal of heat was being generated in a building with a 28' roof height. Included in the operation was an industrial oven that added significantly to the heat.

The heat collected at the top of the building, as there were no fans or other devices to mix the building air and prevent stratification. Temperatures at the ceiling exceeded 125°F.

The return air intakes for each unit were positioned at the roof level. The original design called for the return air ducts to drop much lower, but the idea was rejected by the owner as a cost-saving measure. The technician, after measuring the return air temperature, checked the superheat at each compressor in multiple units. The lowest superheat reading exceeded 50°F (28°C). Several exceeded 60°F (33°C), due to the extreme heat load.

In industrial buildings like this one, there is little or no reason to attempt to cool the entire volume of the building. The people and primary equipment are on the ground. Although it was costly with production in progress, the technician convinced the client to lower both the return and supply ducts down to the 16' level. Heat that rose above this level was intentionally allowed to stratify and remain. Not only did the compressors begin to survive and produce more effective cooling, but the operating cost also fell dramatically.

Many compressors are protected from overheating by an internal or external thermal limit switch. Internal limit switches also provide some protection for three-phase compressors. When a three-phase compressor tries to start with only two phases of power provided (single-phasing), it becomes hot very quickly. The thermal limit switch, also referred to as a *thermal overload*, breaks the electrical circuit.

Without other forms of protection against a single-phase condition, this cycle can repeat itself many times. After several cycles, the entire mass of the compressor becomes very hot. It can take some time for it to cool after a technician arrives.

If a compressor overheats, check for a high superheat. This is often related to a refrigerant undercharge. A high compression ratio can also cause overheating. For example, a system with a TXV may have a condenser coil that is badly clogged with debris. Although the TXV will try to compensate and the superheat may not be excessively high, the compression ratio may be extremely high due to the increase in high-side pressure. The compression ratio increases as the differential between the high and low side of the system grows.

On Site

How Much Oil?

One concern for technicians during a compressor change is related to the amount of oil in the system. A system may have been operating under unusual conditions, such as undercharged or at reduced capacity, for a long time before the compressor failure. As a result, much of the oil can be trapped in the refrigerant circuit. Some oil always remains distributed in the piping, coils, and accessories.

With a semi-hermetic compressor, a sight glass shows how much oil is in the crankcase. With hermetic compressors, it is much more difficult to tell. A hermetic compressor can be drained after removal to determine how much it contains, but all the oil may not come out.

On systems that experience multiple compressor failures, the volume of oil in the system becomes an issue. Unfortunately, there is no foolproof way of determining exactly how much oil is in the circuit except by tracking how much went in and how much was taken away. When an oil sight glass is present, monitor a new compressor at full load long enough to ensure that oil floodback and slugging does not occur, due to trapped oil that returns.

Note that the ideal oil level visible in the sight glass differs between manufacturers and compressor models.

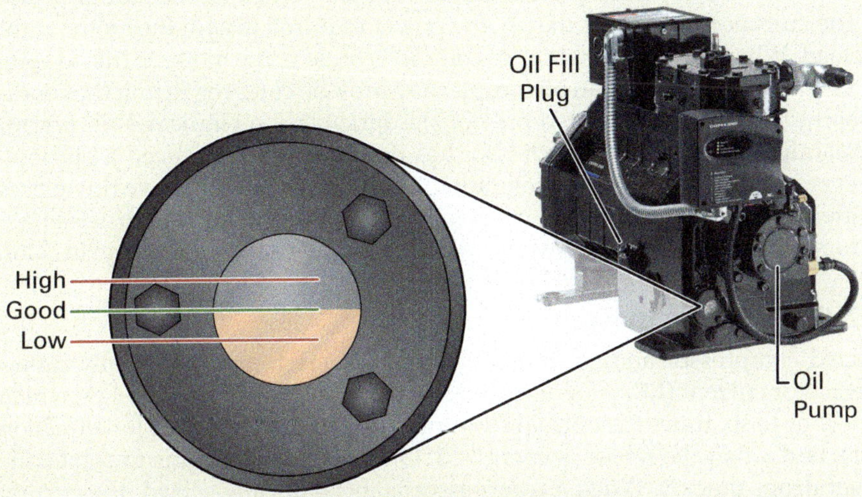

Source: Courtesy of White-Rodgers by Copeland (right)

2.2.0 Compressor Mechanical Problems

Compressors do experience mechanical problems of their own. Some failures are the result of mechanical flaws or normal wear.

A visual examination of an oil sample may provide clues regarding the condition of the compressor. As noted previously, dark, black oil indicates significant overheating, assuming the compressor has not already experienced a burnout. However, oil with gray or silver coloration, or oil that appears to have tiny pieces of glitter in it, indicates internal mechanical wear. Minute metal particles flaking from piston rings, cylinder walls, or bearings will cause this change in color.

2.2.1 Valves and Piston Rings

In a reciprocating compressor, valves and piston rings separate the high-pressure and low-pressure sides. If the pistons or their related mechanical parts are damaged, the compressor must be replaced. Valves can be field-replaced in semi-hermetic and open-drive compressors. Fortunately, pistons and piston rings are far less likely to be damaged than the valves.

Suction and discharge pressure tests can help determine the condition of the valves. Suction valves are damaged more often than discharge valves. Damaged valves or rings may be a problem if the suction or discharge pressures do not fall/rise as anticipated. Note that this assumes the system is properly charged and under normal load.

The compressor's running current can also be a clue to its mechanical condition. If the running current under a normal load is considerably less than normal, and the suction/discharge pressures are also abnormal, faulty valves and/or piston rings should be suspected. During this type of test, ensure that there are no active compressor capacity-control devices in operation that could cause misleading results.

A scroll compressor may exhibit the same combination of symptoms that indicate valve problems in a reciprocating compressor. However, a scroll compressor has no valves. When the same symptoms are encountered in a scroll compressor, it indicates damage to one or both scroll plates.

Occasionally, a compressor will appear to do absolutely nothing when it is energized. There may be little or no obvious sound coming from it. A check of the refrigeration gauges shows no activity. Yet voltage is being applied to the compressor terminals, and no safety device is, or has been, activated. The operating current will be measurable, but tiny when compared to normal.

These are the symptoms of a compressor with a broken crankshaft or piston rod. The current measured is merely the power required to spin the motor without a load. When the crankshaft or piston rod is broken, the motor is free to spin.

A visual examination of an oil sample may provide clues regarding the condition of the compressor. As noted previously, dark, black oil indicates significant overheating, assuming the compressor has not already experienced a burnout. However, oil with gray or silver coloration, or oil that appears to have tiny pieces of glitter in it, indicates internal mechanical wear. Minute metal particles flaking from piston rings, cylinder walls, or bearings will cause this change in color.

2.2.2 Seized Compressors

A seized compressor audibly hums but will not start, and it typically draws locked rotor current (LRA) for at least a few seconds. Normally, the LRA, which can be four to six times the normal running current, is approached for a fraction of a second when the motor is energized. Once the motor begins to rotate, the current drops quickly. When a compressor is mechanically seized, the current will remain at or near its LRA value until a safety device intervenes.

A seized compressor is a mechanical condition. However, there are also electrical problems that can prevent the compressor motor from rotating. Before condemning or attempting to start what appears to be a seized compressor, check for the following conditions:

- Low supply voltage or missing phase
- Three-phase compressor contactor not closing on all poles (single-phasing)
- Defective or open start relay (if present), preventing the start capacitor from assisting
- Start and/or run capacitor open or weak
- System pressures not balanced at startup

If the supply voltage is within ±10% of the motor's nameplate rating, but a single-phase PSC motor fails to start, a *hard start kit* may be needed to increase the motor starting torque to correct the problem. A hard start kit refers to a start capacitor that provides a capacitance boost to the motor as it starts. These kits are not designed to compensate for low voltage conditions or similar problems.

If electrical tests prove a PSC compressor has no open, shorted, or grounded windings, and if the bulleted factors listed at the beginning of this section are not an issue, a technician can attach a hard start kit to a PSC compressor (*Figure 14*). Connect the kit to the compressor using the wiring diagrams provided. The normally closed current relay included in the kit disengages the capacitor from the circuit once the compressor is rotating.

Figure 14 Hard start kit.
Source: RectorSeal, LLC

Once started, the compressor should continue to run at its normal speed. Allow the system to run long enough for the pressures to stabilize and check all the appropriate operating characteristics. After shutting the system down, allow the operating pressures to equalize. Then try restarting the compressor again without the kit attached. If the compressor starts the first time and starts on subsequent attempts, the hard start kit can remain or be removed. If the compressor will not start without the kit, it can be permanently installed. Alternatively, a correctly sized start capacitor and start relay for the compressor can be installed.

Start kits designed for use with most compressors are available from HVACR distributors. Note that each kit has a range of compressor sizes that it is designed to work with, such 1 to 3 tons.

When a compressor is changed, a different run or start capacitor rating may be required. This is sometimes overlooked when compressors are replaced. The new compressor may be equipped with a different motor than the previous model. Every time a compressor or motor is changed, even when there is no change in capacitance required, the capacitor is best replaced. The terms of many manufacturer compressor warranties require the installation of a new run capacitor when a compressor is replaced.

CAUTION

Hard start kits should never be applied to inverter-driven compressors, such as those used in most ductless systems.

CAUTION

Always use the correct start kit specified for the compressor capacity or horsepower. Failure to use the appropriate start kit may damage the compressor motor.

Three-Phase Compressors

Seized three-phase motors can sometimes be started by temporarily reversing or interchanging two of the leads. After the motor has started and is running, turn off the power immediately, then reconnect the compressor leads to previous state. Although reciprocating compressors can typically operate in either direction, scroll compressors cannot.

CAUTION

Three-phase scroll compressors should only be reversed for several seconds. While rotating in the wrong direction, no pressure differential will develop, and the compressor will likely be noisy. The compressor may be permanently damaged if left rotating in the wrong direction.

2.2.3 Scroll Compressors

Similar to reciprocating compressors, scroll compressors sometimes require start-assist components. However, the need for a hard start kit is far less likely since there is less friction between mechanical parts. Scroll compressors require less energy to place in motion than reciprocating compressors.

Because some scroll compressors can rotate backwards momentarily at shutdown, it is important to employ a short-cycle timer. Short-cycle timers provide the compressor the time required for pressures to equalize and its rotation to stop completely.

Unlike reciprocating compressors, scroll compressors cannot pump when their rotation is reversed. It is important to remember that three-phase scroll compressors must be properly phased at startup. Single-phase compressors should rotate in the correct direction regardless of polarity.

2.2.4 Rotary Compressors

Rotary compressors sometimes become *oil locked*. This is a condition in which the rotor will not turn due to the tight clearances inside. Only a thin film of oil typically coats the surfaces to prevent leakage from the high-pressure side to the low-pressure side. Oil lock can occur in new compressors as well as old.

To resolve an oil-locked compressor, try to start the motor several times. If this does not work, install a hard start kit. If the kit does not help the compressor start, it must be replaced.

Another problem unique to rotary compressors is that of stuck vanes. When a vane sticks, there is no pumping action, and the motor current is low. The rotor is turning, but the vane(s) is not moving out to create compression pockets.

NOTE

This method of testing a rotary compressor typically requires refrigerant recovery and brazing processes, making it labor intensive. However, if the compressor does not begin to pump, the compressor or equipment will need to be replaced. If it does work, then the effort required to reconnect the piping, evacuate, and recharge the system is considered worthwhile.

Before replacing the compressor, try disconnecting the suction line from the compressor shell. Use a heat gun to warm the tubing and inside the suction inlet. The heat may cause the stuck vane to release. Before reconnecting the suction line, momentarily start the compressor with the palm of a heavily gloved hand over the suction fitting. If the vane is no longer stuck, the compressor will pull on the palm of your hand. This indicates that the vane is now moving and creating suction pressure.

Worn vanes are generally unable to develop sufficient pressure differential for the system to function properly. Since rotary compressors are usually hermetic, worn vanes can only be resolved by compressor replacement.

2.3.0 Compressor Replacement

The procedures required to replace a compressor differ depending on whether the cause of the compressor failure is electrical or mechanical.

The procedures in this section describe the replacement of a typical hermetic compressor. Always consult the system and replacement compressor manufacturer's service literature for the specific system being serviced and the compressor being used before replacing a compressor.

The following basic guidelines apply when replacing a compressor following a mechanical failure:

WARNING!

Be sure all sources of energy to the equipment are turned off before beginning a compressor replacement. Open, lock, and tag disconnects. Use extreme caution around pressurized or hot components. Follow all safety instructions labeled on the equipment and provided in the manufacturer's service manual. Failure to do so may result in personal injury.

Step 1 Test the new compressor for open windings, grounds, and shorts before the task begins. Compare the nameplates of the two compressors and ensure the new compressor is the correct type and voltage, although the model numbers will not always match.

Step 2 Turn off the equipment. Lockout and tag all sources of power and energy.

Step 3 Recover the refrigerant, regardless of its condition. Recycle or reclaim the refrigerant as its condition dictates.

Step 4 Disconnect all wiring from the compressor and tag the leads as necessary.

WARNING!

Danger exists if the compressor terminal assembly is physically damaged while the system is pressurized. The terminal assembly, sealed to the compressor shell, may weaken or crack under severe electrical or mechanical stress. When making voltage, current, or continuity checks on a hermetic or semi-hermetic compressor in a pressurized system, always make measurements at terminal boards and points away from the compressor terminals. Disconnect the wiring from the compressor terminals after the refrigerant has been recovered.

Step 5 Use a tubing cutter to cut the refrigerant tubing connected to the compressor. Do not use a hacksaw; it will introduce copper chips into the system. Tape the open lines in the equipment to prevent dirt or moisture from entering while work continues.

WARNING!

Do not use a torch to disconnect the tubing from the compressor. Use a tubing cutter.

Step 6 Remove the mounting bolts, and then remove the old compressor. Save any vibration isolators for reinstallation. Be sure to get help lifting and moving the compressor to avoid injury.

Step 7 Unpack the new compressor and read the literature that accompanies it. Replacement compressors often have new features that need to be acknowledged, such as changes in the run capacitor size.

Step 8 Install the vibration isolators on the feet of the new compressor.

Step 9 Mount the new compressor in the equipment and bolt it down. Remember that the mounting bolts are not usually tightened all the way, except when the equipment is being prepared for shipment. The bolts should be tightened until the heads nearly contact the rubber vibration isolators or the top of the foot. This allows the compressor to float on its isolators.

Step 10 Measure, cut, and fit new piping components for the compressor suction and discharge lines.

Step 11 Braze the system refrigerant lines to the compressor. A nitrogen purge should be maintained to prevent copper oxides and prevent oil ignition. Clean the joints after they cool.

Step 12 Remove the existing liquid line filter-drier and replace it with one with the proper capacity for the system. Oversizing the filter-drier for this type of failure is not typically necessary.

Step 13 Leak test the system and apply proper evacuation techniques.

Step 14 While the vacuum pump is running, use the time to consult the equipment wiring diagram and connect all the disconnected wiring. Replace the contactor if it has been stressed or damaged and replace the capacitor(s) if present.

Step 15 Recharge the system with the correct type and weight of refrigerant. If the charge is not weighed in, ensure enough is added to start the system while the remaining charge is added as it runs.

Step 16 Start the system and allow it to run to stabilize the system pressures. Refer to the equipment manufacturer's service instructions and follow them to adjust the refrigerant charge. Check and record all the important operating information.

On Site

Liquid Refrigerant Charging

When adding refrigerant following a compressor replacement, charging with liquid from the cylinder is usually required today, for a variety of reasons. With blend refrigerants, liquid charging is always a necessity. Using a fitting with an orifice in the charging hose between the gauge manifold and the system connection, as shown here, ensures that the liquid vaporizes as it enters, and at a flowrate the compressor can handle.

Source: Ritchie Engineering Co, Inc./ YELLOW JACKET

NOTE

Make sure that only one liquid line drier is installed in the system. Remember that manufacturers may place one inside the condensing unit.

2.4.0 Compressor Replacement Following an Electrical Failure

Compressor electrical failures related to open windings do not usually create a significant problem in the refrigerant circuit. Sometimes the internal safety device fails, opening the windings permanently. However, when an internal path to ground is formed, arcing occurs, and an astonishing amount of contaminants can be produced very quickly as the winding insulation burns.

When the windings are shorted to ground, it is usually referred to as a burnout. The term refers to the breakdown of the motor winding insulation due to the arcing and extreme temperatures generated at the moment of failure. Burnouts can be mild or severe.

A mild burnout usually occurs suddenly, causing the motor to stop quickly before a lot of contaminants are created and pumped out into the refrigerant circuit. Mild burnouts often occur as the compressor starts. If the compressor does not rotate, or only makes a few rotations before the windings fail, it is not pumping as the event occurs.

Another factor in a mild burnout is the overcurrent protection device(s). If the fuse or circuit breaker opens quickly, the amount of time for damage to be done is significantly reduced.

Severe burnouts develop over a longer period. The compressor is operating and moving refrigerant as the event takes place. If the process begins slowly and the overcurrent protection device is slow to respond, lots of contaminants are produced and pumped throughout the refrigerant circuit. This is sometimes referred to as a *running burn*. More aggressive actions to clean up the refrigerant circuit are required.

Before replacing a compressor that has experienced a burnout, try to determine if the event was mild or severe. Check the system for acid, using one of the many kits available. Acid/moisture test kits can typically be connected to a system service port to obtain a sample.

Oil test kits require a sample of the system oil. For systems with hermetic compressors, an oil sample may be difficult to obtain without removing the compressor.

Another way to determine the severity of the burnout is to open the suction and discharge lines and check for carbon by running a clean swab inside the lines. If you find a significant amount of carbon in the discharge line but none beyond the condenser coil, perhaps it was mild. During a running burn, if the system has a filter-drier, most of the carbon stopped there. If there is no filter-drier, then the carbon and contaminants can easily continue throughout the circuit.

You'll have to judge the severity by how much carbon you see. Carbon deposits in the suction line usually indicate a severe burnout. In a semi-hermetic compressor, the easiest way to tell the severity of the burnout is to remove a cylinder head and look for the carbon.

2.4.1 Mild Burnout Procedures

Since most contaminants produced by a mild burnout remain within the compressor itself, the procedure for changing the compressor is the same as that used when there is a mechanical failure. An oversized liquid line filter-drier may be chosen to provide extra surface area to capture particles and contaminants. Triple-evacuate the system before charging it with refrigerant.

During the compressor replacement process, consider installing an oil sample trap for future oil testing. The liquid line filter-drier should be replaced after several hours of operating time to permanently remove what it has captured from the circuit.

2.4.2 Severe Burnout Procedures

Servicing a system after a severe burnout requires a thorough cleanup of the refrigerant circuit to remove all harmful contaminants. In many cases, compressor replacement is only the beginning of the work to be done. A proper system cleanup can be a time-consuming and very costly process.

Successive compressor failures can often be traced to improper cleanup. The following is an overview of the procedure for replacing a compressor after a severe burnout. It is essential to follow the system and/or compressor manufacturer's guidelines for replacing a compressor under these conditions. Note that the compressor warranty may depend on following manufacturer-specified steps that may differ from this example.

Step 1 Recover the system refrigerant and remove the compressor. This refrigerant should be sent away for reclamation, as it is unlikely that any field recycling process can successfully remove the contaminants.

Step 2 Remove the liquid line filter-drier.

Step 3 Flush the system piping in the opposite direction from normal refrigerant flow, using a refrigerant circuit flushing kit (*Figure 15*). The indoor and outdoor coils should also be flushed. While purging the system with nitrogen won't hurt, it will not remove contaminants trapped in piping or coils.

Step 4 Remove and clean, or replace, the metering device(s), receiver, accumulator, and oil separator, if equipped.

Step 5 Install the new compressor. Use a nitrogen purge while brazing.

Step 6 Install the liquid line filter-drier. For sealed-core driers, use one size larger than usual. For larger systems, it is best to install a replaceable-core filter-drier shell (*Figure 16*) so that the core can be replaced as needed once the system is returned to service. The large cores also provide far more surface area than sealed filter-driers.

- Install isolation ball valves on each side of the filter-drier with an access valve to recover refrigerant from this section of the line. The access valve will also be used to evacuate this section and the new filter-drier when it is replaced. This step eliminates the need to recover the entire system charge to change the filter-drier later. Many replaceable-core filter-drier shells have a fitting in the cap for the installation of an access valve.

NOTE

Remember that a reversible filter-drier is required when it is placed in the liquid line of a heat pump.

Figure 15 Refrigerant circuit flushing kit.
Source: RectorSeal, LLC

(A) Filter-Drier Shell

(B) Replaceable Core

Figure 16 Replaceable-core filter-drier shell and core.
Source: Courtesy of Danfoss

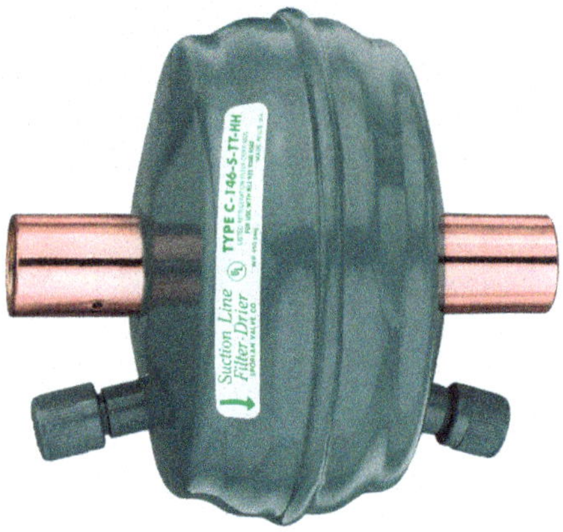

Figure 17 Suction line filter-drier.
Source: Courtesy of Sporlan Division – Parker Hannifin Corporation

NOTE

Even during a running burn, a limited amount of contamination typically reaches the suction line. Suction line filter-driers can often be removed completely after several hours of system operating time. If a replaceable-core shell is used, the cores can be removed and replaced with simple filter inserts or be left empty. When the pressure drop is not increasing over hours of run time, remove the filter-drier assembly or core.

NOTE

When working with semi-hermetic, also change the oil after a few hours of run time. This is especially important if the filter-driers are becoming fouled, indicating a large volume of contaminants are present.

Step 7 Add a suction line filter-drier with input and output pressure taps (*Figure 17*). The pressure tap(s) on a suction drier allow for the pressure drop to be tested while the system is in operation. A replaceable-core suction filter-drier may be required, depending on the capacity of the system. Suction line filter-driers should be selected for a maximum pressure drop of 3 psi in a permanent installation, and 8 psi for temporary use.

- Check the pressure drop after startup to provide a reference for future pressure drop measurements. Suction line filter-driers must be larger than liquid-line models due to the lower density of the vapor stream. Suction line filter-driers should not be left in a system indefinitely. Check the pressure drop periodically and remove or replace the drier or core as needed.

Step 8 Triple-evacuate the system.

Step 9 Recharge the system with new refrigerant.

Step 10 Start the system and adjust the charge as needed. Immediately check the pressure drop through the suction line filter-drier on startup to provide a baseline. If possible, also check the pressure drop through the liquid-line filter-drier. Check the pressure drop periodically over the first few hours of operation. Shut down the system and change the filter-driers when the pressure drop shows a significant increase above the baseline pressure drop.

Step 11 After a few hours have elapsed, stop the system and test for acid to make sure the system is clean. If the test is negative, replace the liquid line filter-drier if it has not been replaced already. Remove the suction line filter-drier or drier cores from the system.

- If acid or moisture is still present in the system, change the oil (semi-hermetic compressors only) and repeat *Step 10* and *Step 11* as necessary to achieve a clean system.
- Never leave a suction line filter-drier in a system for an extended period. For an extra measure of caution, a suction line filter-drier or core can be left in the circuit, but only if the pressure loss through the clean filter remains below 3 psi. Even then, note that this additional pressure drop affects system energy use.

On Site

Drier Cores

A wide variety of drier cores are offered by various manufacturers for specific uses. Some are designed to capture and hold more water than a normal core; others are designed specifically for POE oils; still others are designed to do a better job of capturing acids. Always consult the filter-drier manufacturer's specifications and literature to determine the proper size and best core type for your specific needs.

Drier cores are fragile and should be handled with care. If subjected to vibration and impacts, the cores can chip or crack, even inside their packaging.

Source: Courtesy of Sporlan Division – Parker Hannifin Corporation

2.4.3 Changing Oil

After a semi-hermetic compressor has been changed due to a severe burnout, the oil will likely need to be changed one or more times. The refrigeration oil, because it is always in a liquid state, tends to capture and carry particles, acids, and other contaminants that enter the system. Although it would be good to change the oil in a hermetic compressor as well, it is not usually practical.

Before removing oil from a semi-hermetic compressor, test a sample of the oil for the presence of contamination. Use an acid test kit and follow the test kit manufacturer's instructions to determine the acid content. The test provides a baseline for future comparison as the cleanup process continues.

Depending on the compressor model, the oil can be removed one of two ways: draining through an oil drain in the sump or pumping it out through an oil fill port (*Figure 18*). Some compressors do not have an accessible drain plug. For these compressors, the oil fill port must be used. Note that draining oil through the plug can be a messy affair unless a fitting and valve have been installed.

Figure 18 Semi-hermetic compressor crankcase ports.
Source: Courtesy of White-Rodgers by Copeland

Remove the oil from a semi-hermetic compressor with a drain plug using the following general guidelines:

WARNING!

When removing oil, wear rubber gloves and eye protection to prevent possible injury. Contaminated refrigerant oil may contain acid. Do not allow contact with your skin or eyes as severe burns may result. Also ensure that the crankcase pressure has been relieved before opening a drain or fill port.

Step 1 If possible, run the compressor fully loaded for some time to maximize oil return from the system. Then close the suction service valve and reduce the crankcase pressure to 0 psig as the compressor runs. Note that the low-pressure switch may have to be temporarily bypassed. Otherwise, a refrigerant recovery process is required to reduce the crankcase pressure to 0 psig.

Step 2 Stop the compressor and isolate it from the system by closing the discharge service valve.

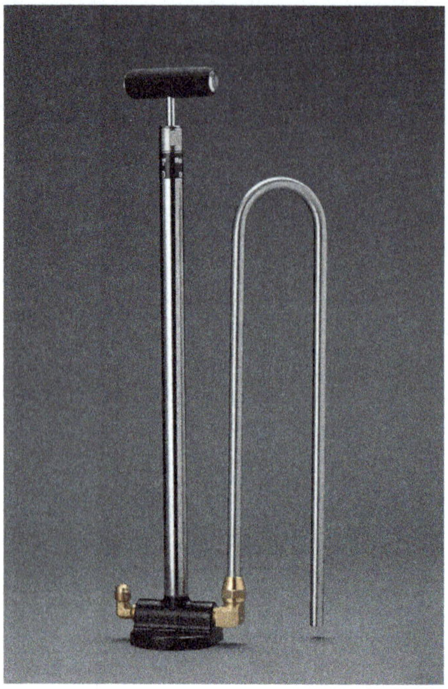

Figure 19 Oil pump.
Source: Ritchie Engineering Co, Inc./YELLOW JACKET

CAUTION

POE oil readily absorbs water vapor when exposed to air. Containers of POE oil should not be used if previously opened to the atmosphere for an extended period. Never store POE in an open container.

NOTE

The correct sight glass oil level varies from one compressor manufacturer to another. Refer to the manufacturer's instructions. Most manual oil pumps can add oil to the crankcase while it is pressurized, but pumping the compressor down to a lower suction, as previously described, is recommended.

Step 3 Loosen the drain plug until the oil begins to seep out around the plug threads. Removing the plug too quickly will allow the oil to drain uncontrollably. Drain the oil into a suitable container so that it is properly captured for disposal. After the oil stops draining, reinstall the plug.

When there is no accessible drain plug, the oil must be removed through the oil fill port. Remove the oil through an oil fill port using the following general guidelines:

Step 1 If possible, run the compressor fully loaded. Then, close the suction service valve and reduce the crankcase pressure to 0 psig. Remember that the low-pressure switch may have to be temporarily bypassed. Alternatively, isolate the compressor and recover the refrigerant from the compressor.

Step 2 Stop the compressor and isolate it from the system by closing the discharge service valve.

Step 3 Remove the oil fill plug. Use a refrigeration oil pump (*Figure 19*) equipped with flexible tubing to remove the oil from the compressor. Insert the tubing through the oil fill port and pump the oil out, collecting it in a suitable container. Continue until all the oil has been removed.

Before charging oil into the compressor, refer to the manufacturer's instructions for the compressor being serviced. Make sure to use the correct volume and type of oil specified. This is very important; do not use substitutes unless authorized to do so. The new oil must be compatible with the oil currently in the system.

Fresh oil is charged into a compressor after it has been drained. Use a refrigeration oil pump to charge oil, such as the one shown in *Figure 19*. Follow these basic steps to add fresh oil to the compressor:

Step 1 If the compressor has an oil drain plug, make sure it is reinstalled and tight. The oil fill plug should be removed from the fill port. Some compressors may have a special valve in the port that can connect directly to the pump.

Step 2 Mount and secure the oil pump in or next to (depending on the pump style in use) the container of new oil. Lay the end of the charging hose in a suitable container to capture purged oil. Purge the oil pump line by operating the pump until oil appears without air bubbles. Then quickly connect the end of the hose to the compressor fill port.

Step 3 Open the oil port valve (if applicable). While monitoring the oil level in the compressor sight glass, pump oil into the compressor crankcase from the container of oil as needed.

Step 4 Close the service valve and disconnect the pump.

Step 5 Evacuate the compressor to 500 microns or less.

Step 6 Open the suction and discharge service valves, and then restart the compressor.

Step 7 Run the system for about 30 minutes. Full-load operation is necessary to ensure that refrigerant velocities in the piping system are sufficient to entrain and move oil back to the compressor. Then recheck the oil level in the sight glass. Add additional oil if necessary.

2.0.0 Section Review

1. How does slugging differ from floodback?
   a. Floodback can only happen when the compressor starts.
   b. Floodback can only result from an overcharge of refrigerant.
   c. Slugging refers to liquid having entered the compression chamber or cylinder.
   d. Slugging can only happen long after the compressor starts.

2. A compressor hard-start kit _____.
   a. provides a capacitance boost
   b. compensates for a low voltage condition
   c. can only be applied to three-phase compressors
   d. slowly ramps up the AC voltage as it energizes the compressor

3. Which of the following steps in compressor replacement should be done before the others?
   a. Recover the refrigerant; recycle or reclaim as needed.
   b. Install the vibration isolators on the feet of the new compressor.
   c. Test the new compressor windings and compare the nameplates.
   d. Use a tubing cutter to cut the refrigerant tubing connected to the compressor.

4. When do the worst compressor burnouts typically occur?
   a. Before the compressor is rotating.
   b. While the compressor is shut down.
   c. When the compressor is running and pumping.
   d. When a fuse or circuit breaker opens very quickly.

Module 03210 Review Questions

1. On a properly operating system, what ratio of liquid to vapor would you expect to be exiting the metering device?
   a. 20% liquid, 80% vapor
   b. 30% liquid, 70% vapor
   c. 55% liquid, 45% vapor
   d. 80% liquid, 20% vapor

2. If a refrigerant liquid line is routed through an unusually hot area on its way to the metering device, how might this affect the device?
   a. Subcooling can increase.
   b. Subcooling can decrease.
   c. Superheat can decrease.
   d. Compressor current can decrease.

3. Air contamination in a refrigerant circuit will likely result in which condition?
   a. High suction pressure
   b. Low suction pressure
   c. Low head pressure
   d. High head pressure

4. A system with an expansion valve that is overcharged with refrigerant will exhibit which of the following symptoms?
   a. The subcooling will be high.
   b. The superheat will be high.
   c. The suction pressure will be low.
   d. The evaporator coil will usually freeze.

5. When the suction pressure is low on a system with a fixed metering device, how might you determine that the problem is low evaporator airflow and not a low refrigerant charge?
   a. The compressor will likely be very hot.
   b. The superheat will be low rather than high.
   c. The condenser coil will be frozen.
   d. The evaporator coil will be warm.

6. On a system with a fixed metering device, what are the symptoms of an undersized orifice?
   a. High superheat and low suction pressure
   b. High suction pressure with low superheat
   c. Low superheat and a sweating compressor
   d. Low head pressure and high suction pressure

7. In an HFC-410A system, how much pressure drop must you consider for a liquid line riser?
   a. 1 psi per 10' of rise
   b. 1 psi per foot of rise
   c. 2.3 psi per foot of rise
   d. 0.43 psi per foot of rise
8. Which of the following conditions can lead to expansion valve hunting?
   a. An undersized TXV
   b. Subcooling below 20°F
   c. A very light evaporator load
   d. Superheat set to maintain 10°F
9. What causes oil to foam in a compressor crankcase when it starts?
   a. Foaming is common when synthetic oil is in use.
   b. The movement of refrigerant vapor through the crankcase.
   c. The condensing of refrigerant as the pressure suddenly increases.
   d. The flashing of liquid refrigerant as the crankcase pressure decreases.
10. Which of the following may be used to stop the migration of refrigerant to the compressor crankcase during the off cycle?
    a. A TXV
    b. Liquid-line solenoid
    c. A fixed metering device
    d. Evaporator pressure regulator
11. A compressor regularly experiences flooded starts. Which of the following is a likely cause?
    a. Not enough oil
    b. Failed crankcase heater
    c. Insufficient evaporator load
    d. An undercharge of refrigerant
12. If a compressor starts but no pressure differential develops and its operating current is very low, which of the following is the most likely problem?
    a. The metering device has failed.
    b. The line voltage is below the minimum.
    c. There is a severe restriction in the refrigerant circuit.
    d. The compressor has a broken crankshaft or piston rod.
13. True or False? Three-phase scroll compressors work in either direction of rotation, but single-phase scroll compressors will only work in one direction.
    a. True
    b. False
14. A running burn is often indicated by ______.
    a. a failed run capacitor
    b. a high suction pressure
    c. a lack of acid in the system
    d. acid and contaminants distributed through the refrigerant circuit
15. When sizing a suction line filter-drier, what is the maximum pressure drop to consider as a permanent installation?
    a. 1 psi pressure drop
    b. 2 psi pressure drop
    c. 3 psi pressure drop
    d. 4 psi pressure drop

Answers to odd-numbered review questions are found in the Module Review Answer Key at the back of this book.

Answers to Section Review Questions

Answer	Section	Objective
Section One		
1. d	1.1.0	1a
2. c	1.2.0	1b
3. a	1.3.2	1c
4. a	1.4.0; *Figure 3*	1d
5. b	1.5.1	1e
6. d	1.6.0	1f
7. c	1.7.1	1g
8. a	1.8.1	1h
Section Two		
1. c	2.1.1; 2.1.2	2a
2. a	2.2.2	2b
3. c	2.3.0	2c
4. c	2.4.0	2d

MODULE 03311

Troubleshooting Heat Pumps

Source: Darkdiamond67/iStock

Objectives

Successful completion of this module prepares you to do the following:

1. Identify the characteristics of heat pumps and describe common heat pump operating modes.
   a. Describe heat pump characteristics and identify different heat pump types.
   b. Describe the common heat pump cooling, heating, and defrost modes of operation.
2. Recount the sequence of operation for the common heat pump operating modes.
   a. Recount the sequence of operation for the cooling mode.
   b. Recount the sequence of operation for the heating mode.
   c. Recount the sequence of operation for the defrost mode.
   d. Recount the sequence of operation for dual-fuel systems.
   e. Describe how to evaluate electronic controls in heat pumps.
3. Explain how to test and troubleshoot various functions and components of heat pump systems.
   a. Explain how to test field and factory wiring.
   b. Explain how to test and troubleshoot heat pump thermostats.
   c. Explain how to test various valves.
   d. Explain how to test defrost controls.

Performance Tasks

Under supervision, you should be able to do the following:

1. Demonstrate the ability to diagnose and identify the proper solution for an instructor-selected mechanical problem in a heat pump.
2. Demonstrate the ability to diagnose and identify the proper solution for an instructor-selected electrical problem in a heat pump.
3. Initiate and evaluate the defrost mode of a heat pump.

Overview

This module reviews the operation of heat pumps and presents methods and procedures for their troubleshooting and repair. Heat pumps contain more controls and have more complex circuits than a cooling-only system. Understanding how they differ from cooling-only systems and what to expect during the heating mode is essential to successful troubleshooting and repair.

Digital Resources for HVACR

SCAN ME

Scan this code using the camera on your phone or mobile device to view the digital resources related to this craft.

1.0.0 Heat Pump Operating Modes

Performance Tasks

There are no Performance Tasks in this section.

Objective

Identify the characteristics of heat pumps and describe common heat-pump operating modes.

a. Describe heat pump characteristics and identify different heat pump types.
b. Describe the common heat pump cooling, heating, and defrost modes of operation.

Servicing a heat pump is no more difficult than servicing any other air conditioning equipment. While a heat pump has different operating modes and many have an auxiliary heat source, heat pump control circuits follow the same principles as the various circuits you've encountered thus far. More circuits do not necessarily increase the difficulty.

Heat pumps currently in use have many variations and continue to grow in popularity. Heat pumps installations in colder areas are increasing rapidly. All heat pumps rely on the same mechanical refrigeration cycle with which you are familiar. While some of the troubleshooting procedures presented in other modules apply to heat pumps, a few are unique to these systems.

1.1.0 Heat Pump Systems

A heat pump is a combination heating and cooling unit. It produces cooling in the same manner as a conventional air conditioner, and then reverses the cycle to produce heat (*Figure 1*). Heat pumps and conventional cooling systems have many components in common.

In addition to typical air conditioning components, a comfort-system heat pump requires a reversing valve as well as additional metering devices and control circuits. Because heat pumps perform both heating and cooling functions, the coils are referred to by location. The condensing coil and evaporator coil are now referred to as the outdoor coil and indoor coil.

The main advantage of a heat pump is that it is more efficient than electric resistance heat. Depending on the operating conditions, a high-efficiency heat pump can provide more than four times as much heat as electric resistance heaters consuming the same amount of power.

Depending on the climate and related utility costs, a heat pump may or may not be as cost effective to operate as a gas furnace. However, improvements in heat pump efficiency and performance continue to be made, and replacing a fossil-fuel system with a heat pump also offers environmental benefits.

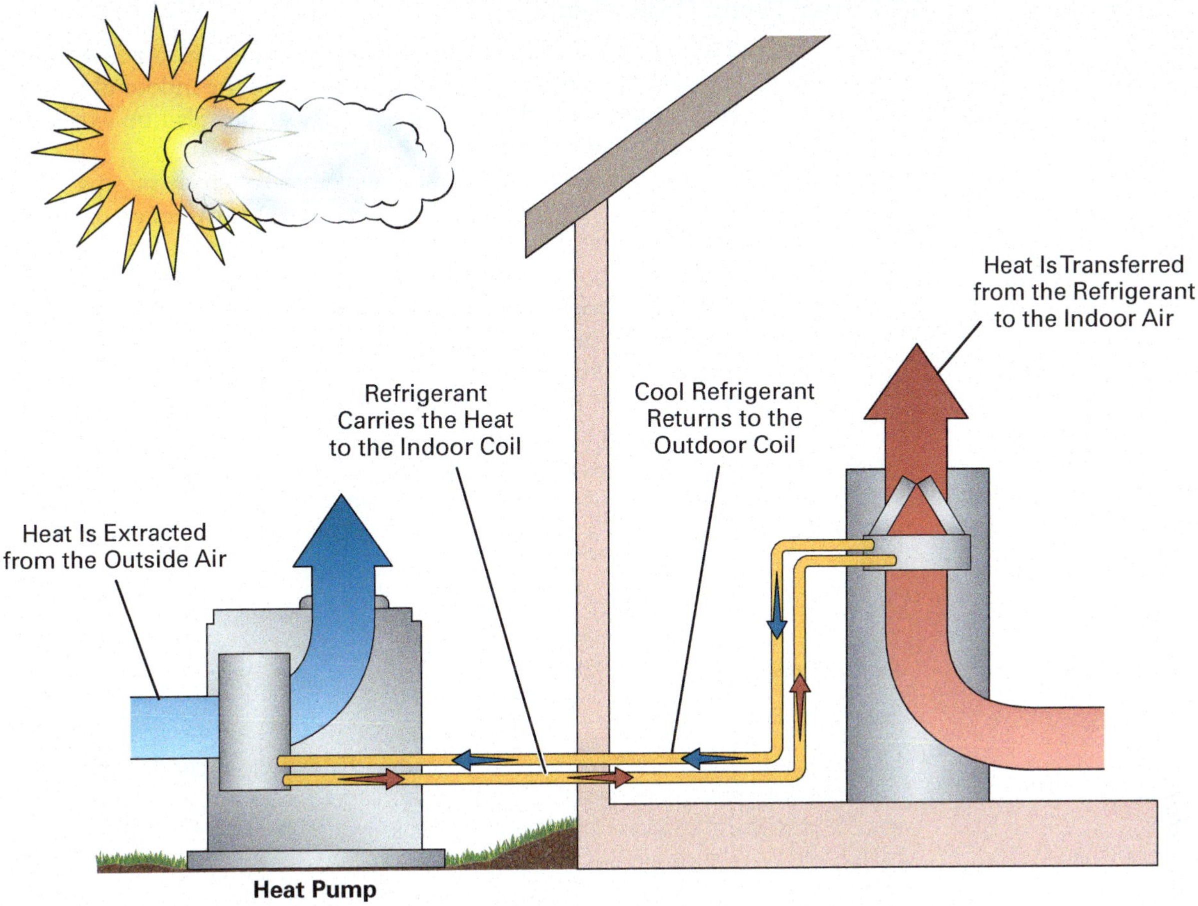

Figure 1 Heat pump operation in the heating mode.

Did You Know?

HVAC and Heat Pump Installation Standards

All HVAC systems should be installed to the highest standards possible to enhance reliability and performance. HVAC equipment and labor is expensive, and users have a right to expect your best efforts.

To provide guidance in this area, the Air Conditioning Contractors of America (ACCA) has developed several equipment installation and commissioning standards. ACCA develops standards as an American National Standards Institute (ANSI) accredited standards developer. These standards address quality installation and commissioning requirements for many equipment types, including heat pumps.

All HVACR technicians and installers should review the following standards and incorporate the guidance into their installation practices:

- ANSI/ACCA Standard 5, *HVAC Quality Installation Specification* — provides the minimum design and installation standards for residential and commercial HVAC applications.
- ANSI/ACCA Standard 9, *HVAC Quality Installation Verification Protocols* — establishes minimum requirements for verifying that installed residential and light commercial HVAC systems meet ANSI/ACCA Standard 5. The standard also provides comprehensive check-lists related to system startup and commissioning.

Note that both standards, as well as other ACCA standards related to residential maintenance practices and system cleanliness, can be downloaded from ACCA at no cost. Visit *www.acca.org/standards/quality* for more information.

1.1.1 Heat Pump Characteristics

Even at very cold temperatures, there is some heat in the air. The air-to-air heat pump extracts heat from the outdoor air and transfers it to the indoor air, using the standard vapor-compression refrigerant cycle. Heat can also be extracted from water using geothermal systems.

The reversing valve is the primary component that sets heat pumps apart from cooling-only systems. The reversing valve reverses the direction of refrigerant flow through the system when the unit is switched to the heating mode. In the heating mode, the outdoor coil becomes the evaporator (absorbs heat from the outside air), and the indoor coil becomes the condenser (transfers the heat gained to the conditioned space).

The control circuits are usually arranged so that the reversing valve is energized in the cooling mode. When de-energized, it defaults to the heating mode. Thus, if there is a reversing valve failure, heating is still available. Heating is considered the most essential mode of operation. A reversing valve or solenoid coil failure results in a loss of cooling. Further, since the defrost mode is a variation of the cooling mode, a reversing valve failure also disables the defrost mode.

NOTE

Not all systems energize the reversing valve in the cooling mode. You will occasionally encounter systems that energize the reversing valve in the heating mode. Do not assume that a system defaults to the heating mode—examine the wiring diagram and sequence of operation. Thermostats can be wired or configured to energize the reversing valve in either mode.

Going Green

Heat Pump Pool Heaters

Heat pumps also make efficient pool heaters. Heat is extracted from the air and transferred to the swimming pool by a refrigerant-to-water heat exchanger. Heat pumps offer an energy-efficient alternative to traditional fossil-fuel pool heaters. Hybrid units that combine gas heating with a heat pump are also available.

Source: cmspic/iStock

Another key difference between cooling-only systems and heat pumps is the arrangement of the metering devices. As shown in *Figure 2*, heat pumps need two metering devices. Note the temperatures and pressures reflect common values for a system operating on HFC-410A refrigerant. The arrangement shown is based on fixed metering devices. Built-in or separate check valves are used to block or allow full refrigerant flow, depending on the mode of operation. Only one metering device is actively metering in either operating mode.

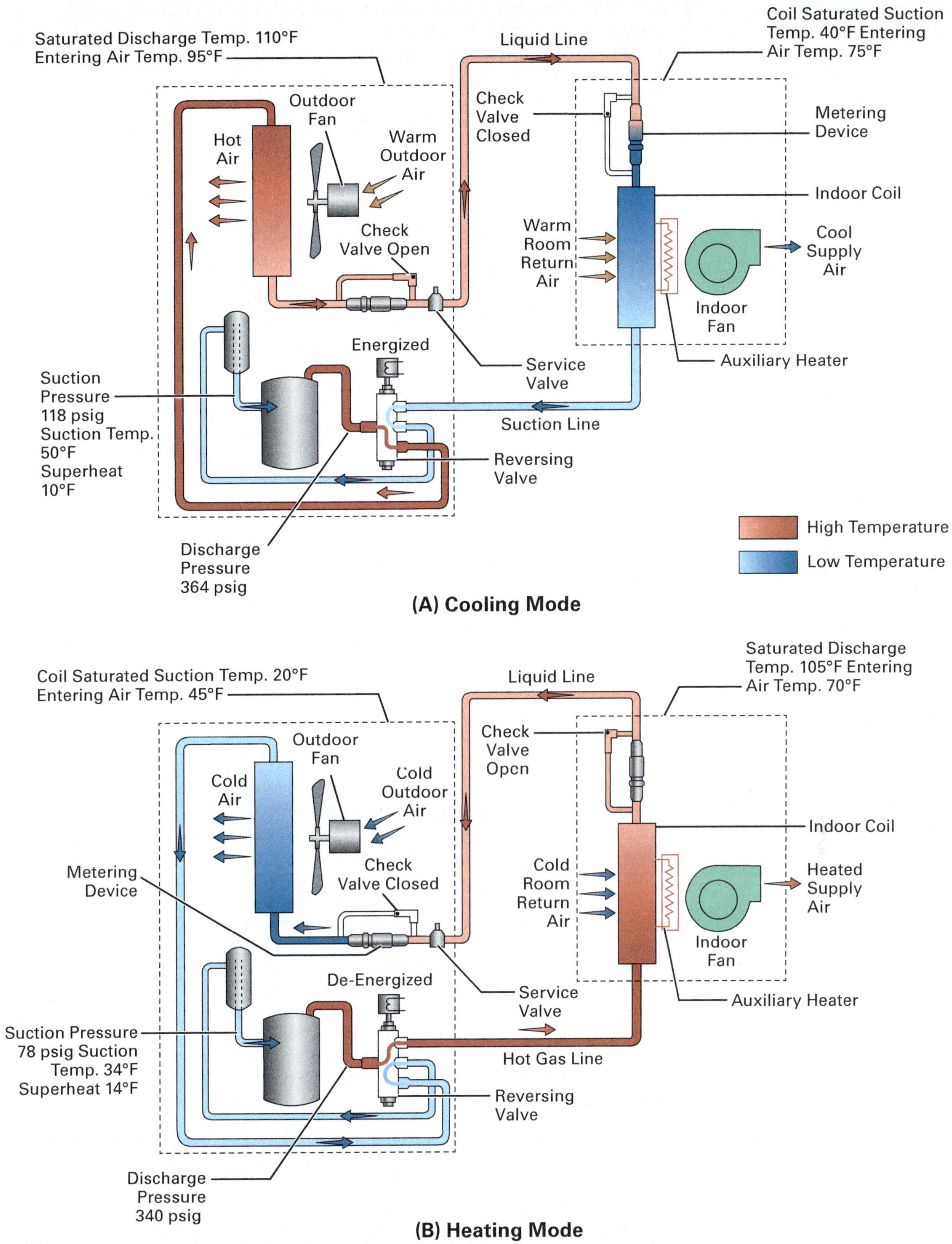

Figure 2 Refrigerant flow in an air-to-air heat pump (HFC-410A).

This arrangement may only be used today for specialized systems with highly consistent loads and operating conditions. When these factors are consistent, a fixed restrictor offers good performance in a very simple form.

In this specific scenario, the restrictor is truly fixed; it does not move back and forth inside its housing like the piston-style fixed-orifice shown in *Figure 3*. Thus, it requires the external check valve shown to manage flow through or around the orifice.

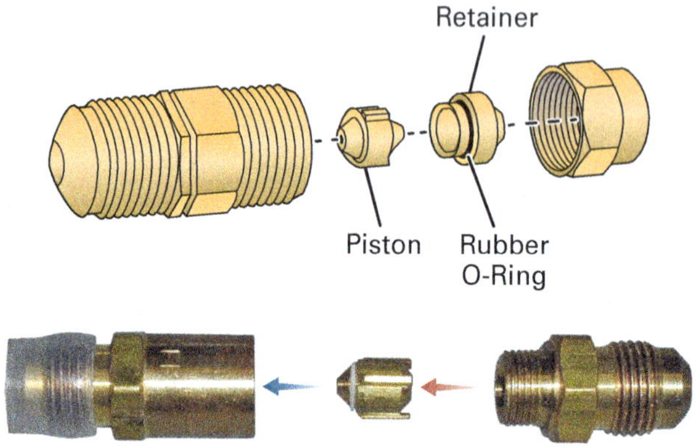

Figure 3 Piston-style, fixed-orifice metering device.

The piston-style orifice was developed to integrate a replaceable fixed orifice with the check valve function in a very simple package. Check valves may occasionally be used with thermostatic expansion valves (TXVs), although many expansion valves today are offered in a bi-flow configuration, with the check valve built in. The valves are designed specifically for heat pump service.

1.1.2 Heat Pump Classifications

Heat sink: A substance or structure that absorbs or disperses heat.

Heat pumps are classified according to their heat source and the medium to which the heat is transferred (**heat sink**). The most common types of heat pumps used for comfort are air-to-air and water-to-air. These heat pumps are commonly referred to as *air-source* and *water-source*, respectively. Heat pumps that transfer heat from air, or water, to water are more often used to heat domestic water, swimming pools, and similar applications.

In a common air-source system, air is the medium used to transfer heat both indoors and outdoors. In the cooling mode, heat is transferred to the outdoor air. In the heating mode, heat is extracted from the outdoor air and transferred indoors.

A water-source system also uses air to heat and cool indoors, but water is used as the heat source for heating and as the heat sink in cooling. The refrigerant is circulated through one side of a heat exchanger, and water is circulated through the other side.

The source of the water varies. Some commercial systems rely on a connected boiler and cooling tower to warm or cool the water in a loop that serves many heat pumps. Residential water-source systems may rely on water drawn from a well and discharged into a second well, or on fluid pumped through loops of tubing buried in the earth. When earth provides the heat source or heat sink, it is referred to as a *geothermal* or *ground-source* system.

The ground and groundwater are more effective and consistent heat sources than air, but they are not always practical resources to access. Installing two wells with some distance between the wells may not be practical for the property. Drilling wells is expensive, as is the trenching and tubing installation required for a closed-loop system. Local codes may also prohibit the reintroduction of the water back into the ground. Because of the limitations and installation cost, air is the predominant heat source for residential and small commercial heat pump installations. However, properly installed water-source systems are very efficient and reliable.

On Site

Geothermal Heat Pumps

Geothermal systems can be installed several different ways, depending on the property and its geology. The illustration on the left shows a vertical closed-loop system. The tubing for closed-loop systems can be installed horizontally or vertically in the ground, or even submerged in a body of water. Open-loop systems use two separate wells, as shown in the diagram on the right.

Because pumping requirements vary from one installation to another, the pump that circulates the fluid is usually an external component. The term *fluid* is used here because the water in closed-loop geothermal systems often needs protection from freezing. A glycol-based antifreeze is typically mixed with water in the loop.

Vertical Closed-Loop System

Open-Loop System (Two Separate Wells)

1.2.0 Heat Pump Operating Modes

Air-source heat pumps have three basic operating modes: cooling, heating, and defrost. Water-source systems do not require a defrost mode.

To provide sufficient heat, air-source heat pumps typically require a supplemental source of heat. The supplemental heat provides additional heating capacity below the heat pump's **balance point** as well as an alternative form of heat if the heat pump fails. When a system thermostat is switched to the Emergency Heat mode, the heat pump is locked out and the alternate heat source is used.

Balance point: The outdoor temperature at which the heating capacity of a heat pump is equal to the heat loss of the structure it serves.

Dual-fuel systems, which typically combine a heat pump with a fossil-fuel heat source, typically have a changeover outdoor temperature that determines which heat source is used. In most cases, the furnace and heat pump can't operate at the same time. Below the balance point, the system usually relies on the secondary source. Above the balance point, the heat pump has sufficient capacity to handle the heating load alone.

1.2.1 Cooling Mode

In most heat pumps, the reversing valve is energized in the cooling mode and de-energized in the heating mode. When the unit is in the cooling mode, the refrigerant flow is the same as that of a cooling-only unit. Although this arrangement is the most common, do not assume all heat pumps are configured this way.

Refer again to *Figure 2A*. Note the refrigerant path through the system. Hot, high-pressure vapor leaving the compressor is routed to the outdoor coil where heat is rejected, and the refrigerant vapor condenses to its liquid state. The check valve downstream of the outdoor coil is open, allowing unrestricted refrigerant flow around the metering device.

The check valve upstream of the indoor coil is closed, forcing the liquid refrigerant to flow through the metering device. The saturated and vaporizing refrigerant flowing through the indoor coil absorbs heat from the conditioned space, eventually exiting the coil as a superheated vapor and returning to the compressor through the reversing valve.

On Site

Dehumidification Mode

Dehumidification mode, also referred to as *dry mode*, is an option for many of today's more sophisticated cooling and heat pump systems. The primary operational difference is that the indoor blower speed is reduced to allow the air to spend more time in contact with the evaporator coil. On systems with expansion valves, the valves can throttle the flow of refrigerant enough to prevent floodback. But this also results in a reduction in the suction pressure and evaporator saturated-suction temperature. This is not a significant problem—it actually enhances dehumidification by increasing the temperature difference between the coil and the return air. The only problem is that the space may be overcooled before the humidity is sufficiently reduced.

On systems with an inverter-driven, variable-speed compressor, the compressor also slows down. This allows dehumidification to continue with less risk of overcooling the area.

1.2.2 Heating Mode

Heat pump operation in the heating mode is shown in *Figure 2B*. The reversing valve has changed its position. The refrigerant leaving the compressor is routed in the opposite direction from that of the cooling mode.

Instead of flowing through the outdoor coil, the hot, high-pressure refrigerant vapor leaving the compressor flows to the indoor coil. When this happens, the indoor coil acts as a condenser. Heat is extracted from the refrigerant because the air in the conditioned space is cooler than the refrigerant in the coil. The check valve downstream of the indoor coil is forced open, allowing refrigerant to bypass the metering device.

The cooled refrigerant then flows to the outdoor coil. The check valve at the outdoor coil is now closed, forcing the refrigerant to flow through the metering device. Acting as an evaporator, the outdoor coil absorbs heat from the warmer outdoor air. Notice the significant differences in the pressures and temperatures at various points in the system compared to the cooling mode.

An accumulator and a crankcase heater, which are optional accessories on cooling units, are required on most heat pumps. The crankcase heater prevents liquid refrigerant from condensing in the crankcase of the compressor, which generally occurs in winter conditions. The accumulator captures any liquid refrigerant returning from the outdoor coil before it reaches the compressor. The accumulator will also capture any liquid that returns from the indoor coil in the cooling mode, but it is far more likely to hold some volume of liquid refrigerant during the heating mode.

The two coils—indoor and outdoor—often have different internal volumes. As a result, less refrigerant is needed in the circuit to support the heating mode, when the indoor coil becomes the condenser. Thus, excess refrigerant gathers in the accumulator. Another reason this occurs is because the temperature difference between the refrigerant and the outdoor temperature is often less than when the indoor coil is acting as the evaporator. Some liquid may leave the outdoor coil as a result. The liquid refrigerant collected slowly vaporizes in the accumulator and eventually returns to the compressor. Oil (as well as a small amount of liquid refrigerant) returns to the compressor through an orifice (*Figure 4*).

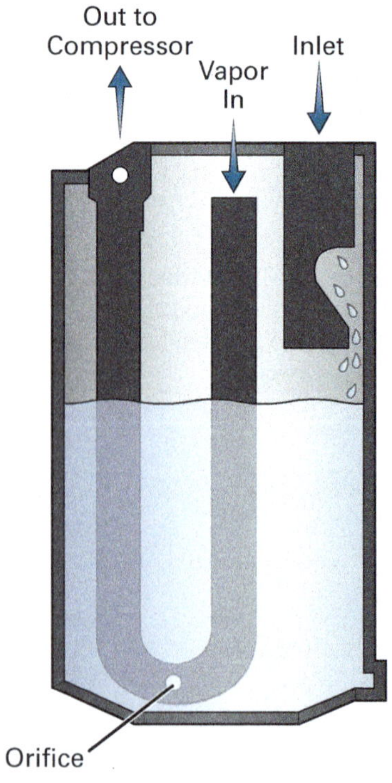

Figure 4 Suction accumulator.

1.2.3 Defrost Mode

When the outdoor temperature or the outdoor coil is below freezing, frost will form on the outdoor coil in the heating mode. The amount of frost that collects is largely dependent on the moisture content of the air. As the frost builds, it restricts airflow and significantly reduces heat transfer. The heat pump shown in *Figure 5* has gone beyond an acceptable level of frost accumulation. Most of what you see on the coil has become solid ice due to limited thawing and refreezing.

The frost is melted by reversing the flow of refrigerant, routing the hot discharge gas from the compressor to the outdoor coil. The defrost mode timing is automated. The initiation and termination of the mode may be based on elapsed time, the outdoor coil temperature, or a combination of the two inputs.

The defrost mode is a slight variation of the cooling mode. The primary difference is that the condenser fan remains off during a defrost cycle. Since the objective is to melt frost and ice, moving air across the outdoor coil would work against that goal. The accumulated frost and cold outdoor air provide a significant heat sink, preventing the head pressure from rising excessively. The condenser fan isn't needed to perform its normal function.

Because the unit is operating in the cooling mode, the supply air temperature indoors can be quite cold. Most systems are supplemented by an electric resistance heat package. The electric heat is energized during a defrost mode to reheat the air as it leaves the indoor coil.

Ductless heat pump systems are rarely equipped with any form of supplemental heat. To avoid blowing cold air into the space, both the indoor and outdoor fans are de-energized during the defrost mode.

Figure 5 Frost and ice on a heat pump.
Source: HVACGuru

1.0.0 Section Review

1. Which of the following is required to help manage the flow of refrigerant in a heat pump when fixed metering devices are used?
   a. Gate valve
   b. Check valve
   c. Reversing valve
   d. Pressure-relief valve

2. True or False? Ductless heat pump systems avoid blowing cold air into a room during defrost by isolating the indoor coil from the refrigerant circuit.
   a. True
   b. False

2.0.0 Heat Pump Operating Sequences

Performance Tasks

1. Demonstrate the ability to diagnose and identify the proper solution for an instructor-selected mechanical problem in a heat pump.
2. Demonstrate the ability to diagnose and identify the proper solution for an instructor-selected electrical problem in a heat pump.
3. Initiate and evaluate the defrost mode of a heat pump.

Objective

Recount the sequence of operation for the common heat pump operating modes.

a. Recount the sequence of operation for the cooling mode.
b. Recount the sequence of operation for the heating mode.
c. Recount the sequence of operation for the defrost mode.
d. Recount the sequence of operation for dual-fuel systems.
e. Describe how to evaluate electronic controls in heat pumps.

A technician troubleshooting a heat pump must understand the operating sequence in every mode. For air-to-air heat pumps, there are five common modes of operation:

- Cooling
- Heating with heat pump alone
- Heating with heat pump plus supplemental heat
- Defrost
- Emergency heat with supplemental heat alone

2.1.0 Cooling Sequence of Operation

Refer to the wiring diagram in *Figure 6*. The following is an example sequence of operation for a heat pump in the cooling mode:

- With the thermostat function switch set to Cool and the room temperature above the setpoint, the thermostat calls for cooling (1). The reversing valve solenoid (2) is energized, shifting the direction of flow for the cooling mode. If the high- and low-pressure switches (as well as any other safety switches in the circuit) are closed, the compressor contactor coil (3) is energized.
- The call for cooling at the thermostat also applies power to the G terminal, energizing the indoor fan relay (4) and starting the indoor fan motor at its high speed (5).
- With the contactor energized, line voltage is applied to the compressor and outdoor fan motor (6).

The unit will continue to operate until the room thermostat is satisfied, de-energizing the various control circuits. The indoor blower often remains running for a predetermined length of time, maximizing the cooling effect of the refrigerant remaining in the indoor coil.

NOTE

To maintain simplicity in the diagrams shown in *Figures 6* through *8*, the internal thermostat circuitry for the Emergency Heat switch is not shown.

NOTE

Many thermostats energize the O terminal (where reversing valve control is connected) as soon as the thermostat function switch is set to the Cool position, regardless of the room temperature. This prevents the reversing valve from shifting during each on and off operating cycle. All thermostats traditionally energize O as soon as the cooling mode is selected, and B when the heating mode is selected.

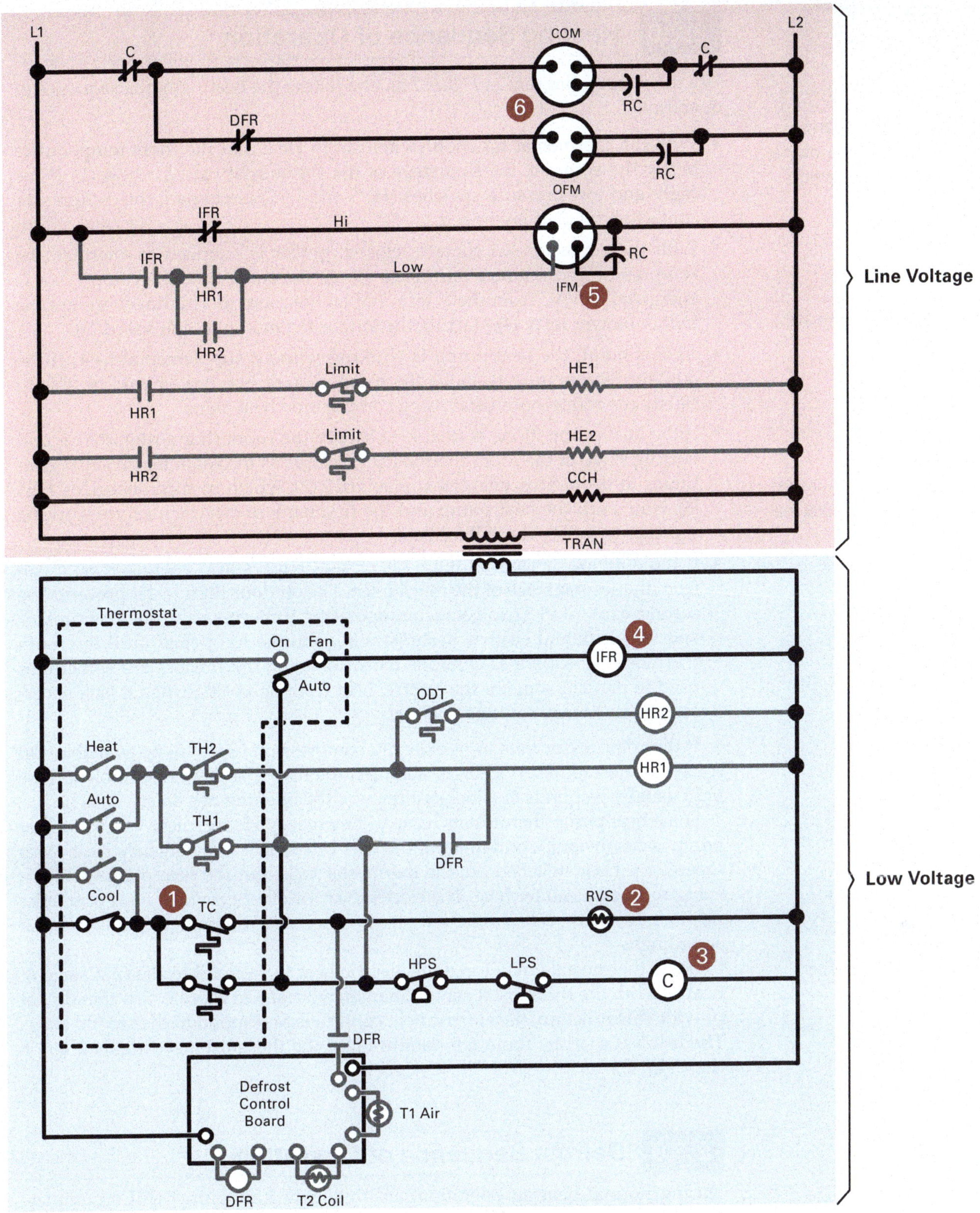

Legend:

C	Compressor Contactor
CCH	Crankcase Heater
COM	Compressor
DFR	Defrost Relay
HE	Heating Element
HPS	High-Pressure Switch
HR	Heater Relay
IFM	Indoor Fan Motor
IFR	Indoor Fan Relay
LPS	Low-Pressure Switch
ODT	Outdoor Thermostat
OFM	Outdoor Fan Motor
RC	Run Capacitor
RVS	Reversing Valve Solenoid
T	Thermistor
TC	Cooling Thermostat
TH	Heating Thermostat
TRAN	Transformer
-------	Mechanical Connection

Figure 6 Example heat pump ladder diagram (cooling sequence).

2.2.0 Heating Sequence of Operation

NOTE

To maintain simplicity in the diagrams shown in *Figures 6* through *8*, the internal thermostat circuitry for the Emergency Heat switch is not shown.

NOTE

The reversing valve is de-energized in the heating mode in this scenario.

NOTE

New cold-climate heat pumps are specifically designed to provide the needed heating capacity without supplemental heat.

Use the wiring diagram in *Figure 7* as you follow the heating-mode sequence of operation:

- With the thermostat function switch set to Heat and the room temperature below the setpoint, the first stage of the thermostat calls for heat (1). If the high- and low-pressure switches (HPS and LPS) are closed, the compressor contactor coil (2) is energized.
- With the contactor (C) contacts closed, power is supplied to energize the compressor and outdoor fan motor (3). At the same time the compressor is energized, 24VAC from the G terminal on the room thermostat energizes the indoor blower relay (4), starting the indoor fan motor at high speed (5).
- At this point, the heat pump is working without supplemental heat. If the outdoor temperature is above the balance point, the unit should be able to satisfy the room thermostat using compression heat alone.
- If the heat pump alone is unable to satisfy the room thermostat, the second heating stage of the room thermostat (6) closes as the room temperature continues to drop. This energizes relay HR1 (7), which in turn energizes HE1 (8). Now both the heat pump and the first bank of electric heat are working together to satisfy the call for heat.
- If the outdoor thermostat (ODT) is closed, HR2 is also energized on a call from the second stage of the thermostat. The outdoor thermostat prevents the second bank of electric heat from operating unnecessarily. The compressor and supplemental electric heater(s) will continue to operate until the room thermostat is satisfied. Outdoor thermostats or thermistors are sometimes used to prevent some of the electric heat from operating when it isn't necessary, as an energy-saving measure.

If the compressor fails to operate, no compression heat will be available. On some thermostats, a service light will alert the user that there is a problem. This light usually responds to a lock-out relay or the opening of a safety switch.

Some heat pump thermostats have an Emergency Heat switch. When the heat pump is clearly faulty or damaged, the user can switch to Emergency Heat. With Emergency Heat switched on, the thermostat locks out the heat pump (will not energize Y) on a call for heat. The electric heat, or other provided form of supplemental heat, then becomes the only heat source and responds to the thermostat accordingly.

Some thermostats may force the electric heat to respond to the first stage of heating with the thermostat set to Emergency Heat. On thermostats that do not provide this function, the electric heat continues to respond to the second stage. The result is that the room temperature lags the thermostat setting by 2°F–3°F (1.1°C–1.7°C).

2.3.0 Defrost Sequence of Operation

During normal heating operation, an air-source heat pump will accumulate frost on the outdoor coil. The frost restricts airflow across the coil, reducing heat transfer.

To remove the accumulated frost, the heat pump must defrost the outdoor coil periodically. This is done by running the unit in the cooling mode, directing the hot refrigerant vapor to the outdoor coil. The method is quite effective for removing frost. However, a very cold supply airstream indoors results. To minimize discomfort, the supply air temperature must be tempered.

The heat pump must be operating in the heating mode for a defrost mode to start; an idle heat pump will not start with the intention of defrosting itself, as the defrost controls do not control compressor operation. This means that the compressor, indoor fan, and outdoor fan are already running when the defrost

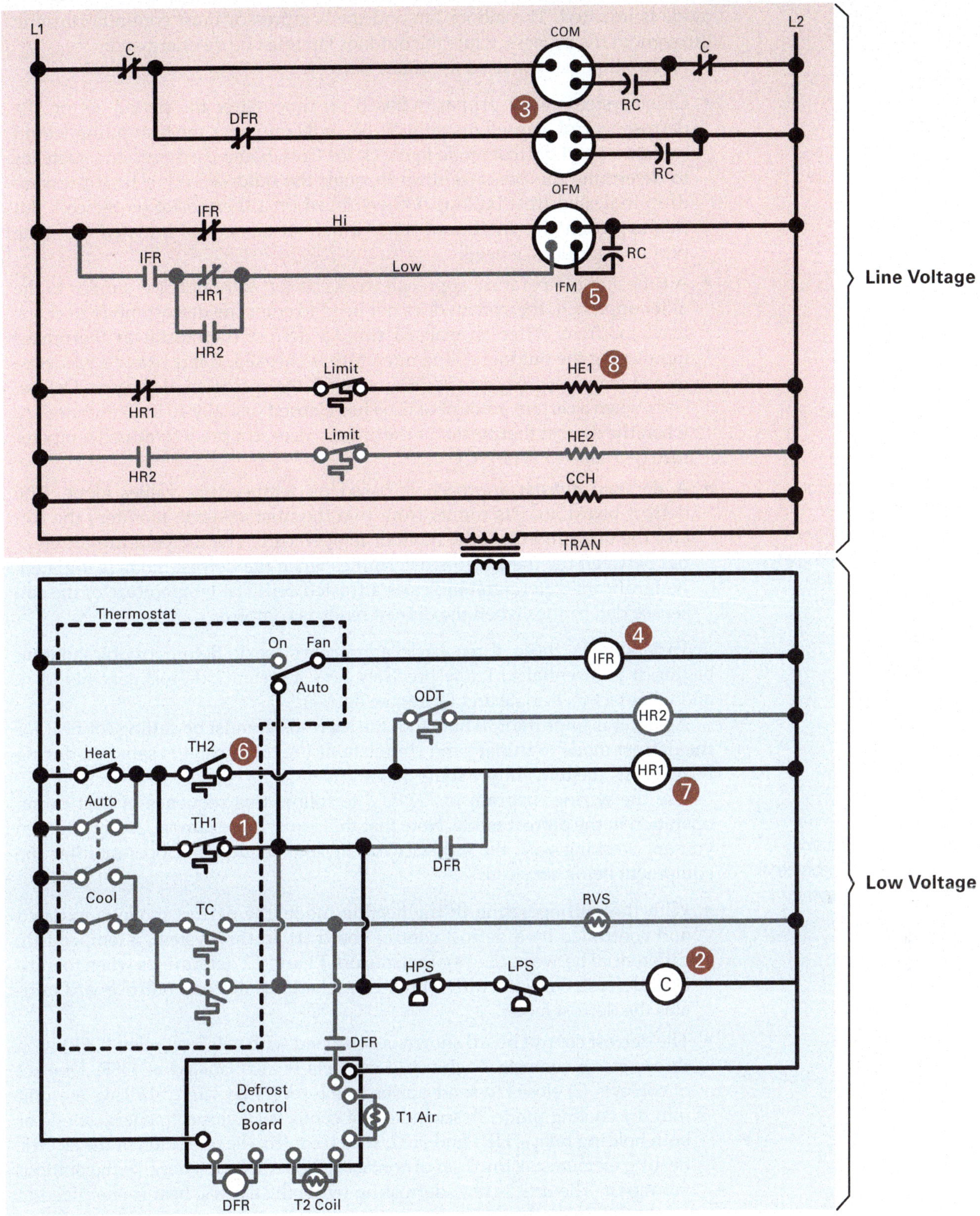

Legend:

C	Compressor Contactor	LPS	Low-Pressure Switch
CCH	Crankcase Heater	ODT	Outdoor Thermostat
COM	Compressor	OFM	Outdoor Fan Motor
DFR	Defrost Relay	RC	Run Capacitor
HE	Heating Element	RVS	Reversing Valve Solenoid
HPS	High-Pressure Switch	T	Thermistor
HR	Heater Relay	TC	Cooling Thermostat
IFM	Indoor Fan Motor	TH	Heating Thermostat
IFR	Indoor Fan Relay	TRAN	Transformer
		-------	Mechanical Connection

Figure 7 Example heat pump ladder diagram (heating sequence).

mode is initiated. The indoor fan and the compressor must continue running throughout the process, while the outdoor fan must be de-energized.

A few strategies are used to initiate defrost:

- Older systems were primarily based on time. After the period set on the defrost control elapsed (usually 30, 60, or 90 minutes), the heat pump would initiate a short defrost mode to check for frost. Some used pressure switches to determine the pressure drop through the outdoor coil; a high pressure drop indicated frost buildup. Others relied on a thermostat on the coil. But this approach was inefficient, since a defrost mode was initiated for some period regardless of need.
- A time and temperature approach works better. Although it is similar to the older approach, the system does not need to enter the defrost mode to check itself for frost. After an elapsed time, a defrost thermostat or thermistor monitoring the outdoor coil temperature is checked. If it is below a predetermined temperature (perhaps 25°F, or –4°C), the defrost mode begins. Defrost ends when a certain amount of time has elapsed (usually 10 to 13 minutes), or when the defrost thermostat or thermistor rises to a predetermined temperature (perhaps 45°F, or 7°C).
- A **demand defrost** approach is based on temperature values alone. The defrost board accepts inputs from two thermistors—one monitors the coil and the other monitors the outdoor temperature. The temperature differential between the thermistors determines when the defrost mode is initiated. A significant differential indicates a frosted coil. The temperature of the coil sensor determines when the defrost mode is complete.

Demand defrost: A defrost control strategy that does not depend on time to determine when defrosting is needed. Temperatures and temperature differentials determine when frost is present, and defrosting is required.

Variations to these three basic approaches exist. Before troubleshooting begins, it is essential to know precisely how a system's defrost controls work and what factors initiate and terminate defrost.

Again, it is important to note that the thermostat must be calling for heat for the defrost mode to initiate and complete. If the thermostat is satisfied during the process, the unit simply shuts down.

NOTE

To maintain simplicity in the diagrams shown in *Figures 6* through *8*, the internal thermostat circuitry for the Emergency Heat switch is not shown.

Use the wiring diagram in *Figure 8* to follow this sequence of heat pump operation in the defrost mode. Note that this is just one example. Always ensure you are working with the correct diagram and sequence of operation for the equipment being serviced:

- With the unit operating in the heating mode, the defrost mode is initiated and controlled by a defrost control board (1). In this system, a temperature differential between the two thermistors T1 and T2 determines when to initiate a defrost. Once the differential is reached, the defrost control board initiates the defrost mode.
- The defrost control board shown is equipped with a defrost relay (DFR) with three sets of contacts. Each set of contacts is also labeled as DFR. One set of contacts (2) closes to send power to the reversing valve, shifting the unit into the cooling mode. A second set of contacts (3) closes to energize one or both holding relays, HR1 and HR2, which control the operation of the electric heating elements. A third set of contacts (4) opens to de-energize the outdoor fan motor. The unit is now defrosting while the electric heat is energized to reheat the very cold air leaving the indoor coil.
- The second stage of electric heat will be energized if the outdoor thermostat (ODT) is closed (5).
- Once the coil temperature at T2 rises above a predetermined temperature, the defrost mode is terminated. To ensure the unit doesn't remain in defrost for an extended period due to a failed thermistor, the board will also terminate defrost if the maximum time allotted has elapsed.

Once the defrost mode terminates, the heat pump returns to operation in the heating mode. The reversing valve is de-energized, the electric heat is turned off (unless the thermostat continues to request it to reach the setpoint), and the outdoor fan motor is energized.

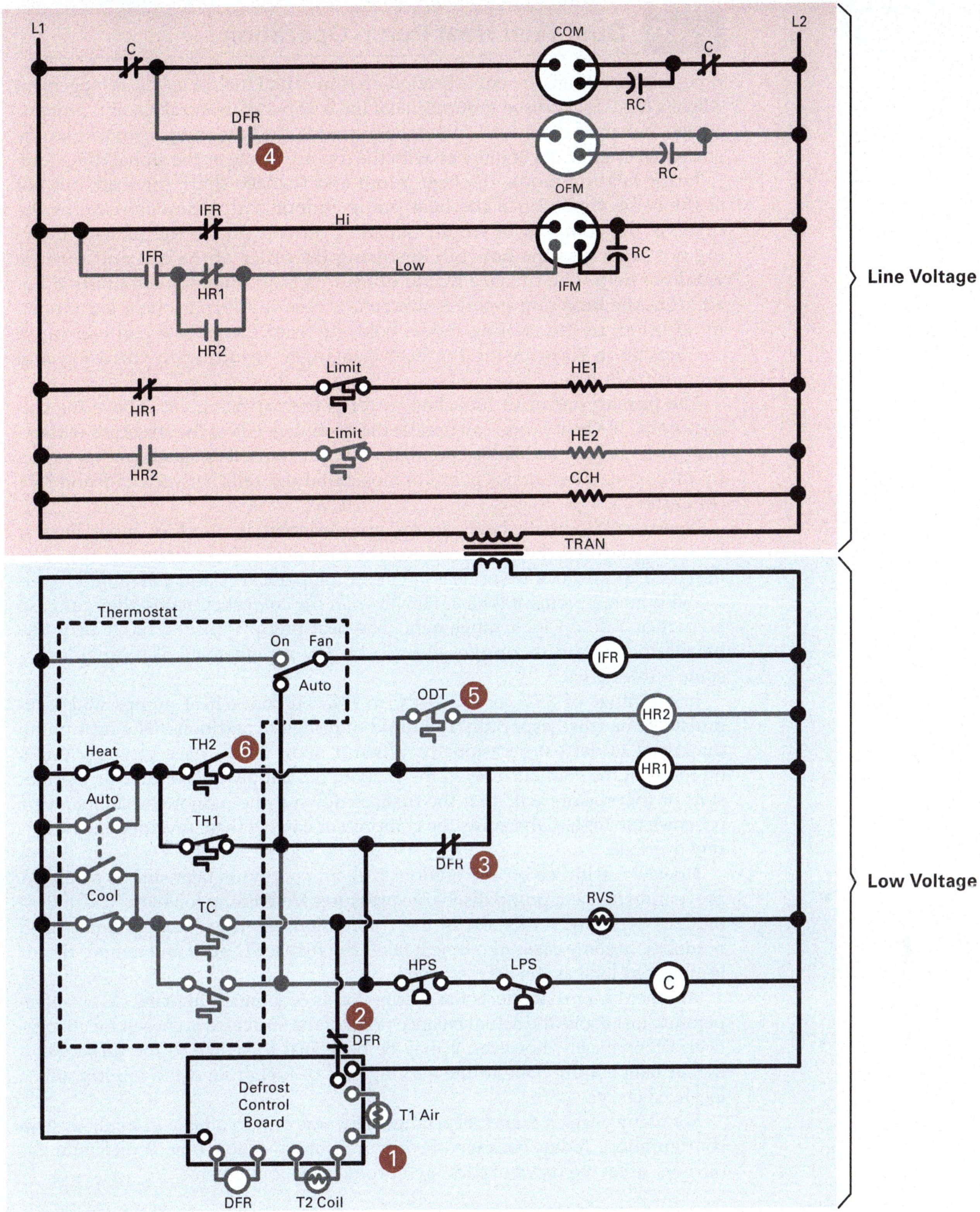

Legend:

C	Compressor Contactor	LPS	Low-Pressure Switch
CCH	Crankcase Heater	ODT	Outdoor Thermostat
COM	Compressor	OFM	Outdoor Fan Motor
DFR	Defrost Relay	RC	Run Capacitor
HE	Heating Element	RVS	Reversing Valve Solenoid
HPS	High-Pressure Switch	T	Thermistor
HR	Heater Relay	TC	Cooling Thermostat
IFM	Indoor Fan Motor	TH	Heating Thermostat
IFR	Indoor Fan Relay	TRAN	Transformer
		-------	Mechanical Connection

Figure 8 Defrost sequence of operation.

2.4.0 Dual-Fuel Heat Pump Operation

A popular use of an air-source heat pump is in a dual fuel, or *hybrid*, arrangement where a fossil fuel is used to supplement the heat pump. Natural gas and propane are the fuels most often used. In the cooling mode, the heat pump operates like any other split-system air conditioner, with the furnace acting as the air handling unit.

In the heating mode, the heat pump and furnace share the load, but not at the same time. Since the heat pump indoor coil is downstream of the furnace, the compressor would quickly overload due to the furnace blowing hot air across the hot coil. Reversing the order of the coil and furnace wouldn't help. The furnace would exceed its high-temperature limits if hot air from the heat pump coil is directed across the furnace heat exchanger. In addition, in the cooling mode, cold air from the indoor coil can cause condensate to form inside the heat exchanger, resulting in corrosion and premature failure.

The primary factor in fuel changeover is the balance point. Above the balance point, the heat pump can handle the heating needs of the structure without assistance. Below the balance point, the furnace, with its greater heating capacity, takes over. The balance point for most buildings falls between 25°F and 35°F (–4°C and 2°C).

Newer cold-climate heat pumps are designed to produce more heat at low temperatures than conventional heat pumps. These units are typically designed to produce more heat at colder temperatures and potentially operate without supplemental heat. However, in the coldest climates, they can also be part of a dual-fuel arrangement. The heat pump typically has a very low balance point, minimizing fossil fuel use. The arrangement also provides for some redundancy.

In the range of 25°F to 45°F (–4°C to 7°C) outdoors, heat pumps will accumulate some frost, especially in humid and foggy conditions. The heat pump may need to defrost occasionally. When it does, the furnace typically cycles on to offset the cold air leaving the indoor coil. Supply and return air thermostats or thermistors will shut the furnace down if the indoor coil temperature becomes too high. Otherwise, the compressor can overheat or experience a current overload.

However, with ice on the outdoor coil, an enormous heat sink is available, preventing the heat pump discharge pressures from rising too high. The defrost mode is typically very short in these conditions, as it isn't exceptionally cold outdoors. In some cases, the time it takes the furnace to start and temper the air is almost as long as the defrost mode.

Another factor that affects the changeover is economics. In many cases, a temperature just above the actual balance point of the structure is chosen for changeover. Economically, however, it may be beneficial to switch to the furnace at a higher temperature. Due to the varying cost of fuel around the country, it is a localized choice.

For many years, a dual-fuel arrangement was only available in a split-system configuration. Today, packaged heat pumps are available that also contain gas burners, allowing users of packaged products to apply them.

Did You Know?

Hybrid Systems and Smart Control

Local utilities often assist homeowners with effective solutions to manage energy use. For example, this Canadian utility offers a program that incorporates a smart thermostat with a hybrid system. The thermostat communicates with algorithms in the cloud that constantly compare the cost of gas versus electricity, enabling it to decide which fuel is the best option in real time.

As a technician, it is important to be aware of such arrangements in your client's control system, as system behavior is less predictable and must be considered in a troubleshooting process.

Hybrid HVAC System with Smart Thermostat

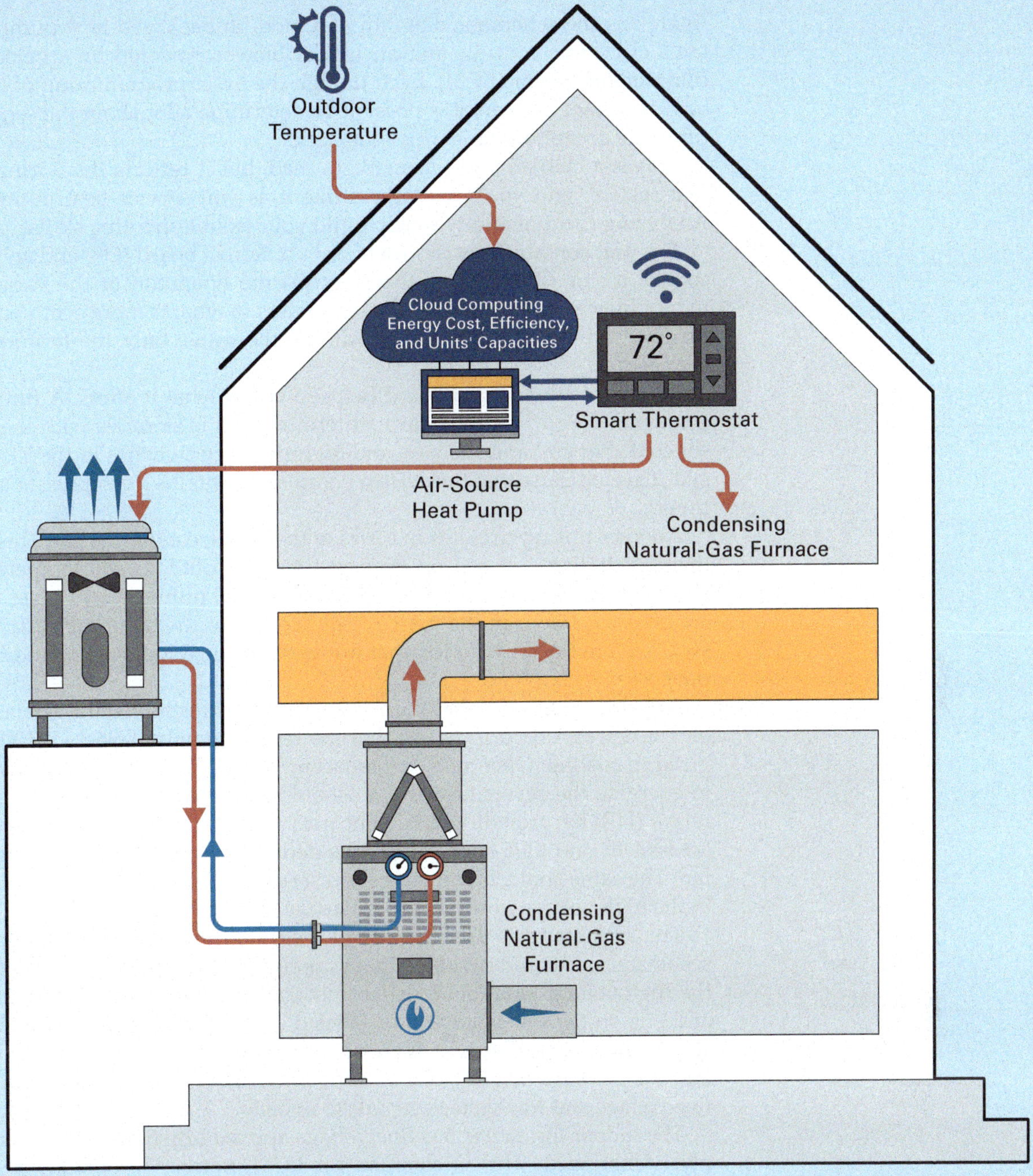

No matter how electricity and gas rates will change in the future, your hybrid system always works in the most economical condition.

We consistently monitor the fluctuation of electricity and gas rates.

We consistently monitor the new government policies and rules placed for protecting the environment.

Weather conditions in your zip code are monitored.

Energy rates for your specific location are monitored.

The efficiency of your HVAC system(s) is determined and considered.

Source: BKR Energy Inc.

2.5.0 Electronic Controls in Heat Pumps

Figure 9 is the schematic diagram of a modern packaged heat pump that contains electronic controls and an indoor blower powered by an electronically commutated motor (ECM). Even though the exact programming of the control boards cannot be seen, it is possible to determine a lot about the sequence and timing of operations from this schematic.

This is a challenging schematic to read, but it reflects the wiring of a current system, and you'll see others like it as you service heat pump systems. Analyzing the diagram will help build your troubleshooting skills.

The unit contains two control boards, a defrost board (DB) and an integrated control motor (ICM) board that controls the operation of the variable-speed indoor blower motor. Technically, this refers to what is more commonly known today as an ECM. ICMs and ECMs are the same; only the terminology has changed.

Looking at the defrost board helps to reveal what it offers. A timing circuit (shown as "logic" on the board, at terminal T2) is in series with the compressor contactor coil and the low- and high-pressure switches in the Y control circuit. This indicates that the defrost board also contains a compressor short-cycle timer.

Another timing circuit is in series with the defrost thermostat. This indicates that the defrost is based on elapsed time and outdoor coil temperature. The elapsed time is field-selectable (30, 60, 90, or 120 minutes) and is set using DIP switches as shown in the bottom right corner of the diagram. Adjacent to the switches are Speed Up pins that are used to force a defrost mode for testing purposes.

A direct connection between the room thermostat cooling contacts and O on the defrost board indicates that the reversing valve solenoid (RVS) is energized in cooling. Open relay contacts on the defrost board close during defrost to energize the reversing valve solenoid and apply power to the electric heat relays (HR) by providing a current path from R to W2. Closed relay contacts across OF1 and OF2 open during the defrost mode to de-energize the outdoor fan. The same contacts can be seen in series with the outdoor fan motor (OFM) in the high-voltage portion (upper part) of the schematic.

The ICM board is used to control the indoor blower motor. Looking at the schematic, one could conclude that four electric motors are in the unit: three on the high-voltage side (outdoor fan motor, compressor, and indoor fan motor), and one on the low-voltage side, labeled as the ICM and controlled through a 16-pin connector. However, there are only three motors. The indoor fan motor and the ICM are actually one and the same. This type of motor requires both line voltage and low-voltage signals to operate.

NOTE

ECM testing and troubleshooting is presented in NCCER Module 03318, *Motor Testing and Troubleshooting*.

The indoor fan motor has line voltage applied any time the unit power supply is energized. This is common for ECM motors. To operate, it must also receive control signals from the ICM board through the 16-pin connector. Thus, the indoor fan motor has two physical connections: one for line voltage (upper part of schematic), and the 16-pin connector (lower part of schematic).

During cooling operation, power on both Y and G informs the ICM board that cooling is requested. The board responds by starting the motor at the programmed speed for cooling. For continuous fan operation, the lack of a Y signal (cooling) and the presence of a G signal (indoor fan operation) cause the ICM board to run at the speed programmed for ventilation. Speed settings are made during the initial system startup.

Always read the manufacturer's literature for the features offered by electronic controls. Training programs available from manufacturers are also very helpful, providing the specifics for a particular product or family of products.

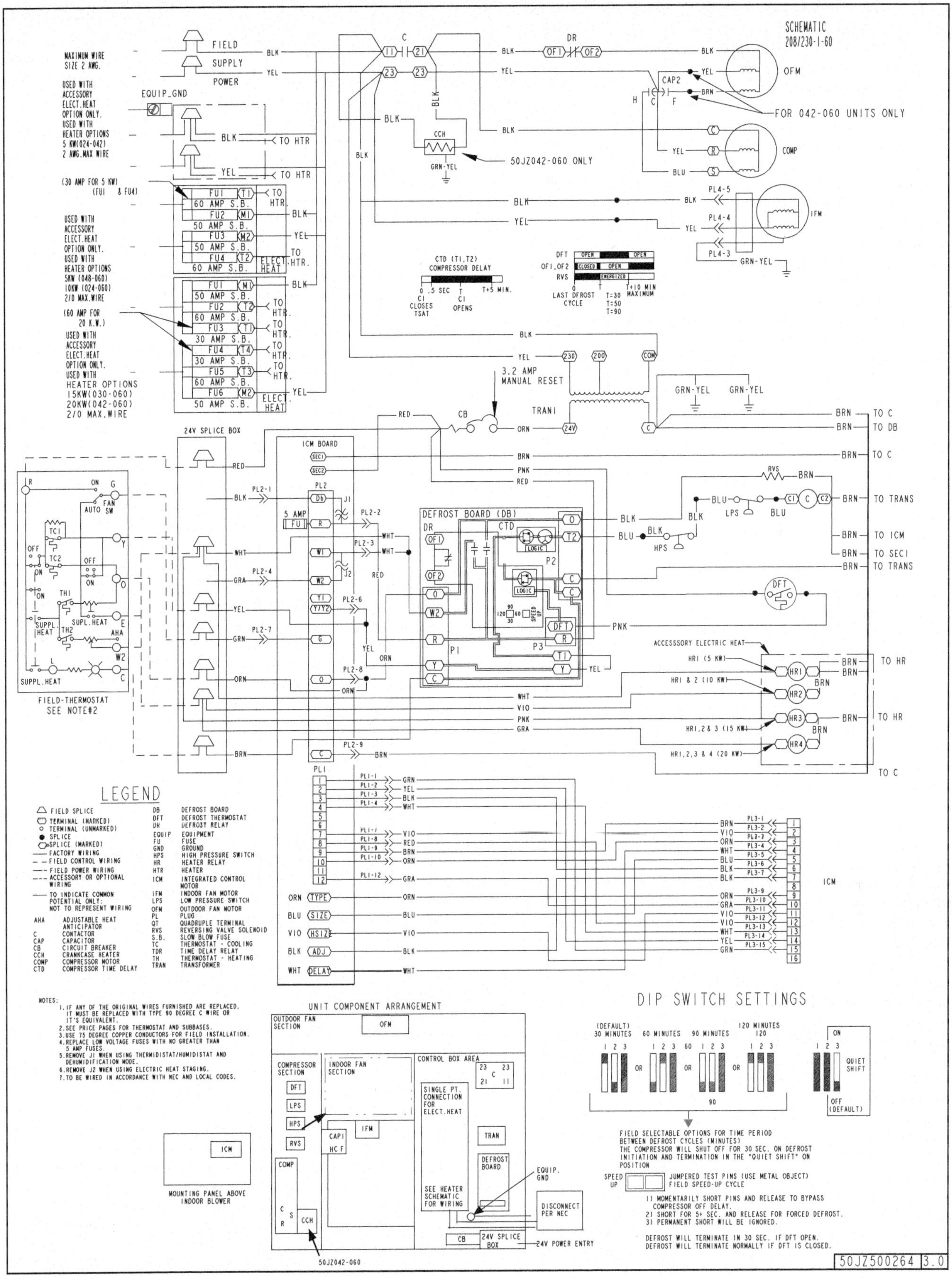

Figure 9 Packaged heat pump schematic.

Source: Carrier Corporation

Technicians are often intimidated when they see the number of components found on electronic control boards, and the diagrams are not as easy to follow as they are for electromechanical controls. HVACR technicians do not need to know how each component on a board works, nor do they need to troubleshoot electronic controls to the component level. What they do need to know is whether a control is functioning correctly by checking for the expected inputs and outputs.

Most electronic controls respond to inputs. If the control has the correct input signals, it responds by generating predetermined output signals. If a technician can confirm the inputs are correct, but the control is not responding with the correct outputs, it is reasonable to assume the board is defective.

Always check the device in question carefully before replacing it. Something as simple as a loose connection or a poorly fitted wiring harness can prevent normal operation. Manufacturers often provide a schedule of input and output signals for an electronic control. One example of this type of information is shown in *Figure 10*. This type of information isn't usually found in the midst of a wiring diagram due to the limited space available.

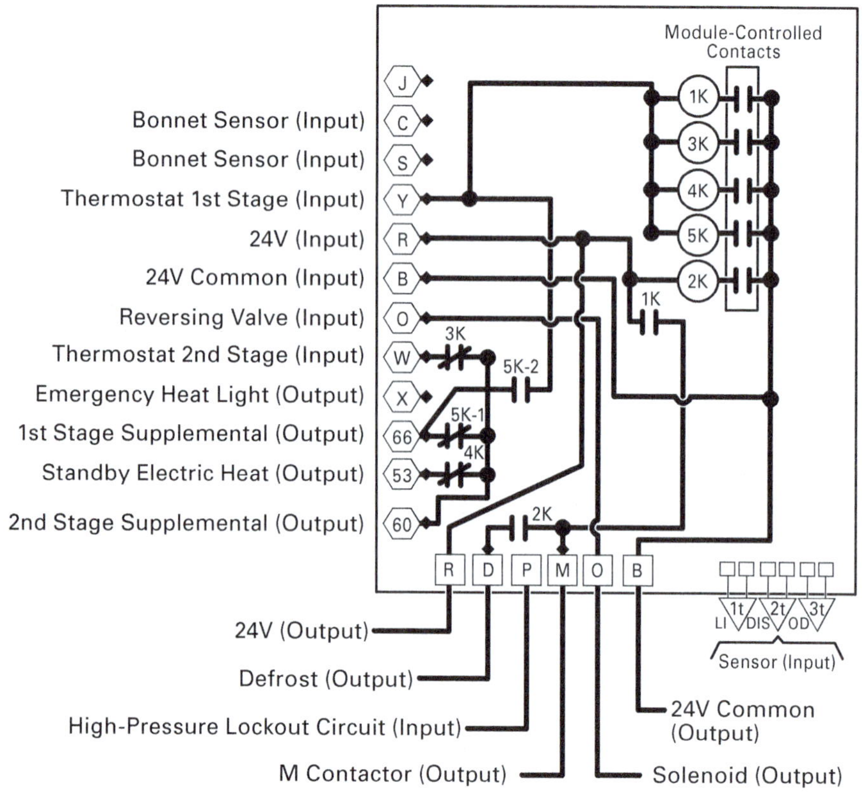

Figure 10 Control module input and output signals.

2.5.1 Variable-Speed Compressors in Heat Pumps

Most heat pumps currently in operation have single-speed compressors. However, variable-speed, inverter-driven compressors have enabled the development of equipment that provides a seasonal energy-efficiency ratio (SEER) of 20+. Variable-speed compressors (*Figure 11*) modulate motor speed to match capacity to demand. These compressors rely on inverter technology for speed control, instead of older multispeed motor technology.

Specific troubleshooting guidance for inverter-drives is outside the scope of this module. The compressor motor is essentially a three-phase AC motor. The motor windings are tested like other three-phase motors by checking the resistance from winding to winding; the resistance should be equal from winding to winding.

Figure 11 Variable-speed, inverter-driven scroll compressor.

Source: Courtesy of Emerson

NOTE

For basic electrical testing of inverter-driven compressors and inverter power modules, see NCCER Module 03318, *Motor Testing and Troubleshooting*.

Testing inverter power modules is best done using an instrument designed specifically for that purpose. Testing the power module with a multimeter is possible but requires detailed information and guidance from the equipment manufacturer.

2.0.0 Section Review

1. When the reversing valve solenoid of a heat pump is energized in cooling, the thermostat prevents the reversing valve from changing position when the thermostat function switch is in the Cool position on each call for cooling by energizing which terminal?
   a. O
   b. W
   c. Y
   d. G

2. A heat pump can manage the heating load without supplemental heat if ______.
   a. an ECM is used to power the indoor blower
   b. the outdoor temperature is above the balance point
   c. ice/frost is never allowed to form on the outdoor coil
   d. the indoor temperature setting is above the balance point

3. In the defrost mode, without the assistance of supplemental heat, a heat pump ______.
   a. will not defrost effectively
   b. still adds some heat indoors
   c. will deliver very cold air indoors
   d. must not de-energize the outdoor fan motor

4. Heat pumps specifically designed for cold climates ______.
   a. do not need to defrost
   b. can still be paired with a fossil-fuel heat source
   c. must be equipped with supplemental heat
   d. can operate at the same time as a connected furnace

5. True or False? An integrated control motor (ICM) and an electronically commutated motor (ECM) operate on different principles.
   a. True
   b. False

3.0.0 Testing and Troubleshooting Heat Pumps

Objective

Explain how to test and troubleshoot various functions and components of heat pump systems.

a. Explain how to test field and factory wiring.
b. Explain how to test and troubleshoot heat pump thermostats.
c. Explain how to test various valves.
d. Explain how to test defrost controls.

Performance Tasks

1. Demonstrate the ability to diagnose and identify the proper solution for an instructor-selected mechanical problem in a heat pump.
2. Demonstrate the ability to diagnose and identify the proper solution for an instructor-selected electrical problem in a heat pump.
3. Initiate and evaluate the defrost mode of a heat pump.

You can now see that heat pump operation is not dramatically different from a common cooling unit. Once the sequence of operation for a heat pump is known, the process of elimination helps isolate a failed component merely by observing

NOTE

Familiarize yourself with different troubleshooting guides offered by manufacturers. Manufacturers also release bulletins periodically regarding field problems that are also helpful.

how the system responds in each mode. These observations can often be made without using test equipment. Some examples of this process are described in this section.

What if the compressor and outdoor fan do not start in the cooling mode but both work in the heating and defrost modes? The malfunction might be found in the room thermostat. All the safety control contacts must be closed, or the compressor will not run at all. The compressor contactor and other controls must also be intact. The thermostat would be a good place to begin troubleshooting in this scenario.

What is a possible cause if the unit runs in heating but won't defrost, and it refuses to switch to the cooling mode? A dysfunctional reversing valve solenoid is a possibility because it defaults to the heating mode. To operate in cooling, the reversing valve solenoid must be energized. If the unit runs in both the heating and cooling modes but won't defrost, only the components directly related to the defrost mode need to be considered, such as a defrost thermostat or thermistor.

Many manufacturers provide troubleshooting guidance in their product literature. While the troubleshooting guidance they provide can be helpful, many are either too general or too complicated. When troubleshooting, there is no substitute for a thorough understanding of the operating sequence of the product.

On Site

Restricted Airflow

The supply air temperature of a heat pump in the heating mode is typically lower than that of electric, gas, and oil heating systems, although the volume of air moved is higher. If the air being discharged into the conditioned space from a heat pump is warmer than 120°F (49°C), it could indicate restricted airflow. Otherwise, the supply air temperature will rarely approach this temperature.

A dirty air filter, dirty indoor coil, or even piece of dislodged duct liner can reduce the airflow. This in turn causes an increase in compressor discharge temperature. High temperatures can cause the compressor to overheat and trip on the overload protection. When the supply air temperature is unusually high, measure the static pressure in the system and look for problems related to a low air volume.

3.1.0 Testing Wiring

CAUTION

Before installing any field wiring, consult the manufacturer's literature to ensure the wire is the correct type and appropriate size for the application. Most field-installed control wiring circuits require 18 AWG wire, but the length of the run is always a factor in electrical wiring. Also pay attention to the rolls of thermostat wire you purchase, as many distributors routinely sell thermostat cable smaller than 18 AWG.

The wiring for HVACR equipment is classified as *field wiring* or *factory wiring*. Field wiring is installed and terminated by the installer. Factory wiring refers to the line- and low-voltage wiring that is installed at the point of manufacture.

3.1.1 Field Wiring

Field wiring must meet the specifications for the job as well as the equipment manufacturer. 18 AWG solid thermostat cable is commonly required as a minimum. Generally, if the run is longer than 100', 16 AWG copper is needed to prevent excessive voltage drop.

It is very rare that control circuit loads require a wire size larger than 16 AWG. Remember though, that voltage drop due to wire sizing will not be evident until the circuit is under load. The voltage will be normal when the circuit is unloaded. Once the installation is complete, replacing the low-voltage wiring becomes a difficult and expensive repair for the installer.

Voltage drop in the control circuit can cause relays and contactors to hum and/or fail to move when energized. This is due to the low strength of the magnetic field created by the coil. When the control voltage is low under load, check the current load and evaluate the wire size. When the control voltage is too low when there is no load, check the primary voltage at the transformer and ensure the transformer is performing as expected. If the transformer primary voltage is low, the secondary voltage will be low as well.

Also look for loose connections and poor terminations. The additional resistance caused by a poor connection can cause a voltage drop downstream. If the terminations appear sound, disconnect each end of a wire in a specific circuit and measure the resistance with an ohmmeter. Excessive resistance often indicates loose connections or corroded terminals; resistance through a wire should be virtually non-existent.

3.1.2 Factory Wiring

Factory wiring in the unit can be checked in the same manner as field wiring. When a defect or flaw in the wiring is suspected, check the field wiring first. Factory connections are generally of better quality because they are completed under ideal conditions using precise and consistent processes. Over time, however, all terminations deteriorate.

Factory wiring damage and failures are often the result of wires contacting sharp and/or vibrating metal components, such as the cabinet or refrigerant piping. Factory wiring can also be damaged by rodents chewing on the wires.

When replacing factory wiring and terminations, use the same wire size and ensure the hardware used is appropriate for the wire size and current load.

Did You Know?

Field Wiring Problems

Field wiring problems are most often found at connection points. This is often because HVACR installers may fail to use the correct devices or procedures to join the wires. For example, it is common to encounter low-voltage thermostat wires that have been simply twisted together and taped. A correctly sized wire nut should be used to make this type of connection.

On Site

Aluminum and Copper Wire Connections

Joining aluminum and copper wires requires special techniques and connectors. The dissimilar materials oxidize, causing the joint to overheat and fail.

For this reason, many HVACR equipment manufacturers prohibit the connection of aluminum wires to their products. Even some products designated for making aluminum-to-copper terminations, particularly wire nuts, aren't always successful.

Products designed for Al/Cu connections are purple or have purple markings to help in identification. The terminal block shown here is a good example. Once the connections are made, the plastic box is closed, minimizing contact with water and fresh air that hasten corrosion.

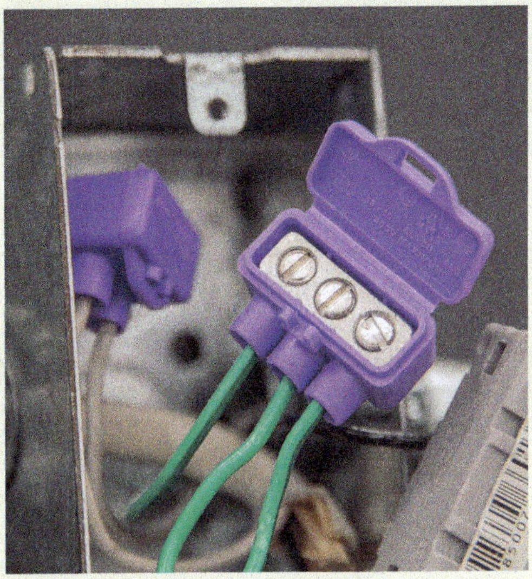

Source: nVent KING INNOVATION

3.2.0 Testing Heat Pump Thermostats

Heat pump thermostats should be checked to ensure that they are a correct match for the system. However, many of today's digital thermostats are flexible enough to handle a variety of system configurations.

Thermostats should not be mounted in a location where extraneous cooling or heating sources will affect it. Although digital thermostats are not affected by an unlevel condition, they should be mounted level for appearance. The wall penetration behind it should be plugged to prevent drafts inside the wall from affecting the accuracy of the thermostat. A clean, level, and functional thermostat installation is the mark of a professional.

Digital thermostats (*Figure 12*) offer precise temperature control, programmability, and application flexibility. The installer configures the thermostat for the desired mode of operation. Some thermostats are set up by answering a series of questions about forms of auxiliary heat, the number of heating and/or cooling stages, and in what mode the reversing valve is energized.

Figure 12 Wi-Fi enabled programmable heat pump thermostat.

Unlike older electromechanical thermostats, digital thermostats are electronic devices and must be powered to operate. This can be done with batteries or with 24VAC available from the control circuit. When the control circuit powers the thermostat, both R and C from the 24VAC control transformer are connected to the thermostat.

There are a few other unique options regarding the use of AC power. Older thermostats rarely required a power source to operate, and a Common wire may not be available in the cable. Installing additional control wires is time-consuming and expensive. This complicates thermostat retrofits.

To address the problem, manufacturers offer thermostats that are powered through *power stealing*. The thermostat "steals" power from the control circuits and stores it in a capacitor. A battery may be provided to maintain the programming during power outages, and it may provide operating power if the capacitor doesn't have enough charge for some reason.

Regardless of the means used to power the thermostat itself, many thermostats have batteries that maintain the programming for some time if the power is interrupted. Wi-Fi thermostats without batteries may store programming in the cloud, although they often lose the time setting during a lengthy power outage.

Before checking a thermostat, evaluate and determine the function controlled by each terminal. The terminal designations are fairly standardized in the HVAC industry, but variations will be found. There are also generally accepted wire colors associated with thermostat wiring. However, the wire colors used are influenced by the number of conductors and colors available in the thermostat cable. Always confirm the wire color connected to each terminal, rather than assume the wires follow the generally accepted color scheme.

Generally, heat pump thermostat terminals are related to the following circuits or functions. However, remember that variations will be encountered, especially as additional brands and models enter the market:

- *Terminal C* — Wire color is typically black or brown. It is the 24VAC common lead from the control transformer secondary winding.
- *Terminal R* — Wire color is red. It is the other 24VAC power lead from the control transformer secondary winding. Two R terminals may be on a thermostat, marked as R_C (or simply R) and R_H. Older systems may have had two control transformers in operation—one serving the heating equipment and one serving the cooling equipment. Separating the R terminals keeps the two circuits apart. When a single transformer is used, a jumper is placed across the two R terminals.
- *Terminal G* — Wire color is green. It energizes the indoor fan control circuit.
- *Terminal Y* — Wire color is yellow. It energizes the compressor control circuit. The Y terminal may be followed by a number (Y_1, Y_2, etc.) to indicate the cooling stage it controls.
- *Terminal W* — Wire color is white. It usually controls the supplementary heat sources. The W terminal is often followed by a number (W_1, W_2, etc.) to indicate the heating stage it controls. In this scenario, Y is used to start the compressor in both heating and cooling, while the W terminals are used to control the supplemental heat sources. In conventional systems, Y is not used for any purpose in the heating mode.
- *Terminal O* or *B* — Wire color is usually orange or blue, depending on which terminal is in use; O and B are very rarely, if ever, used together. These terminals are energized when the heating or cooling mode is selected. When cooling is the selected mode of operation, O is consistently energized. B is energized when the heating mode is selected. The terminals are commonly used to control power to the reversing valve. As you have learned, most systems use O, because the reversing valve is usually energized in the cooling mode.
- *Terminal S* — Wire color is optional. S terminals are points of connection for external temperature sensors.
- *Terminal E* or *X* — Wire color is optional. The E and/or X terminals are generally active when the Emergency Heat mode is selected.
- *Terminal L* — Wire color is optional. L powers the Emergency Heat LED light on the thermostat and is used on systems that energize the light remotely from the indoor unit. Otherwise, the icon is simply powered by the thermostat internally when Emergency Heat is selected.

Another common characteristic of digital thermostats is a configuration menu (*Figure 13*). The configuration menu can have many possible entries and options, each of which can significantly affect the behavior and performance of the thermostat. It is essential to check the configuration menu, especially when a system doesn't appear to be broken but simply doesn't perform as expected.

3.2.1 Communicating Thermostats

A growing number of thermostats operate using a communication bus rather than through common 24VAC wiring terminals. The Carrier Infinity™ system, for example, uses a two- or four-wire approach to control wiring. The wiring terminals are designated as A, B, C, and D. In the four-wire product, the A and B wires are used for communication signals to travel back and forth, while the C and D terminals provide 24VAC power to the thermostat. The primary equipment must be designed to work with a communication bus.

Troubleshooting these units requires a different strategy than troubleshooting standard controls. When problems are encountered, be sure to acquire the service information for the specific model in use.

Configuration Menu						
Screen Reference Number	SS1 MS2	HP1 HP2	Press Key	Displayed Factory (Default)	Press [+] or [◁] to Select from Listed Options	Comments
1	●	●	[+]	MS 2	HP 1, HP 2, SS 1	Selects Multi-Stage (MS2, No Heat Pump), Heat Pump 1 (HP1, 1 compressor), Heat Pump 2 (HP2, 2 compressor or 2 speed compressor), or Single Stage
2	●	●	[+]	(GAS)	ELE	GAS setting: furnace controls blower ELE setting: thermostat controls blower
3		●	[+]	0b (0)	b	Selects Reversing Valve (This item is only to appear if HP1 or HP2 is selected above.)
4	●	●	[+]	Days, **(7) P**	**5** or **0**	Programs per week (7 days, 5-1-1 days or non-programmable)
5	●		[+]	Cool-Off-Heat-Auto	Cool-Off-Heat, Heat Off, Heat, Cool-Off, Auto Off	System switch configuration in non heat pump mode
		●	[+]	Cool-Off-Heat-Em-Auto	Cool-Off-Heat-Em, Off-Em-Auto	System switch configuration, heat pump mode
6	●	●	[+]	(On) **E**	**OFF**	Selects Energy Management Recovery. Not available if 4 is 0
7	●	●	[+]	(FA) Heat, Cr	SL	Selects Adjustable Anticipation, cycle rate, Heat
8	●	●	[+]	(FA) Cool, Cr	SL	Selects Adjustable Anticipation, cycle rate, Cool
9		●	[+]	(FA) Cr/AU, Em	SL	Selects Adjustable Anticipation, cycle rate auxiliary. (This item is only to appear if HP1 or HP2 is selected above).
10	●	●	[+]	(OFF) **SC**	On	Selects Stage Cycle Completion On (OFF).
11	●	●	[+]	(OFF) **CL**	On	Selects Compressor Lockout
12	●	●	[+]	(On) **dL**	OFF	Selects Continuous Display backlight
13	●	●	[+]	0 (Temperature)	5, LO to 5, HI	Selects Adjustable Ambient Temperature Display [range -5 (LO) to +5 (HI)]
14	●	●	[+]	**°F**	**°C**	Selects °F/°C Display
15	●	●	[+]	(On) **b**	OFF	Selects audible Beeper On/Off
16	●	●	[+]	(On) **dS**	OFF	Selects Daylight Saving Time calculation
17	●	●	[+]	(On) Heat, **AS**	OFF	Selects Automatic Schedule for comfort temperature Programming, heat mode (not available if 4 is 0).
18	●	●	[+]	(On) Cool, **AS**	OFF	Selects Automatic Schedule for comfort temperature Programming, cool mode (not available if 4 is 0).
19	●	●	[+]	(OFF) CS	On	Selects Cool Savings Feature On or Off
20	●	●	[+]	**CS** Cool Savings (3)	1-2-3-4-5-6	Selects amount of Cool Savings adjustment
21	●	●	[+]	(OFF) CO	On	Selects Compressor Optimization (not available on earlier models)

Figure 13 Example of a thermostat configuration menu.
Source: Courtesy of Emerson

Communicating models interact with the equipment at a deeper level than common digital thermostats. Digital thermostats, even those that are Wi-Fi enabled, do not provide real-time information about the system. Most of the information provided comes from information gathered by the thermostat. One example of this deeper connection is the display of fault codes when problems occur.

Table 1 shows a small sample of the fault codes that might display. This information is made available to the thermostat from the equipment solid-state control board(s). Each product is different however, and using the proper documentation is necessary when troubleshooting.

3.2.2 Room Thermostat Troubleshooting Sequence

One way to determine if a room thermostat is faulty is to remove it from its sub-base and use jumper wires across the terminals to simulate circuit closures. If a function works as expected using a jumper, but the thermostat fails to energize the terminal when it should, the thermostat is the likely problem.

The digital display can be helpful in this process. Most will display messages such as "Cooling On" or display a snowflake when cooling is called for. Seeing the display but noting that the Y terminal is not being energized is a clear sign that the thermostat is faulty.

TABLE 1 Example Fault Codes for a Communicating Thermostat

Code	Fault
Code 14	Blower motor lockout
Code 15	Gas heat lockout
Code 16	Furnace ignition lockout
Code 18	Abnormal flame signal response
Code 21	Internal control fuse open
Code 25	No low voltage source
Code 30	Contactor coil shorted
Code 33	Failure of high-stage cooling (three times)
Code 34	Failure of low-stage cooling (three times)
Code 40	High-pressure switch open
Code 41	Low-pressure switch open

Before condemning a thermostat, verify that all switch positions are correct during testing, and that the installer programming has been completed properly.

The R terminal is the switched leg of control power. Common provides the second leg of power to energize a 24VAC component, but it is never switched. Common is routed directly to each device. Thus, connecting R to any other terminal (except Common, which results in a direct short) energizes that function.

With the 24VAC transformer energized, place jumpers across the following subbase terminals and observe the response:

- R to G energizes the indoor fan circuit.
- R to Y energizes the compressor contactor.
- R to O or B energizes the reversing valve.
- R to W should energize the supplemental heat. Note that some systems require a set of contacts on the indoor fan relay to be closed before the supplemental heat can be energized.
- R to X should enable the emergency heat circuit. Again, note that some systems require a set of contacts on the indoor fan relay to be closed before the supplemental heat can be energized.

If a component fails to function with the jumper in place, the problem is not in the thermostat, although it could be in the wiring between the thermostat and unit. If a circuit or component does function with the jumper, but fails to function with the thermostat installed, the problem likely lies in the thermostat or its configuration.

Thermostat troubleshooting can also be done at the furnace or fan coil unit. Look for 24VAC between the Common terminal and any other terminal that should be energized by the thermostat. You can also use the jumper wire to connect R to other terminals here, but this does not test the wiring between the thermostat and the indoor or packaged unit.

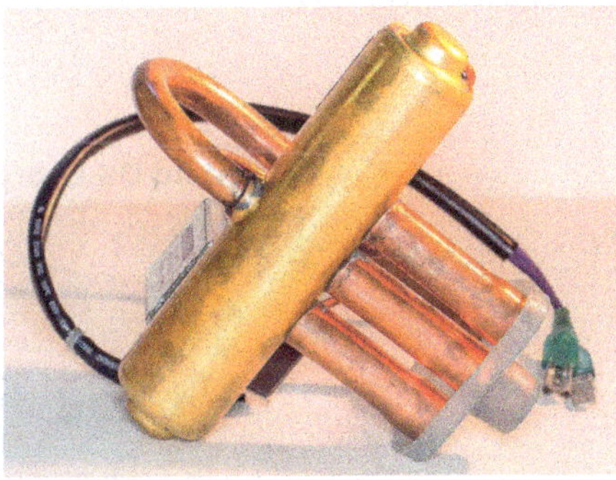

Figure 14 Reversing valve.
Source: Darkdiamond67/Shutterstock

3.3.0 Testing Reversing Valves

A reversing valve (*Figure 14*) is a **pilot-operated valve** with a slide mechanism that directs the refrigerant flow to different paths, depending on the operating mode. *Figure 15* and *Figure 16* show the path of the refrigerant in the cooling and heating modes, respectively. In the defrost mode, the valve is positioned the same as in the cooling mode, directing the hot discharge gas to the outdoor coil.

Pilot-operated valve: A valve that is positioned by a second, smaller valve that ports fluid pressure to the main valve to cause movement. Larger valves and those operating against high differential pressures use this strategy due to the limitations in the magnetic field strength of solenoid coils.

Basic reversing valve inspection and testing includes the following:

Step 1 Inspect the valve body, solenoid-operated pilot valve, and capillary lines for dents, deep scratches, or cracks. Dents in the body will cause the valve to stick. Damaged capillary lines may prevent refrigerant from reaching the main valve piston (also referred to as a *shuttle*).

Step 2 Considering the operating mode, ensure the solenoid coil is energized or de-energized. If the compressor is not running, you may be able to hear a slight

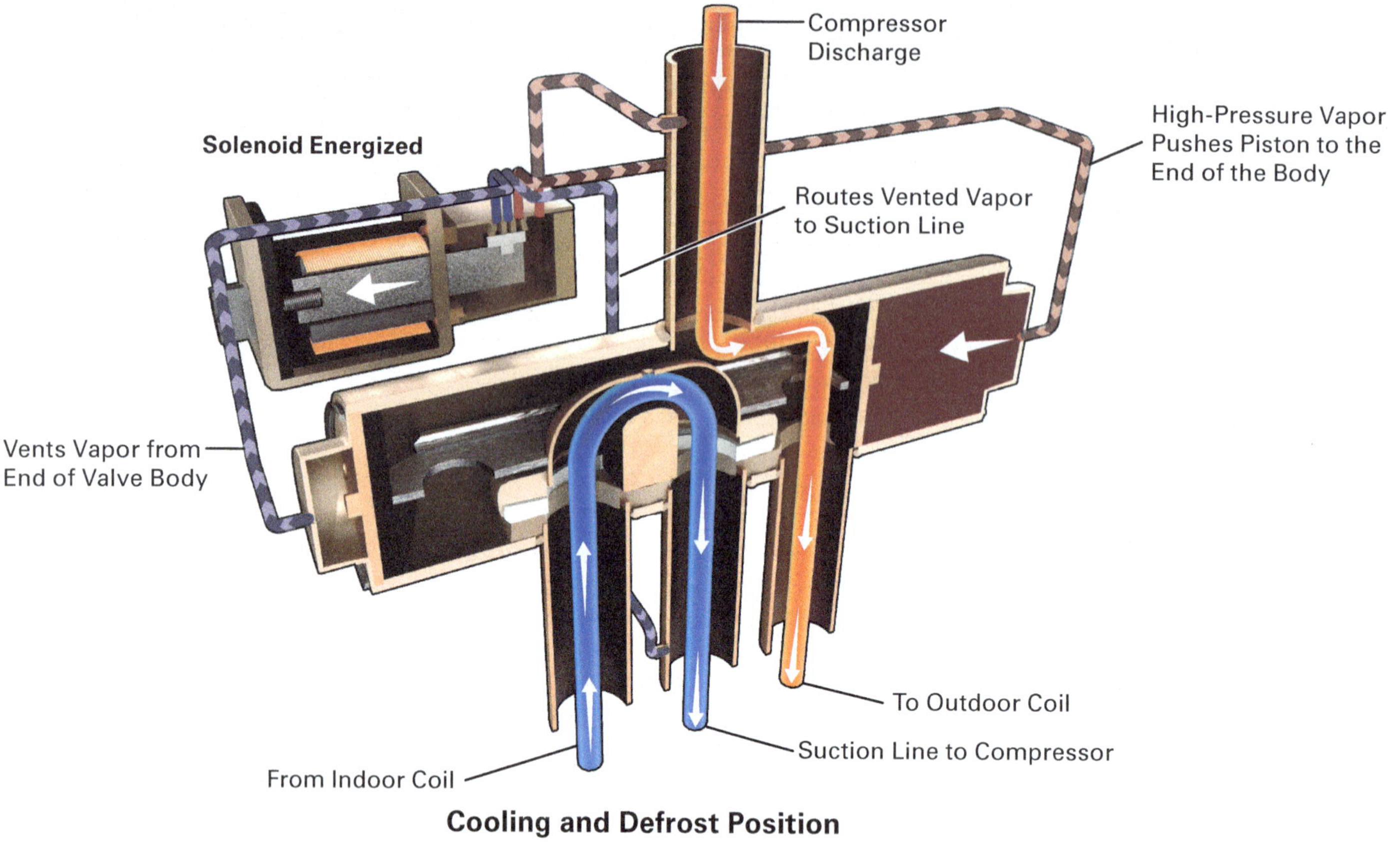

Figure 15 Reversing valve refrigerant flow—cooling and defrost modes.

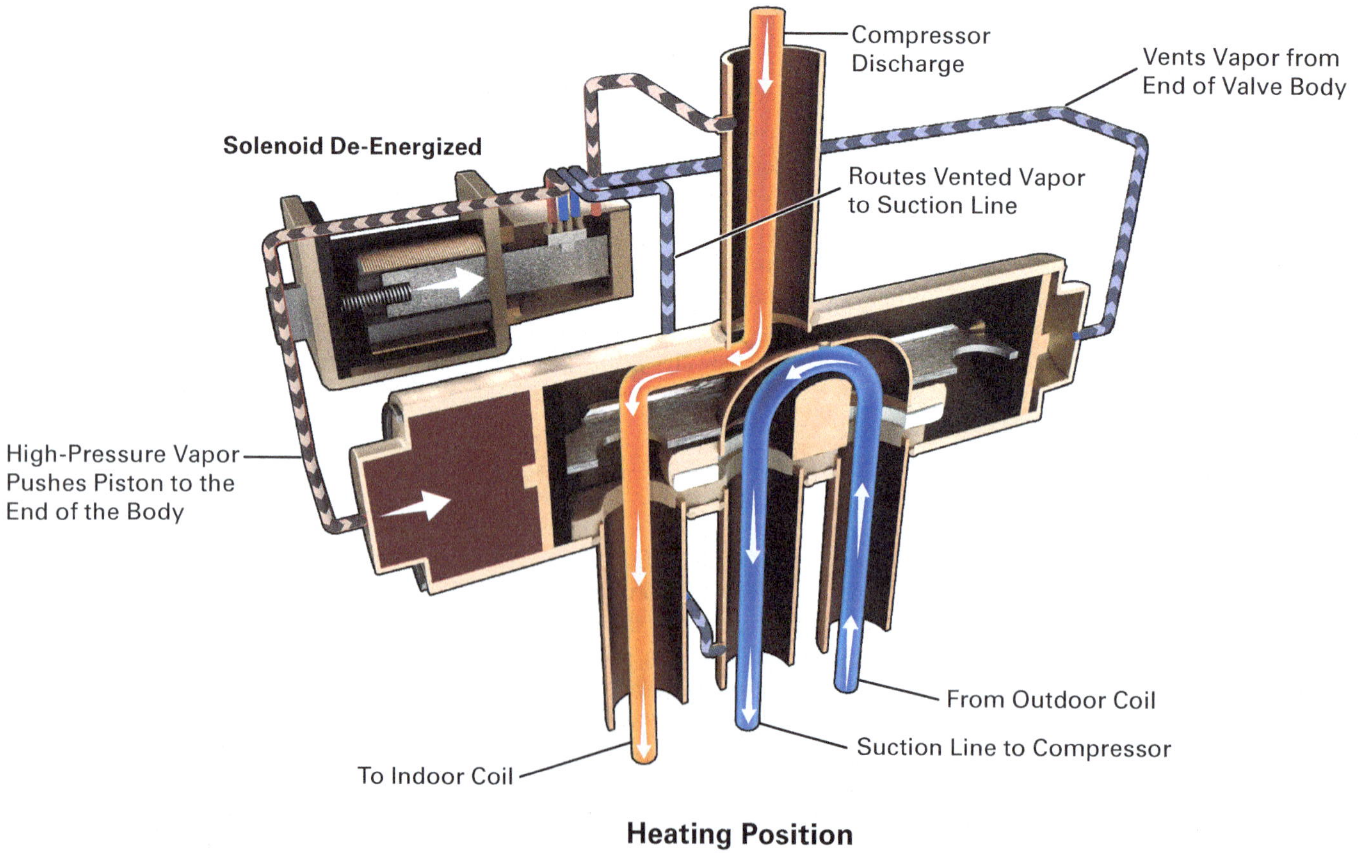

Figure 16 Reversing valve refrigerant flow—heating mode.

click as the pilot valve changes position. When the compressor is running, there should be an obvious "whooshing" sound when the main valve changes position. If the valve is not being energized or de-energized as it should, the problem is related to the controls or interconnecting wiring.

Step 3 Check the system pressure differential with the compressor running. Remember that a reversing valve is a pilot-operated valve. For a reversing valve to operate, a minimum pressure differential is required. While some valves will shift when the differential is as low as 20 psig, others may require a 100-psig differential. Most reversing valve manufacturers do not guarantee the movement of the valve below a 100-psig differential.

CAUTION

While reversing valves appear to be rugged, the valves can easily suffer from mishandling. Handle reversing valves with care. Even small dents or dings can result in binding and/or internal leakage. When replacing a reversing valve, take the necessary steps to avoid overheating the internal seals and parts during soldering or brazing.

A reversing valve may also be tested for internal leakage by measuring line temperatures at strategic locations. If it is leaking internally, the flow will be from the high-pressure side to the low-pressure side, which is the suction line.

Figure 17 shows the measurement points for testing. To avoid error, temperature readings should be made just a few inches away from the valve body. The temperature probes may need to be insulated for accuracy. Note that an infrared thermometer is not the best choice for this test. A contact thermometer or sensor is recommended.

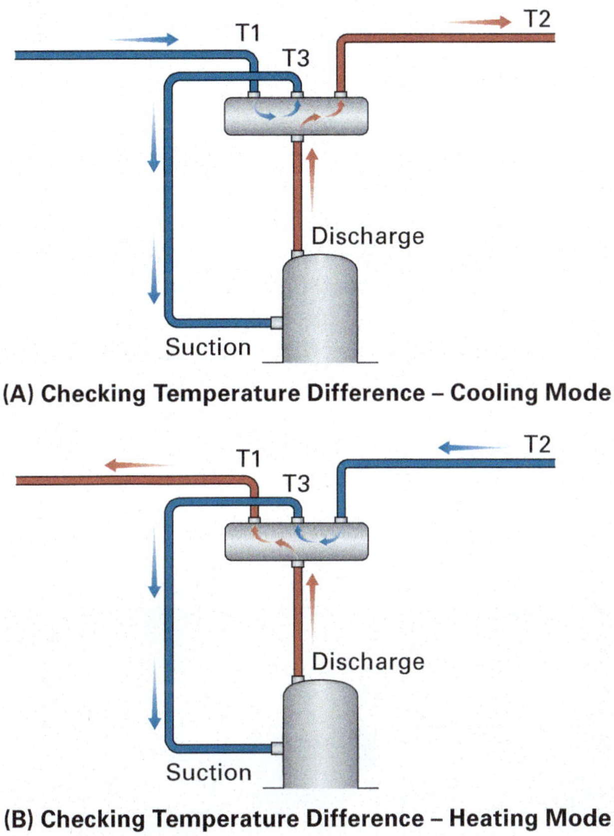

Figure 17 Reversing valve temperature testing.

Refer to *Figure 17* and use the following steps to test the valve for internal leakage:

Step 1 Set the thermostat to start the system in the cooling mode.

Step 2 Measure the temperature at points T1, T2, and T3.

Step 3 If the temperature difference between the measurements at T2 and T3 is greater than 3°F (1.7°C), the valve is leaking internally.

Step 4 Set the thermostat to start the system in the heating mode.

Step 5 If the difference between T1 and T3 is greater than 3°F (1.7°C), the valve is leaking internally.

Reversing valves cannot be repaired in the field. If the valve is leaking internally, it must be replaced. A failed solenoid coil is easily replaced.

CAUTION

When replacing a reversing valve, it is crucial to use a valve that is properly rated for the application and the refrigerant in use.

On Site

Keeping It Cool

The concern of overheating of refrigerant piping components during soldering or brazing has always been an issue. Many components that must be soldered or brazed into the piping system are sensitive to the extreme heat. Some components can be disassembled before the installation, leaving only the body exposed to the heat. However, disassembly is not possible with reversing valves.

Various products are available to use as a heat sink to protect the component. Gels and pastes are common forms. The products readily absorb massive amounts of heat, impeding its movement through the metal.

Rather than bathe the component in the product, apply it in a thick ring around the tubing between the heat source and the component. The product should be applied with good contact and without voids that would allow the heat to simply pass it by. It is also effective to wrap a wet cloth around the main body of the reversing valve and then apply a heat sink in the proper locations, between the joint and the wet rag.

3.4.0 Troubleshooting Defrost Controls

Most defrost controls today are solid-state printed circuit boards that are replaced when they fail. Field repair is not an option.

For testing and troubleshooting, most defrost boards provide a speed-up feature that allows the technician to initiate and observe the defrost mode at any time. Waiting until the system initiates the defrost mode on its own could take hours, assuming the needed conditions are even present.

The most popular method is to use a coin or the tip of a screwdriver to jump across two specific pins or terminals on the board (*Figure 18*). The two pins must be continuously shorted for preset time, such as 10 seconds. Once the feature is initiated, the system enters the defrost mode, allowing the technician to observe the sequence.

Figure 18 Defrost speed-up terminals.
Source: HVACGuru

It is important to ensure the defrost controls are properly set up before assuming there is a fault somewhere. Time/temperature defrost controls allow the installer to select the time between defrost attempts based on the local climate. For example, in cold areas with consistent fog or drizzle, a short interval is selected. In warmer, drier climates, a longer interval is acceptable.

Figure 19 shows a yellow jumper wire permanently attached to the board, which is placed over one of three pins to complete a circuit. The pins on this board are labeled 30, 60, and 90 minutes. What is being set is the amount of compressor run time required before the defrost mode initiates. On warmer days, a heat pump may go for hours between defrost attempts before accumulating the set amount of run time.

Figure 19 Compressor run-time selection for defrost control.
Source: HVACGuru

With a time/temperature cycle, the selected runtime must pass before the defrost mode can begin. At the end of the period, the system will enter defrost if the coil temperature is below a fixed value. The temperature setting of the coil thermostat or thermistor is usually around 30°F, but it does vary.

If the coil is not cold enough to indicate the presence of frost, the unit continues operating in the heating mode, and the timing period begins again. If the temperature is below the set value, the defrost mode begins and continues until either the coil temperature rises above the set temperature or a maximum amount of time elapses (usually 10–13 minutes).

The demand defrost strategy uses two separate thermistors, with no relationship to time. One thermistor monitors the outdoor coil temperature at a strategic location. The second thermistor monitors the ambient temperature. When the outdoor temperature is above a preset level (50°F, for example) the defrost mode is never initiated. Below that temperature, the coil temperature is evaluated along with the outdoor temperature. The defrost control board decides when the conditions indicate that frost is present, and then it initiates the defrost mode in response.

3.4.1 Evaluating Inputs and Outputs

Defrost controls respond to power at various terminals and other inputs. If a defrost control has the correct inputs, it responds by generating relay contact movement or outputs. If you confirm the inputs are correct, but the defrost control is not responding with the correct outputs, it must be replaced.

However, if the inputs and outputs are normal and the component being controlled (such as a reversing valve) is not responding, the problem is more likely with the controlled device or interconnecting wiring.

Typical defrost control board inputs include the following:

- *24VAC power* — If the defrost control does not have power, it will not operate. Onboard functions often require DC power that is rectified from the incoming AC.
- *Temperature input (binary/switch closure)* — With time/temperature defrost systems, the temperature input usually takes the form of an open or closed set of contacts in an outdoor coil thermostat. These thermostats close on a fall; closed contacts indicate a defrost is needed. When those same contacts open during defrost, it usually signifies the outdoor coil temperature is warm enough to skip or terminate the defrost mode. The contact position can be verified with a voltmeter (power on) or an ohmmeter (power off). Note that the contacts will not close at the same temperature at which they open. The temperatures and/or the differential can often be read on the side of the thermostat or be determined from the manufacturer's literature.
- *Temperature input (electronic)* — If a heat pump uses one or more thermistors to sense system temperatures, the input signal(s) is in the form of a resistance that varies with temperature. Thermistors can be tested with an ohmmeter and compared to a manufacturer's chart. The board programming and logic determines when a defrost is needed and when it can be terminated.

NOTE

Thermistor testing is presented in NCCER Module 03317, *Control Circuit Troubleshooting*.

Typical defrost control outputs include the following:

- *Reversing valve control* — The 24VAC reversing valve solenoid is energized by the defrost control. When the defrost mode is initiated, the reversing valve is either energized or de-energized, depending on whether the valve is energized in the heating or cooling mode.
- *Supplemental electric heat* — Systems with a supplemental form of heat use that source to temper the cold air leaving the indoor coil during the defrost mode. When electric heat is used, the substantial current required to operate the heating elements does not flow through the defrost control. A relay or contactor controlled by the defrost control carries the current.
- *Condenser fan motor* — The outdoor fan motor must de-energized during the defrost mode. Relay contacts on the defrost control open to de-energize the motor. The contacts close again when defrosting is complete.

3.4.2 Defrost Troubleshooting Tips

Several things can interfere with the defrost mode. Common defrost problems and possible solutions include the following:

- *The defrost mode activates but does not readily melt the frost* — Ensure the outdoor fan is shutting down during defrost. If not, force a defrost test on the board and check the relay contacts that control the outdoor fan. A strong wind blowing across the outdoor coil can also be a factor, but this would likely be an intermittent issue rather than a consistent one. An incorrect refrigerant charge can also affect defrost performance; check the refrigerant charge according to the manufacturer's instructions. Although hot vapor still leaves the compressor when the refrigerant charge is low, remember that it is primarily the change of state from vapor to liquid that transfers most of the heat to the frost. When there is not enough refrigerant in the system, less heat is transferred.

- *Cold air blown into the conditioned space during defrost* — This problem indicates that the electric heat or furnace is not cycling on. In a system with supplemental electric heat, if the heater works when staged on by the thermostat, it means the problem is likely related to the defrost control board. If the electric heater does not work in either situation, check the heating relays, heating element, and the thermal protection devices. Remember that a fusible link, like a fuse, opens and never recloses. Ductless systems are not equipped with any form of heat to temper the air. Many cycle the indoor fan off during the defrost mode.
- *The defrost mode activates when there is no frost on the coil or defrost does not terminate after the frost is clearly gone* — With a time/temperature defrost scheme, this condition proves that nothing is wrong with the timing or the defrost mode would never start. This problem can be caused by a defrost thermostat that has failed closed. The system will then initiate a defrost at the end of every timing cycle, continuing for the full amount of time allowed by the defrost control (usually 10 minutes). A defective thermistor will also cause the same problem.
- *Ice tends to remain on the lower part of the outdoor coil* — The operation of the defrost controls may appear normal. However, this problem can be caused by problems with the defrost thermostat or thermistors. If the thermistors are not in the proper location or not making good surface contact, the defrost control board may terminate defrost before the task is done. One of the most likely causes, though, is a low refrigerant charge. If the refrigerant charge is low, the hot vapor entering the outdoor coil condenses very quickly, delivering all its latent heat to the upper portion of the coil. Since most of the heat that melts the ice is the latent heat from the change of state, not enough heat is left in the refrigerant to thaw the remainder of the coil.

Because dual-fuel systems usually disable the heat pump at or near the balance point, significant frost buildup is rare. Therefore, defrosting is infrequent, and the temperature is usually near or above freezing. Some dual-fuel systems do start the furnace to temper the indoor air during the defrost mode. If cold air is blowing into the conditioned space with a dual-fuel system and the burner doesn't start, ensure the furnace operates properly with the outdoor temperature below the balance point. If it does, the problem is likely related to the defrost relay, defrost control board, or the interconnecting wiring. If the furnace does not operate in its normal modes, then attention should be focused on the furnace controls.

Also remember that it takes some time for a modern furnace to ignite and warm up. Thus, during the early stages of defrosting, some cold air may be generated before it is tempered.

Depending on how the outdoor coil is circuited, the lower portion is often the last part of the coil to thaw. In some cases, it may be because the system does not defrost often enough. If all aspects of the defrost mode are normal and the refrigerant charge is correct, try increasing the frequency of the cycles on the defrost control board. For demand defrost controls, ensure the thermistors are in the correct location and the coil sensor is firmly mounted and in contact with the surface.

3.0.0 Section Review

1. If a heat pump outdoor unit is located 150' from the air handling unit, the control circuit wire size will likely be _____.
   a. 24 AWG
   b. 22 AWG
   c. 18 AWG
   d. 16 AWG

2. To what does *power stealing* refer?
   a. A method used to power a defrost control board
   b. Sharing a single power source for a heat pump and a furnace
   c. A method used to power a thermostat without a Common wire
   d. A reduction in compressor current when the electric heaters are energized

3. When testing a reversing valve for internal leakage, above what temperature differential would you consider the valve to be leaking?
   a. 1°F
   b. 3°F
   c. 5°F
   d. 10°F

4. A demand defrost approach incorporates which of the following?
   a. Two thermistors
   b. A pressure differential switch
   c. A variable-speed outdoor fan motor
   d. A timer to establish the defrost frequency

Module 03311 Review Questions

1. If a heat pump works normally in the heating mode but will not provide cooling or a defrost mode, which of the following components or their related controls is *most likely* defective?
   a. Compressor
   b. Reversing valve
   c. Room thermostat
   d. Compressor contactor

2. Beyond the reversing valve, what is the other major difference between a heat pump and a cooling-only system?
   a. Heat pumps require two metering devices.
   b. Only cooling systems require a filter-drier.
   c. A heat pump requires a head-pressure control.
   d. Heat pumps only use fixed-orifice metering devices.

3. True or False? Heat pumps do not need to defrost above the balance point.
   a. True
   b. False

4. Generally, in the cooling mode, the reversing valve solenoid is _____.
   a. bypassed
   b. energized
   c. back-seated
   d. de-energized

5. How does oil trapped in the accumulator exit and return to the compressor?
   a. The oil vaporizes during the defrost mode.
   b. The accumulator contains an oil separator.
   c. The oil becomes entrained in the leaving vapor.
   d. The oil leaves through a small orifice in the dip tube.

6. How do ductless heat pumps address the problem of blowing cold air indoors during the defrost mode?
   a. The indoor fan is turned off during defrost.
   b. Ductless heat pumps do not require a defrost mode.
   c. The cold air is directed outdoors through a motorized damper.
   d. Small electric heating elements are mounted to the face of the indoor coil.

7. If a heat pump blows hot air into the conditioned space on a call for cooling, which of the following is a possible cause?
   a. A failed defrost timer.
   b. An excessive refrigerant charge.
   c. A failed reversing valve solenoid.
   d. The outdoor fan continues to run.

8. Which of the following terminals does a thermostat traditionally energize when it is switched to the heating mode?
   a. O
   b. G
   c. S
   d. B

9. Which of the following conditions is *not* common nor appropriate for the defrost mode?
   a. The outdoor fan shuts off.
   b. The compressor shuts off.
   c. An electric heater is energized.
   d. The indoor fan continues to run.

10. In a dual-fuel heat pump system, when can the compressor and furnace burner operate at the same time?
    a. During the defrost mode.
    b. Whenever the thermostat calls for both stages of heat.
    c. Any time there is a call for heat below the balance point.
    d. The heat pump and burner can never operate at the same time.

11. A low control voltage measured *only* when a circuit is under load is most likely caused by ______.
    a. a blown fuse
    b. undersized control wiring
    c. low ambient temperatures
    d. an incorrect transformer primary voltage

12. Thermostats with two R terminals are designed to accommodate ______.
    a. dual-fuel systems only
    b. heat pumps with multiple stages of heat
    c. the need for more room to connect all the needed R wires
    d. systems with separate transformers for heating and cooling

13. What function do terminals marked S on a thermostat subbase serve?
    a. Connection to the local network.
    b. Connection of external occupancy sensors.
    c. Connection of external temperature sensors.
    d. Control of the reversing valve when it is energized in heating.

14. Although reversing valves often change position when exposed to lower pressure differentials, manufacturers typically specify that a reversing valve is operating as expected if it moves at a ______.
    a. 20-psig differential
    b. 75-psig differential
    c. 100-psig differential
    d. 150-psig differential

15. The function of a reversing valve will be affected by all the following, *except* ______.
    a. an open solenoid coil
    b. a dent in the valve body
    c. excessive frost buildup on the coil
    d. a low refrigerant pressure differential

Answers to odd-numbered review questions are found in the Module Review Answer Key at the back of this book.

Answers to Section Review Questions

Answer	Section	Objective
Section One		
1. b	1.1.1	1a
2. b	1.2.3	1b
Section Two		
1. a	2.1.0	2a
2. b	2.2.0	2b
3. c	2.3.0	2c
4. b	2.4.0	2d
5. b	2.5.0	2e
Section Three		
1. d	3.1.1	3a
2. c	3.2.0	3b
3. b	3.3.0	3c
4. a	3.4.0	3d

MODULE 03209

Troubleshooting Gas Heating

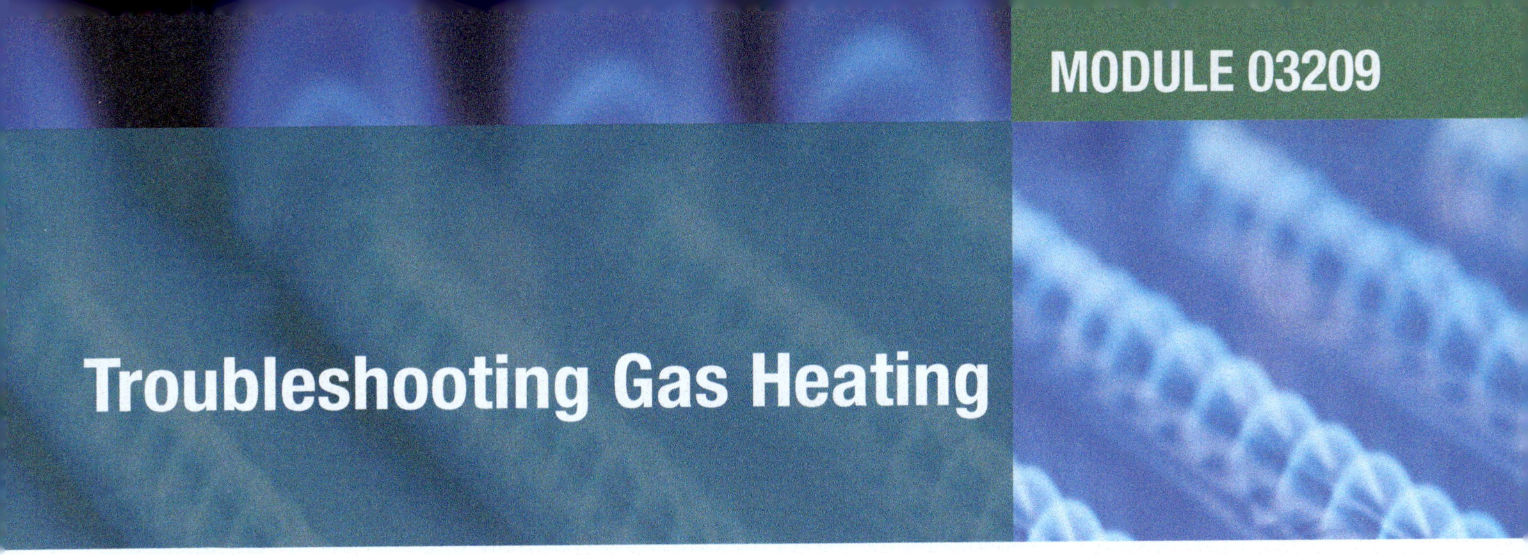

Source: Naumoid/iStock/Getty Images

Objectives

Successful completion of this module prepares you to do the following:

1. Identify and explain how to troubleshoot various gas-heating equipment faults.
   a. Describe typical sequences of operation for gas-fired heating equipment.
   b. Explain how to test ignition controls and other combustion-related components.
2. Identify and outline the operation of various infrared gas heaters.
   a. Identify various types of infrared gas heaters.
   b. Outline the unique operating characteristics of infrared gas heaters.
3. Explain how to conduct a combustion analysis on a gas furnace.
   a. Identify relevant combustion byproducts and describe combustion analysis equipment.
   b. Explain how to conduct a combustion analysis and how to interpret the results.

Performance Tasks

Under supervision, you should be able to do the following:

1. Using the proper tools and wiring diagrams, isolate and correct instructor-selected malfunctions in a gas heating system.
2. Complete a combustion analysis on a gas furnace or gas boiler.

Overview

Gas furnaces have advanced a great deal in the last 25 years in the pursuit of safety and higher efficiency. Today's furnaces are controlled by printed circuit boards that integrate the various safety controls with the operating and ignition controls. HVACR service technicians must understand how to test the various inputs and outputs to identify problems. Technicians must also learn to recognize combustion and venting problems and conduct a combustion analysis to ensure a furnace is operating efficiently.

NCCER Industry-Recognized Credentials

If you are training through an NCCER-accredited sponsor, you may be eligible for credentials from NCCER. The ID number for this module is 03209. Note that this module may have been used in other NCCER curricula and may apply to other level completions. Contact NCCER at 1.888.622.3720 or go to **www.nccer.org** for more information.

You can also show off your industry-recognized credentials online with NCCER's digital credentials. Transform your knowledge, skills, and achievements into credentials that you can share across social media platforms, send to your network, and add to your resume. For more information, visit **www.nccer.org**.

Digital Resources for HVACR

Scan this code using the camera on your phone or mobile device to view the digital resources related to this craft.

SCAN ME

1.0.0 Gas-Heating System Operation

Performance Task

1. Using the proper tools and wiring diagrams, isolate and correct instructor-selected malfunctions in a gas heating system.

Objective

Identify and explain how to troubleshoot various gas-heating equipment faults.

a. Describe typical sequences of operation for gas-fired heating equipment.
b. Explain how to test ignition controls and other combustion-related components.

To troubleshoot a gas-fired heating system, a technician must consider three separate but related subsystems: the controls, the air or water flow, and the combustion process. The ability to read circuit diagrams, use test equipment, correct combustion-related problems, and manipulate the air or water flow are necessary skills.

1.1.0 Gas-Heating Operating Sequences

Control circuits for gas-fired systems vary by manufacturer and furnace complexity. Condensing furnaces and furnaces with multiple stages of heating require more control. Although the control systems have been standardized to some degree, variations will be found.

This module provides examples of circuit diagrams from different furnace manufacturers, along with the operating sequences. Studying these circuit diagrams will reinforce and expand your knowledge, helping you analyze and resolve problems more quickly in the future. Knowing what to expect from a given furnace is an essential part of troubleshooting.

On Site

Indoor Comfort

What is indoor comfort? To most consumers, comfort means hot air coming from the registers in the winter and cold air coming from them in the summer. True indoor comfort today, though, is more than just hot or cold air.

Today's equipment manufacturers are taking advantage of new technologies that allow them to provide improved temperature and humidity control, improved air quality, and energy efficiency levels that were only dreamed of years ago. Advances include the widespread use of electronic controls, capacity control using staged or modulating gas valves and inverter-driven compressors, and the use of variable-speed blower motors.

Technicians must have a working knowledge of these new features when servicing modern gas heating systems to ensure their services result in maximum comfort and operating efficiency for their customers.

NOTE

Refer to Appendix 03209A for a sample troubleshooting flowchart for gas furnaces.

Troubleshooting the combustion process in gas heating equipment begins with observing the ignition sequence and comparing it to what circuit diagrams and service manuals tell you to expect. Careful observation and evaluation help identify the source of a problem quickly with minimal testing.

1.1.1 Microprocessor-Based Furnace Controls

Microprocessor-based operating and ignition controls are now common in HVACR products. Modern gas heating systems typically use a combination of electromechanical and electronic controls. Electromechanical controls include various switches and relays while printed circuit boards with microprocessors provide logic to the control process. Manufacturers use microprocessor-based controls in their equipment to increase safety, reliability, and energy efficiency.

Unlike electromechanical control systems, it is not possible to trace the individual circuits on a printed circuit board (*Figure 1*). Troubleshooting these boards requires testing for the proper inputs and outputs. When a control board is not receiving a proper input, it isn't likely to respond with a proper output. All too often, functional control boards are misdiagnosed due to the failure of an input. When all the inputs are correct and arrive on time, the control board should provide the appropriate outputs.

Figure 1 Gas furnace printed-circuit control board.

Figure 2 shows the wiring diagram for a gas forced-air furnace. A connection diagram is shown on the left while a schematic diagram is shown on the right. Having both types of diagrams allows the technician to see the control structure from two points of view.

Highlights of the operating sequence for a gas-fired, induced-draft, forced-air furnace including the following:

- Heating operation is initiated by a call for heat from the room thermostat. The thermostat closes a connection between R and W. W initiates the heating mode in fossil-fuel equipment. The control board typically responds by starting the induced draft motor (IDM) through a board-mounted relay. The IDM may run a short time before any other action takes place, allowing time for it to purge the heat exchanger of any gases.
- Before the microprocessor ignites the burner, it will first verify that each safety control is in its correct position. For example, a microprocessor like the one in *Figure 2* must sense a completed circuit through the combustion safety controls. These safety controls often include temperature limit switches (LS1 and LS2), a flame rollout switch (FRS) that opens if flames roll out into the control vestibule, and an optional low gas pressure switch (LGPS). In addition, the presence of a draft in the vent (due to the IDM running) must be proven by a pressure switch that monitors the pressure differential between the heat exchanger and atmospheric pressure (PRS). This normally open (NO) pressure switch must be closed.

If these conditions are met, the microprocessor will ignite the burner:

- With the hot surface igniter (HSI) and gas valve energized, the microprocessor looks for a current signal from the flame rod to verify that a flame is present. If so, the ignition process ends while the burner continues operating, warming the heat exchanger.
    - If the microprocessor does not sense the presence of a flame, its logic usually allows for additional attempts to ignite the burner. After each attempt, the microprocessor imposes a short delay before ignition begins, giving the IDM time to purge the heat exchanger of any remaining fuel gas. Once the maximum programmed number of attempts has occurred without success, the microprocessor locks out and a fault indication is provided through a displayed code or a flashing LED. The system is then disabled until the power supply is manually reset. Note that systems using LP gas may have different lockout schemes, since LP vapors are heavier than air and tend to fall rather than readily rise out of the heat exchanger.

NOTE

One electronic furnace control on the market will attempt ignition as many as 33 times over a 15-minute period before it locks out. Other controls may only allow three or four attempts before a control lockout occurs.

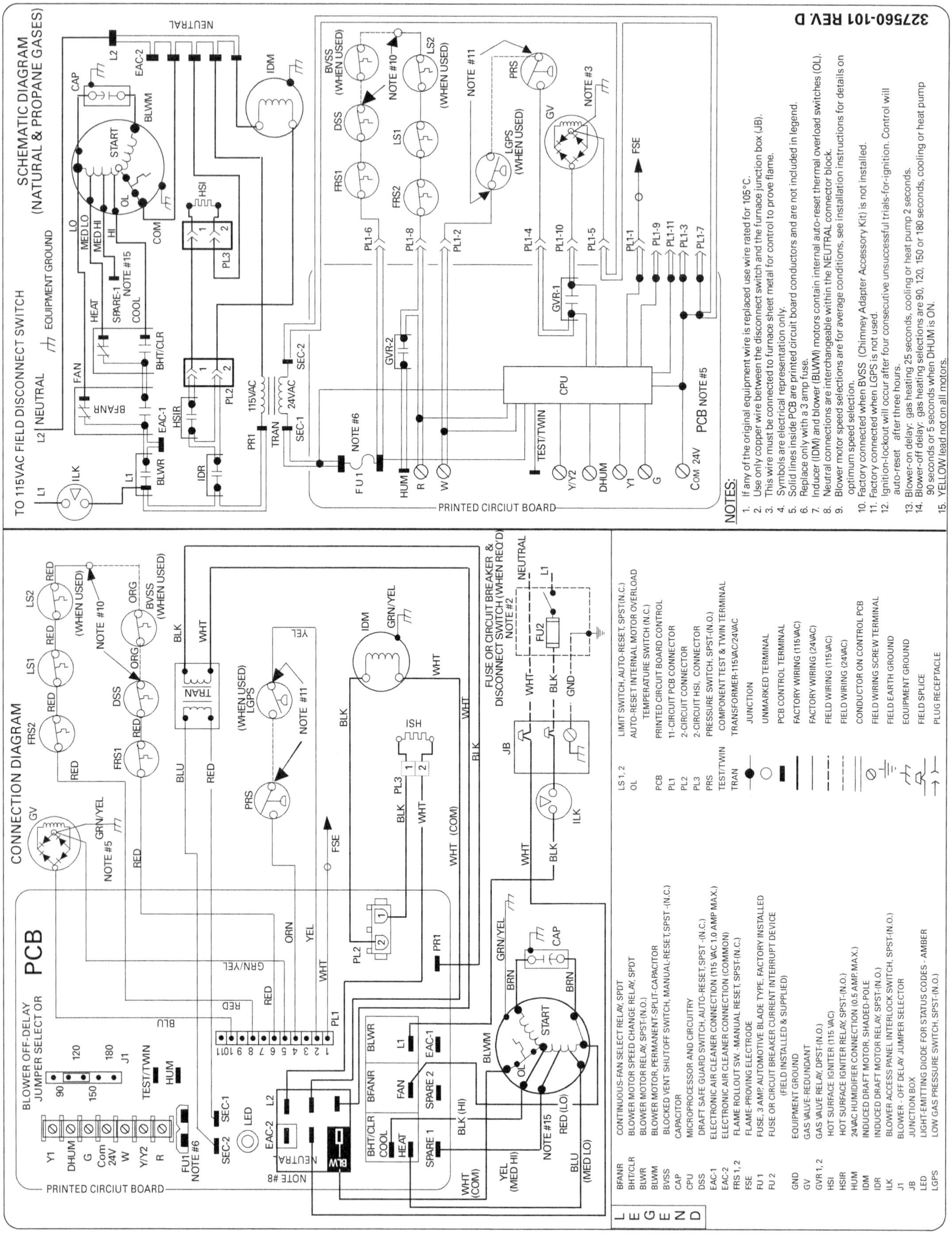

Figure 2 Gas furnace connection and schematic diagrams.
Source: Carrier Corporation

WARNING!

LP gas, also known as *propane*, is heavier than air. While natural gas tends to rise and dissipate easily, LP gas tends to collect in low places and remain there. Use extreme caution when dealing with LP gas leaks or when purging gas lines.

- Once the existence of a stable flame is proven, the microprocessor logic is ready to start the indoor blower (BLWM). However, before the blower starts, a short time delay allows the heat exchanger time to warm up. This delay helps prevent cool air from being circulated in the conditioned space. At the end of the delay, the indoor blower starts. Burner and blower operation continue until the thermostat is satisfied or a safety control opens.

The operating sequence just described is common to many current gas furnaces. Differences between systems are generally related to special features that can be programmed into the microprocessor. Some of these features include the following:

- Furnace blowers run for a period after the wall thermostat is satisfied and combustion has stopped. This allows the extraction of heat remaining in the heat exchanger. This also prevents the high-limit switch from opening due to residual heat. The off-delay period may be fixed, or the microprocessor may automatically lengthen the off-delay if the high-limit switch opens during the cool-down.
- Variable-speed blowers driven by electronically commutated motors (ECMs) are common in today's furnaces. ECMs are controlled by signals from the furnace's printed circuit board rather than directly through a simple blower relay.
- Most microprocessor-controlled systems can identify common faults. Many automatic diagnostics include a failure message to indicate that the control board itself has failed. This must be checked before the board is replaced because other problems, such as an intermittent power loss, might also cause this message.
- Some gas furnaces provide more than one stage of heating to maximize efficiency and comfort. The microprocessor in these systems may automatically adjust the length and ratio of the high-fire and low-fire cycles to optimize the multistage feature.

NOTE

Testing and troubleshooting ECMs is presented in NCCER Module 03318, *Motor Testing and Troubleshooting*.

Think About It

Circuit Diagram Interpretation Exercises

Test your knowledge of microprocessor-based heating controls with these exercises. Use the diagram in *Figure 2* as a reference.

1. Which component on the control board controls blower speed in the heating and cooling modes?
2. If the draft inducer (IDM) runs and the igniter (HSI) energizes, but the gas valve (GV) does not energize, what should you do next?

Integrated Diagnostic Aids

Electronic furnace controls (*Figure 3*) are usually equipped with a diagnostic aid in the form of multiple or single light-emitting diodes (LEDs). Controllers with a single LED only energize the LED when the ignition sequence is interrupted by a fault. The LED flashes a sequence corresponding to the type of failure detected. For example, for the furnace control shown in *Figure 3*, two flashes mean the pressure switch is stuck closed, while three flashes mean the pressure switch is stuck open. Diagnostic codes and light patterns differ among furnaces. A table showing the meaning of the diagnostic codes is usually posted on the wiring diagram or on the control enclosure.

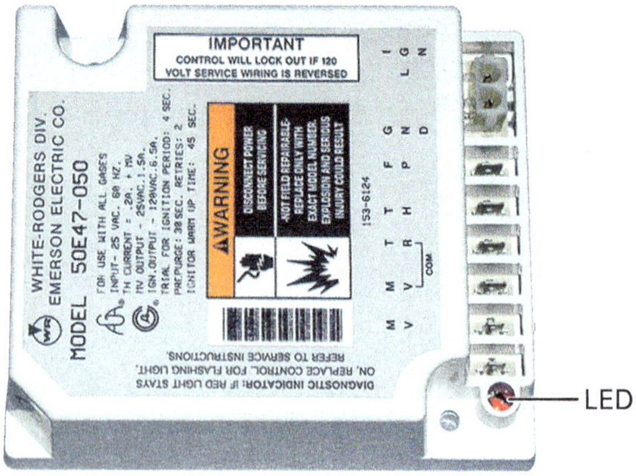

Diagnostic Indicator Flash Codes

1 Flash = System Lockouts (Retries of Lockouts Exceeded)
2 Flashes = Pressure Switch Stuck Closed
3 Flashes = Pressure Switch Stuck Open
4 Flashes = Open High-Temperature Limit Switch
5 Flashes = Flame Rollout Sensed
6 Flashes = 115 VAC Power Reversed/Improper Ground
7 Flashes = Low Flame Sensor Signal

Continuous Flash – Flame Sensed 0.5 Sec. without Gas Valve
Continuous On – Internal Control Failure

LED On = Failed Controller
LED Flashing = See Flash Codes
Controller with Minimal Diagnostics

Figure 3 Example diagnostic codes.
Source: Courtesy of Emerson Climate Technologies

Diagnostic codes are helpful, but codes cannot replace proper troubleshooting to verify the condition. For example, if the diagnostic codes indicate a pressure switch is stuck open, the problem could simply be a loose wire. The microprocessor has no means of determining why the circuit is open. Before condemning the pressure switch, check the wiring carefully and use your voltmeter to monitor the switch as the induced draft motor runs.

To properly diagnose an ignition sequence failure, the sequence must be observed in progress to identify at what point the sequence stops. Then, input and output voltages at the board can be monitored at the appropriate moments.

Did You Know?

Electronic Ignition Systems

The development of electronic ignition for gas furnaces was a major step for all furnaces, but it may have been even more significant for packaged units. Because packaged units are exposed to wind and have very short flue vents, standing pilot flames were often blown out. To restart the unit, the pilot flame would have to be relit manually.

1.1.2 Induced-Draft Furnaces

The 80% AFUE gas furnace shown in *Figure 4* is a typical induced-draft gas furnace. Ignition is accomplished with an HSI. The burner flame is monitored by a flame rod integrated into the HSI, and the operating sequence is managed by a sealed, solid-state controller containing miniature relays. This furnace is similar to the one represented by the wiring diagram in *Figure 3*.

A detailed sequence of operation for the furnace follows. Refer to *Figure 3* as you follow the sequence:

1. The furnace control board, labeled as the PCB, is powered through terminals Common and R. Power to the R terminal flows through the 3A fuse (FU1). A 24VAC signal from the room thermostat is applied to the W terminal of the control board when the thermostat heating contacts close.
2. The PCB performs a self-test and then verifies that the pressure switch (PRS) contacts are in the NO position before energizing the induced draft motor (IDM). The IDM will run for 15 seconds to purge the heat exchanger of any gases. The change in pressure in the heat exchanger causes the pressure switch PRS contacts to close. This proves that the induced draft motor is running and that a draft has been established through the heat exchanger.
3. At the end of the purge period, the HSI is energized. After the igniter warms up, the gas valve relay (GVR) contacts close, applying power to the gas valve (GV). The igniter ignites the gas/air mixture flowing to the burners.

(A) Top Half of Furnace

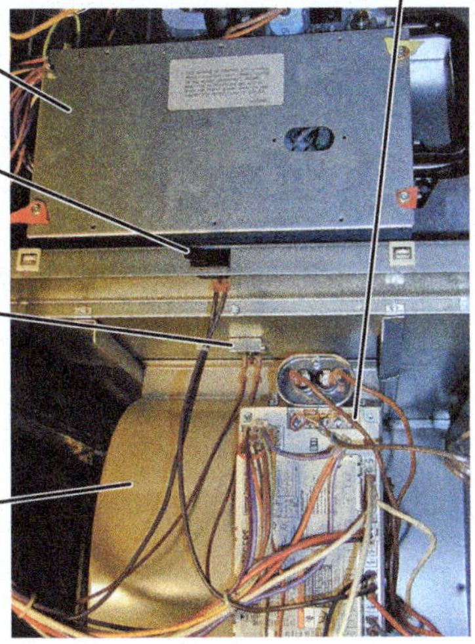

(B) Bottom Half of Furnace

Figure 4 Typical 80% AFUE induced-draft gas furnace.

4. The flame rod (FSE) has two seconds to prove that an adequate burner flame is present. The HSI is de-energized as soon as a flame is proven.
   - Failure to sense a flame will cause the control board to de-energize the gas valve and restart the ignition cycle. If additional attempts fail, the control board will eventually lock out and display a diagnostic code. Interrupting the line voltage to the furnace or the control power to the module will reset the control.
5. Assuming the burner flame is proven, the blower motor (BLWM) is energized 25 seconds after the gas valve opens. The blower motor operates at the selected HEAT speed. The motor speed for heating is field-selectable to meet the needs of the air distribution system.
6. Burner and blower operation continue until the thermostat is satisfied and the W terminal at the thermostat is de-energized. The gas valve closes, and the flame is extinguished. The induced draft motor will run for a short period to purge the remaining combustion byproducts from the heat exchanger. The indoor blower motor continues to run for up to 180 seconds. The blower-off delay time is field-selectable.

On Site

Modulating Gas Valves

Modulating gas valves can provide very precise levels of indoor comfort. These valves may be combined with a variable-speed blower motor that matches a specific indoor air volume to the gas valve position.

A modulating gas valve can typically modulate between 40% and 100% gas flow, with some able to modulate the flow in 1% increments. A microprocessor-based furnace control positions the gas valve based on values such as the indoor temperature and the current deviation from setpoint.

Modulating gas valves are positioned by either 0–10VDC or 4–20mA signals produced by the control board. Signals at the low end of either signal type close the valve; higher values cause it to open. These advanced valves also provide feedback to the board to indicate the valve's current position.

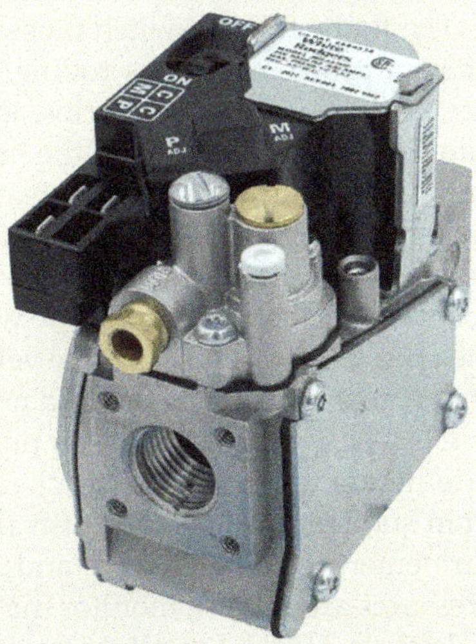

Source: Courtesy of White-Rodgers by Copeland

Limit Switch Activation

Limit switches are essential safety devices to prevent overheating. A flame rollout limit switch is positioned in the burner vestibule. Another limit switch monitors the temperature around the heat exchanger (*Figure 5*).

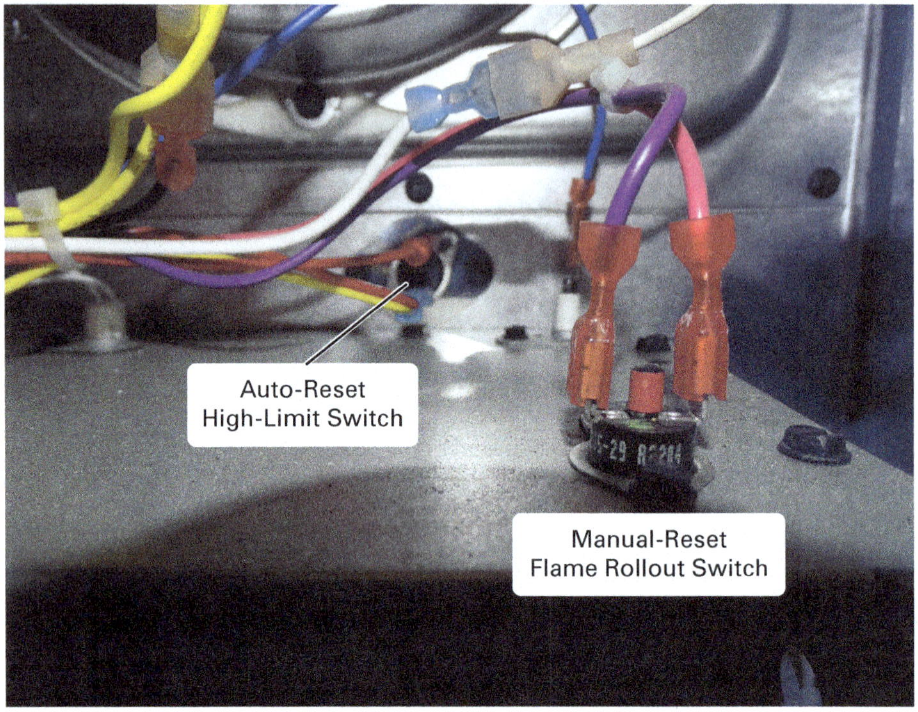

Figure 5 Induced-draft furnace limit switches.

Temperature rise: The difference between the return air temperature and the supply air temperature across the heat exchanger. The acceptable temperature rise for each furnace is listed on the furnace nameplate.

The primary high-limit switch, positioned in the heat exchanger, typically resets to the closed position automatically when it cools off. If the control opens repeatedly, check the **temperature rise** across the furnace. A higher-than-normal temperature rise can indicate blocked filters or some other obstruction in the duct. It may also be due to the blower speed being set too low, or too may registers and diffusers are closed. The switch may be defective, opening at a temperature below its factory setpoint, but failure is rare. Consider it the least likely problem until all other possibilities have been considered.

It is also important to note that a user won't always be aware that the high-limit switch is opening. Since it recloses automatically, the furnace will restart after the switch returns to its closed position. If the system maintains reasonable comfort, the user may not notice the cycles.

The flame rollout switch must be manually reset, as it represents a more significant problem. Flames rolling out into the combustion vestibule can start a fire. If the rollout switch trips, the cause must be corrected. Flames may roll into the furnace vestibule if the heat exchanger or vent is blocked, or if the draft inducer blower wheel is loose or deteriorated. However, the draft pressure switch (*Figure 6*) should prevent that from happening by proving the presence of a draft. If the draft inducer pressure switch has failed closed (another rare occurrence), the control module will not be aware of these conditions. Note that the pressure switches used are very sensitive and operate at very low pressures. Draft pressure switches should not be adjusted in the field.

Delayed ignition is another possible cause of flame rollout. If a gas/air mixture builds up in the heat exchanger before ignition takes place, the pressure from sudden ignition can push flames into the vestibule. Check the gas pressure entering the gas valve and in the manifold during operation. If the gas pressures are correct, monitor the ignition process carefully to determine why it is delayed. The HSI or pilot assembly may not be positioned correctly in relation to the gas flow into the combustion chamber. Check the burners for obstructions and the gas orifices for damage that would affect gas flow.

Figure 6 Draft pressure switch.
Source: HVACGuru

On Site

Door Interlock Switch

Door interlock switches have traditionally been used to disconnect power when the door of a cabinet containing electrical circuits is opened. This helps to minimize the potential for electrical shock.

On gas furnaces, the door interlock switch serves a second purpose: it prevents furnace operation if the blower door is removed. Ensuring the door is in place during furnace operation prevents harmful combustion byproducts from being drawn into the duct system by the blower.

During troubleshooting procedures, it is acceptable to temporarily tape the door interlock switch closed. Make sure the switch is working when the troubleshooting procedure is complete.

1.1.3 Condensing Furnaces

A condensing furnace (*Figure 7*) has both a primary and a secondary heat exchanger. The ability of these units to extract additional heat from flue gases produces AFUEs that reach as high as 98%. Because the secondary heat exchanger extracts most of the remaining heat from the flue gases, moisture containing acids condenses and collects on the secondary heat exchanger. This is an important factor, as it means condensate collection and disposal must happen year-round. The acidity of the condensate in the heating season complicates its disposal.

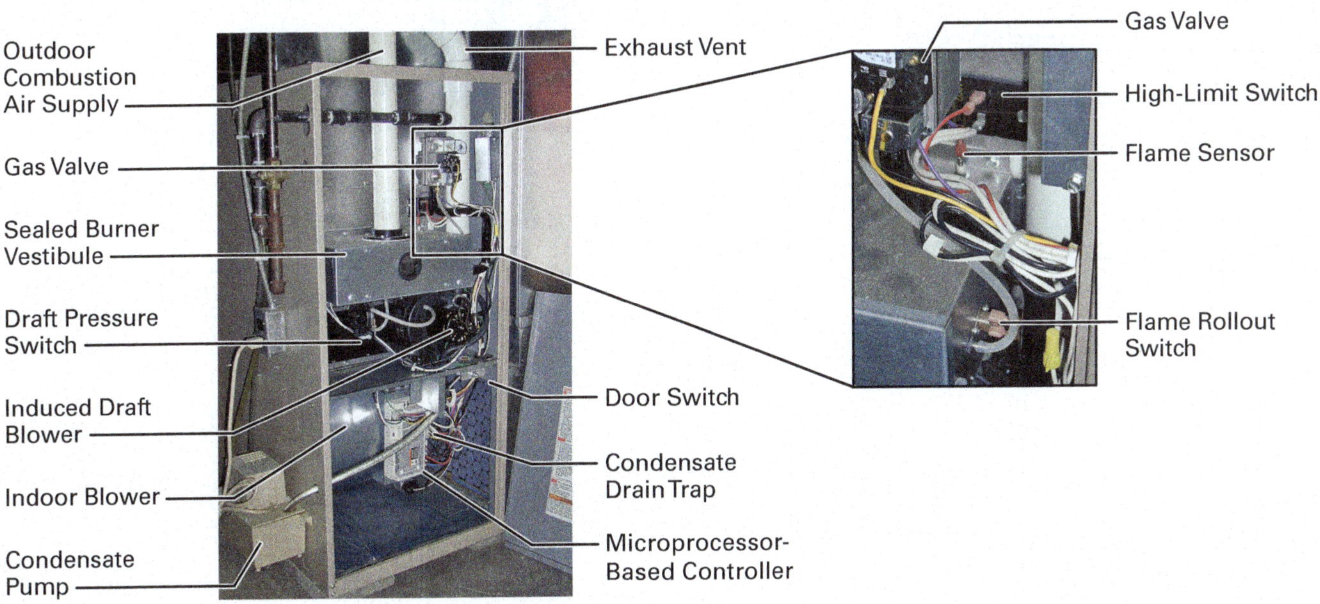

Figure 7 Typical 90% AFUE condensing gas furnace.

Acidic condensate is drained from the secondary heat exchanger and routed through a condensate drain trap to an approved drain, either directly or using a condensate pump. Some local codes require that the condensate pass through an acid neutralizer before it is routed to a drain.

The typical sequence of operation, along with the combustion devices, controller, and monitoring devices, is essentially the same as an induced-draft furnace. However, the condensate pump serving the furnace may be equipped with a float switch that opens if the pump fails to operate. The switch is wired into the furnace control circuit as an external safety device to shut down the furnace. This additional control would not be found in furnace control diagrams.

Condensing furnaces typically draw combustion air from outside the conditioned space. For this reason, the combustion chamber is sealed to avoid drawing room air into the combustion process. The flame characteristics are observed through an observation port.

Most manufacturers do not recommend operating condensing furnaces with return air temperatures below 55°F (13°C) for prolonged periods. Condensation may occur on the primary heat exchanger, causing it to corrode and fail prematurely. There is no provision to capture or collect condensate in the primary heat exchanger—only in the secondary heat exchanger. The owners of condensing furnaces should be instructed to set the room temperature thermostat no lower than 60°F (16°C) if the building will be unoccupied for an extended period.

1.1.4 Gas-Fired Boiler Operation

The sequence of operation for a gas-fired boiler differs from that of a forced-air furnace. A typical boiler control circuit is shown in *Figure 8*.

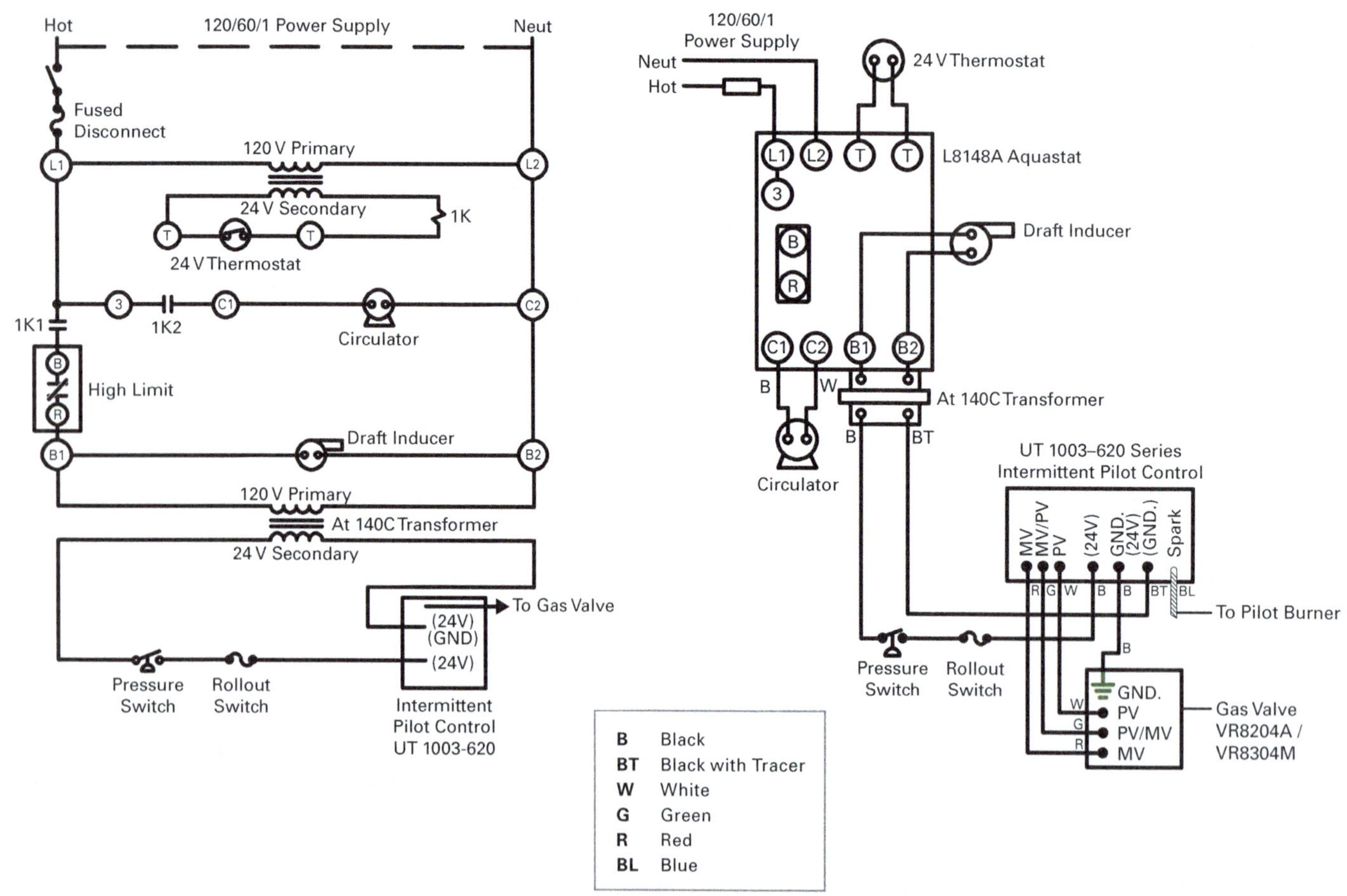

Figure 8 Gas-fired boiler control circuit.
Source: Carrier Corporation

The sequence of operation for the gas-fired boiler shown in *Figure 8* is described as follows:

1. A call for heat from the room thermostat closes the circuit between the two T terminals on the aquastat, energizing the 1K relay. What is not shown on the right-hand connection diagram is that there is a small control transformer inside the aquastat enclosure. However, it is represented in the ladder diagram; it is the transformer shown near the top of the ladder diagram. One leg of the control circuit is broken through the 24V thermostat shown. This completes the circuit to the 1K relay coil. Having both a connection diagram and a schematic or ladder diagram is very helpful.
2. The 1K relay contacts 1K1 and 1K2 close when the relay is energized by the thermostat. Assuming the high-limit switch is closed, 120VAC power is applied to the circulator pump and the draft inducer.
3. The pump and the draft inducer motor start. If there is no vent blockage, the draft pressure switch closes, indicating proper airflow through the combustion chamber. With the rollout safety switch also closed, 24VAC power is applied to the intermittent pilot control.
4. The intermittent pilot control opens the pilot gas valve and simultaneously generates a high-voltage arc to light the pilot flame.
5. When an adequate pilot flame is proven to the pilot control by a flame rod, the ignition arc stops.
6. The main gas valve opens, and the pilot flame ignites the main burners and operates until it satisfies the aquastat setting.
7. If the water temperature exceeds the high-limit setpoint, the line voltage circuit serving the primary of the main control transformer (labeled as AT-140C) is broken. The loss of control voltage from the transformer de-energizes all the combustion controls. The circulator will continue to run because the 1K relay remains energized. This also serves to cool the overheated water. When the water temperature drops below the high-limit setpoint, burner operation will resume, assuming the limit resets automatically and the room thermostat is still calling for heat. Otherwise, the high-limit control will require a manual reset.
8. If the draft is insufficient, the draft pressure switch will open and de-energize the pilot control and gas valve. The induced-draft fan remains powered as long as the room thermostat continues to call for heat and the high-limit switch contacts remain closed. If the pressure switch recloses, burner operation resumes. If the induced-draft fan is found to be running but no combustion is taking place, the draft pressure switch and rollout switch should be checked first.
9. When the room thermostat is satisfied, the 1K relay de-energizes, opening the 1K1 and 1K2 contacts to de-energize the burner, circulator pump, and draft inducer operation.

On Site

The Incredible Shrinking Furnace

As late as the 1980s, commonly available upflow gas furnaces were up to 60" (152 cm) tall. Due to improvements in heat exchanger technology and the use of induced draft systems, many manufacturers now offer upflow furnaces that are less than 36" (91 cm) tall.

1.2.0 Troubleshooting Combustion-Related Controls and Components

Some systems use spark igniters to light a pilot flame when the thermostat calls for heat. The pilot flame then ignites the main burner when the gas valve opens. *Pilotless systems*, which are common today, use either an HSI or a direct spark igniter to directly ignite the main burner. As the name implies, there is no pilot flame in a pilotless system.

A spark ignition system has a flame rod that provides a signal to prove a flame has been established whenever the pilot or burner lights. The flame rod provides the signal through **flame rectification**. In essence, the flame rod

Flame rectification: A process in which a small alternating current is rectified to a direct current through the exposure of a sensing rod to a flame, resulting in a signal that proves the existence of a flame.

(*Figure 9*) ensures that the burners do not continue to receive fuel if a flame isn't established quickly. This process involves rectifying an AC current sent to the flame rod into a DC current. The current produced is very small, measured in microamps (μA). The flame itself acts as a conductor, allowing the current to pass through it.

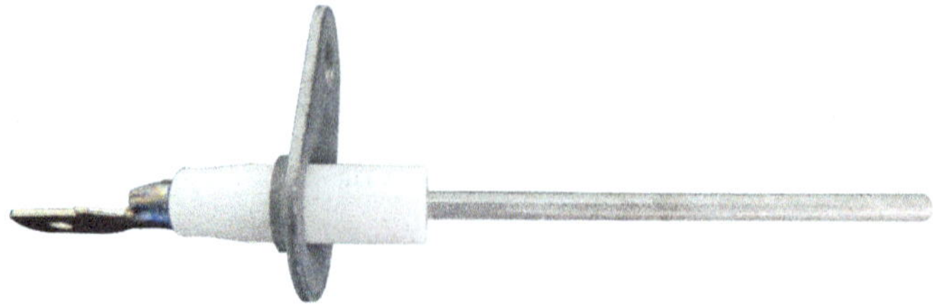

Figure 9 Typical flame rod.

Some ignition assemblies use the same rod that is used for the ignition arc as a flame rod. Others, especially systems that rely on a pilot flame, have a separate flame rod. When an HSI is used, the flame rod may stand alone and be visible in the burner area. However, some HSIs incorporate the flame rod into the HSI assembly. Regardless of the arrangement, if the flame rod is not producing the appropriate current, the gas supply will shut off. The current produced is usually in the range of 2 to 6μA.

When troubleshooting spark igniters or HSIs in an electronically controlled system, keep in mind that the system will lock out the ignition circuit if it fails to establish and maintain a flame after some number of attempts. Usually, the control or power source must be cycled off and on to reset the controls.

Remember that ignition devices require a reliable ground connection to operate. If the ignition controls are poorly grounded, an ignition arc can't be established.

1.2.1 Spark Ignition Systems

Spark ignition systems may rely on a pilot flame or light the burner directly. *Figure 10* shows a typical spark-ignition pilot assembly. You should be able to see and hear when the spark igniter is working. If you can hear the spark, but the gas is not igniting, the ceramic insulator surrounding the electrode may be cracked. This allows the spark to be misdirected toward another grounded surface.

Figure 10 Typical spark-ignition pilot assembly.
Source: BostonHeatingSupply.com

Misalignment may also be causing the ignition arc to jump across a shorter gap to another grounded surface where it isn't properly exposed to the gas-air mixture. If there is no spark at all, the spark gap may need to be adjusted according to the manufacturer's specifications. If the gap is correct but there is still no spark, or the spark is weak or intermittent, the problem is most likely in the ignition module or the high-voltage ignition wire. For example, a cut in the ignition wire will allow it to arc to any nearby grounded surface.

Ignition control testing must be done carefully due to the high voltage produced by the igniter. With the furnace de-energized, disconnect the ignition cable from the pilot assembly. While holding the insulated portion of the cable with a pair of properly insulated pliers, hold it within $\frac{1}{4}$" of a paint-free, grounded surface on the furnace cabinet. Power the ignition control and see if an arc is produced.

WARNING!

Take precautions when troubleshooting spark igniters. The high-voltage arc operates in the range of 10,000 to 20,000 volts to allow it to cross a substantial gap. Do not touch the spark igniter with your fingers or a screwdriver or attempt to disconnect the igniter without first de-energizing the control.

Also inspect the ignition lead. This wire is much like a spark plug ignition wire in an automobile. It is very flexible and heavily insulated, and often has a boot around the connection at the ignition control. Ensure the connection at each end is sound and that the insulation is intact. If you suspect a break, test the wire for continuity with an ohmmeter.

If there is a strong spark but no pilot flame, the pilot orifice may be blocked. During the summer months, small insects may build webs or cocoons around the orifice. The coil of the pilot valve, part of the main gas valve, may have failed as well. However, gas valves are one of the most reliable components in the furnace.

1.2.2 Hot Surface Ignition Systems

HSIs (*Figure 11*) may wear out or crack after several years of use. Many HSIs are constructed of a material called *silicon carbide*. When replacing an HSI, consider replacing it with a more durable and reliable type made from *silicon nitride*. Silicon nitride HSIs tend to last longer than silicon carbide and heat up more quickly. However, the product must be compatible with the furnace and the ignition control module.

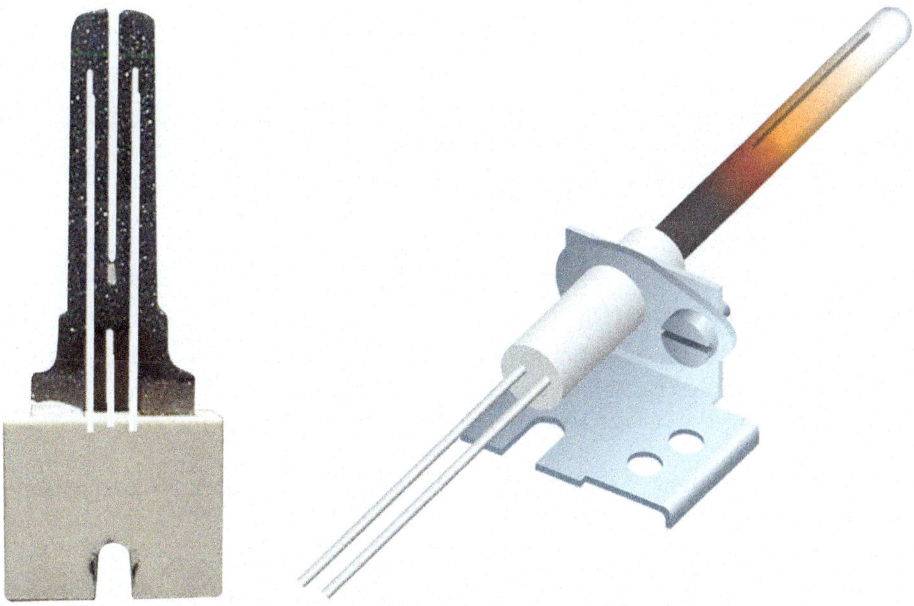

(A) Silicon Carbide Ignitor

(B) Silicon Nitride Ignitor

Figure 11 Hot surface igniters (HSIs).
Source: (B) Courtesy of Emerson Climate Technologies

HSIs have a high electrical resistance. When electrical power is applied, the HSIs heat up rapidly, becoming very hot and glowing red (*Figure 12*). Most HSIs have a specified time required to reach the target ignition temperature. The warmup time generally ranges from 10 to 20 seconds. The silicon-based material is very brittle and can be cracked or broken if handled carelessly. If a bright white line is visible anywhere on an operating HSI element, the element is cracked, but it hasn't failed completely. Replace a cracked HSI to ensure reliability.

Figure 12 HSI energized.
Source: JJ Gouin/iStock/Getty Images

If an HSI fails to glow during an ignition sequence and the required preceding have occurred, then one or more of the following is likely:

- No power is being supplied to the HSI.
- The HSI element is open (no continuity).
- The resistance of the HSI is insufficient to heat it to the needed temperature.

WARNING!

Testing an HSI involves measuring 120VAC power. Use caution to avoid contacting the HSI element, any live wires, or bare electrical contacts when making voltage measurements. Make sure the 120VAC power to the equipment is turned off before disconnecting and connecting wiring.

Testing for Power to the HSI

Observe the HSI element without removing it from the burner section. If the element is not visibly cracked or damaged, monitor the 120VAC power applied to the HSI during the ignition sequence. This can be done at the connector as shown in *Figure 13*.

Cycle the equipment through the ignition sequence while monitoring the HSI voltage. If the proper voltage is detected, the HSI itself will be tested next. If multiple failures of the HSI have occurred, make sure that the voltage does not exceed 125VAC. Excessive voltage can significantly reduce the life of an HSI.

If no power is found at the inline connector, check the required inputs to the ignition control. If the inputs are present but the control will not energize the HSI, the ignition control is likely defective.

Testing the HSI

The HSI can be tested for continuity and to determine its resistance value. If power is applied but it does not glow at all, it is likely open.

The HSI may glow faintly but fail to reach a temperature hot enough to ignite the gas. This often indicates the resistance of the HSI has dropped below its required range. The resistance may begin to drop after many cycles. The resistance is measured with an ohmmeter and must conform to the manufacturer's

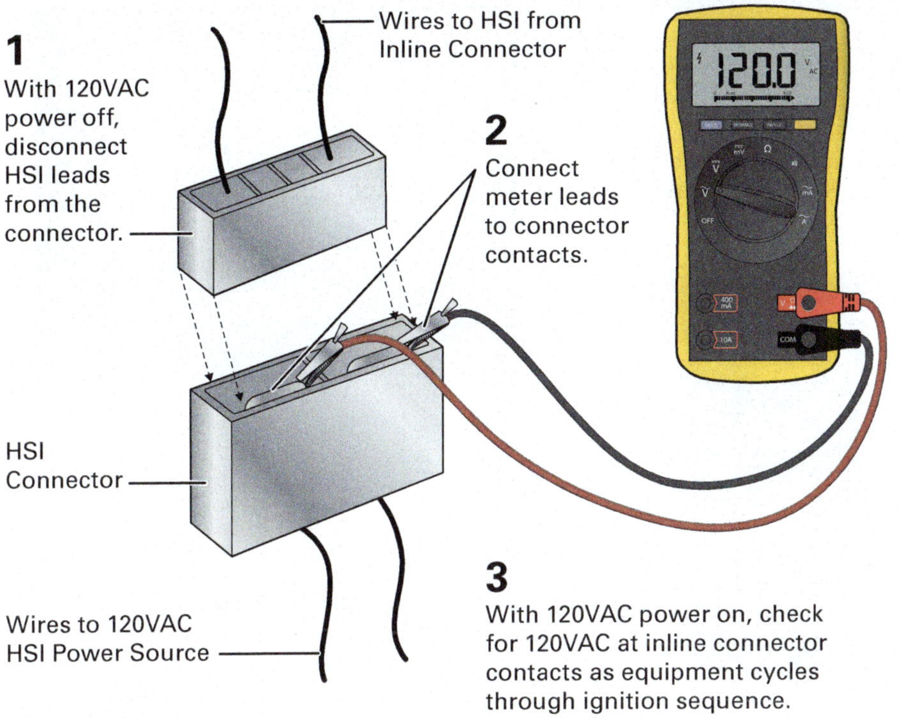

Figure 13 Checking HSI source voltage.

specifications. If there is no continuity measured, the HSI is clearly defective and must be replaced.

A common acceptable resistance range is 40–100 ohms (Ω) at room temperature, but do not rely on these values. The measured resistance is highly dependent on the HSI temperature. To determine the condition of the HSI, you will need to know the required resistance value when it is cold. *Cold* in this case refers to room temperature. A hot HSI will have a much higher resistance than one at room temperature. Thus, testing the resistance while it is hot or even warm cannot be done unless you have a temperature/resistance chart for that specific HSI. The required resistance varies among products, so it is essential to have the right chart. If the resistance is outside the required range, replace the HSI.

Note that some sophisticated ignition controls monitor the HSI and will not energize the gas valve unless the resistance is within the proper range.

1.2.3 Flame Rod Testing

The flame rod may be mounted on the pilot assembly (if equipped) or mounted along the front of the heat exchanger where it can sense the burner flame. The furnace ignition control senses the tiny DC current it produces.

Figure 14 shows how to test a flame rod. A multimeter capable of reading DC microamps must be connected in series between the flame rod and the ignition control. With the power restored, the thermostat is set to call for heat. Once the pilot flame or the burners ignite, the current reading should be visible on the meter display. A common range for the DC signal is 2–6μA. However, refer to the manufacturer's specifications for the correct value.

If the current reading is absent or too low, it could be caused by one of the following problems:

- The ignition module is not properly grounded. Note that the DC signal current travels back to the ignition control through the ground connection.
- The flame rod is poorly positioned in the flame.
- The flame rod has developed oxidation.
- The ignition control is defective.

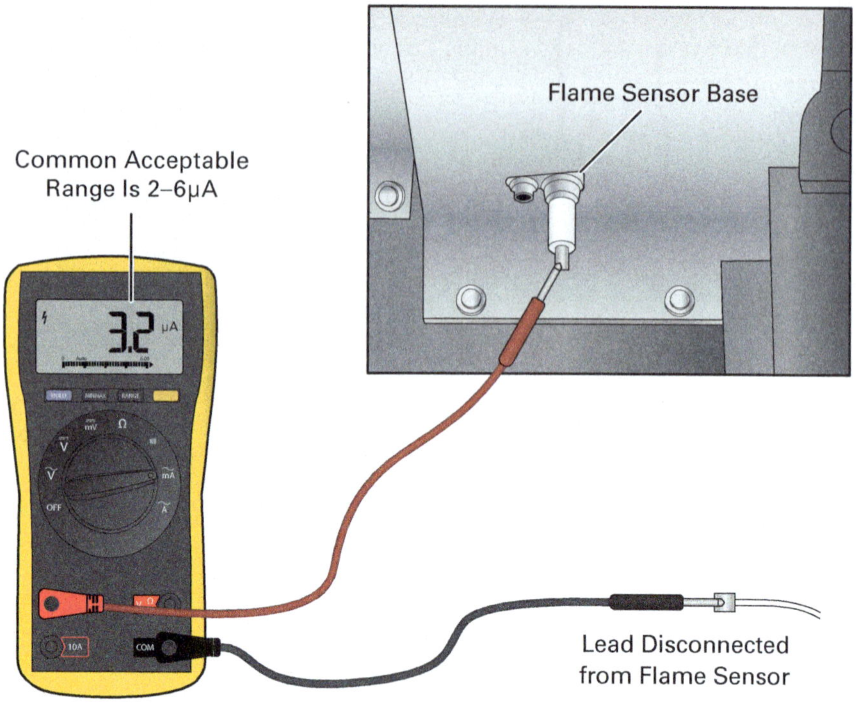

Figure 14 Flame rod test.

Ensure that the ground connections are sound. Consult the manufacturer's specifications for the proper position of the flame rod.

Over time, oxidation can collect on the surface of the flame rod, impeding current flow. Cleaning the rod with fine steel wool removes the oxidation.

Flame rods are quite durable and rarely fail. Remember that the ignition control must send an AC signal to the rod, which is then rectified to DC. Without that output, the flame rod can't conduct the signal current.

1.2.4 HSI-Integrated Flame Rod Testing

The flame rod is often integrated with an HSI in direct-ignition systems. The ignition control changes its outputs to the HSI as ignition progresses. Power is sent to the igniter for a prescribed time to ignite the flame. After a set time delay to allow a flame to be established, the control opens the circuit to one of the leads, then sends the AC signal to be rectified through the remaining lead.

As the HSI ages, oxidation forms on the surface of the element. This is normal, but it reduces the strength of the flame-proving DC signal to the controller. If the burner ignites, but immediately or intermittently extinguishes before the thermostat is satisfied, a weak or fluctuating flame signal is the most likely cause. Replace the HSI if it appears to be excessively oxidized.

Whenever problems with the flame-proving system exist, check the quality of the ground circuit. Ensure there is steady continuity between the burners, the flame carryover tube, the equipment cabinet, and ground. Also ensure the HSI is positioned so that it is well immersed in the flame. The tip should be fully engulfed in the flame, protruding in about $\frac{3}{4}$" to 1".

1.2.5 Induced Draft Systems

Induced-draft furnaces have more restrictive heat exchangers that extract more heat from the flue gases. The induced-draft fan overcomes the friction losses, so the byproducts of combustion are properly vented.

The induced-draft fan is primarily designed to overcome the restriction of the heat exchanger alone. Thus, the flue gases exiting the heat exchanger should be at or near a neutral pressure. Natural drafts established in the vent then assist the travel of flue gases through the vent piping. However, in condensing furnaces, the induced-draft fan does create a positive pressure in the vent piping. Remember that condensing furnaces have a smaller vent size, and the vent lines often travel horizontally. In addition, the combustion air is drawn in through a second pipe. More mechanical assistance is therefore required to ensure the needed volume of air moves through the heat exchanger and vent system.

In most units, the first event that occurs after a call for heat is the startup of the induced-draft fan. A pressure switch is typically used to prove that the inducer is running and producing a sufficient pressure differential. If the inducer runs but the pressure switch doesn't close, look for restrictions in the combustion air supply or vent piping. Also ensure that the induced-draft fan blower wheel is intact and turning at the correct speed.

Due to the high level of sensitivity, it is best to avoid making field adjustments to the draft pressure switch. Although many pressure switches appear to be adjustable, they are typically factory preset and carefully calibrated. This is often indicated by a small drop of colored cement or paint placed on the threads of the adjusting screw. Field adjustment of the switch may result in unsafe operation. If a pressure switch is not operating properly, it is best to replace it with a manufacturer-provided part with the correct setting.

1.0.0 Section Review

1. If the microprocessor of an ignition control does not sense the presence of a flame and makes additional attempts to ignite the burner, for what reason is a short time delay applied before the process restarts?
   a. To allow the microprocessor time to cool off.
   b. To allow the HSI or ignition electrode time to cool off.
   c. To allow the gas pressure in the manifold to return to normal.
   d. To allow the induced-draft fan to purge any fuel gas from the heat exchanger.

2. What type of feedback does an ignition control receive from a flame rod?
   a. Filtered AC voltage
   b. Rectified DC current
   c. Variable 4–20mA signal
   d. Variable 0–20VDC signal

2.0.0 Infrared Gas-Fired Heaters

Objective

Identify and outline the operation of various infrared gas heaters.

a. Identify various types of infrared gas heaters.
b. Outline the unique operating characteristics of infrared gas heaters.

Performance Tasks

There are no Performance Tasks in this section.

Infrared heaters are used in buildings with a high volume of air exchange due to consistent external door operation and in high-bay buildings where forced-air heating is impractical. Infrared heaters heat objects rather than air.

Standard forced-air heating systems heat air that then moves about a space and warms the people and objects. When loading doors such as those found in a warehouse are opened, much of the heated air escapes and the process must begin again. In a high-bay facility, the heated air rises and remains near the ceiling. It is difficult to encourage this air to move down to the occupants.

Since infrared heaters heat nearby objects, little heat is lost when a large door is opened, as the objects continue to absorb and radiate heat back to the space. Infrared heaters are ideal for commercial applications in factories, foundries, large workshops, vehicle repair garages, warehouses, aircraft hangers, and sports arenas. Smaller versions are available for use in residential and light commercial applications. Even when cold air rushes into the space through open doors, the occupants can still feel the warmth from infrared heaters. You will also find these heaters in outdoor spaces such as patios and dining areas.

Remember that the effectiveness of infrared heat relies on a "line-of-sight" principle. People and objects must be within view of the heater to feel the effects.

2.1.0 Infrared Heater Types

Infrared heaters are identified as either *low-intensity* or *high-intensity*. Low-intensity heaters have a long steel tube that acts as the combustion chamber and heat exchanger. These heaters usually require a draft inducer to overcome the resistance of the tube. Low-intensity heaters are typically vented outside the space, like traditional furnaces.

High-intensity heaters have been in use since the 1950s. These heaters are usually mounted high on a ceiling, use an open flame to heat a ceramic surface, and are not typically vented. In other words, the byproducts of combustion are released in the space. Due to the large volume of air inside the space and the rate of fresh air exchange, the combustion byproducts are not considered a problem.

2.1.1 Low-Intensity Infrared Heaters

Figure 15 shows typical gas-fired, tubular, low-intensity infrared heaters. The tubes can be straight, L-shaped, or U-shaped to provide more heating capacity and coverage in the same space.

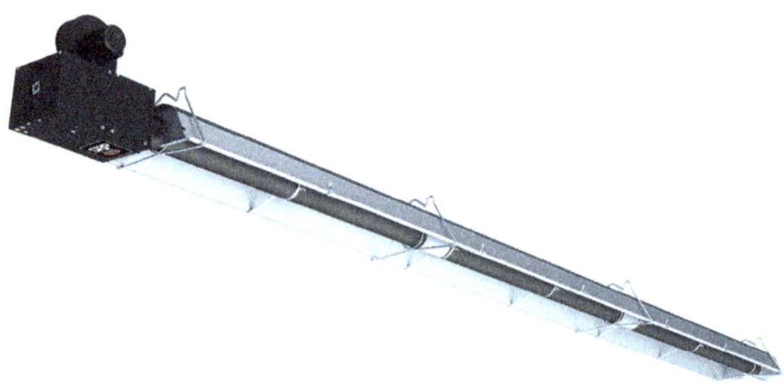

(A) Straight Configuration

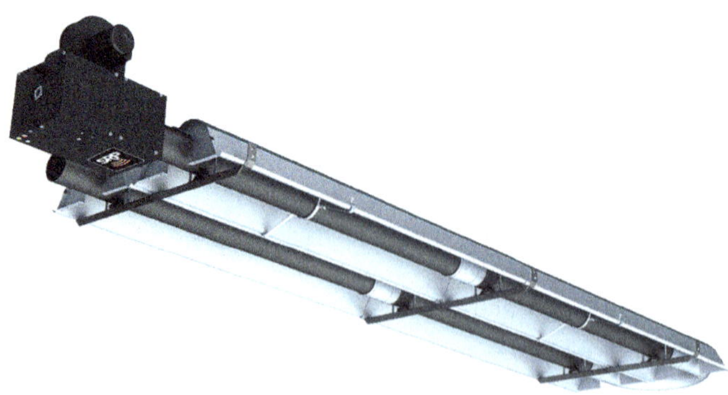

(B) U-Shaped Configuration

Figure 15 Low-intensity infrared heaters.
Source: Superior Radiant Products

These low-intensity units are mounted relatively high in a building. The gas flame is directed from a burner assembly through a heat-treated aluminum or steel tube, heating the tube. The tube's hot outer surface produces infrared radiant energy focused by a polished aluminum or stainless-steel reflector onto the floor, occupants, and other surfaces beneath the unit. The effect is similar to the sun's infrared rays. This creates a comfortable environment free of drafts.

The design of the reflector is very important. Ideally, the reflector will reflect the infrared light down and away from the heated tube. Any light reflected toward the tube leads to inefficiency.

Although some low-intensity tubular heaters are available in a non-vented design, many require venting (*Figure 16*). Most models also require some form of mechanical assistance for venting (induced draft) due to the resistance of the long tubular heat exchanger.

Figure 16 Vented low-intensity heaters.
Source: Courtesy of Detroit Radiant Products Company

If more than one heater needs to be connected to a single vent, a negative pressure can be created by a single vacuum pump located at the vent's exit to the outdoors. Think of the vacuum pump as an induced-draft fan placed at the end of the tubing run. Multiple burners can be situated in the system layout, all depending on the single vacuum pump to create the draft. The burners can be vented vertically through the roof or horizontally through a wall. Thus, infrared heating systems can be classified as either positive or negative pressure systems. Negative-pressure systems allow for complex layouts in a variety of patterns.

In a non-vented unit, combustion byproducts are discharged into the heated space, adding additional heat. Non-vented heaters are permitted in areas with a high infiltration rate, meaning there is a lot of fresh air consistently entering the space. Building exhaust fans can also be installed and operated to dramatically increase the rate of fresh air entry.

The burner assembly operates in much the same way as the burner in an induced-draft gas furnace. The ignition and safety devices used in low-intensity infrared gas-fired heaters are similar to those used in induced-draft furnaces but rearranged to suit the application. Although true condensing infrared heaters aren't practical, some models are still efficient enough to use PVC or special plastic venting materials.

Burner capacity modulation is also available, allowing the heaters to track the heating demand more accurately. In cold weather, cycling the units on and off is very noticeable because the air is not being heated. Thus, when the units cycle off, it is much like a warm sun disappearing behind a cloud on a cold day. Modulation allows more consistent operation without overheating the space or its occupants.

2.1.2 High-Intensity Infrared Heaters

Reverberator: A wire or corrugated metal screen used to cover the ceramic grids in a gas-fired radiant heater to boost the radiant heat output by converting convective heat to radiant heat.

Non-vented, high-intensity heating units (*Figure 17*) are typically used as spot or space heaters. The units operate by burning gas between two porous ceramic grids. The grids warm to about 1,600°F (870°C) and produce intense infrared radiant energy. In some cases, the grids are covered by an optional wire screen, called a **reverberator**, which increases the infrared radiation.

WARNING!

By design, infrared heaters have exposed surfaces that reach extremely high temperatures. To avoid serious personal injury, never touch an infrared heater assembly before checking the temperature of the surface.

Figure 17 High-intensity infrared heater.
Source: Superior Radiant Products

High-intensity units are often mounted at a specified angle (*Figure 18*). Polished reflectors around the grids focus the radiant energy to the desired location. Like non-vented, low-intensity units, a large volume of fresh air must move through the space to dilute the byproducts of combustion. These units operate more like traditional natural-draft gas furnaces, with no need for an induced-draft fan.

Figure 18 High-intensity heater serving an ice hockey arena.
Source: Courtesy of Detroit Radiant Products Company

On Site

Electric Infrared Heaters

Electric infrared heaters are popular for heating occupants in small indoor and protected outdoor areas. These heaters offer an option when gas and combustion ventilation are not available or practical. Electric infrared heaters produce quiet and clean heat in a compact package. They can be portable, suspended from a ceiling, or mounted on a wall. Electric units may even be used in automative painting and detailing applications.

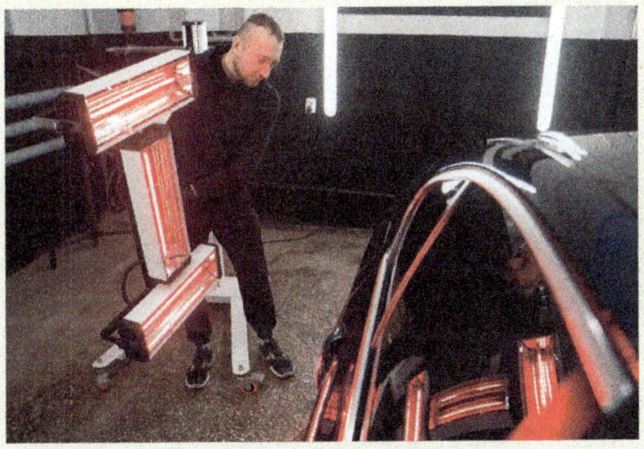

Source: Andrii Borodai/iStock/Getty Images

2.2.0 Infrared Heater Operation

Space temperature control with gas-fired infrared heaters differs somewhat from conventional temperature control. A simple On/Off switch can often provide adequate control. For example, a worker at a desk on an unheated loading dock can turn the heat on when the area is occupied, then turn off the heat when the occupants leave. Since these units do not heat air, trying to maintain a specific air temperature is not always practical.

Low-voltage and line-voltage thermostats are often used with infrared heaters, providing some semblance of control. However, in practice, the thermostats are often used like an On/Off switch.

Figure 19 shows a wiring diagram for a single-stage, low-intensity tubular heater controlled by a thermostat. The sequence of operation is as follows:

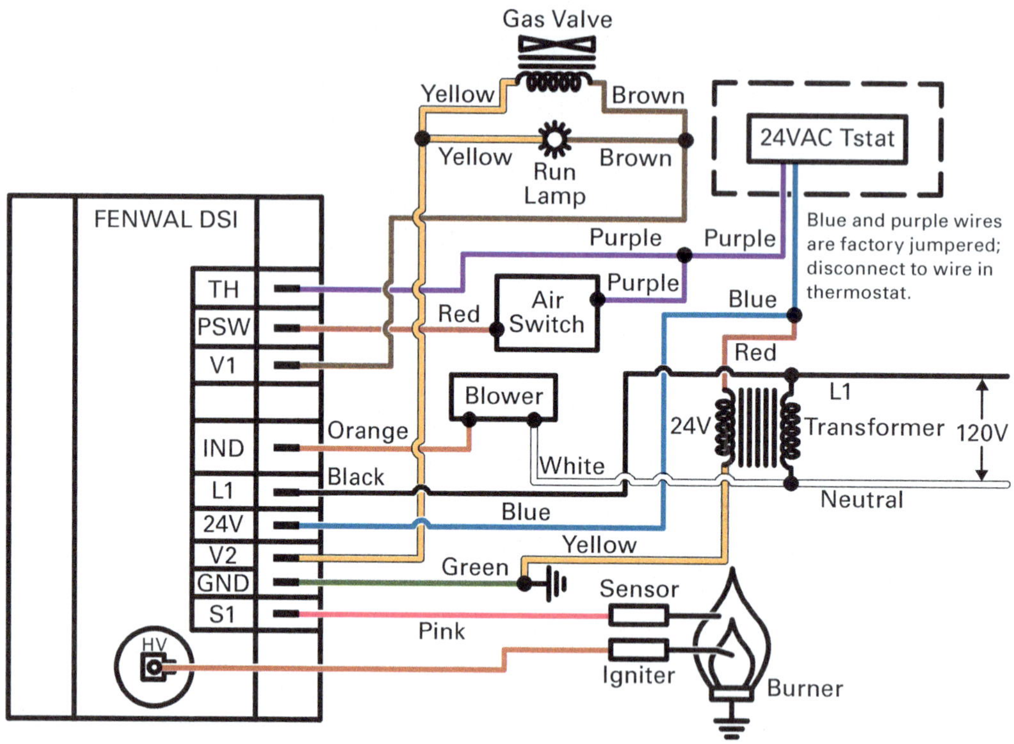

(A) 24VAC Thermostat

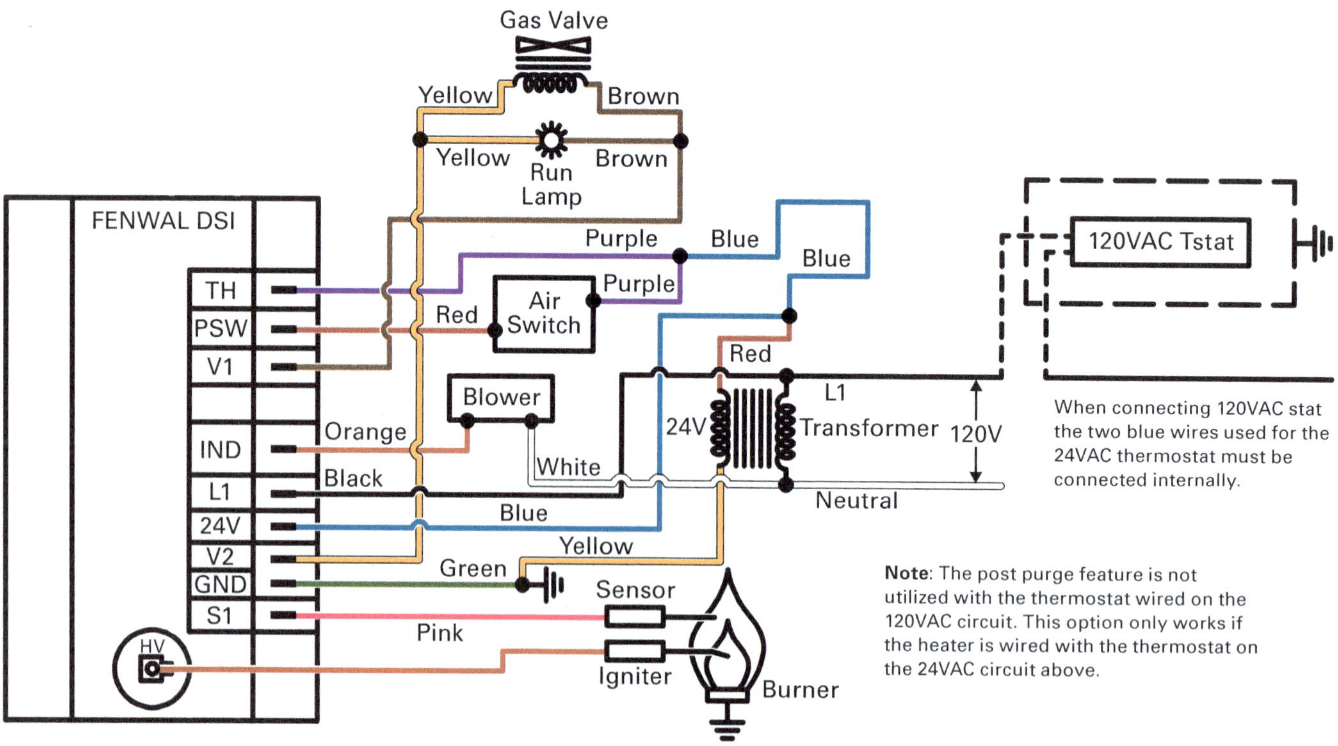

(B) 120VAC Thermostat

Figure 19 Low-intensity infrared heater wiring diagram.

Source: Superior Radiant Products

WARNING!

Infrared heaters are typically mounted high in an area. Maintenance or repair of these units will require the use of a ladder, scaffold, or suitable personnel lift. To prevent serious personal injury, remember to use a personal fall arrest system (PFAS).

1. On a call-for-heat, the induced-draft fan (BLOWER) is energized by the ignition control. When it reaches operating speed, the pressure switch (AIR SWITCH) closes and to prove a draft has been established.
2. After a 30-second pre-ignition purge by the induced-draft fan, the ignition module energizes the igniter electrode. Simultaneously, the gas valve is energized for about 15 seconds to try and establish a flame. The pre-ignition purge ensures that no unburned gas is in the heat exchanger that could explode on ignition.
3. Once burner ignition occurs, the flame rod (SENSOR) rectifies the signal applied and returns the tiny current that proves ignition. This signal to the ignition module allows the gas valve to remain energized and the igniter electrode is de-energized.
4. If no flame signal is returned, the gas valve closes. Before another ignition trial begins, a 30-second purge is used to clear any unburned gas from the tube. Following the purge, another ignition trial begins. After three trials without success, the module locks out for roughly one hour. After the hour passes, the module will repeat the three-attempt cycle. The one-hour delay can be circumvented by cycling power to the burner off and back on again. To ensure the ignition module resets, the power must remain off for a minimum of five seconds, allowing its capacitors time to fully discharge.
5. If a 24VAC thermostat is used for temperature control, the induced-draft fan will run for 30 seconds after the thermostat is satisfied to purge combustion byproducts from the unit. With this specific product, this feature is de-activated when a line-voltage thermostat is used for temperature control.

NOTE

For this specific piece of equipment, when LP gas is used as the fuel, only one ignition trial is allowed. The ignition module locks out after the first failed attempt.

Although this sequence of operation applies to only one specific single-stage infrared heater, all similar models will operate in a similar manner. More sophisticated units can provide staged or modulated heating capacity.

Multiple infrared heaters may be controlled from a central multizone control panel (*Figure 20*). Panels typically include digitally programmed time switches that can be set up according to a variety of heater control scenarios. The controls are designed to integrate with popular communication protocols used with building management systems and wireless connectivity. The controls also provide real-time equipment monitoring and inputs from web-enabled devices.

Figure 20 Multizone infrared-heating control panel.
Source: Superior Radiant Products

2.0.0 Section Review

1. The ceramic grid in a high-intensity infrared heater operates at a temperature of about _____.
   a. 500°F (260°C)
   b. 1,200°F (650°C)
   c. 1,600°F (870°C)
   d. 2,000°F (1,090°C)

2. What purpose does a pre-ignition purge period serve?
   a. It proves that the draft pressure switch is working.
   b. It ensures no unburned gas is in the heat exchanger.
   c. It provides time for an HSI to warm up to the ignition temperature.
   d. It allows time for the flame rectification circuit to develop and return a signal.

3.0.0 Gas Combustion Analysis

Performance Task	Objective
2. Complete a combustion analysis on a gas furnace or gas boiler.	Explain how to conduct a combustion analysis on a gas furnace. a. Identify relevant combustion byproducts and describe combustion analysis equipment. b. Explain how to conduct a combustion analysis and how to interpret the results.

Figure 21 Inshot gas burner.
Source: pricelessphoto/iStock/Getty Images

A combustion analysis on a gas furnace or boiler can determine how efficiently the furnace combustion process is operating and ensure that the most harmful combustion byproducts are minimized. Given a sufficient supply of combustion air, a good draft, and the proper gas pressure, today's inshot burners (*Figure 21*) are virtually foolproof. Inshot burner designs offer little or no means of adjustment that affects the combustion process. However, combustion analysis is still a valuable tool that can reveal the presence of a combustion problem and provides clients with peace of mind. In other words, the analysis results are more informative than diagnostic in nature.

A combustion analysis is done by measuring the flue gas temperature and quantifying the gases it contains using specialized instruments. These figures are then used by the instrument to calculate combustion efficiency.

Use the listed AFUE of the equipment as a guide for expected efficiency. However, expect the combustion efficiency results on new and properly tuned equipment to be slightly lower than the AFUE number, which is determined under specific laboratory operating conditions.

The terms *AFUE* and *combustion efficiency* are terms that are often used interchangeably. While similar, the meanings do differ. When the AFUE is determined for a furnace or boiler, the product is tested in a laboratory where conditions are tightly controlled, and the furnace is cycled on and off under part-load conditions. A field combustion analysis is done under real-world conditions with many variables that can change the results, such as altitude, heat content of the local gas supply, and combustion air temperature.

For this reason, it is unlikely that combustion analysis results will match the AFUE rating of the furnace. Many field tests have shown that combustion analysis results are lower than the posted AFUE of a product, but the field results may also be higher. However, the slight difference between the numbers does not represent a problem.

3.1.0 Combustion Byproducts

Air is composed of roughly 21% oxygen (O_2), 78% nitrogen (N), and 1% other gases (*Figure 22*). In complete combustion, the O_2 in the air combines with hydrocarbons in the fuel to produce the combustion products carbon dioxide

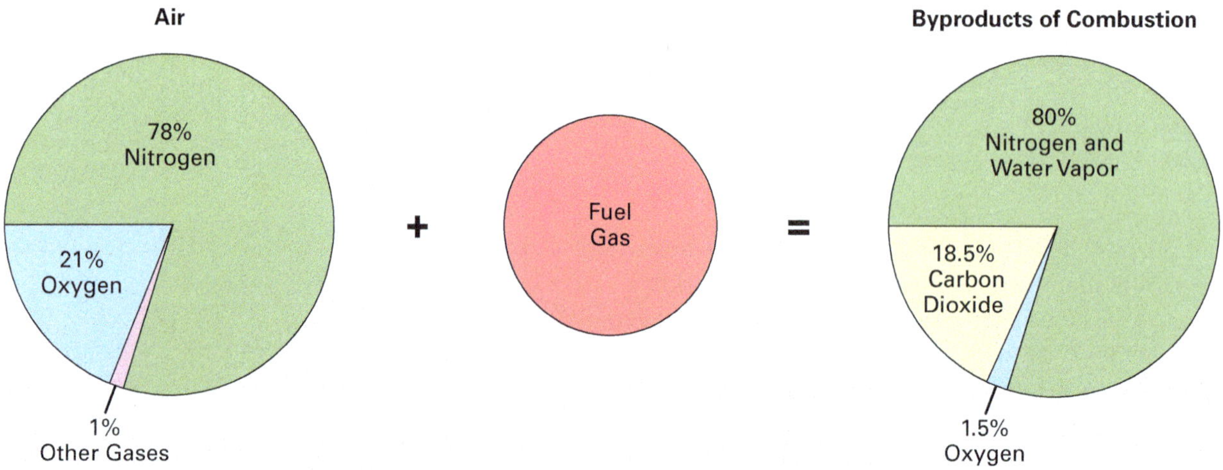

Figure 22 Byproducts of combustion.

(CO_2) and water vapor (H_2O). In fact, the byproducts of combustion from a furnace also contain some O_2 because more air is supplied to the combustion process than is consumed. This excess air helps minimize the development of toxic carbon monoxide (CO).

To summarize, the byproducts of combustion from a gas furnace are expected to include CO_2, water vapor, and a small amount of O_2. Nitrogen from the air is not required for combustion, so most of it passes through the burner. However, the heat of the combustion process can create small amounts of nitrogen oxides (NOx), which are considered pollutants.

Some states and regional pollution control authorities require periodic combustion testing of gas-fired heating equipment to determine the levels of CO, CO_2, and NOx being exhausted to the atmosphere. NOx compounds represent a group of highly reactive gases. Many of the compounds are both colorless and odorless. However, one common pollutant, nitrogen dioxide (NO_2), along with smoke and other particulates in the air, is often seen as a reddish-brown smog in urban areas.

NOx pollutants form when fuel is burned at high temperatures. The primary sources of atmospheric NOx are motor vehicles, electric utilities, and power generation facilities that burn fossil fuels. NOx contributes to *acid rain* and is known to degrade surface water quality. NOx also has a much higher global warming potential (GWP) than CO_2, which is the standard used to rate the GWP of refrigerants and other substances.

NOTE

Another common nitrogen-oxide compound is nitrous oxide (N_2O), which is not a product of combustion nor is it included in the NOx category of air pollutants. It is a relatively inert compound commonly used as a propellant for aerosol products, oxidizers in rocket fuels, and even as a mild anesthetic. In the latter case, it is often referred to as *laughing gas*.

Going Green

Low-NOx Furnaces

NOx byproducts are considered an air pollutant, and NOx emissions are subject to regulation in many parts of the country. For example, residential gas furnaces and gas water heaters installed in California must be designed to reduce NOx emissions.

Since NOx byproducts are produced at high temperatures, one solution to reduce these emissions is to reduce the temperature of the burner flame. This is often done by positioning metal rods, metal screens, or metal tabs into the path of the flame. The metal acts as a heat sink, cooling the flames so that fewer pollutants are produced.

In the United States, all major furnace manufacturers currently offer low-NOx furnaces. However, California currently plans to eliminate all new gas appliances by 2030, and other states are considering a similar path forward.

The flue gases of a properly operating gas burner will have the following characteristics:

- The flue gas temperature must be high enough to assist in its movement through the vent and prevent condensation inside, but a higher temperature contributes to efficiency loss. Although you should always check the manufacturer's documentation for accurate information, an 80% AFUE induced-draft furnace typically has a **net stack temperature** of 300°F to 400°F (~150°C to 200°C). The net stack temperature is the difference between the measured flue gas temperature and the entering air temperature of the combustion air.
- The CO_2 and O_2 levels will be at acceptable levels per the combustion analyzer, indicating that enough excess combustion air is being supplied to the burners.
- Only a small volume of CO should be present in the flue gas. Modern furnaces are designed to emit a CO volume of less than 50 parts per million (ppm). It is generally accepted that anything near or over 100 ppm of **air-free CO** is cause for concern and investigation. The presence of a significant amount of CO is an indication of incomplete combustion. If the CO content is outside the normal range, make sure the furnace is receiving

Net stack temperature: The measured stack (vent) temperature minus the entering air temperature of the combustion air.

Air-free CO: A measure of CO volume in flue gases that accounts for the dilution caused by air and other gases. The air-free CO measurement from a combustion analyzer will always be higher than a standard CO reading, and air-free CO is the value of interest during a combustion analysis.

adequate combustion air, the venting system is operating properly, and the manifold gas pressure is correct.

- Traces of nitrous and sulfur oxides may be present. It is important to be familiar with the regulations regarding NOx emissions in your area and respond accordingly.

3.1.1 Combustion Analysis Equipment

The electronic combustion analyzer shown in *Figure 23* is one of many combustion analyzers currently available. The kit typically contains multifunction sampling probes that plug into the analyzer. Results are available in real time on the display. Graphs can also be produced from this data. The analysis results can be saved and/or printed. Many are also capable of wireless communication and are remotely accessible via an app.

Before using a combustion analyzer, read the instruction manual thoroughly and become familiar with its various features. Most analyzers require specific steps to be taken before an analysis begins.

Figure 23 Combustion analyzer.
Source: Courtesy of Fieldpiece Instruments

3.2.0 Conducting a Combustion Analysis

Before conducting a combustion analysis on a gas-fired furnace or boiler, look for and resolve obvious problems that might affect the analysis or furnace performance. Things to look for during a preliminary inspection include the following:

- Plugged or dirty supply air filters
- Flaws in the venting system

- Soot buildup in the combustion chamber
- Restricted or blocked combustion air intakes

WARNING!

Use caution when conducting a combustion analysis. Flue piping can reach temperatures exceeding 400°F (204°C). Remember that the analyzer probe will be at this temperature when it is withdrawn. Avoid touching the probe and do not set it down on flammable materials.

Prepare the equipment for the combustion analysis by drilling a sampling hole in the vent pipe to sample the flue gas and measure the flue gas temperature. *Figure 24* shows the location of the sampling hole(s).

Induced-draft and condensing furnaces should have the sample hole drilled about 6" to 8" above the vent connector on the furnace. Condensing furnaces should have two sample holes; one in the vent pipe as it leaves the furnace and one on the combustion air intake pipe where it enters the furnace. The hole in the combustion air intake allows for the measurement of the combustion air supply temperature. Alternatively, this can be measured outdoors at the intake. The combustion air entering a condensing furnace can be very cold because it typically comes directly from outdoors.

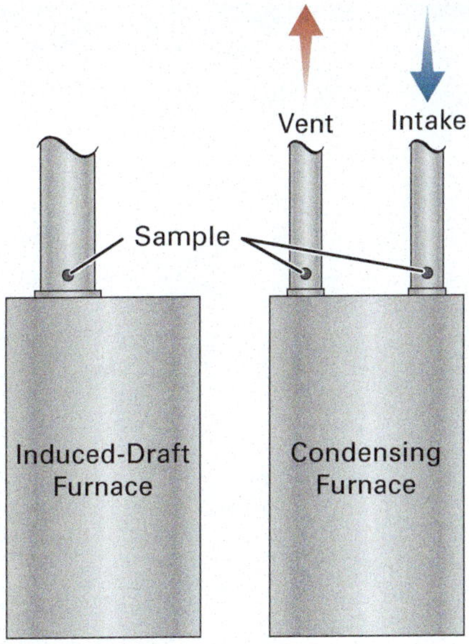

Figure 24 Sampling hole locations.

NOTE

After completing the analysis, seal all sampling holes with a suitable metal or silicone plug. Do not rely on tape.

3.2.1 Using the Combustion Analyzer

Electronic devices quickly measure the temperature and content of the flue gases and calculate burner efficiency. It is essential to read the analyzer's operating and maintenance manual thoroughly before conducting an analysis. Analyzers are complex and sensitive instruments that must be used as directed for accurate and reliable results. You must also maintain the analyzer as directed.

Combustion analyzers typically require time to warm up and complete a calibration routine. This may require a trip outdoors in fresh air, where the analyzer will also measure ambient CO concentration and calibrate the CO sensor. The analyzer will also require you to select the fuel and adjust for the altitude if needed.

At some point, you will also need to measure the temperature of the combustion air entering the furnace. For a condensing furnace, the entering air temperature can be very cold since it is typically drawn from outdoors. For a standard induced-draft furnace, the air comes from the room where the furnace is located. Many analyzers allow this temperature to be entered manually. The temperature of the combustion air entering the furnace is used by the device to determine the net stack temperature.

Once the analyzer is ready, you can initiate a combustion analysis (*Figure 25*). Insert the probe so that the tip is positioned midway in the vent and is secured in place by the cone-shaped lock.

Allow the furnace time to warm up and stabilize to ensure accurate and consistent readings. As the furnace starts, the readings will change rapidly. The analyzer will display the readings in real time (*Figure 26*).

It is best to insert the analyzer into the flue gases before the indoor blower starts. Radical changes in the O_2 content when the indoor blower starts usually indicate that the heat exchanger is compromised and air is leaking in, affecting the flue gas readings.

With the burner stabilized and the readings consistent, print and/or save the results. Equipment owners often like to see the results, and you'll need them for comparison to future results.

CAUTION

If the CO and/or O_2 levels suddenly rise after the indoor blower starts, it is a sign that the heat exchanger is leaking, and room air is entering the system. Run the analysis again to confirm the results before conducting a thorough inspection of the heat exchanger.

Figure 25 Conducting a combustion analysis.
Source: Courtesy of Fieldpiece Instruments

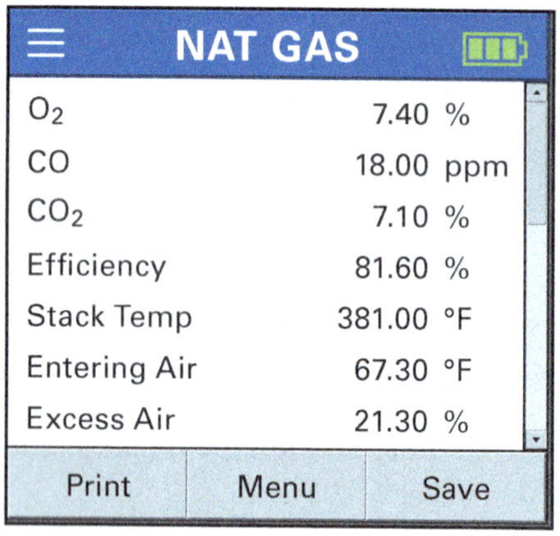

Figure 26 Sample combustion analyzer display.

3.2.2 Analyzing the Results

When the combustion analysis is complete, the results should be reviewed, and corrective actions taken to solve any problems. Problems the analysis may reveal, along with suggested corrective actions, include the following:

- *Excessive O_2 readings after the indoor blower starts* — This is generally an indication that the heat exchanger is leaking. A thorough heat exchanger inspection is necessary in this case. Note that not all heat exchanger cracks will leak. The cracks still represent a hazard, as they will only become larger over time due to expansion and contraction as the furnace temperature changes. The operating pressure in the heat exchanger of an induced-draft furnace is always negative. Thus, air leaks into the heat exchanger rather than flue gases leaking out. The positive pressure applied around the heat exchanger by the indoor blower increases the volume of air leaking into the heat exchanger. This results in a significant increase in O_2 volume in the flue gases.

WARNING!

Never attempt to repair a heat exchanger. When a compromised heat exchanger is discovered, the furnace should be locked out and tagged until the furnace or heat exchanger is replaced.

- *High flue gas temperatures* — A high flue gas temperature may be the result of an overfired burner, or an excessive temperature rise across the heat exchanger. Check the manifold gas pressure to ensure it is within the manufacturer's specifications. Excess gas pressure creates more heat. A high temperature rise across the heat exchanger is usually caused by insufficient indoor air flow. Check the blower speed and ensure the air filters are clean.
- *Lower than normal CO_2 and O_2 levels* — This may indicate the burner is not receiving adequate combustion air. Look for and correct the combustion air deficiency. On a condensing furnace, one thing to look for is a blockage in the combustion air piping.
- *NOx emissions detected* — Electronic combustion analyzers can often detect other gases that may be present in flue gas, including NOx. Many states and localities have limits on NOx emissions. A combustion analysis can determine whether these emissions are within acceptable limits. Although excess air helps ensure complete combustion, too much excess air can cause a rise in NOx emissions. Low-NOx furnaces typically have a lower flame temperature. Note that this may also serve to increase CO emissions.
- *High CO readings* — High CO readings are usually an indication that there isn't enough combustion air. That's the first thing to look for—anything that may be restricting the furnace's source of air. However, other issues can also increase the CO. Make sure all the burners are clean and properly aligned with the heat exchanger. When the flames strike a surface rather than moving cleanly into the heat exchanger, more CO and impurities develop. Also remember that other sources of CO can be encountered around buildings, especially parking garages. If CO is already in the combustion air, the volume can only increase as it passes through the burner.

As always, the best and most reliable source of information regarding combustion analysis and the results is the equipment manufacturer.

3.0.0 Section Review

1. What volume of CO is generally accepted as the threshold of concern in a modern gas-furnace combustion analysis?
   a. 50 ppm CO
   b. 50 ppm air-free CO
   c. 100 ppm CO
   d. 100 ppm air-free CO

2. Which of the following is a possible cause for a high flue gas temperature?
   a. Excessive gas pressure
   b. Leaking heat exchanger
   c. Excessive indoor air flow
   d. Low combustion air temperature

Module 03209 Review Questions

1. Pilotless heating systems use either an HSI or a(n) _____.
   a. indirect spark igniter
   b. bimetal igniter
   c. thermocouple igniter
   d. direct spark igniter

2. What is the typical signal produced by a flame rod?
   a. 2 to 6μA
   b. 30 to 70Ω
   c. 0 to 5VDC
   d. 4 to 20mA

3. Which of the following is a problem with a spark igniter that you might expect to encounter?
   a. Faulty cold junction
   b. Loss of resistance
   c. Grounded condition
   d. Cracked ceramic insulator

4. The more durable HSIs are made from ______.
   a. nitrous oxide
   b. carbon fiber
   c. silicon carbide
   d. silicon nitride

5. How does oxidation on a flame rod affect it?
   a. It prevents corrosion of the rod.
   b. It enhances its thermal conductivity.
   c. It impedes current flow.
   d. It causes it to glow more brightly.

6. In most induced-draft furnaces, what is the first event that occurs after a call for heat?
   a. The pilot gas valve opens.
   b. The induced-draft fan is energized.
   c. Power is sent to the igniter or HSI.
   d. The indoor blower motor is energized.

7. Upon what principle does the effectiveness of infrared heat rely?
   a. The conductivity of the air
   b. The movement of air
   c. The line-of-sight principle
   d. Heat stratification

8. What is the term for an optional wire screen covering the grids of high-intensity infrared heaters that increases the radiating effect?
   a. Radiator
   b. Reflector
   c. Wiremold
   d. Reverberator

9. True or False? Adjusting the primary airflow entering an inshot burner is a common practice when conducting a combustion analysis.
   a. True
   b. False

10. What is the typical range of the net stack temperature for an 80% induced-draft furnace?
   a. 175°F to 275°F (~80°C to 135°C)
   b. 300°F to 400°F (~150°C to 200°C)
   c. 425°F to 575°F (~220°C to 300°C)
   d. 600°F to 750°F (~315°C to 400°C)

Answers to odd-numbered review questions are found in the Module Review Answer Key at the back of this book.

Answers to Section Review Questions

Answer	Section	Objective
Section One		
1. d	1.1.1	1a
2. b	1.2.0	1b
Section Two		
1. c	2.1.2	2a
2. b	2.2.0	2b
Section Three		
1. d	3.1.0	3a
2. a	3.2.2	3b

Troubleshooting Oil Heating

Source: antos777/iStock

Objectives

Successful completion of this module prepares you to do the following:

1. Identify and describe the primary components and operation of oil-fired furnaces.
   a. Describe a basic oil-fired heating system.
   b. Identify the primary components and operation of pressure-type oil burners.
   c. Identify and describe common oil-furnace safety controls.
   d. Describe fuel oil supply systems.
2. Explain how to perform periodic maintenance on oil-fired heating equipment.
   a. Explain how to perform basic maintenance procedures on an oil-fired system.
   b. Explain how to perform a combustion analysis.
3. Explain how to troubleshoot a typical oil-fired heating system.
   a. Describe troubleshooting procedures for common oil heating problems.
   b. Describe troubleshooting procedures for oil furnace control devices.

Performance Tasks

Under supervision, you should be able to do the following:

1. Remove and replace an oil burner pump.
2. Conduct a complete combustion analysis, including smoke test and draft.
3. Remove and replace an oil burner nozzle and set the electrode position.
4. Test a cad cell flame detector.

Overview

Oil-fired furnaces and boilers are often used in homes in northern sections of the United States that have limited access to other reliable fuels. Oil-fired heating systems have provided reliable comfort to homes and businesses for many years, and are relatively simple to maintain and repair. To ensure reliability, oil furnaces do require annual maintenance, so it is important to know how to service these furnaces if that is within your scope of work.

Digital Resources for HVACR

Scan this code using the camera on your phone or mobile device to view the digital resources related to this craft.

SCAN ME

NCCER Industry-Recognized Credentials

If you are training through an NCCER-accredited sponsor, you may be eligible for credentials from NCCER. The ID number for this module is 03310. Note that this module may have been used in other NCCER curricula and may apply to other level completions. Contact NCCER at 1.888.622.3720 or go to **www.nccer.org** for more information.

You can also show off your industry-recognized credentials online with NCCER's digital credentials. Transform your knowledge, skills, and achievements into credentials that you can share across social media platforms, send to your network, and add to your resume. For more information, visit **www.nccer.org**.

1.0.0 Oil-Fired Heating Systems

Performance Tasks

There are no Performance Tasks in this section.

Objective

Identify and describe the primary components and operation of oil-fired furnaces.

a. Describe a basic oil-fired heating system.
b. Identify the primary components and operation of pressure-type oil burners.
c. Identify and describe common oil-furnace safety controls.
d. Describe fuel oil supply systems.

Annual Fuel Utilization Efficiency (AFUE): An industry standard used to define furnace combustion efficiency.

Weatherized furnace: A furnace that has been well insulated and has a weather-resistant case as specified by government standards.

Oil furnaces are designed to burn fuel oil, which is stored on site in a storage tank. The fuel oil is mixed with air, sprayed as a fine mist into the furnace's combustion chamber, then ignited by a continuous electric arc. Heat produced in the combustion chamber heats up the furnace's heat exchanger. Air returned from the conditioned space is directed over the heat exchanger by a blower in the furnace unit. Heat is then transferred to that air and recirculated by the duct system.

A typical oil furnace is shown in *Figure 1*. New high-efficiency condensing oil furnaces can achieve an average **Annual Fuel Utilization Efficiency (AFUE)** of up to 98%. However, these models represent a very small percentage of the market. The US Department of Energy has established minimum AFUE ratings for oil furnaces of 83% for **weatherized furnaces** and 78% for non-weatherized furnaces. Most oil furnaces on the market today have an AFUE of roughly 86%.

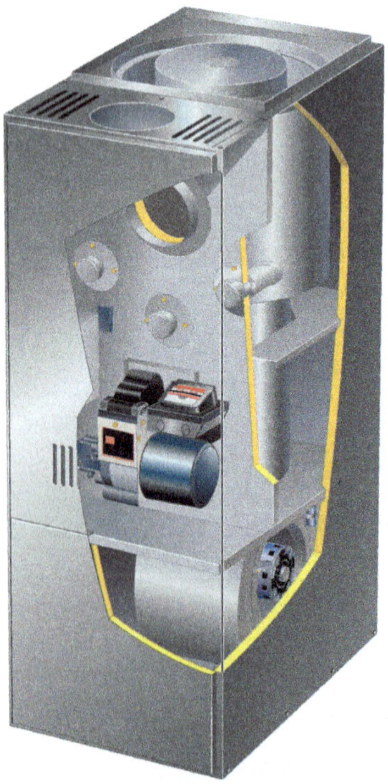

Figure 1 Oil furnace.
Source: Courtesy of Lennox Industries, Inc.

Going Green

Heating with Biodiesel

Biodiesel is a fuel made from new or recovered vegetable oil or animal fat. Experiments with blends containing 80% heating oil and 20% biodiesel in oil furnaces have proven successful. In addition, modifications of the furnaces have not been necessary.

Further experiments by the US Department of Energy have shown that a 70/30 blend can be used with no change in heating performance. One of the main benefits of using such a blend is the reduction of carbon emissions. In addition, biodiesel, unlike fossil fuel, is a readily renewable resource.

1.1.0 Oil-Fired Heating System Basics

In a typical oil-fired furnace, the oil burner receives oil from a storage tank on the premises (*Figure 2*). The oil burner sprays oil through a nozzle where it immediately mixes with air that contains the needed oxygen. Electrodes located at the point where the oil spray enters the combustion chamber provide a continuous high-voltage arc that ignites the oil.

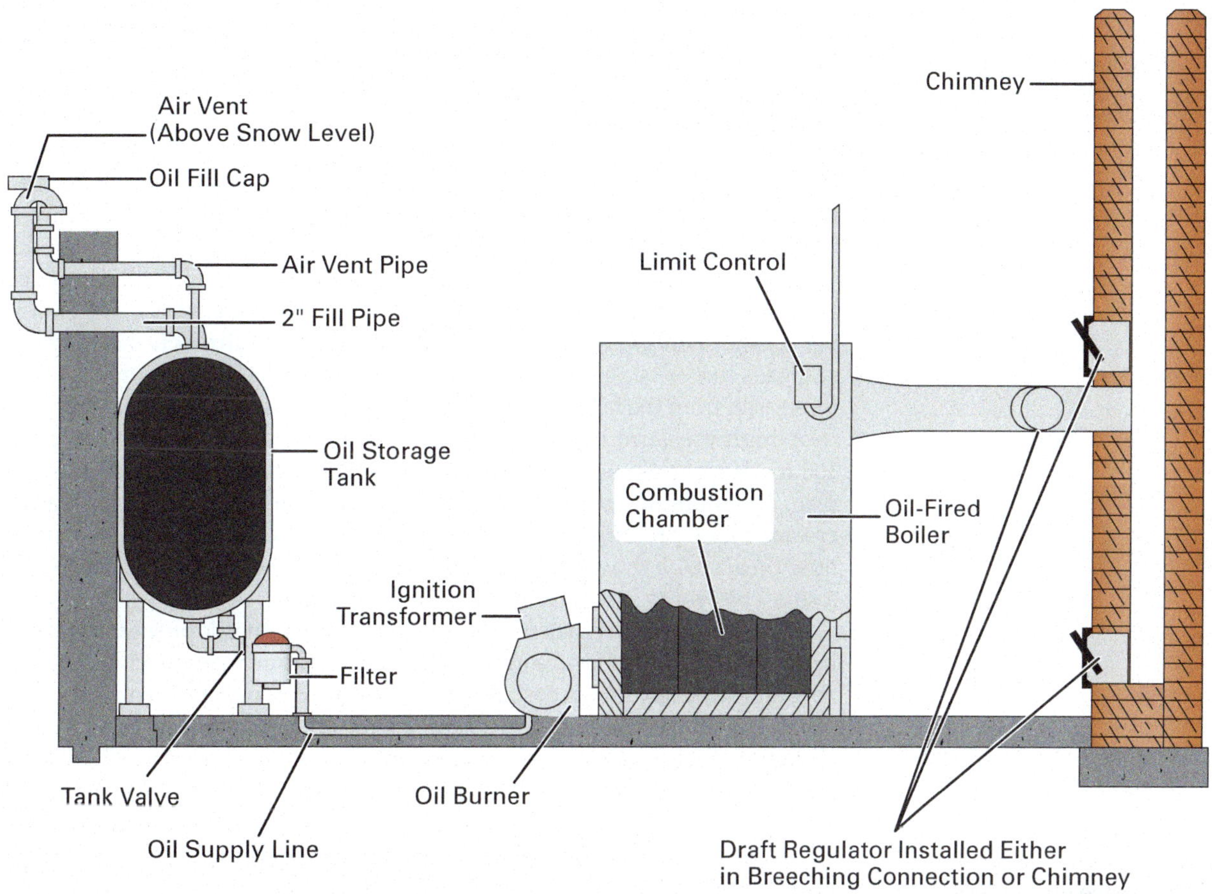

Figure 2 An oil-fired heating system.

1.1.1 Combustion Chamber

The combustion chamber protects the heat exchanger from heat damage and reflects heat back toward the burner. The reflected heat enhances combustion. The flame shape must fit the combustion chamber and the nozzle must be located at the correct height above the chamber floor. Flaws in furnace design or deficiencies in the burner can result in incomplete combustion. Incomplete combustion can produce **soot** and carbon monoxide (CO) as well as waste fuel.

Three types of materials are commonly used for combustion chamber construction: metal, typically stainless steel; insulating fire brick; and molded ceramic. In small furnaces such as those used in residential applications, the combustion chamber is built into the furnace. For large commercial and industrial applications, the combustion chamber may be constructed on site. These large chambers are made of fire brick or pre-cast ceramic **refractory** material. If pre-cast materials do not fit the job, a combustion chamber may be formed and poured on site using ceramic materials designed for this purpose.

A moldable refractory material, referred to as *wet pack*, is often used to construct or re-line combustion chambers. This material is moist when it is purchased and is shipped in sealed plastic bags. Because it is moist, it can be easily molded into the desired shape. Exposure to air causes the wet pack to dry into a rigid structure. The material has poor heat transfer characteristics, making it an excellent insulator that protects the combustion chamber structure from excessive heat.

Soot: Fine carbon particles resulting from incomplete combustion. An accumulation of soot on heat exchangers will reduce combustion efficiency and result in increased carbon monoxide production.

Refractory: Materials designed to withstand extremely high temperatures and used to line or fabricate combustion chambers.

1.1.2 Heat Exchanger

The heat exchanger of an oil-fired furnace has primary and secondary heating surfaces. The primary heating surface surrounds the flame, whereas the secondary heating surface is often a series of steel sections over which the hot combustion gases pass before exiting the furnace.

The main elements of an oil-fired system include the following:

- Oil burner assembly
- Controls
- Fuel supply system

1.2.0 Oil Burners

Pressure-type oil burners, also referred to as *gun-type burners*, force the oil through a nozzle and then spray it into an airflow nozzle. Pressure-type oil burners can be broken down into low-pressure and high-pressure models.

In a low-pressure gun-type burner, which may be found in older furnaces, the oil and primary air are mixed before reaching the nozzle. At a pressure of 1 to 15 pounds per square inch gauged (psig), the mixture is forced through the nozzle where it **atomizes** as it exits. Secondary combustion air is drawn into the spray mixture after it is released from the nozzle. An electric arc is used to light the combustible mixture.

Atomizes: Reduces a liquid to a very fine spray.

A high-pressure burner forces oil through the nozzle under a pressure of 100 to 150 psig. Some burners may be designed for higher pressures, up to 200 psig. The higher pressure breaks the oil into a finer mist. The atomized oil spray creates a low-pressure area as it leaves the nozzle. This causes secondary combustion air to be drawn into the oil spray. Primary combustion air is supplied by a small blower that is part of the burner assembly.

The high-pressure burner is the most common domestic burner today. It is simple in construction and effective in operation. The parts are mass-produced, readily available, and relatively low in cost. Few low-pressure oil burners remain in service today.

Oil burners introduce the atomized oil and air mixture to the combustion chamber. An ignition transformer establishes an arc across a set of electrodes to ignite the fuel-air mixture. The **primary control** monitors the process, disabling and locking the burner out if there is a flame failure. Lockouts must be manually reset. A **cad cell** is commonly used to sense the light of the flame to verify ignition occurred, and that combustion continues.

Primary control: A combination switching and safety control that responds to signals from the thermostat and other safety controls.

Cad cell: A light-sensitive device (photocell) in which the resistance reacts to changes in the amount of light. In an oil furnace, it acts as a flame detector.

An oil burner assembly is shown in *Figure 3*. Its primary components include the motor, blower wheel, oil pump, ignition transformer, and oil nozzle (*Figure 4*).

Figure 3 Typical oil burner.
Source: Courtesy of RW Beckett Corp.

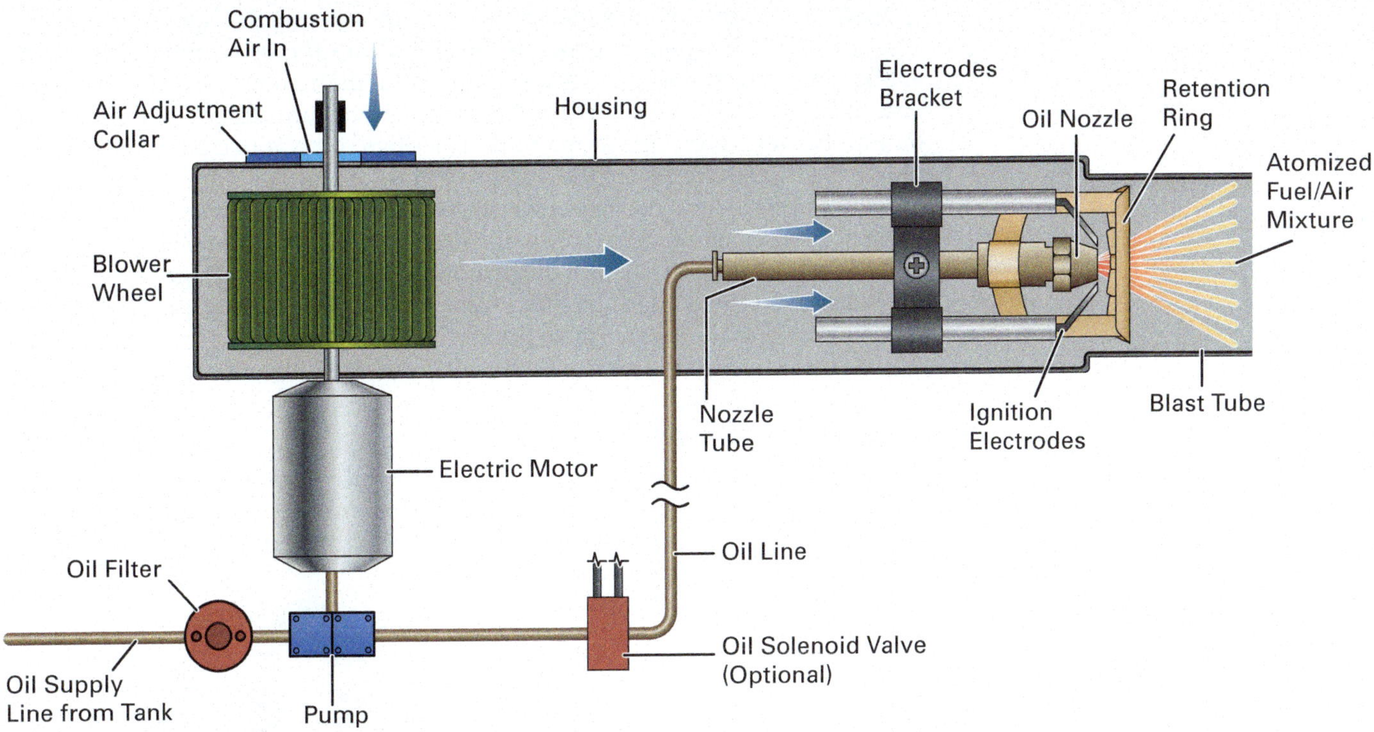

Figure 4 Oil burner components.

The burner cycles on or off in accordance with the demands of the room thermostat. Grade 2 fuel oil is the most commonly used fuel.

1.2.1 Power Assembly

The power assembly of an oil burner consists of the motor, blower wheel, and oil pump (*Figure 5*). A single motor drives both the fan and oil pump. The blower wheel, mounted on the motor shaft, forces air through the blast tube to provide combustion air. The inlet to the fan has an adjustable opening so that the air volume can be manually controlled. The oil pump, coupled to the motor shaft, draws oil from the storage tank and delivers it to the nozzle. The fuel/air mixture is ignited by the electric arc formed between two electrodes in the nozzle assembly.

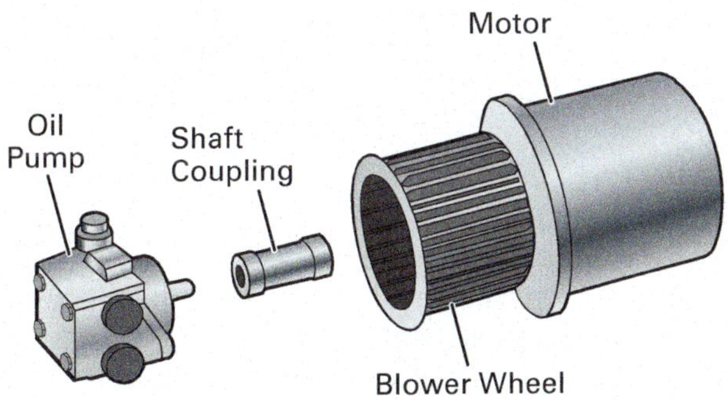

Figure 5 Power assembly.

The pressure-regulating valve inside the oil pump has an adjustable spring that permits adjustment. The pump can typically deliver more oil than the system can use. The excess oil is returned to the tank through a second oil line or is bypassed inside the pump. A two-pipe system allows for the excess fuel oil to return to the tank, while a single-pipe system does not. The choice depends on several factors, including the oil tank location and distance from the burner. On some burners, an oil solenoid valve stops the flow of oil as soon as the primary control stops the burner, thereby preventing fuel oil from dripping into the combustion chamber.

> **Did You Know?**
>
> **Oil Pressure**
>
> For many years, it was commonly accepted that the oil pump outlet pressure should be 100 psig on a high-pressure burner. Today, some manufacturers specify higher oil pressures. Always check the manufacturer's literature before adjusting the pump outlet pressure.

> **NOTE**
>
> Do not confuse single- and two-pipe operation with single- and two-stage pump operation.

Most oil pumps are designed to work with both single-pipe and two-pipe systems. A bypass plug is generally inserted into the oil pump for two-pipe applications. Most bypass plugs are simply a small threaded plug with an Allen head. The plug closes off an internal passage in the pump, forcing the excess fuel oil into the return oil line to the tank. For single-pipe applications, the bypass plug must be removed from the pump.

Oil pumps are designed for single-stage or two-stage operation. A single-stage pump is used when the oil supply is above the level of the burner and the oil is fed to the pump by gravity. The two-stage unit is used when the storage tank is below the burner and the oil must be lifted some distance. The first stage draws the oil to the pump and the second stage boosts the pressure to the nozzle. The vacuum on the suction side of the pump should not be less than 15" of water column (w.c.).

1.2.2 Nozzle Assembly

The nozzle assembly (*Figure 6*) consists of the oil nozzle tube and the nozzle. The nozzle controls the oil flow volume and atomizes the fuel. The oil passes through an integral strainer before flowing through the nozzle orifice (*Figure 7*).

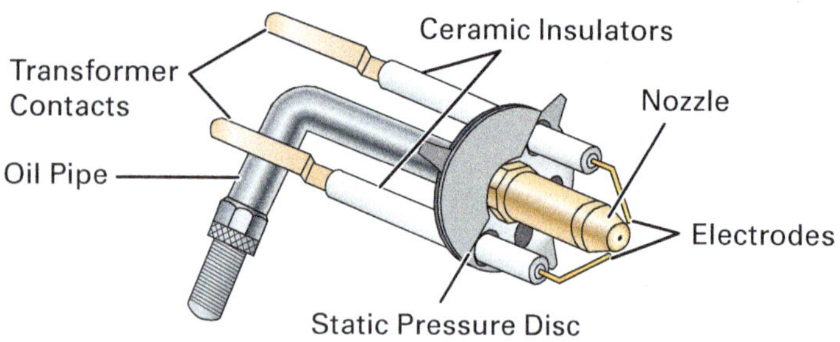

Figure 6 Nozzle assembly.

Figure 7 Oil nozzle.
Source: Courtesy of Delavan Spray Technologies, an R.W. Beckett Company

Figure 8 Retention ring.
Source: Courtesy of Carlin Combustion Technology

The nozzle orifice is precisely shaped and sized to deliver a specific amount of fuel in a specific pattern suitable for the application. The oil spray then passes through slots cut into the *retention ring*, as shown in *Figure 8*. The pattern in the retention ring causes the oil and air mixture to spin while entering the combustion chamber.

Nozzles are selected by the manufacturer to provide the correct oil volume and spray pattern for a given oil burner. Due to the precise nature of the nozzle orifice, dirty or defective nozzles are usually replaced rather than cleaned or repaired. If it is necessary to replace a nozzle, the replacement should be the same size and pattern as the original.

Nozzle spray patterns vary with the type of application. The shape and angle of the spray are determined by the nozzle. The three basic spray patterns are hollow, semi-hollow, and solid (*Figure 9*). The hollow and semi-hollow patterns are commonly found on domestic burners because they are more effective when applied to modern combustion chambers.

The angle of the spray must also correspond to the combustion chamber (*Figure 10*). An angle of 70 degrees to 90 degrees is usually best for square or round chambers. An angle of 30 degrees to 60 degrees is best suited for long, narrow

Figure 9 Nozzle patterns.

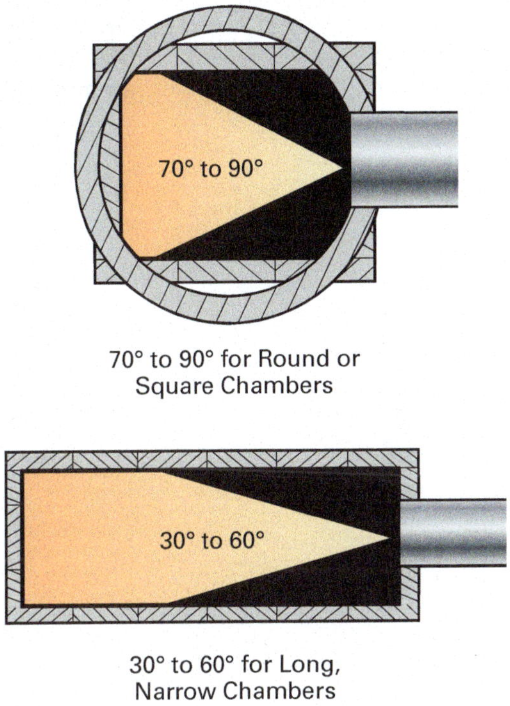

Figure 10 Spray angles.

chambers. Again, it is important to use the spray pattern and nozzle angle recommended by the manufacturer for proper performance. An incorrect nozzle will not provide ideal performance and could result in damage to the furnace.

Nozzle Identification

Oil burner nozzles are sized in gallons per hour (gph). This refers to the volume of oil at a given pressure, usually 100 psig. The nozzle's capacity and type are stamped on the nozzle body. Manufacturers use different letters or combinations of letters to identify the spray pattern. For example, Delavan® Type A nozzles are a hollow-cone design, while Type B nozzles produce a solid cone (*Figure 11*). When replacing a nozzle, consult the manufacturer's catalog information and cross reference to another brand if necessary.

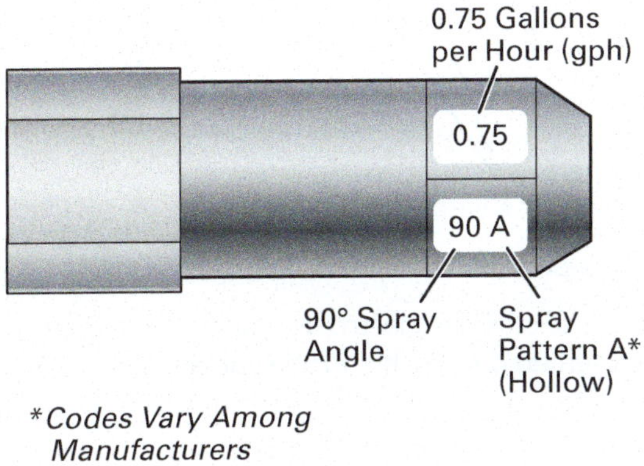

Figure 11 Oil burner nozzle characteristics.

The oil burner nozzle flow rate determines the burner capacity. For example, one gallon of Grade 2 fuel oil has a heat content of approximately 140,000 Btus. Thus, if using a 0.5-gallon nozzle, the burner input would be 70,000 Btus per hour (Btuh). The flow rate is related to furnace input and not the output. The heating output will be lower and is determined by the combustion efficiency of the burner.

1.2.3 Ignition System

The ignition system for high-pressure burners consists of a step-up transformer (*Figure 12*) connected to two electrodes. The transformer supplies a high voltage that causes an arc to develop between the two electrodes, igniting the oil and air mixture.

Figure 12 Ignition transformer.
Source: France/A Scott Fetzer Company

Ceramic insulators surround the metal electrodes (*Figure 13*). The ignition transformer increases the voltage from 120V to about 10,000V at a current of about 20 milliamperes (mA). The ceramic insulators prevent arcing from the electrode to other parts of the burner. A 10,000V potential can cross a gap of roughly $^3/_8$", or 1 centimeter (cm), between conductive surfaces. Low current reduces wear on the electrode tips.

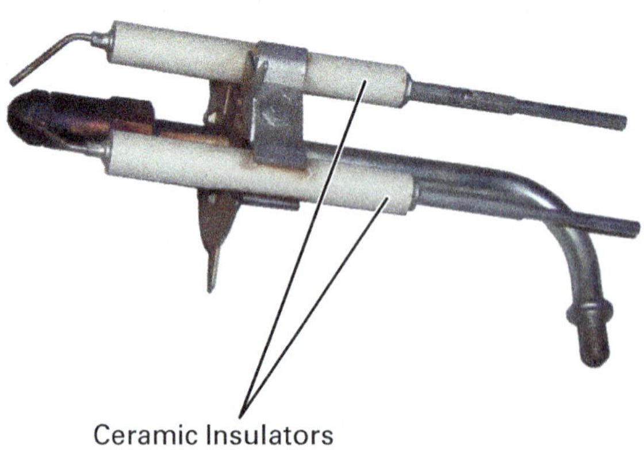

Figure 13 Insulated electrodes mounted on a nozzle assembly.

It is important to make sure the electrodes are positioned correctly (*Figure 14*). If the electrodes are not centered in the blast tube, the flame may be distorted, causing soot to form on the nozzle or retention ring. Accumulated soot distorts the flame and blocks the passages through the retention ring. Ignition problems also result when the electrodes are set too far above or behind the oil spray. The proper settings are established by the equipment manufacturer based on the chosen nozzle.

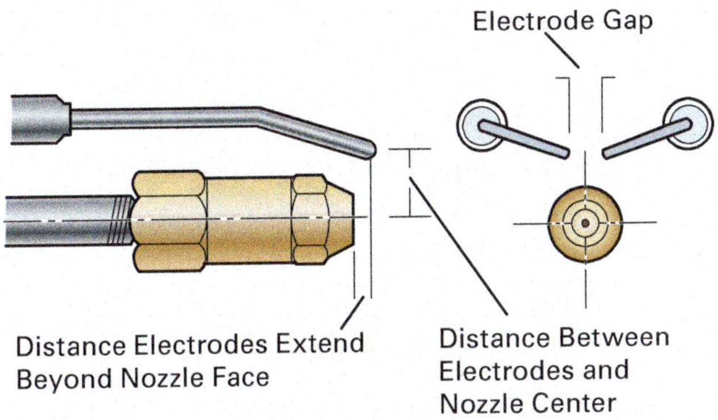

Figure 14 Electrode position measurements.

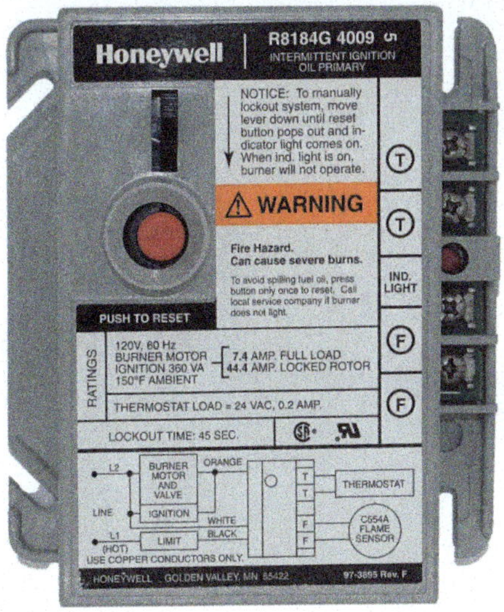

Figure 15 Primary control for an oil burner.
Source: HVACGuru

1.3.0 Oil Burner Controls

All residential oil burners are equipped with a control device known as the *primary control* (*Figure 15*). The primary control provides the 24VAC control voltage source, and the room thermostat is connected to the primary control through the T terminals on the front of the primary control.

The primary control is usually mounted on the burner assembly itself, as shown in *Figure 3*. The primary control manages burner operation and energizes the burner motor and ignition transformer in response to the thermostat. As a safety control, the primary control disables the motor and transformer if the burner does not ignite or the flame goes out. External safety controls, such as the furnace or boiler high-limit switch, are typically wired into the thermostat circuit. In an oil furnace, the high-limit switch is typically located in the plenum that surrounds the heat exchanger.

> **NOTE**
>
> Some furnaces may have the high-limit switch and other external safety controls connected to the line voltage power source serving the primary control.

In modern furnaces, a cadmium sulfide *photoelectric* cell, or *cad cell* (*Figure 16*), is used to detect the presence of a flame. The cad cell is sensitive to light and is installed in the burner assembly on a line of sight with the burner flame. The light of the flame causes the cad cell to generate a small voltage, proving that a flame is established. If the cad cell does not produce a voltage when the primary control has energized the burner motor, a lockout occurs, and the burner shuts off.

Older furnaces may have a stack switch (*Figure 17*) in place of a cad cell. The stack switch is installed in the vent connector. This switch senses heat and disables the burner if heat is not detected in the stack. Many stack switches also combine the function of a primary control and burner ignition proof into a single component. These combined switches are more accurately referred to as *stack controls* or *stack relays*.

1.4.0 Oil Supply System

Oil furnaces are supplied from either indoor or outdoor oil tanks. It is important to make sure that oil tanks are properly installed and maintained. Oil leaks cause environmental damage.

Outdoor oil tanks are often buried. Because many underground oil tanks have leaked in the past, strict regulations currently govern their placement, construction, and alarm systems. Aboveground tanks over a certain size may also require special containment arrangements to capture leaking oil. An oil tank installation should not be undertaken without a clear understanding of local and national codes.

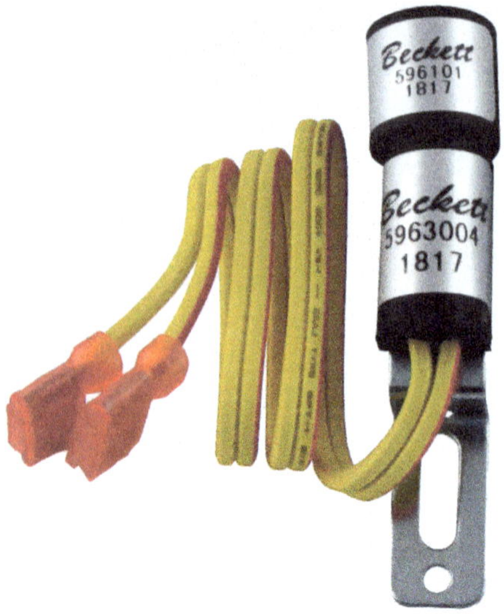

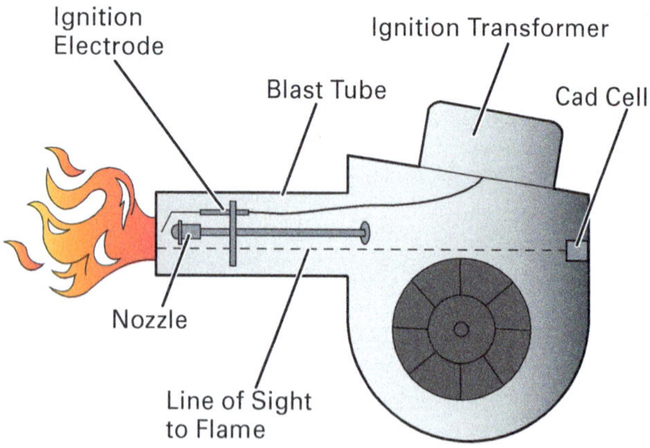

Figure 16 Cad cell.

Source: Courtesy of RW Beckett Corp. (top)

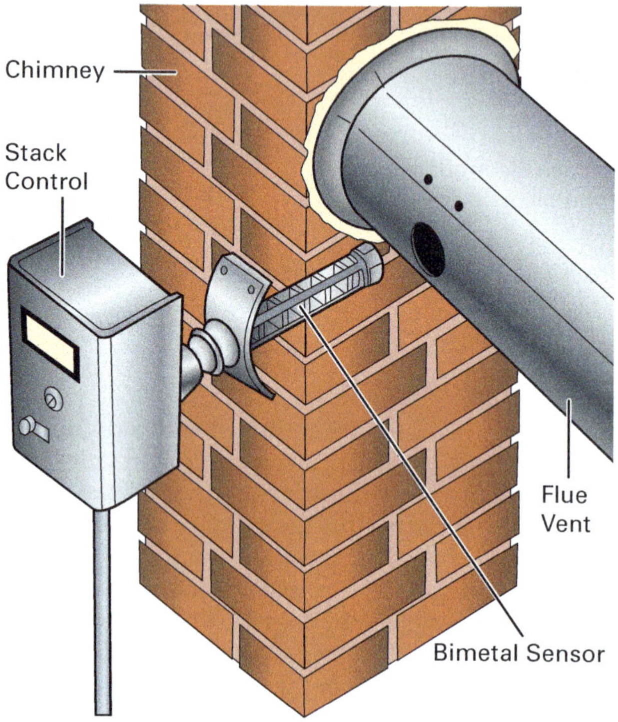

Figure 17 Stack switch.

1.4.1 Fuel Oil

Fuel oils are rated according to their heat content in Btus per gallon and the specific gravity as listed by the American Petroleum Institute (API). The six common grades of oil are described as follows:

- *Grade 1* — A light-grade distillate for use in vaporizing-type oil burners.
- *Grade 2* — A heavier distillate for use in domestic pressure-type oil burners.
- *Grade 4* — A light residue or heavy distillate used for higher-pressure commercial oil burners.
- *Grade 5 (light)* — A medium-weight, residual-type fuel used in commercial oil burners that are specifically designed around it.
- *Grade 5 (heavy)* — A residual-type fuel for commercial oil burners. Usually requires preheating.
- *Grade 6* — A very heavy residual fuel used for industrial burners. Preheating is necessary in the tank to permit pumping. Additional preheating is needed at the burner to convert the fuel into a fine spray or atomize it. Grade 6 is also known as Bunker C.

Grade 2 fuel oil is the most common grade used in domestic oil burners. Grade 1 fuel oil is also referred to as *kerosene*. Kerosene has a slightly lower heat content than Grade 2 oil. However, in northern climates, Grade 1 may be used if the oil is to be stored in an external tank. Grade 2 fuel oil thickens when exposed to extreme cold, which makes it harder to pump from the storage tank to the burner. If the fuel oil is too thick, it cannot be easily atomized at the burner.

1.4.2 Piping Systems

Piping connections to the oil pump can either be single-pipe or two-pipe. In the single-pipe system, only one pipe runs between the storage tank and the burner. In the two-pipe system, two pipes run between the tank and the burner. One pipe carries the oil supply; the second pipe carries bypassed oil back to the tank. Oil pumps typically pump a great deal more oil than the nozzle requires; any excess oil is bypassed back to the tank. Oil lines are typically connected using flare fittings; compression fittings should never be used on fuel oil piping.

The oil supply line must be purged of air for a solid column of oil to enter the pump. While single-pipe systems generally need to be primed and the air purged from the lines, two-pipe systems prime easily and tend to remain free of air.

Various oil supply installations are presented here. National, state, and local codes covering oil furnace installations, especially underground tanks, are under constant review and are subject to change. Consult the manufacturer's installation instructions, as well as the applicable codes, before beginning an installation.

Figure 18 shows a single-pipe system with the supply tank in the same space as the burner. Use the single-pipe system only when the burner is located on the same level or above the bottom of the oil supply tank.

CAUTION

If an underground tank is used, make sure it meets current construction standards. The US Environmental Protection Agency (EPA) has published special requirements for commercial tank installations that differ from residential installations.

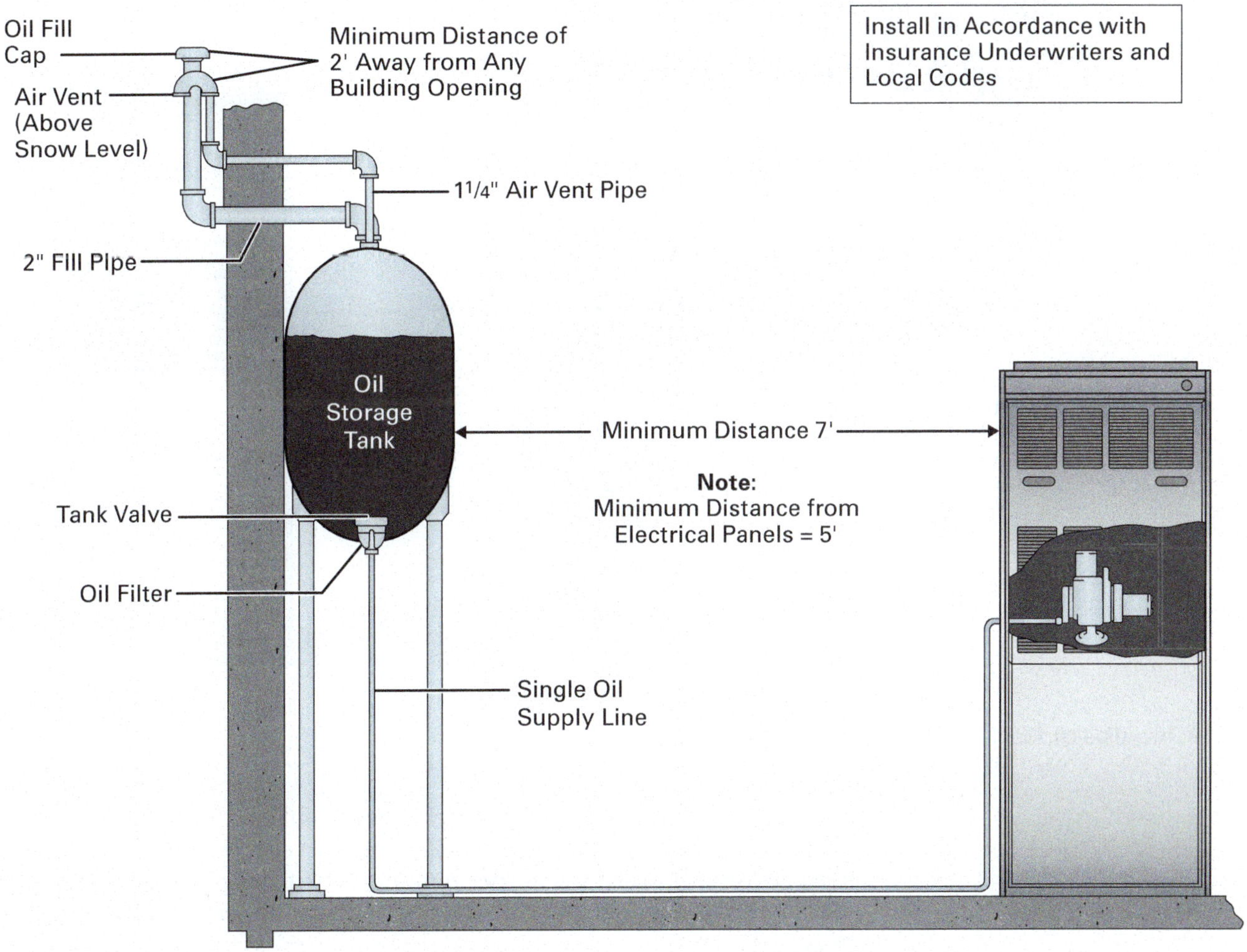

Figure 18 Single-pipe oil supply with an indoor tank.

Single-pipe systems often depend primarily on gravity to feed oil to the burner. The pump used on pressure-type burners creates a vacuum to assist in moving the oil, but not enough to lift the oil very far. In gravity-fed installations, a solenoid oil valve at the tank outlet may be needed to help prevent oil leakage from a damaged oil line.

The piping on a single-pipe system must be free of leaks. Air cannot be allowed to enter the oil piping after the supply line has been purged. If air does enter through a leak, the system will likely need to be primed again. When the oil supply is interrupted, the flame goes out and the primary control will lock out the burner.

Figure 19 depicts a two-pipe system with the burner located below or on the same level as the supply tank. However, oil must first be lifted out of the tank to a height above the burner. Because the oil must be lifted out of the tank, a two-pipe system is more reliable even though the oil is level with or slightly above the burner.

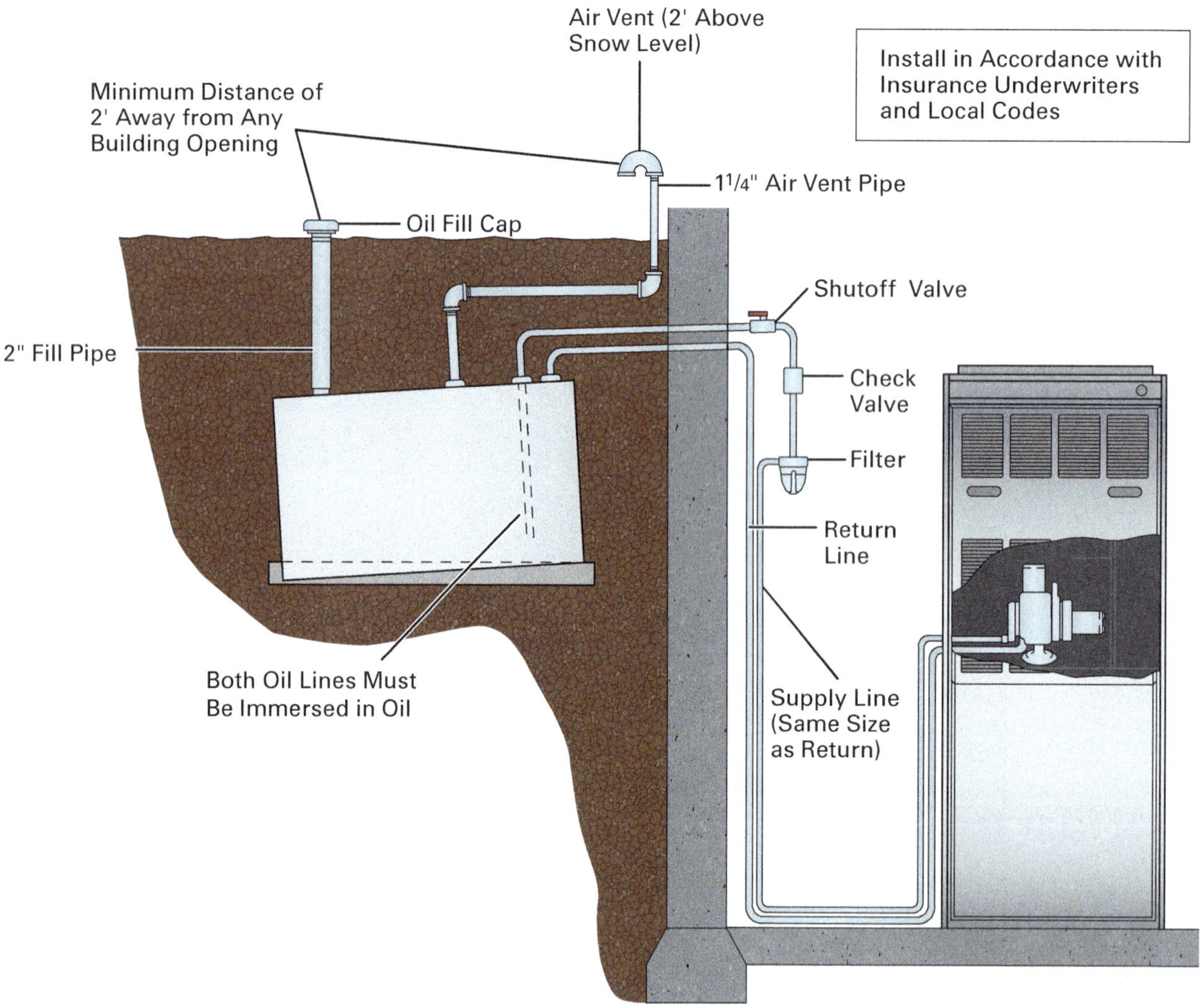

Figure 19 Two-pipe oil supply with burner below tank level.

In this arrangement, a brass check valve should be installed in the supply piping as close to the tank as possible. This prevents the oil from falling back toward the tank when the burner pump is not running. A valve in the supply piping to manually shut off the oil supply for burner service or removal is also required.

Figure 20 shows the two-pipe system with the burner located above the tank when a lift of less than 15' is required. A check valve and globe shutoff valve should be included in this piping network.

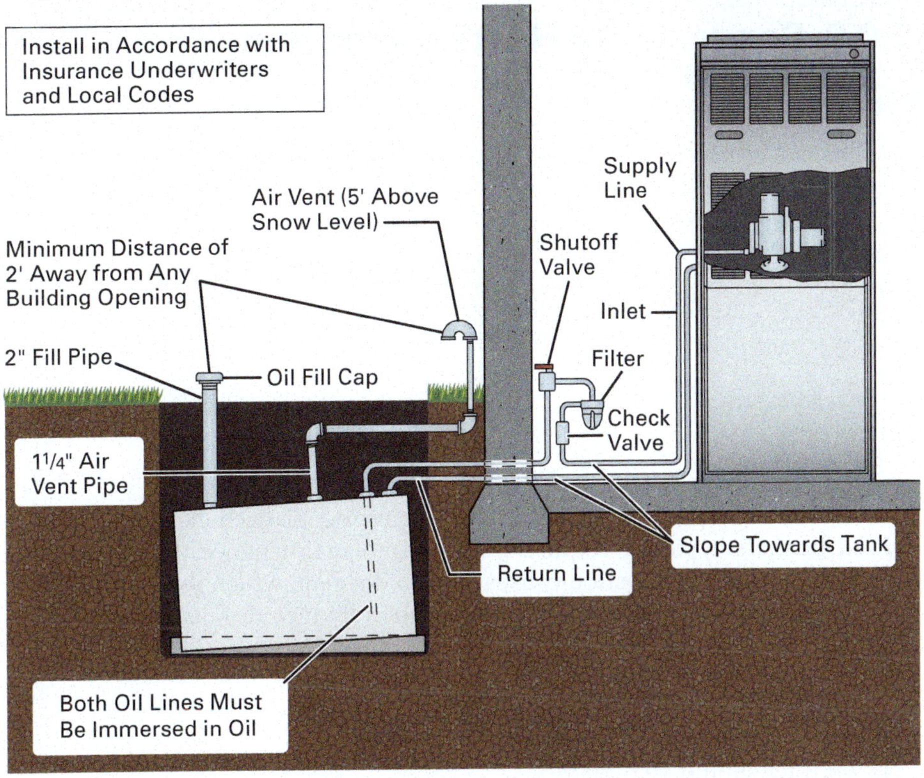

Figure 20 Two-pipe oil supply—burner above tank level.

Did You Know?

Moisture in a Fuel Storage Tank

How does water get into a fuel storage tank? Most of it gets in through condensation of water vapor on the interior walls of the tank. Because the tank is vented, moisture-laden air can move into the tank naturally. Due to the range of conditions, outdoor tanks are more susceptible to the problem.

Keeping the fuel tank full throughout the year can help prevent condensation. Accumulated moisture can be removed by inserting a rigid pipe or hose through the top of the tank until it touches the bottom. The water can then be pumped or siphoned from the bottom, since it has a higher specific gravity than the oil.

1.0.0 Section Review

1. True or False? Incomplete combustion can produce soot and carbon monoxide.
   a. True
   b. False

2. Which of the following devices supply combustion air in an oil-fired furnace?
   a. Natural draft
   b. The furnace indoor blower
   c. Vanes in the oil pump
   d. Burner blower wheel

3. Where is the high-limit switch typically found in an oil furnace?
   a. Combustion chamber
   b. In the plenum surrounding the heat exchanger
   c. In the air discharge plenum
   d. In the oil burner assembly

4. Grade 1 fuel oil is also known as _____.
   a. kerosene
   b. gasoline
   c. Bunker C
   d. propane

2.0.0 Servicing Oil Heating Systems

Performance Tasks

1. Remove and replace an oil burner pump.
2. Conduct a complete combustion analysis, including smoke test and draft.
3. Remove and replace an oil burner nozzle and set the electrode position.

Objective

Explain how to perform periodic maintenance on oil-fired heating equipment.

a. Explain how to perform basic maintenance procedures on an oil-fired system.

b. Explain how to perform a combustion analysis.

Oil furnaces should be serviced at the beginning of each heating season. Oil-fired burners typically require a little more maintenance than gas-fired systems. Inefficient combustion causes soot to develop, which results in a loss of efficiency. Annual furnace maintenance generally includes tasks such as the following:

- Air filters should be cleaned or replaced.
- The blower and heat exchanger should be cleaned.
- The oil filter should be replaced.
- The burner nozzle should be replaced, and the electrode position checked for accuracy.

The following items are commonly needed when servicing oil furnaces:

- Multimeter
- Combustion analyzer
- Oil pressure gauge kit
- Digital thermometer
- Vacuum cleaner suitable for soot
- Electrode gauge
- Flashlight
- Hand mirror
- Nozzle wrench
- Wire brush
- Bucket

2.1.0 Servicing Procedures for Oil-Fired Systems

In addition to the specific safety precautions provided in the product literature, the following general safety precautions should be followed when servicing an oil-fired system:

- Unless it is necessary to have the power on for a particular procedure, always de-energize and lock out the power supply before servicing the equipment. If it is necessary to check voltage or current, be sure to keep one hand outside the unit and avoid touching wiring connections.

- Do not use compression fittings on oil piping. Use flare fittings to prevent air and oil leaks in the fuel lines. The best flare fittings for oil piping have a long shoulder (*Figure 21*).
- Wear protective respiratory equipment when cleaning the soot out of a furnace.
- Do not try to clear the oil burner nozzle. It should be replaced.

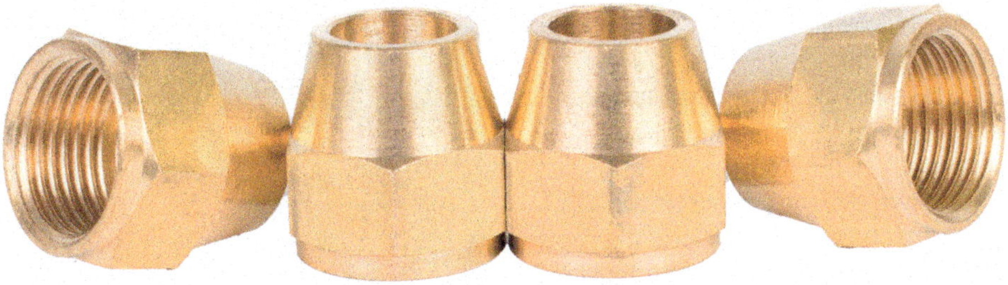

Figure 21 Long flare nuts.
Source: ISARA SUKSARN/Shutterstock ID: 392676778

Before performing any maintenance, check with the customer and listen to any complaints. Make note of any comment that would indicate a possible problem. Then proceed as follows:

WARNING!

Never fire an oil furnace if oil has collected in the combustion chamber and formed a puddle. This can occur when the user has used the reset button multiple times without a successful restart. Excess oil creates a high risk of explosion or fire. Turn off the oil supply, then pump out or soak up the oil before firing the furnace.

Step 1 With the furnace de-energized, check and clean the blower compartment and blower motor. Replace the air filter.

Step 2 Check the electrical wiring for loose connections, insulation breakdown, and corrosion or dirt on the terminals.

Step 3 Inspect the combustion chamber using a hand mirror and flashlight. Look for carbon buildup, holes in the combustion chamber, and erosion of the blast tube and retention ring. If the combustion chamber is lined with refractory, check for damage or deterioration.

Step 4 Check for soot in the heat exchanger:
- Disconnect the flue vent pipe.
- Use a flashlight or trouble light to look inside for carbon buildup. It is not necessary to clean the heat exchanger unless there is an excessive amount of carbon or scale on the heat exchanger surfaces.
- If the heat exchanger needs cleaning, refer to the manufacturer's service manual for the proper cleaning procedure.

Step 5 Check the oil piping:
- Check all oil supply lines between the supply tank and burner for dents, rust, kinks, and loose connections. Oil spots on the equipment or the floor indicate a leak.
- Ensure all the oil piping is properly supported.

Step 6 Check the thermostat for proper operation following the inspection by ensuring the furnace responds to a call for heat and shuts down when the thermostat is satisfied.

On Site

Soot Removal

Certain vacuum cleaners are designed specifically for soot removal. While these vacuums look similar to a shop vacuum, there are significant differences. Soot vacuums have a metal canister that provides a tight seal to prevent soot from escaping. A special filter is used to capture the soot. Soot vacuums are also, in general, more powerful than shop vacuums.

Several chemical products are available that assist in cleaning the soot-coated surfaces of an oil furnace. Always follow both the furnace and chemical manufacturers' instructions and precautions when using soot-removal products.

Source: RectorSeal, LLC

2.1.1 Changing the Oil Filter

Figure 22 Fuel oil filter element.

The oil filter (*Figure 22*) should be changed during a maintenance visit. The following is the basic procedure for changing the filter:

Step 1 Check the filter housing for the name and type of filter element. Ensure you have the correct element for the housing.

Step 2 Place a container below the filter housing to catch any spilled fuel oil.

Step 3 Close the fuel line valve and unscrew the housing mounting screw on the top. Gently pry the top loose from the casting, remove the filter, and place it in a bucket.

Step 4 Clean the inside of the filter housing with a clean, soft cloth.

Step 5 Remove the ring gasket at the top of the filter housing. It may stick to the casting. Scrape away any filter gasket particles with a wide-blade putty knife and then clean the casting and the top of the housing with a clean cloth.

Step 6 Replace the filter element and fill the cartridge with fresh oil.

Step 7 Install a new ring gasket, replace the top, and install the hardware. Do not to overtighten.

Step 8 If the oil tank is above the level of the burner, open the fuel line valve and bleed air from the filter housing through its bleed port by gravity. This shortens the priming process.

Step 9 Connect a plastic hose to the bleed port of the oil pump. Route the end of the hose into a suitable container to capture the oil and start the burner. Crack open the bleed port with a small wrench (roughly one-quarter turn). When oil starts to discharge steadily, close the bleed port.

Step 10 After closing the bleed valve and the burner has begun to receive a steady flow of oil, it should ignite quickly.

2.1.2 Checking the Primary Safety Control Circuit

While the burner is operating, disconnect the cad cell leads at the primary control. The burner should stop in less than 90 seconds and the primary control should lock out. If it does not, the primary control is likely defective and unreliable. Clean the lens of the cad cell during the maintenance procedure.

If the burner locks out while the combustion process appears stable, the cad cell needs to be tested. Cad cell testing is covered in the next section related to troubleshooting.

2.1.3 Checking the High-Limit Safety Switch

Prevent the blower from operating by disconnecting the motor wiring from the blower relay. Initiate a call for heat at the thermostat and place a suitable thermometer in the heat exchanger plenum. The safety limit control should shut down the furnace at or near 200°F (93°C) after a couple of minutes of operation. If the furnace continues to operate beyond the high-limit switch setting, shut it off immediately and replace the high-limit safety switch.

2.1.4 Checking the Temperature Rise

Temperature rise is the difference in temperature between the return and supply airstreams. It reflects the amount of heat transferred to the air as it passes over the heat exchanger. The temperature rise range is specified by the manufacturer and found on the furnace label. To determine the temperature rise, measure and compare the temperature of the two airstreams.

If the temperature rise is out of tolerance, it can often be corrected by adjusting the furnace blower speed. Note however, that this is also a symptom of a furnace being over- or under-fired due to an incorrect nozzle size.

To determine the temperature rise, complete the following steps:

Step 1 Make sure all the furnace access doors are secured and the air filters are clean. Initiate a call for heat at the thermostat.

Step 2 Drill holes in the supply and return ductwork (*Figure 23*) and insert a temperature probe into each hole. The thermometer in the supply duct must be located out of the line of sight of the furnace heat exchanger to prevent radiant heat from affecting the reading.

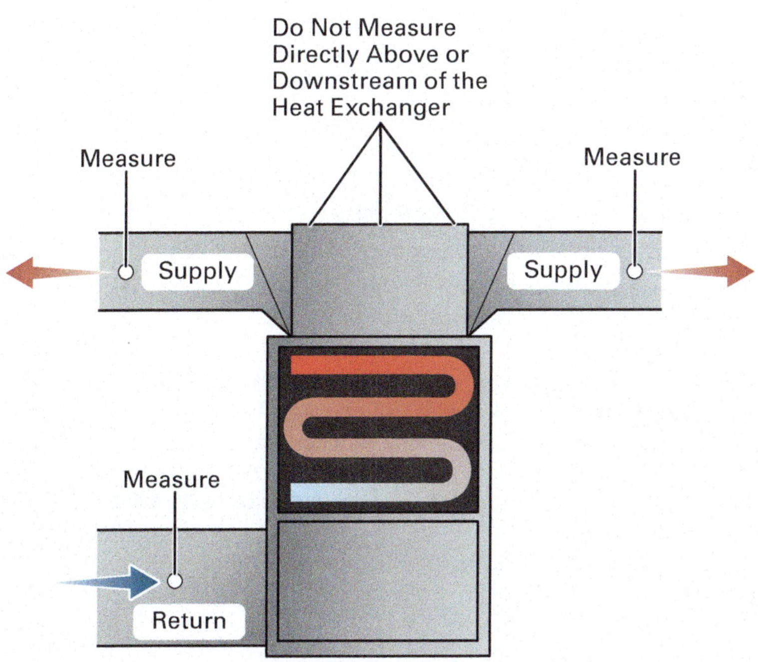

Figure 23 Hole locations for measuring temperature rise.

Step 3 Subtract the return air temperature from the supply air temperature. If the resulting temperature rise is below the specified range, reduce the blower speed. If it is above the specified range, make sure there are no airflow restrictions and increase the blower speed if possible.

Step 4 If minor adjustments in blower speed do not cure the problem, remove the nozzle assembly from the burner and check the nozzle size. Ensure that it is within the operating range of the furnace.

2.1.5 Servicing the Oil Burner

The oil burner assembly (*Figure 24*) is the primary active component in an oil-fired system. It must be carefully inspected and serviced.

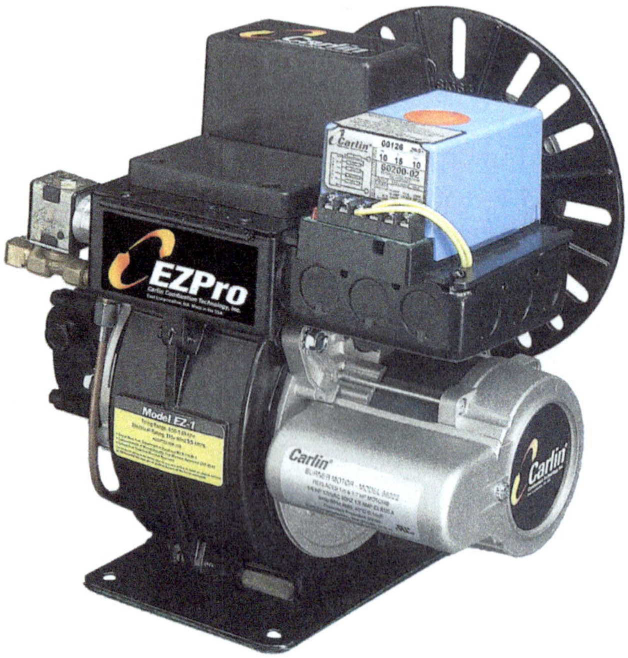

Figure 24 Oil burner assembly.
Source: Carlin Combustion Technology, Inc.

Inspect and service the burner as follows:

Step 1 Turn off the power.

Step 2 Disconnect all wiring and oil piping to the burner.

Step 3 Remove the burner assembly, being careful not to damage any of the components.

Step 4 Access and clean the blower wheel with a stiff brush.

Step 5 Clean the blast tube with a soft cloth or disposable towels.

Step 6 Remove, clean, and inspect the electrodes. Look closely for cracks in the ceramic insulators. Use fine steel wool for cleaning the metal electrodes; do not use sandpaper.

Step 7 Remove and replace the oil burner nozzle. Nozzles are inexpensive and are best replaced as part of periodic maintenance. Make sure the nozzle installed is the correct one for the equipment by checking the furnace or boiler data plate; do not refer to the data plate on the burner assembly alone, as the burner will accept a wide range of nozzle sizes. The furnace or boiler is the controlling factor for the nozzle size.

Step 8 Replace the nozzle using the nozzle wrench designed for that purpose (*Figure 25*). Do not overtighten it. The nozzle wrench consists of two sockets, with the outer socket larger than the second socket on the inside. Slip the wrench over the nozzle where the outer socket seats on the hexagonal section of the oil tube. The inner socket is then pushed over the nozzle, and the nozzle is loosened with minimal effort by rotating the handle attached to the inner socket.

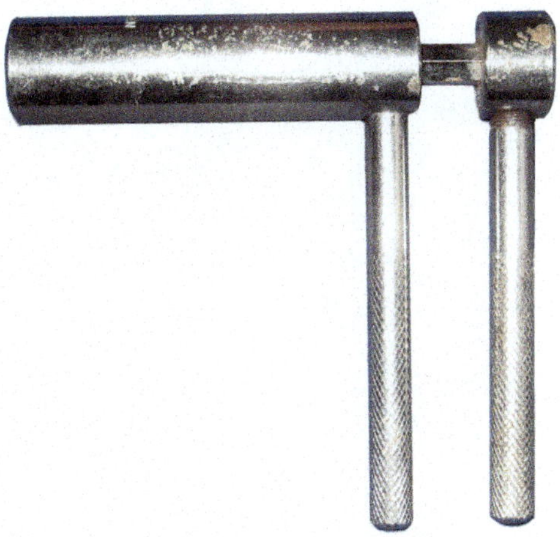

Figure 25 Nozzle wrench.

Step 9 Reinstall the electrodes and check the position of the electrode tips. Electrode insulators must be clean and free of cracks and carbon buildup. Clean or replace as necessary. Adjust the gap and spacing from the nozzle as specified by the oil burner manufacturer. A simple tool used for this purpose is shown in *Figure 26*. Additional information regarding electrode positioning follows this step list.

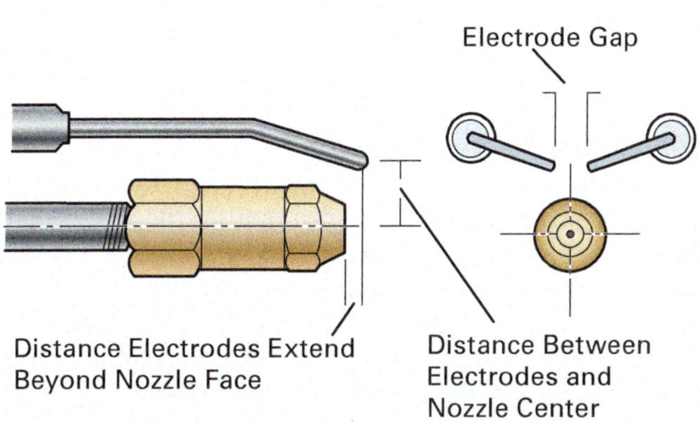

Figure 26 Electrode gauge used to set the gap and height.

Going Green

Multifuel System

A multifuel furnace can burn coal, wood, or oil. When the cost of oil is high, wood or coal might be the most affordable option. When there is no one available to stoke the furnace, fuel oil can be used instead.

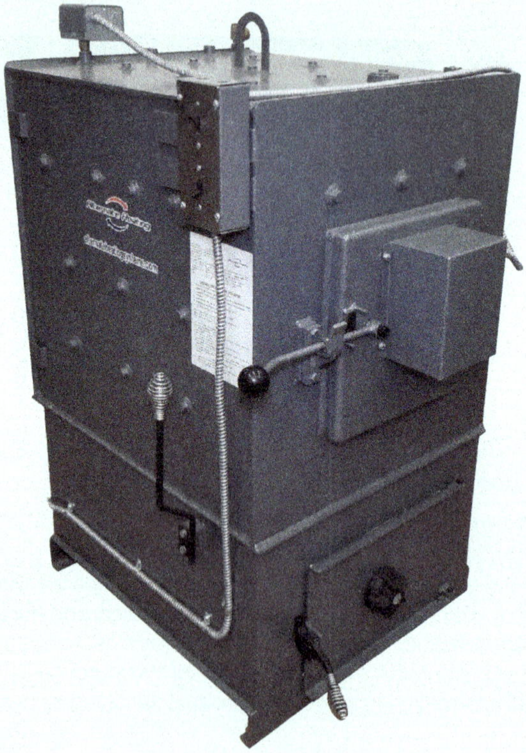

Source: Courtesy of Alternate Heating Systems

On Site

Disposing of Waste Oil

Waste oil cannot simply be thrown away or poured on the ground because it contaminates the environment. Waste oil should be disposed of at an appropriate recycling facility or service station.

Step 10 Inspect the cad cell flame detector and clean the lens. The cad cell must see light clearly to generate sufficient voltage. If the face of the cad cell becomes coated with soot or a film of oily dirt, the light will be obscured. Clean the face of the cad cell with a soft cloth or cotton swab to remove any debris.

Step 11 Check the oil pump coupling. Replace or service it as necessary.

Step 12 Remove and clean the oil pump screen.

Step 13 Reinstall the burner assembly and complete an operational test.

Step 14 Perform a combustion efficiency check.

Accurately setting the position of the electrode tips can be challenging without the correct tool. The angle of the nozzle spray pattern determines how far forward the electrodes are positioned in front of it. Nozzles with narrow angles require the electrode tips to be positioned further forward, ensuring the arc is at the edge of the oil spray. Wider angles require the electrode tips to be moved further back. The distance between the two tips, and the vertical distance of the tips above the nozzle, are also crucial to proper ignition.

The electrode gauge pictured in *Figure 26* simplifies the process of positioning the electrodes. The possible spray angles for nozzles are molded into the tool. Simply locate the angle of the nozzle stamped around the tool, then slip the tool over the nozzle tip with the selected angle pointing up. Position the tips of the electrodes in the tool's grooves. This properly positions the electrodes in all three dimensions. Once the electrode mount clamp is tightened to secure the electrodes, slide the tool off the nozzle.

2.2.0 Measuring Combustion Efficiency

Combustion efficiency is a measure of how well an appliance is burning its fuel. The ideal, of course, would be 100% efficiency, which means all the energy available in the fuel is being used to produce usable heat. Unfortunately, some losses in the process always occur. Combustion efficiency affects the operating cost, so it is important to keep the equipment operating at peak efficiency.

Combustion efficiency is determined by measuring the temperature of the flue gases and how much oxygen and CO_2 are present.

Draft is also an important factor when determining combustion efficiency. Insufficient draft through the venting system can cause dangerous carbon monoxide (CO) to back up into occupied spaces or overheat the combustion chamber. Too much draft will reduce the combustion efficiency.

Efficient combustion produces the least amount of smoke. The smoke produced by burning the fuel contains soot, which is deposited on the surfaces of the heat exchanger, combustion chamber, and vent system. Heavy soot deposits on the heat exchanger and vent system cause a reduction in heat transfer efficiency and restrict air movement through the vent.

A combustion analyzer (*Figure 27*) is essential test equipment for technicians servicing gas and oil furnaces. Some analyzers have optional printers that can be connected to the device with a USB cable or by wireless connection. Others can communicate wirelessly with smartphones and test data can be saved for further review.

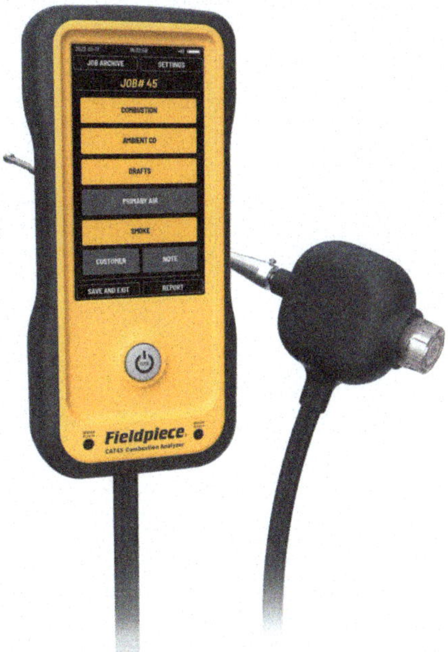

Figure 27 Electronic combustion analyzer.
Source: Courtesy of Fieldpiece Instruments

The instrument simultaneously calculates and displays combustion efficiency, flue gas temperature, CO_2 content, O_2 content, nitrous oxide (NOx), and other values. The analyzer performs combustion calculations for a variety of fuels, including natural gas, propane, coal, and several grades of fuel oil. Smoke density is the only relevant combustion parameter not measured.

The combustion analyzer shown in *Figure 27* is equipped with a probe assembly. The probe is inserted into an opening in the flue vent and connected to the analyzer.

2.2.1 Smoke Spot Testing

The combustion analyzer shown in *Figure 27* cannot test the density of the smoke. However, it does use the information you gather using a separate device to test the smoke. The results of the smoke test are manually entered into the device.

A **smoke spot test** should be performed before the rest of the analysis. The air shutter on the burner is then adjusted to obtain the proper smoke density before combustion testing begins. Since the position of the air shutter affects the analysis, it is important to set it at what appears to offer the best smoke result. The test is performed using a smoke spot tester and a comparison chart (*Figure 28*).

Smoke spot test: A test used to determine the density of smoke in the flue gases.

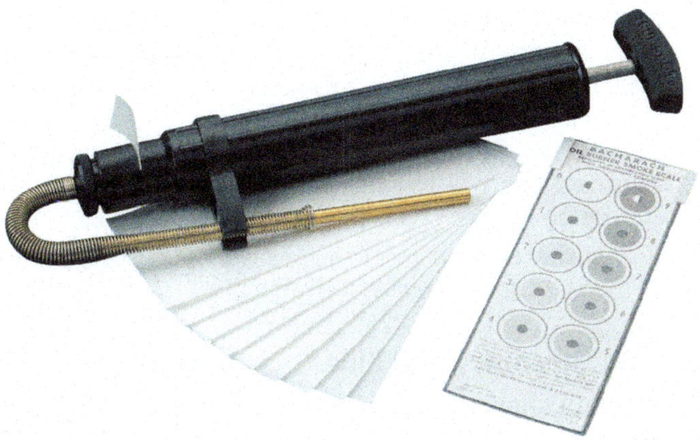

Figure 28 Smoke spot test kit.
Source: Used with permission from MSA

To test the density of the smoke in the flue gases, follow the steps listed below:

Step 1 Drill a test hole in the flue vent upstream of the equipment draft diverter and about 8" to 12" from the vent connector to the equipment, as shown in *Figure 29*. The opening should be at least 8" ahead of the draft diverter. This opening will also be used for combustion efficiency testing. Most probes will fit into a ¼" hole.

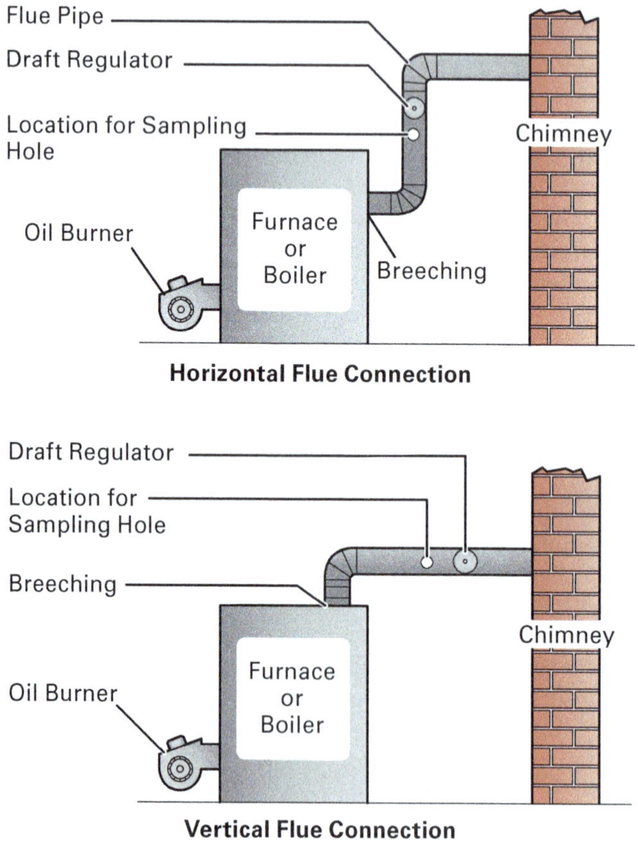

Figure 29 Combustion testing hole locations.

Step 2 Operate the furnace for 10 to 15 minutes before performing the smoke spot test. This ensures that all temperatures in the system have stabilized.

Step 3 Before starting the test, insert a small piece of test paper into the tester, in accordance with the manufacturer's instructions.

Step 4 Insert the probe of the smoke spot tester into the hole in the vent and pump the handle 10 full strokes, pumping at 1- to 2-second intervals. The gases pass through the filter paper inserted into the chamber, typically leaving a black circle on the paper. The darkness of the circle indicates the level of smoke in the gases.

Step 5 Compare the smudge on the paper to the chart that comes with the tool. A No. 1 smoke is standard for conventional burners, while a No. 0 smoke is standard for a **flame-retention burner**. Charts typically show examples up to No. 9 smoke.

Flame-retention burner: A high-efficiency oil burner with a flame retention ring surrounding the oil nozzle.

Step 6 Adjust the air shutter on the burner to achieve the desired smoke level, per the equipment manufacturer's recommendations.

2.2.2 Combustion Efficiency and Flue Gas Testing

Before testing for combustion efficiency and flue gases, make sure the blower compartment door is secured. The furnace should be operated for a minimum of 10 minutes before testing.

Follow the procedure listed below to test the combustion efficiency:

Step 1 Set up the combustion analyzer in accordance with the manufacturer's instructions. The probe assembly on the test set may contain a water trap and filter assembly. Before using the device, make sure the trap is dry and empty, and replace the filter. Water cannot be allowed to enter the instrument. The test instrument may also contain oxygen, CO, and CO_2 sensors that require periodic replacement. Again, follow the manufacturer's instructions.

Step 2 Insert the probe so that the tip is in the center of the vent pipe and allow time for the readings to stabilize.

Step 3 If required, enter the results of the smoke test. Although many analyzers do not test the smoke density directly, the results of an independent smoke test may be manually entered.

Step 4 Record or print the test results and compare them to the manufacturer's requirements. The values will vary, depending on the type of burner and the AFUE rating of the furnace, so it is important to make this comparison. Typical values for a flame-retention burner are as follows, but these values should not be considered as the standard:

- Oxygen: 3% to 5%
- CO_2: 10% to 12.5%
- CO: <50 ppm
- Stack temperature: 400°F to 450°F
- Vent draft: 0.02 to 0.04" w.c.
- Overfire draft: Within 0.02" w.c. of vent draft

Step 5 If the combustion efficiency is not within the required tolerances as established by the manufacturer, adjust the air shutter on the burner to obtain the correct value. If this value cannot be achieved by air adjustment without significantly affecting the smoke setting, additional troubleshooting and inspection is needed to isolate the cause.

Step 6 Take a negative draft reading through the inspection door (over the flame). A difference between this reading and the reading from the flue of more than 0.02" w.c. indicates a dirty or restricted heat exchanger. Adjustments in the draft regulator may also be required.

Step 7 If all is well, clean up the work area and seal the test holes in the vent, and complete the service call.

If the required CO_2 level cannot be reached without exceeding an acceptable smoke number, poor mixing of air and fuel is likely. This could be caused by an improper nozzle spray pattern for the burner. Also check the blast tube and flame retention head for obstructions.

Also look for an unbalanced burner flame. If the flame appears unbalanced, the fuel input may be too low. Check both the nozzle size and oil pressure. If the nozzle is partially plugged or defective, replace it with another as recommended by the manufacturer. Never try to repair or increase the size of a nozzle orifice. The machined opening is easily damaged or distorted, ruining the pattern.

2.0.0 Section Review

1. Which of the following should occur if the cad cell leads are disconnected from the primary control while the furnace is running?
   a. The high-temperature limit switch should open after one minute.
   b. The primary control should lock out and the burner should stop.
   c. The burner should continue running.
   d. The indoor blower should shut off immediately.

2. How many times should the handle be pumped before drawing a sample of flue gases for smoke analysis testing?
   a. Once
   b. Twice
   c. Five times
   d. Ten times

3.0.0 Oil Furnace Troubleshooting

Performance Task

4. Test a cad cell flame detector.

Objective

Explain how to troubleshoot a typical oil-fired heating system.

a. Describe troubleshooting procedures for common oil heating problems.

b. Describe troubleshooting procedures for oil furnace control devices.

On a service call, always listen carefully to the customer's description of the problem. A wise technician can often narrow down the possibilities before the conversation is done.

3.1.0 Troubleshooting Common Oil Heating Problems

This section provides a troubleshooting guide for common oil heating problems. The following checks should be made if the customer says that the furnace is not producing enough heat or is not operating at all:

Step 1 Check for power at the furnace. If voltage is present, de-energize the furnace for the remaining checks. If no voltage is present, check the circuit breaker or fuse serving the furnace.

Step 2 Check the burner motor fuse (if equipped).

Step 3 Ensure the oil supply is adequate and all manually operated oil valves are open.

Step 4 Check the limit switches to ensure the contacts are closed.

Step 5 Inspect the combustion chamber for excess fuel oil.

WARNING!

Make sure the combustion chamber is free of excess oil before starting the burner. Failure to do so could result in an explosion.

Step 6 Set the thermostat to call for heat. Reset the burner primary control, if necessary. Observe it as the ignition sequence begins. If the burner operates as expected, observe its operation until it satisfies the thermostat, several times if necessary to ensure its reliability.

If the burner and indoor blower do not operate properly, note the point at which it fails. Use the troubleshooting chart in *Appendix 03310A* to determine which parts of the oil heating system are most likely to be at fault. *Appendix 03310A* shows the most likely causes of common oil heating system failures. Once you are aware of the possible causes for the problem, troubleshooting can begin.

3.1.1 Oil Burner Troubleshooting

The oil burner nozzle is an occasional source of problems. The following conditions can be traced to a worn or damaged nozzle:

- No flame (partial or complete blockage)
- Flame asymmetrical in size and shape
- Flame veering to one side of the combustion chamber (*Figure 30*)
- Low carbon dioxide reading in flue gases (less than 8%)
- Delayed ignition
- Oil odors

If there is any doubt about the nozzle, replace it. If the nozzle strainer is clogged or dirty, determine the cause. The fuel oil filter and oil pump strainer should capture debris before it reaches the nozzle. The oil supply may also be contaminated with water.

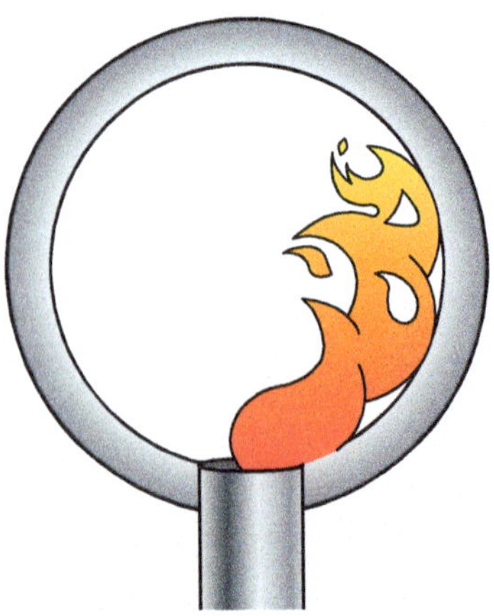

Figure 30 Improper burner flame.

If a varnish-like substance has formed on the nozzle, it may indicate overheating. Three major causes of overheating are as follows:

- The burner is overfired (nozzle too large for the furnace).
- Ignition is occurring too close to the nozzle.
- The nozzle is positioned too far forward.

To determine if a burner is overfired, check the furnace nameplate to find the heating input or nozzle size. Then, check the nozzle and find the flow rate stamped on the nozzle body. If the nozzle size is not shown on the nameplate, do the math to determine the heating capacity of the nozzle. When working with Grade 2 fuel oil, the heating input is calculated at 144,000 Btus per gallon. A nozzle that produces a heating input above that shown on the nameplate is overfired.

If the flame ignites too close to the nozzle, check the oil pressure. A small increase in oil pressure may solve the problem. The nozzle may also be positioned too far forward. Check the manufacturer's specifications and position the nozzle accordingly.

Did You Know?

Oil Burner Nozzle Wear

Oil burner nozzles can wear out over time, but they do survive many hours of operation. The abrasive action of microscopic particles in the oil, which is forced at high pressures through a tiny orifice, can erode the inside of the orifice. In time, this abrasion can distort the orifice by creating tiny grooves. This type of wear is referred to as *wire drawing.*

3.1.2 Fuel Supply Troubleshooting

If an oil pump is working properly, the pressure should be steady. If the needle of an attached gauge pulsates, check for a fouled fuel filter or pump strainer and for air leaks in the fuel supply lines.

A vacuum gauge connected to the inlet side of a single-stage pump should read less than 10" of vacuum. On a two-stage pump, the gauge should read less than 15" of vacuum. If a lower pressure is measured, check for kinks or restrictions in the fuel supply lines and a plugged fuel filter. Also listen for a whining sound at the pump, which can indicate a restriction of some sort on the suction side of the pump.

Oil supply lines in residential installations generally use $^3/_8$" outside diameter (OD) soft copper tubing. To prevent oil and air leaks in fuel lines, use flare fittings on copper tubing connections. Compression fittings should never be used on copper fuel lines.

Remember that water can condense in the fuel tank. If enough water collects in the bottom of the tank, it can make its way into the fuel line serving the burner. The water can cause corrosion and harm the pump, in addition to causing ignition and flame stability problems. Fuel tanks should be periodically checked for the presence of water. If water is found, it must be drained or pumped from the tank.

Going Green

Disposing of Oil Storage Tanks

Oil storage tanks for heating systems, particularly those for residential systems, may not be covered by the same federal regulations as industrial underground storage tanks, but they still contain potentially hazardous fuel oil residue. Always check and abide by all applicable federal, state, and local codes when disposing of oil storage tanks.

3.1.3 Odors

To find the source of a fuel oil odor, proceed as follows:

Step 1 Check for leaks in the oil supply and return piping. Tighten or replace connections as necessary.

Step 2 Check the fuel tank vent. Correct venting problems where necessary and securely tighten or seal the vent connections.

Step 3 Check for seepage around the fill or vent connections if an oil film is visible on the tank.

Step 4 Check for seepage past the oil pump cover gasket or the pump shaft. If necessary, replace the pump.

Step 5 Check for loose oil line connections at the burner and repair as necessary.

Step 6 Check for leaks around the fuel filter housing. Leaks anywhere on the oil line leading from the tank to the oil pump not only cause odors, but also cause a loss of prime. If air can enter the line during the off cycle, gravity will cause the oil to fall down the line.

3.1.4 Flue Vents and Chimneys

The flue vent piping and chimney can also become a source of problems. The proper design and installation of the vent system can eliminate many potential problems.

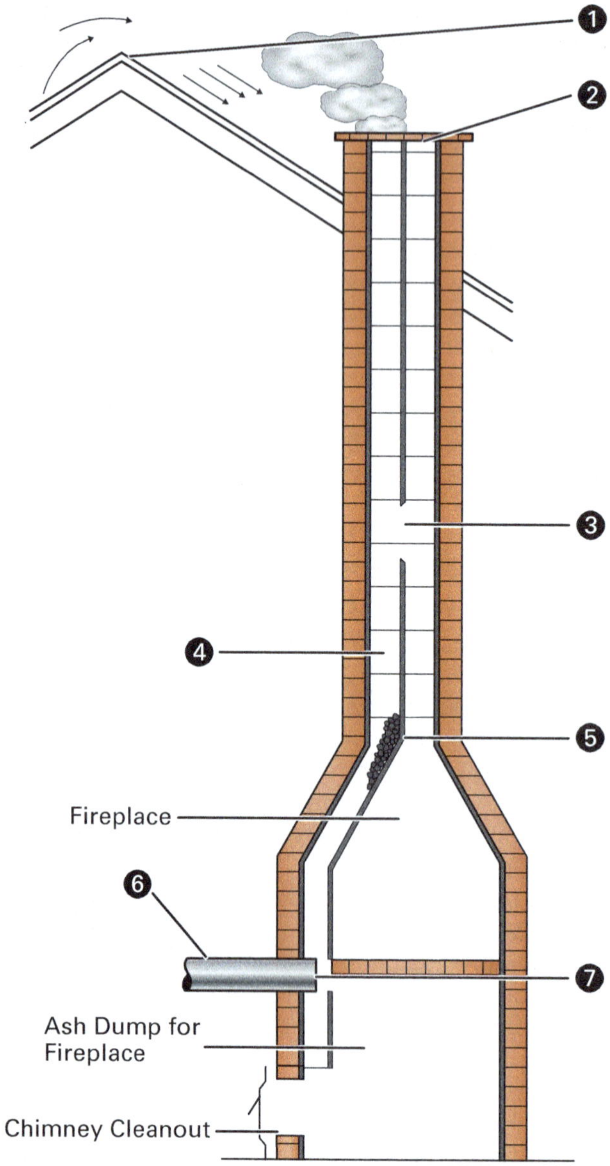

Figure 31 Common chimney problems.

Refer to *Figure 31* and the numbered list here for a review of potential venting issues:

Downdraft: A situation in which outdoor air is forced down a chimney, preventing proper draft. A downdraft can occur if the top of the chimney does not extend above the roof line.

Coping: The upper masonry course of a chimney designed to keep water out.

1. A **downdraft** might occur if the top of the chimney is lower than surrounding objects. The chimney should exceed the height of any obstructions within 30' or as required by local building codes.
2. A restricted chimney opening can be caused by problems with the chimney cap, ventilator, or **coping**.
3. In masonry chimneys, a break in the lining or tile can allow pieces to fall away and block the chimney. A breech between a chimney with two passages can allow flue gases to enter the building through a fireplace or other connected appliance.
4. An oversized chimney might not be able to establish and maintain an adequate draft.
5. Soot from poor combustion can collect at narrow openings and changes of direction in the vent piping. Obstructions in the vent piping and chimney can be located by using a bright light and a mirror.

6. Loose or open joints in the vent connector can allow the flue gases to leak into the room, especially during start up when the vent system is cold, and a reliable draft has not yet developed.
7. A vent connector that extends too far into the chimney can interfere with the flow of flue gases.

3.1.5 Draft Regulator

A draft regulator (*Figure 32*) consists of a small door on the side of the flue pipe. It is hinged near the center and controlled by adjustable weights. A draft regulator is necessary to help manage the volume of room air drawn into the vent system downstream of the combustion process.

Figure 32 Draft regulator.
Source: Field Controls

The draft regulator is barometric in nature, meaning that it responds to changes in pressure inside the flue pipe. When there is little or no draft through the vent, the weight on the front keeps the damper closed. As the draft increases and a negative pressure develops behind the damper, it opens slightly to admit more air.

The position of the weight is adjusted to control the amount of air the regulator admits as the draft increases. Without a draft regulator in place, all air exiting the furnace must come through the combustion chamber, wasting a great deal of heat and interfering with the combustion process.

If the draft in the vent is excessive, too much heat is lost. If the draft is insufficient, incomplete combustion results. The draft regulator should maintain a consistent draft over the fire, typically 0.01 to 0.03" w.c.

On Site

Chimney Liners

Correcting masonry-related chimney problems is much easier today due to advances in technology. If a chimney is oversized or is improperly lined, a flexible chimney liner kit rated for use with fuel oils can be installed to correct the problem.

3.2.0 Control Troubleshooting

The ability to effectively troubleshoot oil furnace problems requires an understanding of the controls as well as the ability to interpret the schematic and/or wiring diagrams provided by the manufacturer. The control system for an oil furnace is relatively simple. *Figure 33* provides an overview of the external control system.

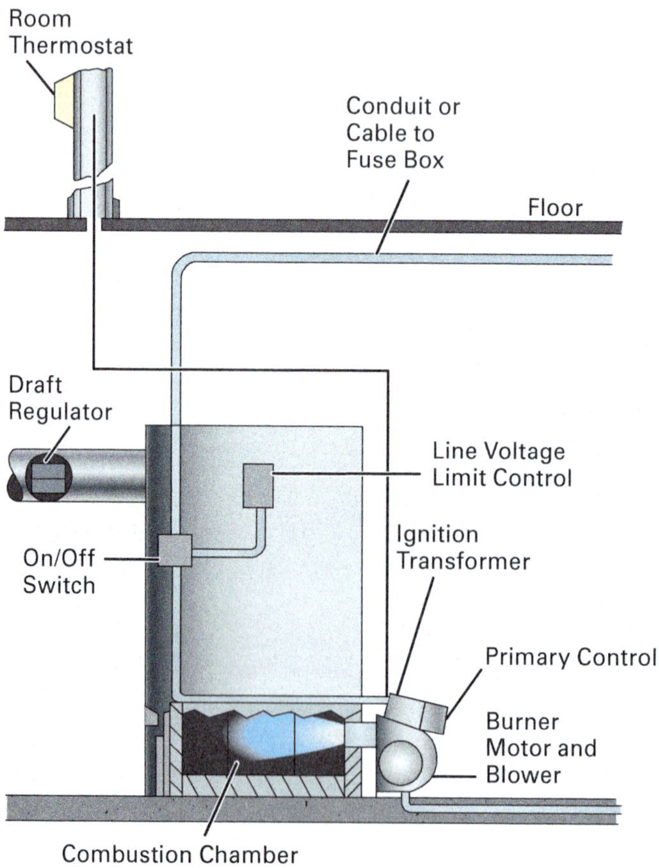

Figure 33 Oil furnace external control system.

Figure 34 is a schematic diagram for an oil furnace and add-on air conditioner. This schematic is for an older primary control that may still be found in service. The operating sequence is shown on the diagram. *Figure 35* is the connection diagram for the same furnace. This is a unique case where the primary control is providing the 24VAC control voltage for the entire system.

Before troubleshooting the furnace controls, make sure the following conditions are in order:

- The ignition electrodes are properly positioned for the nozzle in use.
- Contact between the ignition transformer terminals and the electrodes is firm and reliable.
- The oil burner nozzle is clean and correctly sized.
- The oil pressure is adequate.
- The oil supply is adequate. Many service calls are related to a lack of fuel in the tank.

Older systems used a temperature switch in the furnace plenum to start the indoor blower after the furnace reached a preset temperature. The blower control was often integrated with the high-limit safety switch. When combined, these controls were referred to as *combination fan/limit controls*. Some furnaces

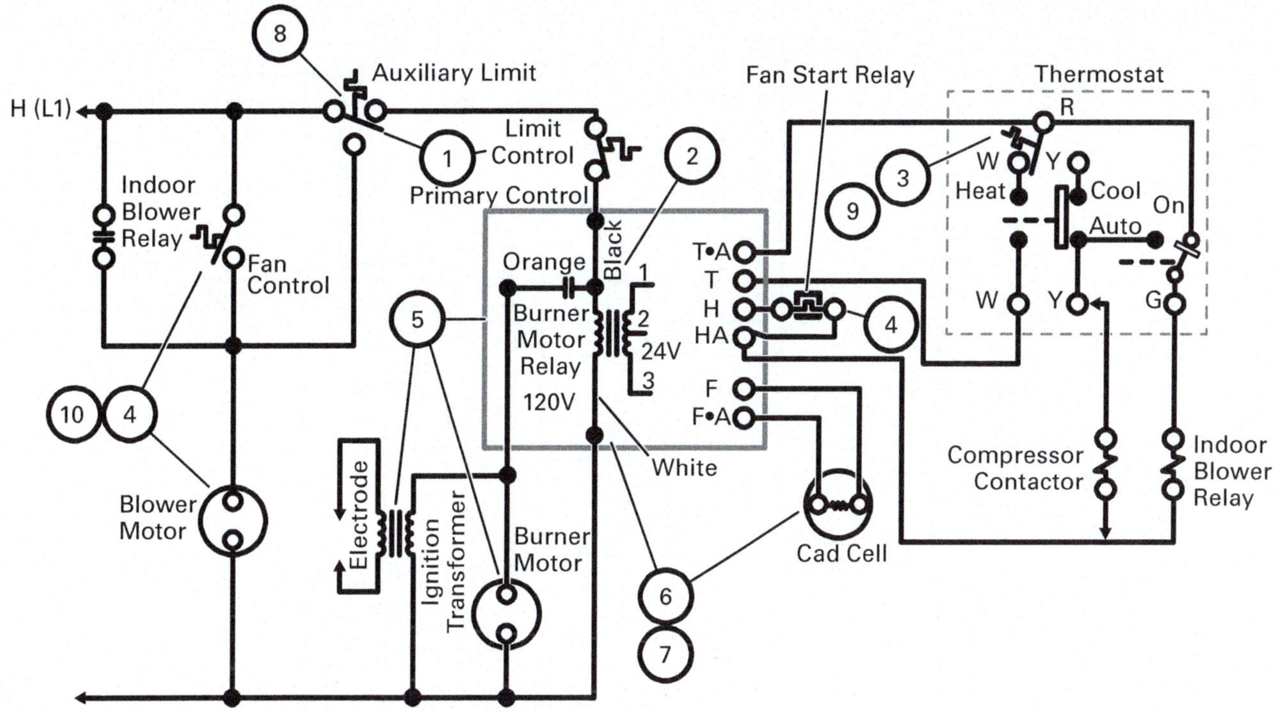

1. Line potential feeds through the secondary limit and limit controls to power primary control.
2. The primary control provides 24-volt control circuit.
3. The thermostat closes the circuit between R and W to call for heat.
4. The fan control heater is energized through primary control. After a short period, the heater provides sufficient heat to close the fan contacts. This energizes the blower motor.
5. The primary control simultaneously energizes the burner motor and ignition transformer at the oil burner. The burner motor operates the oil pump and combustion blower to feed air and oil vapor into the combustion chamber. The fuel mixture should ignite with the spark furnished by the ignition transformer.
6. If combustion does not take place within the time specified on the primary control, as detected by the cad cell, the primary control locks itself out.
7. Should a flame failure occur during the run cycle, the primary control locks itself out in response to the cad cell.
8. The secondary limit opens at temperatures above setpoint. This de-energizes the primary control but still allows blower motor operation through the NO contacts.
9. As the heating demand is satisfied, the thermostat heating contacts open. This de-energizes the oil burner circuits.
10. The blower motor continues running until furnace temperature drops below fan control

Figure 34 Oil furnace schematic diagram.

operating today still use this type of control. Otherwise, a fan relay equipped with a time delay feature, much like those used in gas furnaces, is used to control the indoor blower.

3.2.1 Primary Control Troubleshooting

The primary control shown in *Figure 36* is a modern control and safety device. The module contains a switching circuit that energizes the burner motor and the ignition transformer on a call for heat. Although the two low-voltage thermostat terminals are visible on top, the line voltage connections are on the bottom (*Figure 37*).

The primary control allows a set time to elapse before it de-energizes and locks out the burner, giving the burner time to ignite and stabilize. The timing of a primary control is not adjustable. Different primary controls have different lockout times. Burner manufacturers select a primary control that is design specific. If a flame is established within the prescribed time, burner operation continues until the thermostat is satisfied.

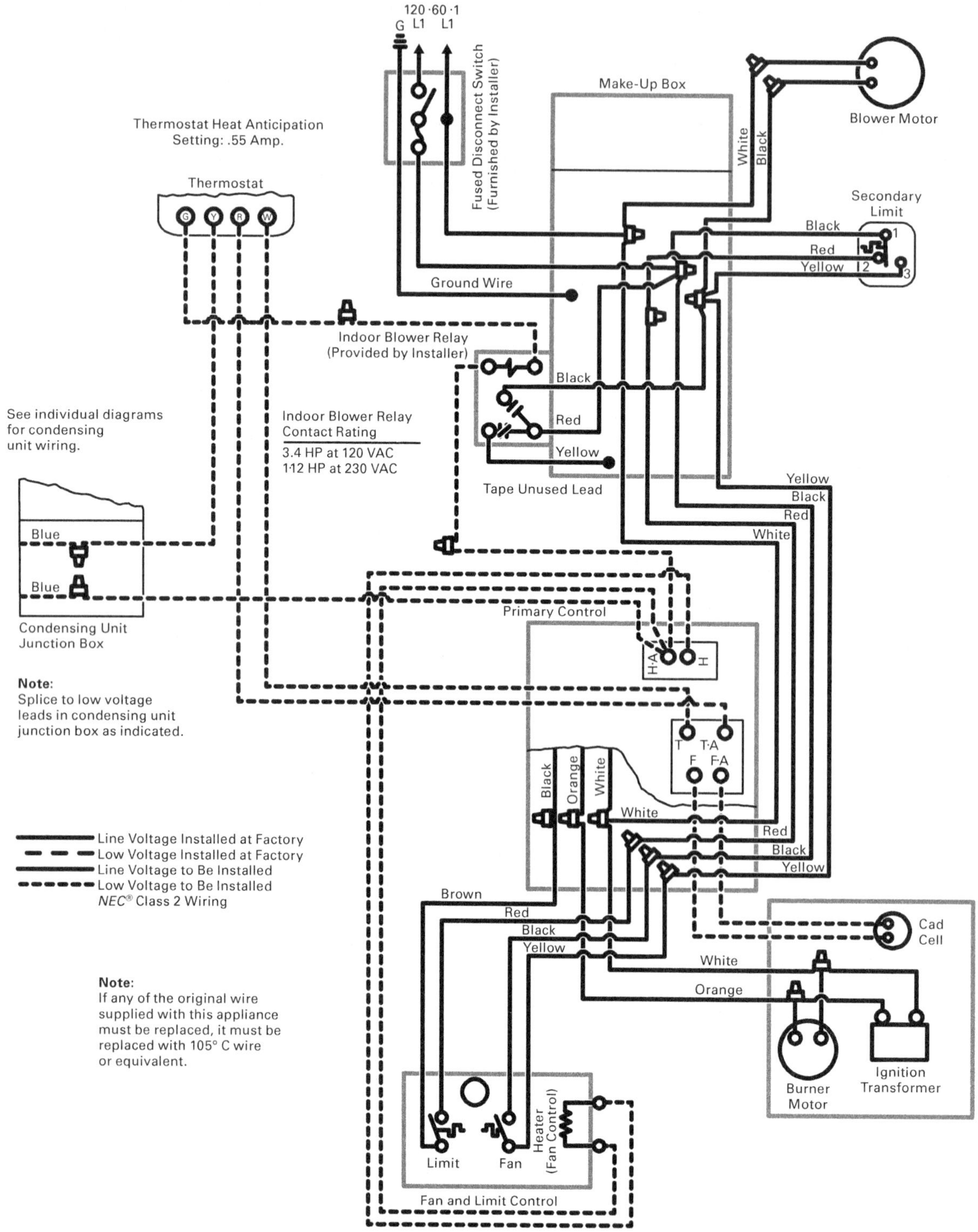

Figure 35 Oil furnace connection diagram.

The following describes the operation of this primary control. Refer to the connection diagram shown in *Figure 38*:

- The room thermostat closes its contacts when a drop in space temperature occurs, signaling the burner to start.
- The primary control sends power to the burner motor, ignition transformer, and oil solenoid valve (if equipped). The burner motor starts, providing combustion air and driving the oil pump. The ignition transformer establishes an arc across the electrodes. The oil solenoid valve is energized after a short delay, allowing the blower time to reach its normal operating speed.

Figure 36 Beckett Genysis® primary control.
Source: HVACGuru

Figure 37 Beckett Genysis® line-voltage terminals.
Source: HVACGuru

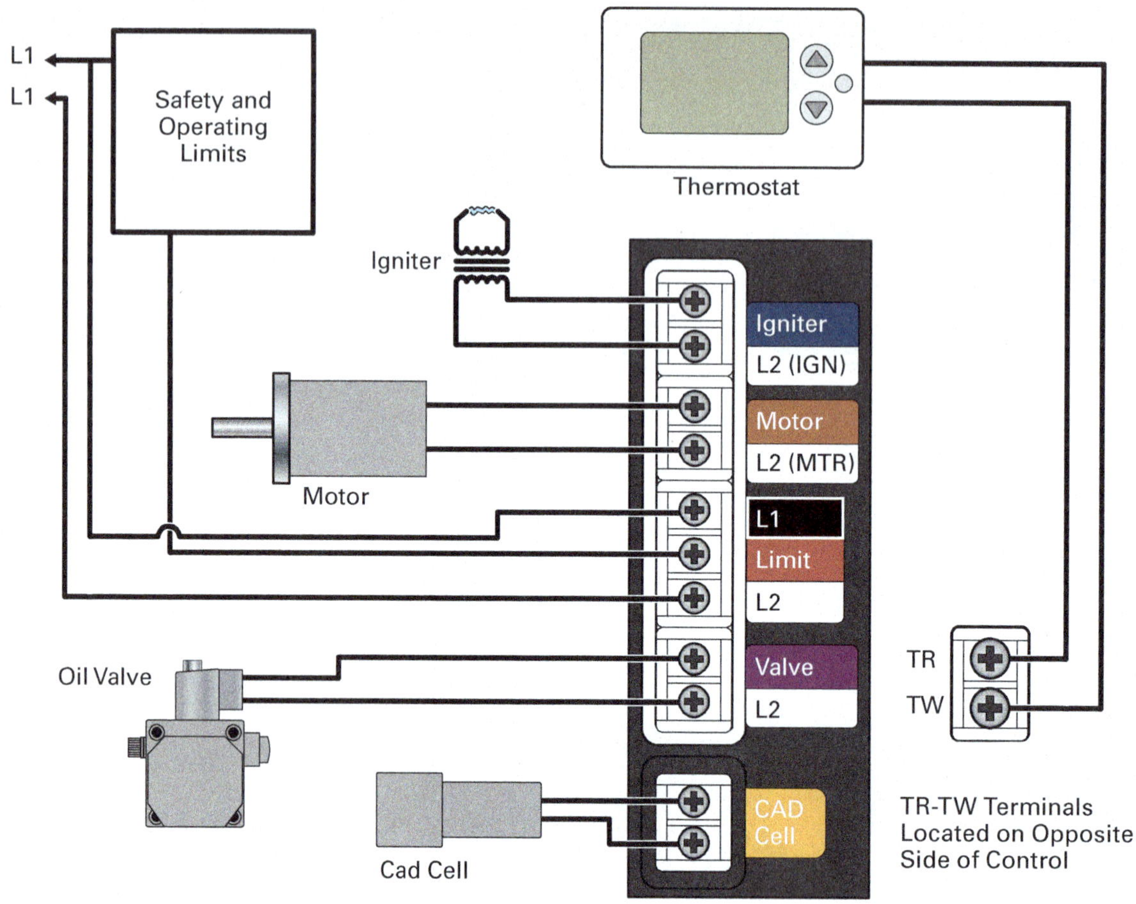

Figure 38 Beckett Genysis® connection diagram.
Source: Courtesy of RW Beckett Corp.

- The primary control waits until the factory-set time delay has elapsed. If the cad cell is not sending sufficient voltage to the primary control to confirm a flame at that time, the primary control de-energizes the burner motor, ignition transformer, and oil solenoid valve. The primary control locks out and can be reset by pushing the Reset button for >1 second. This is referred to as a *soft lockout*.
- If the flame is lost (insufficient voltage from the cad cell) while the burner is firing, the primary control shuts down the burner. After a one-minute delay, it attempts to restart the burner. The control will only attempt a restart a few times before a *hard lockout* occurs. A hard lockout reset requires the Reset button to be pressed for >15 seconds. This helps prevent the user from repeatedly attempting to reset the burner and filling the combustion chamber with oil.
- If the high-limit safety switch(es) open(s), the primary control immediately de-energizes the burner.

Testing the primary control and cad cell must often be done with the system energized. Observe all electrical safety precautions when working with energized equipment.

The equipment required for checking the primary control and cad cell include the following:

- Multimeter
- 1,500Ω resistor
- Several sets of insulated jumper wires with alligator clips

To troubleshoot a primary control if the burner locks out, proceed as follows:

Step 1 Make sure all safety/limit switches are closed. Confirm line voltage is present at the L1 and L2 terminals of the primary control. Then de-energize the burner.

Step 2 Disconnect the room thermostat from terminals TR and TW. Place a jumper wire across TR and TW.

Step 3 Disconnect the cad cell leads from the primary control. Connect one end of a 1,500-ohm (Ω) resistor to one of the cad cell terminals. This resistor will simulate a cad cell that senses light when it is later placed into the circuit. Be prepared to connect the other end of the resistor, via a jumper wire, to the other cad cell terminal.

Step 4 Energize the furnace and press the Reset button on the primary control. Quickly (within 5 to 10 seconds) connect the other end of the 1,500Ω resistor to the open cad cell terminal.

Step 5 If the primary control locks out after the ignition trial period has elapsed, it indicates that the primary control is defective. If in doubt, wait five minutes and repeat the test.

If the primary control does not lock out with the resistor in place, allowing the burner to continue running, the primary control is functioning properly. The cad cell may be the problem.

WARNING!

Never replace or bypass a cad cell with a resistor, even temporarily. The cad cell is an essential safety device for oil burner operation. Bypassing it can lead to fires and explosions that result in serious property damage and physical injury.

3.2.2 Cad Cell Troubleshooting

The cad cell flame detector consists of a light-sensitive cell and a socket with a mounting bracket and lead wires. It is mounted in the burner and connected to the primary control as shown in *Figure 38*. The position of the cad cell is critical. It must be positioned so that it can detect the light from the flame.

When checking the cad cell resistance with an ohmmeter, the resistance should range from 300Ω to 1,600Ω when a stable flame is present. With the cad cell lens covered (no light at all), the resistance should be roughly 100,000Ω. This is high enough to block any current flow through the circuit. If the cad cell resistance is higher than 1,600Ω with the burner operating properly and a stable flame, the cad cell may need cleaning or re-aligning on its target.

To test a cad cell, proceed as follows:

Step 1 De-energize the furnace and burner. Disconnect at least one lead of the cad cell from the primary control terminals. Use a multimeter to check the resistance through the cad cell. In the dark, as it should be, the resistance could be as high as 100,000Ω. If a small amount of light is visible to it, the resistance will likely be lower, between 25,000Ω and 100,000Ω.

Step 2 Reconnect the cad cell to the terminals. Disconnect at least one lead of the room thermostat from the TR/TW terminals and place a jumper wire across the TR and TW terminals.

Step 3 Energize the furnace. With the jumper wire across the thermostat terminals, the burner should start quickly.

Step 4 After the burner ignites the fuel, quickly disconnect both leads of the cad cell from the terminals. Remember the burner will stop after a short delay. Measure the resistance through the cad cell before reconnecting it.

Step 5 With the burner operating, the resistance should not exceed 1,600Ω. If the resistance is higher than 1,600Ω, the cad cell may be damaged or misaligned. A resistance over 1,400Ω is cause for concern, as it may cause nuisance lockouts.

NOTE

You can install the 1,500Ω resistor across the cad cell terminals, if necessary, to keep the burner running while you monitor the resistance of the cad cell.

3.2.3 Ignition Component Testing

The ignition system consists of the ignition transformer and the electrodes. Perform the following steps to troubleshoot the ignition system:

WARNING!

Follow the manufacturer's instructions and be extremely careful when performing this procedure. The ignition transformer produces 10,000V at a current sufficient to cause a severe shock.

Step 1 De-energize the furnace and burner.

Step 2 Shut off the fuel supply.

Step 3 Disconnect the blower motor leads from the primary control.

Step 4 Connect a jumper wire across TR and TW or go to the room thermostat and set it to call for heat.

Step 5 Connect a voltmeter capable of measuring high voltage across the ignition transformer output (secondary) terminals and energize the furnace. The voltage should be at least 10,000V. If the voltage is low, replace the transformer.

Note that the ignition transformer can be tested in the same manner as other transformers for open windings and grounds.

The ignition transformer must make good contact with the contacts of the electrodes. The contacts mounted to each electrode are typically flexible and can be bent or curved to ensure good contact is made with the transformer terminals.

If the electrode insulators are cracked or chipped, an arc may be established between the electrode and any grounded portion of the burner assembly. If the ignition transformer is working properly, yet no arc is observed at the end of the electrodes, check their position and gap and inspect for flaws in the insulators.

3.2.4 Solenoid Oil Valves

Solenoid oil valves can be mounted directly on the oil pump or in the discharge line of the pump. This valve controls fuel flow to the burner. A solenoid oil valve can help maintain the prime in the oil piping when the tank is below the burner.

Today's primary controls typically energize the valve after a short delay, allowing time for the burner motor to start and reach its normal operating speed. The valve is spring-loaded and closes immediately when de-energized. Some models have an integral thermistor that delays opening until the burner motor reaches full speed. As shown in *Figure 38*, the valve has its own terminals on the primary control. If the burner does not have an oil solenoid valve, the terminals remain open. It is best to slide insulated terminals over the connections to prevent electric shocks when troubleshooting with the primary control energized.

Oil solenoid valves offer reliable service and rarely fail. A multimeter is used to test the valve coil for continuity and grounds. Although the valve can also fail mechanically, it is extremely rare.

A failure usually results in the valve not opening, preventing fuel from reaching the burner. To check the solenoid oil valve, proceed as follows:

Step 1 As the valve energizes and de-energizes, listen for an audible click as it opens and closes.

Step 2 If the valve does not open when the thermostat calls for heat:

- Ensure the oil pump is producing normal pressure.
- Check the power supply to the valve. If voltage is not present with the primary control having already started the burner motor, the primary control is likely the problem.
- If voltage is present, de-energize the circuit and test the solenoid coil with an ohmmeter. A failed coil can sometimes be replaced, depending on the valve construction. Note that the solenoid valve on modern burners may be integrated into the oil pump (*Figure 39*).

Figure 39 Beckett Cleancut® oil pump with integral solenoid valve.
Source: Courtesy of RW Beckett Corp.

3.0.0 Section Review

1. True or False? Compression fittings are best used to connect oil supply piping.
   a. True
   b. False

2. In older oil furnaces, the indoor blower was typically controlled by which device?
   a. Primary control
   b. Stack switch
   c. Combination fan/limit control
   d. Solid-state fan relay

Module 03310 Review Questions

1. The AFUE rating of a non-weatherized oil furnace must be at least ______.
   a. 65%
   b. 78%
   c. 92%
   d. 100%

2. A small blower in a high-pressure oil burner is used to ______.
   a. prevent overheating
   b. cool the oil
   c. atomize the oil
   d. provide combustion air

3. An oil burner's input, in Btus, is determined by the ______.
   a. type of fuel
   b. nozzle flow rate
   c. position of the electrodes
   d. size of the ignition transformer

4. In addition to atomizing the fuel, what else does the oil nozzle do?
   a. The nozzle modulates the oil supply to control capacity.
   b. A nozzle sensor detects light from a flame.
   c. A nozzle sensor detects the heat of the flame.
   d. The nozzle controls the oil flow volume.

5. Which of the following receives a signal from the thermostat and controls power to the burner motor and transformer?
   a. Cad cell
   b. Limit switch
   c. Primary control
   d. Master cylinder

6. Which device is most used in oil furnaces to verify a continuous flame?
   a. Flame rectifier
   b. Thermopile
   c. Cad cell
   d. Hot surface igniter

7. When the fuel oil must be lifted from a buried tank outdoors, where should a check valve be installed in the supply piping?
   a. Next to the oil nozzle
   b. As close to the tank as possible
   c. Above the ceramic insulators
   d. Next to the burner

8. When is a single-pipe oil system best used?
   a. When the burner is located on the same level or above the bottom of the oil supply tank.
   b. When the oil supply level is located below the burner.
   c. When the vertical distance from the tank to the furnace is more than 15'.
   d. When the vertical distance from the tank to the furnace is more than 30'.

9. An electrode gauge is used to ______.
   a. measure the temperature of the electrodes
   b. measure the diameter of the electrodes
   c. set the position of the electrode tips
   d. check the condition of the electrodes

10. What does a smoke spot tester measure?
    a. The amount of smoke in the combustion chamber.
    b. The density of smoke in the flue gases.
    c. The amount of smoke in the heat exchanger.
    d. The toxicity levels of smoke in the combustion chamber.

11. Where should the test hole for smoke spot and combustion testing be drilled?
    a. As close as possible to the furnace.
    b. Between the draft regulator and the chimney.
    c. At least two flue pipe diameters from the furnace breeching.
    d. 8" to 12" from the vent connection to the equipment.

12. If the burner flame is veering off to one side of the combustion chamber, the most likely cause is ______.
    a. an electrode gap that is too wide
    b. a worn or damaged nozzle
    c. weak ignition transformer
    d. loose oil pump coupling

13. What can a pulsating needle on an oil pump pressure gauge indicate?
    a. A defective cad cell
    b. A cracked heat exchanger
    c. A clogged fuel filter or strainer
    d. A defective nozzle

14. Which of the following is the possible cause of a downdraft in a chimney?
    a. The top of the chimney being too high.
    b. The top of the chimney being lower than surrounding objects.
    c. An obstruction in the chimney.
    d. A crack in the chimney lining.

15. When measuring cad cell resistance with an ohmmeter while a stable flame is present, the resistance should range from ______.
    a. 10,000Ω to 20,000Ω
    b. 2,000Ω to 8,000Ω
    c. 300Ω to 1,600Ω
    d. 25Ω to 75Ω

Answers to odd-numbered review questions are found in the Module Review Answer Key at the back of this book.

Answers to Section Review Questions

Answer	Section	Objective
Section One		
1. a	1.1.1	1a
2. d	1.2.1	1b
3. b	1.3.0	1c
4. a	1.4.1	1d
Section Two		
1. b	2.1.2	2a
2. d	2.2.1	2b
Section Three		
1. b	3.1.2	3a
2. c	3.2.0	3b

MODULE 03312

Troubleshooting Accessories

Source: Marvin Samuel Tolentino Pineda/iStock

Objectives

Successful completion of this module prepares you to do the following:

1. Explain how to troubleshoot common residential HVAC system accessories.
    a. State how to approach the troubleshooting process.
    b. Explain how to troubleshoot residential humidifiers.
    c. Explain how to troubleshoot electronic air cleaners.
    d. Explain how to troubleshoot ultraviolet lighting devices.
2. Explain how to troubleshoot economizers and recovery ventilators.
    a. Describe the operation and troubleshooting of economizers.
    b. Describe the operation and troubleshooting of recovery ventilators.

Performance Task

Under supervision, you should be able to do the following:

1. Using the proper tools and circuit diagrams, isolate and resolve problems in instructor-selected accessories.

Overview

HVAC systems, especially in cold, dry climates, are often equipped with humidifiers to maintain a healthy moisture level indoors. Electronic air cleaners may also be installed in both residential and commercial environments. Economizers, which take advantage of cool outside air to provide free cooling indoors, are common in commercial rooftop systems. Energy and heat recovery ventilators are another important indoor air quality (IAQ) accessory that are applied in both residential and commercial systems. Servicing HVAC accessories requires an understanding of how the devices affect the indoor environment and are integrated with the primary system.

NCCER Industry-Recognized Credentials

If you are training through an NCCER-accredited sponsor, you may be eligible for credentials from NCCER. The ID number for this module is 03312. Note that this module may have been used in other NCCER curricula and may apply to other level completions. Contact NCCER at 1.888.622.3720 or go to **www.nccer.org** for more information.

You can also show off your industry-recognized credentials online with NCCER's digital credentials. Transform your knowledge, skills, and achievements into credentials that you can share across social media platforms, send to your network, and add to your resume. For more information, visit **www.nccer.org**.

Digital Resources for HVACR

SCAN ME

Scan this code using the camera on your phone or mobile device to view the digital resources related to this craft.

1.0.0 HVAC Accessory Troubleshooting

Performance Task

1. Using the proper tools and circuit diagrams, isolate and resolve problems in instructor-selected accessories.

Objective

Explain how to troubleshoot common residential HVAC system accessories.

a. State how to approach the troubleshooting process.
b. Explain how to troubleshoot residential humidifiers.
c. Explain how to troubleshoot electronic air cleaners.
d. Explain how to troubleshoot ultraviolet lighting devices.

HVAC system accessory troubleshooting uses the same approach as troubleshooting the primary equipment. It requires close examination of the equipment and an analysis of the symptoms to eliminate possibilities before using your tools and instruments to identify a specific problem. Following standard practices makes troubleshooting accessories easier and eliminates the need to troubleshoot by trial and error.

By using the manufacturer's manuals, wiring diagrams, and troubleshooting aids, the cause of a problem can be isolated quickly. A technician must actively listen to the owner's complaint, observe the system in operation, then identify and pursue the most probable cause of the problem.

Many accessories can be applied to HVAC systems. This module covers some of the most common accessories encountered in residential and light commercial systems, including the following:

- Humidifiers
- Electronic air cleaners
- Ultraviolet (UV) lighting
- Air-side economizers
- Energy and heat recovery ventilators

1.1.0 Approach to Troubleshooting

Whether you are troubleshooting an electrical or mechanical problem, a logical, structured approach to troubleshooting is essential. You must have the skills to test specific control devices and components. A good set of service tools and the ability to read electrical wiring diagrams are needed. With the right tools and skills, guessing the cause of a problem is not necessary, although you will speculate based on the symptoms to determine where to begin testing.

Before you begin, it is suggested that you review the material found in NCCER Module 03317, *Control Circuit Troubleshooting*. It describes a systematic approach to control troubleshooting and presents many of the skills needed to successfully complete this module.

1.1.1 Electrical Troubleshooting

Electrical troubleshooting skills relevant to accessories are the same as those needed for the primary equipment, including the following:

- Isolating a faulty circuit initially by eliminating those that are clearly functional
- Isolating a faulty component, such as a relay or thermistor
- Single-phase and three-phase voltage and current measurement
- Fuse and circuit breaker testing
- Resistive and inductive load testing
- Control transformer testing

Many HVAC system accessories are equipped with an electronic control board or module. When troubleshooting these devices, remember how to best test them. Control boards and modules accept inputs and react accordingly in the form of outputs. You must know what to expect the device to do in response. When the correct inputs are in place, but the outputs are missing, then the device itself is the likely problem. Otherwise, the problem is related to an input or an external device that isn't responding to an output.

1.1.2 Troubleshooting Aids

The first step in troubleshooting accessories is to consult the product literature. The manufacturer's literature typically describes the mechanical and electrical operation of the equipment, including its sequence of operation.

Service literature almost always includes troubleshooting aids such as wiring diagrams and troubleshooting charts. These aids will help to isolate an equipment problem to the specific failed component.

Onboard diagnostics that are sometimes integrated into control boards and modules are also considered a troubleshooting aid. Make use of the diagnostic capabilities when they are available. If the legend of fault codes is not displayed on the device itself, the product literature will provide the definition.

1.1.3 Sequence of Operation

Knowing the sequence of operation of any accessory is necessary to successfully troubleshoot a problem. If you don't know what to expect from the equipment, it's virtually impossible to diagnose a problem with any confidence.

Although there are exceptions, different brands of the same equipment usually follow a similar sequence of operation. For example, in gas forced-air furnaces, the burner ignition sequence usually begins with the induced draft fan, regardless of brand. Similar accessories will usually operate in a similar manner, regardless of brand.

Remember that many accessories require system airflow to function. Duct-mounted humidifiers and electronic air cleaners (EACs) are two examples. Even a residential fan-powered humidifier relies on system airflow to distribute the water vapor; the humidifier fan merely encourages airflow across the wetted media. For both humidifiers and EACs, a component, such as a pressure switch, proves airflow is present before the accessory is energized.

The sequence of operation may be presented in several ways. One way is a list of steps in a normally occurring sequence (*Figure* 1). One step must be completed before the next step can begin. Sometimes the operating sequence is presented graphically, using a flowchart or a spreadsheet to show events and relationships to other events over time (*Figure* 2). Both operating sequences shown are related to steam humidifiers.

1.2.0 Troubleshooting Humidifiers

Humidifiers introduce water vapor into the conditioned air supply or directly into a space. Humidifiers must be connected to a potable water supply, and most also require a drain connection.

Humidifiers must be carefully installed and mounted as directed in the manufacturer's installation instructions. Some humidifiers have a small fan to move air across a wetted element, although the power consumption of the fan motor is typically low. Steam humidifiers with an electric heating element usually require a more substantial power supply.

Humidifiers are mechanically reliable. Most humidifier problems can be prevented with proper periodic maintenance. Humidifiers collect minerals contained in the water, left behind as the water evaporates. These minerals collect and harden on wetted media and in the reservoir.

Operation	Unit Activity
Power is provided or restored to the unit:	• Humidifier controller is energized. • LEDs three times; On LED remains lit. • All sensors activated.
Water level sensor determines the steam chamber is empty:	• Following short delay, the fill valve opens and the Fill LED illuminates in unison. • Water flows into the steam chamber for up to 14 minutes, depending upon chamber size. • Water level sensor determines chamber is full. The HPC-450i de-energizes the fill valve and related LED.
On a call for humidification by the control device:	• A DC signal is completed at the control, applying DC power to terminal H1 on the humidifier. The signal will range from 1 to 10 VDC based on the ΔRH between the new setpoint and the measured space RH. • The fan relay will energize in 5–7 seconds. • The HPC-450i will monitor for closed circuits in the fan proving switch and the high limit switch. • If circuits remain closed, the STEAM LED will illuminate and remain on as long as the demand for humidity is present. • The STEAM LED will de-energize if the monitored circuits open or the DC demand signal falls below 1 VDC. • After a 20-second delay, the contactor will close to energize the electric elements. • In roughly 9 minutes, the water in the steam chamber will boil and steam will be generated. Steam enters the dispersion hose and enters the manifold. **WARNING:** Steam chamber and hose will be very hot!
As humidification continues, the water level in the steam chamber will fall:	• Fill sequence as noted above begins.
The demand for humidity ceases:	• Steam LED cycles off. • The electric elements are now de-energized. • The fan continues to run for 4 minutes, and then the fan relay is de-energized by the humidifier.

Figure 1 Example sequence of operation for a steam humidifier.

Power On	Draining Cycle			Filling Cycle	Boiling Cycle		
1	2	3	4	5	6	7	8
Up to 8 sec	Up to 3 min	Up to 5 min	Up to 2 min	Up to 2 min	Up to 5 sec	Up to 6 min	Up to 20 sec
	Up to 10 minutes total for drain cycle				1 steam cycle		

Figure 2 Graphical representation of an operating sequence.

The float, float valve orifice, and valve seat may also need periodic cleaning to remove mineral deposits. Any strainers in the water supply line should also be cleaned. Plastic or metal strainers (*Figure 3*) are generally found at the water inlet to the humidifier, or inside a solenoid valve.

Algae and bacterial growth also cause problems in reservoirs. Chemical treatments of various types can be used to reduce bacterial growth. Mold and bacteria growing in the reservoir are distributed throughout the conditioned space, potentially causing significant problems for the occupants.

Most manufacturers recommend that the water reservoir be drained and the wetted humidifier components cleaned bi-monthly during the heating season. In hard water areas, more frequent cleaning may be necessary. When the humidifier won't be in use for months at a time, the water supply should be turned off and the reservoir drained and cleaned.

Figure 4 is a simple humidifier control circuit. This bypass humidifier does not have an integral fan; it relies exclusively on the HVAC system blower to move across the wetted media. The media is attached to a rotating drum that is partially submerged in the water reservoir.

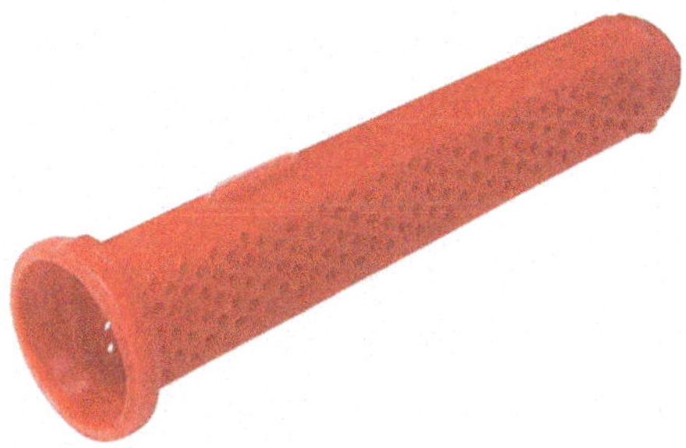

Figure 3 Inline humidifier strainer.
Source: AprilAire

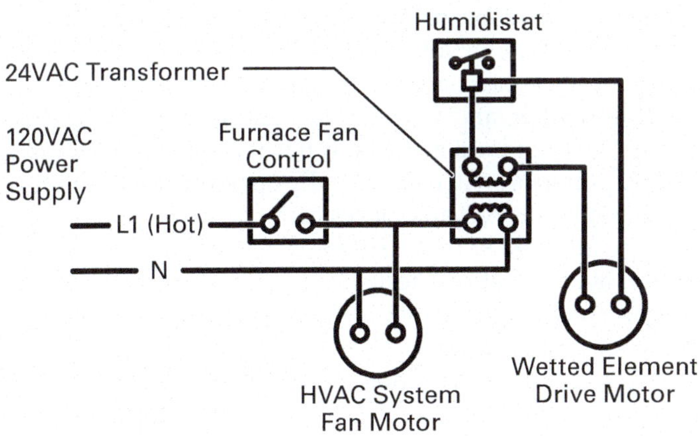

Figure 4 Basic humidifier control schematic.

The furnace blower motor and a 24VAC transformer are connected in parallel and are simultaneously energized when the furnace fan control closes. Note that this transformer is dedicated to the humidifier only; a separate transformer serves the primary equipment. A humidistat in the low-voltage circuit is in series with a small 24VAC gear motor in the humidifier. The motor is geared to rotate the media slowly.

In *Figure 4*, the humidifier gear motor will only rotate the wetted media if the blower motor and transformer are energized, and the humidistat is closed. Sometimes a 24VAC solenoid will be connected in parallel with the gear motor, allowing water to flow to the humidifier. Both the motor and the solenoid valve are powered by a low-voltage power supply.

Figure 5 is a variation of the circuit shown in *Figure 4*. This is a residential fan-powered humidifier with a water solenoid valve. Both the humidifier fan and the solenoid valve are powered by the 120VAC line voltage (wired in parallel), controlled by the contacts of relay R. In this circuit, the humidistat contacts close to power the coil of relay R. When energized, the relay contacts energize the humidifier fan and solenoid valve. These humidifiers require line voltage because the fan motor requires more power than the gear motor of a bypass humidifier.

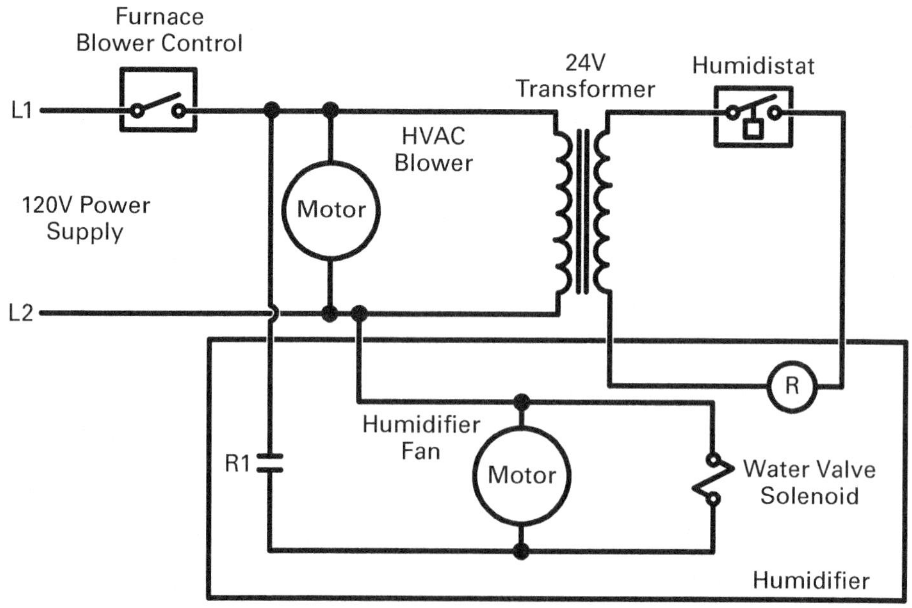

Figure 5 Schematic of a typical fan-powered humidifier circuit.

Figure 6 is a schematic for a three-phase infrared humidifier. An infrared humidifier does not depend on heat supplied by the HVAC equipment to evaporate water. Instead, infrared lamps heat a pan of water to vaporize it. The infrared lamps are powered by line voltage and energized by a contactor controlled by a timer. The primary control and safety devices are connected in series. The humidistat, airflow switch, and high-limit aquastat must all be closed for the lamps to operate. The high-limit aquastat is normally closed (NC) and opens on a rise. The airflow pressure switch is normally open (NO) and closes on a rise in pressure. The humidistat is NO and closes on a fall.

Troubleshooting all three of these different circuits involves checking for the presence of power at key locations; testing circuits for open and closed switches, either by continuity or voltage testing; and testing the electrical integrity of various control and load components.

The following is a list of common problems associated with residential humidifiers:

- *Humidifier not running but not providing moisture* — This is usually caused by an electrical issue. Assuming the primary HVAC equipment is powered and operational, ensure that any additional transformers related to the humidifier are powered and functional. Then check the humidistat circuit and any other humidifier control components. Once it is confirmed that power is available and the controls are working as they should, test important load components, such as the humidifier fan motor. A failed water solenoid valve (if equipped) can also prevent humidification. It's also not unusual to find that the water shutoff valve is closed, especially at the beginning of the heating season.

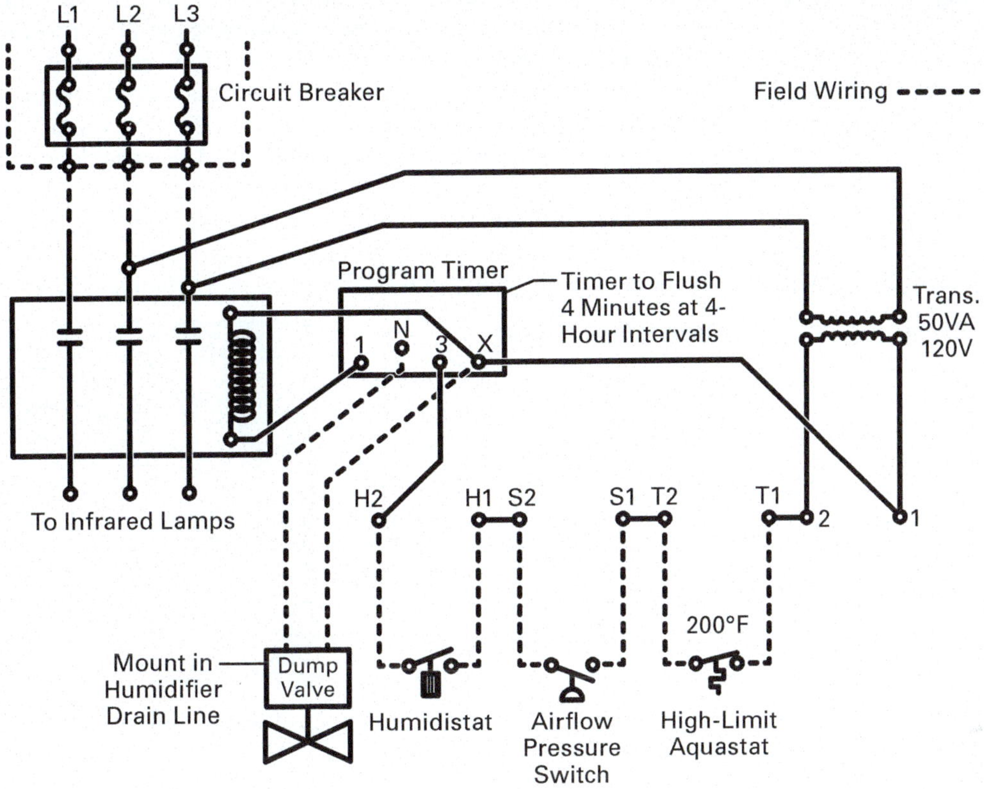

Figure 6 Three-phase infrared humidifier schematic.

- *Consistent water overflow* — Check for a dirty, misadjusted, or faulty float. If the humidifier is equipped with a solenoid valve, disassemble it, and check for dirt or scale on the valve plunger and seat. Also ensure the drain line from the humidifier is not blocked or restricted.
- *Unusual noise* — If the unit is equipped with a solenoid valve, the valve may make a clicking sound as it opens and closes, or it may hum consistently when energized. Both the clicking sounds and the humming sound can be amplified by sheet metal ductwork. One manufacturer supplies a rectifier kit that can be installed on the terminals of the solenoid. The rectifier converts AC to DC to help eliminate the hum.
- *Low or high relative humidity* — Traditional humidity controls are not known for laboratory precision. Wall-mounted humidistats may differ by several percent from handheld measuring instruments that are typically better calibrated and more accurate. The humidity in other areas can differ significantly from the humidity at the control location. Precision is not a significant concern in most applications, although many humidistats offer a calibration routine or an offset in the display to improve accuracy for the user. It is best to check humidistat accuracy using a sling or digital psychrometer.

When the humidity is too high or too low, first check the humidistat setpoint (*Figure 7*). Electromechanical humidistats have been used for many years. These humidistats contain a material that expands and contracts relative to the moisture level in the air. That material is connected to switch linkage in the humidistat. The calibration of electromechanical humidistats has always been difficult to maintain.

Figure 8 shows a digital humidistat that senses moisture levels electronically. Digital humidistats typically offer more options, such as diagnostics, and maintain their calibration better than electromechanical models. For example, the humidistat shown in *Figure 8* will automatically start the system fan periodically for a short time. This ensures the air doesn't remain stagnant for too long, and it ensures that the humidistat is surrounded by air that represents conditions through the structure. Sophisticated models may also be equipped with an outside-air temperature sensor and the ability to lower the humidity setpoint in

Did You Know?

Humidifier Selection

When mineral deposits and bacterial growth are problems, select a humidifier design that minimizes these problems. For example, many bypass and fan-powered humidifiers lack a water reservoir. Water flows through the humidifier only when it is active; excess water drains away immediately. Eliminating the reservoir reduces bacterial growth, while draining excess water helps flush out mineral deposits.

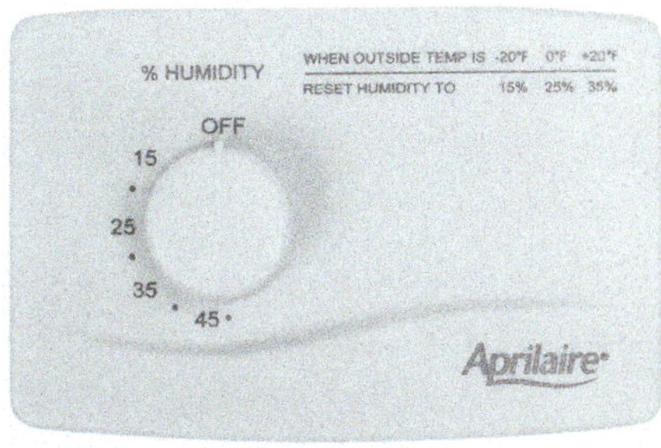

Figure 7 Humidistat.
Source: AprilAire

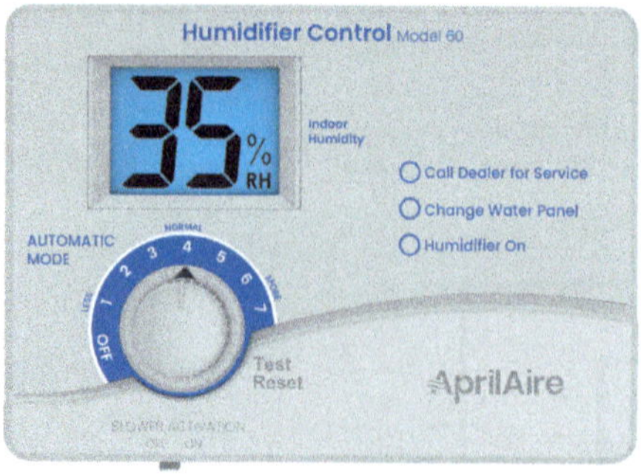

Figure 8 Digital humidistat.
Source: AprilAire

response to colder outdoor temperatures. This helps avoid condensation on the windows and other sensitive locations. Many digital thermostats contain a digital humidistat, eliminating the need for a separate wall-mounted control.

Smart thermostats with built-in humidistats can also control dehumidification. Better HVAC systems offer a dehumidification mode. This arrangement typically relies on compressor operation and a variable-speed, electronically commutated motor (ECM) to reduce the airflow across the indoor coil. This maximizes dehumidification while minimizing cooling.

On Site

Bypass Humidifiers

A bypass humidifier is normally installed on the supply duct with the bypass duct connecting to the return duct. However, the humidifier can operate just as effectively when installed on the return duct with the bypass connecting to the supply plenum. This makes bypass humidifiers a bit more versatile than other types of humidifiers. The bypass duct is typically closed in the summer months.

Beyond the humidification equipment, there are other reasons why it can be difficult to maintain the humidity setpoint. Excessive air infiltration from poorly sealed windows and doors, exhaust fans running consistently, and open fireplace dampers all increase the number of air changes in a space. In this case, the humidity being added is quickly finding its way outdoors.

In some cases, an oversized furnace can also cause the humidity to remain low. Bypass and fan-powered humidifiers, for example, rely on warm air from the heating system to moisturize and distribute the air. When a furnace is too large, it may not operate long enough for the humidifier to keep up with the demand. In this case, it may be necessary to set the blower for continuous operation so that the humidifier can run as needed. However, it is important to note that the capacity of an evaporative humidifier is substantially reduced when only room temperature air is being circulated, as opposed to heated air.

On Site

Saddle Valves

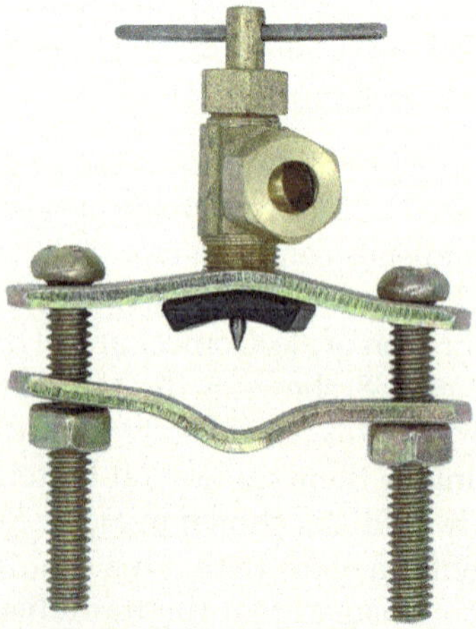

To obtain water for a humidifier, a nearby water line is often tapped using a self-piercing *saddle valve*. This valve is clamped around a copper water line and contains a piece of rubber that surrounds the opening created, acting as a gasket to prevent water leakage.

The water line is tapped by turning the valve handle clockwise. This forces a sharp steel point into and through the copper tube. Once the tube is pierced, the steel point seats itself against the copper to shut off water flow. Flow begins when the handle is turned counterclockwise to retract it. One or two turns should allow sufficient flow.

Some local codes do not allow the use of saddle valves. The valves do tend to leak over time, as the rubber seal dries out and hardens. Consider the condition of the valve during inspections, as a leaking valve can do a lot of water damage before being discovered.

Going Green

Humidifiers and Hot Water

Some bypass and fan-powered humidifier installers tap into a hot-water line instead of a cold-water line for the humidifier. However, some manufacturers prohibit this practice as their components may not be rated for hot water exposure.

Hot water does vaporize more quickly in the humidifier, but the energy used to heat the water increases the operating cost. However, it takes time for the hot water to reach the humidifier, and cold water is being adding to the humidifier reservoir in the interim. In the end, the practice has little or no value in most situations.

1.3.0 Troubleshooting Electronic Air Cleaners

EACs (*Figure 9*) remove airborne particles and odors from the air circulated to the conditioned space. EACs can be stand-alone units or may be mounted in the return-air side of a duct system. HVACR technicians typically service only duct-mounted models.

EACs create an electrostatic field that applies a positive or negative charge on the particles being carried along in the airstream. The charged particles are removed from the airstream by drawing them toward oppositely charged collector plates.

EACs are powered by a high-voltage, solid-state power transformer. The voltage produced by the transformer, which can range from 6,000 to 10,000 VDC, is used to charge particles in the airstream as they enter the filter assembly. *Figure 10* illustrates typical EAC operation.

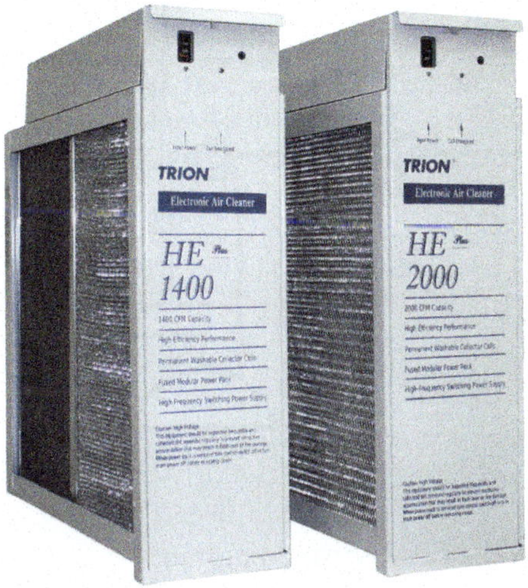

Figure 9 Duct-mounted electronic air cleaners (EACs).
Source: Trion Indoor Air Quality

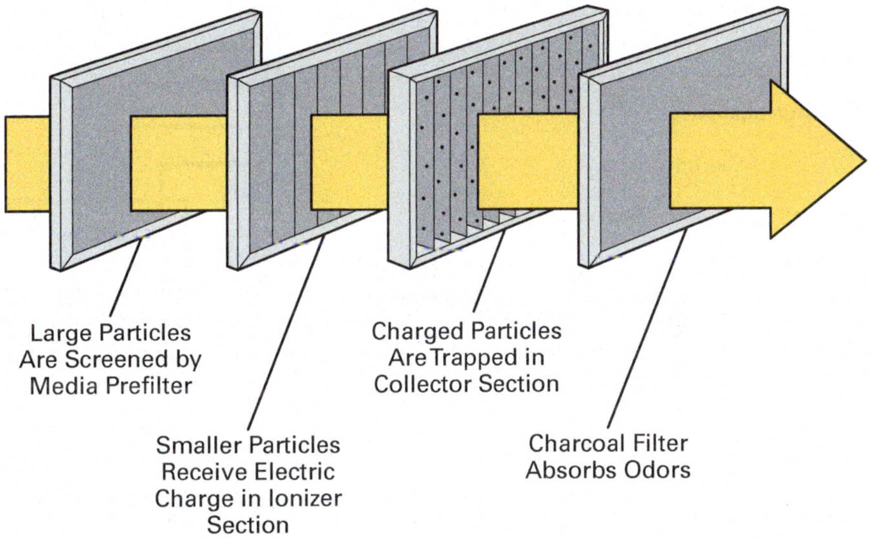

Figure 10 Electronic air cleaner operation.

The components through which the air travels include the prefilter, ionizer section, collector section, and charcoal filter section. As the air enters the EAC, larger particles are trapped by the media prefilter. Smaller particles pass through the prefilter into the ionizing section.

The ionizing section is a grid consisting of fine tungsten wires. The wires are connected to the high-voltage DC power supply. The electrostatic field created by the high voltage applied to the wires charges the passing particles.

The collector section consists of a series of vertical, parallel, equally spaced collector plates. The plates are also connected to the high-voltage DC power supply. The even-numbered plates have a positive charge while the odd-numbered plates have a negative charge. As charged particles flow between the collector plates, they are attracted to and held on oppositely charged plates. The air may then pass through a charcoal post-filter section. Charcoal has the unique ability to **adsorb** odors. Not all EACs are equipped with a charcoal post-filter.

Adsorb: To attract and hold gas or liquid molecules on the surface of a solid.

Figure 11 shows an example electrical diagram for a typical EAC. In the example shown, the high-voltage power supply is made from discrete components such as diodes, capacitors, and resistors. In newer modules, the power supply may be a module that is fully enclosed in a non-repairable, sealed assembly.

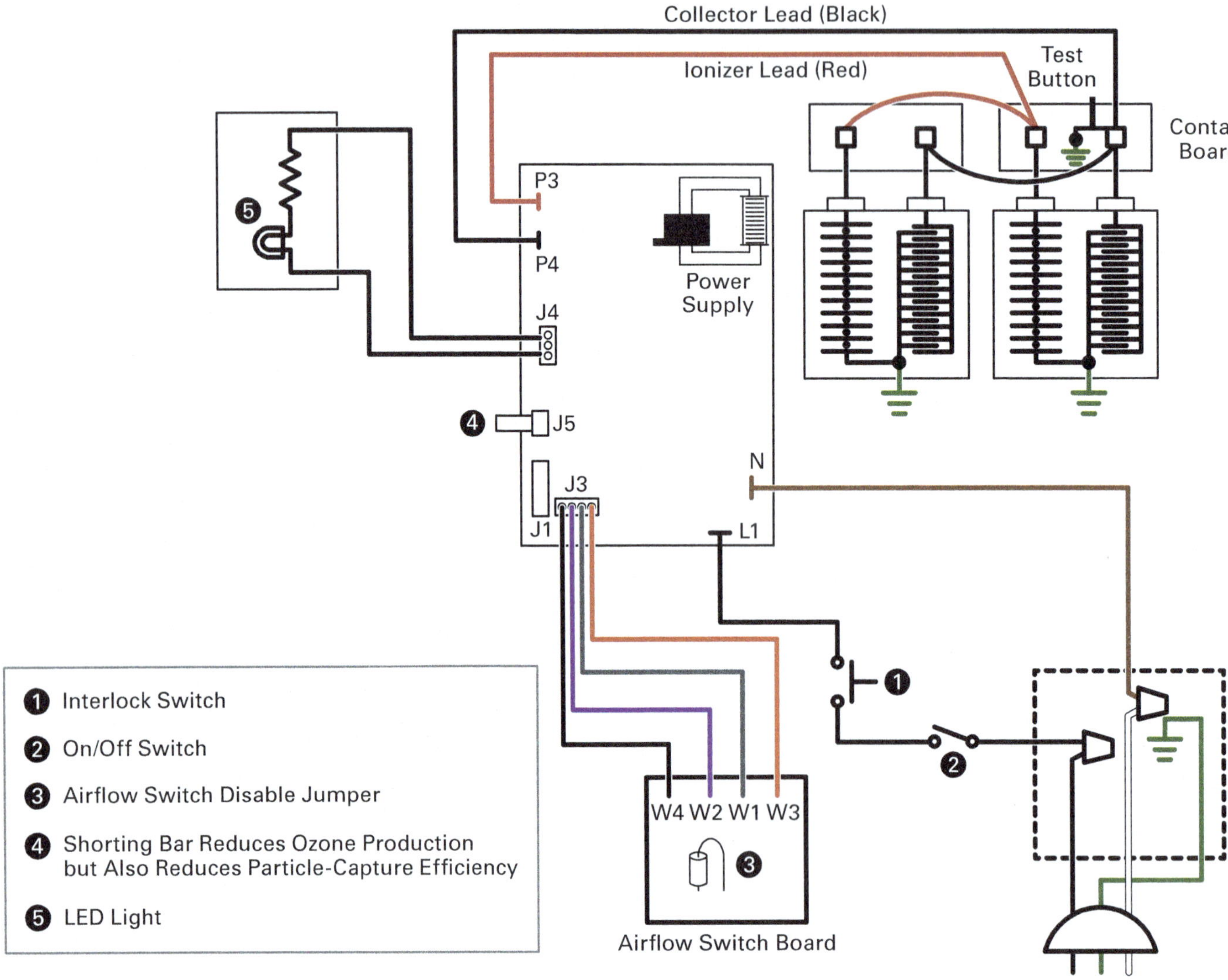

Figure 11 Example EAC schematic.

Here are some helpful comments about the EAC wiring diagram:

- The interlock switch shown in the power connection to L1 is a safety device. It automatically de-energizes the EAC power supply when the cabinet is opened for service or inspection. Its integrity is easily tested with a meter or by simply ensuring that removing the cover de-energizes the unit.
- Many EACs are equipped with an airflow-detection sensor or pressure switch that proves airflow before the power supply energizes the ionizing wires and collector plates. The airflow switch disabling jumper shown bypasses the airflow sensor. This allows the unit to operate any time it is energized. Since most EACs today are energized by the furnace or fan coil unit to which they are connected, using an airflow switch as a primary control is unnecessary.
- The shorting bar is merely a tiny pin jumper on the power supply. Not all EACs are equipped with this option. Removing the jumper reduces ozone production, but also reduces the effectiveness of the unit to a lesser degree.

EAC performance is sensitive to air volume and velocity. When the air velocity is above that recommended by the manufacturer, EACs become far less efficient. When the velocity is below the manufacturer's recommended minimum, excess **ozone** can be generated.

Ozone: A colorless and unstable gas with a pungent odor, formed from oxygen in response to electrical discharges in the atmosphere and UV light. It contains three atoms on its molecule (O_3) as opposed to oxygen with two atoms (O_2). Ozone in the upper atmosphere provides protection from UV radiation, but ground-level ozone is a significant and concerning pollutant.

Note the model numbers of the EACs shown in *Figure 9*. One is a Model HE1400, and the other is Model HE2000. The model numbers, in this case, refer to the volume of air the EACs are rated to handle—1,400 cfm and 2,000 cfm, respectively. If an EAC is operating but doesn't collect as much debris as expected, or if the smell of ozone is clearly present, make sure that the system air volume and velocity through the filter is within the specifications.

EACs are less effective when dirty. The collection plate assembly, often referred to as a *cell*, and prefilter should be cleaned every two to four months. The cells of most models can be safely placed in a dishwasher. The cells can also be soaked in an appropriate detergent and rinsed clean with a hose.

When cleaned regularly and handled with care, many EAC problems can be avoided. Some conditions can be expected to cause an EAC to capture more than usual, such as the following:

- New carpeting installed
- Unsealed, exposed concrete floors
- Nearby construction work
- Seasonal pollen

Begin troubleshooting by performing a visual inspection for damaged or distorted collector plates and broken ionizer wires. This should also be done any time the cells are removed for cleaning. For electrical problems, use the troubleshooting aids in the manufacturer's service literature to isolate the fault.

Figure 12 shows a typical manufacturer's troubleshooting flowchart. Note that the chart calls for a reduced ionizer voltage (7,500 to 8,500VDC) when the cells are in place, compared to the value measured with the cells removed. This is due to the voltage drop that occurs when a load is imposed on the high-voltage transformer. This is also true of the collector voltage; expect roughly the same percentage of reduction.

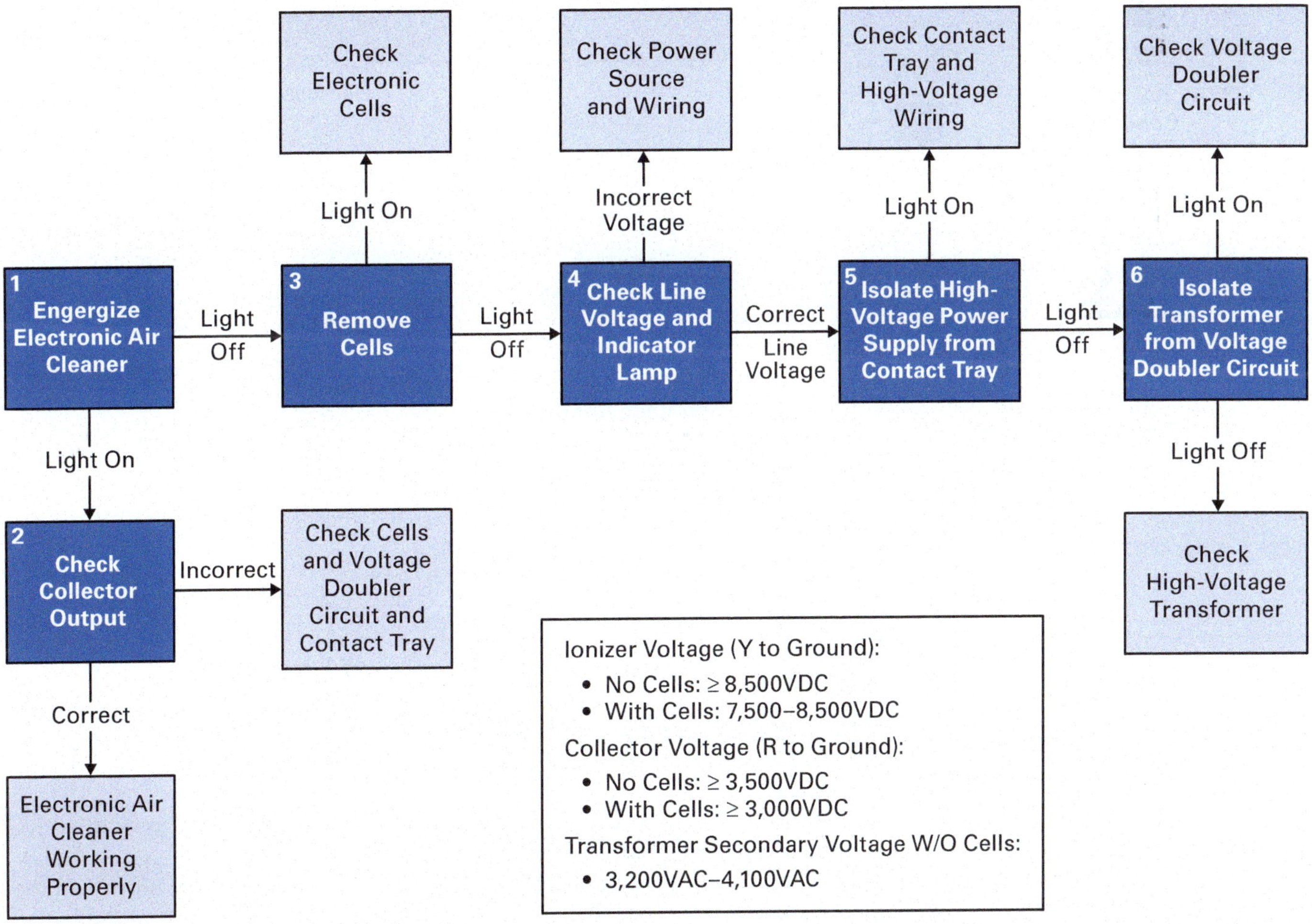

Figure 12 EAC troubleshooting flowchart example.

WARNING!

When making electrical measurements on energized EAC units, ensure the multimeter is rated for the anticipated voltage. Use extreme caution when near the exposed power supply.

On Site

Connecting Accessory Devices

Many HVAC equipment manufacturers prewire their products with sets of terminals to which accessory devices such as humidifiers and EACs may be connected. This eliminates wiring errors and field modification of their product to accommodate the accessory. The furnace control module shown here has line-voltage terminals to power an EAC and 24VAC terminals for a humidifier.

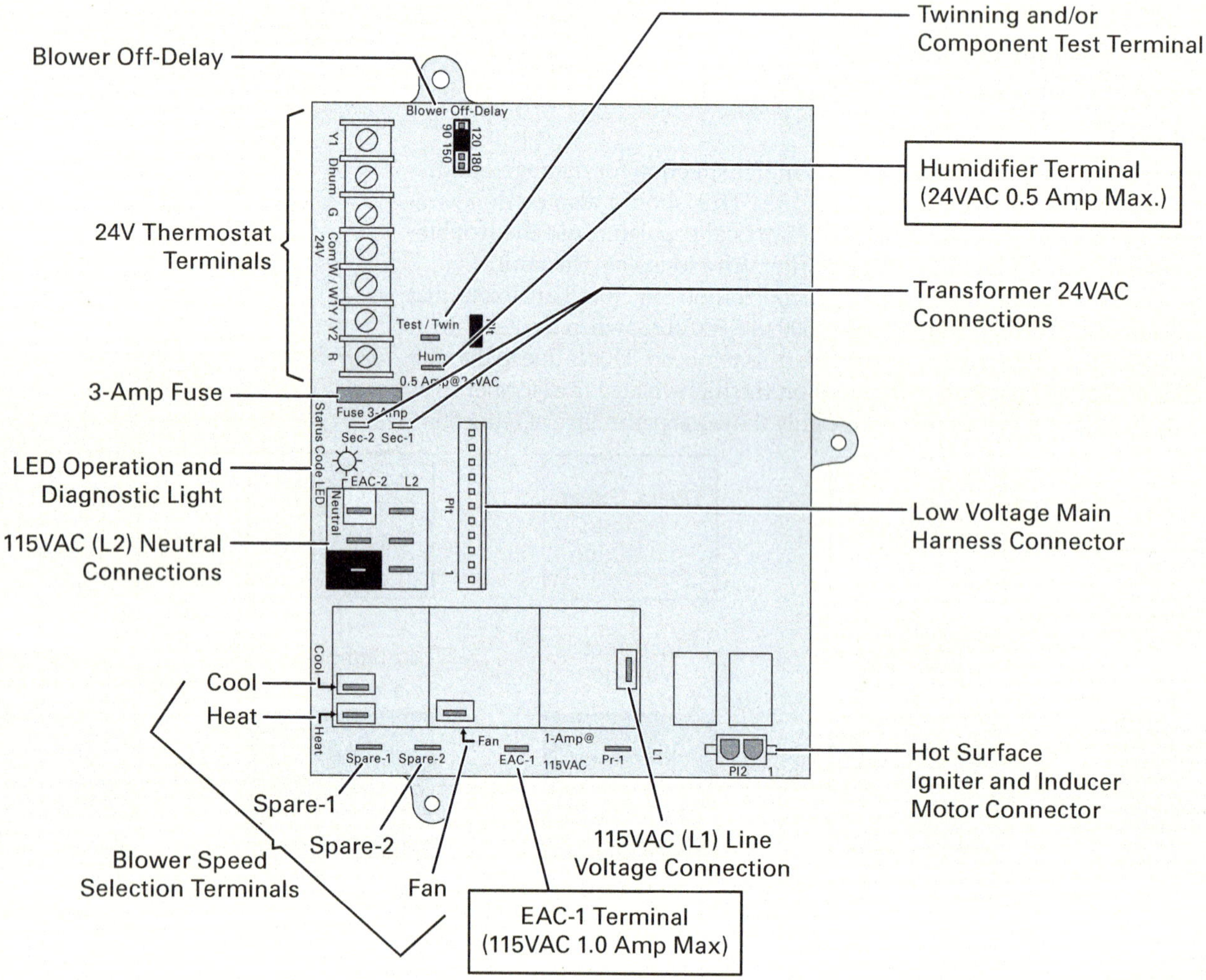

Source: Carrier Corporation

1.4.0 Troubleshooting Ultraviolet Lamps

Ultraviolet (UV) lamps (*Figure 13*) are used to control microbial growth on evaporator coils, drain pans, and duct insulation. UV light can also kill airborne pathogens.

Two different approaches are used to apply UV light. One very effective approach constantly bathes a fixed object, such as an evaporator coil, with UV light. The other approach focuses on the air, with the lamps positioned in the moving airstream. Since microbes in a moving airstream move past the lamps quickly, exposure to the UV light is limited. This makes this approach less effective but still valid.

Figure 13 Ultraviolet (UV) lamp and power supply.
Source: American Air & Water, Inc.

WARNING!

UV light can cause serious and permanent damage to the eyes and skin. Appropriate PPE to protect yourself from exposure includes UV-rated eyewear, UV face shields, long-sleeved and tightly woven clothing, and gloves that cover the hands up to the shirt cuff. When servicing UV lamps, review and follow all manufacturer safety warnings.

Troubleshooting UV lamps is not complicated. If an entire assembly is not working, ensure that power is present. Some assemblies have a cord and are plugged into a standard 120VAC receptacle. Check the individual lamps for physical damage and make sure no debris is in the lamp socket.

UV lamps must be replaced after a certain number of hours of operation. UV light output diminishes over time, although it may not be obvious visually. Consult the manufacturer's instructions for information about the life span of the lamps and what replacement intervals ensure effectiveness.

WARNING!

Conventional UV lamps contain mercury, a hazardous substance. If a lamp breaks during handling, be cautious when cleaning up the debris and dispose of it according to local regulations. Always wear lightweight gloves when handling UV lamps.

1.0.0 Section Review

1. True or False? A fan-powered humidifier mounted on a forced-air furnace can function independently, without the system blower running.
   a. True
   b. False

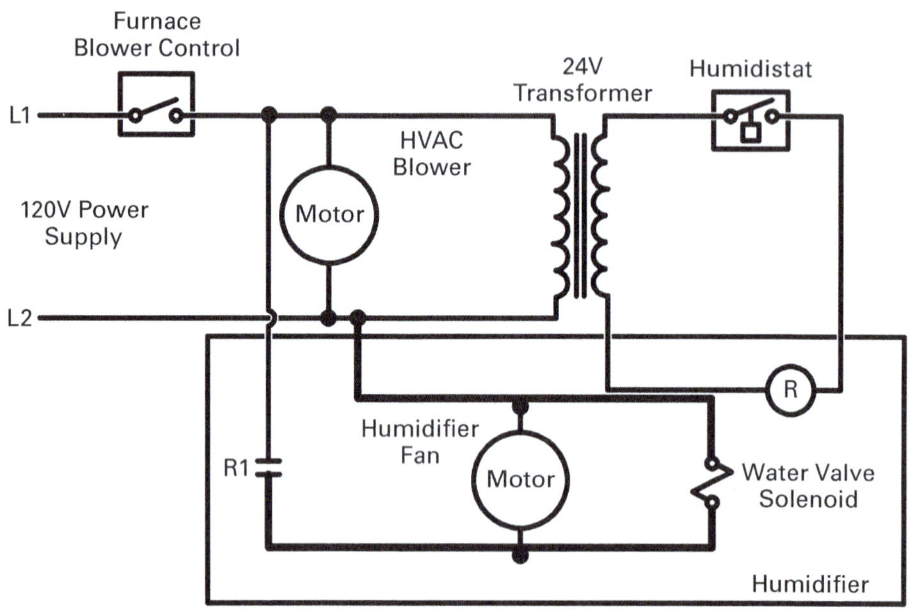

Figure SR01

2. Refer to *Figure SR01*. What will the result be if the HVAC blower motor fails?
   a. The 24VAC transformer will not be energized, disabling the humidifier completely.
   b. The humidifier fan and water solenoid will continue to operate when the humidistat closes, but the moist air won't be distributed.
   c. Relay R will be energized when the humidistat closes, but no line-voltage power will be available to the humidifier fan and water solenoid.
   d. The 24VAC transformer will not be energized, but the humidifier fan and water solenoid will be energized as long as the furnace remains energized.

3. Which condition is indicated by an excessive amount of ozone produced by an EAC?
   a. Low air velocity
   b. High air velocity
   c. Low line voltage
   d. Very dry room air

4. Which hazardous substance do conventional UV lamps contain?
   a. Arsenic
   b. Mercury
   c. Cadmium
   d. Sulfuric acid

2.0.0 Outside-Air Accessory Troubleshooting

Objective

Explain how to troubleshoot economizers and recovery ventilators.

a. Describe the operation and troubleshooting of economizers.

b. Describe the operation and troubleshooting of recovery ventilators.

Performance Task

1. Using the proper tools and circuit diagrams, isolate and resolve problems in instructor-selected accessories.

When the inside of a building is warmer than the outside, it makes sense to use outside air for cooling. It's often why we open windows—to invite cooler air in. The problem with the window is that the air volume isn't controlled, and the air often carries dust, pollen, and other debris.

HVAC units equipped with economizers can introduce outside air in a controlled manner, and in far greater volume than natural air currents can achieve. The equipment can also filter the air and ensure that the outside air conditions are suitable for effective cooling.

Recovery ventilators also bring in outside air, but for a different purpose. The ventilators are used to satisfy the requirement for fresh air indoors to maintain good indoor air quality (IAQ). In addition, recovery ventilators exchange inside air for outside air, transferring heat from one airstream to the other and recovering that energy.

2.1.0 Economizer Operation

An economizer (*Figure 14*) is an accessory for commercial HVAC units that reduce operating cost by using outside air for cooling (**free cooling**). Any time cooling can be obtained without operating the compressors, a great deal of energy is saved.

Free cooling: A mode of economizer operation when the system compressor is turned off and the indoor fan alone is used to bring outside air into the building as the sole source of cooling.

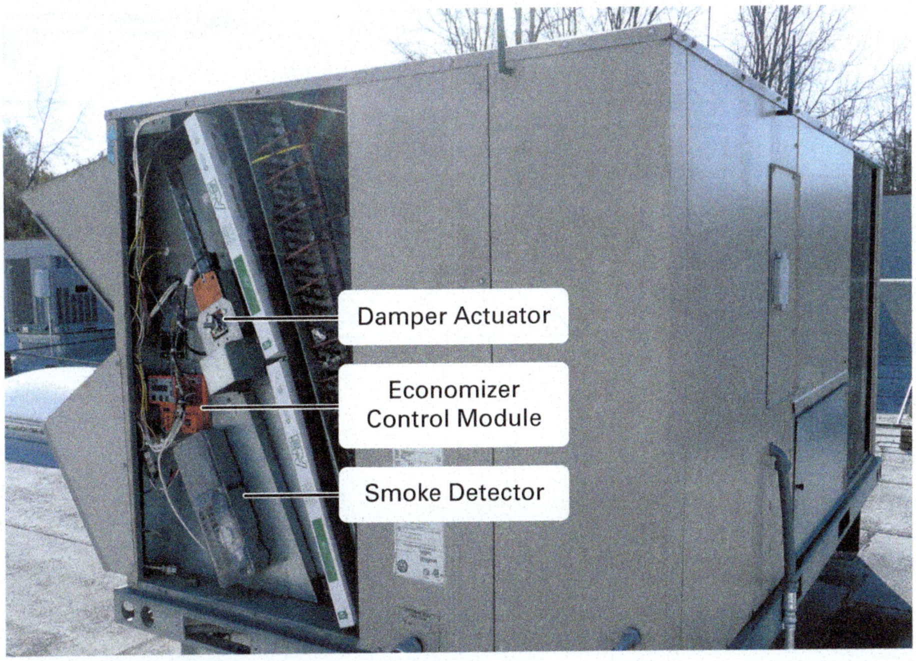

Figure 14 Rooftop unit with an economizer.
Source: Belimo Aircontrols

Figure 15 shows the basic components of an economizer package. An economizer system consists of an economizer control module, related sensors, one or more damper actuators, and damper assemblies. The dampers are not part of an economizer kit because dampers must be made to fit each piece of equipment. Operable dampers are required on the outside air duct or intake and in the return air duct. Economizer kits can be field installed, but most of the kits are sold to equipment manufacturers for factory installation. *Figure 16* shows a schematic for a basic economizer package.

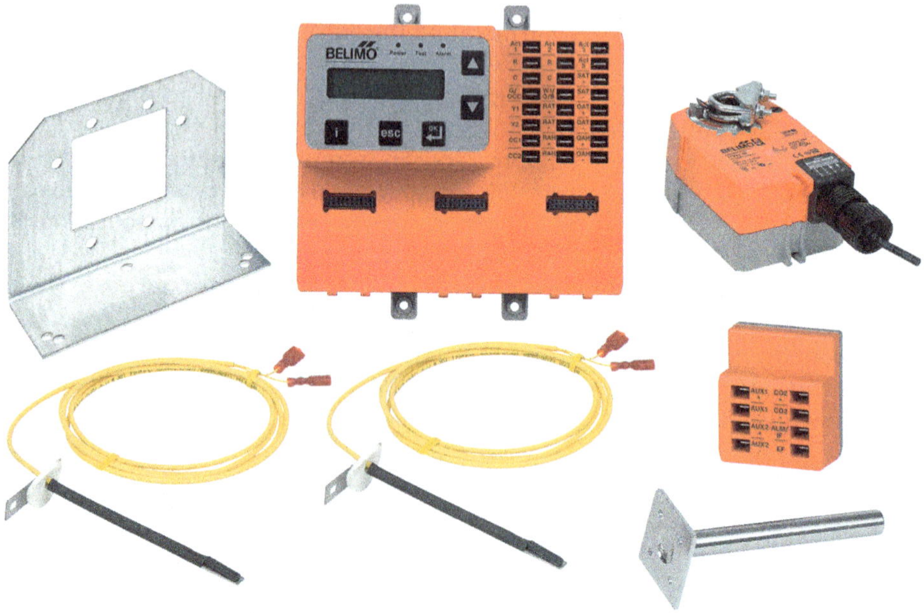

Figure 15 Basic economizer kit.
Source: Belimo Aircontrols

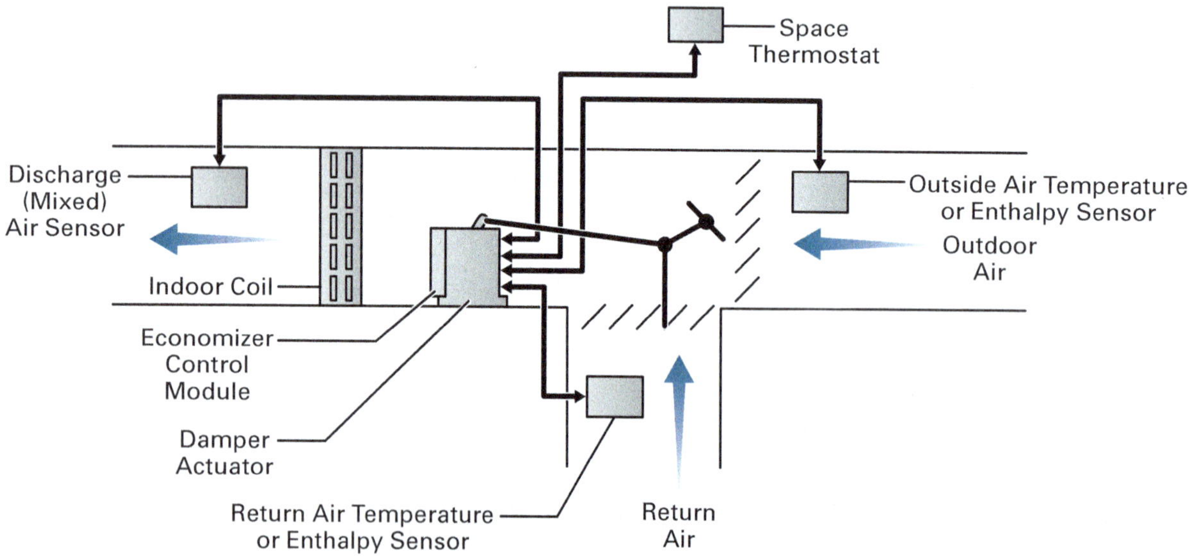

Figure 16 Economizer schematic.

Inputs to the economizer control module come from multiple sources. Primary inputs include the following:

- The room thermostat or building management system (BMS)
- An outside air temperature or **enthalpy** sensor
- A return air temperature or enthalpy sensor
- A discharge, or mixed, air sensor

Enthalpy: A measure of the total heat contained in an air source, accounting for both the dry-bulb temperature and the volume of water content in the sample. Moist air, for example, contains more heat than dry air at the same temperature. Enthalpy is measured in units of Btus per pound of air (Btu/lb air).

2.1.1 Economizer Types

Three basic types or economizers will be encountered in the field:

- Dry-bulb economizers
- Enthalpy economizers
- Integrated economizers

The *dry-bulb economizer* (*Figure 17*) offers the simplest arrangement. When the outside air temperature (OAT) falls below a user-specified setpoint (58°F, for example), the economizer module activates a control relay. The control relay (CR) locks out mechanical cooling (the compressor and condenser fans). Space cooling is then accomplished by modulating the volume of outside air that is mixed with the return airstream. The resulting discharge, or mixed, air temperature (MAT) must be kept above a reasonable temperature to avoid discomfort. Blowing 35°F air into a room, for example, will result in a lot of complaints.

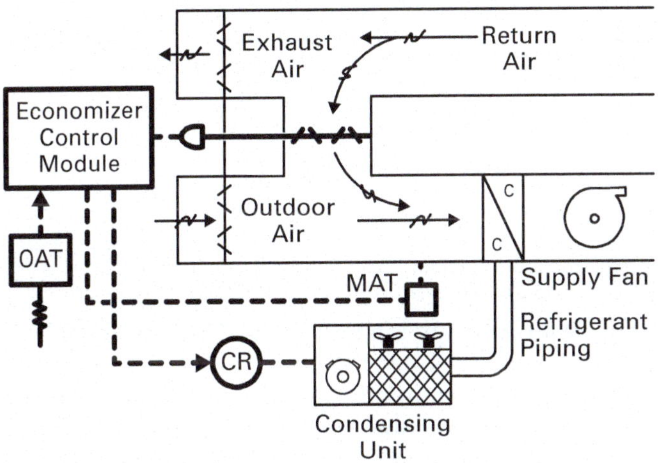

Figure 17 Dry-bulb economizer.

Since the common discharge air temperature from a mechanical cooling system is around 55°F, that is a common target setpoint for the economizer discharge air temperature. As the outside air temperature falls below 55°F, more return air must be mixed with the outside air to maintain the 55°F discharge air temperature setpoint. If the outside air temperature and discharge air temperature setpoint are the same, then the control module will close the return air damper completely and operate on 100% outside air.

The weakness in the dry-bulb approach is that the sensors and control module have no way of knowing how wet the outside air is. Consistently moving a large volume of wet air into a building can create more IAQ problems than it solves. Moist air also contains more heat than dry air, increasing discomfort in summer and increasing the load on the mechanical system when it takes over. Thus, dry-bulb economizers are only common in areas with limited rainfall and low humidity.

An *enthalpy economizer* uses an outside-air enthalpy sensor (HOA) rather than a temperature sensor. Because a measurement of enthalpy accounts for the volume of moisture in the air, bringing in air that carries too much moisture and the heat it contains can be avoided. When the outside air enthalpy is below the setpoint (24.7 Btu/lb is a common setpoint), the control relay locks out the compressor. The return air and outside air dampers are then modulated to produce the desired mixed air condition for the supply airstream. If the outside air temperature is too cold to send into the building directly, the return air is mixed with outside air to warm it up.

An *integrated economizer* (*Figure 18*) monitors the enthalpy of both the outdoor air and the return air. The control module calculates the enthalpy of both the outside air and the return airstream. Below a pre-determined outside air enthalpy (24.7 Btu/lb in this example), the economizer can provide free cooling with the compressor locked out, just like the other economizer types.

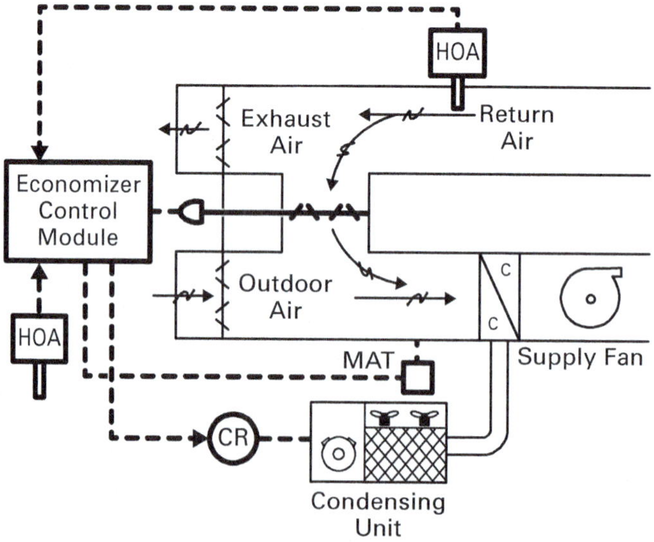

Figure 18 Integrated economizer.

When the outside air enthalpy is above the 24.7 Btu/lb setpoint, an enthalpy economizer controller switches back to the compressor, and the outside air dampers are driven to their minimum position to provide the needed fresh air. However, an integrated economizer compares the enthalpy of the two airstreams—outside air and return air—and determines which airstream contains the least amount of heat. When the outdoor air enthalpy is less than the return air enthalpy, outside air is used in place of return air to reduce the mechanical cooling load. The term *integrated* reflects the integration of mechanical cooling with free cooling.

Going Green

Economizers Need Your Help!

According to studies conducted by economizer system manufacturers, up to 70% of installed economizers are not functioning properly. In most cases, it is because the system was left in the factory-default configuration or was otherwise set up improperly. A tremendous amount of energy is being wasted as a result.

Economizer failures often go undetected for years. As long as the mechanical cooling works, users rarely notice that something is wrong. As you provide periodic maintenance on units equipped with an economizer package, examine the setup, and ensure the economizer is working as it should.

Demand Control Ventilation

Technically, demand control ventilation (DCV) is not an economizer mode. However, since many of today's economizers are being equipped with this option, it is worthy of discussion.

You'll recall that commercial buildings require a certain amount of outside air to be introduced to maintain good IAQ and provide fresh oxygen for the occupants. However, rather than measure the percentage of oxygen in the building, the need for ventilation is usually based on the carbon dioxide (CO_2) content of the indoor air. As the CO_2 volume increases, more outside air is brought in to purge it from the space.

The same dampers and damper actuators used by the economizer must also be used to manipulate the outside and return airstreams to ensure proper ventilation is achieved. Thus, the economizer system is also related to the ventilation requirements. By interlocking an indoor CO_2 sensor with the economizer, the ventilation rate can be monitored and controlled.

When the economizer is active and introducing fresh air for cooling purposes, the ventilation rate is well above the minimum necessary. At all other times, the economizer module monitors the CO_2 sensor and increases the outside air volume to match the human occupancy load.

2.1.2 Relief Air Dampers

In *Figure 17* and *Figure 18*, you'll note a set of exhaust dampers is shown. When a system operates using 100% outside air (or any high percentage of outside air), the building pressure rises. As the pressure indoors increases, the amount of air that can be pushed into the space is reduced. Think of it as blowing up a paper bag. At some point, you are unable to blow more air into the bag, other than replacing any air that happens to be leaking out.

A building may be designed to accommodate the introduction of fresh air and be equipped with a means for excess air to escape. In many cases, however, the HVAC unit must accommodate the pressure increase by providing an escape path. Exhaust dampers, often referred to as *relief air dampers*, can be added to the unit (*Figure 19*) to relieve the excess pressure.

Figure 19 Rooftop unit with economizer and relief air dampers.

There are two primary approaches to relieving the air pressure in a building:

- The simplest type of relief damper is a *barometric damper*. A barometric damper is not motorized and has no springs to hold it or return it to any position. The blades move freely and easily. As the building pressure increases, the dampers begin to open and relieve the pressure. When the building air pressure is less than or equal to the barometric pressure outdoors, the dampers remain closed.

Figure 20 Rooftop unit relief-air hood with fans installed.
Source: MicroMetl Corporation

- Simply opening a path for building air to escape outdoors isn't always enough. In cases where a more aggressive approach is needed to reduce the building pressure, a powered relief-air fan is used. Relief fans are usually propeller-type fans because there is no significant friction loss to overcome. *Figure 20* shows a hood containing two fans that can be attached over the relief-air opening and barometric dampers. The barometric dampers remain in place, preventing air from entering the building through the opening when a negative pressure develops indoors.

Relief-air fans are often single-speed models, responding to an output from the economizer control module. The fan(s) may be energized when the outside-air damper opens to a specific percentage or more. For more precise control over the building pressure, variable-speed fans can be used. A variable-frequency drive (VFD) controls the motor speed based on the pressure in the building. A pressure transducer provides the needed input to the VFD.

2.1.3 Troubleshooting Economizers

Before troubleshooting an economizer, consult the specific manufacturer's service literature before proceeding. Although the functional concepts will be similar across brands, the controls they use can differ significantly.

Economizers use the dampers to manage the return air source. The economizer control module responds to its programming and temperature and/or enthalpy sensors provide essential inputs. The electronic controls and sensors tend to be very reliable and are not as prone to problems or failure as the electromechanical components, such as the dampers, damper actuators, and interconnecting linkage. Unrealistic setpoints and flaws in the wiring connections are two of the more common problems encountered.

Use the following procedure to confirm an economizer is operating normally:

- *Check for power* — Ensure that the economizer control module is energized. Most modules operate on 24VAC power.
- *Check temperature sensors* — Thermistors are used to monitor temperature in most modern economizers. Thermistors are easily tested for accuracy by comparing the resistance to a matching temperature chart as outlined in NCCER Module 03317, *Control Circuit Troubleshooting*. Ensure that all temperature and enthalpy sensors are secured in the precise location outlined by the manufacturer.
- *Check enthalpy sensors* — Enthalpy sensors are more specialized than thermistors, and the correct type must be used with a given economizer package. For example, one brand of economizer controls uses on enthalpy sensor that produces a 4–20mA signal, while another uses an enthalpy sensor that produces a 1–10VDC signal. *Figure 21* shows an example of a chart comparing current output to combinations of temperature and relative humidity. Obtain the correct chart for the sensor in use and measure the temperature and relative humidity of the air to which it is exposed. Then measure the current or voltage and compare the value to the chart.
- *Check motorized damper operation* — Damper actuators are typically spring return, meaning that a spring pulls the actuator back to its default position when it is de-energized. The outside-air damper usually closes while the return air damper may move to its fully open position. De-energize the rooftop unit at the disconnect switch and watch the two dampers. Restoring power should return the outside-air damper to its minimum position, per the economizer setting. With the damper in its minimum position, vary the setting on the control module that controls the minimum position. If the damper responds, restore the original setting. It is also important to ensure that the mechanical operation of the damper and linkages is smooth and does not bind.
- *Check economizer operation* — Set the thermostat to simulate a call for first stage cooling or disconnect the thermostat and use a jumper wire at the unit terminal board. If the outside air temperature or enthalpy is low enough, the

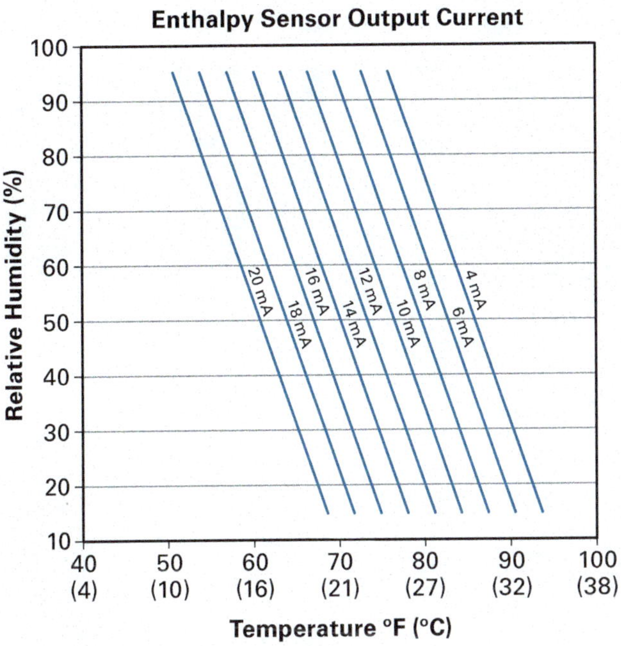

Figure 21 Example enthalpy-sensor output chart.

outside air damper and return air damper should react to provide free cooling, and the indoor blower will start. In warm weather, you may have to fool one or more sensors by applying an ice pack.

The dampers must function properly for an economizer to be effective. Leaky dampers waste energy. Check damper assemblies for the following:

- Make sure that the actuator and the damper assembly are securely mounted and don't move as the damper is being positioned. Check the condition of any caulking applied around the frame to prevent air leaks.
- Ensure the actuator drive is secured to the damper shaft so that it fully opens or closes at the end of the actuator's travel in each direction. Damper actuators have a specified range of rotation, such as 135 degrees. Ensure the motor's range is compatible with the requirements of the damper. The range is typically adjusted on the actuator itself. Actuators are also rated in terms of torque, such as 180 inch-pounds (in-lb).
- Ensure the damper blades close tightly. Look for damaged or bent blades or blade gaskets (if equipped) and replace as needed. Also check any seals around the edges.
- Check the blade bushings or bearings for wear and corrosion and replace any as required. Lubricate as needed with a light oil.

2.2.0 Recovery Ventilators

Energy recovery ventilators (ERVs) and heat recovery ventilators (HRVs) draw fresh outside air into occupied spaces (*Figure 22*). At the same time, recovery ventilators exhaust stale indoor air. The volumes of air entering and leaving are nearly the same so that the pressure indoors remains stable. Recovery ventilators also recover much of the energy from the outgoing indoor air and transfer it to the fresh incoming air.

ERV and HRV units are similar in construction. At the heart of both units is a special heat exchanger. The heat exchanger is referred to as a *recovery core*. As an ERV replaces inside air with outside air, it transfers both sensible and latent heat

Figure 22 Energy recovery ventilator (ERV).
Source: Image courtesy of RenewAire LLC

between the two airstreams. This allows it to remove both heat and moisture from the incoming air during the cooling season. An HRV is designed to transfer only sensible heat between the entering and leaving airstreams.

ERVs can better condition the incoming airstream and save more energy by removing moisture as well as heat. For this reason, ERVs are used in most areas of the United States and some parts of Canada. Since HRVs cannot transfer moisture and the latent heat it contains in the summer, the burden of dehumidifying the air is transferred to the HVAC system.

A recovery ventilator can be installed as a standalone unit with its own ductwork. However, many are integrated into a forced-air heating/cooling system. Remember that, despite the transfer of some energy between the two airstreams, the introduction of outside air increases the heating cooling load. It is important to ensure that the outside air load is considered in the load calculations for a building. The increase in the cooling and heating load is substantial.

The recovery ventilator can be controlled independently by a simple wall-mounted control. When it is integrated into a central HVAC system, a blower interlock relay ties the operation of the recovery ventilator to the HVAC system blower.

When the recovery ventilator is in operation, the interlock relay is energized, causing the HVAC system blower to start as well. Typically, a recovery ventilator is set up to bring in slightly more air than it exhausts. This helps maintain a slightly positive pressure in the space. If the space is at a negative pressure, uncontrolled and unconditioned outside air is drawn into the space through every crack and crevice.

CAUTION

Some recovery cores contain cardboard structural components and cannot be washed with water. Follow the manufacturer's cleaning and maintenance instructions carefully for the specific model being serviced to avoid damage.

The recovery ventilator air filters must be replaced periodically. It is important the filters fit well to effectively protect the core. The recovery core must be vacuumed periodically to remove dust and dirt. These tasks should be performed at intervals recommended by the ventilator manufacturer.

2.2.1 Recovery Ventilator Operation

When a recovery ventilator is started up, the two airstreams must be balanced. The amount of fresh air needed for the space is usually determined based on American Society of Heating, Refrigerating, and Air Conditioning Engineers

(ASHRAE) Standard 62.2, *Ventilation and Acceptable Indoor Air Quality in Low-Rise Residential Buildings*. Many recovery ventilator manufacturers offer apps or calculators based on the ASHRAE standard to help contractors quickly determine the volume of fresh air required in a residence.

For example, using the calculator on one website requires the user to input the size of a home in square feet and the number of bedrooms. Entering 2,600 ft^2 and three bedrooms returns a solution of 108 cubic feet per minute (cfm) of outside air. This requirement applies around the clock. To maintain a slight positive pressure indoors, the exhaust air volume can be set slightly lower, such as 95 cfm.

Many recovery ventilators also offer a *boost mode*. Boost mode increases the fan speeds to accommodate a specific need for more outside air. For example, in the evening when a family may be taking showers and operating bathroom exhaust fans, boost mode can be initiated for a period to increase the outside air volume and maintain a positive pressure indoors.

The outside air and exhaust air volumes are set by taking pressure readings, calculating the pressure drop, and comparing the readings to a chart such as the example shown in *Table 1*. Pressure ports are permanently installed in the unit where a manometer can be connected for the measurements.

TABLE 1 Example Recovery Ventilator Airflow Chart

Measured Pressure Drop	Airflow MERV 8 Filtration	Airflow MERV 13 Filtration
0.1" w.c.	27 cfm	23 cfm
0.2" w.c.	46 cfm	38 cfm
0.3" w.c.	64 cfm	53 cfm
0.4" w.c.	82 cfm	69 cfm
0.5" w.c.	101 cfm	84 cfm
0.6" w.c.	119 cfm	99 cfm
0.7" w.c.	139 cfm	116 cfm

Air pressure ports should be located near the return air, exhaust air, outside air, and supply air inlets (*Figure 23*). Compare the readings taken at the return air and exhaust air ports. The difference between the two is the pressure differential (pressure drop) used for the chart. You can also connect the two manometer ports at the same time, which will read the pressure drop between the two locations directly. Then use the same process to determine the pressure differential between the outside air and supply air ports.

Potentiometers are provided to adjust the speed of the electronically commutated motors (ECMs) and control air volume. Once the pressure drop matches the desired air volume on the chart, the setting is complete. The manometer can remain connected while the adjustments are made. Boost-mode airflow is established the same way, but at higher airflow rates. In *Table 1*, note that more efficient air filters do reduce the air volume somewhat.

Potentiometers: Three-terminal variable resistors used to change the resistance in a circuit. The recovery ventilator interprets the resistance and determines the speed output signal sent to the motors.

NOTE

A discussion of air filter efficiencies and MERV ratings can be found in NCCER Module 03204, *Air Quality Equipment*.

A wiring diagram for a small ERV is shown in *Figure 24*. The power source is connected to two terminals on the ERV control module. A door switch (Door Kill) is wired into the power circuit to de-energize the unit when the access door is removed.

Low-voltage (24VAC) wiring is connected to the four terminals on the left side of the module. The 24VAC and High terminals provide a switching circuit to initiate Boost Mode. The switch can be a simple manual switch, or another type of device, such as an occupancy sensor.

A damper can be added into the ERV duct to prevent natural air movement when the system is off. The damper can be connected one of two ways. When it is connected to the 24VAC and COM (Common) terminals, the damper is driven open anytime the unit is energized. This is preferred when the unit will

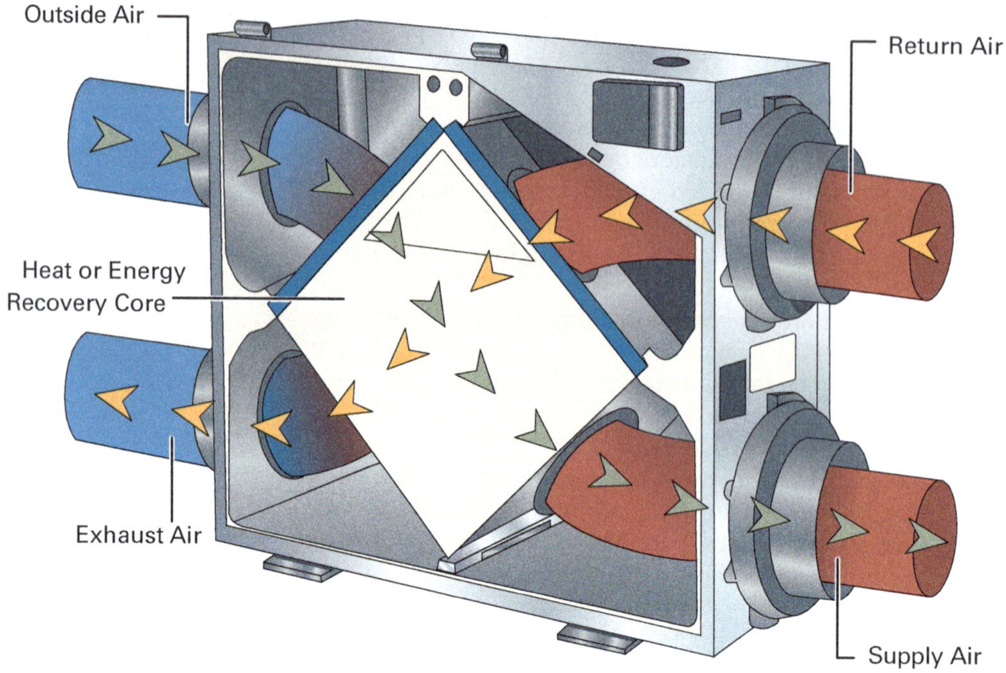

Figure 23 Recovery ventilator airflow pattern.

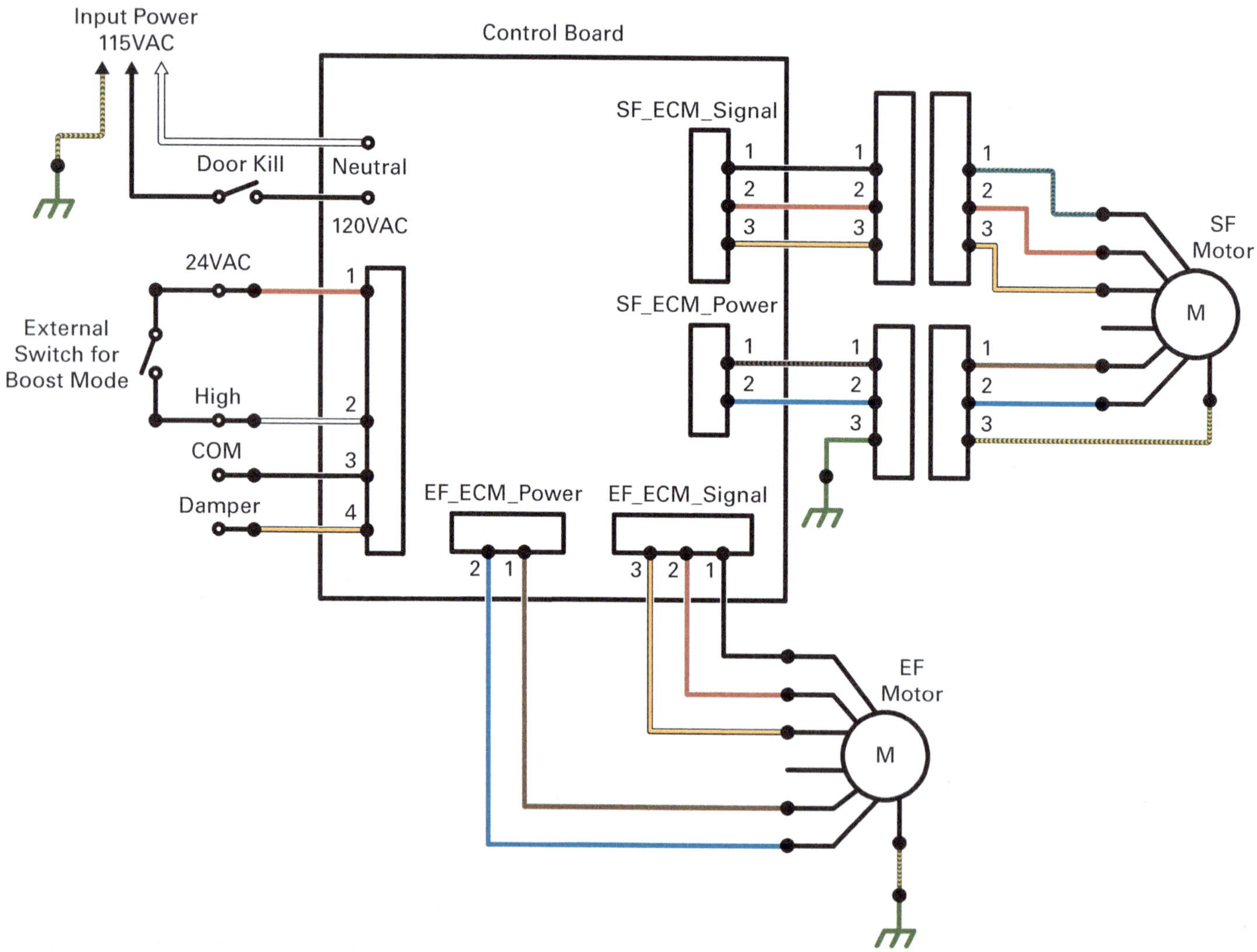

Figure 24 ERV wiring diagram.

Source: Image Courtesy of RenewAire LLC

run continuously. Alternatively, when usage is intermittent, the damper can be connected to the DAMPER (Damper) and COM terminals to operate when the ERV starts.

The two ECMs are labeled as SF (Supply Fan) and EF (Exhaust Fan). You'll recall that ECMs have an attached control module that requires both a power source and signals from a controller to initiate and control the speed of rotation. Two separate connectors for each motor provide the power source and the signal inputs.

2.2.2 Recovery Ventilator Troubleshooting

Before troubleshooting a recovery ventilator, ensure that the proper maintenance has been performed. The air filters should be reasonably clean. The core of the unit must also be vacuumed periodically, even though the air is being filtered before it reaches the core. Dirt and debris on the core significantly impede airflow and heat transfer.

It is also a good idea to check the connected ducts and ensure the ducts are intact and routed in a manner that supports good airflow.

Remember to begin every troubleshooting process with the proper literature and wiring diagrams in hand. Also, recovery ventilators may offer both a low-speed and a high-speed (Boost) operating mode. Thus, a failure can occur in one mode that does not affect the other mode.

- If the ventilator does not operate in either mode, ensure the unit has power. If power is available at the main terminals, ensure power is also available at the connectors for each ECM. Also ensure the door switch is closing when the access door is properly installed.
- Ensure that the potentiometers are adjusted to an appropriate position, rather than turned all the way down to the stop. Also ensure that the connectors to the motors are properly seated and secure.
- If you hear the fans running but there is little or no airflow, ensure that any motorized dampers are opening as they should. However, it is also important to remember that a small unit may only move about the same volume of air as a bathroom exhaust fan. Don't expect a substantial airstream.
- If the unit operates at high speed but not low speed, check the operation of any external control that initiates and enables ventilator operation.
- If the unit operates at low speed but not high speed, place a jumper across the 24VAC and HIGH terminals. If the unit starts at high speed, the problem lies with the external control device or the interconnecting wiring.
- If one motor operates but the other does not, test the non-working motor in the same manner as you would test other ECMs. If neither motor operates when the external inputs to the control module are correct and power is available, the problem is likely in the control module itself.

NOTE

ECM testing and troubleshooting is presented in NCCER Module 03318, *Motor Testing and Troubleshooting*.

When a new unit is started, it's a good idea to record the final settings both on your service report and on a document or sticker that remains with the unit. After troubleshooting or repairs are conducted, be sure to return the unit to the same airflow settings.

2.2.3 Frost Control

Under the right conditions of temperature and humidity, frost can develop on the ventilator core. This significantly impedes airflow and must be dealt with in many cases.

Because ERVs transfer both sensible and latent heat, a solution to frost accumulation is rarely required. ERVs may experience some frosting when the outside air is well below freezing. However, HRVs are far more likely to develop frost, with frost often developing when the outdoor temperature is around 25°F and lower. This is another reason ERVs are typically a better choice than an HRV.

To prevent frost from developing in HRVs, the outside air can be preheated to a temperature just above freezing. This approach does reduce the heat transfer efficiency of the unit, but no defrost cycles are required. Electric heating elements in the outside airstream is a common choice for preheating. Other approaches may be used in large commercial systems.

A more common approach is to periodically defrost an HRV unit by temporarily stopping the cold air from entering while the air normally exhausted outdoors is recycled back to the space. This allows warm room air to melt the frost, but it also means that no outside air is entering during the cycle, nor is any stale air being exhausted.

If frost becomes a problem in an HRV, ensure that you are familiar with the manufacturer's defrost or frost prevention strategy and proceed as needed.

2.0.0 Section Review

1. How does an integrated economizer differ from an enthalpy economizer?
   a. Integrated economizers never require relief air dampers.
   b. Integrated economizers require the user to manually select outside air for cooling.
   c. An integrated economizer only monitors the dry-bulb temperature of the airstreams.
   d. An integrated economizer can use outside air and mechanical cooling at the same time.

2. The core of a heat recovery ventilator (HRV) transfers ______.
   a. latent heat only
   b. no usable energy
   c. sensible heat only
   d. sensible and latent heat

Module 03312 Review Questions

1. Most humidifier problems are caused by ______.
   a. overheating
   b. faulty controls
   c. chemical attack
   d. a lack of maintenance

2. A residential fan-powered humidifier often has a water solenoid valve that ______.
   a. increases the water pressure
   b. opens whenever the fan turns off
   c. is wired in parallel with the fan motor
   d. must open before the humidifier fan will start

3. Many residential humidifiers today are primarily controlled by a(n) ______.
   a. digital thermostat
   b. integral humidistat
   c. differential humidistat
   d. duct-mounted humidistat

4. The capacity of an evaporative humidifier is substantially affected by ______.
   a. the applied voltage
   b. system static pressure
   c. type of air filtration used
   d. the supply air temperature

5. In an electronic air cleaner, to which components is the output of the high-voltage transformer applied?
   a. Ionizer section
   b. Collector section
   c. Both the collector and ionizer sections
   d. Prefilter, collector, and ionizer sections

6. What often happens when an electronic air cleaner operates with a lower air velocity than its manufacturer recommends?
   a. Excess ozone is generated
   b. Warping of the collector plates
   c. Overheating of the ionizer wires
   d. Overheating of the high-voltage transformer

7. Which hazardous substance does an infrared lamp contain that requires it to be disposed of according to local regulations?
   a. Ozone
   b. Lithium
   c. Mercury
   d. Cadmium

8. An enthalpy economizer uses a mixed/discharge air sensor to control the position of _____.
   a. only the outside air damper during free cooling
   b. only the outside air damper during mechanical cooling
   c. both the outside air and return air dampers during free cooling
   d. both the outside air and return air dampers under all operating conditions

9. When an economizer is in operation to provide free cooling, the simplest method to relieve excess pressure in the building is to add a _____.
   a. manual damper
   b. barometric damper
   c. powered parallel-blade damper
   d. powered opposed-blade damper

10. The air volume settings through both sides of a recovery ventilator are made based on what measurement?
    a. Core velocity
    b. Damper position
    c. Air pressure drop
    d. Fan rotational speed

Answers to odd-numbered review questions are found in the Module Review Answer Key at the back of this book.

Answers to Section Review Questions

Answer	Section	Objective
Section One		
1. b	1.1.3	1a
2. b	1.2.0	1b
3. a	1.3.0	1c
4. b	1.4.0	1d
Section Two		
1. d	2.1.1	2a
2. c	2.2.0	2b

MODULE 03213

Sheet Metal Duct Systems

Source: zvirni/iStock

Objectives

Successful completion of this module prepares you to do the following:

1. Identify and describe common types of sheet metal.
   a. Identify and describe various types of steel sheet metal.
   b. Identify and describe various sheet metal alloys.
2. Identify various methods of joining sheet metal duct.
   a. Identify various types of duct seams.
   b. Describe various methods of connecting sections of duct.
3. Describe methods used to suspend and support sheet metal duct.
   a. Describe methods used to suspend sheet metal duct.
   b. Describe methods used to support sheet metal duct.
4. Describe methods used to insulate and attenuate sheet metal duct.
   a. Describe duct liner materials and the installation process.
   b. Describe external duct wrap and the installation process.
5. Identify various sheet metal duct accessories and describe the installation requirements.
   a. Identify and describe the installation of various types of dampers.
   b. Identify and describe the installation of duct takeoffs and access doors.
6. Identify types of flexible duct and describe the installation process.
   a. Identify types of flexible duct.
   b. Explain how flexible duct is connected and supported.

Performance Tasks

This is a knowledge-based module. There are no Performance Tasks.

Overview

HVAC ductwork can be made of sheet metal or fiberglass ductboard. Because the interior surfaces of sheet metal ducts are smooth, metal offers less resistance to airflow than fiberglass ductboard. However, in most cases, metal duct must be insulated to reduce noise transmission, minimize heat transfer, and prevent condensation. Working with sheet metal ductwork requires a familiarity with various metals as well as with the unique methods used to join, support, and insulate the duct.

Digital Resources for HVACR

SCAN ME

Scan this code using the camera on your phone or mobile device to view the digital resources related to this craft.

NCCER Industry-Recognized Credentials

If you are training through an NCCER-accredited sponsor, you may be eligible for credentials from NCCER. The ID number for this module is 03213. Note that this module may have been used in other NCCER curricula and may apply to other level completions. Contact NCCER at 1.888.622.3720 or go to **www.nccer.org** for more information.

You can also show off your industry-recognized credentials online with NCCER's digital credentials. Transform your knowledge, skills, and achievements into credentials that you can share across social media platforms, send to your network, and add to your resume. For more information, visit **www.nccer.org**.

1.0.0 Introduction to Sheet Metal

Performance Tasks

There are no Performance Tasks in this section.

Objective

Identify and describe common types of sheet metal.

a. Identify and describe various types of steel sheet metal.
b. Identify and describe various sheet metal alloys.

Sheet metal ductwork and fiberglass ductboard can serve the same purpose in most applications. Many instances occur, however, when sheet metal is both preferred and/or required. Sheet metal ductwork offers a longer life than fiberglass ductboard and is less likely to be damaged. Sheet metal is also used for outdoor applications, and it may be required in healthcare or similar facilities due to fire safety codes.

Standards published by the Sheet Metal and Air Conditioning Contractors National Association (SMACNA) provide HVAC installers with guidelines for proper construction, support, and insulation of ductwork. In many instances, sheet metal ductwork must be insulated to reduce noise transmission and prevent heat loss or heat gain. Therefore, it is important for anyone installing sheet metal ductwork to understand the proper techniques for insulating this type of ductwork.

1.1.0 Steel and Other Metals

Steel is a combination of the metallic element, iron, and the non-metallic element, carbon. Though iron is the most abundant of all metals, it is rarely found in its pure state naturally. It is usually found in combination with oxygen and other elements in the form of ore mixed with rocks, clay, and sand.

Alloy: A metal mixed with other metals or elements to create strength and durability.

Metals are classified as either pure metals or **alloys**. A pure metal is one that is not combined with any other metal. Iron, aluminum, copper, lead, tin, zinc, and gold are examples of pure metals. Pure metals are often too soft and lack sufficient strength to be used in construction applications, so other elements are added to create alloys. Alloys are engineered to be stronger and generally have other structural advantages necessary in certain construction applications.

Sheet metal can be divided into three basic groups:

- Basic metals
- Coated metals
- Alloy metals

Basic and coated metals are presented in this section. Alloy metals are presented in the section that follows.

Sheet steel: An uncoated sheet of steel with a bluish-black surface; commonly referred to as *black iron* in the trade.

Cold-rolled steel: Metal that has been formed by rolling at room temperature, usually to obtain an improved surface or higher tensile strength.

1.1.1 Basic Metals

The basic metals used in sheet metal ductwork are **sheet steel**, **cold-rolled steel**, hot-rolled steel, and copper. Sheet steel is an uncoated sheet with a bluish-black surface. Cold-rolled steel is silver-gray and has a smoother surface than other sheet products. The name refers to the steel mill process used to make it. The steel is compressed and rolled into the desired shape and thickness after it has cooled, giving it an improved surface finish and/or higher **tensile strength**.

Tensile strength: The resistance of a material against rupture when placed under tension.

Hot-rolled steel is formed as it poured out of a smelting furnace. Copper sheets usually contain a small amount of another metal to form an alloy. Copper sheets are commonly used as roofing and guttering materials, but rarely for ductwork due to the extraordinary cost.

Sheet metal comes in various thicknesses. However, sheet metal thickness is generally identified by its *gauge (ga)*. Typical sheet metal thickness ranges from $\frac{1}{64}$" (30 gauge) to $\frac{1}{8}$" (11 ga); the lower the gauge number, the thicker the sheet.

Thickness is measured and verified using a sheet metal gauge (*Figure 1*). Be careful not to confuse this with a wire gauge. When measuring coated metals, you need only account for the base metal, not the coating. The actual gauge of coated metal is one gauge thinner than that indicated by the measuring tool.

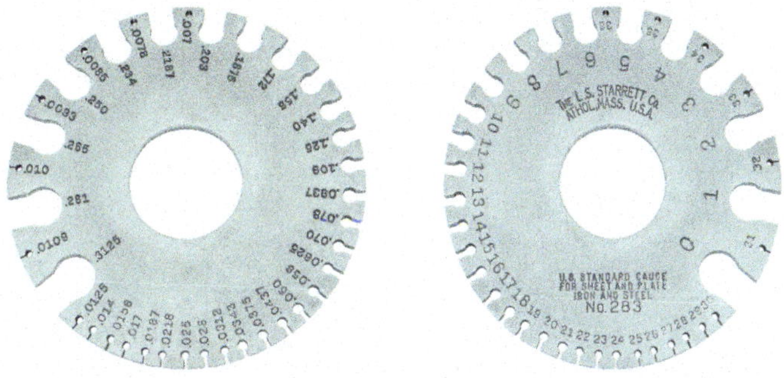

Figure 1 Sheet metal gauge.
Source: Courtesy of The L.S. Starrett Company

1.1.2 Coated Metals

Steel is often coated with zinc to protect it from rust and corrosion. Steel treated in this way is called *galvanized steel* (*Figure 2*). The thickness of the coating on sheet metal may be G60 (0.60 ounces per square foot) or G90 (0.90 ounces per square foot). Some duct applications in areas exposed to corrosive fumes or high amounts of moisture require plastic or fiberglass coatings, as well as other metal coatings.

When the zinc coating is properly applied, galvanizing can protect the base metal from corrosion for 15 to 30 years or more. In addition, the zinc coating provides a surface that helps inhibit the growth of microbes in air distribution systems. Galvanized sheet metals are often used in heating and air conditioning ducts, gutters, and downspouts—all applications that are exposed to common atmospheric corrosion.

Figure 2 Galvanized sheet metal ducts.

On Site

How Metal Is Galvanized

The zinc used to coat steel may be applied by one of two methods:

- Hot dipping
- Electroplating

In the hot dipping method, clean, oxide-free iron or steel is placed into hot, melted zinc. This process bonds the zinc to the surface of the iron or steel.

In electroplating, a pure coating of zinc is applied without heat. This process does not have the soldering effect of the hot-dip process, so it is not suitable for items that must be made watertight. The zinc coating adheres better when hot dipped.

Zinc protects the base metal because the coating is more *electronegative* (charged with negative electricity) than the base metal. The zinc gives the base metal what is called *sacrificial protection*. This means that, if the base metal is scratched, the zinc is slowly consumed instead of the base metal, or sacrificed, so that the iron or steel remains intact. This protection continues as long as the zinc and base metal are in firm contact with one another.

1.2.0 Alloy Metals

The properties of a pure metal can be changed by melting it and mixing in other metals or elements. This process produces an alloy—an entirely new metal that can have different, and more desirable, characteristics than the original elements.

An alloy is generally named after the principal metal in its composition. When metals are added to steel, the resulting product is called a steel alloy. When metals are added to aluminum, the result is an aluminum alloy. Many other alloys are available; some common ones are shown in *Table 1*.

TABLE 1 Common Alloys

Alloy	Components
Brass	Copper and zinc
Bronze	Copper and tin
Stainless Steel	Iron, nickel, and chromium
Steel	Iron and carbon
Solder	Lead and tin

Non-metallic elements can also be part of an alloy. The most common of these elements is carbon. When carbon is added to iron, the result is steel. Thus, steel itself is an alloy.

Did You Know?

Making Stainless Steel

Stainless steel is made in an electric arc furnace. Carbon electrodes make contact with recycled stainless scrap and alloys of chromium or other elements. A current passes through the electrodes, which increases the temperature and melts the scrap and alloys. This molten material then moves to an argon oxygen decarbonization (AOD) vessel. Here, the carbon levels are reduced, and final components are added.

Some of the material is hot-rolled or forged into its final form. Other material is cold-rolled to reduce the sheet thickness or to be made into rods and wire. Most stainless steels receive a final *annealing*—a heat treatment that softens the structure—and *pickling*. Pickling refers to an acid washing process that removes surface scale.

1.2.1 Steel Alloys

Steel alloys are produced when certain other metallic elements are added to plain steel. These other metals include nickel, chromium, manganese, tungsten, and vanadium.

One of the most common alloy steels is stainless steel, which contains between 11% and 26% chromium. The chromium hardens and toughens the steel and makes the grain finer. Chromium also makes the steel less likely to stain (thus the term *stainless*). Sometimes nickel is also added to the mix to produce a type of nickel-chromium stainless steel.

Stainless steel is valued for its strength, toughness, and corrosion resistance. It is often used to make counters and cabinets for hospitals and restaurants. It is also used to make ductwork, particularly in industrial applications.

1.2.2 Aluminum Alloys

Aluminum is the most plentiful metallic element in the Earth's crust. It is the most widely used **non-ferrous** metal, but aluminum is never found in its metallic form in nature.

Non-ferrous: Identifies a metal that contains little or no iron.

Pure aluminum is soft and lacks strength, but it can be combined with copper, magnesium, silicon, manganese, and other elements to form aluminum alloys. These lightweight but strong metals are easily formed and are commonly used for ductwork when steel is too heavy or subject to rapid corrosion. The following are the advantages of aluminum alloys:

- Ductile and malleable
- Corrosion resistant
- Easy to install and maintain
- Recyclable

1.0.0 Section Review

1. Why are pure metals usually a poor choice for sheet metal ducts?
   a. Too expensive
   b. Too soft
   c. Too brittle
   d. Too prone to corrosion

2. Which of the following metals makes stainless steel resistant to stains?
   a. Chromium
   b. Zinc
   c. Copper
   d. Sodium

2.0.0 Joining Sheet Metal

Objective

Identify various methods of joining sheet metal duct.

a. Identify various types of duct seams.

b. Describe various methods of connecting sections of duct.

Performance Tasks

There are no Performance Tasks in this section.

While an airtight duct system might be a desirable goal, absolute air tightness is rarely possible without extreme effort. Seams and joints allow some air to enter and exit the duct, and this is considered when the system is designed. A certain amount of air leakage is expected and is directly related to the following factors:

- Operating pressure
- Square footage of the duct system
- Duct construction
- Sealing

To help keep air leakage at acceptable levels, the duct must be properly sealed at connections, joints, and seams. Installers use approved adhesives, gaskets, or tape systems either alone or in combination to seal the duct. Check the manufacturer's guidelines or follow your shop's standards for choosing and applying the appropriate sealing material.

Duct sealing requirements are divided into three classes (*Table 2*). The duct system designer is responsible for selecting the pressure class and the seal class. This information is usually contained in the job specifications and engineering drawings. Of the three seal classes, Class A duct requires that all seams, connections, and penetrations be sealed, and that the seal be able to withstand the highest pressure. Note, however, that seal classes do not identify the volume of allowable air leakage allowed.

TABLE 2 Pressure Seal Requirements

Class	Pressure Requirements
Class C Duct	Connector joint sealed for 2" water column (w.c.) pressure class and, when specified, for 0.5" w.c. pressure class or for 1" w.c. pressure class.
Class B Duct	Both connector joints and longitudinal seams sealed for 3" w.c. pressure class.
Class A Duct	All connector joints, longitudinal seams, and duct wall penetrations sealed for 4" w.c. pressure class and greater.

2.1.0 Seams

Seams (or *locks*) are used to join the edges of the metal. Seams may be made using mechanical methods or by soldering or welding. The best method for closing a seam depends on the thickness and type of metal, the cost of seam fabrication and assembly, and the equipment available.

Common duct seams include the following:

- Grooved lock seam
- Snap lock seam
- Pittsburgh lock

2.1.1 Grooved Lock Seam

The grooved lock seam is sometimes called the *acme*, or pipe lock (*Figure 3*). This seam consists of two folded edges that are hooked together and locked with a grooving machine or hand groover.

Figure 3 Cross section of a grooved lock seam.

Light-gauge, prefabricated sections of round duct use a similar grooved lock seam. The tension created when the sheet metal is rolled into a round shape forces the two parts of the seam together. The result is a round duct with a single seam.

The hand groover is a hardened steel tool with one end recessed to offset the grooved lock. Hand groovers range in size from $^{3}/_{32}$" to $^{19}/_{32}$" wide. Installers should use a hand groover with a groove that is approximately $^{1}/_{16}$" wider than the width of the seam. Grooved seams are rarely used with metal heavier than 20 ga.

2.1.2 Snap Lock Seam

The snap lock seam consists of a pocket lock and another edge that is inserted into the pocket. This edge is formed using a snap lock punch to raise buttons or bumps that produce a tighter fit inside the pocket. The snap lock seam is widely used in prefabricated sections of light-gauge rectangular ducts (*Figure 4*). The duct section comes in two L-shaped pieces that snap together to form the duct. This process can be done in the field because it does not require any machinery to complete the assembly. Smaller rectangular duct sizes constructed of light-gauge material can be transported in a condensed form (stacked L-shaped sections) and put together at a residential job site. Once the two L-shaped sections are put together, the completed duct section can be joined to other duct sections. This type of assembly is generally not allowed in commercial installations.

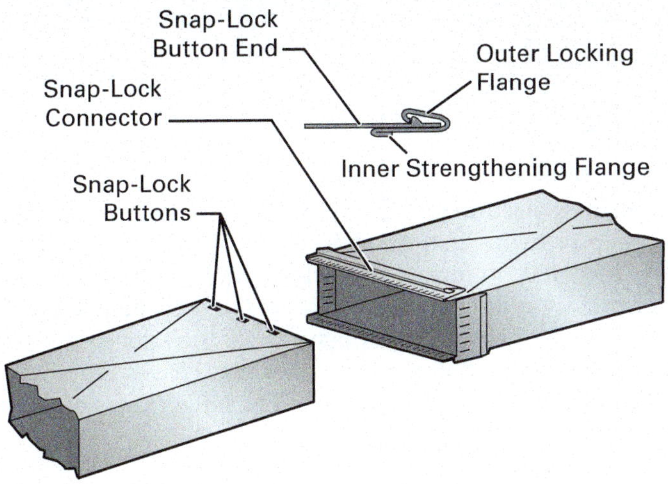

Figure 4 Snap lock seam.

2.1.3 Pittsburgh Lock

The Pittsburgh lock is the most common seam constructed in a sheet metal shop. This seam consists of one edge formed into a pocket and a second edge that is bent and inserted into the pocket. After the bent edge is inserted into the pocket, the pocket edge is folded over to permanently lock the seam shut (*Figure 5*).

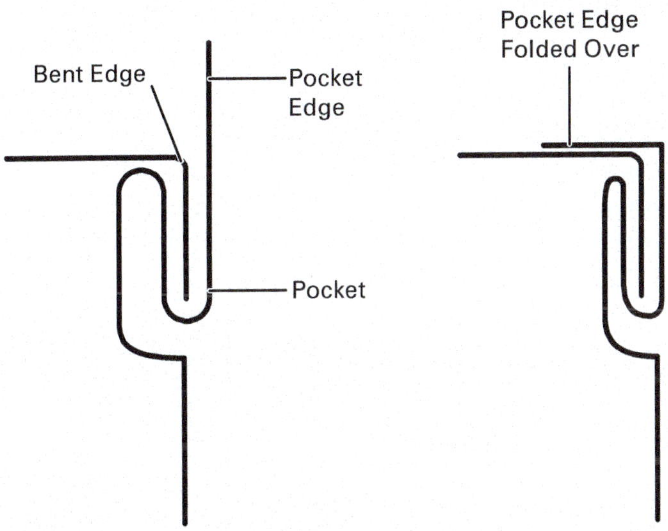

Figure 5 Cross section of a Pittsburgh lock.

A Pittsburgh lock machine can be used to quickly form this type of seam. To form the Pittsburgh lock, metal is inserted in one end of the machine and run through a series of rollers. The metal comes out at the other end with the pocket lock completely formed. If no roll-forming machine is available, installers must form the Pittsburgh lock on a **brake**.

Brake: A sheet metal-working machine, either manually operated or powered, used to bend metal.

2.2.0 Connectors

Connectors are used to join individual pieces of duct. Connectors include the following:

- Drive connector
- S-slip
- Standing seam
- Transverse duct connector

- Standing S
- Standing drive slip
- Ductmate®

2.2.1 Drive Connector

The drive connector is used to join the short edges of two pieces of metal. It is made in a two-step process. First, the short edges of two pieces of metal duct to be joined are turned 180° to form two pockets, each about $\frac{1}{2}$" wide (*Figure 6*). The drive is formed from a separate piece of metal, the edges of which are also turned, forming a sleeve. The drive is pushed over the folded edges of the duct to be connected and driven down with a hammer. Drives are always used to connect the sides (shorter dimension) of duct. The longer sides of the duct, however, must also be connected.

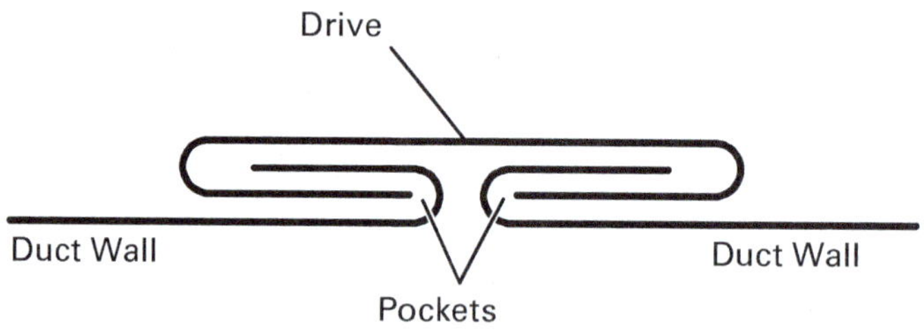

Figure 6 Cross section of a drive connector.

2.2.2 S-Slip

The S-slip is used to connect the two longer sides of duct. The S-slip is an S-shaped connector that forms two pocket locks for metal edges to slip into (*Figure 7*). S-slips are used to connect the top and bottom edges (longest dimension) of ductwork. The S-slip is sometimes also called an S-type connector or flat slip.

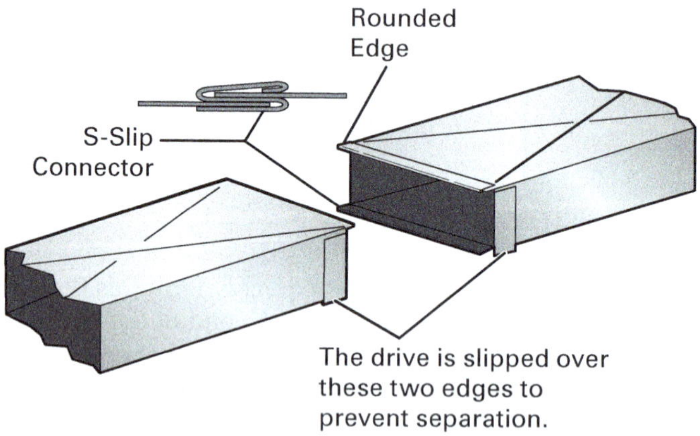

Figure 7 S-slip connector.

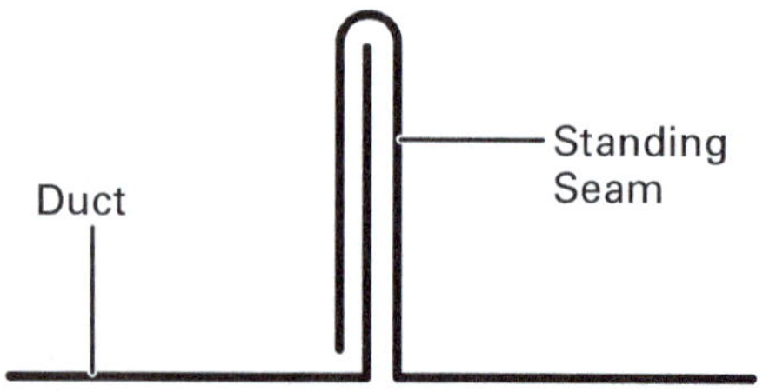

Figure 8 Cross section of a standing seam.

Drive connectors and S-slips must be used together to connect sections of duct end-to-end. The drive pulls the duct sections together, while the slip simply serves to block the path of airflow out of the duct.

2.2.3 Standing Seam

A standing seam eliminates the need for additional reinforcement in a run of duct because it supplies its own reinforcement. This type of lock is easily made and can be used as a cross-seam on larger ducts as a stiffener (*Figure 8*).

2.2.4 Transverse Duct Connector

The transverse duct connector, also called a *pocket lock*, is used to connect two sections of duct. It is used in the same way as the S-slips and drives. The difference is that the finished transverse duct connector is not flush with the duct.

2.2.5 Other Connectors

Three other commonly used duct connectors include the standing S, standing drive slip, and Ductmate®. The standing S is an S-slip with additional reinforcement (*Figure 9*). The standing drive slip is simply a drive connector with this same type of additional reinforcement (*Figure 10*). Both perform the same function as common drive and S-slip connectors but offer greater strength.

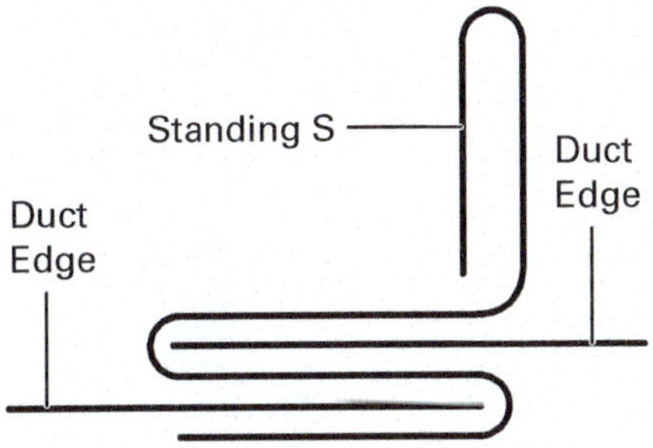

Figure 9 Cross section of a standing S connector.

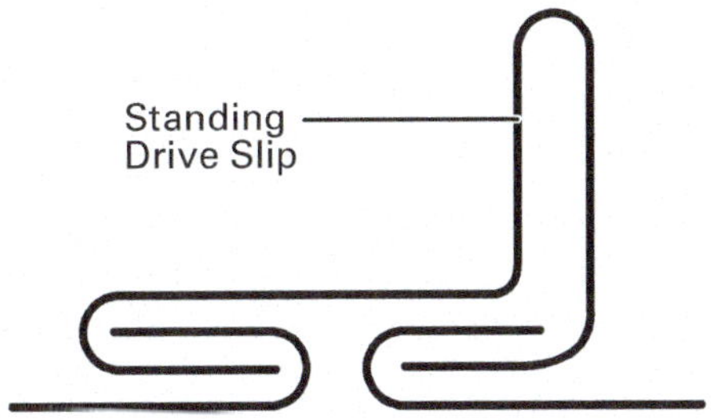

Figure 10 Cross section of a standing drive slip.

Ductmate® is a patented duct connection system used to connect duct and provide greater structural integrity than any of the other options (*Figure 11*). This system is used primarily to prevent air leaks in low- and medium-pressure air systems. It also provides further reinforcement to larger ducts. This system and similar designs are the most popular for joining duct in commercial and industrial applications.

Figure 11 Ductmate® flanges, ready for installation on duct.

Duct that is connected directly to an air handling unit or a furnace may transmit both noise and vibration through the ductwork and into the conditioned space. To minimize this problem, installers often install flexible connectors between the air handler and the duct system (*Figure 12*). The connector is made of fabric such as rubberized canvas. The soft fabric reduces the transmission of noise and vibration into the duct. Flexible connectors are used in both the supply and return air sides of the system. Some codes require that a metal jumper be placed across the flexible connector so that the metal duct system remains grounded, but this is rarely required.

Figure 12 Flexible duct connectors.

On Site

Sheet Metal Shops

HVAC contractors may have their own sheet metal shops, or they may have others do the work. The nature of their business determines whether there is a need to do sheet metal work in-house. Larger mechanical contractors with a focus on commercial work are more likely to have their own shop. The ducts in commercial work tend to be much bigger and of a heavier gauge, and more custom pieces have to be fabricated. This requires specialized equipment and people with the skills to lay out and fabricate custom pieces.

Smaller contractors with a residential focus often have limited or no in-house equipment for metal fabrication. Residential contractors often purchase off-the-shelf standardized sheet metal ducts and fittings. Their need for custom pieces is often limited to making unique adapters or transitions. For example, when replacing a residential furnace, the only custom piece that may be needed is a simple transition from the new furnace's supply air duct flange to an existing warm air plenum. These fairly simple pieces can be made in-house or by an outside sheet metal shop.

2.2.6 Duct Sealing

Proper assembly practices can reduce air leakage from sheet metal duct, but some level of leakage will always occur. In the past, what was considered a small amount of duct leakage was ignored. Today, with higher energy costs, even a small amount of duct leakage adds up to significant energy losses. As a result, significant duct leakage cannot be ignored.

Most codes require that all seams in sheet metal duct be sealed to prevent air leakage. Seams can be sealed with foil tape or a mastic-type sealant (*Figure 13*). When mastic is used, it is applied to exterior duct seams using a brush (*Figure 14*). Sealant can also be applied with a caulking gun or a sprayer.

For mastic duct sealant to successfully seal larger openings, fiberglass mesh tape is often embedded in the mastic (*Figure 15*). Once the first coat of mastic has dried, additional coats are applied. This enables the mastic to bridge the gap better and the mesh helps strengthen the seal, so it is less likely to open.

Figure 13 Mastic-type duct sealant.

Figure 14 Duct joints sealed with mastic.

Figure 15 Fiberglass mesh tape.

Going Green

High-Tech Duct Sealing

It is a fairly simple process to seal a sheet metal duct system. But how do you seal a sheet metal duct system that is already installed? New technologies are available that allow ducts to be sealed from the inside. One such technology, called aerosol sealing, injects small, dry adhesive particles into a pressurized duct system while the cooling coil, fan, and furnace components are blocked off. This isolates these components from the duct system, so they do not become coated with the adhesive sealant. All supply registers and return air grilles are also blocked so that the entire duct system is sealed and isolated.

A fan-powered sealing machine is temporarily connected to the duct system using a large clear plastic duct. Dry adhesive particles are suspended in the air being injected into the duct system. With the duct system plugged, the only place that air can escape is through the leaks in the ducts. As the air passes through leak openings, adhesive particles get stuck and begin to block the openings. Over time, the sealant builds up to seal the leak. The sealing process is monitored in real time using a laptop computer loaded with sealing process software. The manufacturer of this sealing process claims it can seal duct leaks up to $\frac{5}{8}$" in diameter and seal from 70% to 90% of all duct leaks.

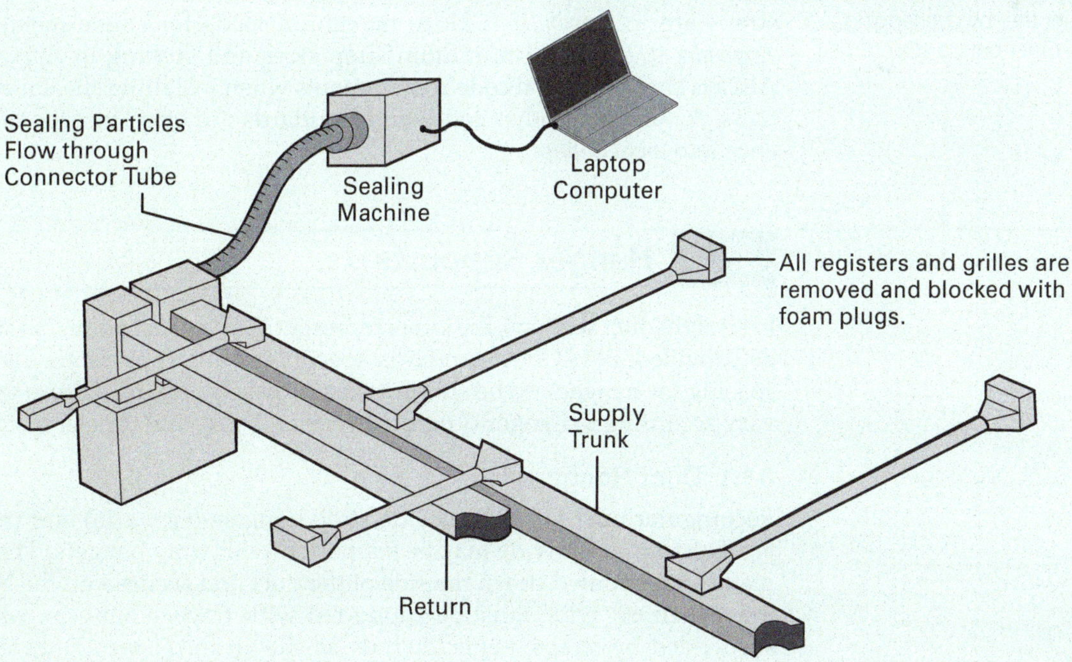

2.0.0 Section Review

1. Grooved lock seams are rarely used in metals thicker than _____.
   a. $^{1}/_{8}$"
   b. 1"
   c. 20 ga
   d. 30 ga

2. Which type of connection provides the greatest structural integrity?
   a. Ductmate®
   b. Standing S
   c. Standing seam
   d. Standing drive slip

3.0.0 Suspending and Supporting Sheet Metal Ducts

Performance Tasks

There are no Performance Tasks in this section.

Objective

Describe methods used to suspend and support sheet metal duct.

a. Describe methods used to suspend sheet metal duct.

b. Describe methods used to support sheet metal duct.

As the ductwork is installed, it must be properly suspended or supported. The hanging system for HVAC ductwork consists of the following components:

- Upper attachment to the structure
- Lower attachment to the duct
- Hanger in between, connecting the duct to the structure

Proper spacing, material, and installation practices must be considered when planning the installation of hangers and supports. The following sections present typical construction code specifications for hanging straight ducts and riser supports. Some specifications for minimum fasteners required on **lap-joined** straps are in *Appendix 03213A*; maximum loads for single metal hangers in *Appendix 03213B*; and minimum strap sizes and spacing in *Appendix 03213C*. Always check the local code requirements when installing sheet metal duct systems. A variety of other codes and standards are often incorporated by reference into local codes.

Lap-joined: A condition in which one piece is joined to another by overlapping one piece with the other before joining.

3.1.0 Hanger Supports

In straight duct sections, the joints represent the weakest points, so support must be provided. SMACNA standards specify the allowable loads and maximum spacing for hangers, so be sure to consult these standards. These specifications vary according to hanger type, duct type and size, and type of installation.

3.1.1 Duct Hangers

Rectangular duct hangers include strap hangers (*Figure 16*) and trapeze hangers. Duct up to 60" wide may be supported with strap hangers. These are sheet metal straps routed down the side of the duct and secured on the bottom. Duct more than 60" wide must be supported with trapeze hangers, which may be suspended by straps, angle, or rods as shown in (*Figure 17*). Stacked trapeze hangers can also be used for stacked ducts (*Figure 18*).

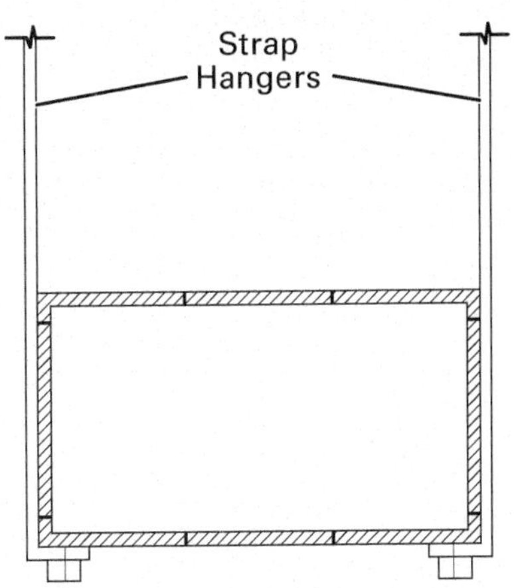

Figure 16 Strap hanger.

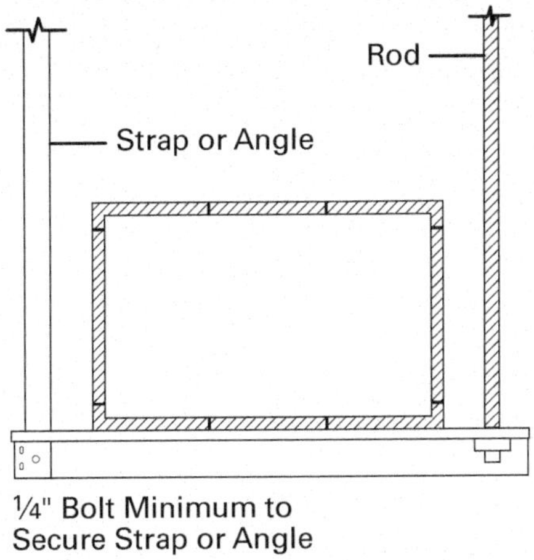

Figure 17 Trapeze hanger.

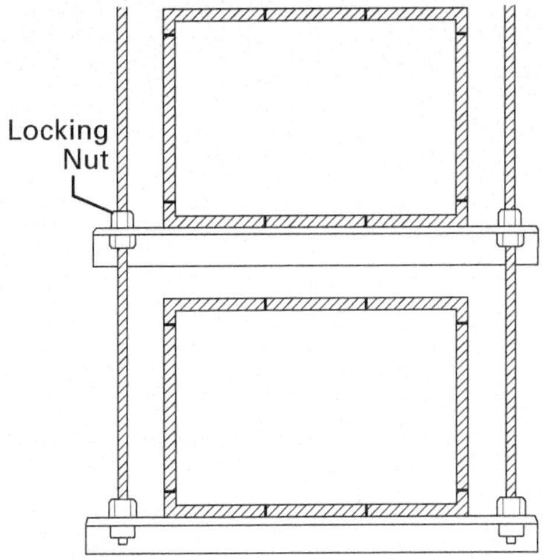

Figure 18 Stacked trapeze hangers.

Make sure that the load does not exceed the load rating of the chosen hanger system. The weight of the duct must be considered when choosing hanger materials. Minimum recommended sizes of hangers for rectangular duct are shown in *Table 3*. Note the dimension provided in the first column (P/2) refers to half of

TABLE 3 Types, Sizes and Spacing of Hangers

Rectangular Metal Duct — 16-Ga Maximum with Allowance for Insulation			
P/2*	**Hanger Pair Spacing**	**Strap Width - Inches × Thickness**	**Wire or Rod**
30" maximum	10' apart	1" × 22 ga	10 ga (0.135")
30" maximum	8' apart	1" × 22 ga	10 ga (0.135")
30" maximum	5' apart	1" × 22 ga	12 ga (0.106")
30" maximum	4' apart	1" × 22 ga	12 ga (0.106")
72" maximum	10' apart	1" × 18 ga	⅜" diameter
72" maximum	8' apart	1" × 20 ga	¼" diameter
72" maximum	5' apart	1" × 22 ga	¼" diameter
72" maximum	4' apart	1" × 22 ga	¼" diameter
96" maximum	10' apart	1" × 16 ga	⅜" diameter
96" maximum	8' apart	1" × 18 ga	⅜" diameter
96" maximum	5' apart	1" × 20 ga	⅜" diameter
96" maximum	4' apart	1" × 22 ga	¼" diameter
120" maximum	10' apart	1½" × 16 ga	½" diameter
120" maximum	8' apart	1" × 16 ga	⅜" diameter
120" maximum	5' apart	1" × 18 ga	⅜" diameter
120" maximum	4' apart	1" × 20 ga	¼" diameter
168" maximum	10' apart	1½" × 16 ga	½" diameter
168" maximum	8' apart	1½" × 16 ga	½" diameter
168" maximum	5' apart	1" × 16 ga	⅜" diameter
168" maximum	4' apart	1" × 18 ga	⅜" diameter
192" maximum	10' apart	No straps allowed	½" diameter
192" maximum	8' apart	1½" × 16 ga	½" diameter
192" maximum	5' apart	1" × 16 ga	⅜" diameter
192" maximum	4' apart	1" × 16 ga	⅜" diameter
193" and greater	Engineering analysis required		

NOTE: The limits are for duct, insulation, and normal reinforcement and trapeze weights. No external loads are allowed in addition to this total weight.
* P = Total perimeter of the rectangular duct. P/2 = ½ of the duct perimeter.

the duct perimeter. For example, a 36" × 18" duct has a perimeter of 108", so P/2 would be 54". Several options are shown for each range of duct sizes.

Angle bracket: An L-shaped metal supporting member used to support vertical risers. Also called angle iron.

In residential and light commercial applications, the ducts are smaller and lighter than those used in commercial applications. For that reason, the support system does not need to be as robust. A simple **angle bracket** (*Figure 19*) is often used to fasten rectangular trunk ducts to the underside of floor joists when ducts are installed in a basement or crawlspace. The rigid nature of the angle brackets allows them to also act as a spacer, keeping the ductwork from contacting the combustible floor joists. Many codes specify that this spacing be a minimum of 1".

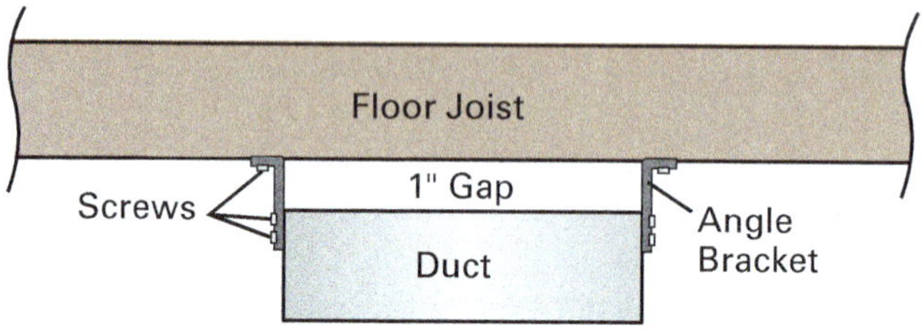

Figure 19 Residential duct hangers.

In residential applications where the ducts are in an attic, these same angle brackets can be used to support the duct slightly above any combustible materials.

Round metal ducts may be used for individual duct runs in residential systems. The round duct branches off from the main rectangular trunk. Round duct runs are typically supported using 1" wide light-gauge sheet metal straps that are attached to floor joists with nails or screws (*Figure 20*). The same metal straps can suspend the round ducts from above.

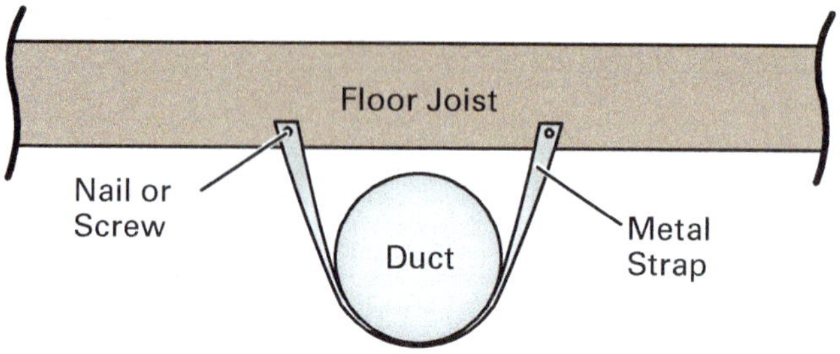

Figure 20 Metal strap supporting round duct.

3.2.0 Riser Supports

Channel: A U-shaped piece of structural steel used as a supporting device.

Rectangular risers (ducts installed vertically) must be supported with angle iron or **channel** secured to the sides of the ducts with welds, bolts, sheet metal screws, or blind rivets. Place riser supports at one- or two-story height intervals, or every 12' to 24'. Some risers are supported from the floor (*Figure 21*).

It may be necessary to install vibration isolators to eliminate noise transmitted through the duct. These isolators reduce noise that can transfer through the riser supports to surrounding walls, floors, or ceilings. These structures sometimes act as sounding boards, amplifying the noise.

Tensile load: The weight required to cause metal to stretch or compress.

Shear load: The amount of weight or pressure that causes an exposed piece of metal to break or shear off.

Some risers are supported from the wall and may be held in place by a band or by angle brackets (*Figure 22*). The allowable load per fastener and the number of fasteners to be used depend on the duct gauge and duct size. Ducts should be located against the wall or a maximum of 2" away from the wall. Each wall anchor must satisfy both the **tensile load** and the **shear load** of the supported member.

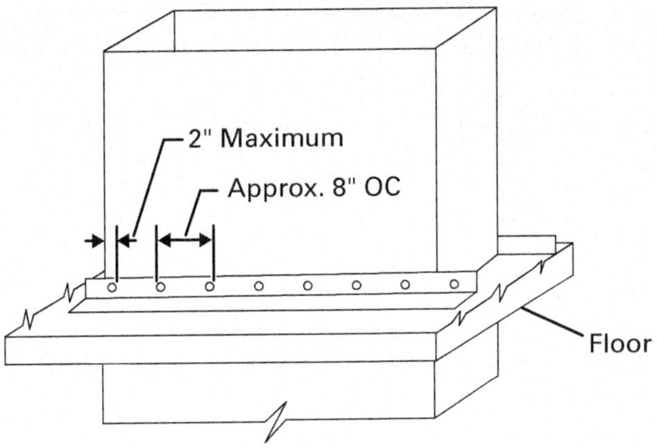

Figure 21 Floor-supported duct risers.

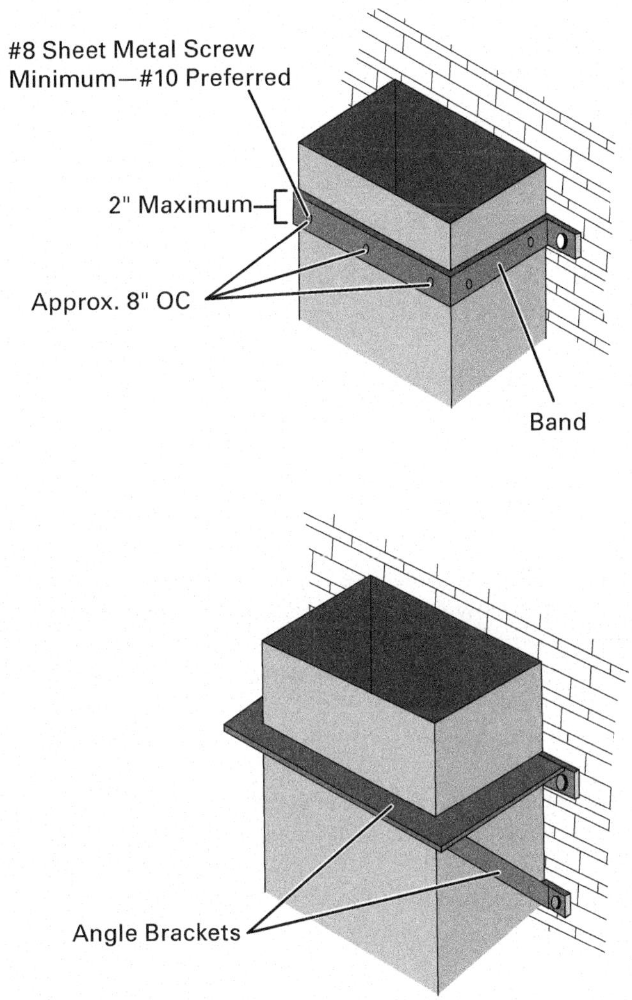

Figure 22 Wall-supported risers.

Going Green

Recycling Metals

A certain amount of scrap will always result when installing a new sheet metal duct system or making changes to an existing system. Recycle leftover scrap metal whenever possible.

3.0.0 Section Review

1. What is the minimum clearance that should be maintained between the trunk duct and combustible materials?
   a. ½"
   b. 1"
   c. 1½"
   d. 2"

2. Which fastening approach can be used to secure rectangular-duct riser support channels?
   a. Nails
   b. Adhesive
   c. Wire
   d. Welds

4.0.0 Duct Insulation and Liner

Performance Tasks

There are no Performance Tasks in this section.

Objective

Describe methods used to insulate and attenuate sheet metal duct.

a. Describe duct liner materials and the installation process.
b. Describe external duct wrap and the installation process.

Vapor retarder: A barrier, such as foil, plastic, or vinyl, placed between insulation and the surrounding air to prevent moisture in the surrounding air from contacting the insulation or the cold surfaces beneath.

Sound attenuation: The reduction in the level of sound that is transmitted from one point to another.

The conditioned air and water that make living and working spaces comfortable may have to travel quite a distance from the air conditioning equipment to get to a room. There is a greater chance of heat loss or heat gain through the duct walls when the duct is longer, because more surface area is exposed. Thermal insulation is required to slow or minimize this movement of heat.

Insulation is required for both hot and cold air distribution systems. In addition to insulation, cold systems require a **vapor retarder** over the insulation to prevent condensation and to protect the insulation from moisture. Insulation also provides **sound attenuation**, reducing the noise generated in an HVAC system.

Thermal insulation is designed to slow the flow of heat energy. It is commonly used to control heat transfer from ductwork to the surrounding area in temperatures ranging from 0°F (–17.7°C) through 3,000°F (1,648°C) and even higher. Thermal insulation has the following characteristics:

- Conserves energy by reducing heat loss or heat gain through ducts, pipes, and buildings
- Controls surface temperatures of equipment for safety and personal comfort
- Allows for temperature control of a piece of equipment or a building
- Prevents condensation on surfaces that have a temperature below the dew point of the surrounding environment
- Helps to control hazardous fire conditions that can be created by grease-laden air in a kitchen exhaust system
- Reduces sound and vibration to some degree

The thermal resistance value, or *R-value*, is a measurement of the ability of insulation to reduce the flow of heat. The American Society of Heating, Refrigerating, and Air-Conditioning Engineers (ASHRAE) has developed a standard that sets minimum recommended R-values for both commercial and residential construction materials. The standard is titled ASHRAE/IES 90.1, *Energy Standard for Sites and Buildings Except Low-Rise Residential Buildings*.

Insulation is made of a variety of porous materials, such as fiberglass, rock or slag wool, calcium silicate, bonded perlite, vermiculite, and ceramics. Insulation may also be made of closed-cell materials, such as cork, foam rubber, polystyrene, and polyurethane. Reflective insulation is another option. It has a smooth metallic or metalized surface applied over rigid foam, plastic film, cardboard, or polyethylene bubbles. The reflective material also serves as a vapor retarder.

WARNING!

Vermiculite insulation encountered in the field is likely contaminated with asbestos. Do not disturb or handle any vermiculite insulation encountered in building attics and walls. For more information about vermiculite insulation and asbestos, refer to the following: *http://www.epa.gov/asbestos*.

Insulation may be flexible, semi-rigid, rigid, or formed-in-place. Flexible and semi-rigid insulation come in blanket or batt form and are available in sheets or rolls (*Figure 23*). A covering or facing material may be fastened to one or both sides. This facing may serve as a reinforcing vapor retarder, reflective surface, or surface finish. Insulation can be applied to the inside or outside of sheet metal ducts.

Figure 23 A variety of insulating materials.
Source: ©CertainTeed

Rigid insulation is available in blocks, boards, or sheets. Rigid insulation for pipes or curved surfaces is supplied in half sections designed to fit standard pipe sizes.

Insulation that is formed in place is available as a liquid or as expandable pellets. This type of insulation is poured or sprayed in place, where it hardens to form rigid or semi-rigid foam insulation. Fibrous materials mixed with binders may also be sprayed in place.

An HVAC system not only moves air, but it can also transmit noise. There are two types of potential noise transmitted through an HVAC duct system:

- System-generated noise
- Occupant-generated noise

The fan, or blower, is the main source of system-generated noise. The more powerful the fan, the noisier it is likely to be. Manufacturers provide a rating, called the *sound power level*, which tells how much noise a fan will typically produce.

Did You Know?

Air Noise

The amount of noise that results from air movement depends on the air velocity and the size of the space through which the air flows. For example, rolling the window up halfway or almost to the top while driving on the highway makes the wind sound higher and louder. In an HVAC system, air produces sound in the same way. The more the airflow is restricted, the higher/louder the sound may be. Any obstructions also play a role in the sound produced.

Going Green

Recycled Material

Shredded bark and ground-up newspapers can be used to make loose-fill insulation out of materials that might otherwise have been thrown away. Recycling such material into useful product helps protect the environment.

As air flows through the ductwork, it produces noise when it changes direction, when it flows through a damper, or when there is a change in the duct size. Under some conditions, air can also produce noise as it flows through straight sections of duct.

Some motors and compressors in an HVAC system transmit vibration and low-frequency noise to the attached structure. Anti-vibration mountings are usually needed between such equipment and the structure. These mountings, which may be made of springs or rubber, help to absorb the vibration, and reduce noise.

Occupant-generated noise can also travel through a duct system. Sheet metal ductwork can act as a speaker tube. For example, if one duct serves two rooms, sound can pass from one room through the duct into the other room. Without proper sound attenuation, this occupant-generated noise, sometimes called *cross-talk*, can get into the ductwork as easily as system-generated noise can get out.

In office areas where private conversations are held and security is an issue, significant consideration must be given to the installation to provide privacy. Acoustics are also an important consideration in facilities, such as libraries, churches, and concert halls.

4.1.0 Fiberglass Duct Liner

Fiberglass duct liner is made of glass fibers bonded with a thermosetting (heat-setting) resin. Two basic forms are available (*Figure 24*):

- Flexible blankets supplied in rolls
- Rigid boards supplied as sheets

(A) Flexible Duct Liner

(B) Rigid Duct Liner

Figure 24 Flexible and rigid fiberglass duct liner.
Source: ©CertainTeed

Fiberglass duct liner is designed to do the following:

- Lessen noise generated by the HVAC system and by building occupants
- Reduce heat loss or gain through sheet metal duct surfaces
- Prevent water vapor condensation on the inside and outside of the duct

Fiberglass duct liner is not recommended for use in the following applications:

- With equipment that does not include automatic temperature controls
- In any application where the operating temperature could exceed 250°F (120°C)
- In kitchen or fume exhaust ducts
- In ducts carrying solids or corrosive gases
- With coal- or wood-fueled equipment
- In any application where the duct may come in contact with water (such as cooling coils, humidifiers, or evaporative coolers), unless the duct is well protected from the water source
- Inside fire damper sleeves
- Near high-temperature heating coils unless the duct is protected from heat radiation
- In systems supplying operating rooms, delivery rooms, recovery rooms, nurseries, isolation rooms, and intensive care units

4.1.1 Required Properties

The American Society for Testing and Materials (ASTM) International has established the required properties for fiberglass duct liner through ASTM C1071, *Standard Specification for Fibrous Glass Duct Lining Insulation*. The standard provides guidance in the following areas:

- Acoustical performance
- Corrosiveness
- Moisture vapor absorption
- Fungi resistance
- Temperature resistance
- Erosion resistance
- Odor emission
- Flame and smoke rating
- Thermal conductivity

4.1.2 Cutting

Fiberglass duct liner may be cut using one of the following methods:

- By hand, using a utility knife or insulation knife. Knives that are specially designed to cut duct liner are available from several manufacturers.
- By machine, using several different types of cutting equipment. Be sure to follow the manufacturer's cutting recommendations for the specific equipment you are using.
- By computer-controlled cutting equipment that can be programmed to cut rectangular or unique shapes (*Figure 25*).

CAUTION

When using mechanical or computer-operated equipment, be sure to follow the manufacturer's operating and safety guidelines.

4.1.3 Adhesives and Mechanical Fasteners

Fiberglass duct liner is fastened inside sheet metal ductwork with special adhesives. These adhesives are also used to repair minor damage to the liner surface and to coat exposed edges of the liner.

Liner adhesives are either solvent-based or water-based and are classified by ASTM according to their flammability when wet and dry. Vapor from

Figure 25 Programmable duct-liner cutting equipment.
Source: Mestek Machinery

solvent-based adhesives can be dangerous. However, solvent-based adhesives generally become tacky and bond more quickly than water-based adhesives. Adhesives can be applied by roller coating, spraying, or brushing.

WARNING!

Vapors from some solvent-based adhesives may be explosive. Never apply solvent-based adhesives near open flames, welding operations, or other potential ignition sources. Always follow the manufacturer's recommendations regarding fire hazards, proper ventilation, and product storage.

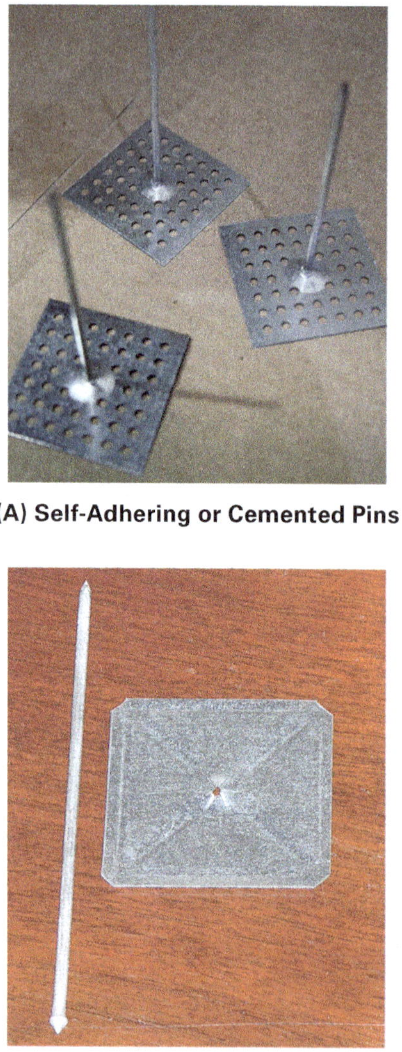

(A) Self-Adhering or Cemented Pins

(B) Weld Pin and Washer

Figure 26 Duct liner stick pins.
Source: (B) Gary J. Bases – BRIL, Inc.

When fastening fibrous duct liner around fittings with tight radiuses, the duct liner may separate from the sheet metal before the adhesive has had time to bond. To solve this problem, you apply additional mechanical fasteners to ensure that the duct liner remains in contact with the sheet metal until the adhesive has cured.

Mechanical fasteners are required in addition to adhesives when installing fibrous duct glass liner (*Figure 26*). Stick pins that help secure the liner to the sheet metal are applied with an adhesive or spot-welded in place.

Welded fasteners either have an integral head or a press-on head. Those with an integral head are installed after the liner or ductboard is in place, as shown in *Figure 27*. The welding equipment used to secure these fasteners must be carefully adjusted to produce a solid weld without burning through the material.

Pins secured with an adhesive come with an adhesive pad on the base, or an adhesive is applied to the metal and the pin is pressed into it. Adhesive pins must always be installed before the liner is placed, then the press-on head is installed to secure the material. Most adhesives lose their gripping power as they age, so this type of fastener may not be appropriate for all applications. Follow the manufacturer's recommendations as well as the job specifications.

When fastening fibrous duct liner around fittings with tight radiuses, the duct liner may separate from the sheet metal before the adhesive has had time to bond. To solve this problem, you must apply additional mechanical fasteners to ensure that the duct liner remains in contact with the sheet metal until the adhesive dries.

Fastener heads should have a minimum area of 0.75 in^2, a minimum thickness of 0.01", and have either a cupped or beveled shape. Fastener heads must not compress the insulation more than $^1/_8$". Excessive compression of insulating materials reduces their ability to suppress heat transfer. Follow the job specifications or the manufacturer's guidance.

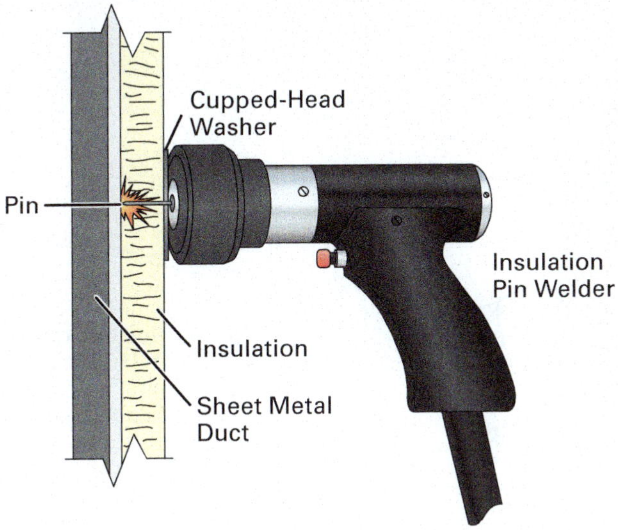

Figure 27 Installing a weld pin.
Source: (bottom) Industrial Construction & Engineering Co.

Duct liner and insulation pins must have the following characteristics:

- Be as corrosion resistant as G60 galvanized steel.
- Be able to indefinitely sustain a 50 lb tensile dead-load test perpendicular to the duct wall.
- Be the correct length for the specified duct liner or insulation thickness.
- Not result in duct leakage.

The correct spacing of the pins is also very important. The pins must be located and spaced regardless of airflow direction as shown in *Figure 28*. The expected velocity of air moving through the duct affects the spacing, per the chart shown in *Figure 28*.

After a lined duct system is installed and has experienced some airflow, conduct a final inspection. The inspection should include checking all registers, grilles, and diffusers to make sure all are clean and free of debris. Any loose pieces of liner may detach from and block the duct. It is important to follow the liner manufacturers' installation instructions. If a failure then occurs, the liability for the failure and any corrective actions taken are generally the responsibility of the manufacturer.

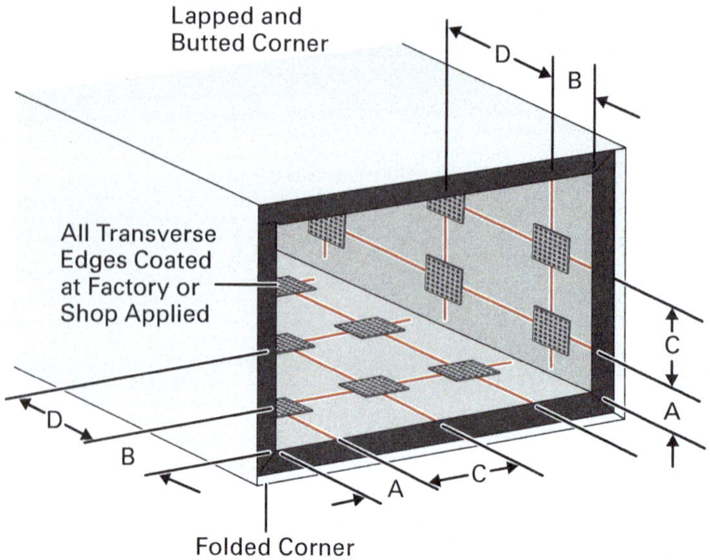

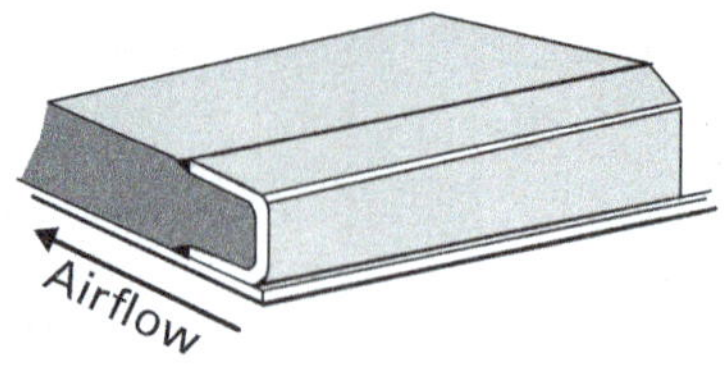

Dimension		Typical Duct Velocity, fpm (m/sec)	
		0–2,500 (0–12.7)	2,501–6,000 (12.7–30.5)
A	From Corners of Duct	4" (100mm)	4" (100mm)
B	From Transverse End of Duct Liner	3" (75mm)	3" (75mm)
C	Across Duct Width, OC, Minimum 1 per Side	12" (300mm)	6" (150mm)
D	Along Duct Length, OC, Minimum 1 per Side	18" (450mm)	16" (400mm)

Figure 28 Duct liner pin spacing.
Source: Industrial Construction & Engineering Co. (bottom)

Figure 29 Duct wrap.
Source: MyrKu/iStock

CAUTION

To prevent fiberglass fibers from transferring to your non-work clothes, wash work clothing separately, then rinse your washing machine thoroughly before using it again. If you have a lot of glass fibers on your clothes, soak them first and then wash the clothes.

4.2.0 Fiberglass Blanket Insulation

Fiberglass blanket insulation for duct is generally referred to as *duct wrap* (*Figure 29*). Duct wrap is typically purchased in rolls. Rigid fiberglass ductboard may also be used to insulate metal duct. While rigid ductboard is easy to apply to straight rectangular duct, it requires significantly more skill and time to properly apply to contoured surfaces.

Fiberglass blanket insulation is made of fine, inorganic glass fibers bonded with a heat-setting resin. Blanket insulation is sold in large rolls that are tightly compressed in a plastic bag, so it may not appear to be as thick as the package indicates. However, the insulation will return to the stated thickness when it is unrolled. It comes in ¾ lb, 1 lb, and 1½ lb densities. Manufacturers often identify the density as Type 75 (¾ lb), Type 100 (1 lb), and Type 150 (1½ lb).

In the case of duct wrap, *density* refers to the weight of the material per cubic foot (ft^3). For example, a density of 1 lb means that 1 ft^3 of insulation weighs 1 lb. The thickness of the insulation ranges from 1" to 3". The length of the standard roll depends on the thickness. The most common lengths are 100' for material ½" thick and 75' for material 2" thick. The standard widths are 48" and 60", which matches the lengths of most straight sheet-metal duct sections.

Manufacturers apply a facing on one side of the insulation blanket. The facing is designed to meet heat, sound, condensation, and fire-resistance requirements. The facing serves as a vapor retarder and provides the fiberglass material with a base to which it can adhere. The vapor retarder is always placed on the exposed side of the insulation.

Facings include foil scrim kraft (FSK), vinyl, and all-service jacket (ASJ). FSK tape is layered aluminum tape reinforced with fiberglass thread embedded between layers. The ASJ jacket adds a white kraft paper layer over the FSK aluminum layers. ASJ jackets are typically used on pipe insulation, but rarely on duct insulation.

All duct wrap products usually have a 2" staple flap along one or both edges to assist in installation (*Figure 30*). The staple flat on the end is created by the installer after a length of insulation is cut from the roll.

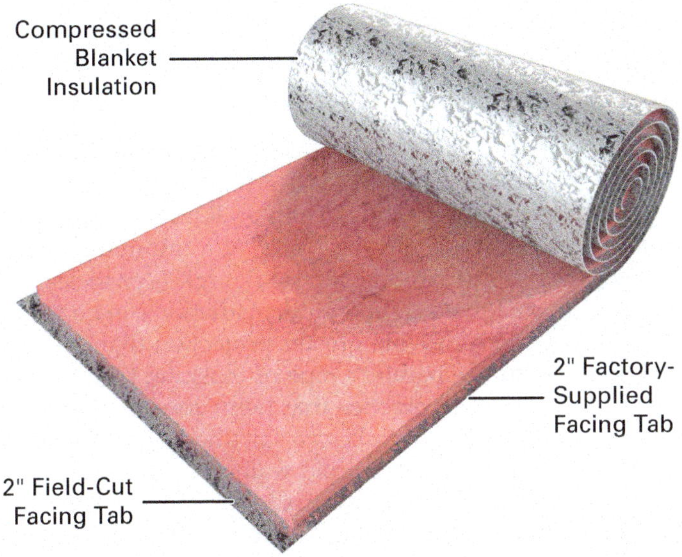

Figure 30 Fiberglass blanket insulation with staple flaps.

In a typical commercial HVAC system, the ducts are insulated after the ductwork is installed, pressure-tested, and inspected, but before the ceiling grid is placed. The trunk or main line of duct should be insulated first and then the branch lines. A simplified drawing of a typical duct system is shown in *Figure 31*.

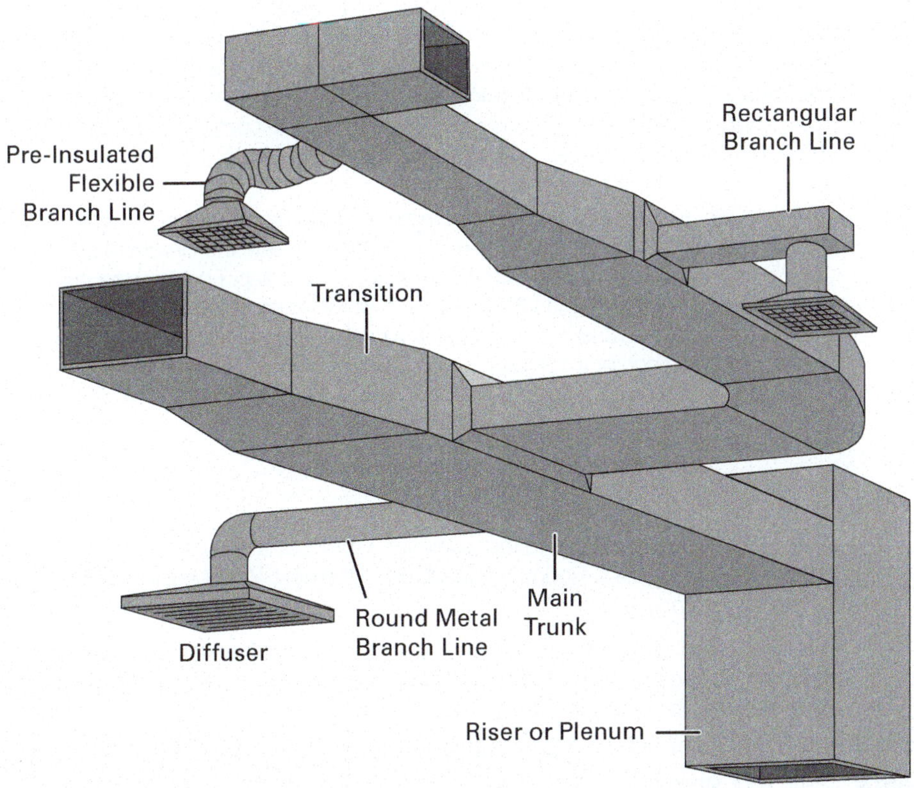

Figure 31 A typical duct system.

Parts of some duct systems may be lined internally. Lined duct may or may not require external insulation. The job specifications will determine if one or both is required. External insulation is often omitted when the duct is exposed or when the exterior must be cleaned periodically. In these cases, the duct liner may be thicker or constructed of rigid fiberglass ductboard instead.

4.2.1 Measuring and Cutting

CAUTION

Always wear proper personal protective equipment (PPE), long pants, and long sleeves when working with fiberglass to prevent irritation to the eyes and skin. Proper PPE should include eye protection (goggles for maximum protection) and a dust mask at a minimum.

To measure and cut fiberglass blanket insulation, you will need the following:

- Tape measure
- Utility knife
- Straightedge
- Large piece of heavy cardboard

Because fiberglass may irritate your eyes and skin, wear safety goggles, gloves, long pants, a long-sleeved shirt, and a hat. In addition, the National Institute for Occupational Safety and Health (NIOSH) recommends that a NIOSH-rated and approved respirator be worn. At a minimum, a simple dust mask should always be used to avoid breathing fiberglass particles.

The tiny glass fibers contained in fiberglass insulation can be very irritating to the skin and eyes. Goggles provide greater protection than standard safety glasses. It is a good idea to use painter's tape or masking tape to seal off shirt sleeves and pants legs. Another practice that helps protect the skin is to rub cornstarch onto exposed areas of skin before starting the work. Upon completion, rinse the skin or shower in cool or cold water. Warm water should not be used because it opens the pores and may increase the discomfort.

To measure and cut fiberglass blanket insulation for rectangular, round, and oval duct, follow these steps:

Step 1 Choose an area in which to measure and cut the insulation. It should be out of the way of foot traffic but near the installation. Place heavy cardboard or plywood under the insulation to protect both the floor and the knife blade.

Step 2 Roll out several feet of insulation with the facing side down. Save the plastic bag the roll came in to collect scrap insulation as you cut.

Step 3 Measure the distance around a straight section of duct. For rectangular duct, when applying $1\frac{1}{2}$" insulation, add 8"; add 10" for 2" insulation. The added length is required to accommodate the extra travel around the corners without crushing the insulation, including an additional 2" to create the staple flap. When the insulation is applied, the corners are expected to be compressed about 50%. An example of how to measure a $1\frac{1}{2}$" thick blanket to fit a 12" × 6" rectangular duct is shown in *Figure 32*.

- For round duct, first determine the diameter of the duct. Then add twice the thickness of the insulation to the diameter and calculate the circumference of a circle that size. Finally, add 2" to create a staple flap. For example, assume you are insulating an 8" round duct with 2" duct wrap. Double the insulation thickness and add it to the diameter, which results in 12". Calculate the circumference of a 12" circle using $C = 2\pi r$:

$$C = 2\pi r$$
$$C = 2 \times 3.14 \times 6"$$
$$C = 37.68"$$

- Add 2" for the staple flap to the result, for a total cut length of 39.68" for an 8" round duct. Round the result up to the next $\frac{1}{4}$" for easier measurements.

Step 4 Check the end of the roll to ensure that it is straight and square. Some ends may become damaged or separated in packing. Use the utility knife and straightedge to cut off any damaged portion.

Step 5 Measure and mark both edges of the unrolled insulation (*Figure 33*).

Step 6 Align your straightedge with the marks and cut from the far edge to the near edge. Make a clean cut through both the fiberglass batt and the facing.

Did You Know?

How Fiberglass Insulation Works

Heat naturally moves toward colder areas by conduction or convection. For example, a stainless-steel pan is an excellent conductor. It allows heat to travel freely from the burner to the food in the pan.

The glass fibers in fiberglass insulation, on the other hand, are poor conductors. Fiberglass is an excellent material for controlling heat loss. Fiberglass insulation is made up of fine strands of glass randomly held together in a mat. Tiny pockets of air lie between each of the fibers, and these pockets contribute to the insulating effect.

Heat can travel through these air pockets by convection. But because the glass fibers are randomly placed, no direct path exists. The longer it takes the heat to travel through the mat of fibers, the more effective the insulation is at controlling heat loss. This is why you must not overly compress the insulation blanket—you will shorten the convection path and reduce the insulation's effectiveness.

Cut Length = [(Duct Width + Duct Height) × 2] + (4 × Insulation Thickness) + 2" Staple Flap

Cut Length = [(12" + 6") × 2] + (4 × 1½") + 2"

Cut Length = 36" + 6" + 2"

Cut Length = 44"

Figure 32 How to measure duct insulation.

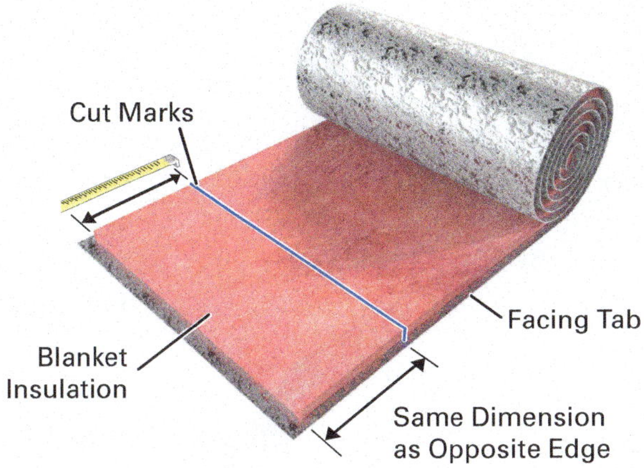

Figure 33 Measuring and marking the cut.

Step 7 Create a facing tab (*Figure 34*). Move the straightedge 2" in from the cut. Cut through the fiberglass but not through the facing material. Remove the 2" wide strip of fiberglass. Use your insulation knife to scrape the fiberglass loose from the facing (*Figure 35*).

Step 8 Place scraps in the plastic bag for proper disposal. The cut piece is now ready for installation.

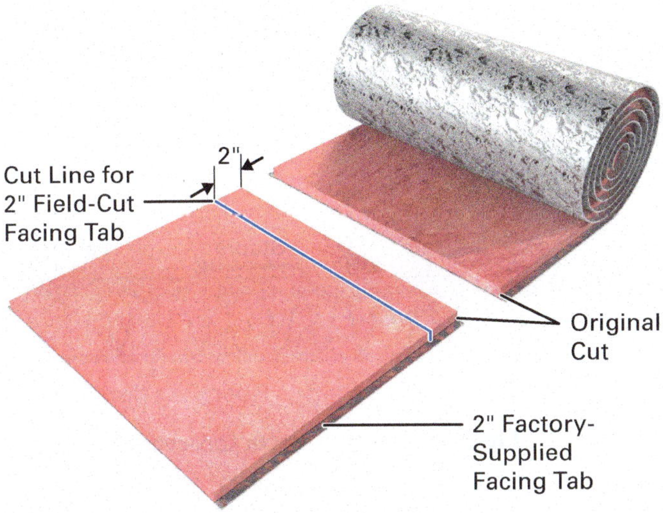

Figure 34 Fiberglass blanket after cutting.

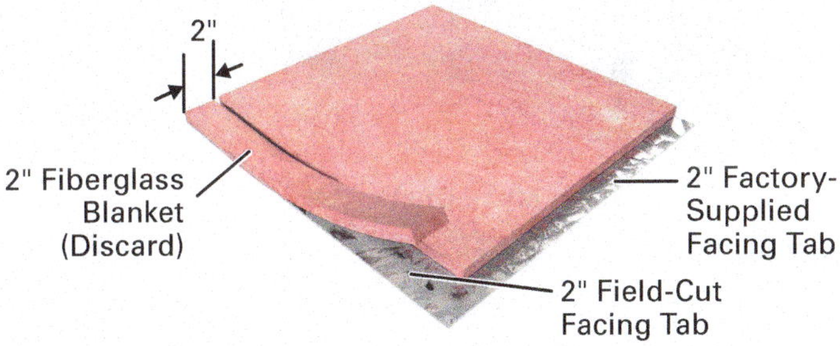

Figure 35 Field-cut facing tab.

4.2.2 Installing Duct Insulation

There are several acceptable methods used to secure fiberglass insulation to ductwork. The most common application is a stapled butt joint. Some specifications, however, require an adhesive to be applied to the duct before the insulation is applied. This is to ensure the fiberglass material is bonded to the duct for a good fit and better thermal performance.

When fabricating a stapled butt joint, use flare-head staples and an outward-clinching staple gun to fasten the ends of the insulation. While the staple gun looks like a normal heavy-duty staple gun, the resulting shape of the staple is different (*Figure 36*) Follow these basic steps to install the insulation around a section of rectangular duct:

Step 1 Place the cut piece of insulation over the top of the duct with the facing tab toward you (*Figure 37*).

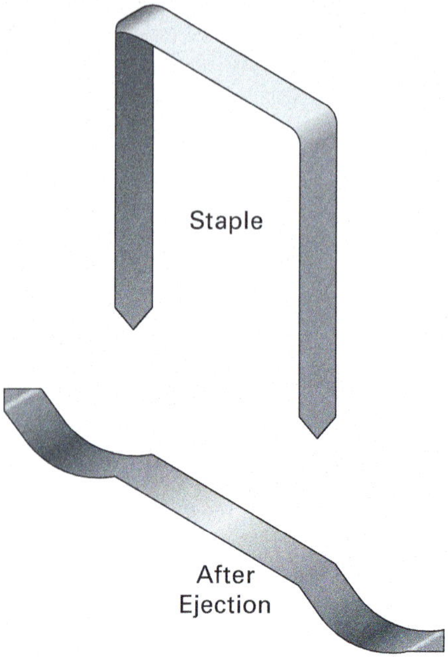

Figure 36 Outward-clinching insulation staple.

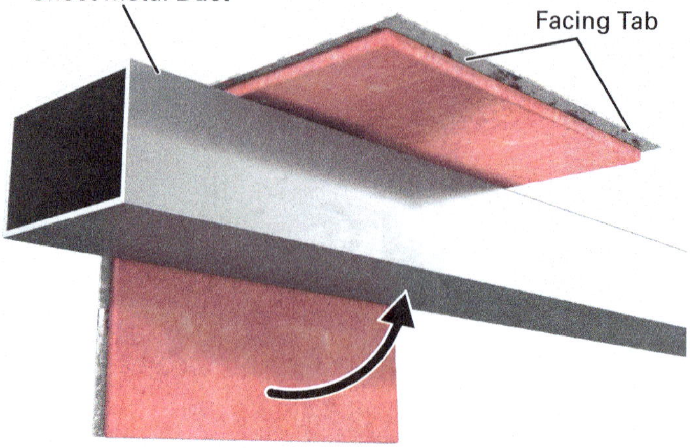

Figure 37 Applying the insulation to the duct.

Step 2 Hold the top end of the blanket and reach under the duct for the other end.

Step 3 Pull the end of the blanket snugly around the duct. Do not compress the blanket more than 50% at the corners; you have allowed for that much in your calculations. Check for proper fit.

Step 4 Hold the facing tab down snugly over the blanket underneath, with the surfaces of the fiberglass butted together. Check the underside of the duct to ensure the insulation is snug. Staple along the middle of the facing tab.

Step 5 Staple every 2" to 4" as required to securely fasten the facing tab. Staple from the center of the piece out toward the edges (*Figure 38*). The staples must have an outward clinch but not bent far enough to penetrate the facing (*Figure 39*).

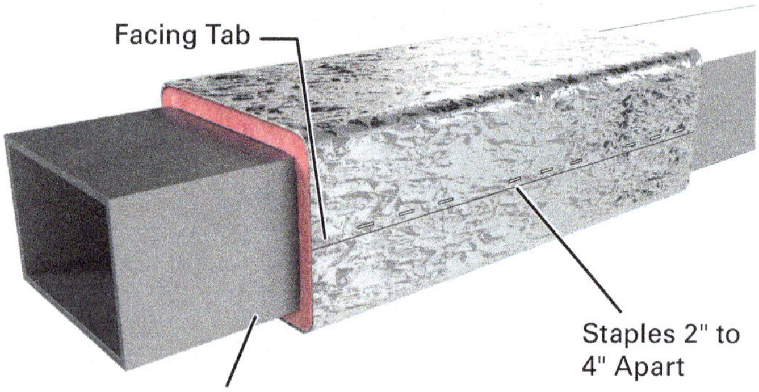

Figure 38 Stapling the insulation.

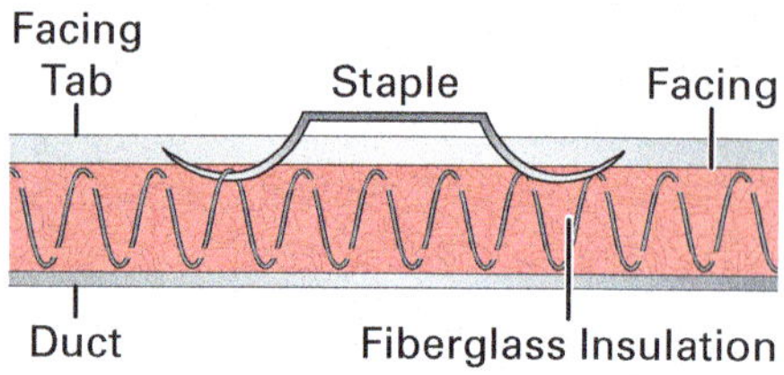

Figure 39 Proper staple application.

Figure 40 FSK aluminum tape.

Step 6 Install the next piece of insulation in the same way, butting it snugly against the first piece. After fastening the second piece with staples at its ends, pull the facing tab on the long side over the first piece of insulation and staple this tab around the duct perimeter.

- When installing blanket insulation around duct hanging straps, split the butt end of the blanket at the hanger to allow the two split pieces to be around the hanger and over the duct. Secure it with staples. Seal the facing split with manufacturer-approved tape, which is typically an FSK aluminum tape (*Figure 40*).
- Ductwork may have projections that stand out $\frac{1}{2}$" to 2" from the surface of the duct. These projections include standing seams, reinforcement flanges, and ribs (*Figure 41*). Install the insulation blanket so that it blouses over the duct and is taut but not significantly compressed at these projections.

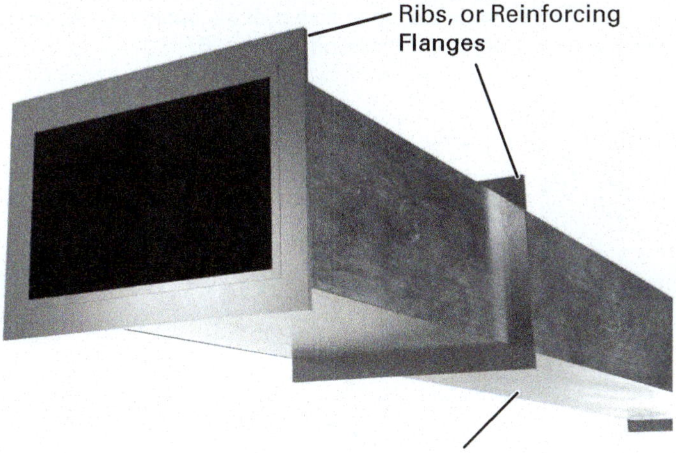

Figure 41 Projections on ductwork.

Step 7 Continue installing sections of blanket until you reach a branch connection, a transition, or other fitting. Do not cut a circular or rectangular piece from the blanket. Instead, split the blanket around the branch to allow it to flare out (*Figure 42*).

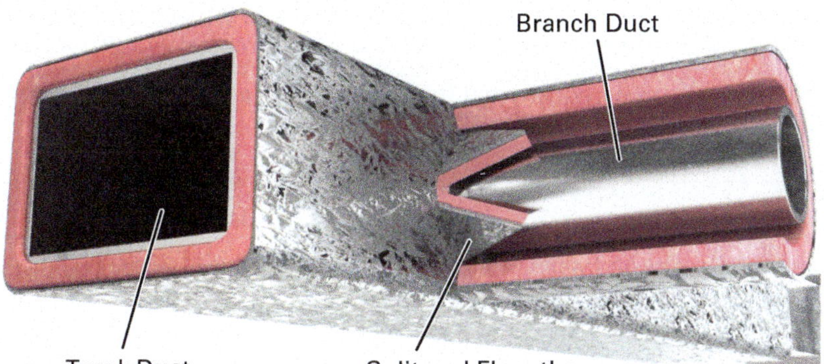

Figure 42 Splitting the insulation around a round branch takeoff.

Step 8 Install the blanket around fittings (offsets, elbows, transitions, etc.) by working from the largest duct to the smaller branches. Wrap the insulation around the largest section of duct first, split it at the corners, tuck it, and secure it by pulling snugly on the smaller sections. *Figure 43* shows how to manage the excess insulation at a transition to a smaller duct size.

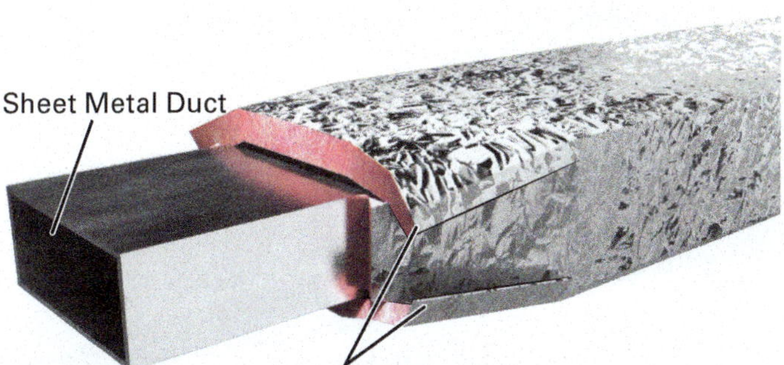

Figure 43 Forming the insulation around a transition to a smaller duct size.

CAUTION

The corners of projections are sharp. Avoid puncturing the insulation facing when applying the insulation. If you puncture the facing, apply a patch of insulation over the puncture and seal it with FSK tape.

Step 9 Install blanket insulation on the branch lines beginning at the trunk line. The insulation should snugly butt or overlap the trunk line insulation. Split the facing tab at the corners to allow the tab to flare over the trunk line insulation. Staple the facing tab to the trunk line insulation (*Figure 44*).

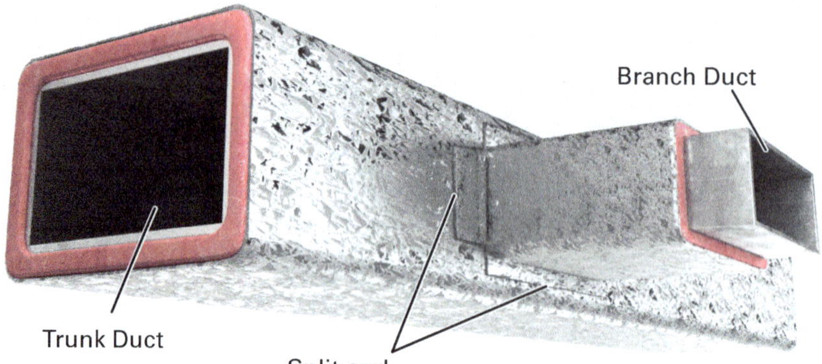

Figure 44 Insulating a rectangular branch duct.

Step 10 When you encounter pre-insulated flexible duct connections, overlap any pre-installed insulation with the duct wrap. Cut, fit, and seal the insulation to the back of each diffuser. Seal all insulation ends at fire dampers, connections, and so forth with manufacturer-approved tape. Make sure the tape overlaps to a metal surface. Always rub the tape firmly to ensure a complete seal.

Job specifications may require that the insulation be attached to the duct with a solvent-based adhesive. To do so, follow these steps:

WARNING!

Solvent-based adhesives and vapors are highly flammable when wet. During application, make sure that there are no nearby sparks or flames.

Step 1 Read the safety precautions on the adhesive label. The precautions may require you to wear a respirator, to work in a well-ventilated area, or both.

Step 2 Make sure no one in your work area is using equipment that produces sparks or flames. Solvent-based adhesives are often highly flammable.

Step 3 Cover finished floors to protect the finish from adhesive drips and splatters.

Step 4 Brush a coat of adhesive on the top edge of the duct.

Step 5 Brush on strips of adhesive spaced about 6" apart around the top, bottom, and sides of the duct. Also apply a strip along both sides of the longitudinal corners (*Figure 45*).

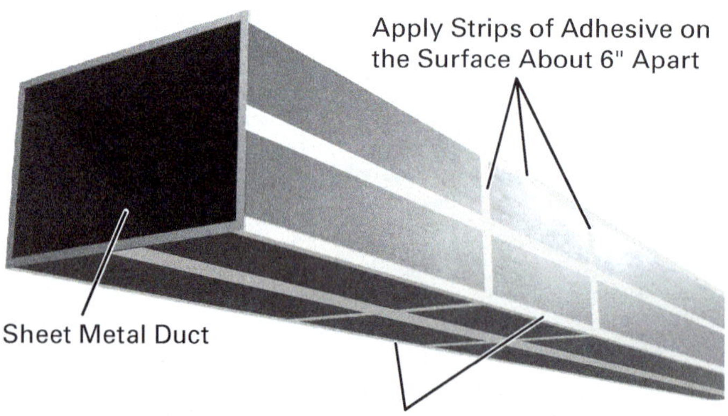

Figure 45 Brushed-on adhesive strips.

Step 6 Place the blanket insulation on the top edge of the duct and press it into place.

Step 7 Pull the blanket snugly around the duct and press it into place.

4.2.3 Sealing

You must seal the seams, joints, and facing tabs on blanket insulation with manufacturer-approved tape or adhesive. Job specifications often require that you also apply vapor-retarding mastic to joints, seams, and staple penetrations. Review the job specifications for sealing requirements and recommended materials.

To apply FSK insulation tape, follow these steps:

Step 1 Measure and cut the amount of tape needed.

Step 2 Peel the protective paper contact strip from the tape.

Step 3 Press the tape down over the staples and the facing joint. Seal any punctures or tears in the facing with tape.

Step 4 Seal the facing-to-metal surfaces at termination points, such as air handling units, diffuser backs, and fire dampers. Rub the tape firmly with a roller or plastic squeegee to ensure a complete seal.

To apply adhesive, follow these steps:

Step 1 Read the label for the manufacturer's instructions regarding application, drying time, and safety precautions. Wear the recommended PPE.

Step 2 Brush the adhesive under the facing tab.

Step 3 Rub or roll the tabs firmly to ensure a complete seal.

Step 4 Staple the blanket insulation or apply a quick-drying contact adhesive.

To use vapor retarder mastic, follow these steps:

Step 1 Read and follow the manufacturer's instructions and safety precautions. Wear the recommended PPE.

Step 2 Protect finished floors and equipment from drops and splatters of mastic.

Step 3 If the manufacturer specifies glass-fiber mesh as reinforcement, fasten the mesh in place by stapling or gluing according to the manufacturer's instructions.

Step 4 Brush or trowel the mastic over the seams and staples. Apply the mastic to the hardest-to-reach areas first and work from the furthest point to the nearest. Do this to avoid reaching over wet mastic. Cover the mesh completely with mastic.

Step 5 Seal all penetrations of the insulation facing (such as duct hangers) tightly.

4.2.4 Mechanical Fasteners

Mechanical fasteners are used on the surface of ducts to secure insulation. The fasteners are especially important on the bottom of large ducts to prevent insulation sagging. Fasteners include self-adhesive stickpins, cemented stickpins, and weld pins. All these pins are similar to those used to install duct liner, although the length may be different to accommodate thicker material without crushing it.

Job specifications may require you to install these fasteners on the bottom surface of all ducts more than 24" wide. Generally, a row of fasteners spaced every 18" along the centerline of a duct that is 24" to 36" wide is adequate (*Figure 46*). Wider ducts require two or more rows.

If you use an adhesive on the duct surface to cement the insulation to the duct, mechanical fasteners are not usually required. Always check the job specifications and SMACNA standards for proper fastener spacing.

Install weld pins with press-on washers before you apply the insulation to the duct. If the weld pins have integral washers, the pins must be installed through the insulation. Use the pin welder, as shown in *Figure 27*, to spot weld the pin to the metal duct.

WARNING!

Insulation pins have sharp points. Be very careful when applying insulation over pins secured to the duct to avoid serious puncture wounds.

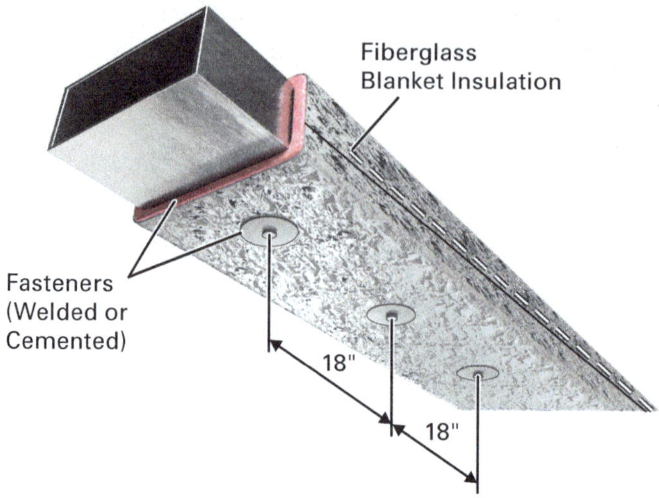

Figure 46 Insulation fastener placement.

Install self-adhering and cemented stickpins before applying the insulation to the duct. For cemented stickpins, use an adhesive designed for this purpose. Many options are available from manufacturers such as 3M®, Tuff-Bond®, and Boss®. Following these steps will help speed production and help avoid a sticky mess:

NOTE

Many adhesive products require storage at standard room temperatures for 24 hours before use and cannot be applied at temperatures below freezing.

Step 1 Cut a piece of cardboard to a manageable size, such as 18" × 18".

Step 2 Stick multiple pins through the cardboard with the anchor bases up (*Figure 47*). Do not attempt to prepare too many fasteners at a time, however, as the fasteners must be applied to the duct within 5–10 minutes.

Step 3 Apply adhesive to the entire base of each fastener. Cover each base with an even coat about $\frac{1}{8}$" thick.

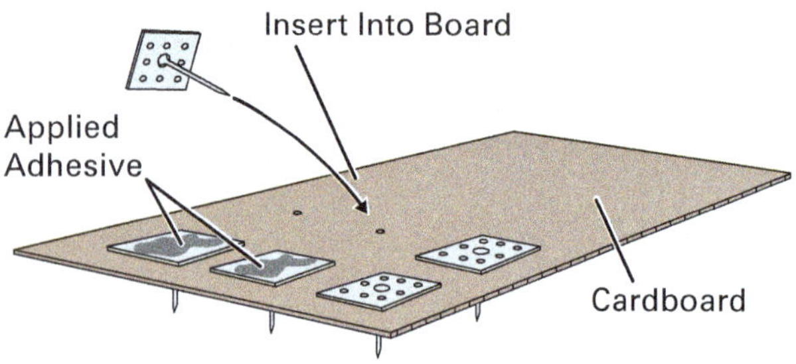

Figure 47 Fasteners pushed through cardboard.

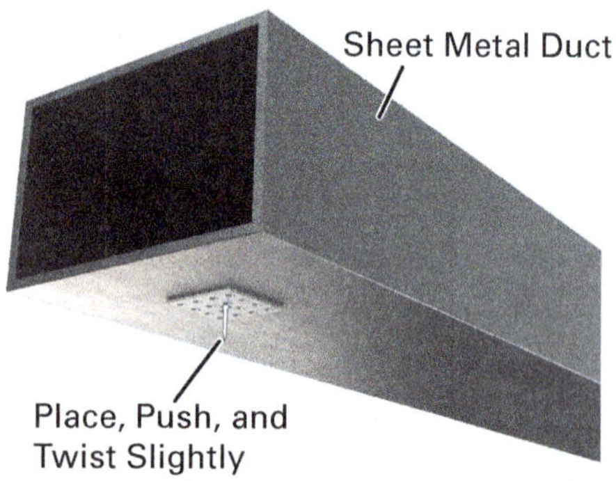

Figure 48 Fastener applied to sheet metal duct.

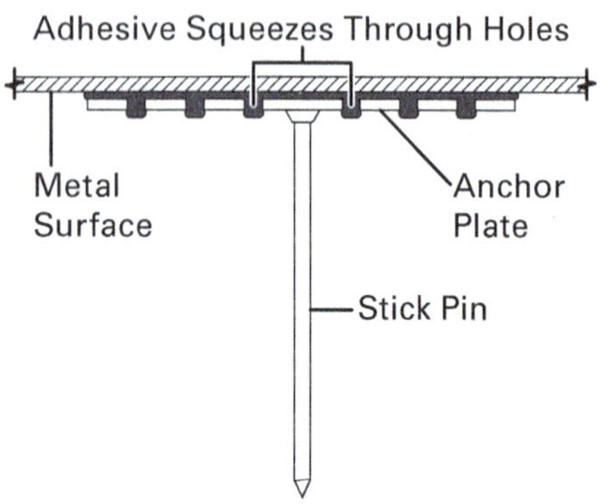

Figure 49 Close-up view of cemented fastener.

Step 4 Place the fastener on the duct surface and press firmly into position with a slight twisting motion (*Figure 48*). Some adhesive should press out through the holes in the base (*Figure 49*).

Step 5 Consult the adhesive instructions to determine how long the adhesive must cure before the insulation is installed.

Apply the insulation after the adhesive has cured. Follow these steps:

Step 1 Carefully wrap the insulation around the duct to avoid puncture wounds. The stick pins must penetrate both the insulation and the facing.

Step 2 Carefully place the washer over the pin. Push the washer down snug against the insulation. Do not push the washer down too far, which would compress the insulation.

Step 3 Cut the protruding pin off flush with the washer using metal nippers.

Step 4 Apply vapor retarder mastic or apply tape over the washer to seal the penetration.

On Site

Bubble Wrap Duct Insulation

Duct insulation is now available in a product that resembles the bubble wrap used to protect items from damage during shipping. The air in the bubbles provides additional insulation value. Both sides of the bubble wrap have a shiny metallic jacket to reduce radiant heat loss.

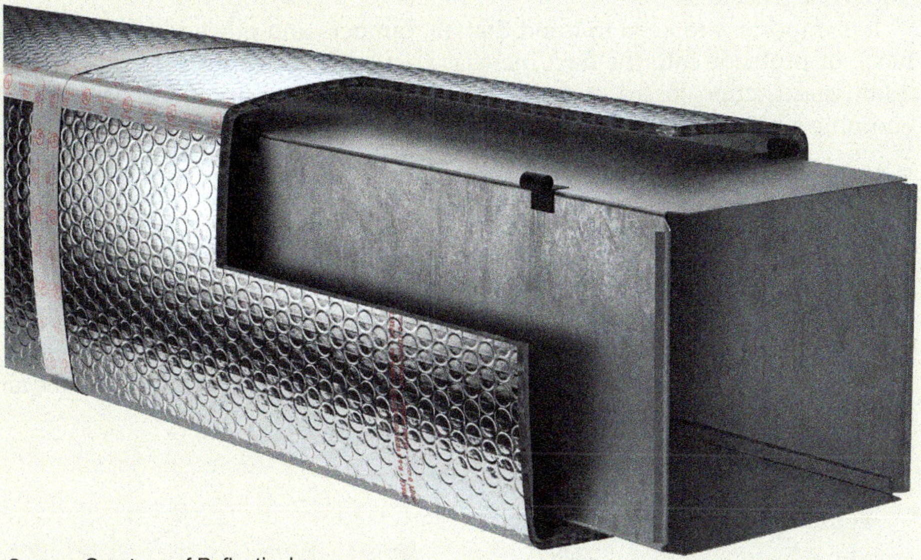

Source: Courtesy of Reflectix, Inc.

4.0.0 Section Review

1. When installing duct liner, stick pin heads must not compress the material more than ______.
   a. 1"
   b. $\frac{1}{2}$"
   c. $\frac{1}{4}$"
   d. $\frac{1}{8}$"

2. How much allowance must be added to the measurement for 2" duct insulation being wrapped around a rectangular sheet metal duct before it is cut, to accommodate the extra travel around the corners and a staple flap?
   a. 6"
   b. 8"
   c. 10"
   d. 12"

5.0.0 Dampers and Access Doors

Objective

Identify various sheet metal duct accessories and describe the installation requirements.

a. Identify and describe the installation of various types of dampers.

b. Identify and describe the installation of duct takeoffs and access doors.

Performance Tasks

There are no Performance Tasks in this section.

The following section describes the installation of dampers, duct takeoffs, and access doors. It is important that HVAC technicians understand the different types of dampers and their unique requirements. Access doors and takeoffs are also common accessories needed in air distribution systems.

5.1.0 Installing Dampers

Dampers are movable blades used to balance airflow by regulating the air volume passing by. Special dampers are also used to control smoke and flame movement through the air distribution system in the event of a fire. Access doors must be installed wherever access to dampers or other components within the ductwork is needed.

It is important to keep in mind that all dampers and other components that block or protrude into the duct increase friction loss and resist airflow. These additional friction losses must be considered when designing a system and choosing a blower package.

5.1.1 Manual Volume Dampers

Manual volume dampers are made of galvanized steel, aluminum, or stainless steel and are round, oval, rectangular, or square. Depending on the size and application, a volume damper may have a single blade or multiple blades (*Figure 50*). Volume dampers are most often found in branch supply air ducts, but can also be used in larger supply and return ducts. An exhaust system with multiple intakes, for example, would use volume dampers at the inlet branch lines to balance the air at the individual intakes.

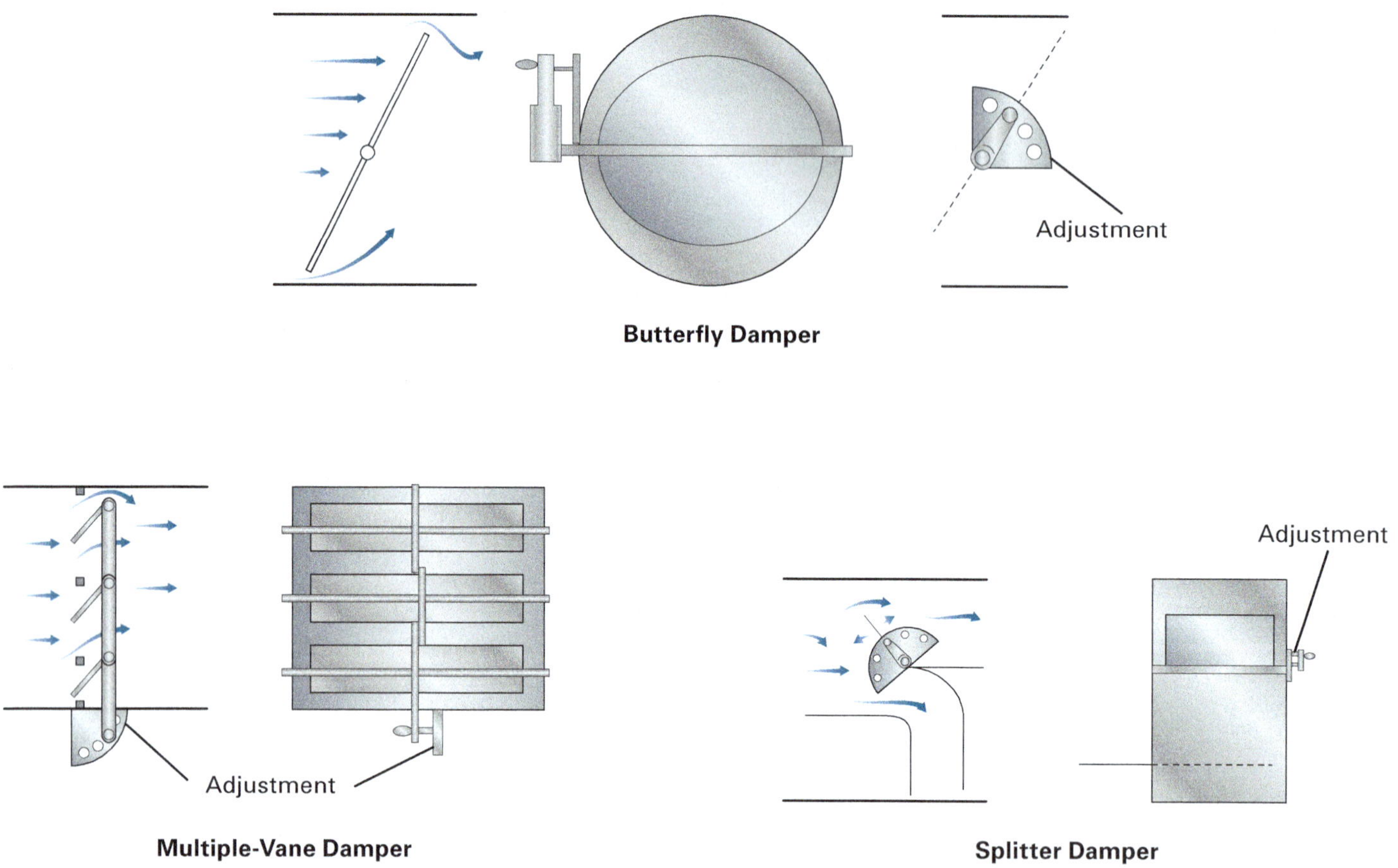

Figure 50 Manual volume dampers.

Multiple-blade dampers are made in two basic types:

- Opposed blade
- Parallel blade

In the opposed-blade damper, each blade moves in a direction opposite to the adjacent blade (*Figure 51*). This design allows for more precise airflow control than can be achieved with a parallel-blade damper.

The parallel-blade damper (*Figure 52*) works like Venetian blinds. This type of damper is a good choice where two-position operation (fully open or fully closed) is required. When the damper must be set to a specific position to achieve a specific air volume, the opposed-blade damper is a better choice.

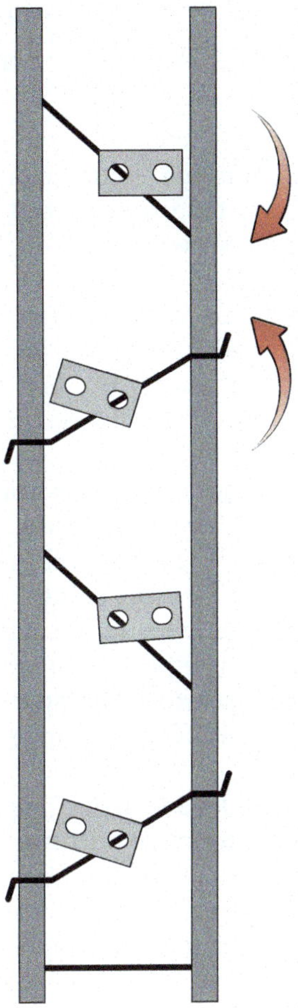

Figure 51 Opposed-blade volume damper.

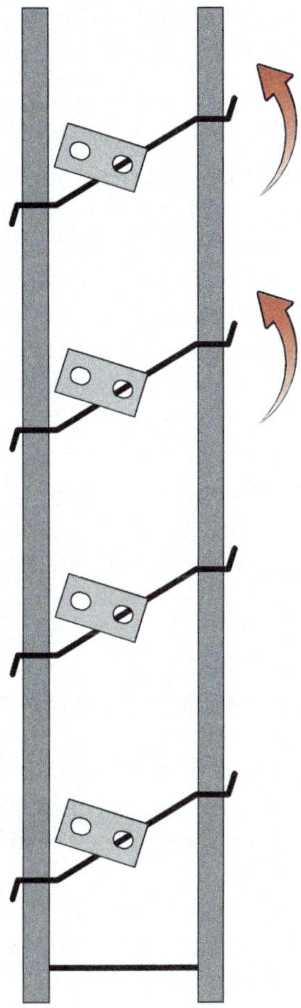

Figure 52 Parallel-blade damper.

Both types of dampers are available in various levels of quality relative to their ability to close tightly. Opposed-blade dampers with blade gaskets are generally able to seal better than models without gaskets. However, not every application requires a high level of performance or extraordinary precision.

Before installing dampers, review the project drawings and specifications to determine the required locations. A butterfly damper will often be an integral part of the takeoff where a branch line attaches to the duct. Always try to orient a butterfly damper so that the quadrant operating shaft is easily accessible and is oriented so that it reflects the position of the damper blade inside.

To install a manual volume damper, follow these steps:

Step 1 Insert the damper into the duct so that you can view the operation of the blades.

Step 2 Install the screws in the frame outside of the travel of the blade while it is opening or closing. The blade must travel freely from the open to the closed position and vice versa.

Step 3 Extend the shaft through the ductwork to mount the damper quadrant with either pop rivets or sheet metal screws.

Step 4 Using a hacksaw, cut a slot in the end of the damper operating shaft. The slot should be parallel to (facing the same direction as) the damper blade. Alternatively, mark the shaft with a permanent marker.

Step 5 Leave the damper in the open position when the installation is complete.

CAUTION

During installation, it is important not to block the damper blade(s) with insulation, hardware, or other material. If one blade of a multiblade damper is blocked, none of the blades can move because the blades are interlocked.

To ensure that the damper is properly installed, use the following checklist:

- Can the damper rotate from fully opened to fully closed and back again, without interference from mounting screws, rivets, duct liner, or other obstructions?
- Does the slot or mark on the damper shaft match the position of the blade(s)?
- Is the damper in the open position with the damper handle locked in place?

Fusible link: A connector designed to melt at a specific temperature to release the spring and allow a fire damper to close.

5.1.2 Fire and Smoke Dampers

Unless specially designed dampers are installed, fire and smoke can easily travel through an air distribution system. This allows the fire to spread, while the smoke can interfere with personnel evacuation or overwhelm the occupants before they can exit the building.

These life-safety dampers include fire dampers, combination fire/smoke dampers, and ceiling radiation dampers. Such dampers are usually located in or near fire-rated walls and fire-rated ceiling assemblies. All fire and smoke dampers must have a UL Solutions (UL) label attached. This label certifies that the damper has been tested and approved for the specified use.

Openings for ducts passing through walls and floors with fire-resistance ratings must be protected by fire dampers as required by local codes. Smoke dampers are used for *smoke containment* or for *smoke control*. Per the NFPA, smoke containment systems "keep smoke from entering specific areas using pressurization and are commonly found in enclosed stairwells." Smoke management systems "maintain tenable environments in the means of egress from large volume spaces or prevent the movement of smoke into surrounding spaces." At locations requiring both fire and smoke dampers, combination dampers that meet the requirements can be used.

Most fire dampers close automatically with a spring-loaded action (*Figure 53*). By closing, the damper restricts the air a fire needs to continue burning and/or prevents the fire from spreading. These spring-loaded dampers are controlled by a **fusible link** (*Figure 54*) that holds the damper open until heat causes it to melt and release the damper. This approach ensures that no electric or mechanical controls can fail at a crucial moment. The fire rating for fire dampers is specified in hours, but the rating does not necessarily match the fire rating of the wall or partition.

A fire damper must remain in place in a wall or floor, even if the ducts or ceiling components on either side of the wall fall. Thus, fire dampers must be installed in a sheet metal sleeve that passes through the wall or floor opening. The wall opening is typically a bit larger than the sleeve since both the sleeve and damper assembly will expand when subjected to intense heat. A common guide is to allow $\frac{1}{8}$" of clearance per linear foot of damper. For example, if the damper is 4' wide, $\frac{1}{2}$" of clearance is required. Stainless-steel damper assemblies often require slightly more clearance, such as $\frac{3}{16}$" per linear foot. It is essential that the job specifications and local codes be followed without deviation.

The fire damper sleeve must be of a minimum specified metal gauge, based on the size of the damper. A 10-ga sheet metal is typically considered the maximum thickness and can be used for any size of damper. The sleeve must extend beyond the wall no more than 6" on each side. Access to the damper through the duct must be provided for inspection and repair. An access door can sometimes be built into the sleeve itself. If an access door is added, the sleeve can then extend beyond the wall up to 16".

Mounting angle (*Figure 55*) is attached to the sleeve or damper frame, but never to the wall. Attaching the mounting angles to the wall would defeat the purpose of enlarging the wall or floor opening for expansion. The framework, sleeve, and damper must be free to expand within the wall opening.

It is important to install the fire damper in the sleeve straight and without any stress applied to its frame. If the damper frame is twisted or distorted when installed, the damper curtain can jam as it tries to close.

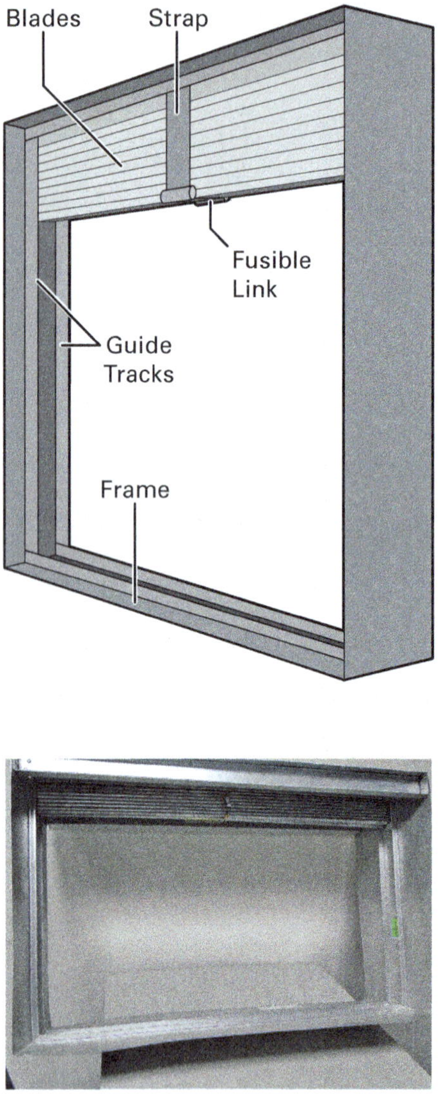

Figure 53 Spring-loaded fire damper.

Figure 54 Fusible links.
Source: Courtesy of Aire Technologies, Inc.

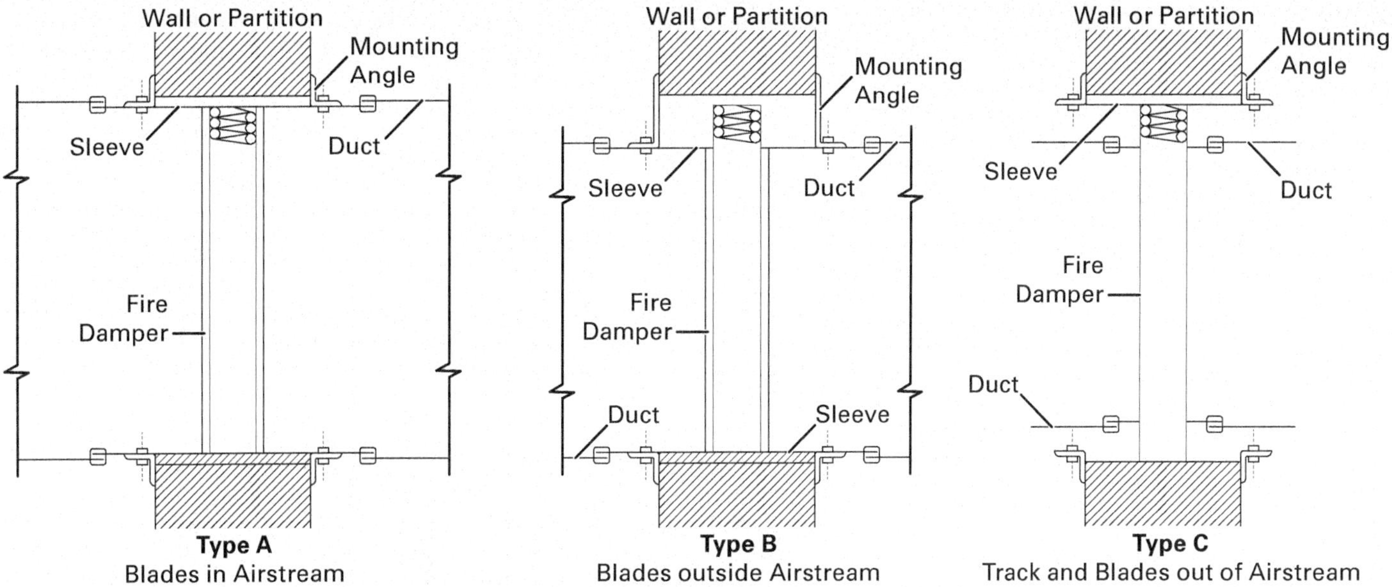

Figure 55 Fire damper mounting styles, using angle.

Depending on the size of the damper, *breakaway connections* may be required between the duct sleeve and the ductwork on either side. This allows the duct to separate from the damper if the duct's primary means of support fails. Several common connection styles qualify as breakaway connections, including the S-slip joint. Ductmate® connections can also be used as long as the corner bolts are left out and a plastic cleat is used to hold the joint together.

A combination fire/smoke damper (*Figure 56*) is a powered damper, with the damper actuator controlled by a heat-sensing device and a building's fire and smoke detection system. This damper is designed to close automatically when an excessive temperature or smoke is sensed. This type of damper must usually be *fail-safe*. In other words, if there is a loss of power, it must fail to the closed position. The actuator is equipped with an internal spring to return the damper to the closed position.

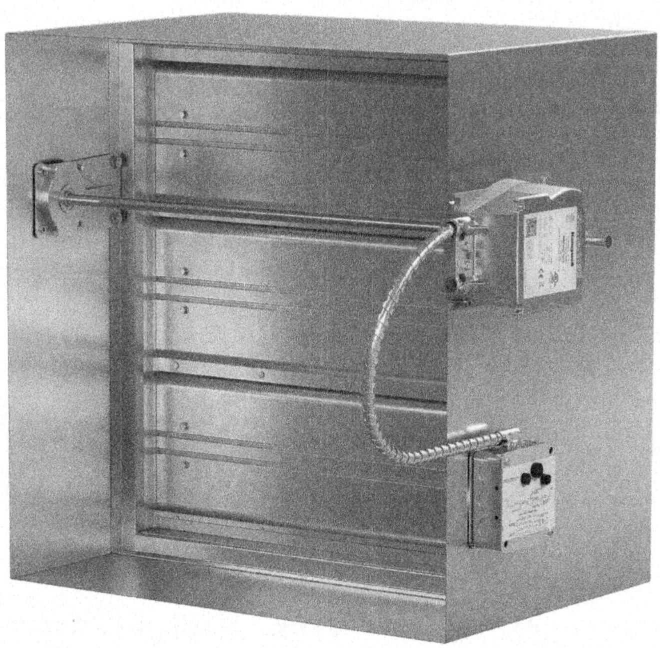

Figure 56 Combination fire/smoke damper.
Source: Image courtesy of Greenheck

Did You Know?

NFPA

The National Fire Protection Association (NFPA) publishes standards for fire prevention in residential and commercial structures. The standards are revised annually. HVAC designers must use the latest edition of the standards to ensure compliance.

A ceiling radiation damper is designed for use in horizontal applications that must be fire-rated. This type of damper can detect heat radiating from the floor below.

Smoke detection systems monitor the smoke density in chimneys or duct systems. These devices provide control over smoke and combination fire/smoke dampers, either directly or indirectly through a control system.

Some duct smoke detection systems use a photoelectric cell (photocell) and beam tube installed in the chimney or duct (*Figure 57*). The tube is long enough to traverse the duct and has a series of holes along its length to allow smoke-filled air to pass through the tube. A signal generated by the photocell reports the density or volume of smoke detected. It is displayed locally and/or the data can be recorded for later review.

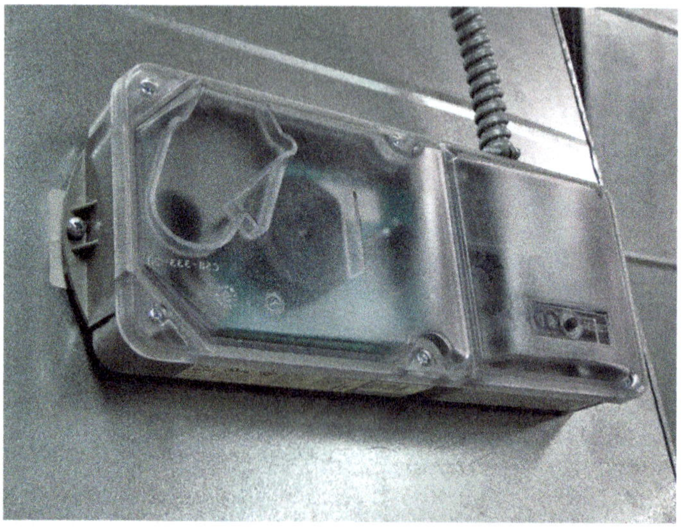

Figure 57 Duct-mounted smoke detector.

On Site

The UL Mark

All fire and smoke dampers must have a UL Solutions (UL) label. UL Solutions is a not-for-profit organization that tests and certifies a wide variety of products for public safety. Founded in 1894, the UL safety mark now appears on more than 22 billion products around the world each year.

If the amount of smoke detected exceeds a preset level, an alarm may accompany other control actions, such as the shutdown of the system and the closure of combination fire/smoke dampers.

The building wall contractor (often the general contractor) usually provides the rough openings for dampers per the plans. Floor and wall openings must be sized according to the damper manufacturer's written instructions, job specifications, and local codes.

Before installing dampers, review the drawings and specifications to determine the locations and installation requirements. Follow the manufacturer's instructions for each type of damper. Because the instructions vary from manufacturer to manufacturer, no specific installation steps are presented here. Never assume that one manufacturer's instructions will apply to a damper made by another manufacturer.

You must also follow the manufacturer's instructions for testing a fire damper. You can test spring-loaded fire dampers by cutting the fusible link that holds the damper in the open position, although the link will need to be replaced. Fusible links can only be used once. The local building inspector will also examine each damper installation and review the manufacturer's installation instructions to verify that the installation has been done correctly.

5.1.3 Motorized Control Dampers

Motorized control dampers are used in HVAC systems to control, or regulate, the mixture of air. These dampers are usually located near the air handler in the mechanical room and modulate the return air, outside air, and exhaust air.

Motorized dampers may be round, oval, or rectangular and may be made of galvanized steel, aluminum, or stainless steel. Depending on the overall size, a motorized control damper may have a single blade, parallel blades, or opposed blades.

Before installing motorized dampers, review the project drawings and specifications to determine the required location. Sometimes, a wall or another duct may block one side of the duct where you need to install the damper. In this case, locate the shaft for the motor on the opposite side.

To install a motorized damper, follow these steps:

Step 1 Insert the damper into the duct so that you can view the operation of the blades.

Step 2 Install the screws into the frame outside of the travel of the blade while it is opening or closing.

Step 3 Make sure the duct liner does not interfere with blade rotation.

Step 4 Note that the actuator may be mounted in the airstream or on the outside of the duct. If the actuator is mounted on the outside of the duct, you must extend the motor's shaft through the duct. The opening for the shaft should be just large enough to accommodate the shaft.

Step 5 Install an access door so that you can install the damper linkage or the motor. This door will also allow for future inspection of the damper.

To ensure that the damper is properly installed, make sure it can open and close freely by hand before attaching the actuator to the linkage.

5.2.0 Access Doors and Takeoffs

Access doors and takeoffs are common accessories that must be installed in sheet metal duct systems. Access doors are somewhat rare in single-family residential installations, but are quite common in larger commercial systems. Takeoffs or starting collars of some sort are required on most systems where the branch lines connect to the main duct. Duct systems that are exposed may have grilles and registers mounted directly on the duct wall, eliminating the need for branch lines.

5.2.1 Access Doors

Once fire and smoke dampers have been installed, access to them is necessary to test the damper installation and to replace the fusible link if it comes apart and the damper closes. A duct access door must be installed at fire damper and combination fire/damper locations. Access doors are often placed near other dampers as well.

Duct access doors are available from many manufacturers. Viewing ports (window-like panels) may be part of the door construction so that damper controls and fusible links can be inspected without opening the door (*Figure 58*). Common framing, fastening, and hinge recommendations are shown in *Figure 59*.

Follow the manufacturer's installation instructions when installing access doors. Observe these general guidelines when installing access doors:

- Measure and cut the opening to the specified size.
- Use the hinge approved for the specific application.
- Make sure that the doors open fully and open out, not in.
- Ensure that nothing will interfere with the door swing. Structural elements, pipe, electrical lines, and conduits cannot be in the way. Note that some of these elements may not yet be installed.
- Make sure the door fits into the opening snugly and that it is level and square with the duct.
- If the access door has a view port, make sure that it is aligned so that you can see the necessary parts of the damper through it.

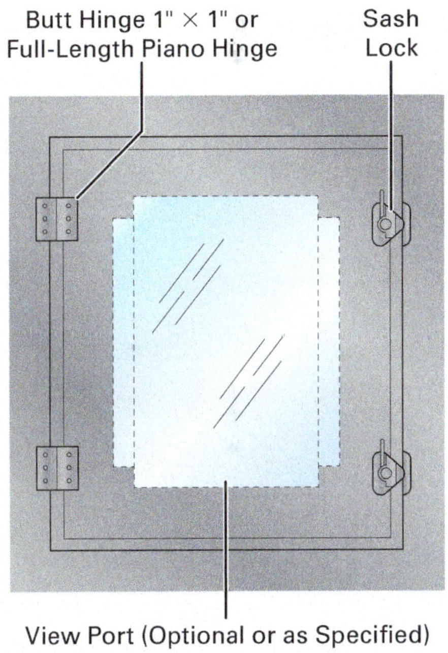

Figure 58 Typical access door.

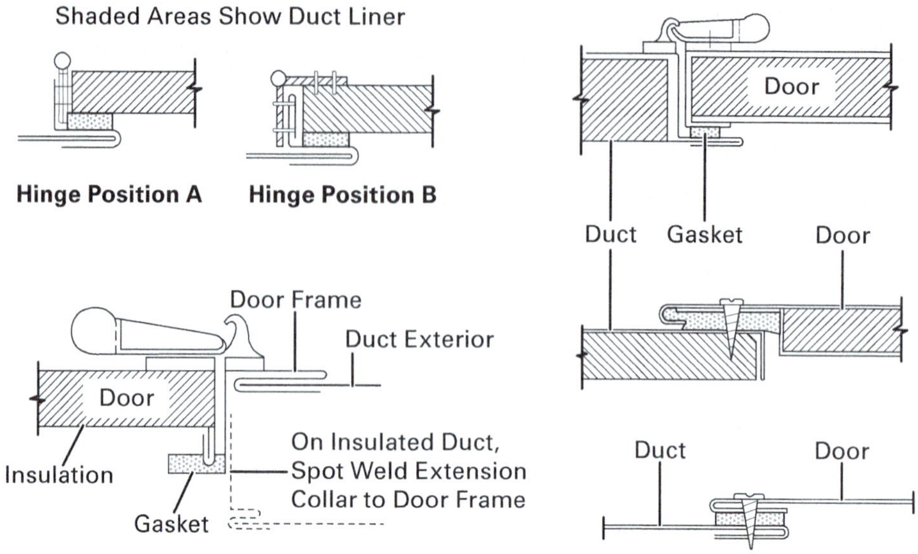

Figure 59 Access door installation methods.

On Site

Special Access Doors

In commercial applications where the duct pressure may be high by design, special access doors that provide pressure relief can be used. Think of them as you would pressure relief valves in a piping system. The access door shown here is designed to blow open to relieve excess pressure in the duct at a specific setting. Otherwise, the duct could rupture in multiple places if the system fan begins to produce more pressure than the duct system can withstand.

5.2.2 Takeoffs

Takeoffs include tap-ins and spin-ins. Tap-ins are square, rectangular, or round fittings that connect branch ducts to the main trunk duct. Spin-ins are round fittings, usually with dampers built in, that are provided for the connection of branch ducts serving the grilles, registers, and diffusers. Round takeoffs are often referred to as starting collars.

To install a takeoff, cut the opening so that the fitting fits into the duct snugly and smoothly. No holes should be in the corners of rectangular openings that would allow significant air leakage. Use a drill to make a small pilot hole and then cut the opening with aviation snips. Circle cutters should be used to make neat openings for round takeoffs. Once the opening has been cut, cut out the duct liner (if present) in the same shape.

On Site

Circle Cutter

Cutting a clean, smooth circle in sheet metal is a simple process with a circle cutter. A drill bit slides into the chuck of the circle cutter, a power drill is attached, and the cutter is set for the proper circle radius. A small hole is then drilled in the center of the desired circle. The center pivot of the tool is inserted into the hole and the drill is started. After the spinning drill bit penetrates the metal, pressure is applied sideways to cut through the metal as the tool pivots.

Source: Courtesy of Malco Tools

A tap-in fitting has tabs cut on one end that allow it to be clinched to the main duct (*Figure 60*). After the appropriate size opening is cut into the duct, the fitting tabs are inserted into the opening and bent over 90° against the inside wall of the duct. Some types require that the tabs be alternated—one inside the duct and the next outside the duct.

To ensure an airtight, professional installation, hammer the tabs smoothly and tightly against the duct wall. The round takeoff shown in *Figure 60A* has tabs. Some square or rectangular fittings have a solid flange, like the one shown in *Figure 60B*.

When the fitting has a solid flange, the curved side of the fitting is placed on the entering-air side to reduce friction loss as the air turns the corner. *Extractors* may sometimes be used at takeoffs as well. An extractor is a section of metal extending into the duct that acts as a scoop, encouraging air to enter a branch line.

A spin-in fitting requires that the opening be cut with some precision. The fitting literally screws into the sheet metal wall. All takeoffs, including spin-ins, should be properly sealed with an appropriate caulking compound or mastic after installation.

To ensure that tap-ins and spin-ins are properly installed, use the following checklist:

- Are the corners neat and free of large openings?
- Are the tabs hammered over neatly, smoothly, and tight against the duct wall?
- Is the fitting installed with the airflow direction in mind?
- Is the damper that may be in the takeoff open and locked in position?
- Has the duct liner been removed?

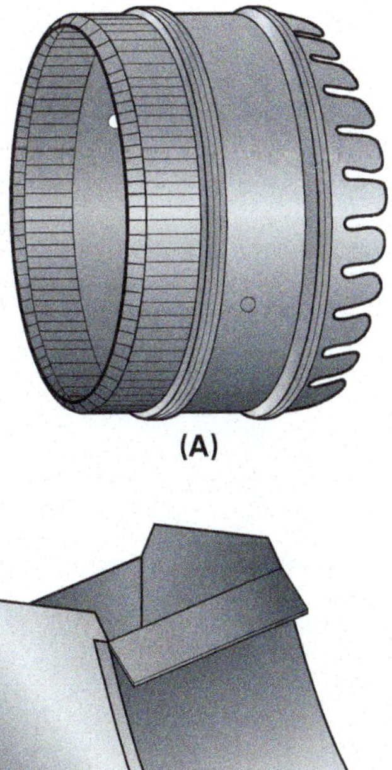

(A)

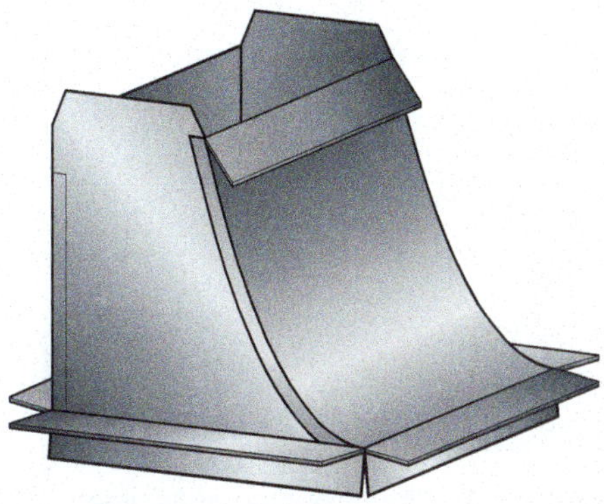

(B)

Figure 60 Takeoff styles.

5.0.0 Section Review

1. Which damper type operates like Venetian blinds?
   a. Opposed blade
   b. Motorized
   c. Manual
   d. Parallel blade

2. What is the function of an extractor in a duct system?
   a. To act as a fusible link.
   b. To encourage airflow to enter a duct branch line.
   c. To dilute smoke or fumes.
   d. To create the opening for takeoff installation.

6.0.0 Flexible Duct

Performance Tasks

There are no Performance Tasks in this section.

Objective

Identify types of flexible duct and describe the installation process.

a. Identify types of flexible duct.
b. Explain how flexible duct is connected and supported.

Flexible duct is used to connect supply and return grilles and registers to the main duct. While it provides great flexibility and eliminates the labor required to fabricate and install metal ducts, it is also easy to overuse. Flexible duct is significantly more resistant to airflow than smooth metal duct.

6.1.0 Flexible Duct Types

Flexible round duct is used for branch lines from low-pressure ducts to mixing boxes, diffusers, and similar terminals. Insulated flexible duct is generally used when the duct is in or passes through an unconditioned space. Flexible duct without insulation may be used in a conditioned space.

Flexible duct is not generally recommended for use on the return side of an air distribution system. As a rule, it does not perform well when the internal pressure is negative, as is the case in return air ducts. However, several flexible duct products are rated to handle negative pressures down to 1" of water column.

Flexible duct is made from a variety of materials. A common core material is a polyester film that is wrapped around a wire helix (*Figure 61*). The core is the internal sleeve through which the air flows. The wire helix gives the duct its round form.

Figure 61 Wire helix core of flexible duct.
Source: Voyagerix/iStock

Ducts used in residential and light commercial applications where the static pressure is low have a thin core material through which the air flows. It is not as strong as the core for ducts that operate at higher pressures.

Flexible duct with a core of corrugated aluminum is available for special applications. The material is available in diameters from 3" up to roughly 20" in diameter, although the smaller and larger sizes may not be consistently stocked. The most common outer jackets are made from a thin metal foil, polyester, or polyethylene.

The insulation on flexible duct must have an R-value (insulation value) that satisfies codes and job specifications. For that reason, duct products of the same nominal (inside diameter) and core construction can have different insulation R-values, as the thickness of the outer blanket of fiberglass insulation differs. The insulation value typically ranges from R-4 to R-8.

The vapor retarder that surrounds the insulation can vary, depending on the application. Ducts designed for light-duty applications may have a vapor retarder made of black polyethylene. For more demanding applications, the vapor retarder may be metalized polyester reinforced with fiberglass mesh, much like FSK. Some flexible duct has metal grommets in the vapor retarder along its length, providing points of hanger attachment.

Due to its flexibility, flexible duct can be dramatically compressed in length for packaging. That makes for a compact package that is easy to handle and transport. Boxes typically contain 25' of duct.

Flexible duct is tested in accordance with UL Standard 181, *UL Standard for Safety Factory-Made Air Ducts and Air Connectors*, to ensure it meets the required specifications. The product is tested for its durability when exposed to tension, **torsion**, and bending. The Air Duct Council (ADC) also tests the properties of flexible duct. These properties include thermal performance, friction loss, acoustical performance, static pressure performance, temperature performance, and air leakage.

Torsion: The twisting of an object along its length by two equal and opposite forces at each end.

6.2.0 Flexible Duct Installation

To connect flexible duct to the main duct, a takeoff must first be installed.

6.2.1 Connecting Flexible Duct

Ensure that tapes, mastics, and non-metallic straps used to connect flexible duct are marked to indicate compliance with UL Standard 181. The following are some general guidelines for installing flexible duct:

- Fully extend the duct material before measuring, cutting, and installing it. The friction loss in flexible duct is much higher when it is compressed. Flexible duct considered to be fully extended is within 4% of its listed length. Thus, a 25' section of duct should be extended to at least 24'. Installing a section of flexible duct that is still 30% compressed quadruples its normal friction loss! Note that there are code limitations regarding the length of flexible duct. *Figure 62* shows how a turn should be made at a diffuser.
- The duct should not be bent sharply around an obstruction. The bend radius should be at least one duct diameter. For example, an 8" duct should have a minimum bend radius of 8".
- Avoid unnecessary bends and/or snaking of the duct. A 45° bend has the same friction loss as 10' of straight duct; a 90° bend imposes the same friction loss as 20' of straight length. Snaking the duct needlessly and leaving it compressed introduces many small changes in direction, increasing the friction loss significantly.
- Avoid allowing the duct to contact pipes, conduit, wiring, lighting fixtures, and similar objects. The duct must remain clear of anything that produces a great deal of heat.

A ductboard collar (*Figure 63A*) may be required for some installations that use insulated flexible duct, especially if it is being connected to fiberglass ductboard. When required, the collar should be the same thickness as the insulation on the duct. The collar insulates the metal takeoff and provides a landing for the outer jacket of the flexible duct. To connect a flexible duct to a takeoff or the connector of an air terminal, use the following procedure:

Step 1 Measure to determine the length of duct required. Allow for any bends or turns.

Step 2 Decompress and stretch the duct out before measuring for the cut. Remember that is should be extended to within 4% of its listed length.

Going Green

Flexible Duct Materials

Flexible ducts are made today with materials that produce little or no volatile organic compound (VOC) emissions and are treated to resist the growth of microbes. The wire, core, and jacket can be made from recycled materials.

CAUTION

Flexible duct with a polyester core should not be used within sight of duct-mounted ultraviolet air purification devices. The UV rays can break down the polyester.

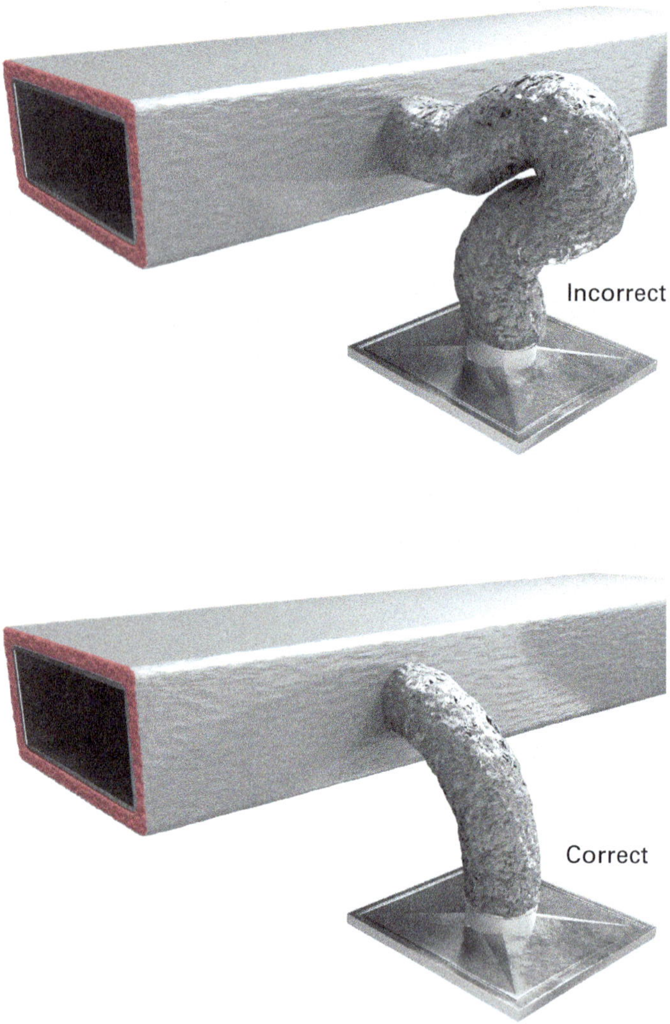

Figure 62 Proper flexible duct fit.

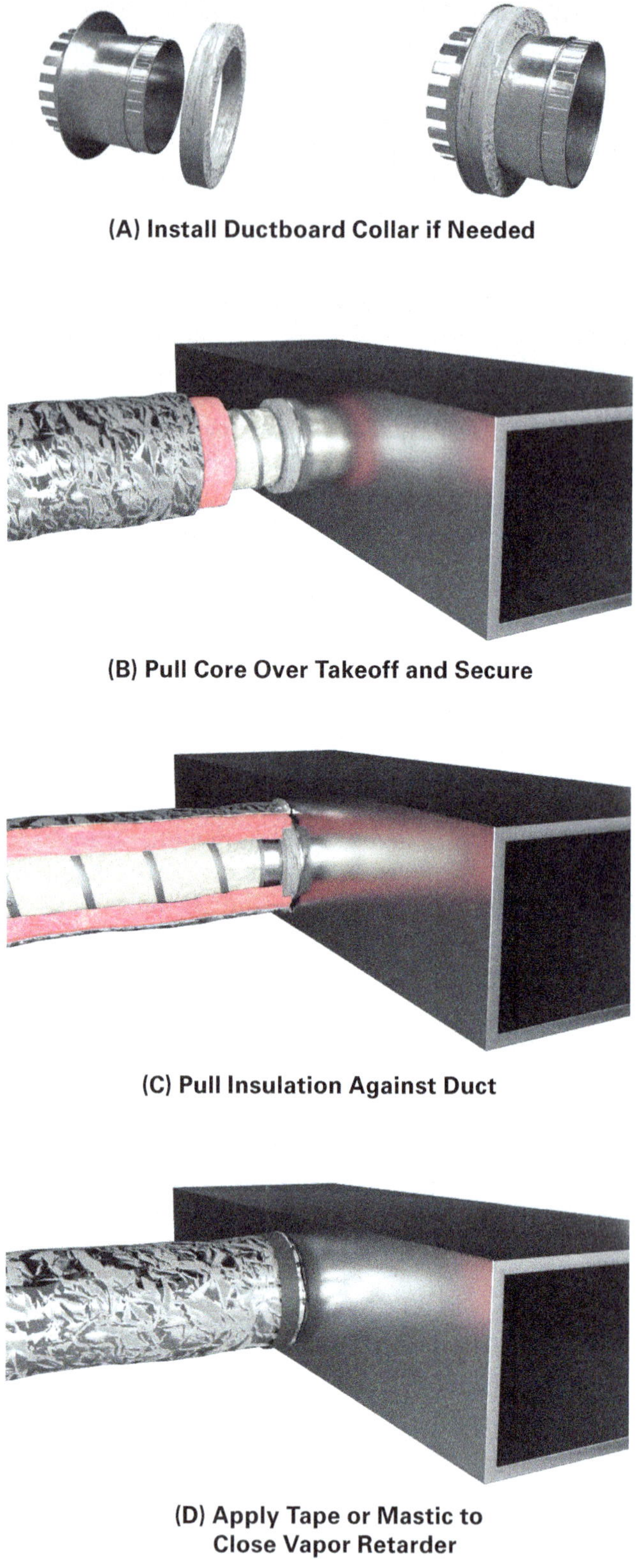

Figure 63 Fabricating a flexible duct joint.

Step 3 Use a sharp utility knife to cut through the outer jacket, insulation, and the core material. Cut the interior wire with electrical lineman pliers or a similar tool.

Step 4 If the duct is insulated, pull back the insulation and outer jacket from the cut end to expose a few inches of the core. Slide the exposed core over the round collar.

Step 5 Attach the core of the flexible duct to the sheet metal collar using duct tape and/or by using a non-metallic locking plastic strap. These are commonly referred to as Panduit® straps. A metallic worm clamp may also be used. If tape is used, wrap the joint at least twice. Some specifications may call for tape to be combined with a metal or plastic strap.

Step 6 If a strap alone is used to make the connection, apply mastic duct sealant to the connection to ensure an airtight seal. The mastic should be applied in a 2" wide strip around the connection.

Step 7 Pull the outer jacket and insulation back over the core and the fitting. Use at least two wraps of tape to seal off the outer jacket to the duct connector. This prevents water vapor from getting between the insulation and the core where it can condense.

6.2.2 Splicing Flexible Ducts

If additional duct length is needed, flexible duct sections can be spliced together using a metal sleeve (*Figure 64*). Each end of the sections to be joined is slid over the ends of the metal sleeve. The metal sleeve should be a minimum of 4" long, allowing 2" of metal for each side of the splice.

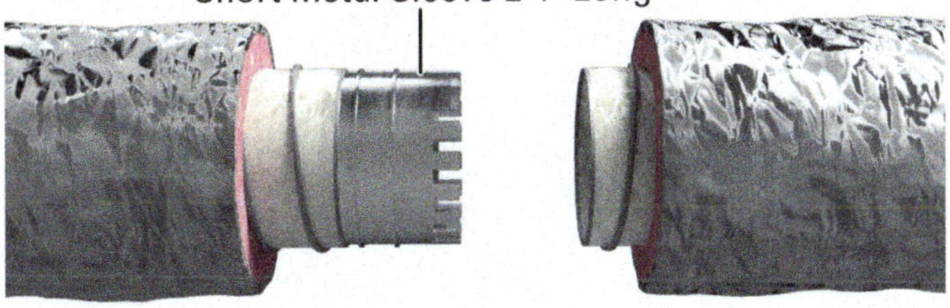

Figure 64 Metal sleeve splice.

Once the sleeve is in place, with the core of both sections pulled over it, the process is identical to making a connection to a duct takeoff.

Any holes in the vapor retarder of a flexible duct must be sealed with tape that meets UL Standard 181. If the core is punctured, resolve the problem by cutting the damaged section of duct away and installing a splice.

6.2.3 Hanging and Supporting Flexible Duct

After a flexible duct run has been connected to the main trunk, it must be well supported to ensure proper airflow to the register or diffuser it serves (*Figure 65*). To hang and support flexible duct, follow these general guidelines. However, follow all local codes to ensure compliance:

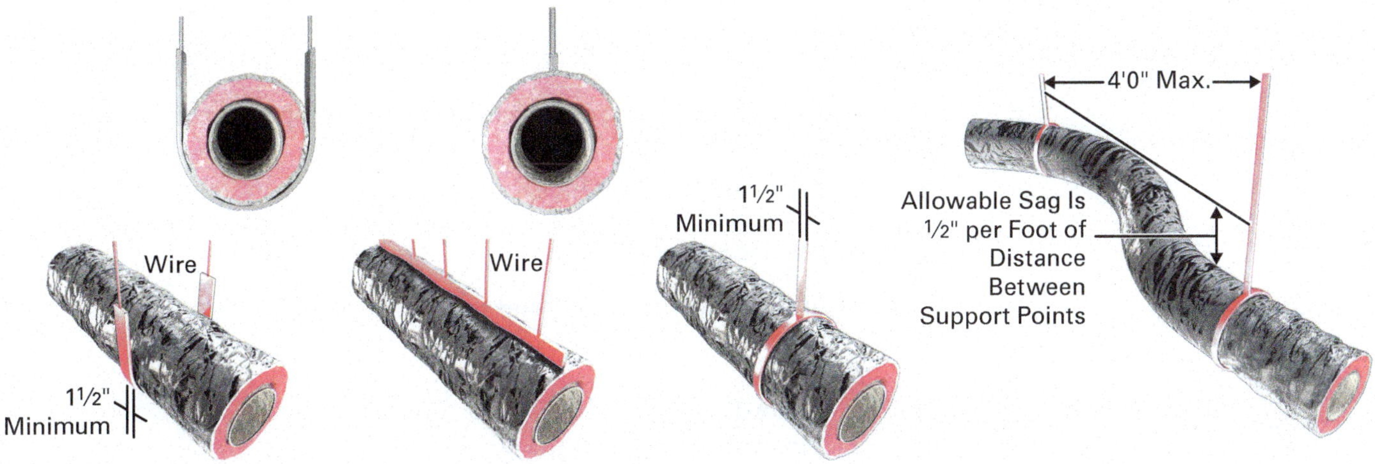

Figure 65 Flexible duct support and suspension options.

- Support the duct at the manufacturer's recommended intervals, but at least every 4'. The maximum permissible sag is $\frac{1}{2}$" per foot of spacing between the supports. Therefore, if the supports are 4' apart, the total allowable sag between them is 2".

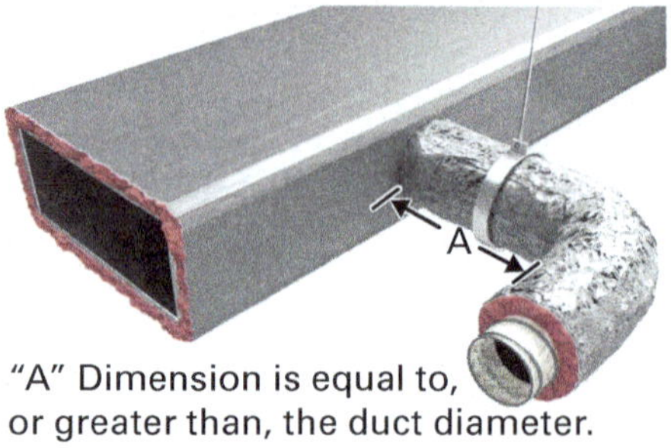

Figure 66 A bend close to the point of connection.

- For long horizontal runs that have sharp bends, use additional supports before and after each bend.
- The material used to suspend the duct from beneath must be wide enough so it does not cause any restriction of its internal diameter. A thin wire, for example, doesn't provide good support. The material should be at least $1\frac{1}{2}$" wide; wider is better.
- Flexible duct can rest on ceiling joists as long as the spacing between supporting points is less than 4'. Most ceiling joists are much closer together than this.
- It is best to allow the duct to extend at least one duct diameter straight away from a connection to the main duct before a bend is allowed (*Figure 66*). A hanger should be placed between the point of connection and the bend.
- If flexible duct is installed vertically, supporting straps should be installed no more than 6' apart (*Figure 67*).

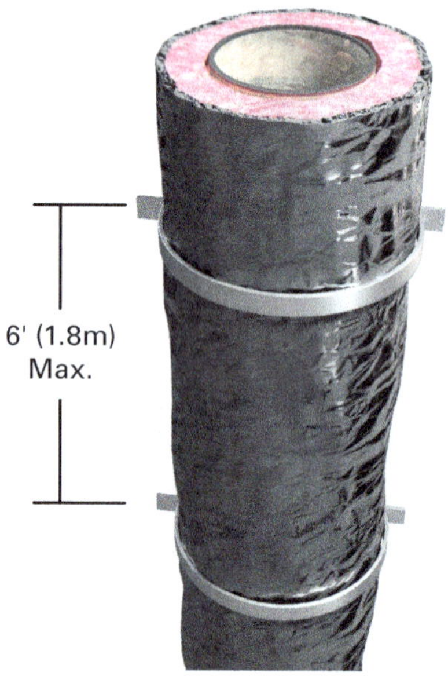

Figure 67 Flexible duct routed vertically.

6.0.0 Section Review

1. Which of the following organizations tests the properties of flexible ducts?
   a. NFPA
   b. Air Diffusion Council
   c. SMACNA
   d. Flexible Duct Association

2. What is the recommended bend radius for a flexible duct?
   a. 4"
   b. 8"
   c. 1 duct diameter
   d. 2 duct diameters

Module 03213 Review Questions

1. A metal that is not combined with any other metal is called a(n) ______.
   a. base metal
   b. alloy
   c. pure metal
   d. natural metal

2. Steel that is coated with zinc is said to be ______.
   a. stainless
   b. galvanized
   c. cold-rolled
   d. hot-rolled

3. What alloy is formed when iron and carbon are combined?
   a. Steel
   b. Cupronickel
   c. Stainless steel
   d. Bronze

4. True or False? Iron is the most plentiful metallic element on Earth.
   a. True
   b. False

5. The duct seal class that requires all joints, seams, and penetrations to be sealed is Class ______.
   a. A
   b. B
   c. C
   d. D

6. Which duct seam design relies on raised bumps or buttons?
   a. Grooved lock seam
   b. Pittsburgh lock
   c. Drive seam
   d. Snap lock seam

7. If no roll-forming machine is available, what machine or tool do installers use to form a Pittsburgh lock?
   a. Press
   b. Snips
   c. Crimpers
   d. Brake

8. Which of the following duct connectors provides its own reinforcement and is also used as a cross-seam for larger ducts?
   a. Transverse
   b. Standing seam
   c. S-slip
   d. Drive

9. True or False? Round duct runs in homes are typically supported by solid flexible wires.
   a. True
   b. False

10. True or False? The term *riser* refers to a duct that runs vertically.
    a. True
    b. False

11. Thermal insulation is designed to ______.
    a. transfer heat by convection
    b. transfer heat by conduction
    c. reduce the flow of heat energy
    d. support the flow of heat energy

12. An appropriate application for fiberglass duct liner is ______.
    a. inside kitchens or fume exhaust ducts
    b. to line fire damper sleeves
    c. excessively moist environments
    d. in ducts that operate at temperatures below 250°F

13. True or False? Adhesives used to fasten fiberglass to the inside of a metal duct are applied with a caulking gun to form a narrow bead.
    a. True
    b. False

14. Which of the following is the standard width of fiberglass insulation used to wrap ducts?
    a. 24"
    b. 36"
    c. 48"
    d. 72"

15. When creating an insulation staple flap, how much fiberglass must be cut away?
    a. 2"
    b. 4"
    c. 6"
    d. 8"

16. How long should a section of 2" insulation be to wrap around a 12" round duct, including the staple flap?
    a. 16"
    b. $39\frac{3}{4}$"
    c. $52\frac{1}{4}$"
    d. 73"

12" duct diameter + (2" insulation thickness × 2) = 16" outer diameter

$$C = 2\pi r$$

$$C = 2 \times 3.14 \times 8\text{" radius}$$

$$C = 50.24\text{"}$$

Add 2" to create a staple flap:

$$50.24\text{"} + 2\text{"} = 52.24\text{", or } 52\tfrac{1}{4}\text{"}$$

17. When attaching fiberglass insulation using the stapled tab method, always use ______.
    a. stainless steel staples
    b. 1" staples
    c. flare-head staples
    d. a pneumatic stapler

18. Manual volume dampers are most often found in ______.
    a. branch ducts
    b. main trunk ducts
    c. plenums
    d. smoke ducts

19. Within duct systems, smoke detection systems monitor smoke ______.
    a. velocity
    b. color
    c. density
    d. conductivity

20. Fire dampers are typically controlled by a(n) ______.
    a. microprocessor
    b. fusible link
    c. heat-sensing coil
    d. internal thermostat

21. When installing tap-in and spin-in takeoffs, you must cut the opening so that ______.
    a. the fitting can extend above the opening by ⅛"
    b. there are no holes at the corners
    c. there is no need for tabs or other fasteners
    d. the damper can move with the airflow

22. What material is used to give flexible duct its form?
    a. Polyester
    b. Fiberglass
    c. Wire
    d. Metalized nylon

23. Which of the following is a common core material in flexible ducts?
    a. Mylar
    b. Polyethylene
    c. Nylon
    d. Polyester

24. True or False? The R-value of the insulation used in flexible ducts can vary.
    a. True
    b. False

25. The minimum length of the metal sleeve used to splice flexible duct sections is ______.
    a. 2"
    b. 4"
    c. 6"
    d. 8"

Answers to odd-numbered review questions are found in the Module Review Answer Key at the back of this book.

Answers to Section Review Questions

Answer	Section	Objective
Section One		
1. b	1.1.0	1a
2. a	1.2.1	1b
Section Two		
1. c	2.1.1	2a
2. a	2.2.5	2b
Section Three		
1. b	3.1.1	3a
2. d	3.2.0	3b
Section Four		
1. d	4.1.3	4a
2. c	4.2.1	4b
Section Five		
1. d	5.1.1	5a
2. b	5.2.2	5b
Section Six		
1. b	6.1.0	6a
2. c	6.2.1	6b

MODULE 03203

Introduction to Hydronic Systems

Source: SerhiiKrot/iStock

Objectives

Successful completion of this module prepares you to do the following:

1. Describe hydronic systems and the principles of closed-system water flow.
   a. Describe the basic properties of water and the significance of its composition.
   b. Describe the relationship between water flow and system pressures.
2. Describe the primary types of hot-water heating systems and system components.
   a. Differentiate between gravity and forced hydronic systems.
   b. Describe various types of boilers.
   c. Identify primary boiler components.
   d. Identify hydronic system components related to air and water management.
3. Identify various hot-water heating piping layouts and the terminal devices used.
   a. Describe the characteristics of one- and two-pipe distribution systems.
   b. Describe hot-water zoning systems.
   c. Identify various hot-water heating system terminal devices.
4. Describe hydronic system pump selection and explain how water flow is monitored and balanced.
   a. Identify devices used to measure and control water flow.
   b. Describe how circulating pumps are selected.
   c. Explain how to measure pressure and water flow in an operating system.

Performance Task

Under supervision, you should be able to do the following:

1. Measure the pressure drop across a balancing valve and determine the flow rate.

Overview

A hydronic system uses water as a medium to provide heating and cooling for comfort or processes. In hydronic comfort-heating systems, water travels from the boiler through pipes to terminal devices where the heat is transferred to the air. This module introduces the basic principles, components, and configurations of hydronic heating systems.

Digital Resources for HVACR

Scan this code using the camera on your phone or mobile device to view the digital resources related to this craft.

SCAN ME

NCCER Industry-Recognized Credentials

If you are training through an NCCER-accredited sponsor, you may be eligible for credentials from NCCER. The ID number for this module is 03203. Note that this module may have been used in other NCCER curricula and may apply to other level completions. Contact NCCER at 1.888.622.3720 or go to **www.nccer.org** for more information.

You can also show off your industry-recognized credentials online with NCCER's digital credentials. Transform your knowledge, skills, and achievements into credentials that you can share across social media platforms, send to your network, and add to your resume. For more information, visit **www.nccer.org**.

1.0.0 Hydronic Heating System Principles

Performance Tasks

There are no Performance Tasks in this section.

Objective

Describe hydronic systems and the principles of closed-system water flow.

a. Describe the basic properties of water and the significance of its composition.
b. Describe the relationship between water flow and system pressures.

Hydronic system: A system that uses water or water-based solutions as a medium to transport heat.

There are essentially two ways to deliver heat from its source of generation to the point where it is needed. A *forced-air system* uses a fan to deliver heated or cooled air through a ductwork system for distribution to the conditioned space. Similarly, a **hydronic system** uses pipes to transport heated or cooled fluid. The fluid used for heating can be either hot water or steam. For cooling, chilled water is used.

Comfort-heating systems often combine both methods. A boiler heats water that is pumped to various terminals, such as hot water coils installed in fan coil and air handling units, and a fan moves air across the coil to deliver heat to the conditioned space. A radiant hot-water system, though, eliminates the fan. A good example of a radiant heating system is a floor heating system, where hot water is circulated through tubing buried in or attached to the floor (*Figure 1*). Many consider radiant floor heating to be the most effective and comfortable form of residential heating.

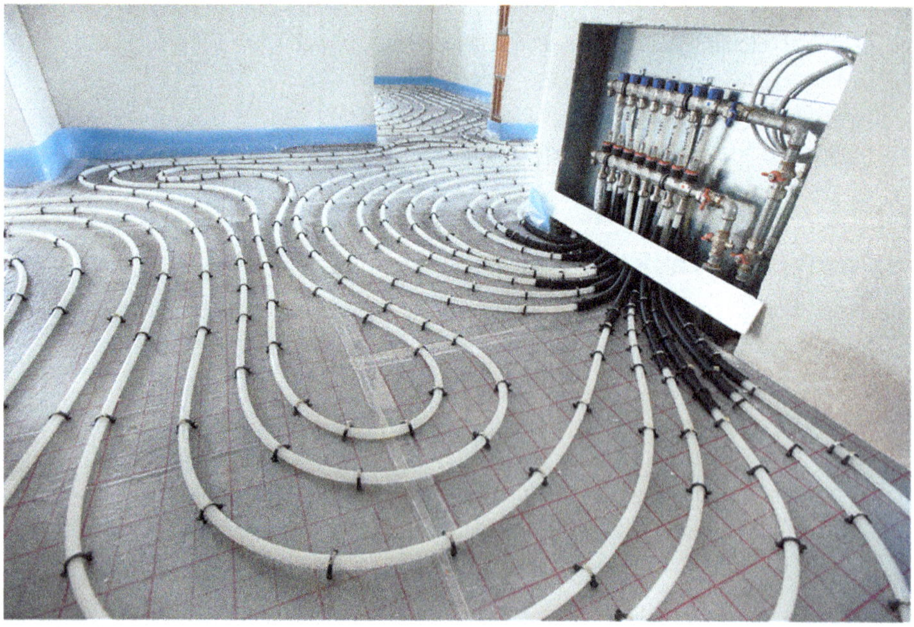

Figure 1 Radiant floor-heating system.
Source: filmfoto/iStock

The term *hydronic* technically refers to all water systems. *Hydronic heating* is the term commonly used when referring to hot-water and steam comfort systems. The term *chilled-water system* refers to a comfort cooling system based on chilled water.

The focus of this module is on hydronic heating systems that use hot water generated by a boiler. For example, a commercial system serving multifamily housing comprised of seven gas-fired modular boilers is shown in *Figure 2*. Steam systems are presented in NCCER Module 03306, *Steam Systems*. Chilled-water systems are presented in NCCER Module 03305, *Commercial Hydronic Systems*.

Figure 2 Commercial boiler room with modular boilers.
Source: SerhiiKrot/iStock/Getty Images

One advantage of hydronic systems over air systems is that heat can be transferred much more efficiently. Water has a higher **specific heat** than air, so it can carry more heat per pound. The specific heat of a substance refers to the amount of heat required to raise the temperature of 1 lb of that substance by 1°F. Water, which is used as the basis for measuring the specific heat of all other substances, has a specific heat of 1.0. Thus, it takes 1 Btu to raise the temperature of 1 lb of water by 1°F. The specific heat of water is roughly four times higher than air.

Specific heat: The amount of heat required to raise the temperature of one pound of substance 1°F. Expressed in units of Btu/lb/°F. At sea level, water has a specific heat of 1 Btu/lb/°F.

1.1.0 Water

Water is a chemical compound composed of the elements oxygen and hydrogen. Water exists in three different states: ice, water, or steam. These represent the solid, liquid, and vapor states of water. At the standard atmospheric pressure found at sea level, water freezes at 32°F (0°C) and boils at 212°F (100°C).

Like most fluids, the weight of water changes as its temperature changes; the higher the temperature, the less it weighs. This is due to expansion and contraction as the molecules move closer or farther apart. For example, 1 cubic foot (ft^3) of water weighs 62.41 lb at 32°F (28.31 kg at 0°C) and 59.82 lb at 212°F (27.13 kg at 100°C). In a closed system, water at different temperatures causes natural circulation by gravity, due to the differences in weight. Colder water naturally moves down, while warmer water rises to the top. This is referred to as **thermal circulation**.

Thermal circulation: The circulation of water in a hydronic system resulting from differences in water temperature and the resulting change in weight.

1.1.1 Corrosion

Corrosion is a major concern in hydronic systems. Water itself is non-corrosive to most materials used in hydronic systems. However, air is also dissolved in fresh water. Although the nitrogen the air contains is relatively inert, the dissolved oxygen will initiate corrosion by combining with iron to create rust.

Corrosion: The breakdown or destruction of a material, especially metal, through chemical reactions. A common form of corrosion is *rust* (iron oxide), which forms when iron combines with oxygen.

Once all the available oxygen has been used in corrosive chemical reactions, corrosion stops. For this reason, the volume of fresh water admitted into any hydronic system must be minimized.

Each time fresh water is admitted into a closed system—following a leak repair, for example—fresh oxygen is also admitted. Corrosion begins again until all the free oxygen has been used. If fresh water is consistently added, the piping system and components will eventually suffer irreparable damage and have a shorter service life.

Open hydronic systems, which are open to the atmosphere at one or more points, allow the water to absorb oxygen from the air continuously. Open systems are rarely used for hydronic heating and cooling systems. However, open systems are commonly used for cooling towers and similar applications where the water transfers heat directly to the air outdoors.

Many hydronic systems use automatic valves to admit fresh water as needed, making up for losses due to leaks or repair activities. Water added to a hydronic system in service is referred to as *makeup water*. Craftworkers are aware of repair activities, and several repairs can sometimes be accomplished at the same time, minimizing the number of times the system is opened.

Leaks, however, can occur in unmonitored areas. For this reason, all hydronic system piping should be inspected periodically. A small leak can result in the constant addition of fresh oxygenated water. On larger systems, water meters are often installed and monitored to see if the volume of makeup water being added is within expectations.

On Site

Black Water

Dissolved oxygen in hydronic systems will react with any exposed iron and steel, causing the water to discolor and darken. The result is referred to as *black water*.

1.1.2 Scale

Water also contains a variety of minerals. The mineral content of local water varies significantly across the country. **Scale** is typically a combination of calcium and magnesium compounds that forms on surfaces in a hydronic system (*Figure 3*).

Scale: Chalk-like mineral deposits, typically a combination of calcium and magnesium compounds, that form and collect on surfaces inside hydronic systems of all types.

Figure 3 Scale on the tube sheet and tubes of a heat exchanger.
Source: RYosha/iStock

Calcium carbonate ($CaCO_3$) is a common component of scale. This substance has a unique characteristic in that it becomes less soluble in water as the temperature increases. In other words, higher temperatures tend to cause it to leave the water and collect as a solid. For that reason, the heaviest deposits of scale are generally found on the warmest surfaces, such as heat exchangers and electric heating elements.

Because scale deposits dramatically impede heat transfer, it is yet another reason to minimize the admittance of fresh water. Chemical additives combine with the minerals to form harmless, non-scale compounds (*Figure 4*). These chemicals are often added to water systems, especially larger ones. They can also reduce free oxygen by using it to produce harmless compounds, thereby locking up the oxygen that would normally react with iron.

NOTE

Always check with the boiler manufacturer before adding chemicals to boiler water. The wrong chemical may damage the boiler and void any warranties. However, many boiler manufacturers not only support, but recommend or require, this approach.

Figure 4 Boiler water treatment chemicals.
Source: Rectorseal, LLC

1.2.0 Water and Pressure

Water moves through a hydronic system due to differences in pressure, flowing from a higher pressure toward a lower pressure. Water at different temperatures within the system will move naturally, as previously described, but our heating needs require the flow rate to be much higher and controllable.

This section presents pressure-related concepts essential to an understanding of hydronic systems.

1.2.1 Pressure Drop

Pressure drop represents a difference in pressure between two points that results from friction in the fluid stream. As the water moves through pipes, heating units, and fittings, it creates friction. Changes of direction, small obstructions, and similar anomalies add additional friction.

Pressure drop: The reduction in pressure between two points in a moving water stream, resulting from friction and other factors.

In a horizontal fluid-filled pipe, assume there is no flow and the pressure is equal at all points. Once flow begins, friction is encountered. The friction increases in direct proportion to the velocity of the fluid. More flow and volume results in more friction.

The change in pressure drop due to a change in flow rate can be calculated using the following formula:

$$\text{Change in pressure drop} = \left(\frac{\textit{new flow rate}}{\textit{initial flow rate}}\right)^2 \times \textit{initial pressure drop}$$

Remember that hydronic flow rates are generally expressed in gallons per minute (gpm). For example, assume that the water flow through a system or device is increased from 5 gpm to 10 gpm at a pressure of 25 psig. The pressure drop at 5 gpm of flow is 5 psig. The new pressure drop is calculated as follows (*Figure 5*):

$$\text{Change in pressure drop} = \left(\frac{10\text{ gpm}}{5\text{ gpm}}\right)^2 \times 5\text{ psig}$$

$$\text{Change in pressure drop} = 2^2 \times 5\text{ psig}$$

$$\text{Change in pressure drop} = 4 \times 5\text{ psig}$$

$$\text{Change in pressure drop} = 20\text{ psig}$$

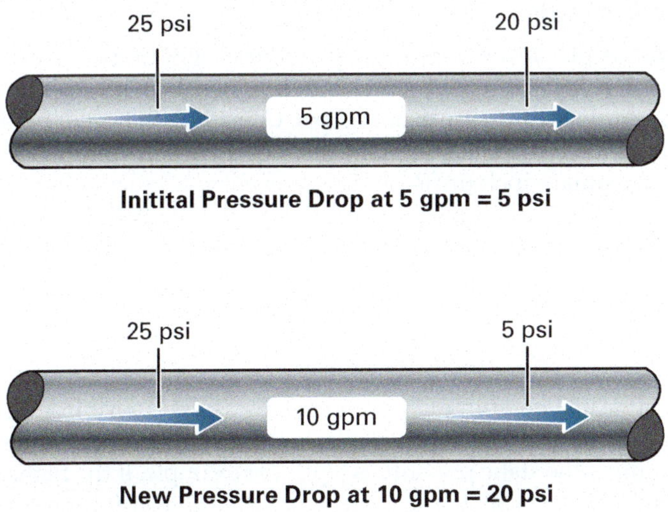

Figure 5 Pressure drop illustrated.

In a hydronic system, additional energy is required to overcome the effects of pressure drop. There must be enough energy to overcome the total pressure drop of the system and still have the required pressure at the farthest reaches of the system. A properly sized circulating pump provides the energy to overcome pressure drop and deliver the needed volume of water.

Manufacturers will provide pressure drop data about their equipment. Charts are readily available to relate the pressure drop through various fittings to an equivalent length of pipe.

Pressure drop in HVACR applications is generally expressed in pounds per square inch (psi) or in feet of water. The two units can be converted as follows (*Figure 6*):

1:psi = 2.31' of water

1' of water = 0.43 psi

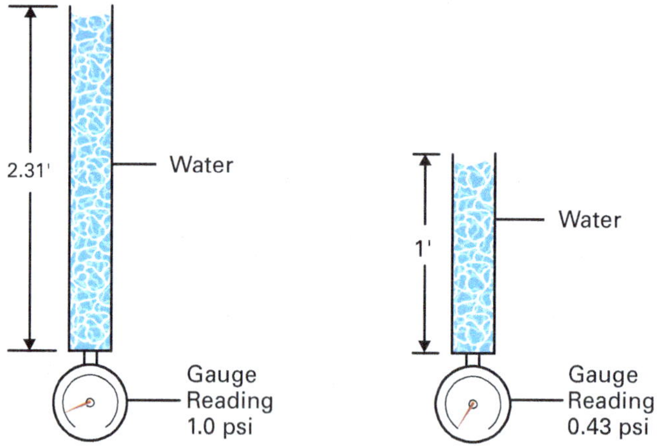

Figure 6 Relationship of feet of water to pounds per square inch.

1.2.2 Head

Head: The height to which a pump can lift water; an indication of the energy that a pump can impart to the fluid.

Head is another measure of pressure, expressed in feet of water. It is often written as *feet of head*, such as 10' head. As shown in *Figure 6*, feet of water can be converted to psi and vice versa.

Head is often used to describe the capacity of a circulating pump. Head indicates how high a pump can lift a column of water. The maximum head of a pump also reflects the maximum pressure drop the pump can overcome to deliver a given volume of fluid.

1.2.3 Static Pressure

Static pressure: Reflects the pressure exerted by the weight of the water that is vertically above the gauge in the system. The static pressure of water is 0.43 pounds per square inch, per foot of water height at standard atmospheric conditions.

Static pressure is pressure created solely by the weight of the water in the system. *Static* means that the water is not moving. Recall that a column of water that is 1' high produces a pressure of 0.43 psi. Consider the *pounds per square inch* unit carefully. What if the cross-sectional area of the water column is 1 square ft (ft^2) instead of 1 square inch (in^2)?

Since there are 144 in^2 in a square foot, we can calculate the weight of the water in a column that is 1' high with a cross-sectional area of 1 ft^2:

$$\text{Weight} = 0.43 \text{ psi} \times 144 \text{ in}^2 \times 1' \text{ height}$$

$$\text{Weight} = 61.92 \text{ lb}$$

Static pressure in a column of water increases as you move from the highest point toward the lowest point of the water column. At the highest point—the top of the water column—the static pressure is 0 psi. For example, if the highest point of a static column of water is 30', the static pressure at various heights are as follows:

- At 30' (highest water level), the pressure exerted is 0 psi.
- At 20' (10' below highest point), the pressure exerted is 4.3 psi (10' × 0.43).
- At 10' (20' below highest point), the pressure exerted is 8.6 psi (20' × 0.43).
- At 0' (lowest water level), the pressure exerted is 12.9 psi (30' × 0.43).

1.0.0 Section Review

1. How many Btus are required to raise the temperature of 0.5 lb of water by 2°F?
   a. 0.5 Btu
   b. 0.75 Btu
   c. 1 Btu
   d. 2 Btus

2. How much pressure per square inch will a column of water 1' high exert?
   a. 0.43 psi
   b. 1.0 psi
   c. 1 psi
   d. 10.0 psi

2.0.0 Hot-Water Heating Systems

Objective

Describe the primary types of hot-water heating systems and system components.

a. Differentiate between gravity and forced hydronic systems.
b. Describe various types of boilers.
c. Identify primary boiler components.
d. Identify hydronic system components related to air and water management.

Performance Tasks

There are no Performance Tasks in this section.

Hot-water comfort systems may be dedicated systems or be segments of much larger hydronic systems used for other purposes. A large system at an industrial facility, for example, may have one or more heat exchangers used to heat water for comfort heating while also supporting a manufacturing process. These large systems will have significantly different piping layouts than the layouts presented in this section.

2.1.0 Gravity and Forced Hot-Water Systems

Hot-water heating systems can be divided into two types: *gravity hot-water systems* and *forced hot-water systems*. Hydronic heating systems can function using gravity and thermal circulation alone. Forced systems that rely on pumps to move the water offer far more flexibility and service.

2.1.1 Gravity Hot-Water Systems

In a gravity hot-water system (*Figure 7*), water in the boiler is heated to the design operating temperature. The circulation of water occurs because of thermal circulation and gravity. No pump is used.

Remember that hot water is lighter than cold water. This is due to expansion as the water temperature increases. When the boiler is operating, the temperature of the water rises and becomes lighter than the rest of the water in the system. This disturbs the system's equilibrium.

As the warmer and lighter water rises upward in the heating main, the cooler and heavier water in the return main moves downward to take its place. Relatively consistent circulation is established. The heat is transferred to rooms through the terminal devices. Cooler water leaving the terminals returns to the boiler, where it is reheated and recirculated.

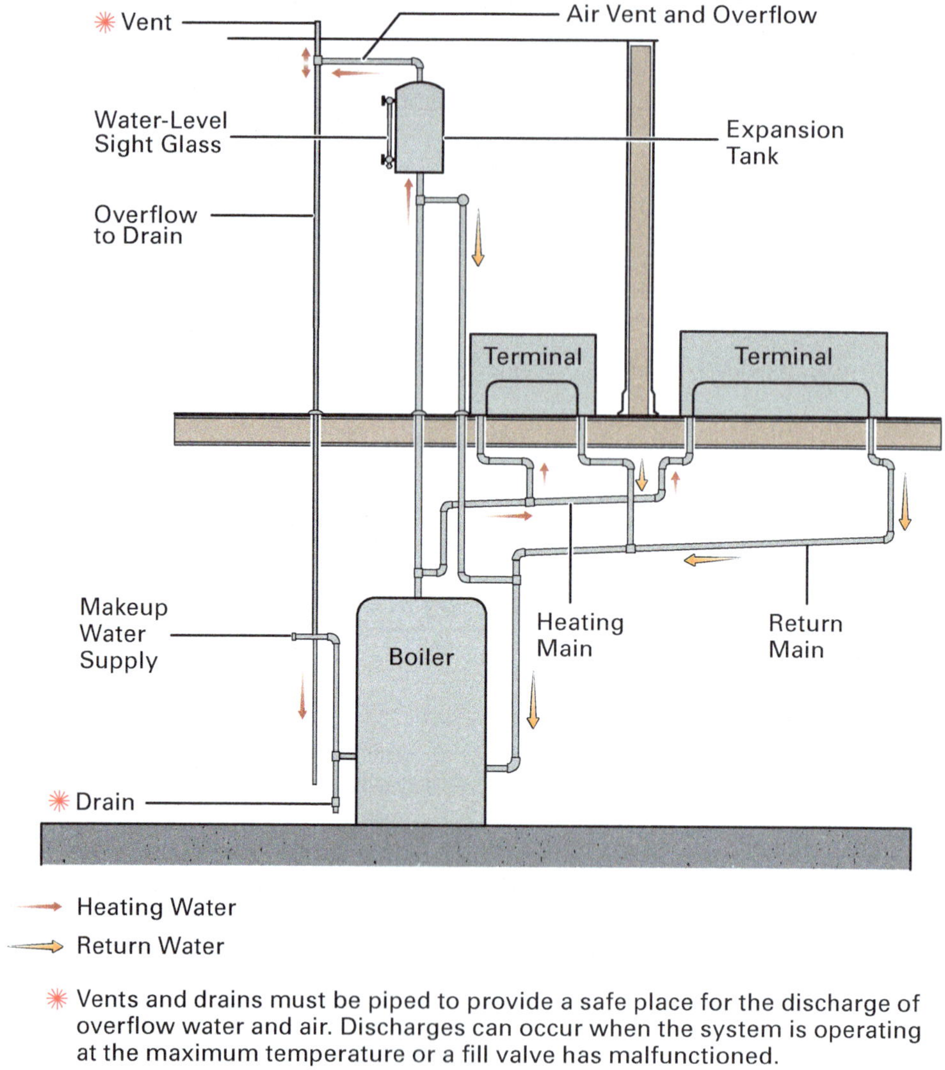

Figure 7 Gravity hot-water system.

Gravity hot-water systems respond slowly to changes in load conditions and are challenging to expand or modify successfully. These systems are both less effective and less efficient than forced systems. Gravity systems are rarely used today, but it is still important to understand thermal circulation in general. While gravity systems rely on thermal circulation, forced hot-water systems employ check valves and other devices to prevent it.

2.1.2 Forced Hot-Water Systems

A schematic diagram for a forced hot-water heating system with two zones is shown in *Figure 8*. Although the layout shown is common, there are many options. Other practical piping layouts will be presented later in this module.

Depending on the type of boiler, the water is heated using fossil fuels or electric heating elements. The burners used for boilers are basically the same as those used in gas and oil furnaces. The combustion or electric heating process is controlled primarily by water temperature, with various safety devices and other components to move the water and transfer heat.

Forced hot-water systems use one or more pumps to move the water through the system piping. The pump(s) is sized to satisfy the system flow requirements.

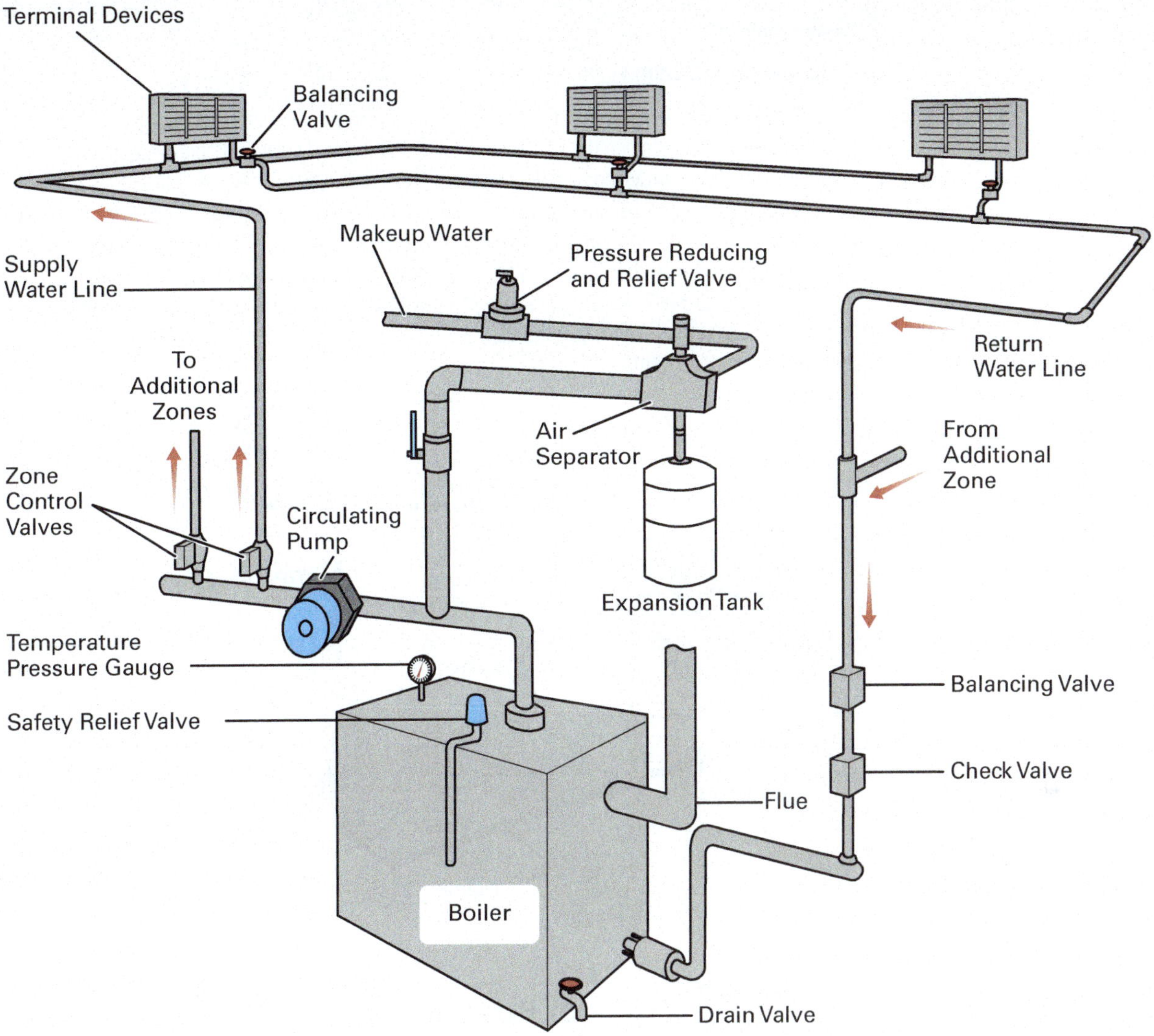

Figure 8 Simplified zoned hot-water heating system.

Buildings are often divided into several zones with independent thermostatic control. When one or more zone thermostats call for heat, the system circulating pump starts. In some cases, a zone valve opens first and a switch in the valve starts the pump. Alternatively, a small circulating pump can be used for each zone.

The pump causes the heated water from the boiler to circulate through a zone control valve and out to the terminal devices in the zone. Boilers that heat water instantaneously require water flow, created by the pump, to enable burner ignition and operation.

Terminal devices transfer the heat contained in the circulating water to the cooler air in the conditioned space. A return line moves the water leaving the terminal devices back to the boiler where the cooled water is reheated.

When the circulating pump is turned off, the check valve located in the return line stops the reverse flow of water due to thermal circulation. Balancing valves add resistance to water flow to limit the volume of water that can flow through individual terminals. Without balancing valves, some terminals get more water than needed while others don't receive enough.

Going Green

Solar Hydronic Systems

Harnessing the energy of the sun to provide heat is not only energy efficient but also good for the environment. However, since the sun does not shine at night, a solar hydronic system is best used to supplement the heat provided by a conventional hydronic heating system. A typical supplemental system consists of a solar collector, circulating pump, water-to-water heat exchanger, controls, and valves.

When the sun shines, an antifreeze solution is pumped through the solar collector where the fluid absorbs heat. The fluid then passes through a water-to-water heat exchanger located in an external storage tank. The heat exchanger transfers heat to the water in the storage tank. Preheating the water means that the boiler's burners do not have to operate as often to maintain the system's water temperature requirements.

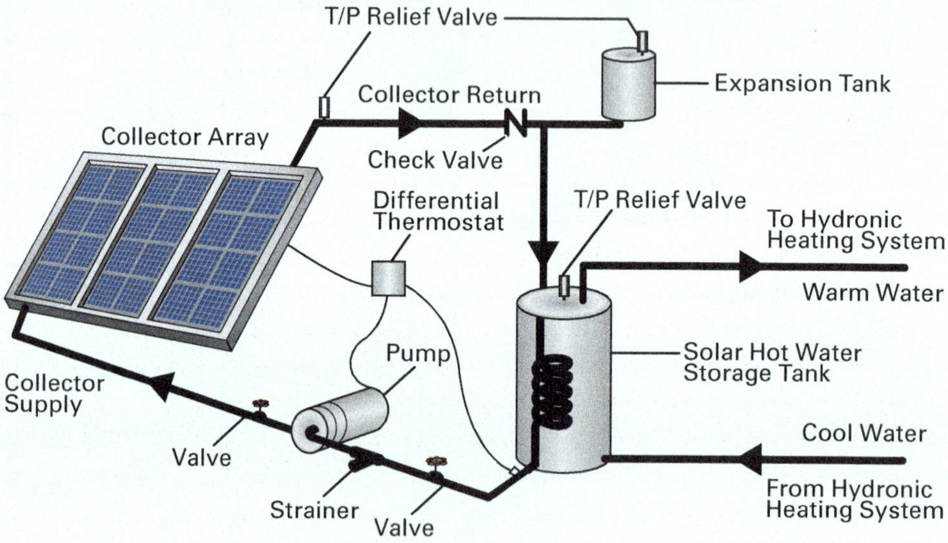

2.2.0 Types of Boilers

Boilers are available in a variety of styles and designs. More sophisticated approaches to combustion are popular as the need for energy efficiency steadily increases.

A fossil-fuel boiler heats water by burning fuels, such as natural gas, propane, and fuel oil, or solid fuels, such as coal or wood. The water may also be heated electrically. Heat pumps can also be used to produce hot water. Some boilers can fire more than one fuel using a *dual-fuel burner.*

In the United States, the residential and light commercial markets are dominated by natural gas in urban and suburban areas, while fuel oil (grade #2) and propane are the primary fuels in rural areas. Coal- and wood-fired boilers are also used in rural areas of the north. Depending on the cost and availability of the solid fuel, these boilers can be less costly to operate and provide a backup heating source. More information about add-on boiler systems is provided in NCCER Module 03409, *Alternative and Specialized Heating and Cooling Systems.*

The burners used on gas- and oil-fired boilers are basically the same as those used with furnaces. The ignition of the gas or oil can be achieved with a standing gas pilot, spark-type ignitor, or hot-surface ignitor (HSI). Standing pilot flames are now rare.

Hot-water boilers and steam boilers are similar with some exceptions. Hot-water operating and safety controls are a little different because a specific water level isn't maintained. Hot-water boilers are entirely filled with water and maintained at a consistent pressure.

Low-pressure boilers are the most widely used. Low-pressure hot-water boilers are constructed with working pressures of up to 160 psi. However, most are designed for a 30 psig (207 kPa) maximum working pressure and operate at 12 to 15 psig (83–103 kPa). Higher pressure settings may be required for multistory buildings.

Low-temperature boilers generally operate at temperature settings below 200°F (93°C). These boilers are often limited to a maximum operating temperature of 250°F (121°C). Medium- and high-pressure boilers can operate at much higher pressures and temperatures. In very large systems, such high pressures and temperatures allow the water to be circulated great distances—from a central heating plant to other buildings, for example—without losing significant heating capacity. It is important to remember that high-temperature and high-pressure systems can be dangerous.

WARNING!

Water at a high temperature and pressure instantly flashes to steam if a small fracture or leak develops in the piping or components. Dissipation of the superheated steam often occurs so rapidly that no steam plume is visible. This is due to its sudden exposure to atmospheric pressures, where water boils at 212°F (100°C). Serious physical harm can result from walking near or moving a hand across the path of the invisible steam. Use extreme caution when servicing or working in the vicinity of these systems.

Boilers are made of cast iron, steel, and/or copper. Many of today's boilers incorporate copper-finned tubes to enhance heat transfer and reduce the volume of water in the boiler. Condensing boilers with efficiencies above 95% are readily available. Some modern boilers resemble the appearance of home appliances, featuring stylish cabinets and colorful displays.

Going Green

Heat Pump Water Heaters

Heat pumps provide heat efficiently. Under the right conditions, a heat pump's energy output can be up to three times its equivalent energy input. Heat pumps are commonly used today to heat swimming pools and supply domestic hot water. Heat pumps can also serve as a heat source in hydronic heating systems.

In these applications, a heat pump's indoor coil (condenser) is a refrigerant-to-water heat exchanger. When the heat pump operates, heat rejected in the heat exchanger preheats the water in the hydronic system. This means that the boiler does not have to operate as long to maintain the water temperature in the system. The control scheme is similar to that used in hybrid air-to-air heat pumps.

Heat-pump water heaters are rapidly gaining in popularity. Earlier models relied on an outdoor condensing unit and an indoor tank and heat exchanger. Many models today are packaged, like the product shown here. While this one is suitable for outdoor installation in a reasonable environment, most must be installed where the temperature remains in the range of 40°F–90°F (4°C–32°C).

Source: Douglas Cliff/iStock

2.2.1 Copper-Finned Tube and Stainless-Steel Boilers

Small, compact boilers with heat exchangers constructed of copper-finned tubes and/or stainless steel are growing in popularity. Older cast iron and steel boilers stored a significant amount of heated water within the boiler to provide hot water on demand. With the introduction of copper-finned tubing (*Figure 9*), the speed of recovery eliminates the need to retain a large volume of heated water. Today's compact condensing boilers do not store any hot water—the water is heated instantaneously and only on demand. This increases the efficiency of the products by reducing heat losses from water in storage.

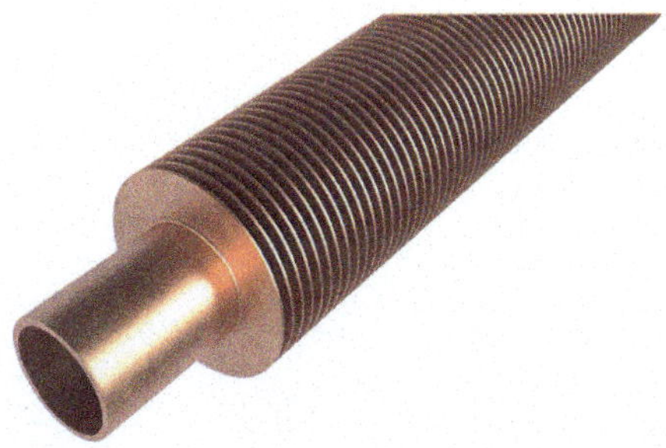

Figure 9 Copper-finned tubing.

Lightweight metals like copper transfer heat quickly and effectively, allowing manufacturers to significantly reduce the size and weight of boilers. With water inside the tubing to act as a heat sink, the copper can withstand a tremendous amount of heat. Older products that occupied 30 ft^3 to 50 ft^3 of space are being replaced with far more efficient units that easily hang on a wall.

Wall-mounted boilers (*Figure 10*) require less space and use space that would otherwise be wasted. These boilers are usually vented through the outside wall just above the location. An example installation of a zoned, wall-mounted, condensing boiler system is shown in *Figure 11*.

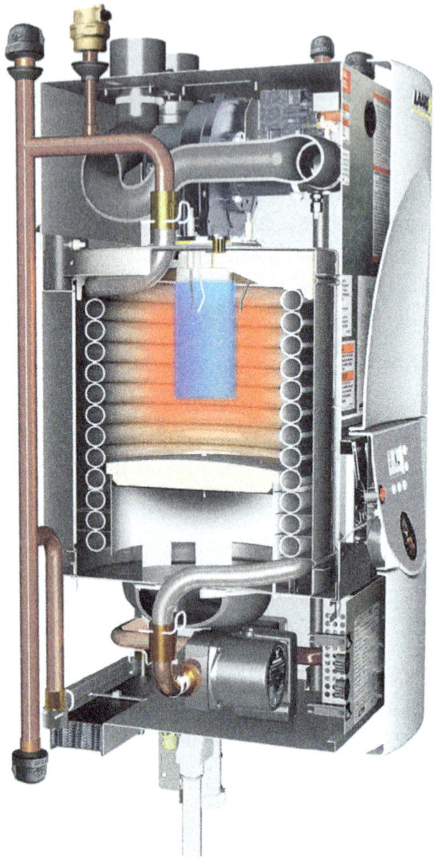

Figure 10 High-efficiency, wall-mounted condensing boiler.
Source: Laars Heating Systems Company

Figure 11 Wall-mounted boiler installation.
Source: matteogirelli/iStock

Commercial applications often require the use of two or more boilers piped in parallel (*Figure 12*). This provides multiple stages of heating to match the demand as well as a level of **redundancy**.

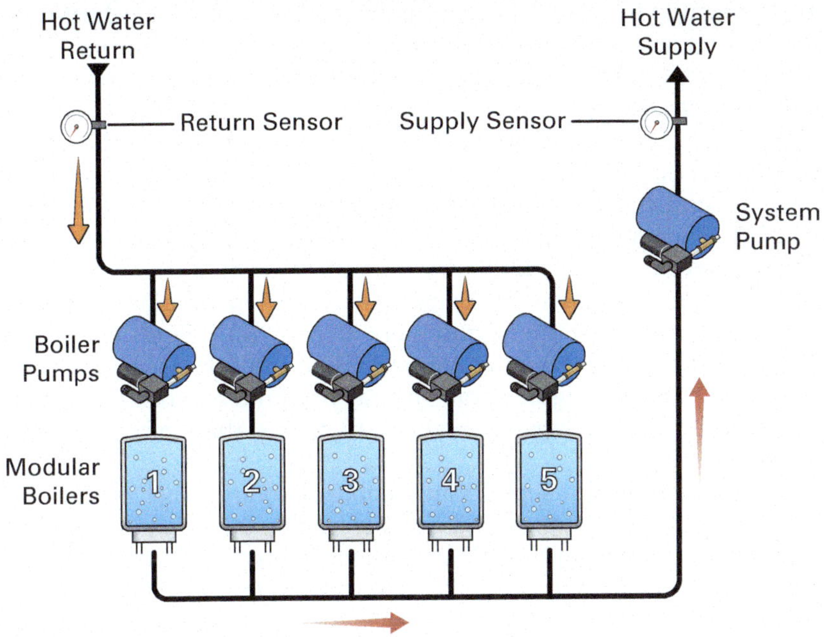

Figure 12 Modular boilers piped in parallel.

Condensing boilers are designed to reach efficiencies above 95%. While smaller condensing boilers are wall-mounted, free-standing models (*Figure 13*) are required for more capacity. These units are typically vented using Schedule 40 PVC pipe due to the low temperature of the flue gas. CPVC can be used in some areas, but it is not preferred. New polypropylene-based materials are on the market and have gained acceptance in many areas. However, Schedule 40 PVC continues to be the most popular material for venting condensing appliances.

Figure 13 Free-standing condensing boiler.
Source: Courtesy of Lochinvar, LLC

Redundancy: In HVACR systems, the provision of equipment to act as a backup to primary equipment, allowing system operation to continue despite a failed unit. An example of 100% redundancy is the installation of two boilers, each sized to handle the complete heating needs of the structure.

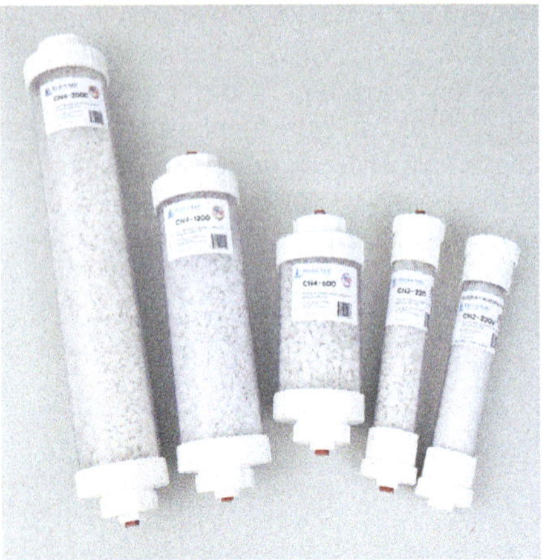

(A) Inline Cartridges

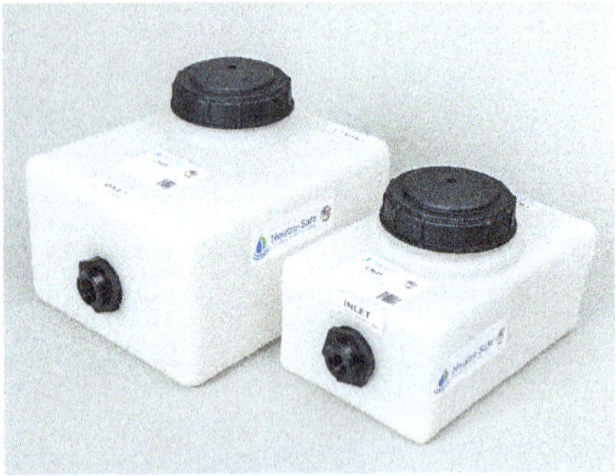

(B) High-Capacity Tanks

Figure 14 Condensate neutralization kits.
Source: Neutrasafe Corporation

Sealed-combustion condensing burners use combustion air from outdoors, rather than using indoor air that has been heated at some cost. When air from inside the structure is used for combustion, it must be replaced by fresh air from outdoors that must be heated to room temperature. The use of draft-inducer fans allows for sidewall venting and other convenient venting options.

Like condensing furnaces, cooling the byproducts of combustion to extract additional heat causes moisture to condense out of the flue gases. This condensate can be quite acidic and must be collected and removed from the unit. The low-pH acid can damage drainage systems and create problems at water treatment plants. Local codes may dictate methods to properly dispose of the condensate from condensing appliances. Some areas require the acidic condensate to be neutralized before it is admitted to a sewage system.

Neutralizers are available for inline installation in tubular form and in high-capacity tanks for larger equipment (*Figure 14*). Blends of materials such as calcite and magnesium oxide neutralize the acid as the condensate flows through the cartridges. Some contractors take a proactive approach and include these neutralizing accessories in all installations of condensing appliances.

Some gas boilers feature sophisticated onboard controls that monitor boiler water conditions, space conditions, and outdoor conditions simultaneously. The firing rate, burner run time, and supply water temperature are controlled based on these conditions. For maximum fuel efficiency, these boilers provide only the water temperature and flow rate required to meet the needs of the conditioned space without waste.

2.2.2 Cast Iron and Steel Boilers

Some boilers are still made from cast iron because it is durable, resistant to thermal shock, and has good heat transfer characteristics. Due to their weight, cast iron boilers are constructed by assembling individual sections (*Figure 15*). The number of sections determines the size and capacity of the boiler. For example, the boiler shown in *Figure 16* can be constructed using from 3 to 12 individual cast-iron sections. A three-section version weighs about 900 lb, while a unit with 12 sections weighs roughly 3,500 lb. These boilers must be field assembled. This particular boiler can accommodate a gas- or oil-fired power burner, which is attached to the front plate.

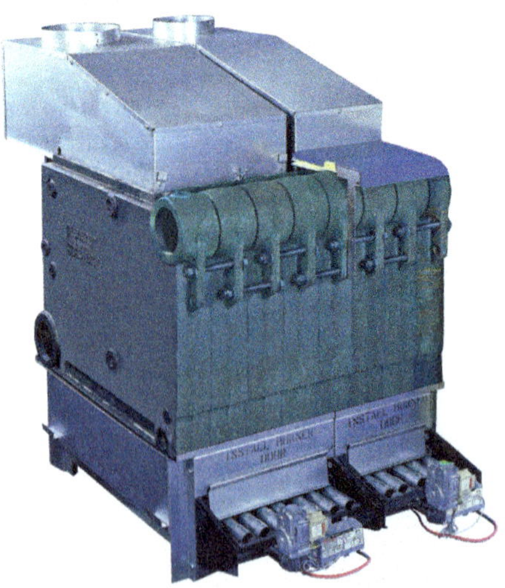

Figure 15 Cast-iron boiler sections assembled.
Source: ECR International, Inc.

Figure 16 Cast-iron sectional boiler.
Source: Burnham Commercial

In a cast iron boiler, the water circulates inside the sections while the flue gases flow around the outside. The heating surface area is large relative to the volume of water in the sections. This allows the water to heat up quickly.

Boilers used for residential and light commercial applications are often supplied fully assembled (packaged). Packaged boilers, other than the compact condensing boilers, are usually made from steel and include the burner, basic controls, and auxiliary equipment (*Figure 17*).

Many boilers used primarily for space heating can be equipped with an internal coil called a *tankless heater* (*Figure 18A*). A tankless heater is a water-to-water heat exchanger used to supply domestic hot water. This isolates the domestic hot-water supply from the water in the heating system. The boiler may run year-round to provide domestic hot water.

Figure 17 Packaged steel boiler.
Source: Burnham Commercial

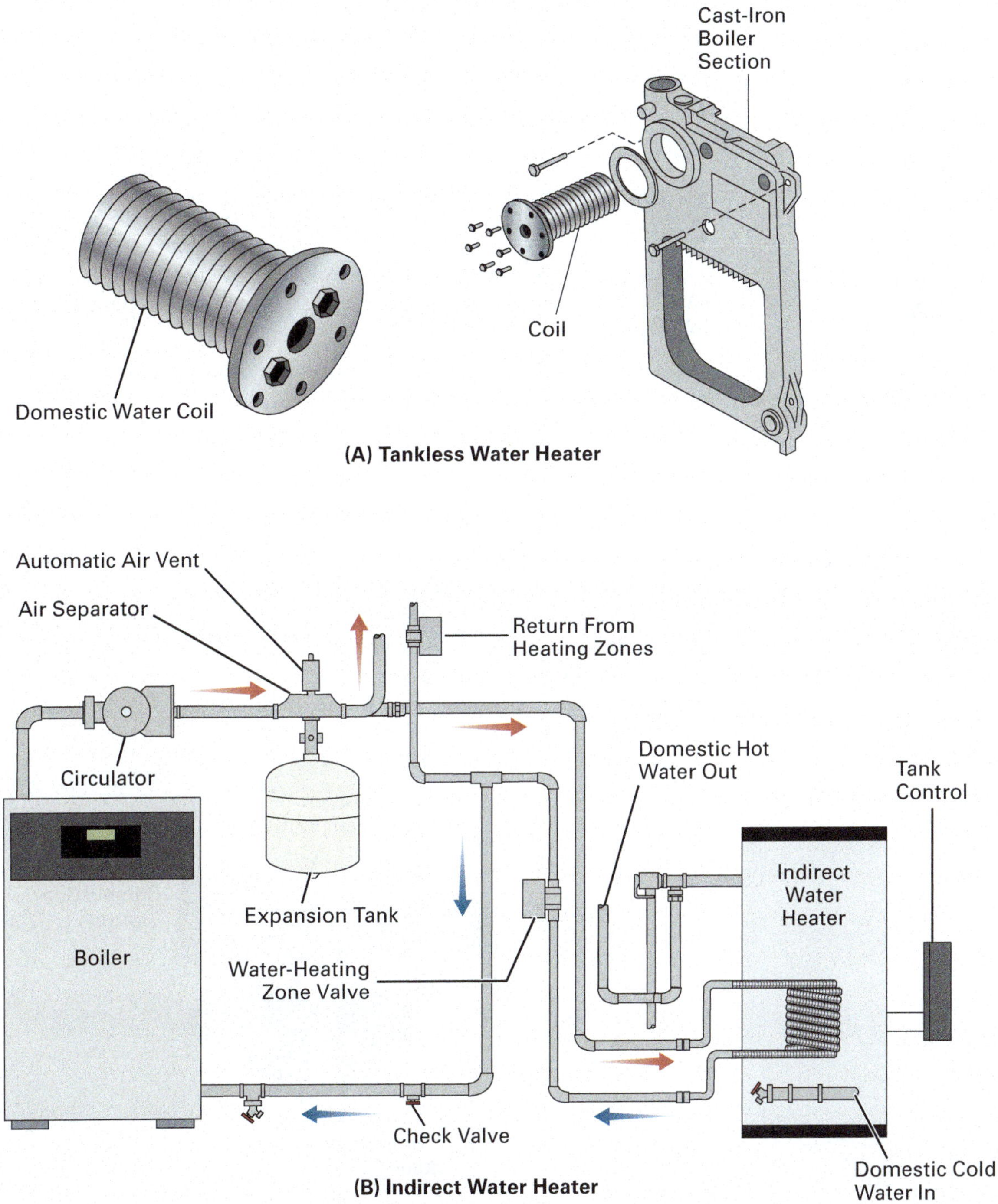

Figure 18 Tankless and indirect water heaters.

Another method of producing domestic hot water is through an indirect water heater (*Figure 18B*). This also relies on a water-to-water heat exchanger, but it is a stand-alone unit connected outside the boiler. It is connected as a separate heating zone with a circulator or zone valve that starts the boiler only when necessary. Heated boiler water flows through the heat exchanger coil in the indirect water heater tank, heating the domestic water contained in the tank. Like boilers with tankless heaters, the boiler in this system must run year-round.

On Site

Dry-Base, Wet-Leg, and Wet-Base Boilers

The terms *dry-base*, *wet-leg*, or *wet-base* are used by manufacturers to describe various boilers. These terms refer to the location of the water-filled sections of the boiler relative to the combustion chamber, or *firebox*. These boilers are defined as follows:

- *Dry-base* — The firebox is located beneath the water vessels.
- *Wet-leg* — The firebox top and sides are enclosed by water vessels.
- *Wet-base* — The firebox is surrounded by water vessels, except for necessary openings.

2.3.0 Primary Boiler Components

A packaged boiler consists of a completely assembled boiler shipped with a basic group of needed components. A packaged boiler typically contains the following primary components:

Aquastat: A thermostatic control for water.

- Hot-water boiler complete with burner and operating controls, such as the **aquastat**
- Combination pressure/temperature gauge
- Circulating pump

Other components needed to complete a system, such as the zone valves and the expansion tank, are field-supplied and not part of the package.

2.3.1 Burners and Combustion Controls

The burners and related safety devices used in gas- and oil-fired boilers are similar to those found in forced-air furnaces. However, the gas burner can either be atmospheric, induced draft, or an in-shot *power burner* (*Figure 19*). Gas-fired power burners operate much like oil burners and can be used to retrofit oil-fired equipment.

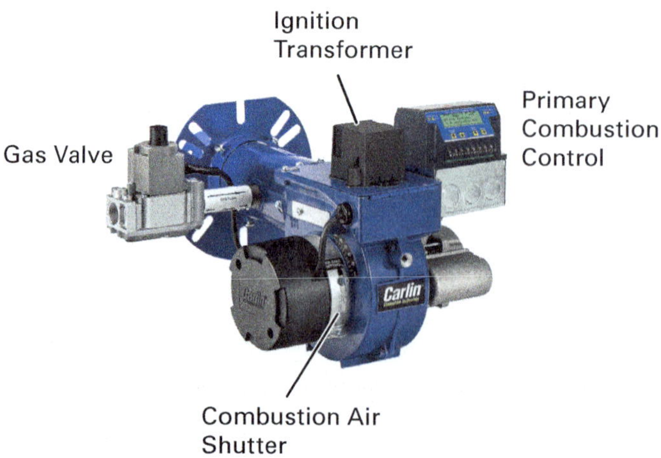

Figure 19 Gas-fired power burner.
Source: Courtesy of Carlin Combustion Technology

2.3.2 Operating and Safety Controls

The aquastat (*Figure 20*) is a control unique to boilers. An aquastat works basically the same as a thermostat except that it is designed to monitor and control water temperature instead of air temperature. While the temperature control shown in *Figure 20* is electronic, many electromechanical aquastats remain in service.

Aquastats may have the sensing element directly or indirectly immersed in the water. Most use a dry well installed in the boiler opening, and then the aquastat bulb or sensor is inserted into the dry well (indirect immersion). This gives the sensing element excellent exposure to the surrounding water while allowing it to be serviced without draining the water.

Aquastats may also be surface mounted to piping (*Figure 21*). It is important to ensure contact between the mating surfaces of the aquastat and the pipe.

Figure 20 Digital aquastat.
Source: Hydrolevel Company

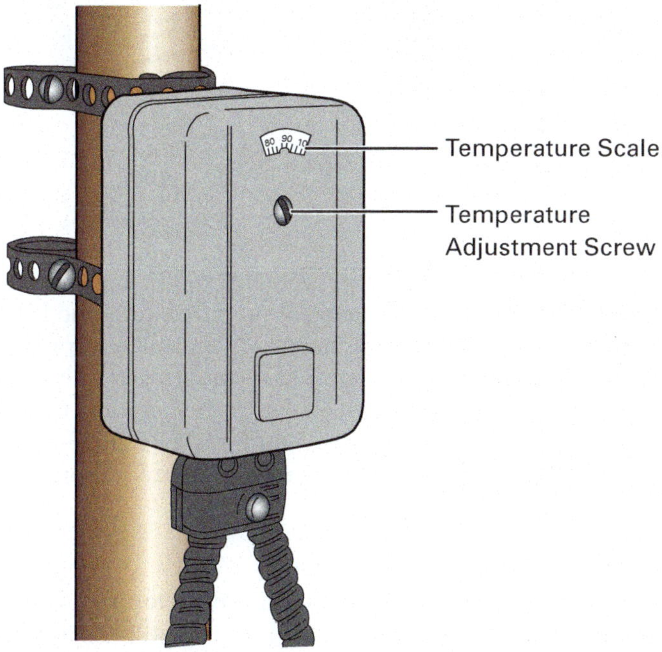

Figure 21 Surface-mounted aquastat.

Aquastats are used both as operating controls and high-limit safety controls. *Direct-acting* aquastats open the contacts on a temperature rise. This type is used mainly for burner control or as a high limit. High-limit controls typically require a manual reset.

Reverse-acting aquastats open the contacts on a temperature fall. This type may be used to stop a pump or fan when the water temperature falls below its setpoint.

Some aquastats also offer low-water cutoff protection for the boiler. Firing a boiler without sufficient water in it can lead to serious damage or fires. Most codes require low-water cutoff devices.

Know the Code

Low-Water Cutoff Controls

The 2021 *International Mechanical Code®* requires both steam and hot-water boilers to have low-water cutoff protection that disables the burner or heating elements when activated. However, the code makes an exception for boilers with coil and finned-tube heat exchangers that require water flow through the appliance to activate the burner. This exempts most wall-mounted models that operate on demand and modulate the burner based on flow rate. Flow switches protect this type of boiler or water heater.

The *IMC®* also requires hot-water boilers to have a temperature gauge and a pressure gauge, or a combination thereof.

Electronic low-water cutoffs that use an electrode placed in the water are common (*Figure 22*). The electrode relies on the conductivity of the water to complete a circuit to ground. If the probe is in the water, the circuit to ground is completed. If it is not, the circuit to ground opens. This provides a reliable means of monitoring the minimum water level. The absence or presence of the circuit to ground can be used to initiate alarms, energize/de-energize relays, and/or start and stop pumps.

Figure 22 Electronic low-water cutoff.
Source: Courtesy of Xylem Inc.

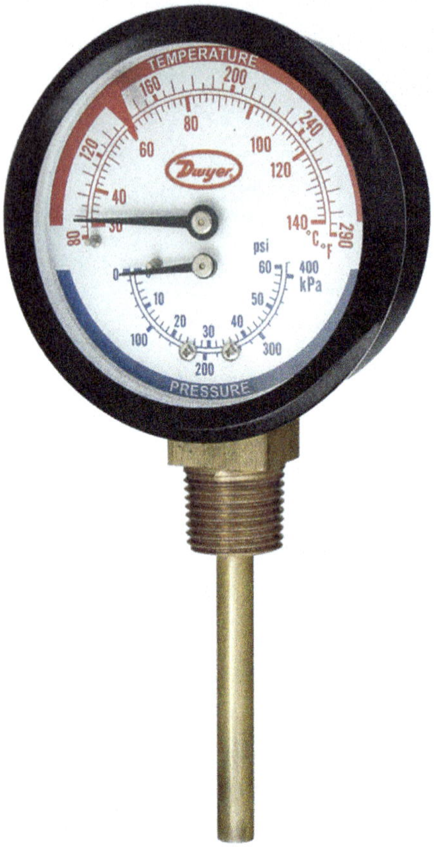

Figure 23 Combination pressure/temperature gauge.
Source: Courtesy of Dwyer Instruments, Inc.

2.3.3 Combination Pressure/Temperature Gauge

Pressure and temperature gauges (*Figure 23*) are often combined into a single device. Larger boilers and systems will have temperature and pressure gauges at other key locations in the boiler room. These gauges typically provide pressure readings from 0 to 50 psi or 0–400 kPa, corresponding to elevations of 0' to 70' (0 m–21 m), and temperatures from 60°F to 320°F (15°C–160°C). Ideally, the boiler's normal operating temperature and pressure fall around the midpoint of the gauge scale.

2.3.4 Circulating Pumps

The circulating pump creates the pressure difference to move hot water from the boiler through the piping, to the terminal units, and back again. It may be referred to as a *booster pump* or *circulator* on smaller systems. Large systems often have two or more base-mounted pumps.

Most circulating pumps are centrifugal pumps, in which the rotating action of an impeller in a spiral housing generates a pressure difference. The pressure difference between its inlet and outlet generates flow. The pressure available and volume of water delivered are functions of the pump size, motor horsepower, and rotational speed.

For smaller systems and systems with individual zone pumps, inline *cartridge pumps* are commonly chosen (*Figure 24*). Base-mounted centrifugal pumps are used in larger systems. Shutoff valves should be installed near the inlet and outlet piping of the pump, so it can be isolated for repair or replacement.

The pump can be installed either in the supply or return piping of the system. The preferred method is to install the pump on the supply side of the system so that the pump is discharging away from the boiler and expansion tank. As always, read the Installation, Operation and Maintenance (IOM) manual, following both the manufacturer's instructions and the project plans and specifications when installing a pump.

Cartridge pumps, such as the one shown in *Figure 24*, have flexible installation requirements and are easy to service. The term *cartridge* refers to the rotating portion of the pump, which includes the *impeller*. Both the cartridge and the motor are easily replaced.

Figure 24 Inline cartridge pump.
Source: Courtesy of Taco, Inc.

On Site

High-Point Air Vents

In large systems, there may be many high points—points at which the piping system rises to its highest level. Air in the system naturally attempts to reach these high points.

When a manual air vent is opened to purge a system or terminal, it is left open until a steady stream of water is obtained. This happens very quickly after a system is full and only a little air remains. However, when filling a new system, it can take some time and multiple visits to the same vent.

Catch the water in a suitable container if it is not piped to a drain. Move from vent to vent in order, opening and closing each vent and collecting the vented water. Repeat this process several times until all the air is purged from the system.

Automatic air vents may also be installed. The outlet of an automatic air vent should be piped to a drain to capture the bits of water it will release.

The pump's location in the system relative to the expansion tank determines whether the pressure generated is added to or subtracted from the system *static pressure*. The static pressure represents the pressure on the system by the water alone, without any pumps running. The point where the expansion tank is connected to the system is the point of no pressure change when the pump is started or stopped. It is recommended that this point be on the suction side of the pump to minimize total pressure.

If the expansion tank connection is made on the discharge side of the pump, the total system pressure must be higher to prevent **cavitation** at the pump inlet. Cavitation results from vapor pockets forming at the pump inlet when the water pressure falls below its vapor pressure for the current water temperature. The vapor pockets that form are essentially bubbles. The bubbles are carried through the pump impeller into an area of higher pressure where the bubbles implode (burst inward). The resulting force can damage the impeller vanes or even the pump housing.

Cavitation: A condition caused by liquids reaching the vapor pressure at the pump inlet, causing bubbles to form that then collapse violently when passing through the pump impeller.

The problem of cavitation can be resolved by maintaining a minimum suction pressure at the pump inlet. The normal operating temperature of the water must be considered when determining the minimum pressure. The minimum suction pressure assigned is referred to as the required *net positive suction head* (NPSH). Symptoms that indicate cavitation is occurring include the following:

- Snapping and crackling noises at the pump inlet (most common indication)
- Pump vibration with no mechanical source
- A reduction in flow
- A drop in the discharge pressure

The circulating pump must be sized to overcome the total system pressure drop and supply the necessary volume of water to all the loads. Pressure drop results from the friction caused by the system piping, fittings, boiler, terminals, and other system components. Thus, the piping system must be sized and designed to minimize pressure drop at the required flow rate. The capacity of a circulating pump is sometimes defined in terms of head pressure. That is just another way of expressing how much pressure drop it can overcome.

The maximum head pressure a pump can develop is its *churn pressure*—the pressure it can develop when there is no flow through the system. It is usually expressed in feet of water.

Manufacturers use *pump curves* to show pump performance, allowing engineers to select a pump and motor combination that meets their needs. *Figure 25* shows an example pump curve for a pump operating at a speed of 1,750 rpm. Note that this curve is for a base-mounted centrifugal pump that can be fitted with four different motor sizes.

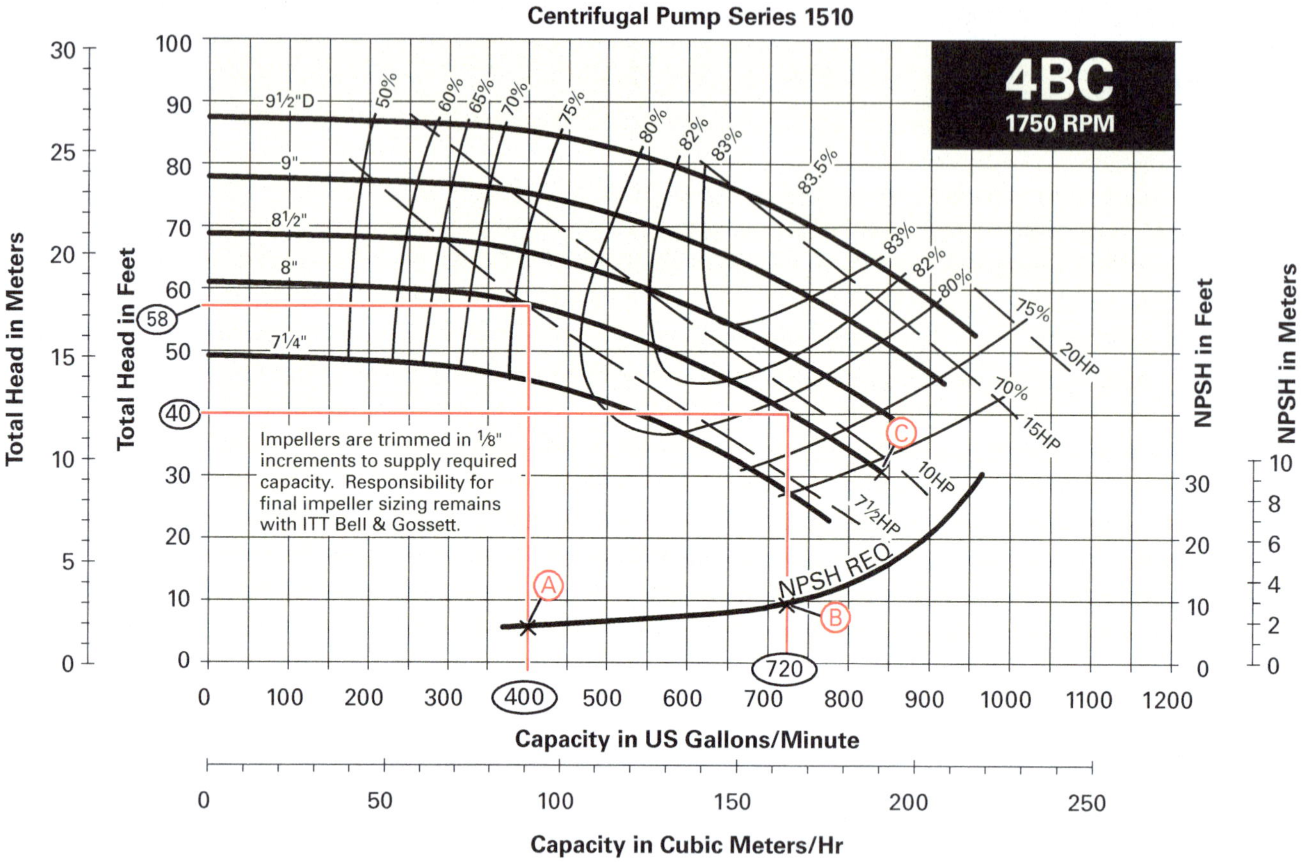

Figure 25 Typical pump curves.
Source: Courtesy of Xylem Inc.

A different set of curves can be acquired for the same pump operating at a different speed. The bottom horizontal scale shows the pump delivery in gallons per minute. The vertical scale shows the head in feet of water.

Examination of the capacity curves for various impeller sizes indicates the following:

- The lower the head (pressure drop), the more water the pump can deliver.
- The lower the volume of water delivered by the pump, the higher the head against which the pump can deliver water.

These points are illustrated by the examples shown on the curves in *Figure 25*. Notice that a pump with an 8" impeller will deliver 400 gpm when operating at a head of 58' (point A on the chart). With a decrease in head to 40', the same pump/motor combination will deliver about 720 gpm (point B on the chart). The maximum flow this pump can deliver is about 840 to 845 gpm (point C).

Figure 25 also plots the curves for pump efficiency and motor horsepower. Examination of these curves shows that at point A, the 8" diameter pump is operating at an efficiency of between 75% and 80% (about 76%) and at a non-overloading motor horsepower of $7^1/_2$ hp. When operating at point B, the same pump is operating at a slightly higher efficiency (about 78%), with an increase of the non-overloading motor horsepower from $7^1/_2$ hp to about 9 hp. Since there is no 9 hp motor, a 10 hp motor would be required.

The NPSHR curve shown in *Figure 25* helps determine the minimal total head in feet that must be maintained above the pump suction inlet to prevent cavitation. Referring to the NPSHR curve for this example, the curves show that when the 8" diameter pump is delivering 400 gpm (point A), the NPSHR is 6'. When the pump is delivering 720 gpm (point B), the NPSHR is 10'.

In some zoned systems, a single circulating pump may be used to provide all water flow. In this case, the pump is wired such that it will start when any zone calls for heat. The pump is sized to handle the total system flow rate at the required head of the zone with the highest resistance.

In other zoned systems, such as the one shown in *Figure 11*, a separate circulating pump serves each zone. Each pump is controlled by its zone thermostat and is sized to handle only the flow rate and head its zone requires.

Here is one way a zone valve can control its dedicated circulating pump:

- A room thermostat in a zone energizes a motorized zone valve, causing it to open.
- The zone valve is equipped with normally open contacts that close when the valve is fully open.
- The closure of the switch contacts completes the control circuit to energize the pump.
- Once the zone is satisfied and the zone valve begins to close, the pump is de-energized.

Single-speed circulator pumps driven by a standard induction motor were common in hydronic heating systems for years. When designing such a system, the pump is selected to handle the correct volume of water under design conditions.

However, design conditions may only exist a few days of the year. On mild winter days, not every zone will be calling for heat at the same time. Thus, the pump tends to be oversized for the job most of the time.

Today, system designers are more likely to use dedicated pumps for each zone. Small circulators using energy-saving electronically commutated motors (ECMs) are now available to maximize performance (*Figure 26*). These ECMs are the same as those used for variable-speed blowers in furnaces and air handlers. ECMs can be used with either approach to zoning.

Figure 26 ECM-driven circulator pump.
Source: Courtesy of Taco, Inc.

An ECM-driven circulator changes its speed to match the conditions in the system. For example, as individual zone valves open and close, the motor changes speed to increase or decrease water volume. Different pump manufacturers use various methods to determine the speed at which the pump should run. Speed changes based on water temperature or system pressure are both common.

In a temperature-based system, the designer selects the water temperature drop between the boiler's supply and return lines (referred to as the delta-T, or ΔT). Temperature sensors on the supply and return lines monitor water temperature and transmit the information to the electronic speed-control module mounted on the motor (*Figure 27*). As the sensors report temperature changes, the motor speeds up or slows down to maintain the selected temperature drop. The result is improved comfort in each zone, less power consumed by the pump, and quieter operation.

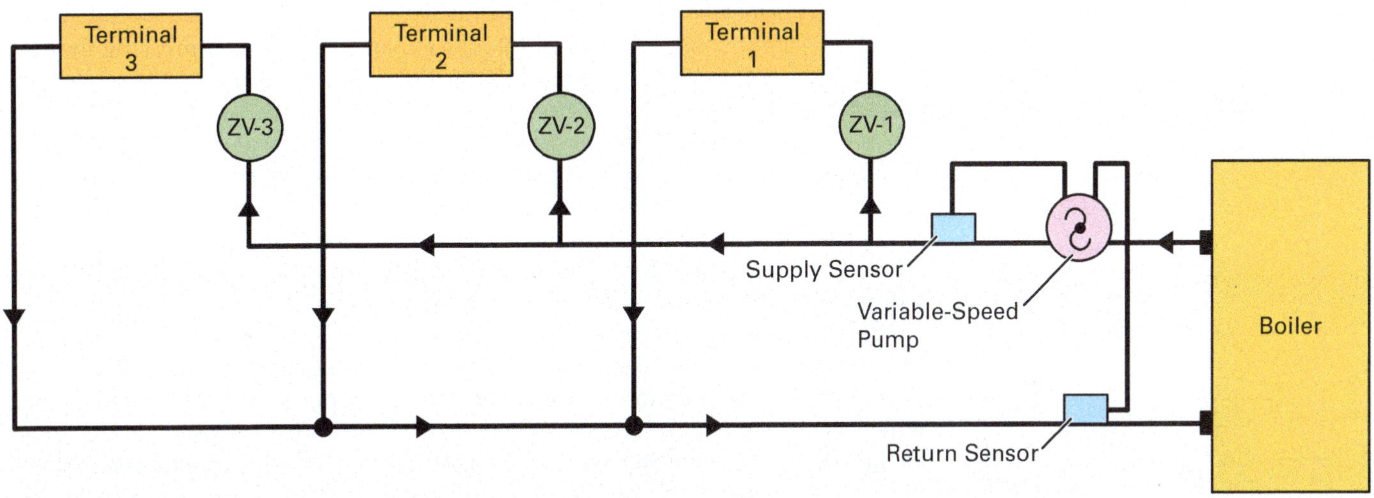

Figure 27 Variable-speed circulator pump controlled by temperature sensors.

Pressure sensors in the system can also control pump speed. The supply and return pressure differential in the system changes as zone valves open and close. The motor speed controller analyzes the pressure inputs on each side of the system and adjusts the pump speed accordingly.

2.4.0 Air and Water Control

Many components in a hydronic system are needed to control air and water. Air must be collected and eliminated from hydronic systems. These components include the following:

- Expansion/compression tanks
- Air-control devices
- Pressure-reducing valves
- Zone valves
- Backflow-preventer valves
- Check valve
- Hand valves
- Two- and three-way valves

2.4.1 Expansion Tanks

An expansion tank (*Figure 28*), also called a *compression tank*, helps maintain the system pressure. It provides a place for the safe expansion of the water as it heats. Otherwise, a system free of air and full of cold water at 15 psig would likely burst at some location when the water is heated to 180°F, due to the resulting pressure increase.

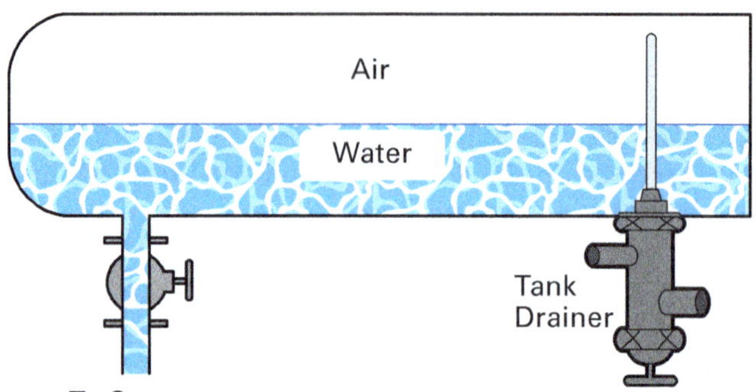

Figure 28 Standard expansion tank.

As the water expands, it pushes into the expansion tank, compressing the air. If the water temperature drops and the water contracts, water can return to the system from the tank.

An expansion tank must be sized to store the excess water that exists when the system is operating at its maximum temperature. In other words, the volume of water in a system and the expected temperature change must be known to calculate the difference in volume. Then the tank's size can be determined, allowing extra volume for a cushion of air. The expansion tank must also help maintain a minimum pressure when the system is cold.

When a heating system is filled with water, free of air pockets, and at its normal operating temperature (perhaps 160°F), the expansion tank should be about one-third to one-half full of water. The upper part of the tank contains air, which acts as a cushion for thermal expansion. The air compresses and expands as the volume of water in the tank changes. This serves to maintain the system pressure at a predetermined level. As mentioned previously, many residential and light commercial heating systems operate at 12 psig to 15 psig.

A traditional expansion tank is shown in *Figure 28*. The air and water are in contact with each other. Larger systems may have two or more tanks in the boiler room. The tanks are fitted with various air control devices to separate air from the water and route it to the expansion tank. However, systems also need air vents at the high points of a system to eliminate most of the air before the system starts. The air control devices at the expansion tank are designed to separate tiny air bubbles that are entrained in the water. Air control devices are presented in the next section.

Traditional expansion tanks are not used as often in newer systems. *Diaphragm expansion tanks* (*Figure 29*) are now common. These tanks contain a flexible diaphragm, or *bladder*, that separates the water from the air. *Figure 30* shows the action of the diaphragm as the water expands. The air pressure must be established manually, or *pre-charged*, while the tank is isolated from the system.

Figure 29 Diaphragm expansion tank.

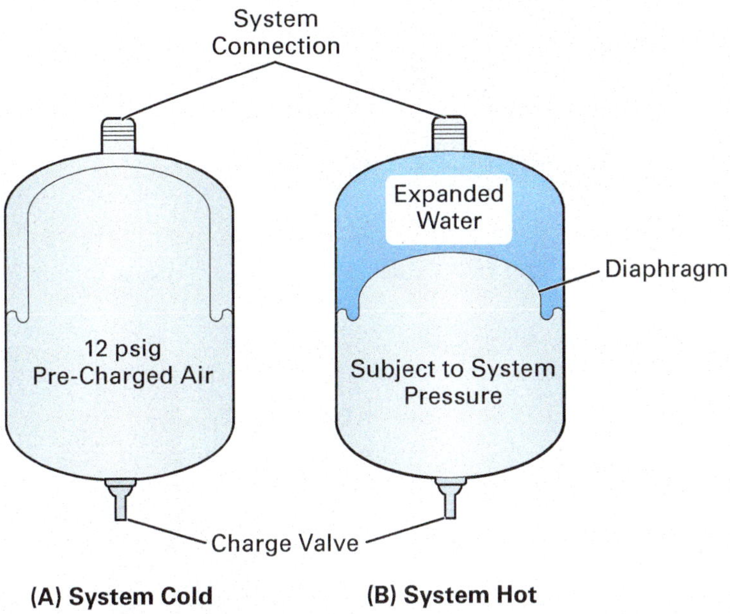

Figure 30 Diaphragm tank operation.

Diaphragm expansion tanks often come pre-charged with air, but the pressure can be adjusted through the charge valve as needed. A simple hand pump can provide the needed pressure. These tanks must be installed according to the manufacturer's instructions to ensure proper operation. Unlike traditional tanks, where entrained air is separated and sent to the top of the tank, the diaphragm prevents that in this type of tank. Any air in the system must be purged to the atmosphere and not allowed to enter the tank(s). One advantage, though, is that a diaphragm tank can often be placed over or under the line to which it is attached. Gravity is not a factor with a diaphragm tank.

Expansion Tank Issues

One sign of a flooded, or *waterlogged*, expansion tank is discharge from the boiler or system relief valve as the boiler is started and warms up. This is a sign that the expanded water has no place to go. Check the tank and drain as necessary, then monitor the system to see what might have caused it to happen.

A leaking diaphragm tank is not rare. When a diaphragm tank fails to hold its air charge, either the air valve or the bladder is leaking. A leaking bladder usually leads to tank replacement, but larger tanks have a replaceable bladder. In *Figure 29*, note the flange at the top of the tank where the system piping connects. This flange can be removed to replace the flexible bladder inside. The tank must be isolated from the system for service.

2.4.2 Air Control Devices

In a closed hot-water heating system, the only air in the system should be the air in the expansion tank. Because water will absorb and entrain air, air separators are necessary to collect air from the water and route it to the expansion tank. Various air control devices are used for this purpose. *Figure 31* shows the two components often used with standard expansion tanks—an air separator that also prevents thermal circulation, and an air charge/fill valve.

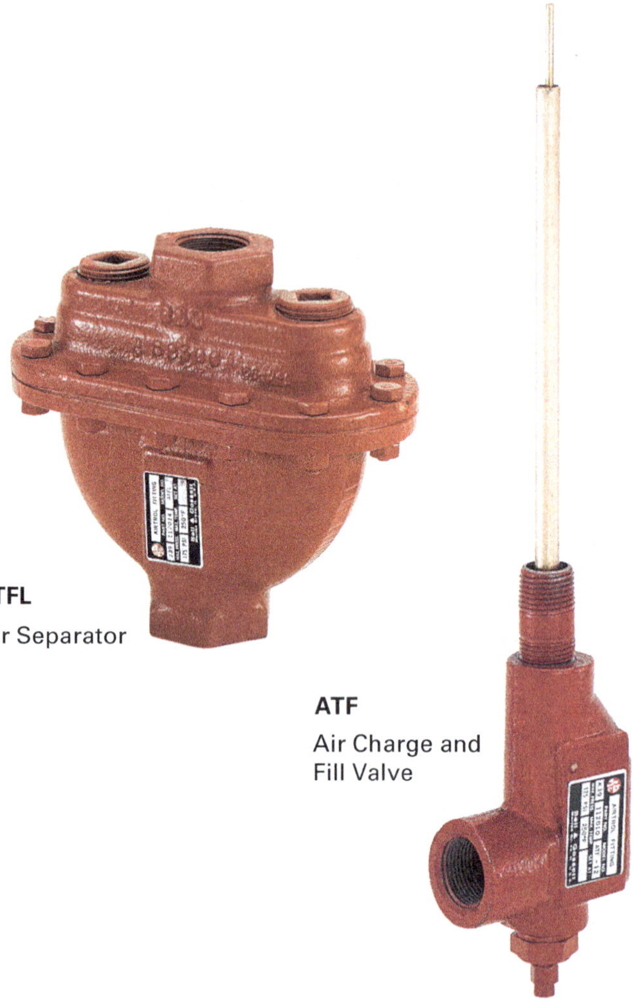

Figure 31 Expansion tank air control devices.
Source: Courtesy of Xylem Inc.

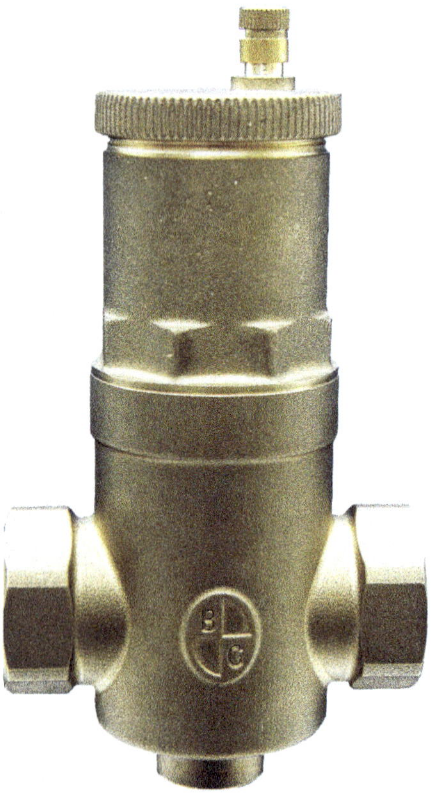

Figure 32 Air separator, or deaerating valve.
Source: Courtesy of Xylem Inc.

The air separator collects the tiniest air bubbles in the water system and routes the air to the top of the expansion tank. It also prevents water from flowing out of the tank as it cools, becoming denser than the water in the system below it. The tank is pre-charged with air and water through the other valve.

Another type of air separator is often used with wall-mounted boilers and diaphragm tanks. The valve shown in *Figure 32* is placed near the hot water outlet of the boiler. It captures the microbubbles in the water as it passes and vents the air to the atmosphere. A common air vent isn't designed to separate air from moving water as these valves do.

On Site

Make-Up Water and Corrosion

Air contains oxygen, which is an essential component of corrosion. A typical closed hot-water heating system contains some microbubbles of air, but heating the water eventually drives the microbubbles out.

If there are leaks in the system that cause fresh water to be introduced consistently, the fresh oxygen it contains initiates corrosion. A properly operating closed hot-water heating system requires very little makeup water during operation. Less makeup water means less oxygen, minimizing corrosion. For this reason, small leaks must not be ignored, even though the water is automatically replaced by the pressure-reducing makeup water valve.

A system without water is full of air. When it is initially filled, large volumes of air collect at the high points of the system. This air must be purged from the system. This can be done with manual or automatic air vents installed at the high points (*Figure 33*). Since a small amount of water may be discharged when air is released from the top, a $\frac{1}{4}$" copper tube is usually attached to the top and routed to an appropriate drain.

Low–Pressure, 1/8" Inlet | High–Pressure, 3/4" Inlet | Medium–Pressure, 3/4" Inlet

Figure 33 Automatic air vents.
Source: Courtesy of Xylem Inc.

On Site

Air Separators for Large Systems

Large hydronic systems require more air separation than a small air separator can handle. The air separator shown is available for pipe sizes 2" to 36". Its centrifugal flow pattern creates a low-pressure vortex in the center that separates both microbubbles and particles from the water.

Source: Courtesy of Xylem Inc.

2.4.3 Pressure-Reducing Valves

Pressure-reducing valves can be used in hydronic systems anywhere it is necessary to reduce a higher water pressure to a lower pressure. However, the valves presented here are used to reduce domestic water pressure to the design pressure of the hydronic system, and automatically admit water as needed to maintain that pressure. A common scenario is a reduction from about 75 psig to 12–15 psig.

Figure 34 Makeup water pressure-reducing valve with fast-fill lever.
Source: Courtesy of Xylem Inc.

The pressure-reducing valve (*Figure 34*) is installed on the makeup water line. The valve automatically replenishes water lost through leaks or from draining the system for service. The valve must maintain the water pressure below the setting of the boiler relief valve. Hot-water hydronic boilers and systems are typically limited to a maximum of 30 psig.

Makeup water pressure-reducing valves have a built-in strainer and a check valve to prevent water from flowing out of the hydronic system back into the domestic water lines. It relies on an adjustable spring that pushes against a diaphragm, acting in opposition to the incoming water pressure. It allows water to pass through the valve when the pressure at the outlet drops below its setting.

The manual fast-fill lever (optional) allows the system to be filled more quickly. Once a positive pressure is measured in the system, the fill lever is released, and the valve controls the flow until the pressure setting is reached.

The spring pressure is manually adjusted to set the desired output pressure level. Typically, these valves are factory-set at 12 psig but are adjustable between 10 psig and 25 psig. The factory setting is appropriate for many residential and light commercial applications. In larger applications, the height of the building requires a higher system pressure.

Pressure Relief Valve

A pressure relief valve (*Figure 35*) protects the boiler and system from excessive pressure. Once activated, the valve opens fully, allowing a significant stream of water to quickly pass through. The construction of a typical hot-water boiler allows for a maximum working pressure of 30 psig.

Figure 35 Pressure relief valve.

The pressure relief valve must be piped to discharge the hot water to a safe location or drain. The drain line must be the same size as the relief valve's outlet port, or larger, and it must not be reduced below the outlet size at any point. The flow away from the valve must not be restricted, and no valve may be installed in the line. If a catastrophic control failure occurs and the burner will not shut down, the temperatures and pressures quickly rise beyond the design limits. In such an event, the valve may discharge both hot water and flash steam.

The installation of pressure relief valves should always follow the manufacturer's directions and conform with local codes. These valves are usually installed in an upright position at the top of the boiler, or in the makeup water line. Combination makeup water/relief valve assemblies are often used on smaller systems (*Figure 36*).

Figure 36 Combination makeup water/relief valve assembly.
Source: Courtesy of Watts

WARNING!

Maintaining proper boiler water pressure is crucial to the safety of nearby personnel, as well as to the boiler itself. Excessive pressure can cause unexpected and explosive failure of components, leading to serious personal injury. Properly selected and maintained pressure relief valves are essential. Ensure that all relief valves are piped to a safe location that does not expose personnel to injury during a discharge.

The pressure setting of a relief valve is not field-adjustable and must not be altered. Adjustments could result in valve failure. Replace a relief valve if it does not open when its pressure rating is exceeded or if it leaks when the boiler is operating normally. Never attempt to field service or repair a critical safety component like a pressure relief valve.

Pressure relief valves require intermittent testing during operation or after long periods of inactivity, according to the manufacturer's instructions. Large relief valves may need to be removed and sent to a certified testing site for periodic reconditioning and testing. Insurance companies often require periodic lab testing and calibration to validate coverage.

On Site

Reusing Boiler Parts

When replacing a boiler, some system components are safe to reuse. However, never reuse a pressure relief valve. Relief valves are relatively inexpensive, vital safety components.

When replacing a boiler, install a new relief valve as specified by the boiler manufacturer. Depending on the manufacturer, the new boiler may come with the relief valve already installed.

2.4.4 Zone Valves

Zone control valves (*Figure 37*) manage the flow of water through each zone. Usually these valves are two-position, thermostatically controlled valves. Zone control valves can be either *two-way* or *three-way* valves. Two-way and three-way simply refer to the number of ports.

Zone control valves may also contain a switch to control the circulator or another load as it opens and closes. Many zone valves are normally open and driven closed so that if the valves fail, the valves fail open to allow flow.

Zone valves can operate by heat or with an electric motor. In a heat-operated valve, a resistance wire around the valve heats the valve's bimetal element when the zone thermostat calls for heat. This causes the bimetal element to expand and slowly open the valve. When the zone thermostat reaches temperature, the bimetal element cools, allowing the valve to close slowly.

Electric-motor operated valves also open and close the valve slowly, using a small motor with gears.

Zone valves open and close slowly to reduce noise and prevent water hammer due to abrupt changes in water velocity. The time required for a typical zone control valve to open and close ranges from about 15 to 30 seconds.

Most zone valves can be opened manually in the event of a power failure. This feature is also useful when troubleshooting a zoned system.

Figure 37 Two-way zone control valve.
Source: Courtesy of Taco, Inc.

2.4.5 Backflow Preventers

A backflow preventer (*Figure 38*) protects the domestic water supply from contaminated boiler water, which would render it unfit for drinking. The valve consists of two inline, independent check valves with an intermediate relief valve. Although the makeup water valve is usually equipped with a check valve, this does not provide sufficient protection for many systems.

A reverse siphoning effect occurs if the inlet pressure to the system falls below the pressure in the boiler system. This can happen if the main water supply to the building is turned off and the pressure relieved for a plumbing repair. The backflow-preventer relief valve opens to permit air to enter and break the siphon.

State and local codes govern the installation and maintenance of backflow preventers, including the craftworker certification to perform these tasks. Most states require certification to service these valves. Always check the state and

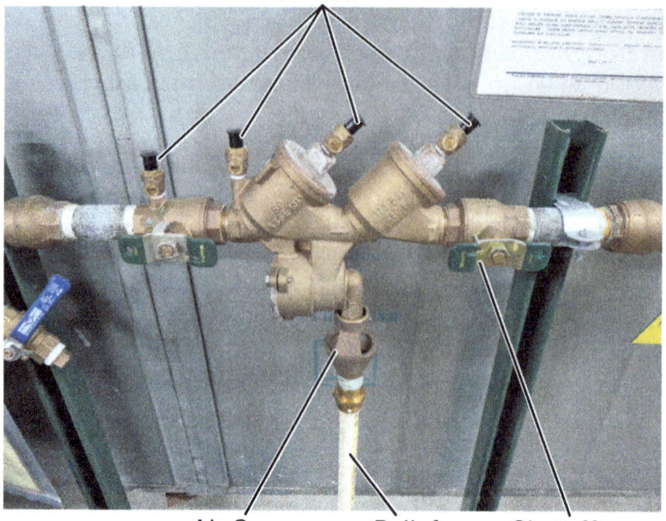

Figure 38 Backflow preventer.

local codes before testing, installing, or working on a backflow preventer. Some states and insurance companies require annual service and testing.

2.4.6 Check Valves

Check valves (*Figure 39*) allow water to flow in only one direction. One common check valve, called a *swing check*, has a swinging flapper that swings open to allow

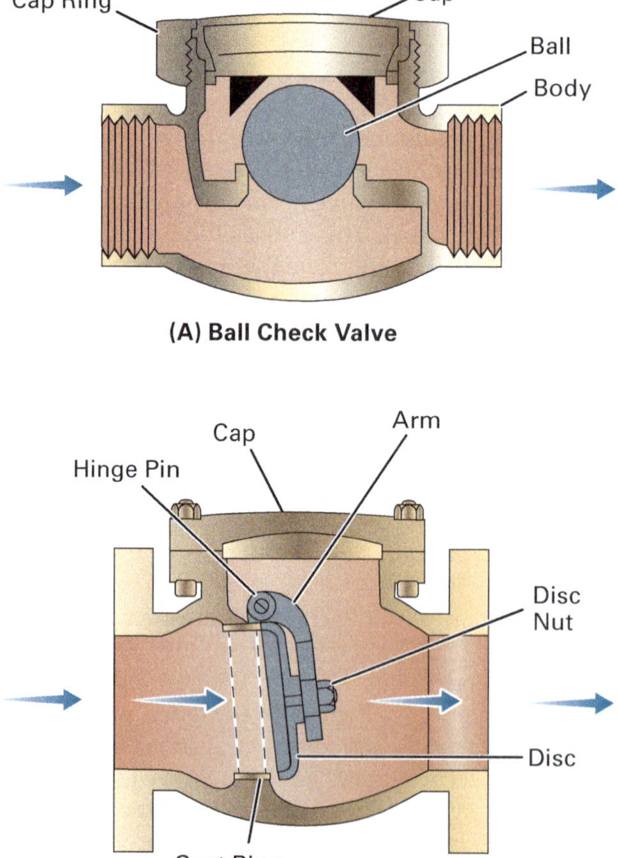

Figure 39 Check valves.

flow in one direction and closes if the flow reverses. This type of check valve must usually be installed in a horizontal line with its hinge at the top. However, some check valves are lightly spring-loaded to help in closing the flapper.

Ball check valves rely on a ball falling into a seat when the direction of flow is wrong. They may also be spring-loaded.

Check valves can be used anywhere in a system where it is necessary to prevent reverse flow. For example, when installed in the piping of a hot-water system, the check valve allows hot water to flow only when the circulating pump is running. When the pump is off, it prevents any backflow caused by gravity or thermal circulation.

2.4.7 Hand Valves

Hand valves generally enable component isolation, but their individual construction characteristics may make one type better suited for an application than another. Hand valves include the following types:

- Gate valves
- Globe valves
- Ball valves
- Angle valves

Gate Valves

Gate valves (*Figure 40*) enable or disable the flow of water. Gate valves have a gate-like disc that slides vertically across the flow path. In its fully open position, the gate is out of the way, and the resistance to water or steamflow is minimal. In its fully closed position, the gate seats tightly and flow stops.

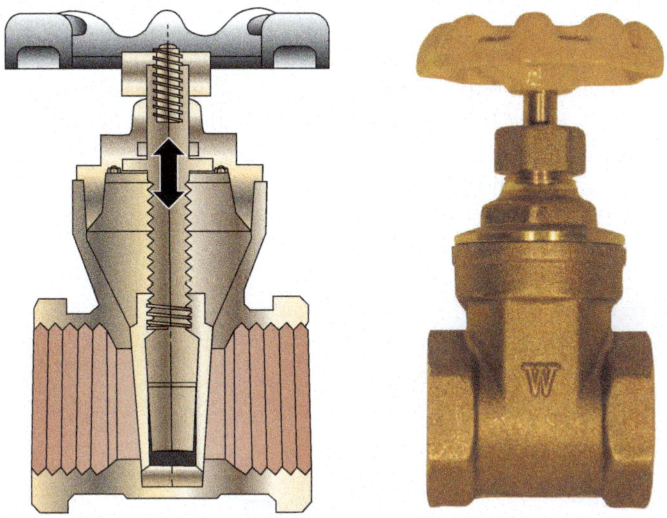

Figure 40 Gate valve.

Gate valves have a straight-through flow pattern. There is no damming effect that might trap sediment inside and interfere with closure. However, these valves are a poor choice for throttling or balancing water flow. Thus, gate valves should either be fully open or fully closed. If opened partway, the turbulence of the flowing water wears the seating surfaces of the valve.

Gates valves do not generally have a preferred direction of flow, meaning that either end can serve as the inlet.

Globe Valves

Globe valves (*Figure 41*) offer a method to adjust flow, within limits. In any given position, though, the volume of flow changes with system pressure variations.

Globe valves are the best type of hand valve to use for throttling water flow in a system. This means that it can be adjusted to some position to reduce flow.

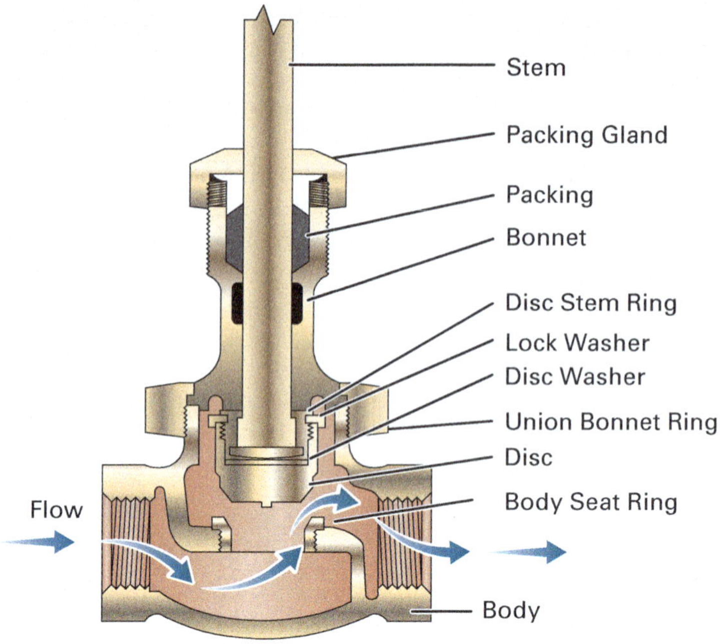

Figure 41 Globe valve.

Flow through a globe valve moves in a Z-pattern that allows throttling without excessive wear on the valve seat.

Globe valves are directional. The incoming flow should always be directed to the bottom of the valve seat as shown in *Figure 41*.

Ball Valves

Ball valves (*Figure 42*) are mainly used for open/closed service. As the name implies, a ball valve has a ball-shaped element with a large opening through its center. The opening is the same size as the nominal size of the valve, providing smooth and unimpeded flow.

Although gate valves and globe valves take a number of revolutions to open and close, ball valves only require a 90-degree turn to move from fully open to

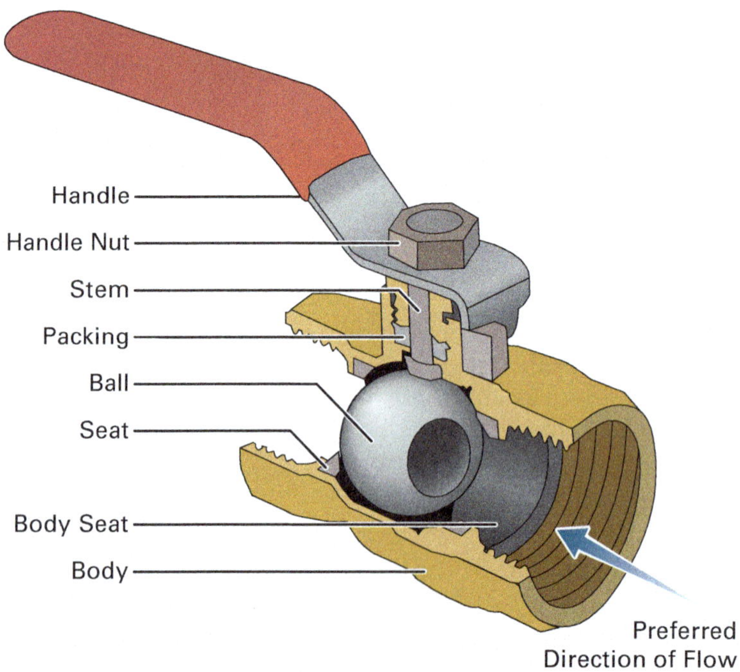

Figure 42 Ball valve.

fully closed. As the ball rotates through the 90-degree arc, the hole moves from being fully aligned with the inlet and outlet ports to being at a right angle to them, blocking flow completely.

Unlike gate valves, ball valves can operate in a partially open/closed position. While throttling is an option, a ball valve is not as efficient at throttling flow as a globe valve, unless the port has a specific shape to enhance it for that purpose. But ball valves are a far better choice for throttling service than gate valves.

Ball valves are easily recognized by the lever-type handle. The position of the lever relative to the centerline of the valve indicates the position of the valve. When the handle is parallel to the line, the valve is fully open. Although either end can serve as the inlet, ball valves should be installed so that the handle points in the direction of flow as shown in *Figure 42*.

Angle Valves

An angle valve (*Figure 43*) is a form of globe valve. Its design features an outlet port that is at a 90-degree angle to its input port. Angle valves are common in various applications where a globe valve and 90-degree elbow would generally be needed. The end opposite the handle should always be used as the inlet.

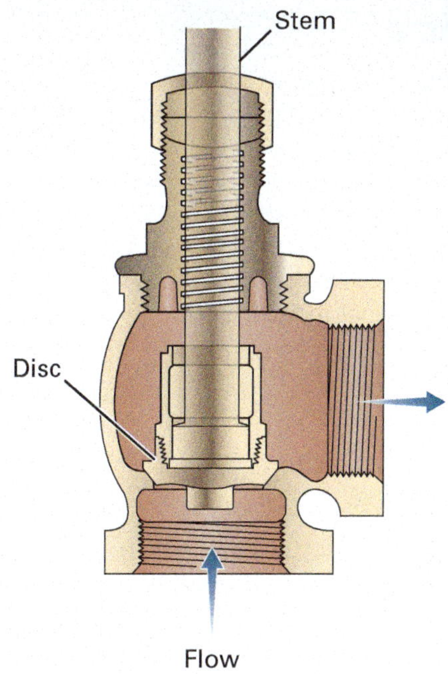

Figure 43 Angle valve.

2.4.8 Two-Way and Three-Way Control Valves

Two-way and three-way control valves control the flow and/or temperature of water, depending on the application. The operation is controlled by a signal applied to the valve's actuator. Valves are commonly modulated to various positions by current or voltage signals. For simple open/close control, the actuator can simply be energized with the correct voltage, or de-energized.

For two-way valves, the water enters the inlet port and exits through the outlet port, as shown in *Figure 44A*.

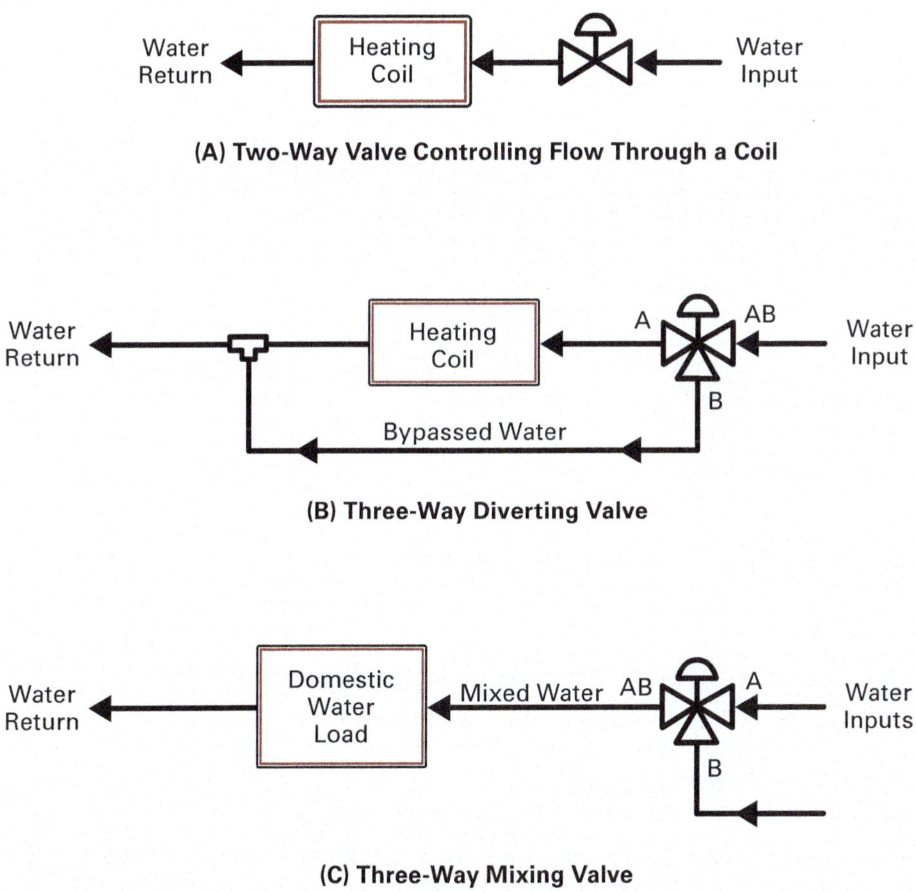

Figure 44 Two-way and three-way valves.

NOTE

Although *Figure 44* depicts the valves installed on the supply-water side of the coil, these valves are often placed on the return-water side of the coil instead. The system designer determines the position. The use of a mixing or diverting valve also depends on the designer and system performance requirements.

Three-way valves can be used to divert one water stream or mix two water streams. Note that, externally, mixing and diverting valves look identical. It is important to note that trying to convert between mixing and diverting duty merely by how the pipes are connected should not be attempted. Diverting and mixing valves are designed and cast specifically for one task or the other.

When a three-way diverting valve is used, it has one inlet and two outlets (*Figure 44B*). It splits a single water stream between two outlets. The position of the valve determines the percentage of the water that goes in either direction. Alternatively, it can direct 100% of the flow to either outlet. Typically, one outlet directs water through a coil or other load. The second outlet provides a bypass around the coil.

A three-way mixing valve has two inlets and one outlet (*Figure 44C*). It mixes two incoming streams of water into a single outgoing stream. The percentage of each incoming stream passing through is determined by the valve's position. The valve can also be positioned to allow 100% of either water stream to pass.

CAUTION

Always ensure that the correct valve type (mixing or diverting) is installed, and that the valve is installed in the specified position. Using the wrong type of valve and/or installing it in the wrong position will impede or prevent proper operation.

Mixing valves are often applied to domestic water systems, mixing hot- and cold-water streams to ensure the delivery of a consistent and safe domestic hot-water temperature. For example, a facility may generate 180°F water for dishwashing but provide a mixed water stream at 130°F to serve lavatories or other loads. Accuracy is very important, as 180°F water can cause significant harm. Diverting valves are more often required in hot-water heating system design.

2.0.0 Section Review

1. True or False? Forced hot-water systems rely on thermal circulation as well as pumps to move water through the system.
   a. True
   b. False

2. What is the maximum working pressure for most low-pressure hot-water boilers?
   a. 10 psig
   b. 12 psig
   c. 30 psig
   d. 45 psig

3. Which of the following is associated with pump cavitation?
   a. A significant increase in discharge pressure
   b. Unusually silent pump operation
   c. Excessive head at the pump outlet
   d. Water vaporizing at the pump inlet

4. What type of valve is best for manually throttling water flow in a hydronic heating system?
   a. Globe valve
   b. Check valve
   c. Gate valve
   d. Zone valve

3.0.0 Hydronic Piping Systems

Objective

Identify various hot-water heating piping layouts and the terminal devices used.

a. Describe the characteristics of one- and two-pipe distribution systems.
b. Describe hot-water zoning systems.
c. Identify various hot-water heating system terminal devices.

Performance Tasks

There are no Performance Tasks in this section.

Water piping systems distribute hot water, chilled water, or both. Once the water reaches the conditioned space, a terminal device transfers heat into or out of the water. This section presents two general classifications of piping systems: one-pipe systems and two-pipe systems.

3.1.0 One- and Two-Pipe Distribution Systems

Although the use of independent terminal devices to deliver or absorb heat is very common, radiant floor heating systems simply use long runs of tubing as an extended terminal device. Since radiant floor heating systems are somewhat unique and can be laid out as either one- or two-pipe systems, these systems are also presented in this section.

3.1.1 One-Pipe Systems

One-pipe water systems are used primarily for hot-water heating purposes. One-pipe systems can be configured as a *series loop* or as a *single loop* system.

A series-loop system has a continuous run of pipe from the supply connection to the return connection. In a series-loop system, all the hot water flows through each heating terminal. Neither the flow rate nor the temperature of the water can be changed without affecting the entire loop.

When the system thermostat calls for heat, the hot water flows from one terminal in the loop to the next. In doing so, it gives up more and more heat as it flows through each terminal. While this design is relatively inexpensive to install, a disadvantage is that rooms receiving the heat first tend to overheat, while those at the end of the loop tend to be cold.

Despite the fact the series-loop systems aren't very practical for a complete system, one or more series loops may be used as branches of a system. Series-loop systems can connect directly to main piping runs, or the loops can run directly to and from the boiler.

The pipe diameter used in a series loop remains the same throughout the loop. The size and length of the series loop piping and the volume of heat to be exchanged are factors in determining the water flow rate, pressure drop, and water temperature drop. *Figure 45* shows a two-circuit series-loop system. The dividing points between the two loops are where the boiler supply and return piping tap into the loop piping. Series-loop systems are common in residential systems.

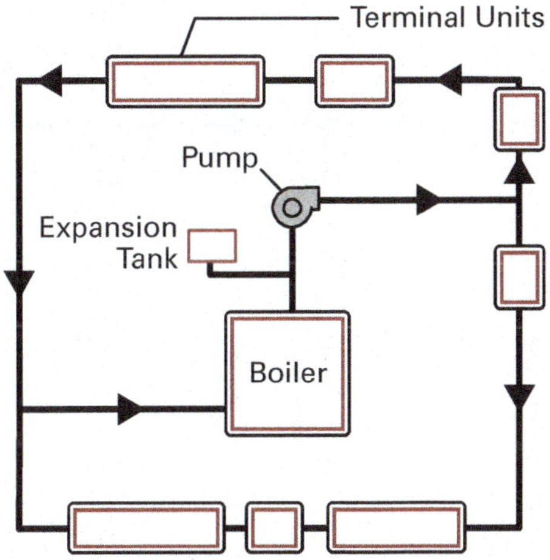

Figure 45 Series-loop, two-circuit hot-water system.

NOTE

Because the focus is on basic system layout, many system components have been omitted from the diagrams in this section.

Single-loop systems (*Figure 46*) place each terminal in its own branch circuit. The piping in each branch is smaller than the main piping. To route hot water through each terminal, a special supply and return tee may be used to connect the terminal to the main piping. The supply tee is a special tee called a *diverting tee*, which is also referred to as a *one-pipe* fitting or a *monoflow* fitting (*Figure 47*). This tee creates a pressure drop in the main flow that forces some of the water flowing in the main to be diverted through the terminal. The rest of the water continues flowing in the main supply line. As the water exits the terminal, it simply joins the rest of the water moving past.

The terminals and diverting tees are matched in design to minimize pressure drop. Some single-loop systems also use diverting tees on the return at each terminal to overcome excessive resistance in piping and terminals.

Because the water returning to the main from each terminal is cooler than the water in the main, the water in the main gets progressively cooler as more terminals empty into it. As a result, water entering each successive terminal in the system will be cooler than the water in the terminal before it. In a well-designed system, the loss of temperature is not enough to cause comfort problems because the volume of water flowing through each terminal is much less than the total amount of water flowing in the main.

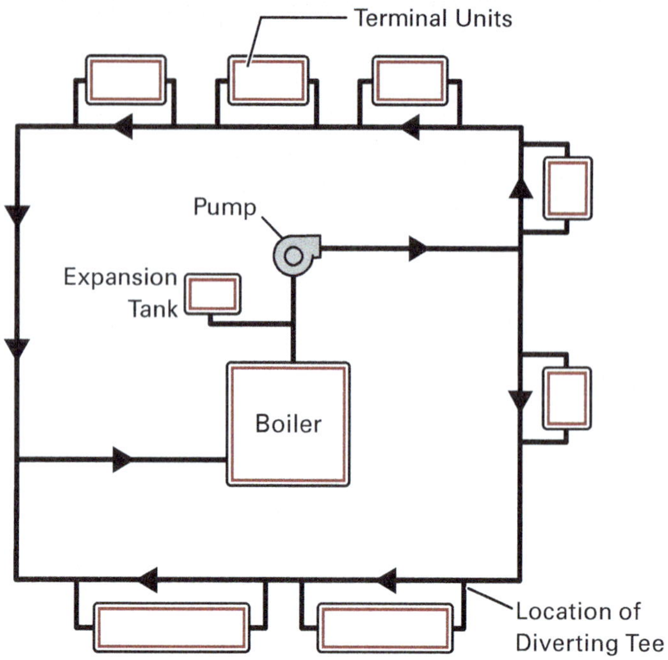

Figure 46 Single-pipe, two-circuit hot-water system.

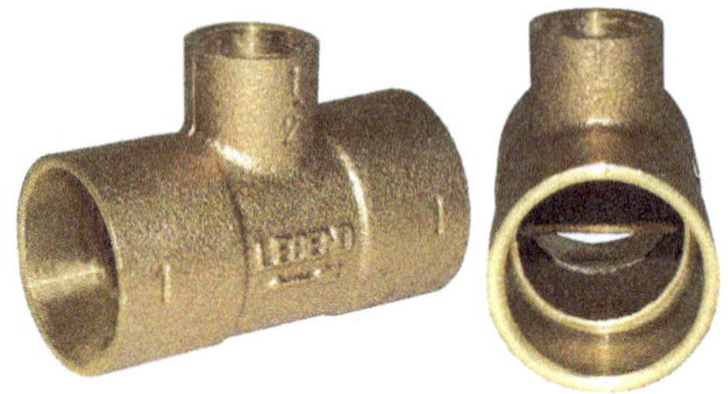

Figure 47 Diverting tee, or monoflow, fitting.
Source: Legend Valve & Fitting Inc

3.1.2 Two-Pipe Systems

Two-pipe systems have terminals connected across separate supply and return mains. Two-pipe systems are classified as either *direct-return* or *reverse-return* systems.

Direct Return Systems

In a direct-return system, the supply and return water flow in opposite directions (*Figure 48*). Also, the return water from each terminal takes the shortest path back to the boiler. This means the water flowing through the terminal nearest to the boiler is the first back to the boiler, because it has the shortest run.

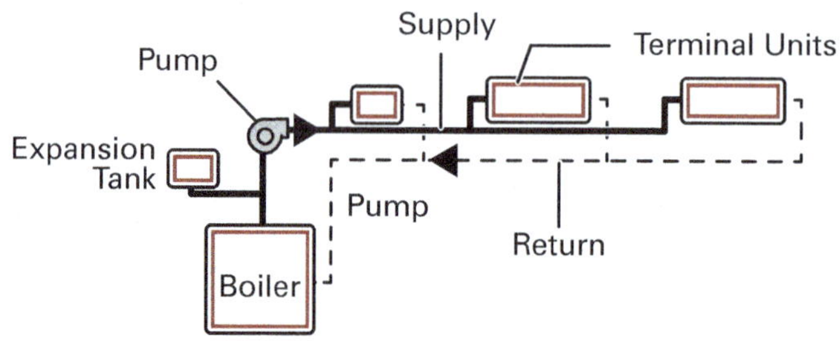

Figure 48 Two-pipe, direct-return system.

The water flowing to the terminal furthest from the boiler has the longest run and is the last to return. This arrangement results in an unequal length of travel for water flow through the terminals. This causes an imbalance in the system pressure drop that often results in poor heat distribution.

To overcome this weakness, balancing valves are normally used on direct-return systems to aid in balancing the system by adding pressure drop to the branches that would otherwise receive too much water. *Figure 49* shows one example of a balancing valve. System balancing is presented in Section 4.0.0 of this module.

Figure 49 Basic balancing valve.
Source: Legend Valve & Fitting Inc

Reverse Return Systems

In the reverse return system, the supply water and return water flow in the same direction through parallel pipe runs (*Figure 50*). In this system, the supply water going to the first terminal is the last to be returned to the boiler and the supply water going to the last terminal is the first to be returned to the boiler.

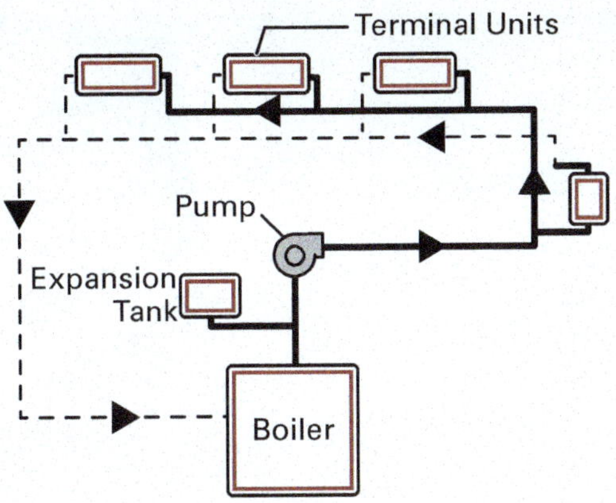

Figure 50 Two-pipe, reverse-return system.

This arrangement equalizes the distance of the supply and return water flow in the system, thus providing a system that is more naturally balanced. For this reason, the reverse return system is more popular than the direct-return system.

3.1.3 Radiant Floor-Heating Systems

Radiant floor heating refers to the use of a heat source in or under a floor. In modern radiant floor-heating systems (*Figure 51*), heated water is pumped through continuous loops of tubing embedded within or fastened beneath the floor. The loops of tubing serve as the terminals.

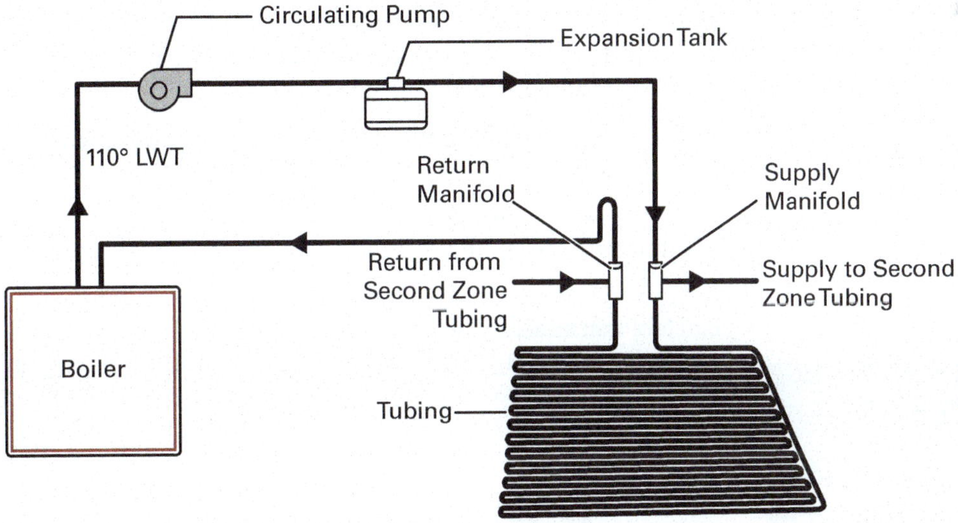

Figure 51 Basic radiant floor heating system.

Figure 52 shows a floor heating system with the supply and return manifolds visible. The red valves identify the supply connections while the blue valves identify return connections. Eight zones are in this system.

The heat from the tubing is absorbed slowly by the surrounding floor materials, and then radiated into the rooms. The result is that the warmth stays near the floor level in an evenly distributed pattern and rises naturally to the occupied zone.

Heating with a floor system is energy efficient and comfortable. Because your feet are in direct contact with the heat source, comfortable room temperatures are achieved at lower thermostat settings. Depending on the application, the water typically only needs to be heated to between 90°F and 110°F (32°C and 43°C), thereby reducing energy costs. Most users feel radiant floor systems provide better comfort than all other types of heating system.

The tubing currently used in radiant floor-heating systems is made from a cross-linked polyethylene material (PEX) or similar polymer-based material. Many types of PEX tubing are available. Hydronic systems require one

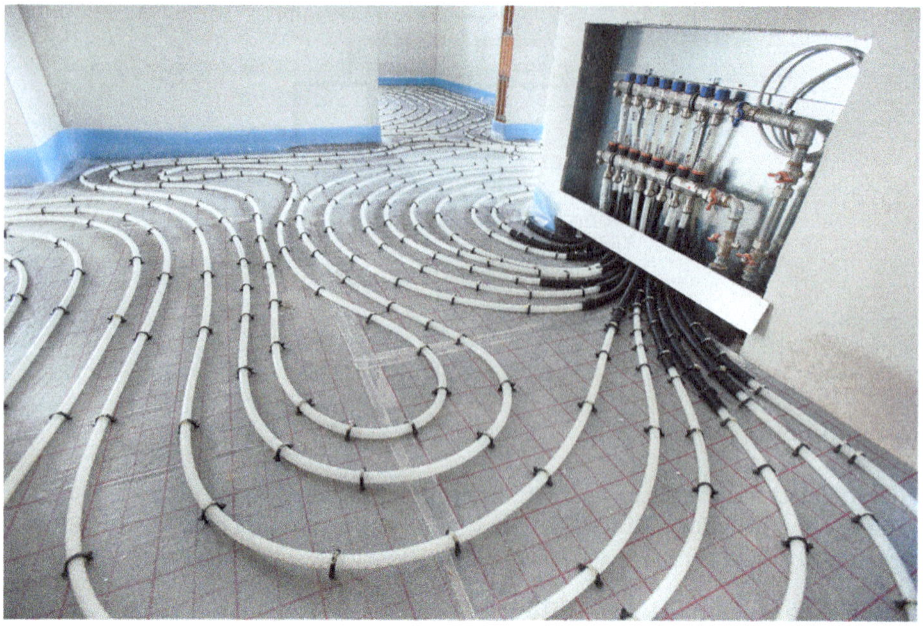

Figure 52 Floor heating system loops and manifolds.
Source: filmfoto/iStock

of several types that are manufactured with an oxygen barrier to prevent the oxygen from penetrating the walls of the tubing. Oxygen-barrier types of PEX include hePEX™ and ThermaPEX®.

The tubing is incorporated into the floors using one of several installation methods. These methods include the following:

- Attaching the tubing to the subfloor and embedding it in a layer of concrete or gypsum
- Mounting the tubing in or below the subfloor
- Attaching the tubing directly to the underside of the subfloor

Figure 53 shows the tubing being stapled in place on the subfloor, using a special tool that attaches a plastic strap to the floor. *Figure 54* shows a special mat that is attached to the subfloor. The tubing is then routed in any pattern desired. Once the tubing is secured, the final layer of flooring (typically concrete) is applied.

Figure 53 Staple-down installation.
Source: Courtesy of Uponor North America

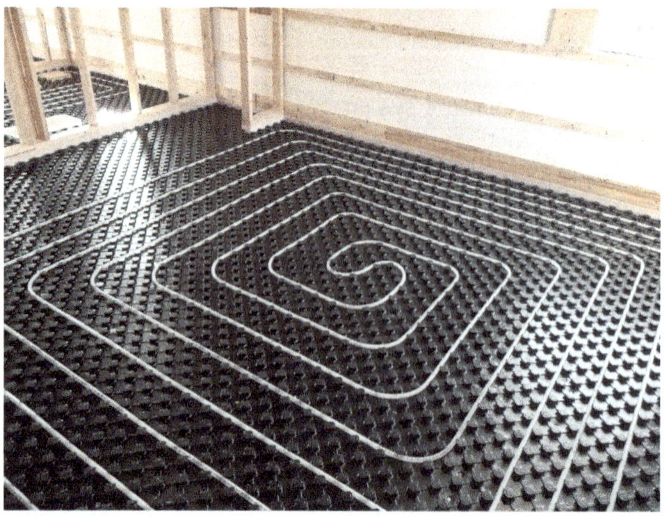

Figure 54 Fast Trak™ mat for tubing installation.
Source: Courtesy of Uponor North America

Because the tubing for a floor heating system must be built into the building's flooring, the installation cost tends to be higher than for other types of heating systems, such as forced-air or baseboard heating. In cold climates, though, it is a worthy investment. Floor heating systems can also be retrofitted to an existing home in some cases.

Around the World

Floor Heating Systems

Floor heating systems have been used for centuries. Evidence shows that around 1300 BC a radiant floor heating system was used in a Turkish king's palace. Over a period ranging from about 80 BC to 324 AD, the Romans used improved versions of these systems in upper-class homes and public baths throughout their empire.

These systems were not hydronic in nature. They typically consisted of tile-lined flues built beneath stone floors, or simple basements beneath the main floor. The floors were warmed by routing the heat from a fire into the flues or basements at one end of the floor. The smoke and heat passed through the flues, then out through a chimney at the opposite end of the floor.

3.2.0 Hot-Water Zoning

To enhance comfort and flexibility, a method of zoning hot-water systems is needed to accommodate different loads throughout a structure. Systems can be easily zoned using multiple circulators and/or zone valves.

Systems using zone valves (*Figure 55*) often have a single pump. Note that an indirect domestic water heater that often uses its own dedicated circulating pump is also shown in the figure. As individual zone thermostats call for heat, the zone valves open and allow water to flow to the zone.

To prevent the pump from *deadheading* (attempting to pump with the flow fully restricted), a differential bypass valve or a three-way valve(s) can be used. The differential bypass valve opens as the pressure difference across it rises to its setpoint, indicating that some or all the two-way zone valves are closed. A three-way zone valve opens its bypass port when the zone temperature is satisfied, acting as a diverting valve. Flow through the terminal device(s) in the zone is either reduced or stopped. Systems equipped with three-way valves at each terminal typically have a very consistent flow rate through the system.

Other systems use multiple circulating pumps, often in addition to a boiler circulator pump (*Figure 56*). Each zone is fitted with its own pump, sized to accommodate the flow rate and head pressure through its own zone piping.

The boiler circulating pump is responsible only for maintaining flow through the boiler loop, rather than through the zones. In this case, both the boiler circulator and the zone circulator pump must operate for the system to function.

3.3.0 Hydronic Heating Terminals

No matter what piping scheme is used, heat in the water must be transferred to the conditioned space. Terminals come in a variety of forms, including the PEX tubing found in floor heating systems. The focus of this section is on terminals used for heating only.

In heating-only systems, the transfer of the heat from the terminals to the space results from a combination of radiation and convection. Normally, the terminals are installed at the points of greatest heat loss, such as under windows, along outside walls, and near door openings.

The following terminals are commonly used in heating-only systems:

- Convectors
- Baseboard units
- Unit heaters
- Fan coil and air handling units

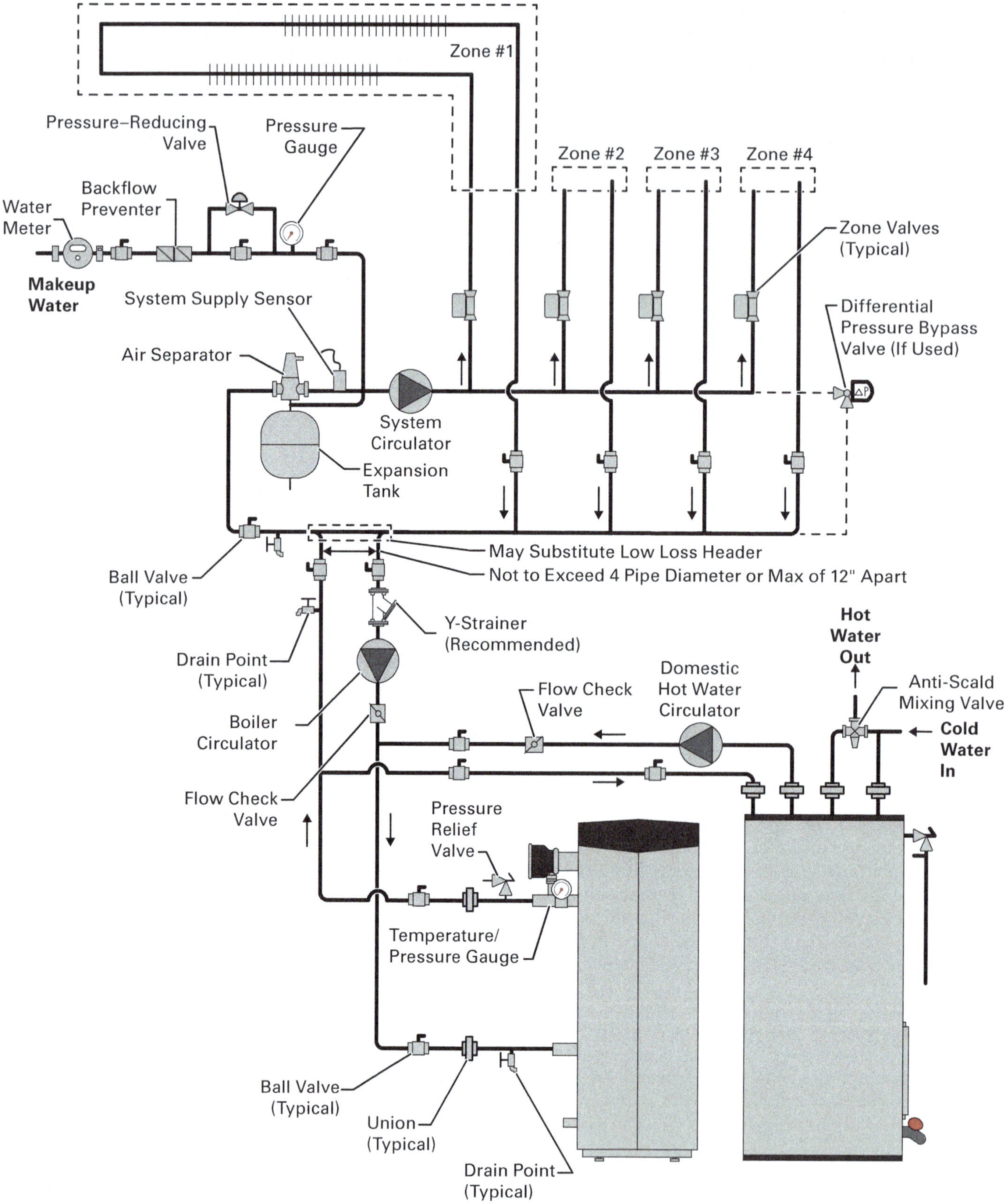

Figure 55 Hot-water system zoned with valves.

Source: Courtesy of Lochinvar, LLC

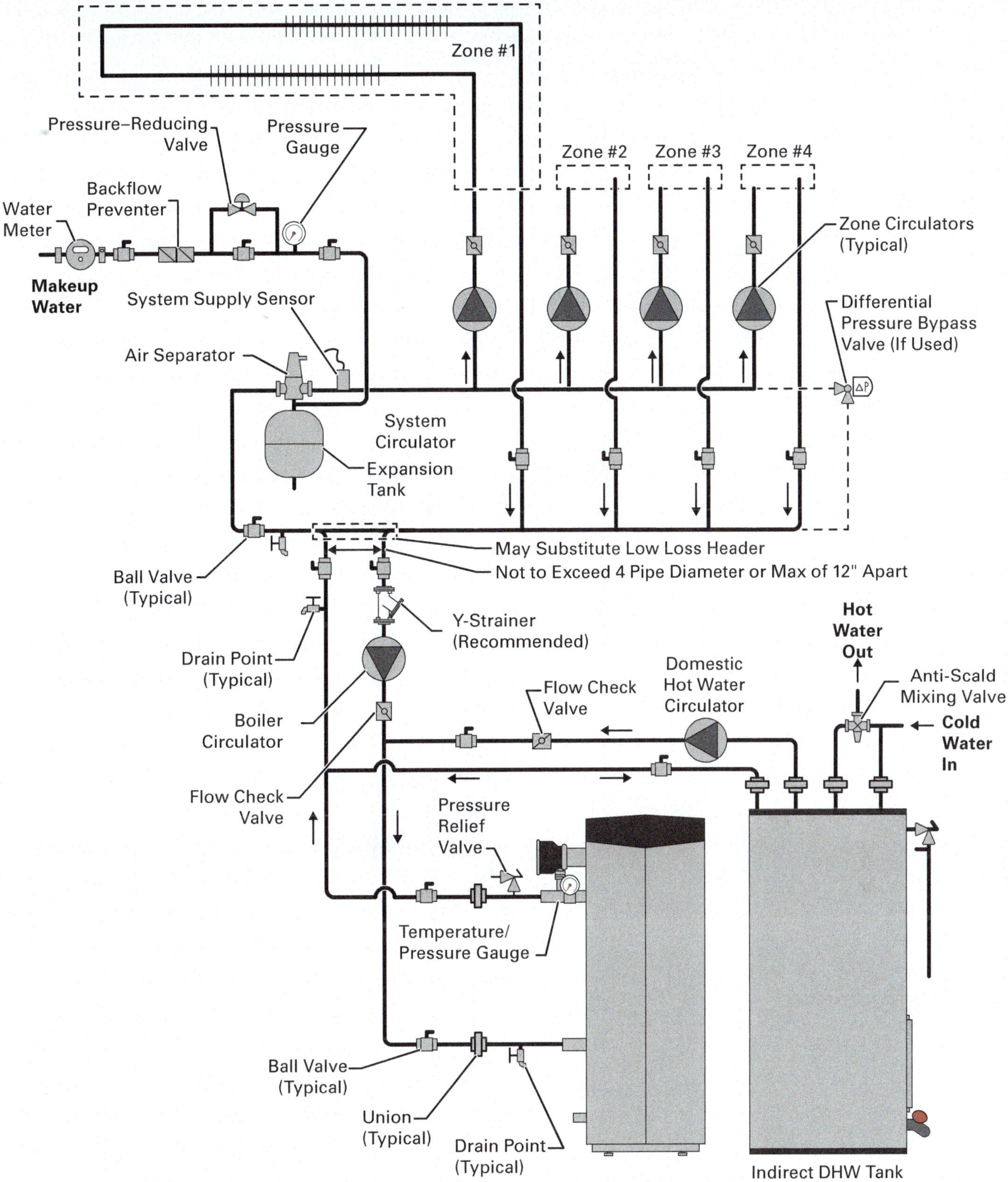

Figure 56 Hot-water system zoned with circulators.
Source: Courtesy of Lochinvar, LLC

A convector (*Figure 57*) has a finned-tube coil(s) mounted in an enclosure designed to enhance the natural, convective flow of air through it. The enclosure can have many shapes. Warm air enters at the bottom, induced to enter by the draft created by the warm coil inside. The rising air passes through the coil and exits the enclosure through a grill at the top.

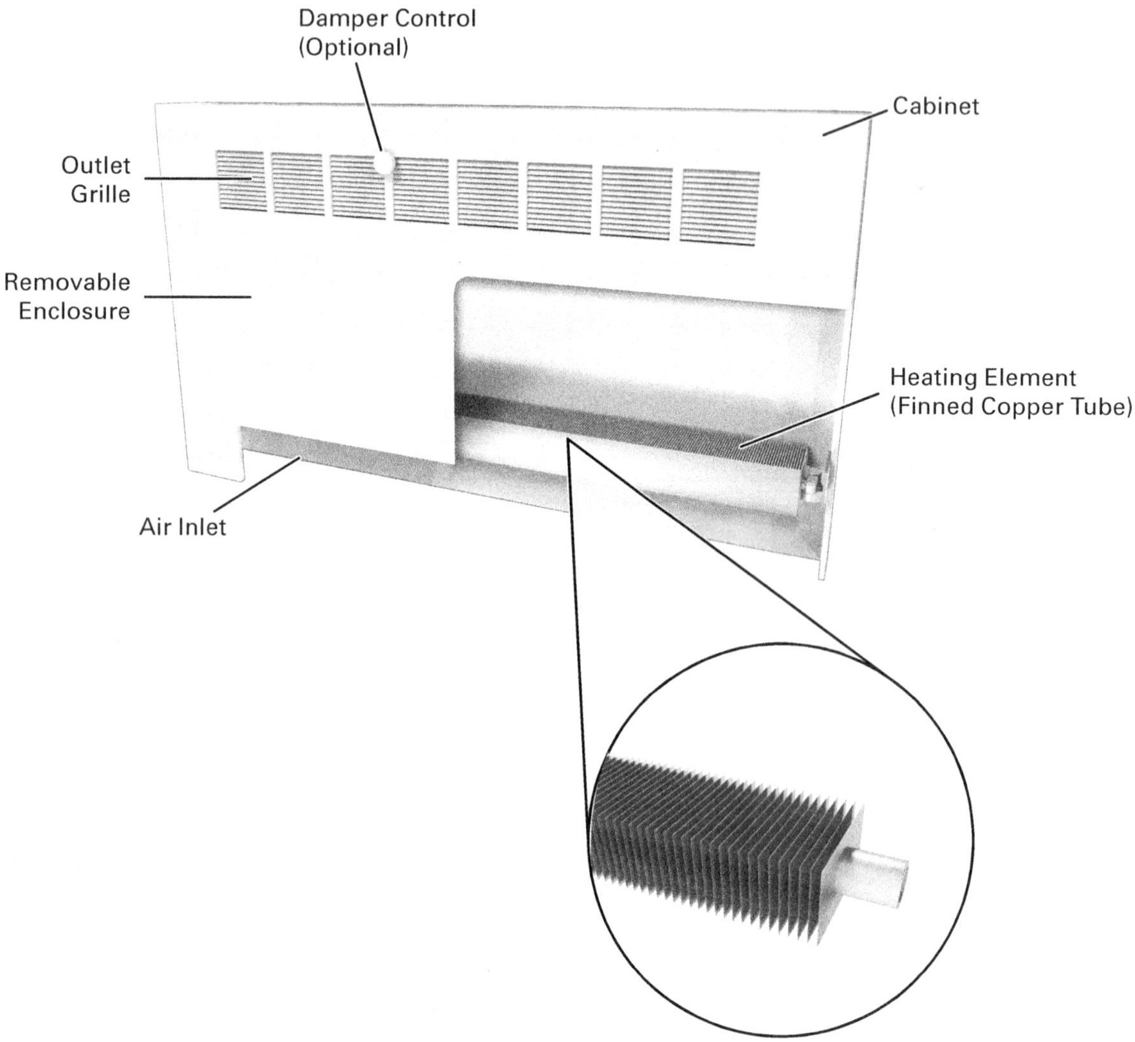

Figure 57 Heating convector.

Baseboard units (*Figure 58*) are mounted on the wall in place of, or over, the usual baseboard. The heating element is the same design as those found in convectors; baseboard units can be thought of as short convectors. However, because baseboard units are significantly shorter, they do not develop as much draft as a convector. Thus, baseboard units have less heating capacity than convectors per foot of length. Baseboard units are usually installed in a continuous run along the outside walls of a room.

A typical unit heater (*Figure 59*) houses a hot-water coil and a circulating fan. The fan blows room air across the heat exchanger. The heat exchanger is made from copper tubing with aluminum fins.

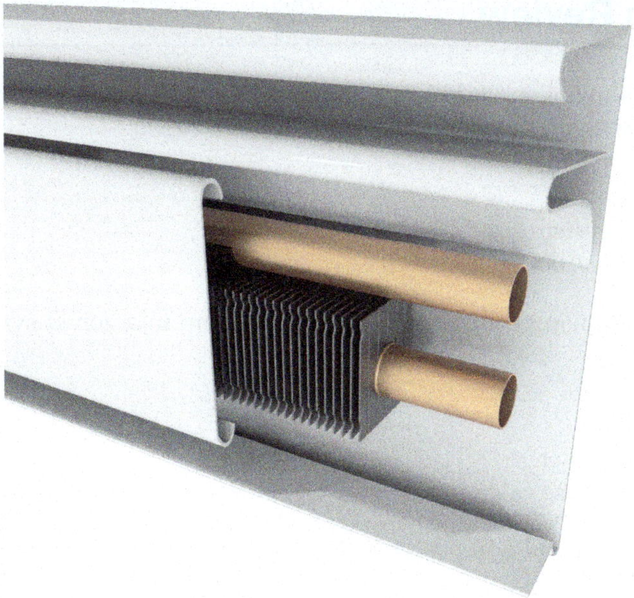

Figure 58 Finned-tube baseboard unit.

Figure 59 Hydronic unit heater.
Source: Courtesy of Beacon Morris/Mestek, Inc.

Unit heaters offer substantial heating capacity for the size and can move heated air a considerable distance. They are used mainly in garages, factories, and warehouses. Both horizontal and vertical units are available. Horizontal units are best used in areas with ceiling heights up to 16'. Vertical units are normally used in areas with higher ceilings. Floor-mounted models that look like convectors but have the advantage of forced airflow can be used in residences and commercial buildings.

Hot-water coils are routinely inserted in fan coil units, air handling units, and variable air-volume (VAV) units (*Figure 60*). Supply and return water connections are made outside the cabinet. Temperature control is achieved by controlling the water flow with a zone valve or similar arrangement.

Figure 60 Hot-water coil in a VAV box.

3.0.0 Section Review

1. Which of the following is a water piping system in which all the hot water flows through each of the terminals?
   a. Single-loop system
   b. Series-loop system
   c. Parallel-loop system
   d. Direct-return system

2. Which of the following components prevent a pump in a zoned system from deadheading?
   a. Check valve
   b. Balancing valve
   c. Backflow preventer
   d. Differential bypass valve

3. True or False? A vertical unit heater is best applied in areas with a ceiling height greater than 16'.
   a. True
   b. False

4.0.0 Water Flow and Balance

Performance Task

1. Measure the pressure drop across a balancing valve and determine the flow rate.

Objective

Describe hydronic system pump selection and explain how water flow is monitored and balanced.

a. Identify devices used to measure and control water flow.
b. Describe how circulating pumps are selected.
c. Explain how to measure pressure and water flow in an operating system.

In hydronic systems, *water balancing* refers to the adjustment of water flow through various parts of the system, especially the terminals. The satisfactory distribution of water depends on a well-designed piping system and properly chosen components. A balanced water system also ensures the correct volume of water is being moved through the boiler or chiller. Too much or too little water flow through the primary equipment can significantly affect performance and lifespan.

After a hydronic system is installed, it must be balanced to ensure it meets design expectations.

4.1.0 Water Flow Measurement and Control Devices

Balancing the water flow through a system requires the adjustment of various balancing devices to add additional head, or pressure drop, to a given terminal or section of piping. Adjustments are made based on measurements taken at various points of the system.

4.1.1 Water Flow Measurement Devices

Larger systems require the flow rate in one or more locations to be accurately monitored. Venturi tubes and orifice plates are commonly used as points to measure the flow in essential locations (*Figure 61*). Both devices place a fixed-area restriction in the pipe to create a predictable pressure drop that can be measured to determine flow rate. The venturi tube is considered the more accurate of the two devices.

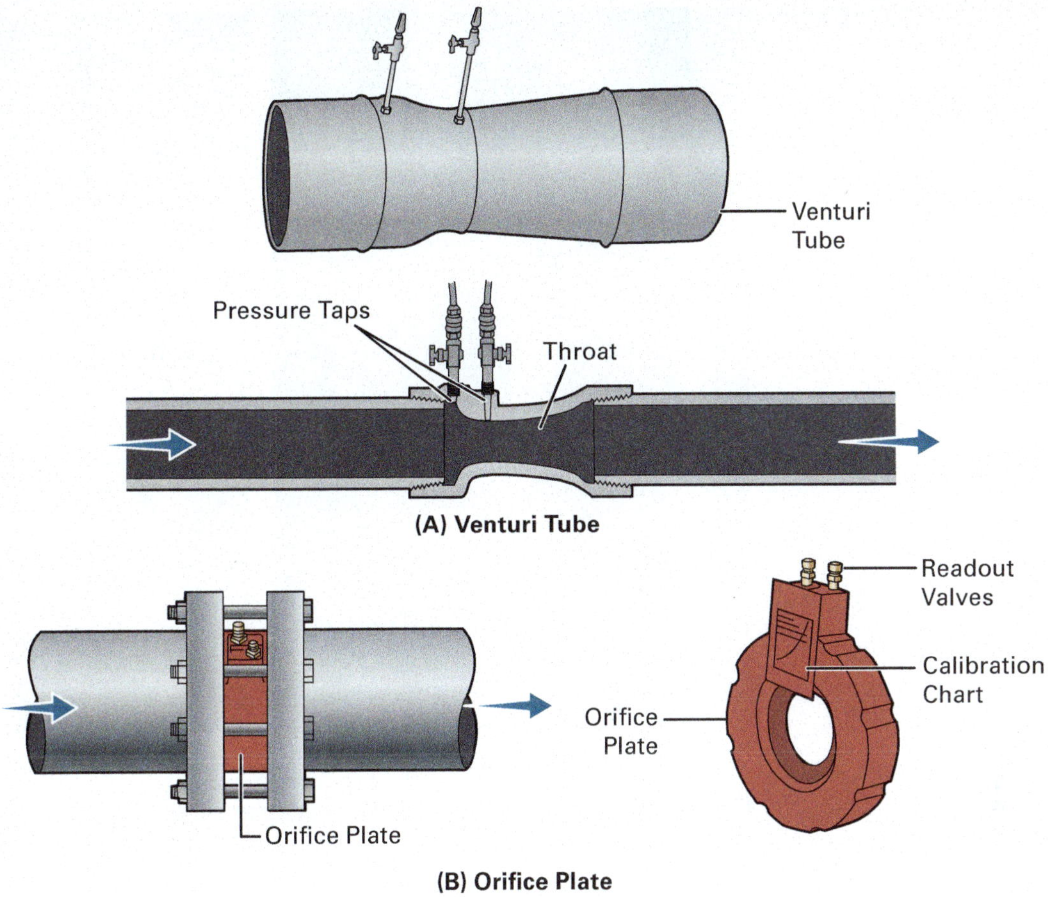

Figure 61 A venturi tube and an orifice plate.

The pressure drop, or *differential pressure,* is measured across the two ports. Using a chart that accompanies the tube or plate, the pressure drop is used to determine the flow rate.

The pressure differential across these devices is measured using a manometer or a differential pressure gauge. Although simple differential pressure gauges are still in use, more advanced instruments are popular (*Figure 62*). A hydronic manometer kit can measure pressure in inches of water, feet of water, pounds per square inch, inches of mercury (in.Hg), and metric values. The differential pressure measurements allow the calculation of the flow rate. Alternatively, the information from the venturi tube or orifice plate can be programmed, allowing the instrument to read the flow rate directly. The data gathered can also be downloaded and saved for later review.

Figure 62 Hydronic manometer kit.
Source: Image courtesy Dwyer Instruments, LLC

4.1.2 Balancing Valves

Figure 63 shows several balancing valves equipped with ports to measure pressure drop. The pressure drop changes as the valve position is changed.

A *triple-duty valve* (*Figure 63A*) is designed for installation at the outlet of a centrifugal pump and provides three functions. In addition to its balancing function, a triple-duty valve serves as a shutoff valve and has a built-in check valve to prevent reverse flow. Similar valves may also be referred to as *multipurpose valves.*

Figure 63B and *Figure 63C* show smaller balancing valves. Note that each valve has a scale to show the setting. In a new installation, the valves are often positioned to a setting determined by the engineer, then set to their final value during the balancing process. Because these valves can also provide shutoff for system service, it is important to note the position so the valve can be reset when the related work is complete.

To read the flow rate through a balancing valve, you must have a chart of the valve's characteristics. The chart (*Figure 64*) shows the flow, in gpm, based on two factors—the setting on the valve scale and the reading taken across the two ports, in feet of water. For example, assume you have a CBV-1 valve, which matches the chart in *Figure 64*. At a valve setting of 10 and a pressure differential of 20' of water, the flow rate is about 12 gpm.

(A) Triple-Duty Valve

(B) Threaded Circuit Setter™

(C) Flanged Circuit Setter™

Figure 63 Balancing valves with flow measurement ports.

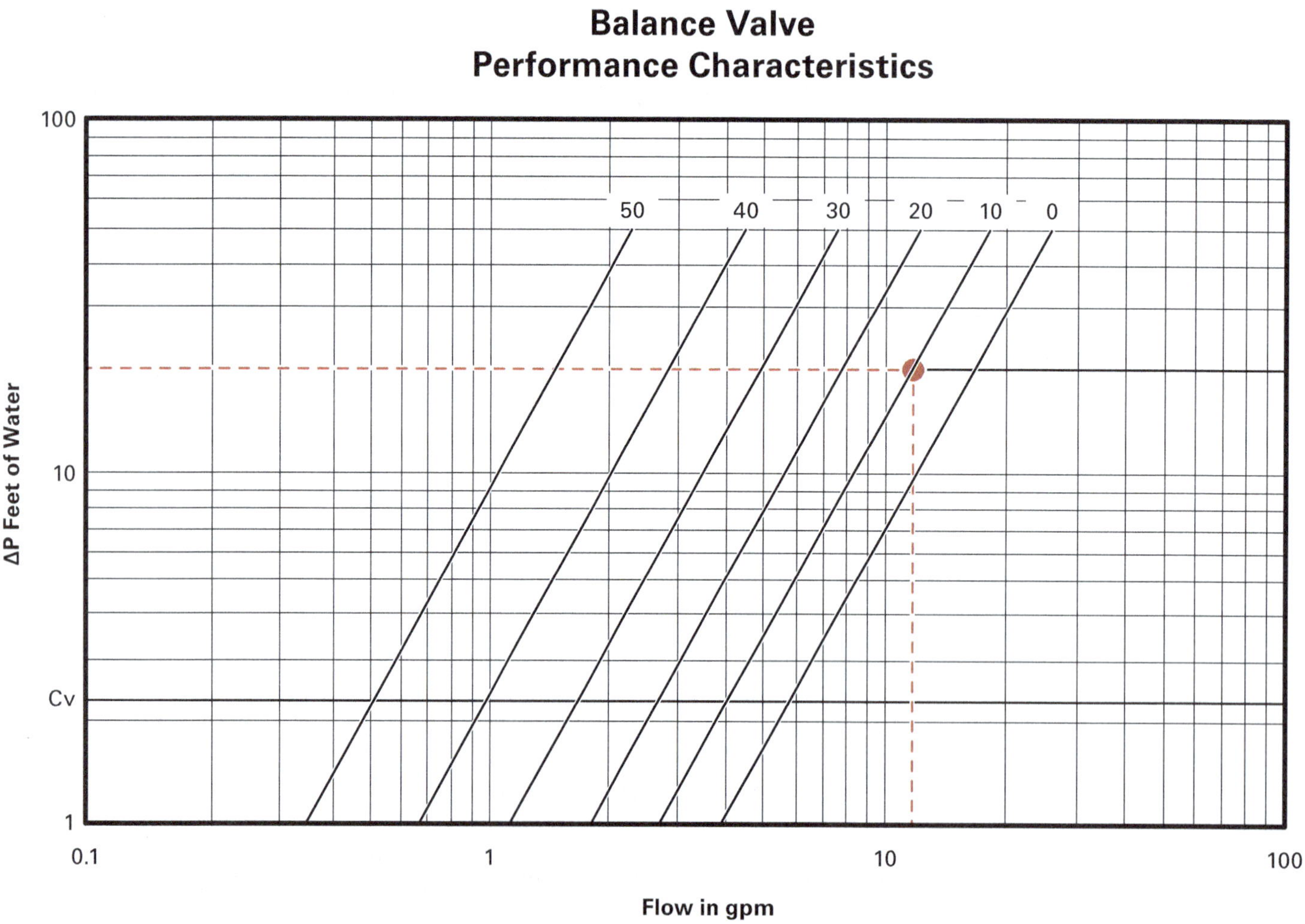

Figure 64 Balancing valve performance chart.

Note that balancing valves like this rely on consistent pressure to remain accurate. When the differential pressure developed by the pump changes, the flow rate through the balancing valves also changes.

Another device offers precise flow control regardless of pressure variations in the hydronic system. A Belimo Pressure-Independent Characterized Control Valve™ (*Figure 65*) is used in commercial hydronic systems with variable-speed pumps. Once the flow rate is set and the valve is energized open, the modulating actuator adjusts the valve position as necessary to ensure the flow rate is maintained even though the inlet pressure may change.

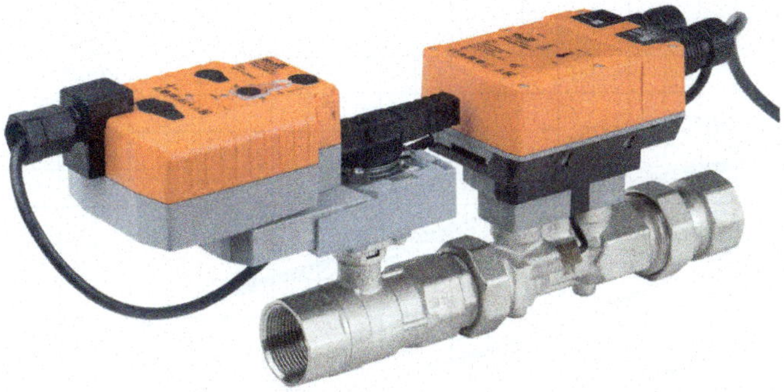

Figure 65 Belimo Pressure-Independent Characterized Control Valve™.
Source: BELIMO Aircontrols (USA), Inc.

4.2.0 System Flow Rates and Pump Selection

The system pump must be able to provide the required volume of water at full load conditions. The total water volume needed, in gpm, is the first piece of information typically needed in the pump selection process. Here is one way the flow rate can be calculated for a system:

$$\text{Water Flow Rate} = \frac{HL}{TD \times 8.33 \times 60}$$

Where:

HL = hourly heat loss in Btuh
TD = design water temperature differential across terminals (°F)
8.33 = weight of one gallon of water, in pounds
60 = number of minutes in one hour

Example 1:

Assume a building's heat loss is 110,000 Btuh. Calculate the conditioned water flow in gallons per minute based on a temperature drop across the terminals of 20°F. This means that the capacity of the terminals might be selected for an entering water temperature of 160°F and a leaving water temperature of 140°F.

Solution:

$$\text{Water Flow Rate} = \frac{HL}{TD \times 8.33 \times 60}$$

$$\text{Water Flow Rate} = \frac{110{,}000\ \text{Btuh}}{20°\text{F} \times 8.33 \times 60}$$

$$\text{Water Flow Rate} = \frac{110{,}000\ \text{Btuh}}{9{,}996}$$

$$\text{Water Flow Rate} = 11.0\ \text{gpm}$$

Once the required flow rate is known, a pump can be selected that will provide the needed flow. As marked on the pump curve in *Figure 66*, a pump with the permissible head of 9.2' at its output is needed to provide the flow rate of 11 gpm.

NOTE

The weight of water changes somewhat as its temperature changes, due to a corresponding change in density. The weight of water at 70°F is 8.33 pounds/gallon. However, it is more accurate to use the weight per gallon associated with the temperature of the water in the system. For example, the weight of water at 140°F—a common water temperature for heating applications—is 8.20 pounds/gallon.

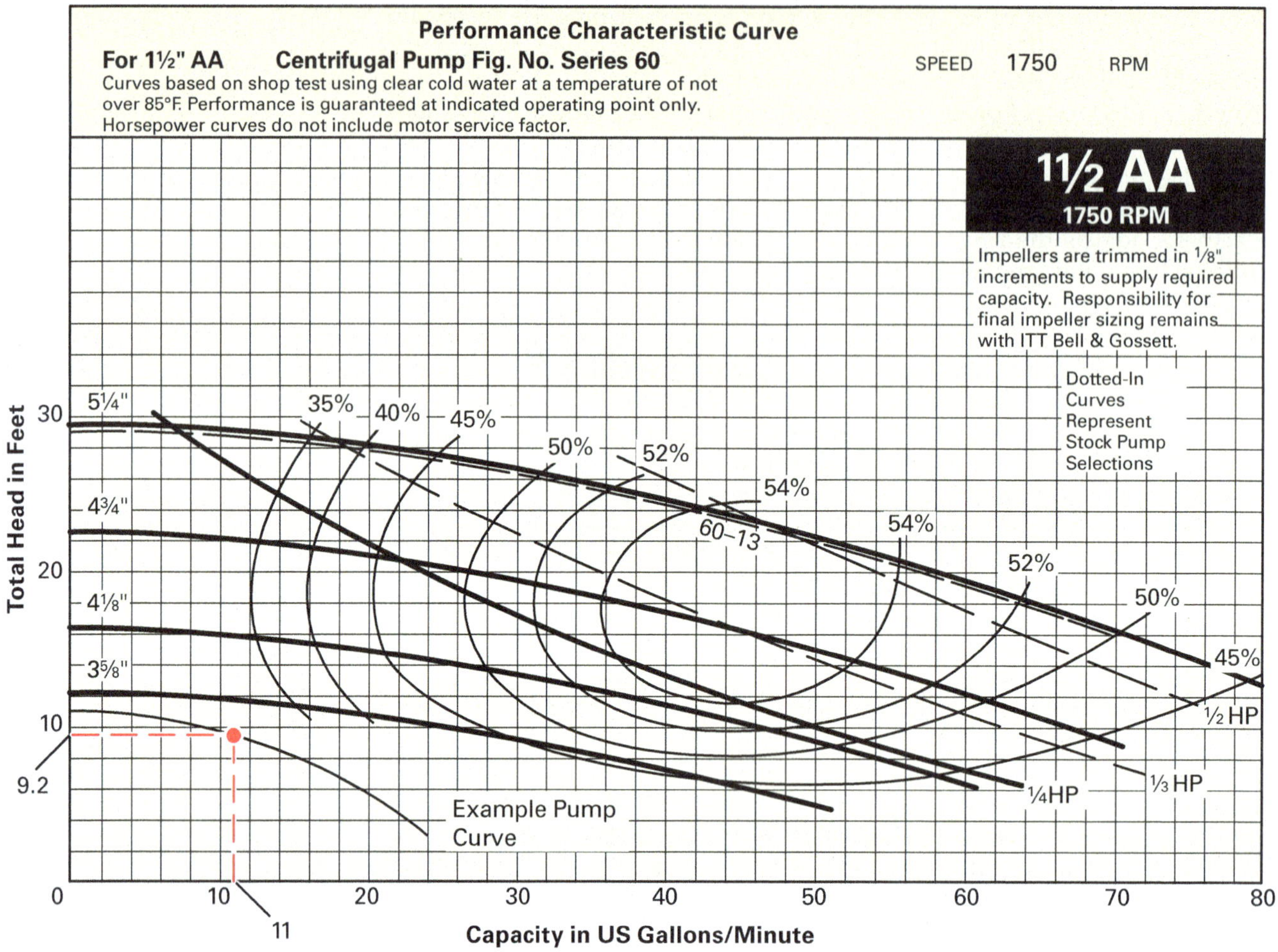

Figure 66 Pump curve (Example 1).
Source: Courtesy of Xylem Inc.

4.2.1 Friction Losses

When water flows, resistance caused by friction results. As in air distribution systems, the components of the system resist the flow of water, and this resistance must be accounted for. Friction losses increase as the water velocity, flow rate, and distance traveled increases. Smaller pipe sizes also increase friction.

Pipe fittings produce more friction loss than straight lengths of pipe due to the turbulence created by the changes in direction and the edges and ridges found at points of connection. This is comparable to what happens with airflow through a duct. Pipe fittings are assigned a value that compares their resistance to flow to a specified length of straight pipe of the same size. This value is known as the equivalent feet of loss. For example, a 90-degree ell of a given size may have an equivalent length of 19'. This means the friction loss as water passes through the elbow is the same as water passing through 19' of straight pipe of the same nominal size.

When sizing a pipe run, the equivalent feet of loss for each fitting is added to the physical length of straight pipe to find the total equivalent length. The friction loss of other system components is determined from the manufacturer's data and added to the system piping losses.

NOTE

Since the result of these calculations is rarely identical to the actual friction loss of the system once the installation is complete, the balancing valves previously introduced are used to add friction in key locations to compensate.

Using the pump head pressure at the required flow rate and the total equivalent length of pipe for the longest run in the system, the friction loss for the circuit can be calculated as follows:

$$\text{Permissable Friction Loss (per 100')} = \frac{\text{Permissable head}}{\text{Equivalent pipe length}} \times 100$$

Example 2:

Assume a head of 9.2' and an equivalent pipe length of 250'. Calculate the friction loss per 100' of pipe.

Solution:

$$\text{Permissable Friction Loss (per 100')} = \frac{\text{Permissable head}}{\text{Equivalent pipe length}} \times 100$$

$$\text{Permissable Friction Loss (per 100')} = \frac{9.2'}{250'} \times 100$$

$$\text{Permissable Friction Loss (per 100')} = 3.68' \text{ per 100' of run}$$

The flow rate and head loss can now be used in conjunction with published tables to determine pipe sizes needed to support the required flow.

If the total friction loss for the circuit exceeds the head that the pump can overcome, some changes in the pipe size may be required. If the total friction loss is very low, pipe sizes in some sections can be reduced to save money. If necessary, corrections can be made by changing some pipe sizes and adding adjustable flow controls or balancing devices. It's important to remember that friction can always be added by a balancing valve, but the friction cannot be reduced once the system is installed.

4.3.0 Determining Pump Pressure, Head Pressure, and Flow

The amount of pressure added by a pump at a given flow rate can be determined by measuring the pressure differential between the pump's suction and discharge ports. Pressure taps used for making such measurements should be located as close to the pump flanges as possible. Both pressure taps should also be at the same height with respect to the pump center line. There should be no gate valves or check valves installed between the readout taps and the pump flanges that will interfere.

The pressure differential measured across a pump can be converted to head, in feet, using the following formula:

$$\text{Head} = (\Delta P \times 144 \text{ in}^2) \div D$$

Where:

ΔP = pressure differential between the suction and discharge ports of the pump, in psi

D = density of the fluid being pumped, in lb per cubic foot (lb/ft^3)

144 in^2 = square inches per square foot

Remember that the density of water and other fluids is dependent on temperature. As the temperature of water increases, its density decreases. The standard density of water measured at sea level and at 68°F is 62.4 lb/ft^3. To find the density of water at other than the standard temperature, refer to tables provided in common references on hydronic systems.

The calculated total pump head in feet and the pump impeller size can be used to determine the amount of water flow through the pump. This is done by plotting the calculated total pump head on the pump's performance curve at zero flow. Then draw a line horizontally to the right to intersect the appropriate pump impeller curve. At this intersection point, draw a line vertically to the bottom of the chart to read the flow in gpm.

Example 3:

Calculate the pump head developed for a measured pump pressure differential of 10.8 psi when pumping water at the standard density of 62.4 lb/ft^3.

Solution:

$$Head = (\triangle P \times 144\ in^2) \div D$$
$$Head = (10.8\ psi \times 144\ in^2) \div 62.4\ lb/ft^3$$
$$Head = 24.92'$$

Using the pump curve in *Figure 67*, determine the flow through a pump with a 5$\frac{1}{4}$" impeller for the calculated total head pressure of 24.92'. As shown in the figure, the intersection of the 24.92' total head horizontal line with the 5$\frac{1}{4}$" pump curve shows that the water flow through the pump is 40 gpm. Per the performance curve, a $\frac{1}{2}$ hp motor is sufficient to drive the pump.

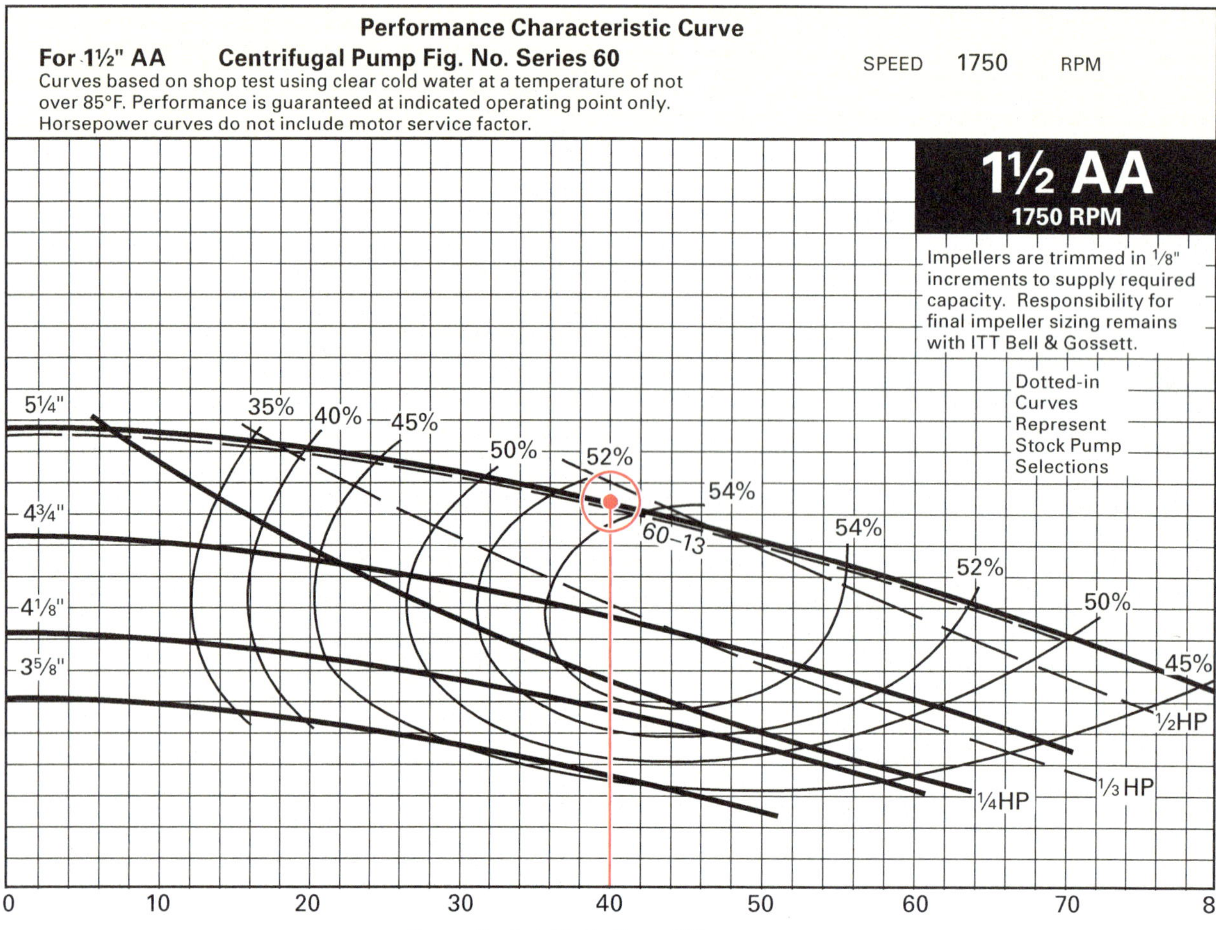

Figure 67 Pump curve (Example 3).
Source: Courtesy of Xylem Inc.

On Site

Hydronic Apps

The websites of many pump manufacturers offer apps that can be used to help select a pump for an application. User inputs to the program typically include the required flow rate, total pump head, desired pump speed, and the type and/or model of pump preferred. Apps for use on mobile devices are also readily available.

Apps are also available for pipe sizing, calculating friction loss and head, and working with balancing valves.

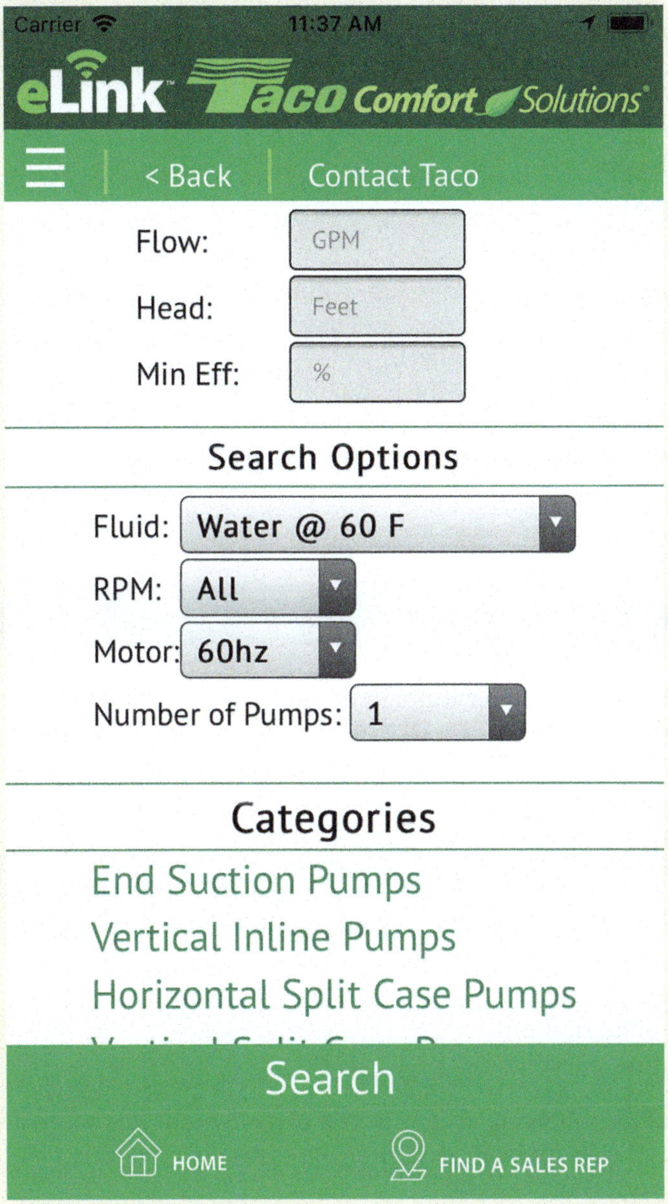

Source: Courtesy of Taco, Inc.

4.0.0 Section Review

1. For the most accurate pressure drop measurement in a water pipe, use a(n) _____.
   a. orifice plate
   b. venturi tube
   c. backflow gauge
   d. high-pressure gauge

2. Assume a building's heat loss is 460,000 Btuh. Calculate the needed hot water flow in gallons per minute based on a temperature drop across the terminals of 25°F.
   a. 11.0 gpm
   b. 27.3 gpm
   c. 36.8 gpm
   d. 55.7 gpm

3. Pressure taps used to measure a pump's pressure differential must be located _____.
   a. right next to the impeller
   b. on the suction side of the pump
   c. on the discharge side of the pump
   d. at the same height relative to the pump center line

Module 03203 Review Questions

1. The movement of water through a forced hydronic heating system is the result of _____.
   a. flow rates
   b. differing densities
   c. pressure differences
   d. temperature differences

2. Pressure drop in a hydronic heating system is caused by _____.
   a. air bubbles
   b. internal friction
   c. lack of lubrication
   d. pressure pulsations

3. Which of the following actions is *most likely* to occur first when a zone thermostat calls for heat?
   a. The boiler burner lights.
   b. The expansion tank drains.
   c. The circulator pump starts.
   d. The burner ignitor warms up.

4. True or False? Low- pressure, low-temperature boilers are typically limited to a maximum operating temperature of 150°F (66°C).
   a. True
   b. False

5. Which of the following statements about wall-mounted condensing boilers is *true*?
   a. Condensing boilers do not require venting.
   b. Wall-mounted condensing boilers heat water instantaneously on demand.
   c. Boiler systems must be supported by at least two boilers piped in parallel.
   d. Unlike condensing furnaces, condensing boilers do not produce acidic condensate.

6. Which of the following is the most common indication of pump cavitation?
   a. Water leaks
   b. Blocked strainers
   c. Overheating zones
   d. Snapping and crackling noises

7. What does a slow-opening zone control valve help prevent?
   a. A cold zone
   b. Water hammer
   c. Pump cavitation
   d. Zone overheating

8. Which term identifies a one-pipe system that has a continuous run of pipe from the supply connection to the return connection, with all the hot water flowing through each terminal?
   a. Series-loop system
   b. Direct-return system
   c. Reverse-return system
   d. Radiant floor-heating system

9. Which term identifies a two-pipe system in which the supply and return water flow in opposite directions?
   a. Series-loop system
   b. Direct-return system
   c. Reverse-return system
   d. Radiant floor-heating system

10. What material is commonly used for the terminal loop in radiant floor-heating systems?
    a. PVC pipe
    b. PEX tubing
    c. Copper tubing
    d. Neoprene tubing

11. A three-way zone valve located at the inlet of a heating terminal acts as a ______.
    a. check valve
    b. mixing valve
    c. diverting valve
    d. pressure-relief valve

12. Which value is determined based on the pressure drop measured across a hydronic measuring device or balancing valve?
    a. Flow rate
    b. Friction loss
    c. Pump efficiency
    d. Temperature difference

13. Which of the following functions is *not* provided by a triple-duty valve?
    a. Check valve
    b. Shutoff valve
    c. Balancing valve
    d. Expansion valve

14. What is the first piece of information typically needed in the pump selection process?
    a. System equivalent length
    b. System discharge pressure
    c. Total water volume needed
    d. Highest point of the system

15. Which of the following factors *increase* friction loss in piping systems?
    a. Increased velocity
    b. Shorter piping runs
    c. Larger pipe diameter
    d. Decreased water flow

Answers to odd-numbered review questions are found in the Module Review Answer Key at the back of this book.

Answers to Section Review Questions

Answer	Section	Objective
Section One		
1. c	1.0.0	1a
2. a	1.2.1; *Figure 6*	1b
Section Two		
1. b	2.1.1	2a
2. c	2.2.0	2b
3. d	2.3.4	2c
4. a	2.4.7	2d
Section Three		
1. b	3.1.1	3a
2. d	3.2.0	3b
3. a	3.3.0	3c
Section Four		
1. b	4.1.1	4a
2. c	4.2.0	4b
3. d	4.3.0	4c

MODULE 03306

Steam Systems

Source: Spirax Sarco Limited

Objectives

Successful completion of this module prepares you to do the following:

1. Describe the properties of water as they relate to steam systems.
   a. Describe the basic properties of water.
   b. Describe the pressure-temperature relationship of water.
2. Describe the basic steam cycle and the primary components related to its operation.
   a. Describe the steam cycle principles of operation.
   b. Identify steam boilers and common accessories.
   c. Identify common steam heat exchangers and terminal devices.
3. Identify common steam system piping arrangements and valves.
   a. Identify common steam system piping arrangements.
   b. Identify common condensate return systems and describe waterside care.
   c. Explain how to size steam and condensate system piping.
   d. Identify steam-based pressure-reducing and thermostatic control valves.
4. Identify and explain how to install and maintain various types of steam traps.
   a. Identify various types of steam traps.
   b. Explain the how to install steam traps.
   c. Explain how to troubleshoot and maintain various steam traps.

Performance Tasks

Under supervision, you should be able to do the following:

1. Perform instructor-selected service or operating procedures on low-pressure steam boilers or systems.
2. Disassemble and maintain instructor-selected steam traps.

Overview

Steam has a wide variety of uses in commercial and industrial sectors. It can carry large quantities of stored energy and be applied in comfort heating, process heating, and even power generation. Most of steam's heat content is stored in its latent form, so it can be transported efficiently at a low cost. Steam has been generated and manipulated as an energy source for many years now. While new technology has enhanced the art of control, some aspects of the topic never change.

Digital Resources for HVACR

Scan this code using the camera on your phone or mobile device to view the digital resources related to this craft.

SCAN ME

NCCER Industry-Recognized Credentials

If you are training through an NCCER-accredited sponsor, you may be eligible for credentials from NCCER. The ID number for this module is 03306. Note that this module may have been used in other NCCER curricula and may apply to other level completions. Contact NCCER at 1.888.622.3720 or go to **www.nccer.org** for more information.

You can also show off your industry-recognized credentials online with NCCER's digital credentials. Transform your knowledge, skills, and achievements into credentials that you can share across social media platforms, send to your network, and add to your resume. For more information, visit **www.nccer.org**.

1.0.0 Steam-Related Fundamentals of Water

Performance Task

1. Perform instructor-selected service or operating procedures on low-pressure steam boilers or systems.

Objective

Describe the properties of water as they relate to steam systems.

a. Describe the basic properties of water.
b. Describe the pressure-temperature relationship of water.

Steam is the most widely used form of energy in industrial facilities throughout the world. Water is plentiful and can be easily converted into steam. The energy stored within water that has been converted to steam can be five to six times greater than the energy contained in water at the same temperature. External energy sources, such as pumps, are not required to move steam from the point of creation to the intended load. This fact alone makes steam a desirable fluid for storing and moving large amounts of energy in a variety of applications.

Did You Know?

Steam Energy

Of the energy needed to produce pulp and paper, 80% comes from steam. Approximately 35% of the energy required for all industrial output in the United States comes from steam.

Source: hxdyl/iStock/Getty Images

1.1.0 Basic Properties of Water

Depending on its heat content and temperature, water can exist in three states: ice (solid), water (liquid), and water vapor (gas, such as steam). *Figure 1* illustrates the changes that occur to one pound of ice when heat is added at sea level. *Figure 2* shows the changing states of water using metric values.

If heat is added to one pound of ice at 0°F, a thermometer will show a rise in temperature until it reaches 32°F. The ice will start changing into water at this point. If heat continues to be added, the thermometer reading will remain fixed at 32°F instead of rising. It will continue to read 32°F until the entire pound of ice has melted.

The increase in temperature from 0°F to 32°F registered by the thermometer is sensible heat. *Sensible heat* is heat that can be sensed by a thermometer or by touch. *Latent heat* is the heat that is added to the ice and causes its change in state from a solid to a liquid but does not register on the thermometer. Latent heat is the heat energy absorbed or rejected when a substance is changing state (solid to liquid, liquid to gas, or vice versa) without a change in the measured temperature.

NOTE

The process discussed here assumes atmospheric pressure conditions. Remember that pressure plays a role in the boiling and freezing points of fluids.

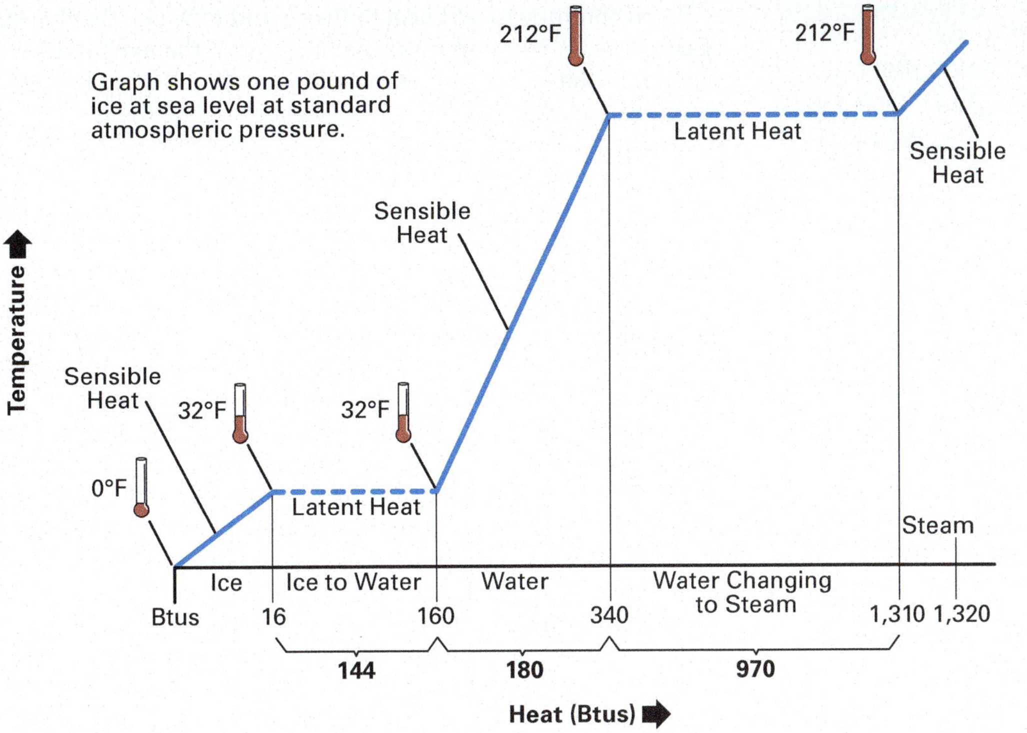

Sensible Heat:
Heat that causes a temperature change and can be sensed by a thermometer or by touch.

Latent Heat:
Heat that produces a change in state without a measurable change in temperature.

Figure 1 Changing states of water—imperial values.

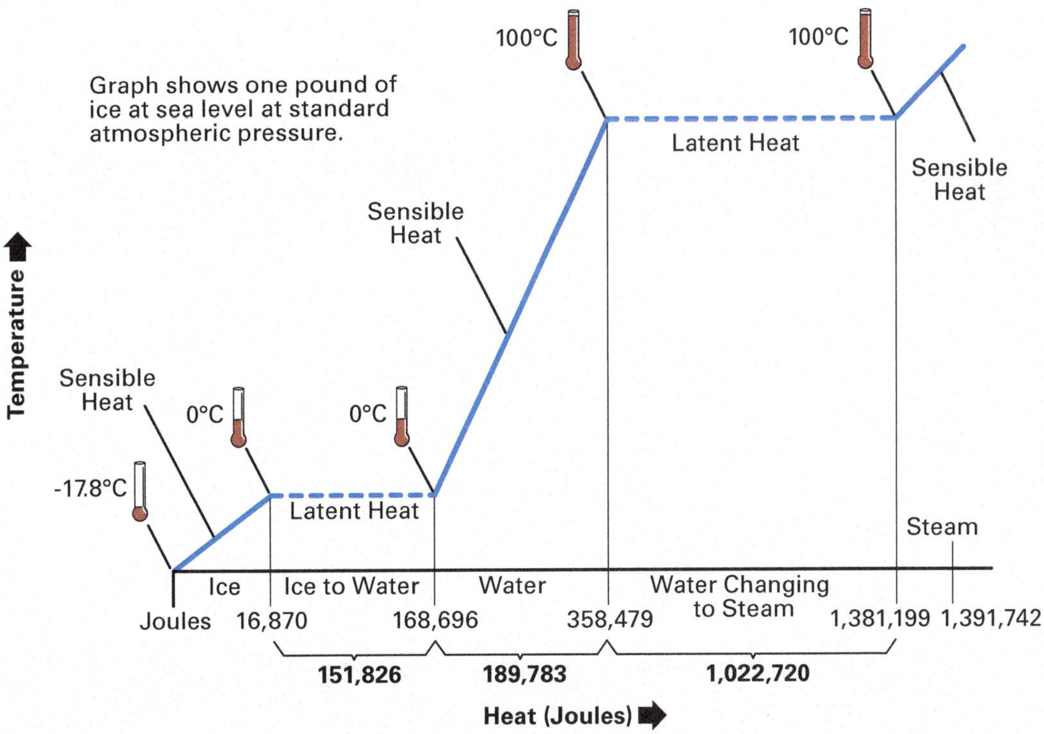

Sensible Heat:
Heat that causes a temperature change and can be sensed by a thermometer or by touch.

Latent Heat:
Heat that produces a change in state without a measurable change in temperature.

Figure 2 Changing states of water—metric values.

Did You Know?

Sensible Heat and Latent Heat

The terms *sensible heat* and *latent heat*, as well as the terms *superheat* and *subcooling*, have the same meaning in the world of steam as in the world of refrigerants.

Subcooling: The reduction in temperature of a liquid after it has completed the change in state from a vapor. Only after the phase change is complete can the liquid begin to decrease below the saturation temperature as additional heat is removed.

Superheat: The increase in temperature of a vapor after it has completed the change of state from a liquid to a vapor. Only after the phase change is complete can the vapor begin to increase beyond the saturation temperature as additional heat is added.

Sensible heat: Heat energy that can be measured by a thermometer or sensed by touch.

Latent heat: The heat energy absorbed or rejected when a substance is in the process of changing state (e.g., solid to liquid, liquid to gas, or the reverse of either change).

If you continue to add heat to the pound of water, the thermometer will once again show an increase in temperature until the temperature reaches 212°F. At this point, the water will start boiling and will change state from water into steam (water vapor). As even more heat is added, the thermometer reading will remain at 212°F until all the water has turned into steam. If you continue to add heat, the thermometer will once again register sensible heat.

As shown in *Figure 1*, it takes more heat to cause a change of state in water than is needed to increase its temperature by one degree. It takes 144 Btus of latent heat to melt a pound of ice before the temperature begins to rise again. This is 144 times as much heat as is needed to raise the temperature of water one degree (specific heat).

It takes 970 Btus of latent heat to vaporize the water to steam. None of the latent heat consumed during the change of state of water registers on the thermometer. Once water reaches the threshold of changing states and begins the change process, the thermometer becomes stable and unchanged until the phase change is complete. This property of latent heat helps to explain why large amounts of energy can be stored and transported through the use of steam.

Figure 3 shows common terms used to describe the addition or removal of latent heat when water is changed to and from solid, liquid, or vapor states. It also shows the terms **subcooling** and **superheat** as they relate to **sensible heat** changes in water. Remember that subcooling and superheating are sensible processes. As water changes state from one form to another, **latent heat** is being exchanged and there is no measurable change in temperature while the transition is in progress.

The terms illustrated in *Figure 3* are further defined as follows:

- *Superheat* — The measurable heat added after all the water has reached its boiling point and changed into steam. For example, the saturation temperature of water at standard atmospheric pressure at sea level (or an absolute pressure of 14.7 psia) is 212°F. If the resulting steam is superheated 10°F, the

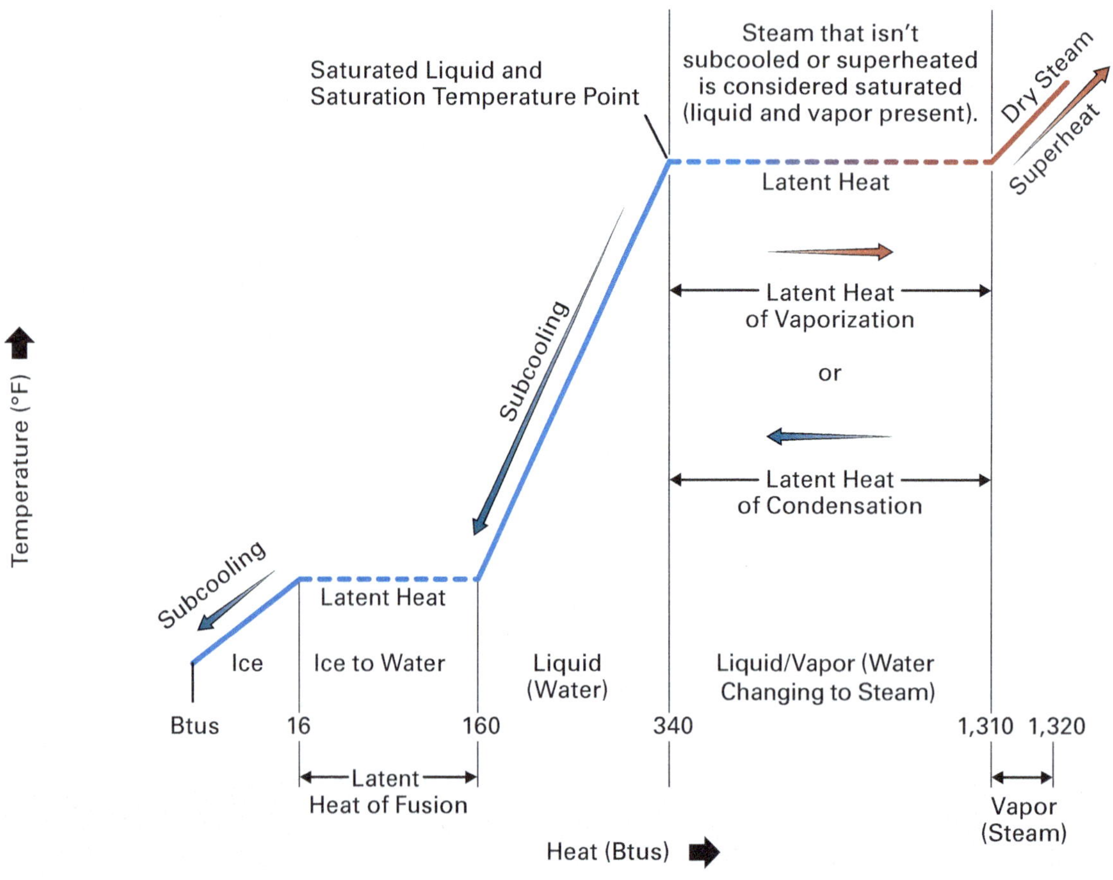

Figure 3 Common terminology related to the change of states.

steam temperature is 222°F. Steam that has been heated above the saturation temperature is called *superheated steam*.

- *Subcooling* — The measurable heat that is removed after all the water has condensed from steam to water. Subcooling is the reverse of superheat, and the subcooling process cannot begin until the change of state from vapor to liquid is complete. For example, the condensing temperature of water at standard atmospheric pressure at sea level (or an absolute pressure of 14.7 psia) is 212°F. If the temperature of the water is cooled to 202°F, it is said to have 10°F of subcooling.
- *Latent heat of fusion* — The heat gained or lost in changing ice to water or water to ice.
- *Latent heat of vaporization* — The heat energy gained in changing water to steam at a given pressure. The temperature at which this occurs is known as the boiling point of water at that pressure. As water absorbs latent heat and the change of state begins, steam rises from the water surface. This steam is referred to as **saturated steam** because liquid remains beneath the vapor in the same space. Whenever both phases—liquid and water—are present in the same space, the substance is said to be *saturated*, or *in saturation*. After all the water has been vaporized and only steam remains, the steam is referred to as **dry steam**.
- *Latent heat of condensation* — The heat removed from steam as it condenses back to water.

Saturated steam: The steam produced and existing at the temperature that corresponds to the boiling temperature of water at the existing pressure (saturation).

Dry steam: The steam that exists after all the water has been vaporized and is moving beyond its saturation temperature (exhibits superheat).

Figures 1 and 2 show that most of the heat content of steam is stored as latent heat. It is this latent heat (970 Btu/lb at sea level) that makes steam so efficient as a means of transferring energy. Steam releases stored energy as it condenses. Large volumes of steam can be transmitted efficiently with little change in its energy content.

Steam properties are dependent on pressure and temperature. This means that the steam system can be controlled by varying either the steam pressure or temperature.

1.2.0 Pressure-Temperature Relationship of Water

Steam tables list pressure-temperature relationships and other properties of saturated steam over a broad range of conditions. *Table 1* shows part of a typical steam table. One entry—the row at 0.0 psig—is for the properties of steam at standard conditions as described above.

The following are descriptions of the information found in *Table 1*:

- Columns A through C show pressure-temperature relationships. Columns A and B show the corresponding gauge (psig) and absolute (psia) pressures.
- For each pressure, there is a corresponding temperature. Column C shows that the saturation temperature of steam at sea level (0 psig/14.7 psia) is 212°F. As steam cools down to these conditions, the latent process of condensation begins while heat continues to be removed. If water is heated up to these conditions, the latent process of vaporization is initiated as more heat is added.
- Column D shows the heat of saturated liquid. This is the amount of sensible heat required to raise the temperature of a pound of water from 32°F to the boiling point at the pressure and temperature shown. It is expressed in Btu/lb.
- Column E shows latent heat. This is the amount of heat required to change a pound of boiling water to steam. It is expressed in Btu/lb. This same amount of heat is released when a pound of steam is condensed back into a pound of water. This heat quantity is different for every pressure-temperature combination.

TABLE 1 Properties of Saturated Steam (abbreviated)

	Column A Gauge Pressure	Column B Absolute Pressure (psia)	Column C Steam Temp. (°F)	Column D Heat of Sat. Liquid (Btu/lb)	Column E Latent Heat (Btu/lb)	Column F Total Heat (Btu/lb)	Column G Specific Volume (cu ft/lb)
Vacuum	29.743	0.08854	32.00	0.00	1075.8	1075.8	3306.00
(in Hg)	29.515	0.2	53.14	21.21	1063.8	1085.0	1526.00
	27.886	1.0	101.74	69.70	1036.3	1106.0	333.60
	19.742	5.0	162.24	130.13	1001.0	1131.1	73.52
	9.562	10.0	193.21	161.17	982.1	1143.3	38.42
	7.536	11.0	197.75	165.73	979.3	1145.0	35.14
	5.490	12.0	201.96	169.96	976.6	1146.6	32.40
	3.454	13.0	205.88	173.91	974.2	1148.1	30.06
	1.418	14.0	209.56	177.61	971.9	1149.5	28.04
psig	0.0	14.696	212.00	180.07	970.0	1150.4	26.80
	1.3	16.0	216.32	184.42	967.6	1152.0	24.75
	2.3	17.0	219.44	187.56	965.5	1153.1	23.39
	5.3	20.0	227.96	196.16	960.1	1156.3	20.09
	10.3	25.0	240.07	208.42	952.1	1160.6	16.30
	15.3	30.0	250.33	218.82	945.3	1164.1	13.75
	20.3	35.0	259.28	227.91	939.7	1167.1	11.90
	25.3	40.0	267.25	236.03	933.7	1169.7	10.50
	30.3	45.0	274.44	243.36	928.6	1172.0	9.40

- Column F shows the total heat of the steam. This is the sum of the sensible heat and the latent heat given in Columns D and E. It is expressed in Btu/lb. It is the total heat in the steam above 32°F.
- Column G shows the specific volume of steam. This is the volume per unit of mass for steam. It is the space that one pound of steam will occupy at the conditions shown. It is expressed in cubic feet per pound. Note that, as pressure and total heat increases, the volume of the steam decreases.

Tables showing the properties of saturated steam can be found in manufacturers' literature, as well as in reference texts for steam system design. These tables assist with determining pressure-temperature relationships. The information contained in steam tables also is used in calculations for designing and sizing steam system components.

1.0.0 Section Review

1. What is the state of steam at atmospheric pressure with a temperature of 240°F?
   a. Saturated
   b. Superheated
   c. Subcooled
   d. Condensed

2. True or False? The specific volume of steam decreases as the pressure increases.
   a. True
   b. False

2.0.0 Basic Steam System Cycle and Operational Components

Objective

Describe the basic steam cycle and the primary components related to its operation.

a. Describe the steam cycle principles of operation.
b. Identify steam boilers and common accessories.
c. Identify common steam heat exchangers and terminal devices.

Performance Task

1. Perform instructor-selected service or operating procedures on low-pressure steam boilers or systems.

The type of equipment used with a steam system depends on the steam volume and pressure required. In addition, steam volume and pressure determine the type of boiler, primary system equipment, and ancillary equipment needed.

2.1.0 Steam Cycle Principles of Operation

Figure 4 shows some basic components commonly found in steam systems. The steam cycle in a closed system begins at the boiler. Water is vaporized in the boiler to generate steam. The initial source of energy used for this purpose typically comes from the combustion of a fossil fuel such as oil, gas, or coal, but electric boilers are also available. Once the water has reached the boiling point and vaporization begins in a closed space, the pressure begins to rise as the steam expands.

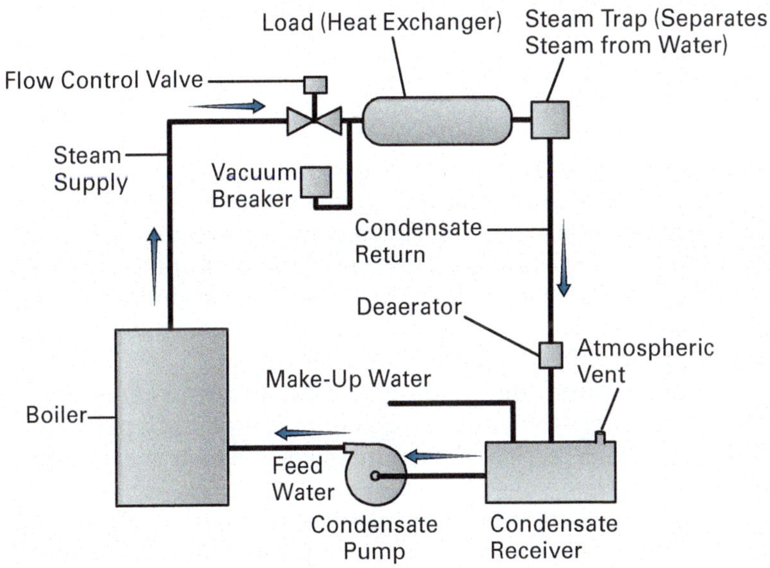

Figure 4 Basic steam system.

The water temperature required for boiling increases due to the change in pressure. As shown in *Table 1*, once the pressure reaches 1.3 psig, the boiling point increases to over 216°F, and it continues to increase as the pressure rises.

The steam in a system is found at the highest temperature and pressure inside the boiler. Steam flows through the supply piping and components as the result of natural movement from an area of higher pressure to an area of lower pressure, much like air through an air distribution system. No pump is needed to transport steam, however.

The control valve is generally an automatic control device. It regulates the flow of steam into the heat exchanger at a rate comparable to the load. The control valve may be controlled by a temperature input, but many other possibilities exist. This ensures that the proper amount of steam is provided to satisfy the needs of the process. In some cases, the control valve may be a pressure regulator that simply controls downstream pressure.

The load or heat exchanger in a steam system may be a steam coil in an HVAC system airstream, a steam-to-domestic water heat exchanger, or perhaps a large vessel where seafood is cooked by the ton. The possibilities are virtually unlimited. The heat exchanger transfers the heat carried in the steam to the load or process, giving up its stored latent heat to the process. Once the latent heat is transferred out, the steam begins to condense. Ideally, only water leaves the heat exchanger, as any steam that leaves represents wasted energy.

The condensing process in the heat exchanger occurs at a lower pressure than that found in the boiler. The pressure at the load largely depends on the pressure losses in the supply piping and the pressure drop across the control valve or regulator.

Steam traps are one of the most important parts of a steam system. The trap in the piping at the output of the heat exchanger holds the steam back until it condenses. Without the steam trap, the live steam would pass through the heat exchanger into the return line, wasting energy and transferring a limited amount of heat.

The trap allows only liquid condensate and/or air to exit the heat exchanger and return to the process. Traps are designed to operate at a specified maximum pressure. Traps will not work properly or could be damaged if placed in a system at higher pressure requirements. The system designer selects the appropriate style and type of trap for the system and considers the design pressure in the selection of every component.

Traps require regular testing and maintenance to ensure proper operation and peak efficiency. Maintaining steam traps helps reduce the chance of failures that interrupt the process. Traps typically require more maintenance and testing than the other system components.

Where elevations permit, steam systems can use a gravity return to route the condensate directly back to the boiler for re-evaporation into steam. However, most systems use gravity flow to drain the condensate into a receiver tank. When the boiler controls call for additional water, a condensate pump cycles on and produces the necessary pressure differential to pump the condensate (feed water) from the tank to the boiler. This method allows gravity to be used to the maximum extent possible, with the condensate pump doing the remaining work. A large and extensive system may have many condensate receivers and pumps.

The condensate receiver and pump generally operate at atmospheric pressure, with the return open to the atmosphere in one or more places. Note the atmospheric vent at the receiver shown in *Figure 4*, which is an advantage to using this type of system.

When steam is applied to heat exchangers that have coils subjected to frigid air or cold water such as in domestic hot-water heaters, the pressure in the heating coil can suddenly drop below atmospheric pressure (vacuum range). In extreme cases, this sudden drop in pressure can cause the tubes in the coil to collapse. This can be prevented by allowing air to enter the heat exchanger through a **vacuum breaker** that senses the drop in pressure. The small amount of air that enters the heat exchanger will be absorbed into the condensate, but most of it will flow in the return line along with the condensate to the receiver. From there it is discharged through the atmospheric vents.

Vacuum breaker: A device used on a steam system to allow air to enter the system in the event a component, vessel, or section of piping falls below atmospheric pressure, thus breaking the vacuum.

Another advantage of using a condensate receiver and pump is that it permits condensate return lines to be located below the water level of the boiler. The pump can compensate for the difference in height. This makes it possible to locate heat exchanger terminals at elevations below the boiler.

2.2.0 Steam Boilers, Controls, and Accessories

Unlike hot-water boiler vessels, steam boiler vessels are not filled with water. Space must be left at the top for steam to occupy. Steam boilers are generally categorized by their working pressure.

Many steam boiler controls and accessories used with steam systems are the same as those used with hot-water boilers. The main difference is that the controls and accessories often require different temperature and pressure ratings.

2.2.1 Steam Boilers

Steam boilers are divided into three broad categories based on the steam pressure generated. *Low-pressure boilers* have a working pressure of up to 15 psig and 250°F. *High-pressure boilers* operate at pressures above 15 psig. It is important to note that these ratings indicate the maximum working pressure of the boiler. For safety purposes, the actual operating pressure of the system must be lower than the maximum allowed.

Some manufacturers and organizations reference a medium-pressure range for steam as well. However, for our purposes, any pressure over 15 psig is considered to be high-pressure steam.

Low-pressure steam boilers (*Figure 5*) are used in a wide variety of applications. HVAC applications include residential and commercial applications in northern climates, but they are used in many industrial applications as well. High-pressure boilers are mainly found in industrial environments, but may also be used in comfort heating applications for the largest facilities, such as college campuses, where long piping runs are required.

(A) Residential Gas-Fired Steam Boiler

(B) Commercial Gas-Fired Steam Boiler

Figure 5 Low-pressure steam boilers.
Source: ECR International, Inc.

Some low-pressure steam boilers are identical in construction to those used for hot water applications. To provide additional safety, boilers rated at a maximum hot-water working pressure of 30 psig will have a reduced maximum rating of 15 psig for steam applications. As shipped, the boiler may only differ in the **trim** provided. However, not every boiler is designed to function in both applications. Keep in mind that steam boilers are not filled with water like a standard boiler. Space must remain at the top for steam to develop and collect.

Trim: External controls and accessories attached to the boiler, such as sight glasses and water feeder controls.

Steam boilers are available in many configurations. Typical designs include the following:

- *Cast-iron sectional boilers* — The cast-iron sections are designed so that the sections can be joined together to create the needed capacity. The sections are joined in the field during the installation process. The boilers shown in *Figure 5* are examples of cast-iron sectional boilers.

- *Steel fire-tube and water-tube boilers* — The heat transfer surface of these boilers is composed of multiple tubes. Water-tube boilers have water passing through the tubes with the heat passing around them. In a fire-tube boiler, the heat and combustion byproducts flow through tubes that are surrounded by water (*Figure 6*).

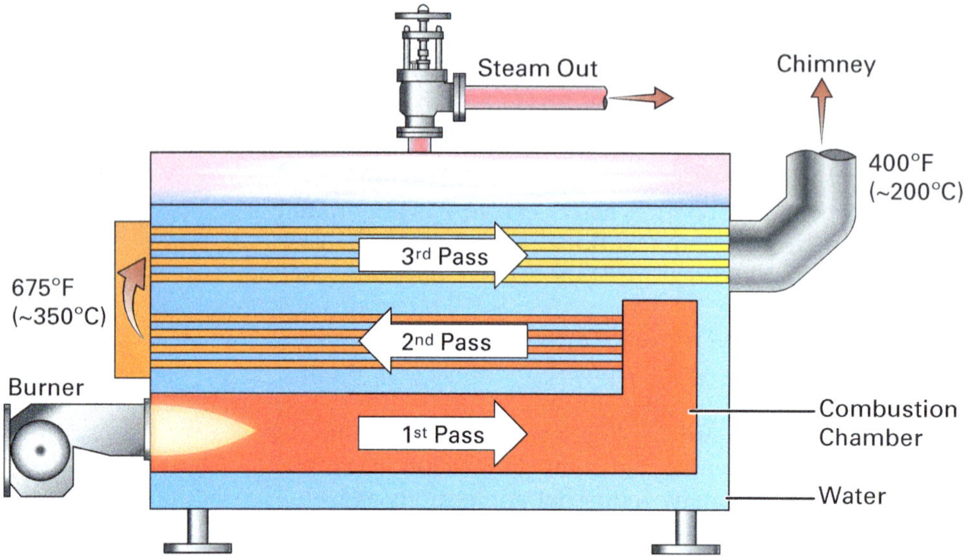

Figure 6 Fire-tube boiler.

- *Steel vertical tubeless boilers* — The heat transfer surfaces are the walls of the vessel inside. Heat and combustion byproducts first pass from the top down through the center of the vessel, then back up and around the outside (*Figure 7*). To maximize efficiency, the hot combustion byproducts may make several passes through the boiler, up and down. This type of boiler features rapid startup, high efficiency, and a small footprint.

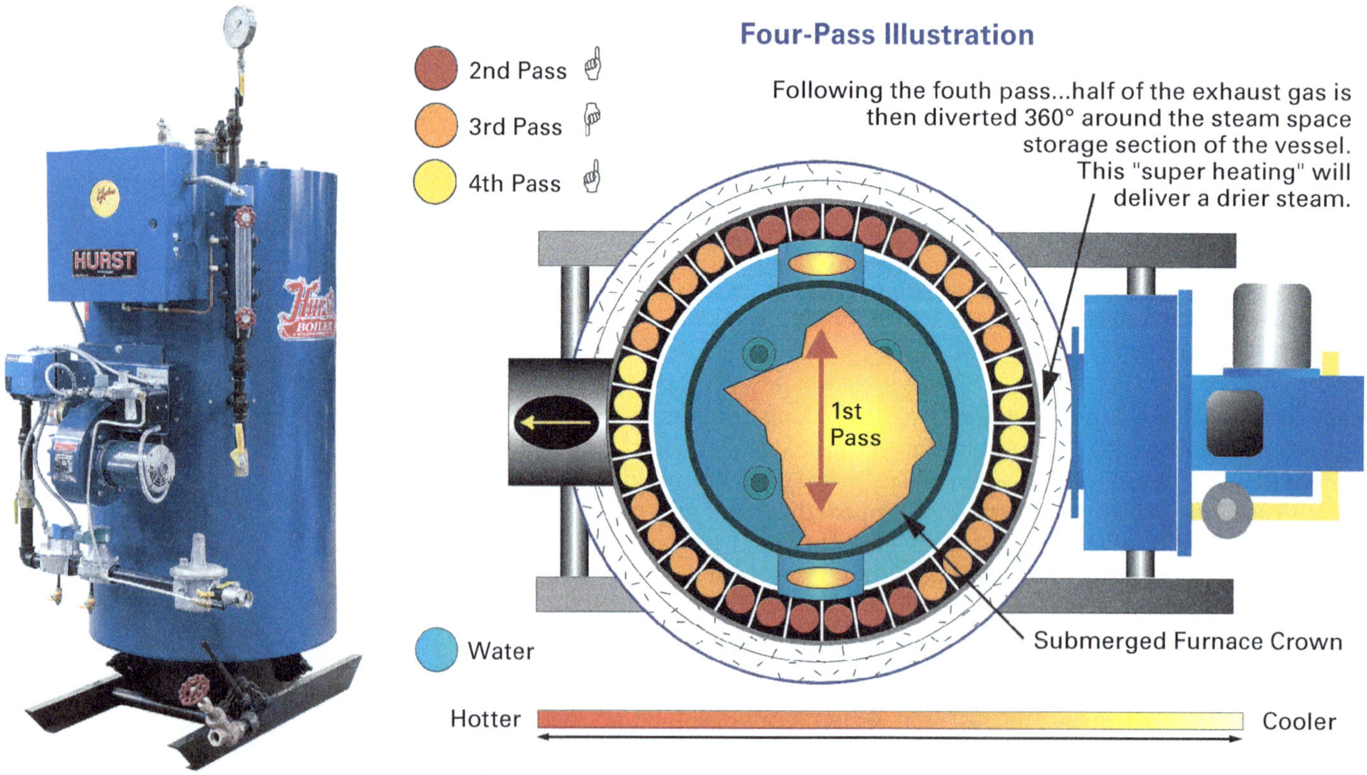

Figure 7 Vertical tubeless steel boiler.
Source: Image courtesy of Hurst Boiler & Welding Co., Inc. of Coolidge, GA

Traditional boilers used in residential and light commercial low-pressure applications are generally of cast-iron sectional or steel one-piece designs. Federal standards require boilers to have a minimum Annual Fuel Utilization Efficiency (AFUE) rating of 80%. Gas-fired boilers earning the EPA's ENERGY STAR designation must have a minimum AFUE rating of 90%, while oil-fired models must achieve 87%.

A standard steam boiler is equipped with the following operating and safety controls, although these devices can be found in a variety of forms:

- Pressure gauge
- Water gauge glass/water column
- Low-water cutoff/water feeder
- High-pressure limit control
- Pressure-relief (safety) valve

2.2.2 Pressure Gauges

Pressure gauges show the pressure within the boiler in pounds per square inch gauge (psig). The gauge is usually installed at the highest point of the boiler's steam space. A *siphon* is generally installed between the boiler and the gauge. Pigtail siphons (*Figure 8*) are very common. The pigtail siphon acts as a water-filled buffer between the live steam and the connected device. Pigtail siphons are always installed vertically and are filled with water before installing the gauge.

Figure 8 Pigtail siphon.
Source: Kele, Inc.

The brass piece shown next to the siphon in *Figure 8* is called a **snubber**. The snubber is also designed to protect a gauge or control device from erratic changes in pressure. It has one or more small orifices that slow down the rate of pressure change on the device, but still allows the pressure to equalize in moments.

Snubber: A simple type of shock absorber used with fluids to suppress, or snub, rapid pressure changes that could damage sensitive instruments and gauges.

Pressure gauges, of course, are not confined to the boiler. Pressure gauges can be found at any point along the steam path, as well as at all connected equipment.

2.2.3 Water Level Indicators

A water gauge glass is used to check the water level in the boiler (*Figure 9*). Although many electronic devices are available today to monitor, report, and control the water level, having a first-hand, physical view of conditions is an essential safety consideration.

The water level is read by cracking open the steam valve at the top of the water gauge, as well as the valve at the bottom. This allows the pressures to equalize with the boiler's water level. A drain valve, or **blowdown** valve, is connected to the bottom of the water gauge. The valve is used to purge the gauge glass of water and any sediment that finds its way in.

Blowdown: To intentionally drain water from a boiler under pressure to reduce the concentration of sediment and impurities.

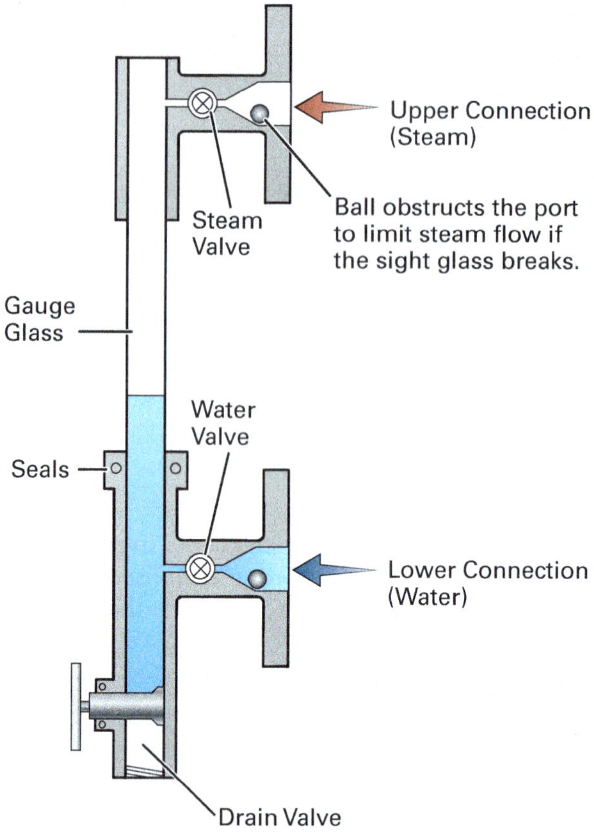

Figure 9 Water gauge glass.

Flash steam: Steam formed instantaneously when hot water is released under pressure to an area of lower pressure.

Use extreme caution when opening a blowdown valve. Steam or high-temperature water can be discharged if the upper and/or lower valves are open. High-temperature water under pressure, when released to atmospheric pressure, creates **flash steam**. The blowdown valve's discharge must always be piped to an appropriate drain, and in a manner that will prevent hot water or flash steam from contacting personnel.

WARNING!

Use extreme caution when using any steam system blowdown or drain valve. The discharge must always be piped to an appropriate drain, and in a manner that will prevent hot water or flash steam from contacting personnel.

On Site

Bleed-Off and Blowdown

The terms *bleed-off* and *blowdown* are often used interchangeably, but both have different meanings in the world of steam.

Bleed-off is generally related to draining water at a trickling rate from a cooling tower or other hydronic loop. This is often done to reduce the concentration of dissolved solids in the remaining water.

Blowdown is related to the removal of dissolved solids and sludge from the water's surface or the bottom of a steam boiler. Thus, blowdown operations can be hazardous, while bleeding water from a system doesn't pose the same kind of hazard.

Know the Code

Water Discharged to a Drainage or Sewage System

Plumbing codes prohibit wastewater that is over 140°F (60°C) from being discharged into a plumbing or drainage system. Water that is discharged from a boiler or similar piece of equipment must be cooled below 140°F before it can be discharged into a system. The water must be transferred to a vessel or tank where it can cool prior to discharge.

On larger boilers, the water gauge may be attached to a water column (*Figure 10*). The water column diverts water or steam from the boiler for application to the water gauge. This helps to stabilize water levels for more accurate water gauge readings. The column also provides a point of attachment for other controls.

Three globe valves are located on the water column shown in *Figure 10*. These small valves are known as *gauge cocks* or *try cocks*. These valves provide a backup method for checking the boiler steam and water levels if the water gauge is damaged. A blowdown valve connected to the bottom of the water column is used to purge the entire column of water and sediment.

In a steaming boiler, the gauge glass often indicates a lower water level than is accurate, as shown in *Figure 11*. This is because the water in the gauge glass is not subject to currents or agitation, is cooler than the water in the boiler, and does not contain rising steam vapor (bubbles).

On Site

Water Columns and ASME Requirements

A water gauge must be mounted on a boiler at a location that meets ASME code requirements. If the system fails and the gauge is not installed at the correct location, the installer may be liable for any damage that occurs.

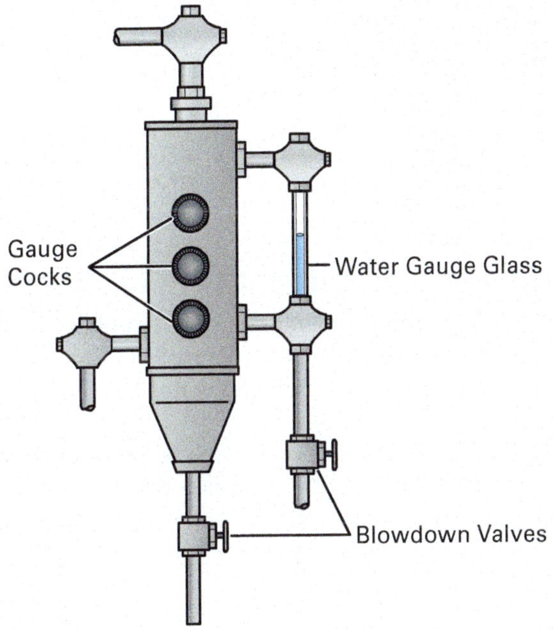

Figure 10 Water column.

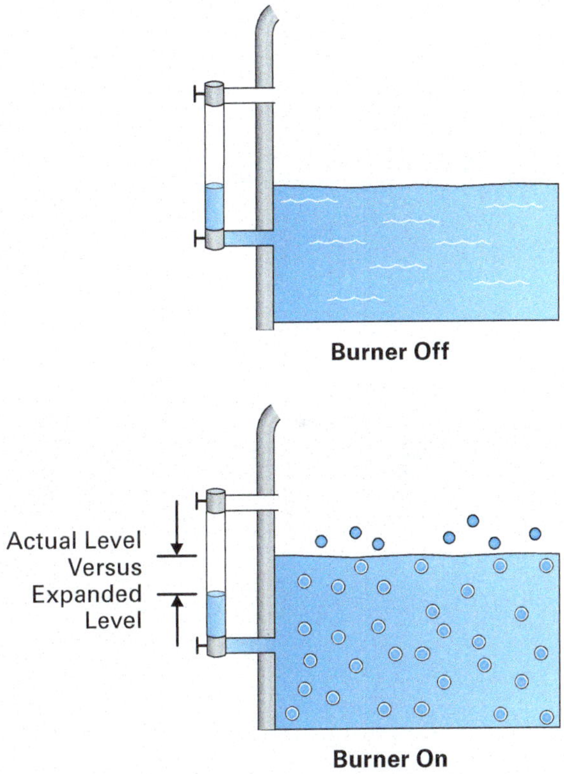

Figure 11 Actual water level versus gauged level.

2.2.4 Low- and High-Water Cutoff and Water Feeder Controls

Low-water cutoff controls protect steam boilers by interrupting power to the heat source when the water falls below the minimum safety level. Water feeders control the flow of makeup water to the boiler.

In systems that effectively collect and return steam condensate to the boiler for reuse, the water feeder uses the condensate reservoir and associated pumps as the primary water source. Fresh water, generally from a municipal water supply, becomes the backup source when condensate levels are insufficient. This type of boiler is generally classified as a steam heating boiler.

Steam boilers are also used in applications where little or no condensate is collected for reuse. In these cases, the water feeder may be quite active, using only the fresh water supply to make up for the lost steam. Boilers used in this manner are classified as steam process boilers.

Typically, low-water cutoff controls for steam boilers use the same methods as those used for hot-water boilers, employing float-operated controls (*Figure 12*). However, while a water boiler must be completely full of water, a steam boiler cannot operate if it is flooded. As a result, both low- and high-water cutoff controls are needed to protect a steam boiler.

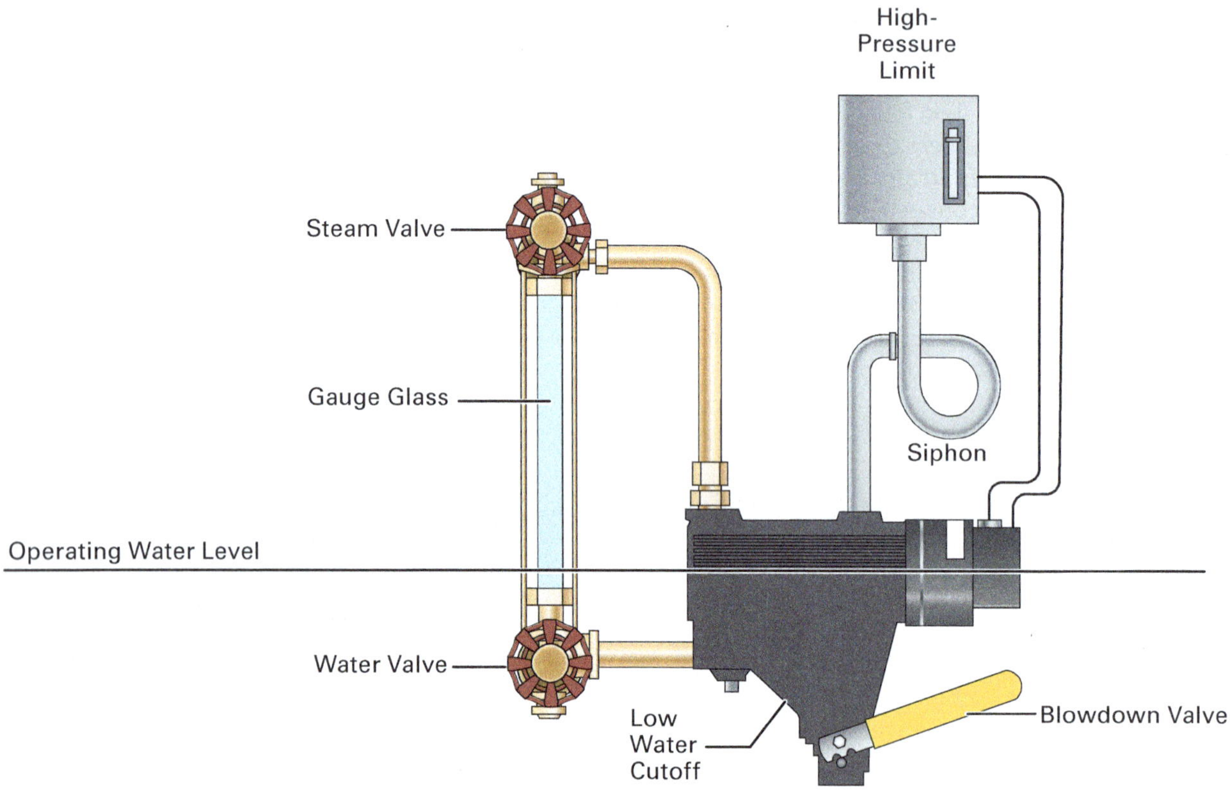

Figure 12 Low-water cutoff.

Figure 13 shows a similar low-water cutoff. However, this model has integral switches that provide an electrical interlock with the control system and the feedwater pump serving the boiler. Thus, it manages the water level as well as serves in a safety role.

Some steam boilers are equipped with an electronic low-water cutoff control (*Figure 14*). A high-water cutoff is not mandatory in most cases while a low-water cutoff is required for safety purposes. A burner firing with little or no water inside the boiler is a serious hazard. Many steam boilers require both a primary and a secondary low-water cutoff.

Figure 13 Low-water cutoff and pump controller.
Source: Courtesy of Xylem Inc.

Figure 14 Electronic low-water cutoff.
Source: Courtesy of Xylem Inc.

Some float-operated and electronic controls can provide low-water and high-water cutoff duty. The control circuit switches have both normally open and normally closed contacts. The switch is wired depending on the desired action. For example, a low-water float control is wired normally open, and the contacts close when the water level is at or above the minimum water level. Conversely, for a high-water cutoff, the switch would be wired normally closed, and the contacts open if the water level is too high.

A marker on the float bowl of the float-operated control used with steam boilers shows the cutoff level for the device. The control is normally installed so that the marker on the float bowl is about ½" higher than the lowest visible point in the water gauge glass.

For some float-operated controls, the operating level is established automatically by fittings on the device that permit its installation in a water column on the boiler. If a low-water condition occurs, float-actuated switch contacts in the control open the electrical circuit to the boiler burner. Note that a little water should still be visible in the gauge glass. The burner cannot restart until the low-water condition is resolved, either manually or automatically.

Figure 15 Electric water feeder.
Source: Courtesy of Xylem Inc.

A water feeder control is used to manage the water level (*Figure 15*). A low water level can usually be attributed to a problem with the feeder, or to a problem with the water source. However, water feeder controls are very reliable when properly maintained and many different kinds of feeders are available. Some feeders use a solenoid valve to admit water from the source, while others use a mechanical float.

Some low-water cutoff controls have a second switch actuated by the float. This switch is used to control the related water feeder. The second switch is set at a higher water level than the cutoff switch and is used to control a solenoid-operated feeder. If the boiler water drops to the operating level, the float closes the switch and energizes the water feeder. This energizes the feeder solenoid valve, allowing makeup water to flow into the boiler. When the boiler water level is restored, the switch opens, and makeup water flow stops.

NOTE

Codes are not the only important factor affecting boiler installation and control schemes. Insurance companies also have specific requirements that may exceed code requirements.

The water feeder control can be combined with the low-water cutoff control into one assembly. These controls are called *combination feeder-cutoff controls.*

A marker on the float bowl casting shows the feeder closing level. Feeders are normally installed so that the marker on the float bowl is about 2" to 2½" below the normal water line.

The selection of low-water cutoff and water feeder devices generally depends on the design system pressure and desired mounting style. Steam boilers over 400,000 Btuh input and/or operating above 15 psig are generally fitted with two low-water cutoff devices for redundancy. The first cutoff (one with the lowest setting) can be a type that resets automatically, while the second cutoff must be a manual reset model.

Water feeders and low-water cutoff controls are essential for proper system operation. These feeders and controls should be disassembled, cleaned, calibrated, and reinstalled at least once a year.

2.2.5 High-Pressure Limit and Other Controls

The high-pressure limit control (*Figure 16*) is activated by a rise in boiler pressure to the setpoint. The high-pressure setpoint should always be lower than the setting of the pressure-relief valve, and within the maximum pressure range of the boiler and any other connected component. High-limit switches generally require a manual reset, due to the serious hazard excessive pressure presents to both personnel and equipment.

Although some high-limit switches use snap-style switches, most use a mercury bulb much like older wall thermostats. If the boiler pressure rises to the setpoint, the mercury rolls to the opposite end of the capsule, opening the control circuit and shutting down the burner.

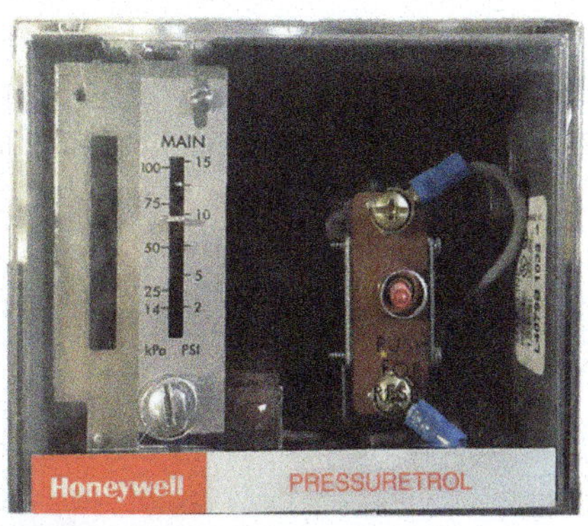

Figure 16 Manual-reset, high-pressure limit control.
Source: Ray Wohlfarth

Pressure switches like these can also provide on-off and sequential control of boilers. The switch is activated when the steam pressure drops below a preset level. This causes automatic feeding of an air/fuel mixture to the furnace, ignition of the burners, and opening of the main air damper(s). As the boiler steam pressure rises, it crosses fixed control points. At each control point, a predetermined mixture of fuel and airflow is selected and applied to the burners. When the boiler pressure reaches the operational pressure limit, the control shuts off the boiler.

2.2.6 Pressure-Relief Valve

A pressure-relief valve protects the boiler and the system from steam pressures exceeding the pressure rating of the boiler or the system (*Figure 17*). It does not operate unless an overpressure condition exists.

Figure 17 Pressure-relief valve.
Source: Anderson Greenwood

A low-pressure steam boiler is built for a maximum working pressure (MWP) of 15 psig. The pressure-relief valve is set to open at the same pressure and discharge steam into the atmosphere. More than one pressure-relief valve must be used on boilers having more than 500 square feet (ft^2) of heating surface area. This is a precaution to guard against the failure of a single valve on a large boiler that can do a lot of damage if the pressure is released catastrophically.

Relief valves should always be installed as directed by the manufacturer with the discharge routed according to local codes. The route and length of the discharge piping should be as short as possible. Relief valves are usually installed in an upright position at the highest point on the steam side of the boiler. No valve or other device can be installed in the discharge line.

The relief valve must be full-sized, with no reductions of its inlet or outlet size. It must be replaced immediately if it does not open when its pressure rating is exceeded, or if it leaks when the boiler is operating at normal pressures. Operation of the boiler with the relief valve leaking can cause a dangerous buildup of scale inside the valve. This can cause the relief valve to stick in the open position should an over-pressurization condition occur.

WARNING!

Unless stated otherwise, the pressure setting of a safety relief valve is not field-adjustable and must not be altered. Incorrect adjustment may cause an explosion, which can result in death or personal injury.

Pressure-relief valves should be checked periodically per the manufacturer's instructions during operation and after long periods of inactivity.

Going Green

Controlling Mercury Hazards

Mercury is a major environmental pollutant that is known to cause cancer. Older pressure switches, like thermostats, contain large amounts of mercury. As mercury switches reach the end of their useful lives, or as boilers are overhauled, the mercury switches are being replaced with other devices, such as snap-action bimetal and electric switches.

The disposal of switches and other devices containing mercury is carefully regulated. Most states have laws governing the storage and disposal of mercury, so it is important to know the applicable regulations in your state.

2.3.0 Steam System Loads

Steam system loads can be manufacturing processes or steam-to-water heat exchangers used for heating potable water or for supplying hot water for space heating. Heating terminals that use steam directly as the medium for space heating are also common loads.

2.3.1 Heat Exchangers and Converters

A heat exchanger is used to transfer heat from one medium to another. It is used in steam systems to generate hot water. *Figure 18* shows a horizontal shell-and-tube heat exchanger, also referred to as a *converter,* used in this manner.

The heat exchanger shown in *Figure 18* consists of tubes encased in a shell. The tubes provide a closed water circuit that allows water to be passed through the converter more than once. The hot water is circulated through the tubes while the steam circulates in the shell surrounding the tubes. There, the heat is transferred from the steam to the water.

Figure 19 shows a typical piping diagram for a heat exchanger used as a hot-water converter in a hot-water heating system. The modulating steam control valve shown is generally thermostatic in nature. It regulates the flow of steam based on the leaving water temperature. More demand for hot water increases the steam flow rate.

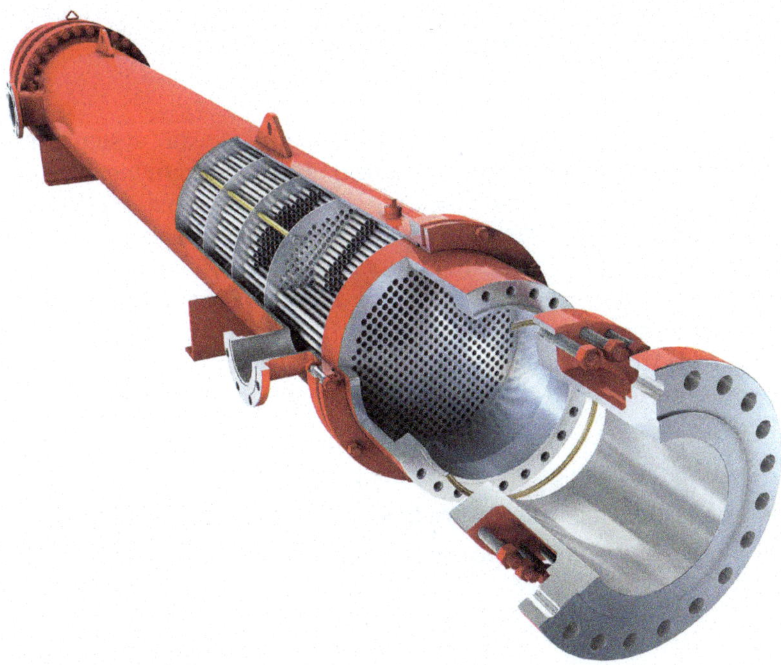

Figure 18 Horizontal shell-and-tube heat exchanger.
Source: API Heat Transfer Inc.

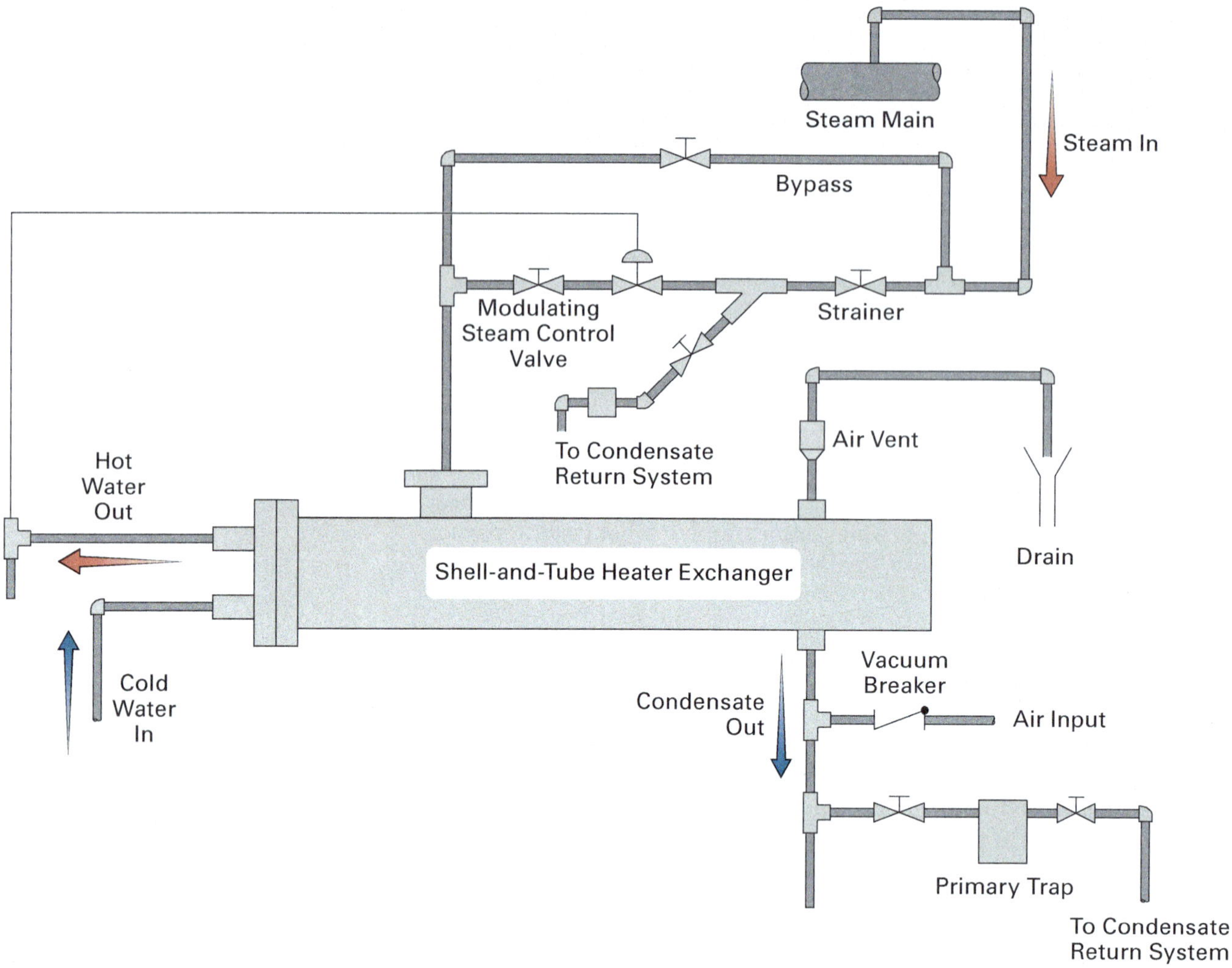

Figure 19 Typical installation of a heat exchanger when used as a converter.

Figure 20 Steam unit heater.
Source: Courtesy of Beacon Morris/Mestek, Inc.

For added safety against contamination, heat exchangers used to heat potable water are of double-walled construction. That is, two walls separate the steam from the potable water. This is especially important when any conditioning chemicals have been added to the steam system.

2.3.2 Terminals

Many styles of air heating terminals can be used in steam systems. *Natural convection* units, which have no mechanical means of moving air, include common radiators and baseboard heating convectors. *Forced convection* units include unit heaters (*Figure 20*), unit ventilators, cabinet heaters, fan coil units, and steam coils found in air handling units.

Steam heating coil designs include standard single-row and multi-pass steam coils, and steam distribution coils. It can be difficult to distinguish between the two types of coils from the outside of a duct.

Single-row coils are used for general heating applications where the following conditions exist:

- The coil will not be subjected to freezing temperatures.
- The flow of steam is not likely to be modulated (consistent flow rate).
- Precise control of the air temperature across the entire face of the coil is not necessary.

Steam distribution coils are generally specified where one or more of these issues are a factor. Distribution coils are of single-row construction but are built with perforated inner tubes that allow the steam to be distributed evenly throughout the coil. Essentially, a perforated tube is inside each visible tube. Some have no perforations in the inner tube, forcing the steam to emerge at the far end of the inner tube and travel back to the coil's condensate header through the outer tube (*Figure 21*). Distribution coils are considered *freeze-resistant*, which is not the same as *freeze-proof*.

Many HVAC systems require steam distribution coils as opposed to standard single-row coils. Systems with outside air intakes that use steam to preheat and temper the air are often exposed to temperatures below freezing. Most HVAC applications also require the flow of steam to be modulated. This makes the steam distribution coil the best overall choice, although it is more expensive due to the unique construction.

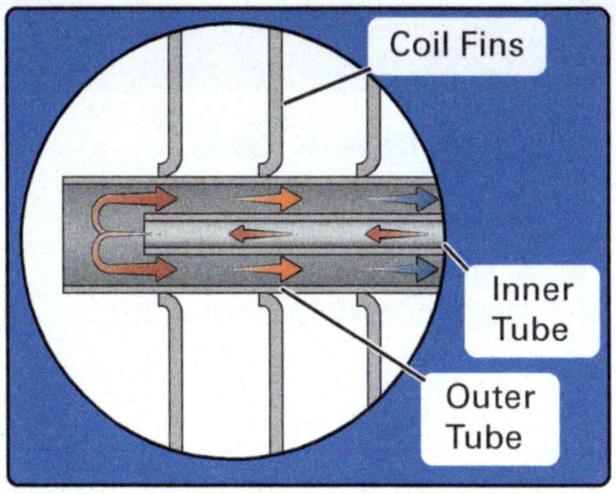

Figure 21 Steam distribution coil flow path.

Did You Know?

Drainable Steam Coils

Draining the condensate from a steam coil helps prevent it from freezing. Some manufacturers offer drainable coils for this reason. However, if the coil is to be out of service for an extended period, it should also be purged with low-pressure air to push the water out. A water and antifreeze mixture can also be added to the coil if it is exposed to freezing temperatures while dormant.

2.0.0 Section Review

1. The steam in a system is found at the highest temperature and pressure ______.
   a. inside a system heat exchanger
   b. at the first control valve
   c. inside the boiler
   d. at the condensate receiver

2. What is the minimum AFUE rating requirement for a gas-boiler to qualify for the EPA's ENERGY STAR logo?
   a. 80%
   b. 85%
   c. 90%
   d. 95%

3. Which of the following is considered a natural convection unit for heat transfer?
   a. Radiator
   b. Unit ventilator
   c. Cabinet heater
   d. Steam coil inside an air handling unit

3.0.0 Steam System Valves and Piping

Performance Task

1. Perform instructor-selected service or operating procedures on low-pressure steam boilers or systems.

Objective

Identify common steam system piping arrangements and valves.

a. Identify common steam system piping arrangements.
b. Identify common condensate return systems and describe waterside care.
c. Explain how to size steam and condensate system piping.
d. Identify steam-based pressure-reducing and thermostatic control valves.

A variety of steam piping systems and valves, as well as support system piping and components, are introduced in the sections that follow.

3.1.0 Steam System Piping

Steam piping systems are either one-pipe or two-pipe systems. The choice is determined by the conditions under which steam is supplied and condensate is returned to the boiler. Each of these piping systems is also classified by the method used to return the condensate (gravity or mechanical flow).

In a simple steam system, condensate can sometimes be returned by gravity alone, but the needed conditions are rare in modern applications. Nearly all steam systems in commercial and industrial applications are two-pipe systems.

3.1.1 One-Pipe Systems

In a one-pipe system, steam flows to, and condensate returns from, the terminal unit via the same pipe. Only one connection to each heating terminal unit is available. One pipe functions as the steam and condensate line.

One-pipe systems return condensate to the boiler by gravity, but some still need a condensate pump. A condensate pump is sometimes needed to lift the condensate from the point where it collects up to the boiler's water level.

One-pipe systems do not rely on steam traps to separate condensate from steam. Air vents are installed at each terminal unit and at the ends of all supply mains. The air vents are used to vent air so the system can fill with steam.

If a one-pipe system does require a condensate pump, air vents are still used at each terminal unit, but steam traps are needed at the end of the steam supply main.

One-pipe systems work best in residential and small commercial buildings. Homes or small buildings with basements often have enough space and height for the needed drainage. One-pipe systems are reliable and require few repairs because of fewer moving parts.

Water hammer: A condition that occurs when hot steam comes into contact with cooled condensate, builds pressure, and pushes the water through the line at high speeds, slamming into valves and other devices. Water hammer can also occur in residential water systems. If the cause is not corrected, water hammer can damage the system.

Water hammer, however, is common in one-pipe systems because the condensate and steam frequently make contact with each other. Condensate is not removed from the system very efficiently. This is especially true during system startup when the piping and components are cold.

Several one-pipe system configurations exist. Steam boilers are partially filled with water to allow room for steam creation and expansion. *Figure 22* shows a basic one-pipe upfed gravity system. The term *upfed* means that the steam supply source is below the level of the heating terminals. Steam flows up a riser to the terminals.

As shown in *Figure 22*, the steam supply main rises from the boiler to a high point, then it is pitched downward to allow any condensate that develops to move on its own. The pitch should be no less than 1" per 20' of run, sloping down toward the condensate collection point. The piping normally remains its full size all the way to the last terminal takeoff.

When the condensate return is above the boiler water line, it is called a *dry return*. In this case, it is only considered to be dry as it is not flooded or filled with water—it carries water, steam, and some air.

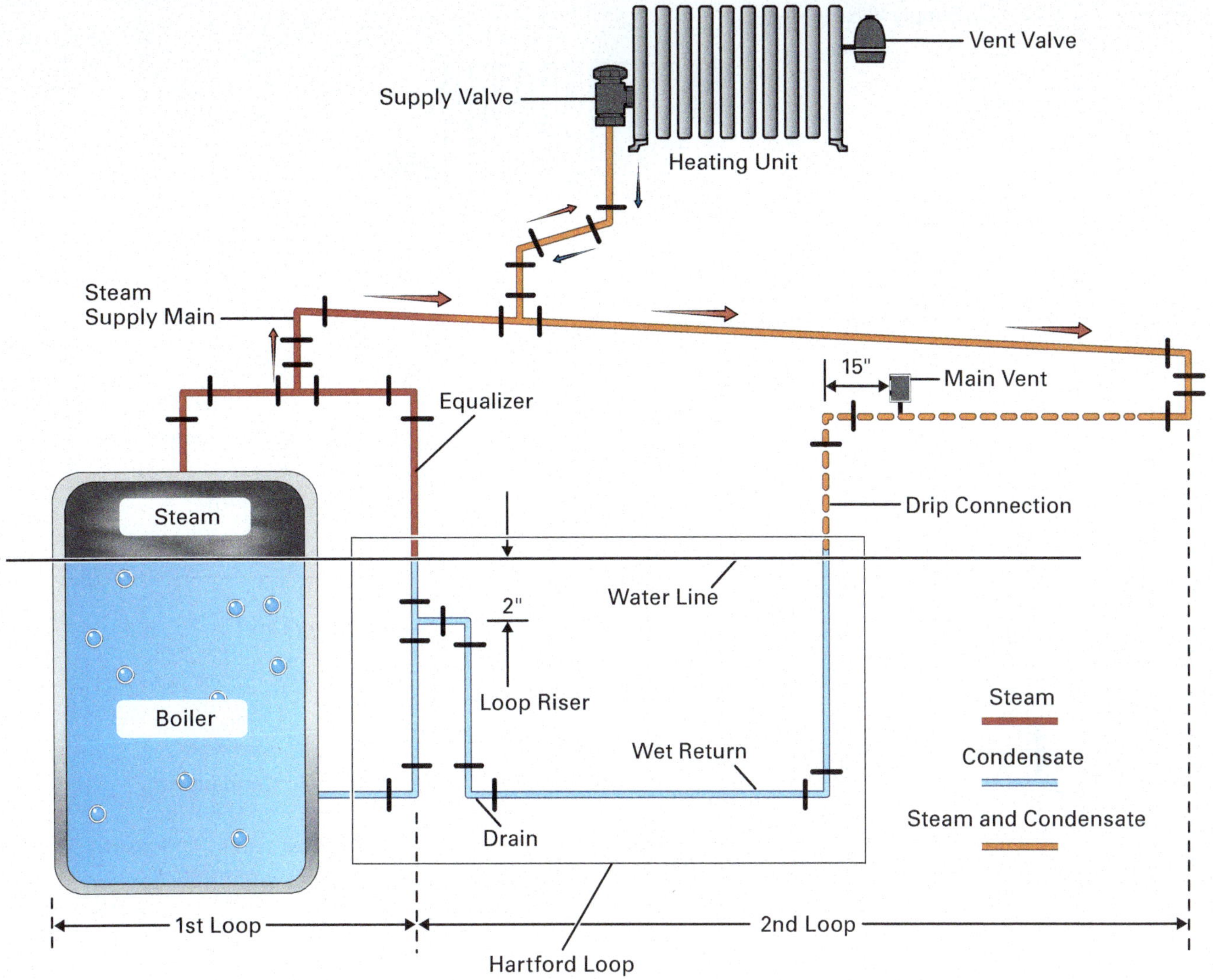

Figure 22 Basic one-pipe upfed system.

The section of return piping installed below the level of the water in the boiler is called the *wet return*. It is normally filled with water and free of any steam. When the system is in operation, the steam and condensate flow in the same direction in the horizontal mains and in opposite directions in the branches and risers. Because the steam and condensate flow in the same direction in the horizontal mains, this type of system is also called a *parallel-flow system*.

The *Hartford loop*, shown in *Figure 22*, is a pressure-balancing loop that equalizes the boiler pressure with the condensate return. This prevents reverse circulation or water leaving the boiler through the return line. The Hartford loop is technically two loops. The first loop is an equalizer line. The second loop is an extension from the steam main through the return and back to a connection at the equalizer loop.

3.1.2 Two-Pipe Systems

The two-pipe system uses separate pipes to supply the steam and return the condensate from the terminals.

A *thermostatic trap* is one type of trap found in steam systems. The traps are installed at the outlet of each terminal. Once the steam has given up its latent heat and condensed, the trap opens to pass the condensate, permitting more steam to enter the terminal. *Float and thermostatic (F&T) steam traps* are often used at the end of each steam main, at the bottom of each riser, and along the horizontal mains. This style offers advantages at these points of the system. *Figure 23* shows a basic two-pipe gravity system.

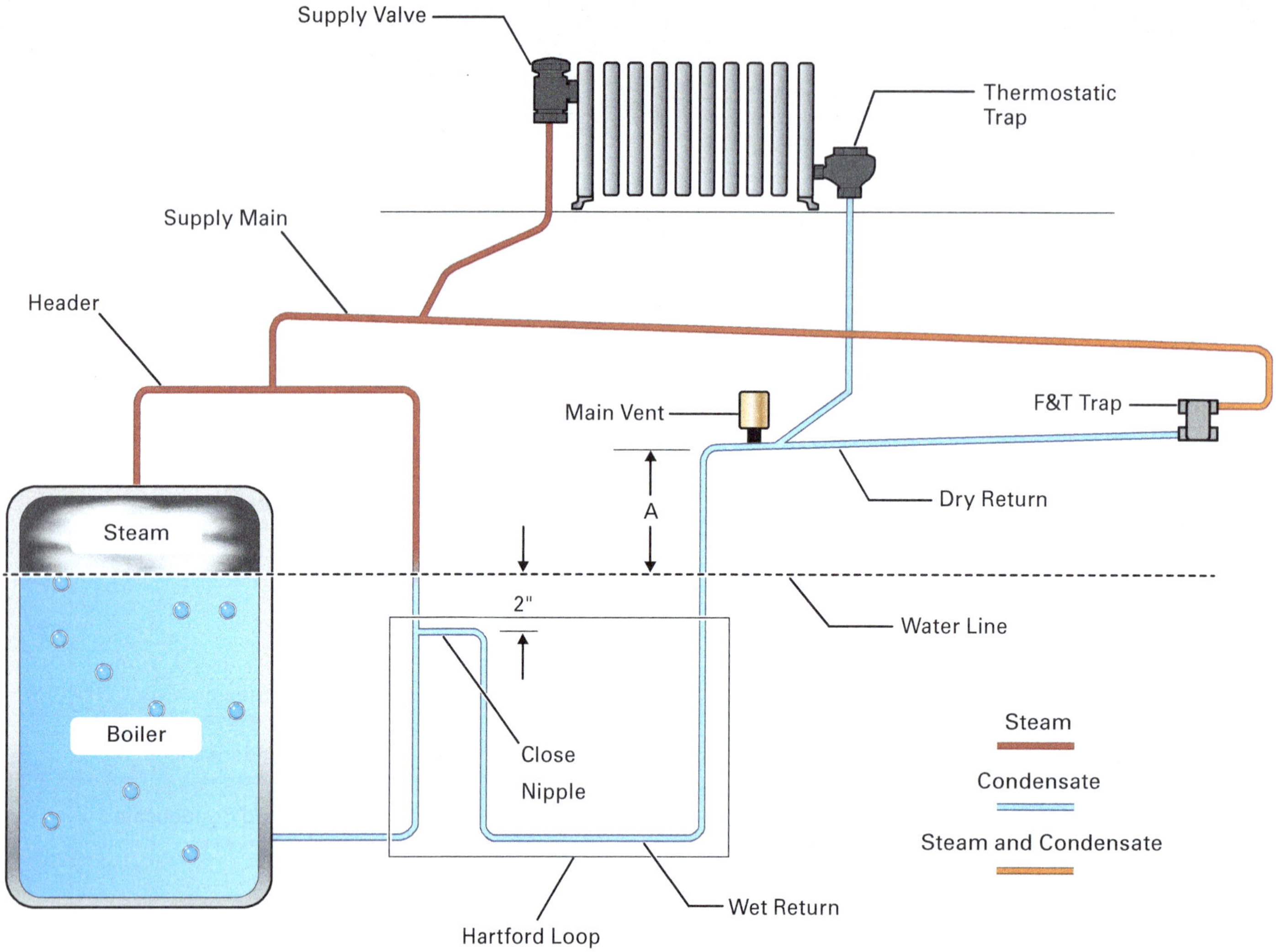

Figure 23 Two-pipe gravity system.

> **NOTE**
>
> Steam traps are covered in detail in NCCER Module 03306, *Section 4.0.0 Steam Traps*.

As steam systems increase in size and complexity, higher pressures are often needed to move the steam along. In addition, gravity alone isn't sufficient to return the condensate to the boiler. A receiver to collect the condensate and a condensate pump are commonly used for this purpose.

Condensate from the return piping system flows by gravity to a receiver. A condensate pump mounted on or near the receiver tank pumps the accumulated condensate into the boiler feedwater piping. *Figure 24* shows a two-pipe steam system with a single condensate receiver and pump.

Note in *Figure 24* that air vents have been eliminated from the design. With a mechanical condensate return, air is released from the system through the open vent pipe on the receiver. The vent opening must be above the boiler water level.

Vacuum Return Systems

A *vacuum return* system (*Figure 25*) is another option to assist the movement of condensate through the system. The system removes air from the condensate lines, leaving the lines below atmospheric pressure. The steam side of the system must remain at a positive pressure, with the traps continuing to separate steam from condensate. It is important to note that the vacuum pump does not actually move the condensate.

Operating with the condensate lines in a vacuum increases the pressure differential between the two sides of the system. This allows the steam to flow more readily and system warmup times are shortened. As a result, steam pressure settings at the boiler can sometimes be lowered for significant fuel savings and improved efficiency. Water hammer is also less likely.

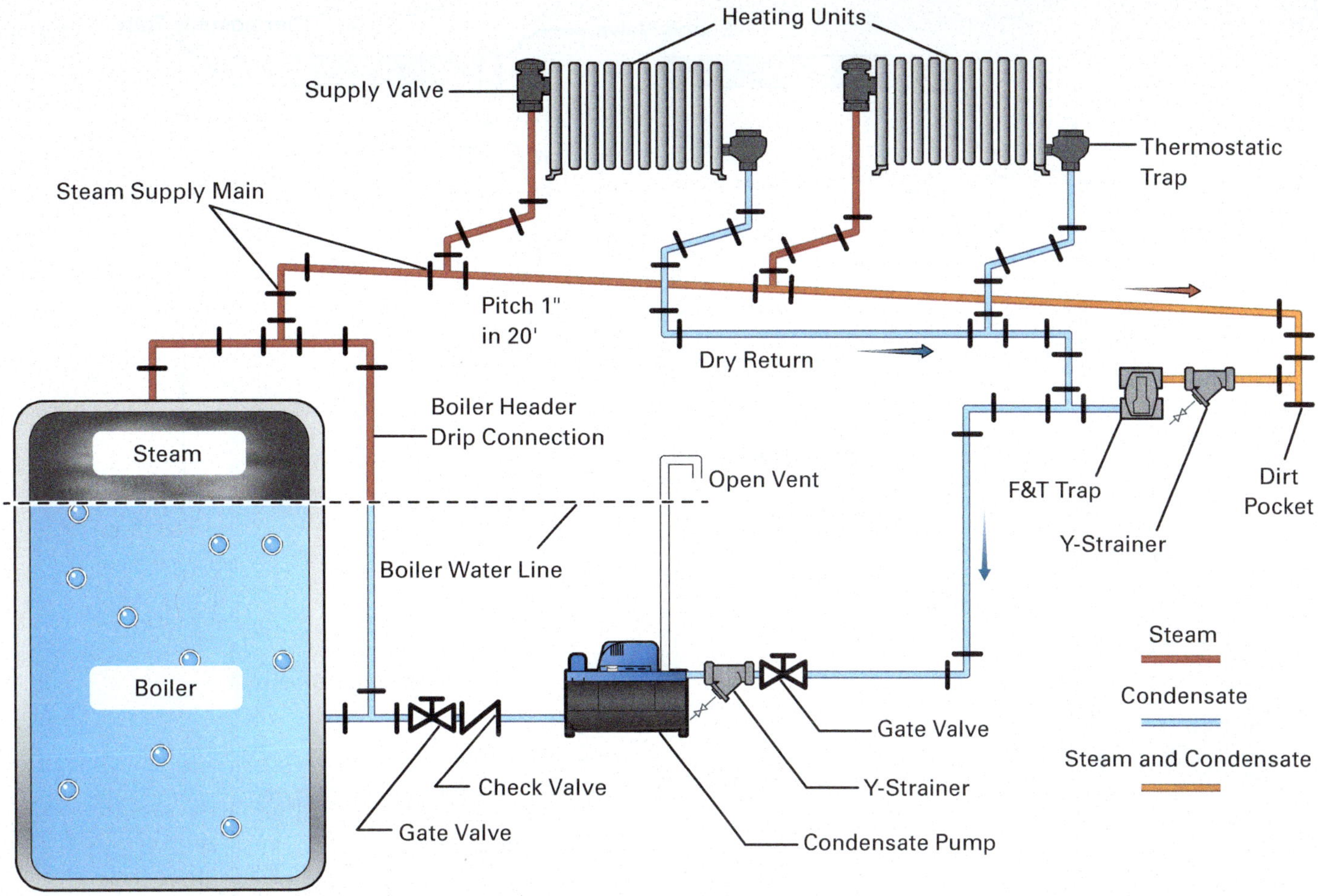

Figure 24 Two-pipe system with mechanical condensate return.

The vacuum pump cycles as needed to maintain a preset vacuum pressure in the condensate lines. However, the vacuum pump can't lift the condensate up to the boiler water level like a condensate pump.

Air leaks in the condensate piping can easily overwork the vacuum pump and prevent the system from reaching the needed vacuum pressure. Because the condensate piping is below atmospheric pressure, leaks aren't as obvious as they would be otherwise. The return piping must be leak-free to maximize the benefits associated with vacuum return and to prevent excessive pump operation.

Remember that water vaporizes to steam at a lower temperature when the water is below atmospheric pressure. For example, at a vacuum of 15" of mercury (Hg), water boils at roughly 180°F (82°C). By reducing the boiling point of the water, greater operating efficiencies are possible. Some vacuum systems, known as *variable vacuum systems*, take advantage of this by creating a vacuum in the supply side of the system, as well as in the condensate return piping. This obviously requires a vacuum pump with more capacity (expressed in cfm). Airtight piping systems are even more important in this scenario.

The vacuum pump controller in a variable vacuum system determines the vacuum level that the pump must maintain. The common range is from 3" to 8" Hg. In a heating system, the controller generally responds to outdoor temperatures. In a production application, it likely responds to steam demand. During mild weather or times of low production use, vacuum levels can be maintained at the lower value of 8" Hg. As the steam demand increases, the vacuum pressure can be reduced to 3" Hg. Hotter steam is then provided to the loads as the boiling point of the water is increased.

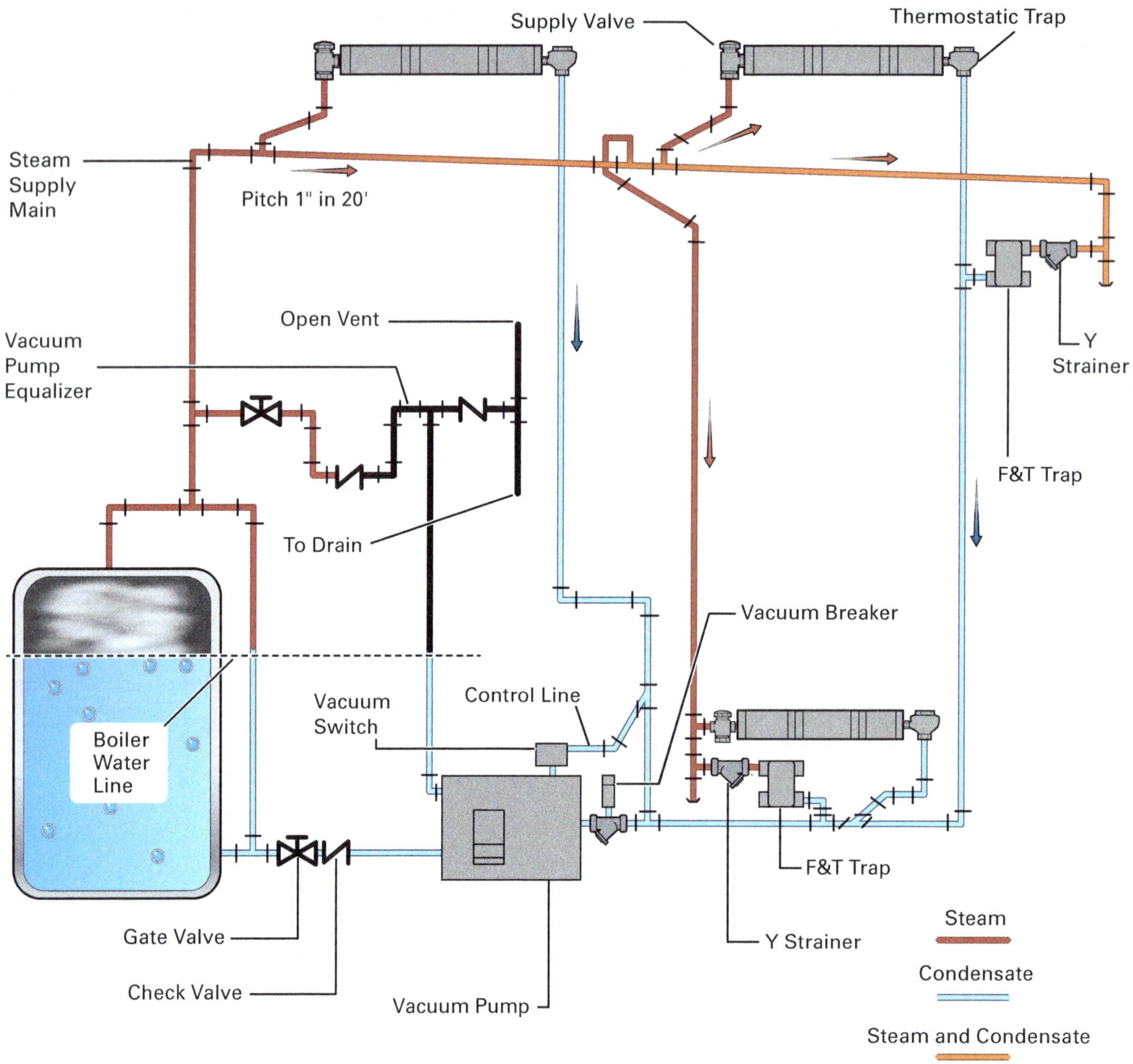

Figure 25 Vacuum return system.

3.2.0 Condensate Return/Feedwater System Components

Condensate is the natural byproduct of a steam system. Once the steam has condensed and given up its latent heat, the condensate must be removed as quickly as possible and returned to the boiler.

The following are the common components in a condensate return/boiler feedwater system:

- Condensate pump and receiver
- Vacuum pump
- Deaerating feedwater heater

3.2.1 Condensate Pump and Receiver

A condensate pump and receiver are used on steam systems to accumulate condensate and return it to the boiler. There are many variations available. Some systems may use a simple condensate receiver equipped with a single pump, while others may use a receiver with two pumps, referred to as a *duplex unit.*

Duplex units (*Figure 26*) may be equipped with an alternator arrangement. In this arrangement, pumps are automatically switched between primary and secondary on a timed basis. This helps balance the operating time of the two pumps. When a primary pump failure occurs, the unit automatically switches pumps to prevent the system from flooding and shutting down.

Figure 26 Duplex condensate receiver.

Figure 27 shows one example of a system with a condensate receiver package. This system uses a feeder pump controller installed at the boiler water line to control the pump on the condensate receiver.

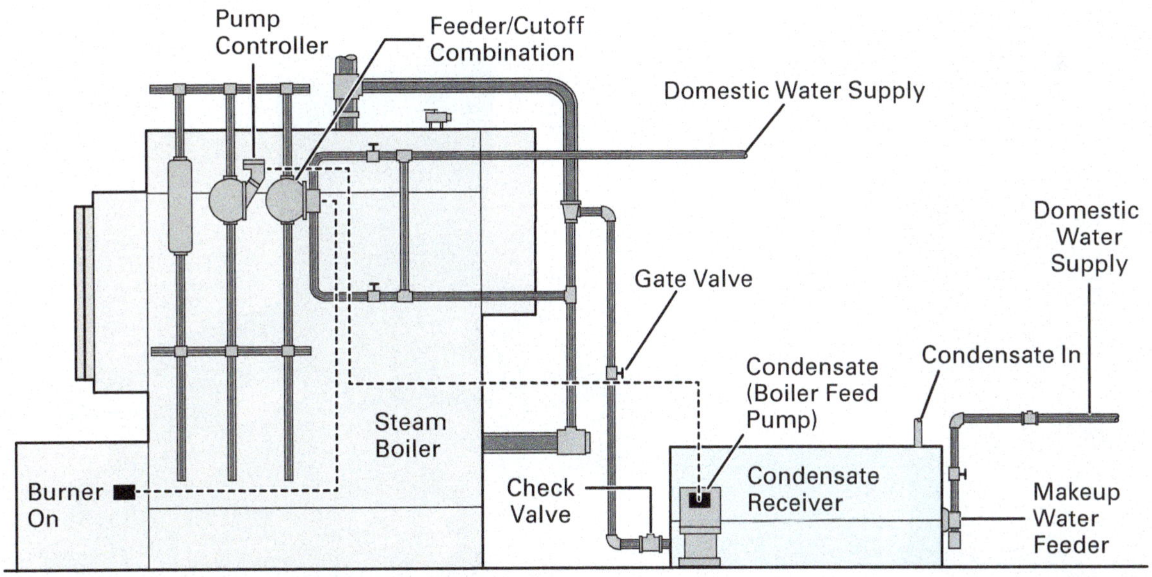

Figure 27 Typical condensate pump and receiver system.

Whenever the pump controller senses a drop in the boiler water level, it starts the condensate pump. Condensate water is pumped from the receiver into the boiler as needed to maintain the boiler water level. A check valve in the condensate pump discharge line prevents boiler water from backing up into the receiver when the pump is off. A gate valve in the line allows the pump to be isolated and serviced without draining the boiler.

The receiver in *Figure 27* is also equipped with a float-actuated makeup water feeder. If the level of condensate water in the receiver drops below the normal operating level, the control float drops and allows fresh water to enter from the domestic water supply. Note that there may also be a fresh water supply available directly to the boiler's water feeder. This source is used for initial filling of the system. Routing makeup water through the condensate receiver warms the incoming water before it enters the boiler.

A combined feeder/low-water cutoff provides boiler protection. In the event of a power failure, condensate pump failure, or a similar problem, this control will use its connected domestic water supply to maintain the boiler water level.

3.2.2 Vacuum Pumps

Prior to the cold start of a two-pipe steam system, the steam mains, load devices, and dry return lines are drained. This leaves the entire system full of air above the boiler's water line. As steam is generated, it must push the air out of the system. The loads begin to warm as the steam flows in and displaces the air.

The load closest to the steam source typically heats first. As air continues to move out of the system through atmospheric and other vents, additional loads receive steam. But it takes time for the system to vent the air and reach normal operating temperatures. Water hammer is also very likely to occur as the system warms up.

Vacuum return systems assist by mechanically removing the air from a system and enhancing condensate flow. By quickly exhausting the air and condensate from the system, the vacuum pump allows the steam to circulate more rapidly, reducing warm-up time and providing quieter system operation. A feed pump is still often required to pump the condensate from the receiver into the boiler.

The vacuum unit and condensate pump are often combined into a single unit, such as the unit pictured in *Figure 28*. However, the vacuum pump package can also be designed to allow the same pump(s) to function in both roles. *Figure 29* shows a duplex unit with pumps that can create a vacuum as well as pump water into the boiler.

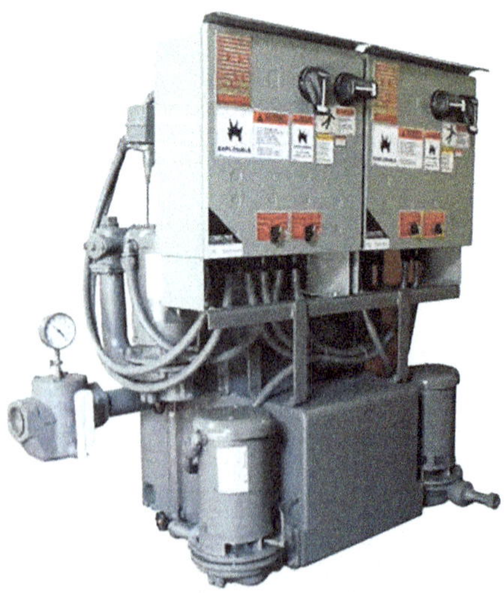

Figure 28 Duplex vacuum-return condensate unit.
Source: Courtesy of Xylem Inc.

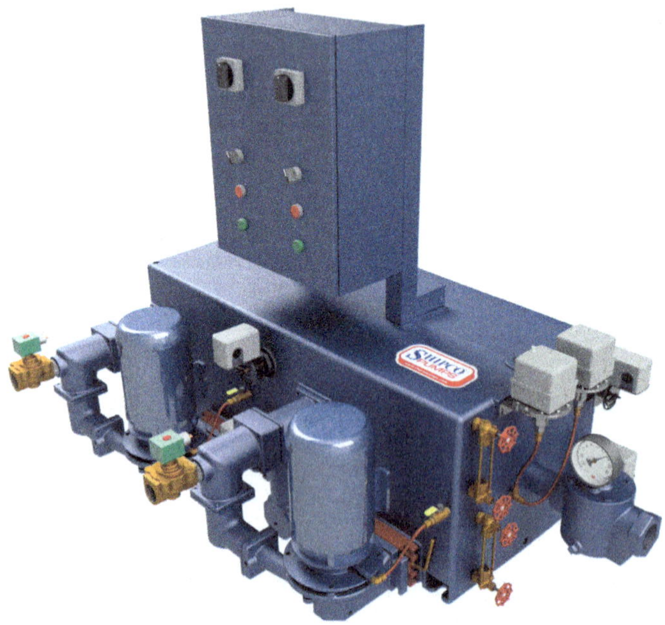

Figure 29 Duplex vacuum-return condensate unit with dual-functioning pumps.
Source: Photo used with permission from Shippensburg Pump Company, Inc., Shippensburg, PA 17257, www.shipcopumps.com

3.2.3 Deaerating Feedwater Heater

In some larger steam systems, the condensate water leaving the receiver does not get pumped directly into the boiler. Instead, it is pumped into a *deaerating feedwater heater,* also called a *deaerator.*

The deaerator (*Figure 30*) uses system steam to preheat the water before it is supplied to the boiler. The purpose is to drive out the dissolved oxygen, but also to deliver feedwater that is closer to the boiler's temperature. Although the oxygen content in the condensate is low, fresh water is mixed with it at the deaerator to keep up with normal steam losses. Such losses are low in heating applications for comfort, but many industrial uses require that some of the steam be released into the atmosphere or process. In large systems, some level of leakage, as well as losses associated with maintenance must also be made up with fresh water.

Figure 30 Stacked feedwater deaerator package.
Source: Photo used with permission from Shippensburg Pump Company, Inc., Shippensburg, PA 17257, www.shipcopumps.com

This is especially important in systems that lose 25% or more of the steam condensate from the circuit during normal operation, or systems that are drained and refilled regularly. Preheating fresh water is necessary because it carries large quantities of dissolved oxygen, and oxygen in the circuit initiates corrosion. Deaeration is also beneficial in high-pressure steam systems.

3.2.4 Flash Tanks

When hot condensate under pressure is released to a lower pressure, part of it vaporizes to form flash steam. This also happens when very hot condensate is discharged into a condensate return line where the pressure is below the saturation pressure of the condensate. Some of the condensate flashes to steam and flows along with the liquid condensate through the return lines back to the boiler. This happens more often in high-pressure steam systems. But many large steam systems must support both high- and low-pressure steam demands.

Flash steam causes an undesirable pressure increase in the condensate lines. *Flash tanks* are used in medium- and high-pressure systems to remove the flash steam from the condensate lines by venting it into a low-pressure steam line for reuse. In some cases, the flash steam is simply released.

The energy content of flash steam is the same as that of live steam. Use of the flash steam in a low-pressure steam main allows the energy to be used rather than wasted. The low-pressure flash steam from a high-pressure industrial process can be used to heat a building, for example.

When flash steam is used as an energy source, the condensate returns are piped to a *flash tank* (*Figure 31*). Flash tanks can be mounted vertically or horizontally. However, vertical mounting provides for better separation of steam and water.

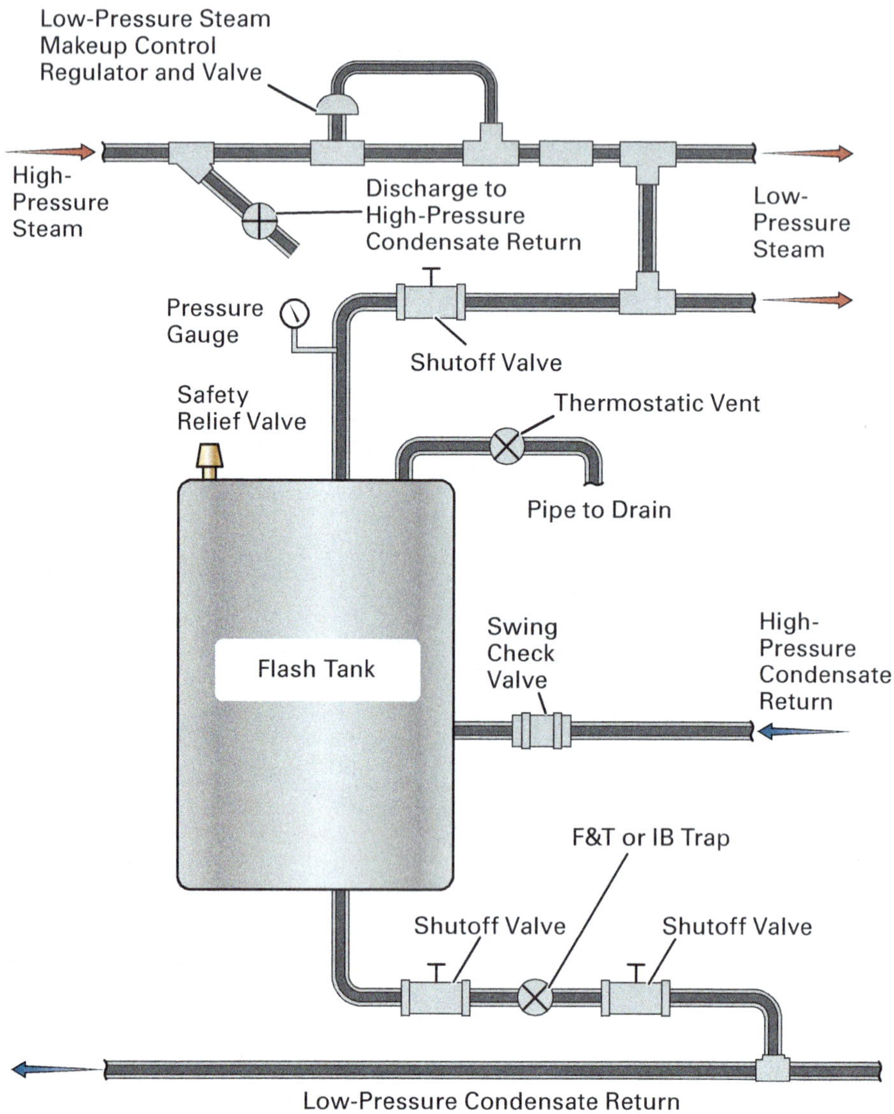

Figure 31 Flash tank piping connections.

When the condensate and any flash steam in the return line reach the tank, the high-pressure condensate is released into the lower pressure of the tank, causing some of it to flash. The flash steam is discharged into the low-pressure steam main. The remaining condensate is pumped back to the boiler or discharged to an appropriate drain.

The demand for steam on the low-pressure main may outpace the volume of flash steam available. When that is the case, a control valve and pressure regulator maintain the needed low-pressure steam volume using high-pressure steam regulated to the lower pressure. This arrangement is shown at the top of *Figure 31*.

Ideally, the demand for low-pressure steam from the flash tank is greater than or equal to the amount of flash steam collected. If it is not, the low-pressure system can be over-pressurized. A safety relief valve is installed at the top of the flash tank to protect the low-pressure system from excessive pressure.

If more than one condensate line feeds into a single tank, a check valve is installed in each line to prevent backflow. The top of the tank should have a thermostatic air vent to vent any air that accumulates. The bottom of the tank should have an inverted bucket or F&T steam trap.

3.2.5 Boiler Skimming and Blowdown

To make up for leaks and natural water consumption related to some processes, makeup water is added to maintain the proper water level in the boiler. This is in addition to any steam condensate that remains in the system. As a result of adding fresh water, sediment, dissolved salts, and similar organic matter gradually accumulate in the boiler. Remember that any water that leaves the system in the form of steam leaves behind all the solids the water contained. Some of these impurities collect as foam on the surface of the boiler water. Heavier material falls to the bottom of the boiler.

All foreign matter must be regularly purged from the boiler. The material is removed through one of two processes. **Skimming** refers to the process of removing surface contaminants. The contaminants are purged from the bottom through the process known as blowdown. Technically, skimming is a form of blowdown, but using the two terms helps avoid confusion. Boilers are equipped with one or more valves for this purpose.

Skimming: Removing lighter-than-water impurities from the surface of the water in a steam boiler.

Skimming valves allow the foam from the surface of the water to flow out. The foam largely consists of oil, grease, and lighter-than-water sediment. Foaming prevents steam bubbles from breaking away from the water's surface and hinders steam production. Not all boilers are equipped with skimming valves. If equipped, the valve is usually found at the normal water level of the boiler when installed.

Small boilers may not have a port for a skimming valve in the right place. *Figure 32* shows how the valve may be connected on a small boiler with no dedicated skimming port. To skim the boiler, the water level must be increased to the level of the valve connection. In other words, the boiler must be flooded.

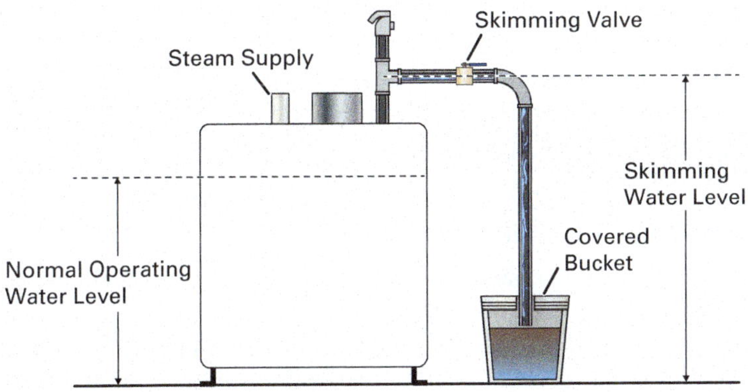

Figure 32 Alternate skimming valve location for small boilers.

Use of the surface blowdown valve to successfully skim off the impurities requires a little skill. Skimming must proceed slowly to encourage the foam and floating impurities to exit. Follow the boiler manufacturer's instructions when skimming the boiler water, as well as the safety guidelines of the facility. The following provides basic guidance for skimming a boiler:

WARNING!

Steam boilers operate at dangerous pressures. When working with or around boilers, follow all applicable manufacturers' instructions, OSHA regulations, and jobsite requirements related to the safe management of steam resources. Always wear the appropriate PPE, including gloves and eye protection.

On Site

Blowdown Separators

Some boilers are equipped with *blowdown separators*. These devices reduce corrosion and scaling by separating flash steam from blowdown water and removing dissolved solids. Blowdown separators also reduce the temperature and pressure of the blowdown water before it is discharged. Blowdown separators are sometimes required for compliance with environmental regulations and municipal plumbing codes.

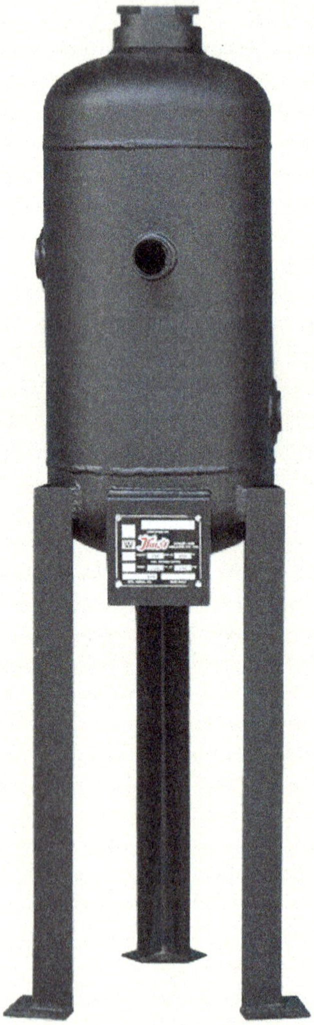

Source: Image courtesy of Hurst Boiler & Welding Co., Inc. of Coolidge, GA

Step 1 If the skimming drain line from the valve is not permanently piped to an appropriate drain, run a temporary drain line from the skimming valve. Alternatively, the water can be drained into a bucket.

Step 2 Operate the boiler until it reaches normal operating pressures and temperatures.

Step 3 Turn off the burner and allow the steam pressure to dissipate or release it very slowly.

Step 4 Manually lift the pressure relief valve to ensure that no pressure remains.

Step 5 Slowly and carefully open the skimming valve. While some water and foam may flow out, it often does not.

Step 6 Use the manual fill valve for the boiler to begin raising the water level slowly. This will cause water to begin flowing from the skimming valve.

Step 7 The skimming process continues until impurities are no longer visible, or until the water is cool enough to handle.

Blowing down a boiler involves draining water from the bottom of the boiler and replacing it with an equal amount of fresh water.

The blowdown valve at the bottom of the boiler is used to drain off the sediment and other impurities that settle out of the water. The impurities accumulate at the lowest point in the boiler. If the boiler is extremely dirty, a thorough blowdown may require that the boiler be completely drained and refilled with water one or more times. This process must be continued until the water discharged from the boiler runs clear.

On smaller systems, the boiler water discharged through the bottom blowdown valve is drained directly into a suitable receptacle where it can cool before entering the building drain. The blowdown water may be piped to a tank where it cools before going into the sewer. Remember that codes limit the wastewater temperature from a steam or hot-water boiler to 140°F (60°C). Boiler blowdown should always be performed as directed in the boiler manufacturer's service instructions. The following is an example of common blowdown procedures:

Step 1 Check the water level in the boiler to ensure it is correct as the burner continues to run.

Step 2 Partially open the bottom blowdown valve. Once the water starts draining, fully open the valve.

Step 3 Remain at the blowdown valve. Monitor the sight gauge during blowdown to make sure that the water level is not dropping faster than it can be made up.

Step 4 When the desired amount of sediment has been drained from the boiler, close the blowdown valve.

3.2.6 Boiler Water Treatment

The water used in a boiler must be kept clean to ensure efficient operation. Never add dirty or rusty water to a boiler.

Hard water can interfere with the efficient operation of a boiler. For this reason, hard water should be chemically treated with water softeners before being used. Scale, corrosion, fouling, and foaming can cause the following boiler problems:

- Scale deposits on heating surfaces increase boiler temperatures and lower operating efficiency.
- Corrosion can damage metal surfaces, resulting in metal fatigue and failure.
- Fouling clogs small passages and pipes with solid materials, thereby restricting circulation and reducing the heat transfer efficiency.
- Foaming results in overheating and can cause water impurities to be carried with the steam into the system.

Various chemical and mechanical methods of water treatment can be used to help reduce impurities. At a minimum, proper water treatment includes blowdown, pH control, and the addition of chemicals to neutralize other contaminants. The specific methods and devices used for water treatment are covered in NCCER Module 03308, *Water Treatment*.

3.3.0 Steam and Condensate Pipe Sizing

Proper steam and condensate line sizing is essential for any steam system to operate efficiently and provide the needed steam volume for each load. To size the piping system, the volume and pressure of the steam required must be known.

In many cases, especially in larger facilities with many steam loads, the steam system pressure is considerably higher than required at any one of the operating loads. This ensures that sufficient pressure remains to satisfy distant loads by compensating for the pressure drop of the piping and accessories.

Maintaining the steam pressure higher than needed downstream provides advantages. The steam pressure can then be reduced using pressure-reducing valves near the inlet to each of the loads. The size of steam distribution mains can be smaller to carry the same amount of energy, thus saving on installation costs. Smaller pipe sizes also reduce piping heat loss. Smaller pipe sizes also mean smaller valves, further reducing the cost.

Undersized piping must also be avoided. Piping that is too small causes excessive pressure drop and increases steam velocity. This increases system noise and promotes the erosion of internal piping surfaces.

It is common for a new load to be added to a steam system long after the initial installation. A piping system that is already undersized or marginal may prevent even a relatively small load from being added without increasing the size of the main steam line. Piping that is too large, however, creates unnecessary heat loss and higher installation costs. But the effects of under-sizing steam piping are far more significant.

Two primary methods are used to size steam piping: sizing by desired velocity and sizing to maintain an acceptable pressure drop. Systems with shorter mains and branch lines may be sized by velocity. Piping runs are most often sized by velocity first, then rechecked to ensure that acceptable pressure drops are not exceeded.

NOTE

Only attempt blowdown when the boiler is operating with a light load. Temporarily suspend the boiler's heating process if necessary to reduce water turbulence and avoid disturbing settled debris.

3.3.1 Velocity Sizing

Steam lines are typically designed for a velocity of 80' to 120' per second (fps), or 4,800' to 7,200' per minute (fpm). Where noise is not an issue and pressure drop is less important (very short piping runs), velocities up to 200 fps/12,000 fpm can be used. Again, these values are not recommended for the reasons stated earlier concerning undersized piping.

Table 2 provides a steam velocity chart for pipe sizing. As mentioned earlier, the volume of steam to be carried, stated in pounds per hour (lb/hr), must be known. The design steam pressure must also be known. Note that this chart is for Schedule 80 pipe (Sch 80). Charts are available for various pipe schedules. High-pressure steam systems require thicker Sch 80 pipe to ensure system integrity.

The following example shows how to determine the correct pipe size:

- A steam system must carry at least 1,500 lb/hr of steam through the steam main. The maximum velocity must not exceed 100 fps, and the design steam pressure is 100 psig. What is the appropriate pipe size for the steam main?

Step 1 Enter the left side of the chart at 100 psig. Since the velocity cannot exceed 100 fps, read across the chart horizontally at the row for 80 fps.

Step 2 Continue across the chart until you reach a value equal to or greater than 1,500 lb/hr. The closest value above the minimum capacity is 1,660 lb/hr.

Step 3 Read up the column to find the steam pipe size. The pipe size is 2".

When troubleshooting steam systems, the sizing chart can be used to check existing conditions. For example, if the system seems too noisy in operation, check the pipe size against the volume of steam being transported to see if velocities exceed the prescribed maximum of 120 fps for quiet operation. This happens when new loads are added to an existing system and the steam pressure is increased to provide the needed additional volume.

TABLE 2 Steam Velocity Chart

Capacity of Sch. 80 Pipe in lb/hr steam

Pressure (psi)	Velocity (ft/sec)	½"	¾"	1"	1¼"	1½"	2"	2½"	3"	4"	5"	6"	8"	10"	12"
	50	12	26	45	70	100	190	280	410	760	1250	1770	3100	5000	7100
5	80	19	45	75	115	170	300	490	710	1250	1800	2700	5200	7600	11000
	120	29	60	110	175	245	460	700	1000	1800	2900	4000	7500	12000	16500
	50	15	35	55	88	130	240	365	550	950	1500	2200	3770	6160	8500
10	80	24	52	95	150	210	380	600	900	1500	2400	3300	5900	9700	13000
	120	35	72	135	210	330	590	850	1250	2200	3400	4800	9000	14400	20500
	50	21	47	82	123	185	320	520	740	1340	1980	2900	5300	8000	11500
20	80	32	70	120	190	260	520	810	1100	1900	3100	4500	8400	13200	18300
	120	50	105	190	300	440	840	1250	1720	3100	4850	6750	13000	19800	28000
	50	26	56	100	160	230	420	650	950	1650	2600	3650	6500	10500	14500
30	80	42	94	155	250	360	655	950	1460	2700	3900	5600	10700	16500	23500
	120	62	130	240	370	570	990	1550	2100	3950	6100	8700	16000	25000	35000
	50	32	75	120	190	260	505	790	1100	1900	3100	4200	8200	12800	18000
40	80	51	110	195	300	445	840	1250	1800	3120	4900	6800	13400	20300	28300
	120	75	160	290	460	660	1100	1900	2700	4700	7500	11000	19400	30500	42500
	50	43	95	160	250	360	650	1000	1470	2700	3900	5700	10700	16500	24000
60	80	65	140	250	400	600	1000	1650	2400	4400	6500	9400	17500	27200	38500
	120	102	240	410	610	950	1660	2600	3800	6500	10300	14700	26400	41000	58000
	50	53	120	215	315	460	870	1300	1900	3200	5200	7000	13700	21200	29500
80	80	85	190	320	500	730	1300	2100	3000	5000	8400	12200	21000	33800	47500
	120	130	290	500	750	1100	1900	3000	4200	7800	12000	17500	30600	51600	71700
	50	63	130	240	360	570	980	1550	2100	4000	6100	8800	16300	26500	35500
100	80	102	240	400	610	950	1660	2550	3700	6400	10200	14600	26000	41000	57300
	120	150	350	600	900	1370	2400	3700	5000	9100	15000	21600	38000	61500	86300
	50	74	160	290	440	660	1100	1850	2600	4600	7000	10500	18600	29200	41000
120	80	120	270	450	710	1030	1800	2800	4150	7200	11600	16500	29200	48000	73800
	120	175	400	680	1060	1520	2850	4300	6500	10700	17500	26000	44300	70200	97700
	50	90	208	340	550	820	1380	2230	3220	5500	8800	12900	22000	35600	50000
150	80	145	320	570	900	1250	2200	3400	4900	8500	14000	20000	35500	57500	79800
	120	215	450	850	1280	1890	3400	5300	7500	13400	20600	30000	55500	85500	120000
	50	110	265	450	680	1020	1780	2800	4120	7100	11500	16300	28500	45300	64000
200	80	180	410	700	1100	1560	2910	4400	6600	11000	18000	26600	46000	72300	100000
	120	250	600	1100	1630	2400	4350	6800	9400	16900	25900	37000	70600	109000	152000

3.3.2 Steam Line Pressure Drop

Steam lines are generally sized based on velocity first, then checked to see if the pressure drop is acceptable. The allowable pressure drop varies widely, based on the system itself. Systems operating at 100 psig serving loads that require 80 psig, for example, cannot be sized for 25 psig allowable pressure drop. A more appropriate choice would be sizing for a 5 to 10 psig loss.

A low-pressure system must be sized for a much lower pressure drop. Even a 5-psig pressure loss is far too much for a 10-psig system. A maximum pressure drop of 0.5 to 1 psig would be more appropriate. Generally, the higher the steam pressure, the greater the pressure drop that can be tolerated in the piping system.

Figure 33 provides an example of a steam pressure drop chart. Note that this chart is designated for Sch 80 pipe. The pressure drop differs for a given volume of flow in other pipe schedules. This is because the inside diameter (ID) of the pipe differs. Sch 80 pipe has a smaller inside diameter than Sch 40 pipe.

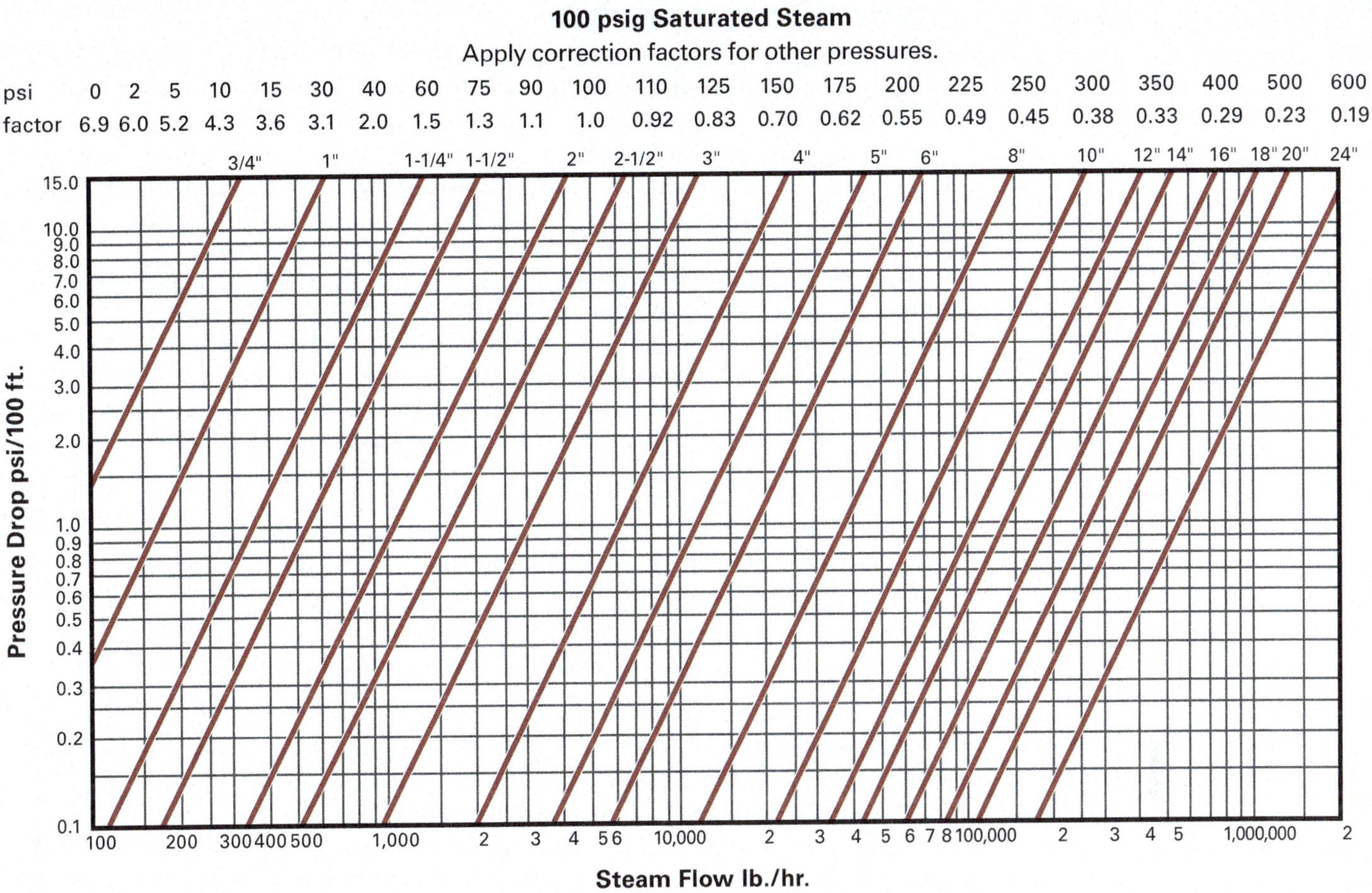

Figure 33 Steam pressure-drop chart.

Also note that the chart is based on a steam pressure of 100 psig. For pressures other than 100 psig, a correction factor must be applied to the value found in the chart.

The correction factors for various pipe sizes are listed in a row at the top of the chart. For example, if a steam system is to operate at 15 psig, then the value read from the chart must be multiplied by 3.6 to determine the correct pressure drop.

The values in the chart provide the pressure drop that exists in 100 **equivalent feet** of pipe. It is important to remember that various fittings are assigned an equivalent length of straight pipe that must be added to the measured straight lengths of pipe.

Equivalent feet: The sum of the measured length of pipe in a system plus the length assigned to fittings and other components that relate the fitting loss to feet of straight pipe. For example, a threaded 1" 90-degree elbow has an equivalent length of 5.2'.

Here is an example problem related to pressure drop. Using the values and resulting pipe size determined in the first example problem, calculate the pressure drop in 75' of pipe (equivalent length):

Step 1 Enter the chart at the bottom, locating the vertical line representing 1,500 lb/hr of steam flow.

Step 2 Follow the line vertically until it intersects with the diagonal line representing 2" pipe.

Step 3 Follow the horizontal line left from this point of intersection and read the resulting pressure drop per 100' of Sch 80 pipe. The result is 2.0 psig of pressure loss per 100' of pipe.

Step 4 To adjust for the actual length of pipe (75' in our example), multiply 2.0 psig by 0.75. The total pressure drop is 1.5 psig in this steam line.

3.3.3 Condensate Line Sizing

Steam condensate can carry as much as 15% of the energy originally used in the process of creating steam. With rising energy costs, it is important that steam systems capture and return the condensate back to the process to preserve this expended energy as well as the water.

Condensate line sizing can be more difficult than sizing steam lines, due to the variety of conditions found in the system. One variation is in the ratio of water to air and steam in the line. A single set of recommendations does not apply. The lines carrying condensate away from the load represent a different set of conditions than those found in condensate lines transporting condensate away from a trap.

Condensate piping systems are classified based on the design pressure. As described earlier in this module, gravity return describes those systems where the condensate line is at or very near atmospheric pressure, with the condensate generally returning to a vessel vented to the atmosphere. *Low-pressure returns* are those operating at a pressure above 0 psig but below 15 psig, while medium-pressure returns are those operating between 15 and 100 psig. Systems that operate with the condensate line at pressures above 100 psig are classified as *high-pressure returns*.

In all cases, the sizing of condensate lines is based on these factors:

- *Pressure* — The difference in pressure from one end of the line to the other must be considered. A significant pressure difference could cause some flash steam to be generated and may help promote flow in the line.
- *Condition of the condensate* — The designer must know whether the condensate is primarily liquid or whether a significant amount of flash steam can be expected.
- *Volume* — The anticipated volume of condensate flow is a very important consideration. The volume can change dramatically as the steam volume moving through the loads changes. During startup, the temperature difference between the load and piping system and the steam supply is at its highest. Far more condensate will be generated during this period as a result.

Condensate must be efficiently drained away from loads through traps, leaving little condensate behind. Generally, no significant pressure drop to consider occurs in the line from the load to the trap, so flash steam resulting from pressure changes is negligible. Also, no pressure differential assists flow because condensate movement is by gravity. However, the volume of the condensate changes dramatically from startup to operation at full load. The volume of condensate is roughly three times greater during startup than at full load. The flow of non-condensables (air) through the system must be accommodated as well.

The pipe size leaving the load must be large enough to ensure good drainage. In most cases, sizing the line for a volume twice that of the expected flow during full-load operation is sufficient to accommodate startup loads without being oversized for normal operating conditions. System designers know the volume of steam that will move through the load, and can determine the resulting volume of condensate based on the steam volume.

Discharge lines leaving traps should rely on gravity as the primary mover wherever possible. A pressure drop occurs at the trap. Flash steam may result from this pressure difference, more significantly at steam pressures above 30 psig. As condensate is discharged from the trap, some small amounts of live steam may also pass through. In the trap discharge line, both liquid and vapor are often moving together. This is known as *two-phase flow*, much the same as the expected flow of liquid and vapor in sewer systems. Undersized drain lines from the trap can dramatically affect the efficient movement of steam through the load.

As a result of heat loss through the piping, condensate will also gather in steam mains, especially during system startup. Steam mains should always be pitched, usually ½" per 10' of run, to allow the condensate to naturally drain toward a collection point. **Drip legs**, also known as *collection legs*, are installed on the bottom of the main along the horizontal run at reasonable intervals. The drip leg should be the same pipe diameter as the main to which it connects. A drip leg should also be installed at the base of all risers in steam mains. A steam trap is then connected at the base of each drip leg to drain the condensate away.

Drip leg: A drain line for condensate that forms naturally in a steam line that is routed to a steam trap for elimination.

Condensate from more than one load is best drained into a common condensate return line. The line must never become completely flooded with condensate, as this will impede the flow of steam throughout the system. It should slope toward the vented receiver tank or flash tank. In most cases, sizing for liquid-only condensate drainage should be for velocities at or below 7 fps.

3.4.0 Valves

The valves used to regulate and direct steam are generally the same as those used in water piping systems. These valves include gate valves, ball valves, globe valves, angle valves, and check valves. Always ensure that the valves used are rated for the pressure and temperature of the system.

3.4.1 Pressure-Reducing Valves

Pressure-reducing valves are used to reduce a higher steam pressure to a lower pressure. A pressure-reducing valve must maintain the pressure downstream consistently, despite small pressure fluctuations in the steam supply or variations in the load downstream.

Several types of pressure-reducing valves are available. *Figure 34* shows a type that controls the output pressure by balancing internal spring pressure acting on a diaphragm and the pressure of the delivered steam. The spring pressure can be adjusted manually to set the desired pressure downstream, but pressure-reducing valves can be automated as well. This allows them to be interfaced with building or process automation systems using a variety of sensor inputs.

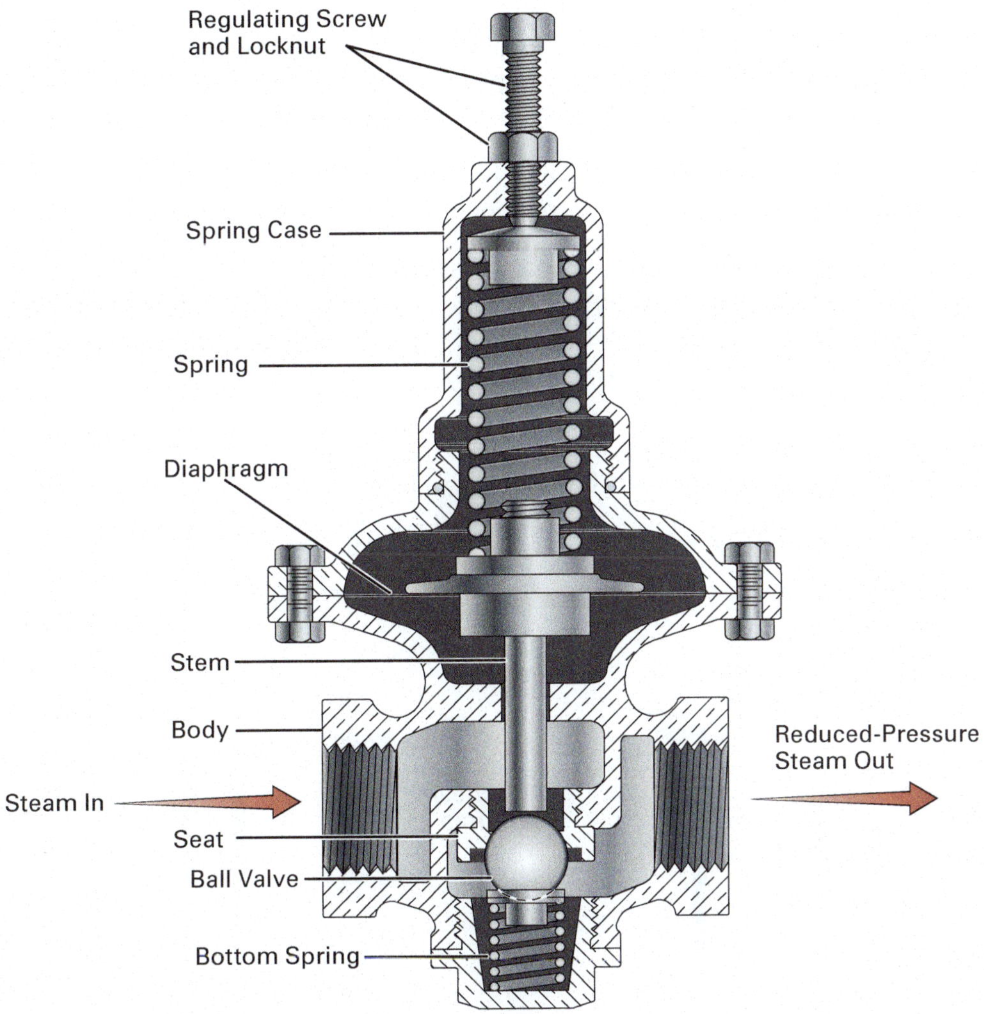

Figure 34 Pressure-reducing valve construction.
Source: Courtesy of Xylem Inc.

Figure 35 shows a typical single-stage pressure regulator. Notice that the pressure regulator in the figure can be bypassed by opening a globe valve. This allows steam to be supplied to the downstream load should the pressure regulator valve malfunction. The gate valve installed on each side of the pressure-relief valve allows the pressure-reducing valve to be isolated for repair without shutting down the system.

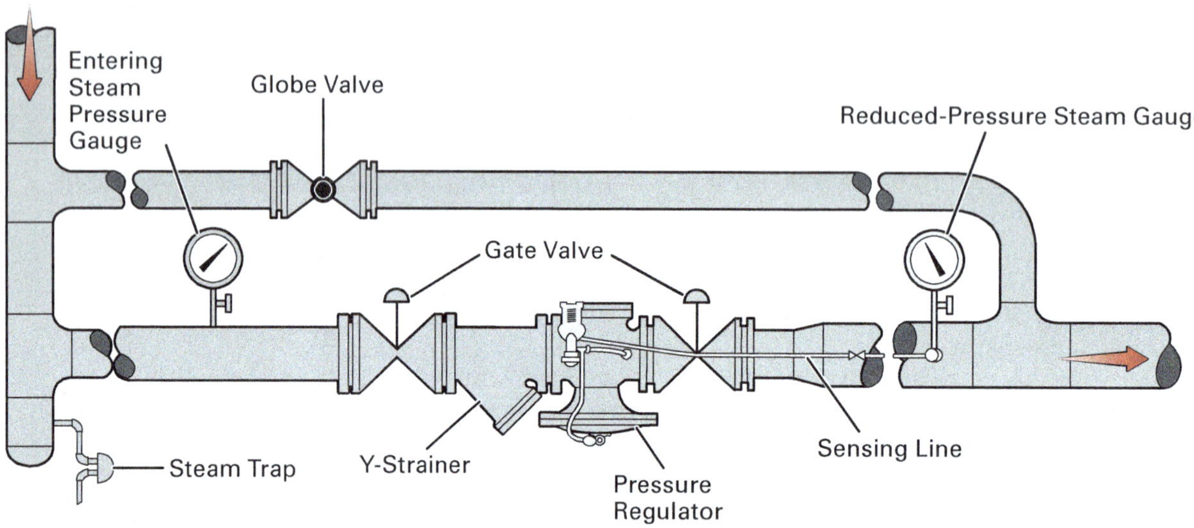

Figure 35 Single-stage pressure-reducing station.

Steam traps are used to prevent the accumulation of condensate on the input side of the pressure-reducing valve. The complete assembly consists of the pressure regulator and related valves and traps. This assembly is called a *pressure-reducing station.*

Large pressure reductions can be accomplished in one or more stages. *Multistage pressure reduction* is commonly used where an intermediate pressure is required or where the total pressure reduction is large. A large pressure reduction is one in which the ratio of downstream pressure to upstream pressure is greater than 50%. This ratio is referred to as the **turndown ratio**. *Figure 36* shows two pressure-reducing valves in series, creating a multistage pressure reduction.

Turndown ratio: In steam systems, the ratio of downstream pressure to upstream pressure as it applies to steam pressure-reducing valves. More than a 50% turndown ratio is generally considered a large pressure reduction.

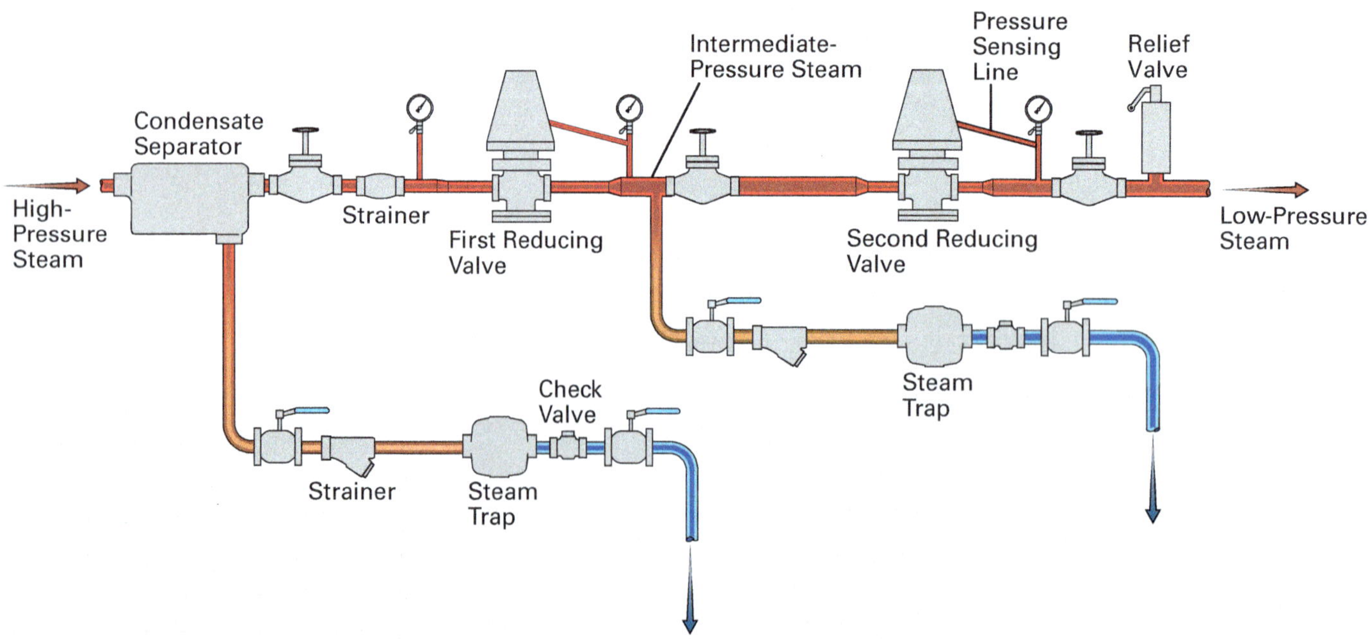

Figure 36 Steam pressure-reducing valves connected in series.

Pressure-reducing valves can also be installed in parallel. This may be done when seasonal or production-related changes in the desired steam pressure are needed. The steam supply is diverted to one valve or the other. This arrangement can also be used to create a low-flow/high-flow scenario.

Figure 37 shows a pressure-reducing station with valves installed in parallel. The valves have different settings that can provide sequenced operation. A smaller regulator valve sized to handle about one-third of the maximum load is connected in parallel with a larger valve sized to handle the remaining two-thirds of the load.

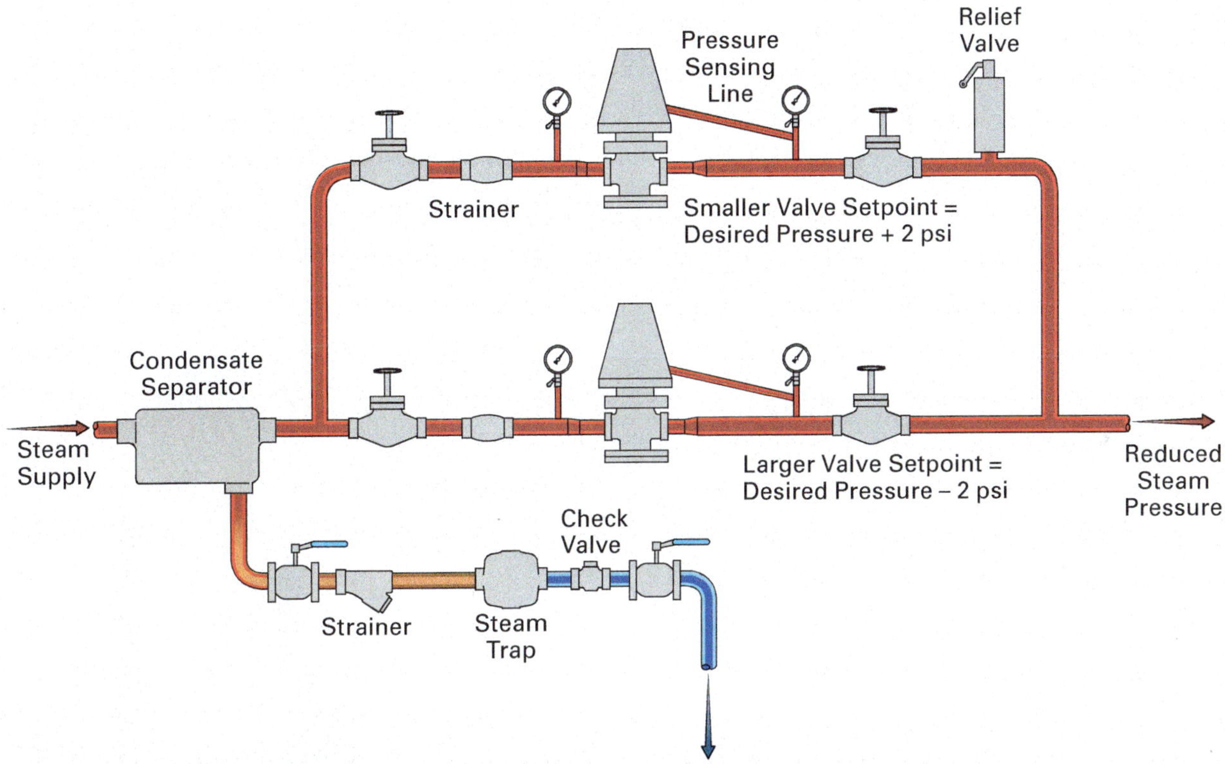

Figure 37 Steam pressure-reducing valves connected in parallel.

The smaller pressure regulator is set for a slightly higher pressure than the larger one. When the load is low, the smaller regulator handles the load. As the load increases to the point where the outlet pressure falls 2 psi below the setpoint, the larger regulator begins to pass steam and provide the additional steam volume.

Factors that should be considered when installing pressure-reducing include the following:

- Follow the manufacturer's guidelines for proper sizing. Oversized valves tend to exceed the desired pressure output and experience greater wear. Undersized valves are unable to pass enough steam to satisfy the downstream load.
- Pressure-reducing valves operate properly in one direction only. Make sure the valve is properly oriented in the line. Most have an arrow on the valve body to indicate the proper direction of flow.
- Many pressure-reducing valves must be installed horizontally. Provisions for condensate drainage from the steam main are usually found just ahead of the valve.
- Pressure gauges are installed near both the inlet and outlet. Valves that use a sensing tube for remote pressure input should also have a gauge installed adjacent to the input tube.

- Dirt and other foreign matter cause most valve failures. Use strainers upstream of the valve. Strainers should be removed and cleaned soon after a new installation has been placed in operation.

3.4.2 Thermostatic Valves

Thermostatic control valves control the flow of steam based on temperature inputs. These valves can be applied in many ways. Possible applications include the following:

- Comfort heating radiation
- Steam water heaters
- Steam-jacketed vessels
- Various component molds
- Tank heating coils

In the simplest applications, a thermostatic valve is used to control the steam flow to a radiator or similar terminal to provide zone temperature control (*Figure 38*). The temperature setting is changed manually. Many industrial thermostatic valves are self-contained, with a liquid-filled temperature-sensing element attached directly to the valve (*Figure 39*).

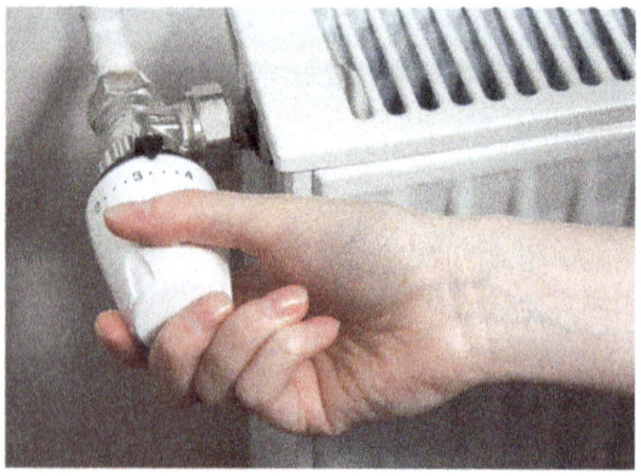

Figure 38 Typical radiator thermostatic control.
Source: morisfoto/iStock/Getty Images

Figure 39 Self-contained thermostatic steam valve.
Source: Courtesy of Danfoss

Various thermostatic elements are available for a given valve, with the selection dependent upon the desired temperature range. As temperature at the sensing element increases, the liquid in the element expands and applies a closing force against the plunger or valve spring. Setpoint adjustments change the spring tension, thereby changing the force required to open or close the valve.

The installation process for thermostatic valves is similar to that of pressure-reducing valves. Although a pressure gauge upstream of the valve is generally recommended, local downstream pressure doesn't usually need to be monitored. Instead, thermometers or temperature sensors should be placed in the controlled medium to provide a quick visual indication of the temperature.

3.0.0 Section Review

1. True or False? Water hammer rarely occurs in one-pipe steam systems because the condensate is very efficiently removed.
   a. True
   b. False

2. Why is a deaerator added to some steam systems?
   a. To drive dissolved oxygen out of fresh feedwater.
   b. To make use of flash steam that often forms in condensate lines.
   c. To lower the temperature of condensate, creating a vacuum in condensate lines.
   d. To remove water from the top of the boiler, providing space for steam at startup.

3. Which of the following should be given primary consideration when sizing the piping of a steam system?
   a. Flash tank capacity
   b. Steam volume and pressure
   c. Condensate line size and lengths
   d. The number of steam traps to be used

4. At what turndown ratio are multiple stages of steam pressure reduction normally considered and applied?
   a. 30%
   b. 50%
   c. 65%
   d. 80%

4.0.0 Steam Traps

Objective

Identify and explain how to install and maintain various types of steam traps.

a. Identify various types of steam traps.
b. Explain how to install steam traps.
c. Explain how to troubleshoot and maintain various steam traps.

Performance Task

2. Disassemble and maintain instructor-selected steam traps.

The steam trap is one of the most important components in steam systems. Steam traps typically require more attention than other parts of the system because they have moving parts exposed to extreme heat. A steam trap acts as a self-actuating drain valve that maintains effective heat transfer in the steam system by consistently purging it of condensate and air. The trap must efficiently discharge the condensate for the system to function properly.

As steam cools, it condenses to its liquid form. Most of this condensation process takes place at the load, such as a heat exchanger where steam is used to heat water (*Figure 40*). However, condensation can also form in the steam piping system. If the condensate is exposed to superheated steam, it can be propelled at a high rate of speed into elbows and other fittings used to change direction. This essentially defines water hammer.

Figure 40 Installed steam traps.

Air is always present during system startup. Air is also dissolved in fresh water used to make up for any losses. In addition, the feedwater may contain carbonates that dissolve and release carbon dioxide (CO_2) gas. Air and CO_2 can cause problems in a steam system.

Air can reduce the efficiency of the steam at the heat transfer surfaces (loads) and encourage corrosion. CO_2 supports the formation of carbonic acid, which is extremely corrosive. Scale and debris found in steam systems move with the condensate to sensitive areas. One of these sensitive areas is the steam trap. Thus, properly installed steam traps typically have strainers to protect them from debris.

4.1.0 Steam Trap Types

The general categories of steam traps are as follows:

- Mechanical
- Thermostatic
- Thermodynamic

Various materials are used to make steam traps including cast iron, stainless steel, forged steel, and cast steel. Steam traps are fitted to the pipe with screwed, flanged, socket-welded, or compression fittings.

4.1.1 Mechanical Steam Traps

Mechanical steam traps respond to the difference in density between steam and condensate. A typical mechanical trap is called an *inverted bucket trap* and has only two moving parts—the valve lever assembly and the bucket (*Figure 41*). This type of trap is often installed in the drain line between the steam-heated unit and the condensate return header.

Prior to initial startup of the load, there is no pressure on the inlet side of the trap and the inverted bucket is in the down position with the discharge valve open. If the trap is properly primed, it has water in it. As the steam moves through the load, the pressure increases, pushing the water out of the trap. As steam enters the trap, it collects in the top of the bucket because it is less dense than the water beneath it. Once steam fills the bucket, the bucket becomes buoyant and floats up, closing the discharge valve. The bucket remains up until the steam cools and condenses. The bucket then loses its buoyancy, falls, and opens

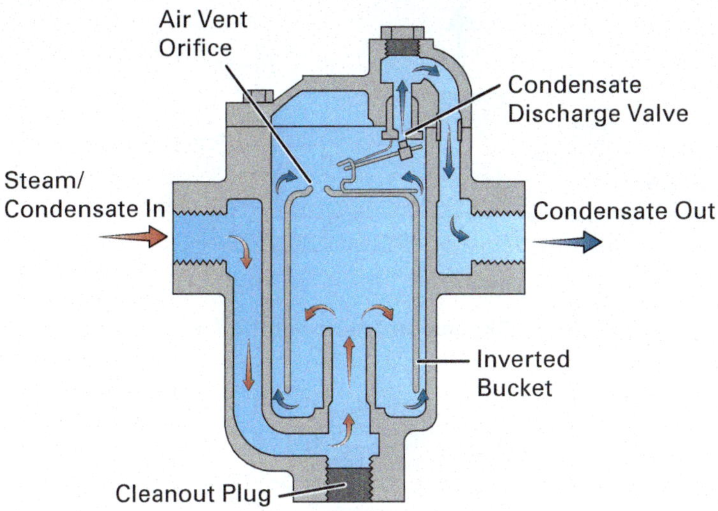

Figure 41 Inverted bucket trap.

the discharge valve again. Condensate is pushed out until enough steam again enters into the bucket to force it up.

Since the bucket relies on the steam in the bucket condensing to reduce its buoyancy, air trapped in the system creates a problem. Since the air does not condense, the bucket remains up and keeps the trap closed while condensate builds up behind it. An air vent is used to allow non-condensable gases to escape from the bucket.

The inverted bucket trap operates intermittently, responding to the accumulation of condensate by repeatedly opening and closing to dispose of the condensate.

Inverted bucket traps offer these advantages:

- Handle high-pressure steam
- Resist damage from pressure surges and water hammer
- Tolerate freezing if the body is made of a ductile material

The disadvantages of inverted bucket traps include a limited air-discharge capacity and a tendency to be noisy. They are not applied where rapid removal of air from a system or load is important.

NOTE

Inverted bucket traps must be partially filled with water to operate. When a new inverted bucket trap is installed or reinstalled after servicing and cleaning, prime it by filling it with water.

4.1.2 Thermostatic Steam Traps

Thermostatic steam traps respond to temperature changes in the steam line. These traps open when the cooler condensate is present and close when hotter steam is present. Four types of thermostatic steam traps are illustrated in *Figure 42*. The following describes the steam traps shown:

- *Liquid-expansion thermostatic trap* — Remains closed until the condensate cools below the steam-system design saturation temperature, which opens the thermostatic element downstream of the valve. This discharges the cool condensate at a constant temperature.
- *Balanced-pressure thermostatic trap* — Opens when condensate cools slightly below the steam saturation temperature at any pressure within the trap's range. The discharge valve is controlled by a liquid-filled bellows. The discharge valve closes when hot condensate vaporizes the liquid in the bellows. It opens when the condensate cools enough to lower the pressure inside the element. Balanced pressure traps are smaller than mechanical traps, open wide when cold, readily purge air, and are unlikely to freeze.
- *Bimetal thermostatic trap* — Relies on bimetal strips that respond to cool condensate by bending to open the discharge valve. The thermostatic bimetal element is set at a temperature that allows it to differentiate between steam and condensate. When steam hits the bimetallic strips, the strips expand and

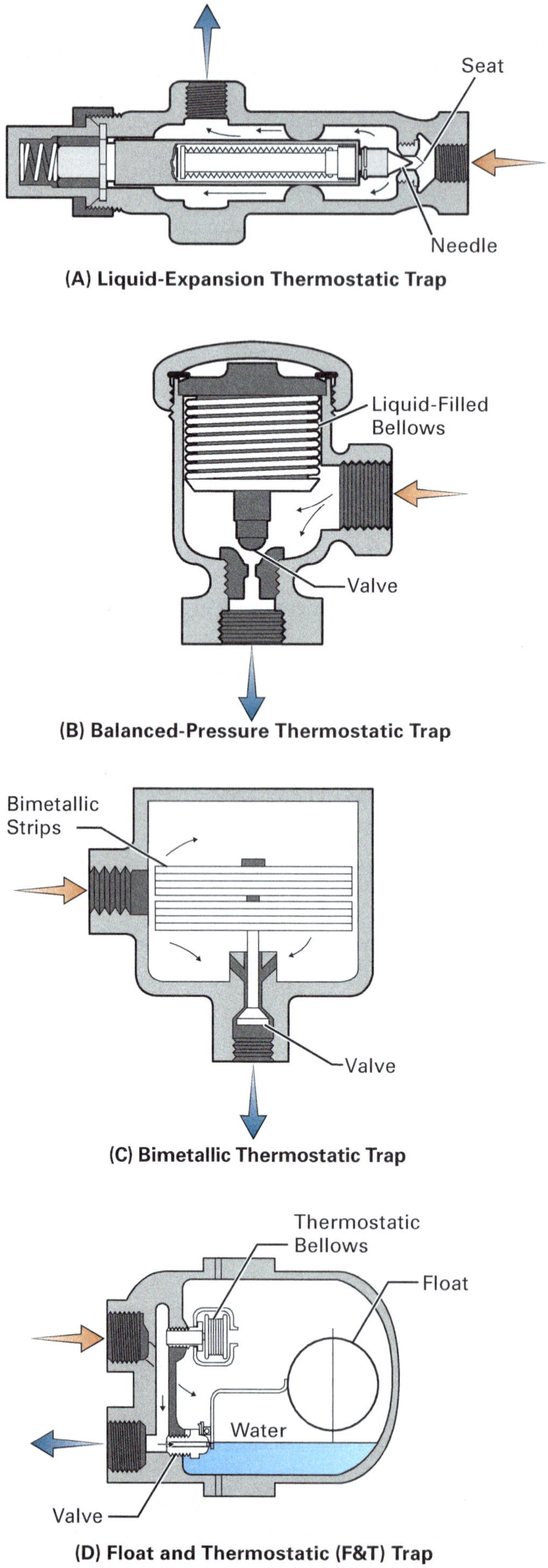

Figure 42 Thermostatic steam trap designs.

close the discharge valve. Considerable cooling is required before the discharge valve reopens. The closing force of some designs varies with the steam pressure, approximating the response of a balanced pressure trap. This trap has a large capacity but responds slowly to process load changes. However, it vents air very well. It is not easily damaged by freezing, water hammer, or corrosion and it can handle the temperatures of superheated steam.

- *Float and thermostatic (F&T) trap* — A float rises when condensate enters and opens the main valve that lies below the water level. At the same time, another discharge valve at the top of the trap, operated by a thermostatic bellows, opens to release air. This trap design vents air well and responds to broad and sudden pressure changes, discharging condensate continuously. In the failure mode, the main valve is closed but the air vent is usually open. F&T traps should always be protected from freezing because they usually hold some water.

4.1.3 Thermodynamic Steam Traps

Thermodynamic steam traps use the heat from hot condensate and steam to control the opening and closing of the trap. The traps operate well at higher temperatures and can be installed in any position because their apparatus does not depend on gravity.

The thermodynamic disc steam trap is widely used. The only moving part in this simple unit is the disc (*Figure 43*). On start-up, steam and air pressure raise the disc. Condensate that arrives, along with air, flows freely through the trap. There are several outlets around the perimeter of the housing, although only one is shown in *Figure 43*.

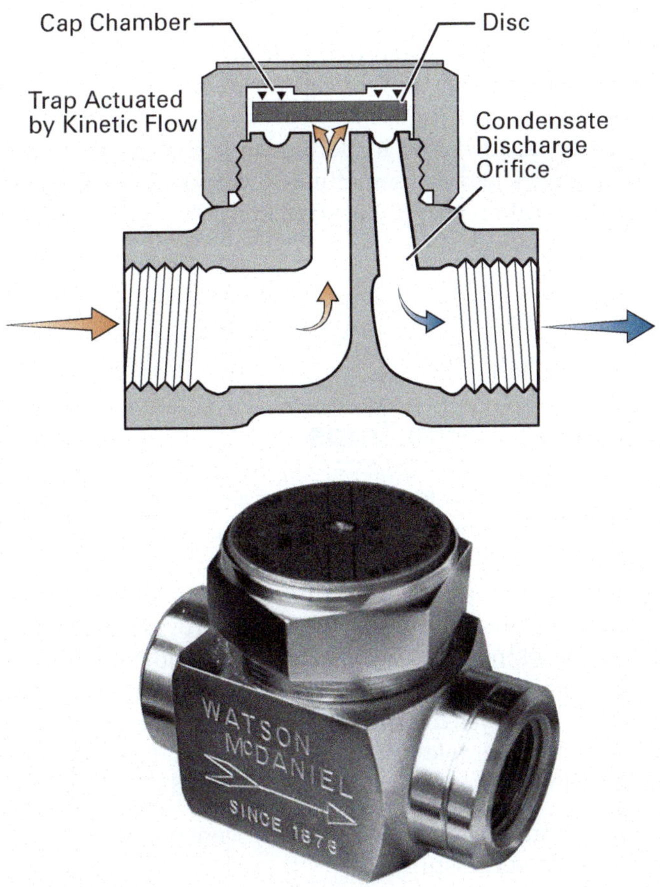

Figure 43 Thermodynamic stainless-steel steam trap.
Source: (Photo) Watson McDaniel Company

Hot condensate entering the chamber beneath the disc experiences a slight pressure loss, and flash steam forms. The flash steam moves rapidly through and into the cap chamber. This results in reduced pressure beneath the disc.

As the disc moves down, the flash steam fills the space above it and becomes trapped, increasing the pressure on the disc until it seats, blocking the flow. As seconds pass, the flash steam trapped above the disc cools and condenses, lowering the pressure above. As fresh condensate enters the chamber beneath the disc, the disc is pushed up, allowing air and condensate to flow out once again. Most thermodynamic traps will remain closed for 20 to 40 seconds per cycle.

The thermodynamic steam trap can withstand freezing temperatures, high pressure, highly superheated steam, and water hammer without damage. The audible clicking of the disc indicates when a discharge cycle occurs, and the sound of disc movement helps in troubleshooting. Thermodynamic steam traps are also very simple to disassemble and maintain.

Thermodynamic traps can be noisy. Also, these traps do not work well if there is only a small pressure difference across the trap. An oversized trap tends to cycle more rapidly and experiences higher wear as a result.

4.1.4 Fixed-Orifice Steam Traps

The fixed-orifice steam trap has a single opening of a predetermined size that passes a predictable volume of condensate at specific pressures. Condensate flow rates, however, rarely remain precise from startup to full load. But the fixed-orifice trap can be used successfully in the right application.

The fixed-orifice trap is merely an orifice with no moving parts. A strainer upstream is essential to keep it from being blocked with debris.

4.1.5 Strainers

A strainer should be installed ahead of the steam trap in most cases. The scale and corrosion in a steam system must be captured before entering a trap, thereby blocking or damaging it.

Strainers must be cleaned and inspected on a regular basis. Strainers are available in many styles and are sometimes incorporated into the steam trap. *Figure 44* shows three different types of strainers.

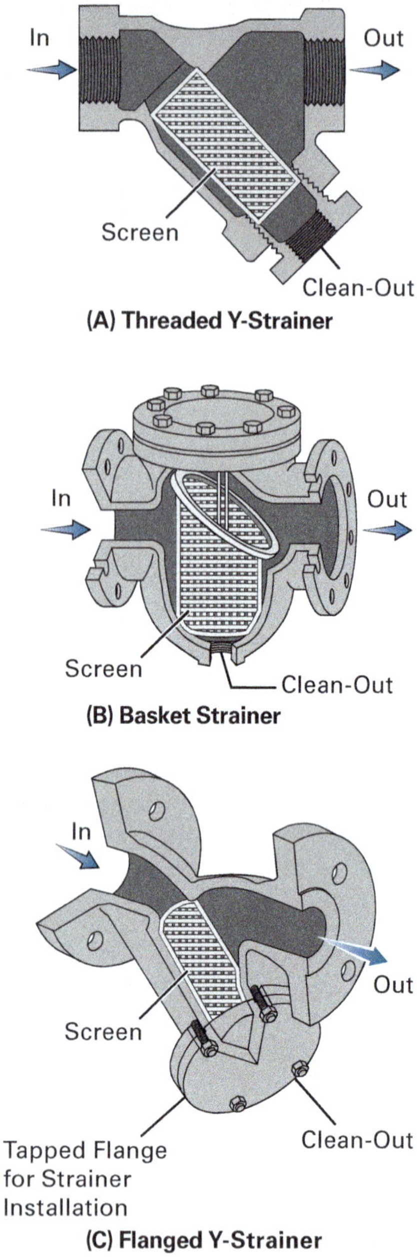

Figure 44 Strainer styles.

4.2.0 Installing Steam Traps

Steam traps are installed at the lowest points of a system where condensate naturally collects. A strainer should be installed immediately upstream of the trap when practical.

Steam trap manufacturers have resources that detail how to service and install their products. Plans and specifications also provide important guidance. When servicing or working with an existing system, do not reduce the quality and specifications of any piping or component without the acknowledgement of a qualified engineer. The system's integrity, as designed, must be maintained.

Figure 45 shows one example of a trap installation. The following provides some general guidelines for steam trap application and installation:

- Provide a separate trap for each piece of equipment or apparatus. Traps rarely serve multiple loads successfully.
- Install an isolation valve ahead of the trap and/or close to the steam header to allow for maintenance or modifications to a live system.

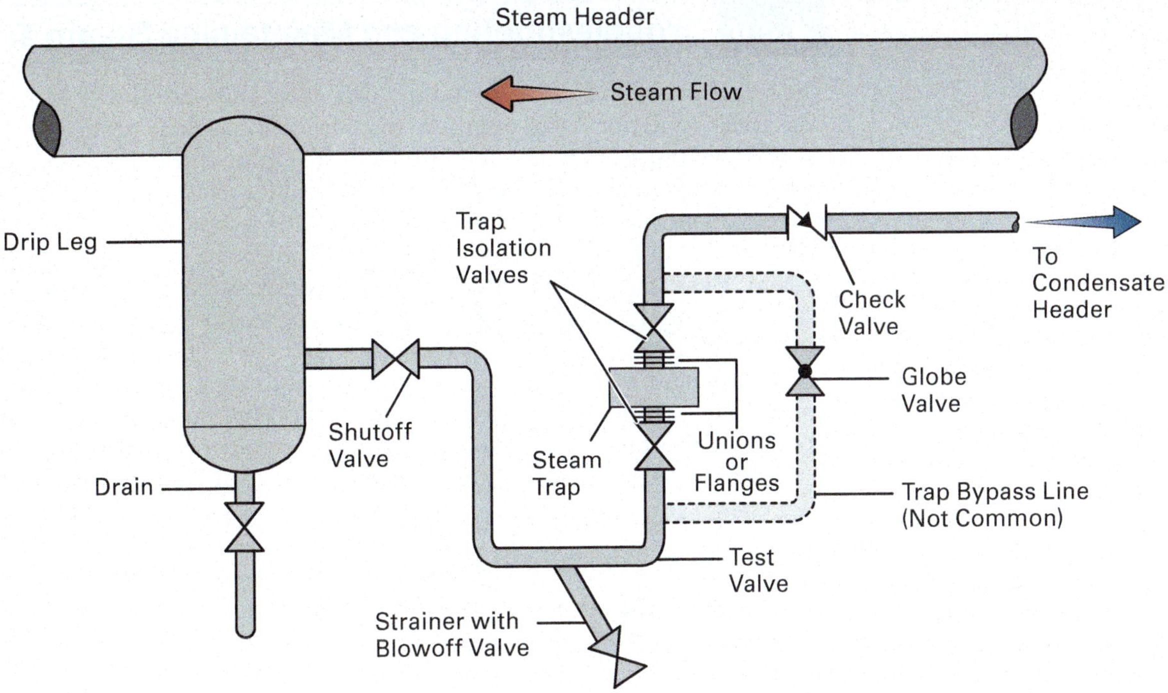

Figure 45 Basic trap installation.

- Install a strainer ahead of each trap to keep rust, dirt, and scale out of the trap. Make sure the strainer is easily accessed for cleaning.
- Provide unions on both sides of the trap for its removal and installation.
- Install a test valve downstream of the trap to allow observation of the discharge when testing.
- Install a check valve downstream to prevent condensate backflow during shutdown or in the event of unusual operating conditions.
- Install a downstream shutoff valve, when practical, where the trap's discharge line joins another line.
- Do not install a bypass around a trap unless it is clearly specified. Bypasses are an additional expense to install and can be left open accidentally, resulting in the loss of steam and decreased efficiency. If a bypass must be installed, a globe valve is used in the bypass line to allow and manually control condensate flow while the primary trap is serviced.

Going Green

Steam Losses

According to the US Department of Energy, steam traps can quietly cause a loss of up to 20% of the steam generated in a system. Of course, losses between systems vary significantly, but the cost of the wasted energy can be enormous.

However, steam traps aren't alone as a source of wasted steam energy. This graphic shows the collection of average losses that occur in a system that has not been optimized through proactive engineering and maintenance programs that directly address these sources. If you work with large steam systems, you'll likely be encouraged to look for opportunities to save energy daily.

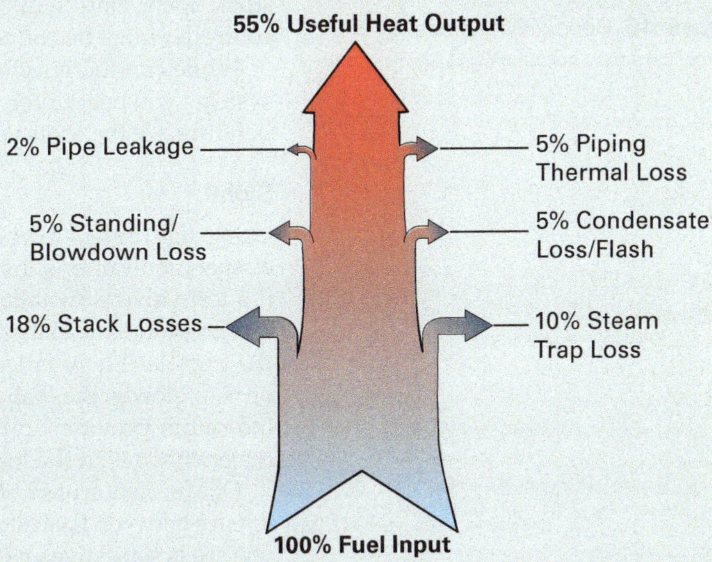

4.3.0 Troubleshooting and Maintaining Steam Traps

The two primary indications of failure in a steam trap are allowing live steam to pass freely (will not close), or failure to allow condensate to pass (will not open). Most steam traps fail in the open position. It is more difficult to sense the problem when this happens because it does not usually affect the performance of the process. Plenty of steam will flow to continue the work.

When a trap fails open, steam flows freely through and into the condensate system. Heat transfer at the load does not suffer, but a great deal of energy is wasted. Normal internal wear of a trap allows steam to leak through in small amounts at first, with the losses increasing along with the wear.

Traps that fail in the closed position are easier to identify because the backup of condensate significantly impacts heat transfer at the connected load. Steam flow is reduced or even stopped altogether. Clogged strainers have the same effect.

Common causes of steam trap failure include the following:

- Scale, rust, or corrosion buildup that prevent the valve from closing
- Defective or damaged valve seat
- Physical damage from severe water hammer
- Foreign material lodged between seat and valve
- Clogged or damaged strainers
- Increased back pressure

4.3.1 Basic Trap Troubleshooting

Always refer to the steam trap manufacturer's service literature for details on how to troubleshoot and repair specific traps. This is always the best place to begin, until you gain experience with a product. Manufacturers' publications and media provide a great deal of information about their products that assist in understanding and troubleshooting their products.

Conductivity: The degree to which a substance conducts electricity. The conductivity of steam and water differs significantly.

Steam trap analysis incorporates four basic tools: **conductivity**, sight, sound, and temperature. General troubleshooting relies primarily on the first three tools.

Conductivity

Figure 46 Conductivity probe.
Source: Lakewood Instruments, LLC

Conductivity can be used to both monitor and troubleshoot steam traps, but a conductivity probe (*Figure 46*) must be in place to do so. The probe is placed strategically inside the trap, or just upstream of the trap where condensate is expected to be. Since water conducts electricity far better than steam, the conductivity can be used to "see" through the walls of the trap or pipe—to see whether condensate or steam is present at the probe location. When a trap is stuck open, there will be steam where it shouldn't be. A compatible display or controller must be connected to the sensor to monitor the conductivity.

While conductivity is a valuable troubleshooting and monitoring tool, most systems you encounter will not be equipped with them. Sight, sound, and temperature are the tools you'll use most often.

Sight

Using sight refers to visually checking for the presence of condensate or steam at specific locations, using available valves. If the condensate line is open, or if a test valve is installed on the condensate line, the condensate and/or steam can be seen as it exits the trap. If live steam is present in the condensate line, the trap has likely failed open. If condensate is not coming from the line at all or very slowly, the trap has likely failed closed. Remember, though, that visual indicators provide limited information if the system is just starting up or if major changes in the load are taking place.

One limitation of sight diagnosis is that the technician must be able to tell the difference between flash steam and live steam. Opening a valve and evaluating what you see is subjective and depends on the skill and experience of the technician.

Most traps are dealing with condensate that is at or near steam temperatures; that is, somewhere above the boiling point of water. When condensate is released from the pressure of the steam system to atmospheric pressure, a certain amount of the condensate may immediately flash to steam. Live steam and flash steam can be distinguished from each other once you have seen it a few times. Live steam emerges as a hard, blue-white straight flow, often with a clear bluish tint at the outlet. Flash steam usually billows, spreads, and dissipates more quickly, as much of the pressure is relieved as soon as the valve is cracked open.

Sound

The use of sound in trap troubleshooting refers to the use of ultrasonic test instruments that allow you to clearly hear activity within the trap. The human ear alone can detect some differences but using an ultrasonic tester (*Figure 47*) allows you to hear every sound inside a pipe or trap clearly.

Ultrasonic probes pick up noise from the steam system and send an audible signal to headphones. Most ultrasonic testers also provide a display that shows the frequency of the sounds. The sound of steam rushing through a small orifice, for example, has a distinctive frequency to it. The technician performing this test must have skill and experience at interpreting the sounds as leaks, condensate discharges, or normal operation. To use an ultrasonic tester with success, you must study the accompanying literature and guidance carefully, and practice with the instrument on various types of traps is essential. All traps do not sound the same when they are operating properly.

Advanced software is now available from manufacturers that analyze the sound and summarize the activity. The software compares the sounds to previously recorded sounds of traps in various states of normal operation and failure. Ultrasonic testers also incorporate temperature measurement (*Figure 48*).

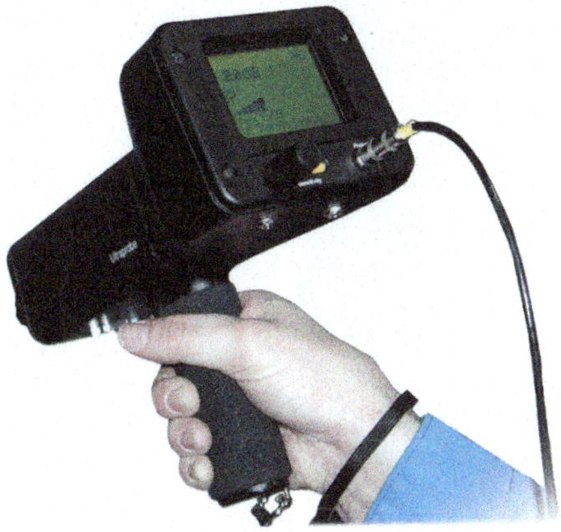

Figure 47 Ultrasonic tester.
Source: Courtesy of UE Systems Inc.

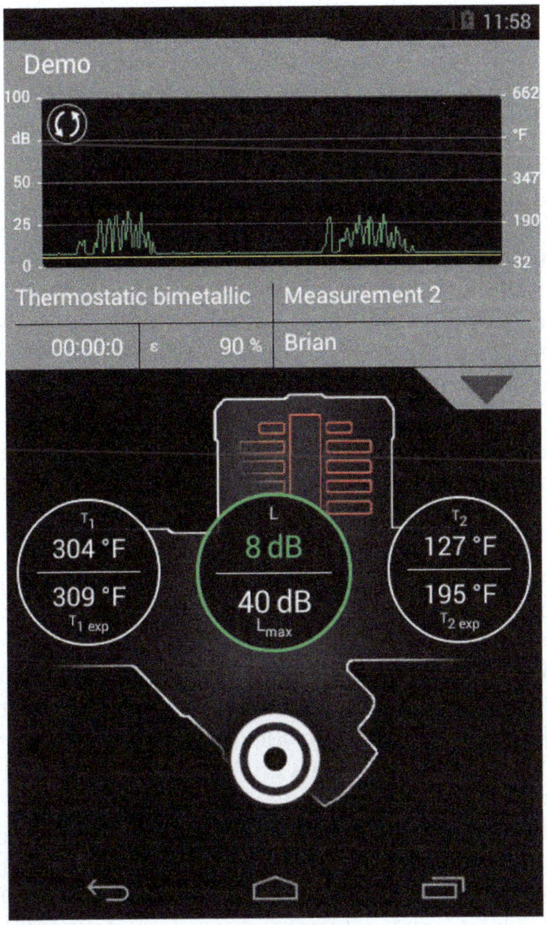

Figure 48 Ultrasonic testing app summary.
Source: SONOTEC

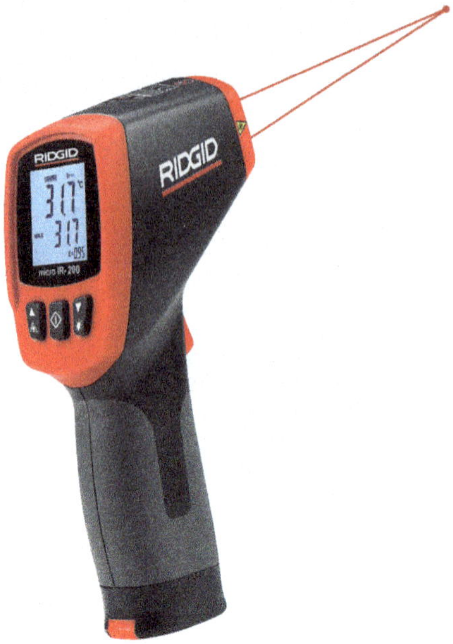

Figure 49 Infrared pyrometer.
Source: Ridge Tool Company

Note that temperature and sound together offer more information than either one alone. Experience and practice are still very valuable, but analytical apps can do a great deal of the work. The apps also help the user learn to distinguish the sounds themselves with practice.

Temperature

Temperature testing is often done with infrared pyrometers (*Figure 49*). These pyrometers are handheld for external readings, either up close or at a distance. To get an accurate measurement with an infrared device you must pay attention to the instrument's *distance-to-spot ratio* (D:S).

The area of temperature measurement for an infrared pyrometer is shaped like a cone. The farther away you are from the measured surface, the larger the area considered in the temperature measurement. For example, if you try to read the temperature of a small pipe from 25' away, the measurement could encompass an area several feet in diameter. This results in inaccurate readings.

Figure 50 shows the D:S ratio for a pyrometer with a 12:1 ratio. At 12", the measured area is 1" in diameter. At 12', the measured area grows to 1'. Thus, it is important to be aware of the ratio and use the device accordingly. Clearly, taking measurements closer to an object provides better results.

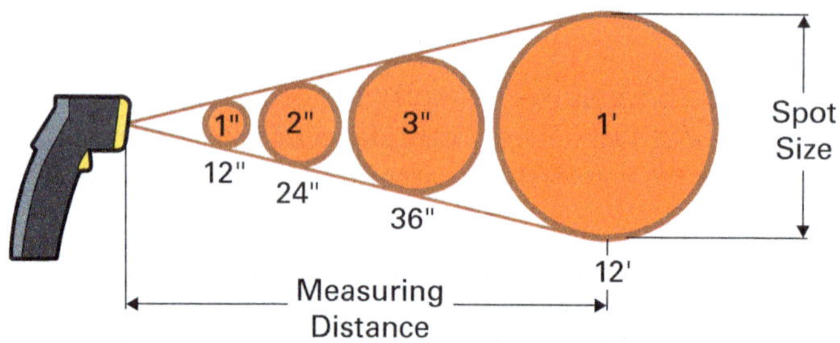

Figure 50 D:S ratio of 12:1 illustrated.

A thermal analysis of a trap should show the inlet as the area with the highest temperature. For functioning traps, the condensate outlet should be cooler than the inlet. Remember, though, that the condensate will frequently be near the steam temperature. You shouldn't expect a huge difference in the two temperatures. However, the condensate outlet should still be at a lower temperature than the steam inlet. If this is not the case, the trap is probably malfunctioning.

Failures are generally specific to the type of steam trap. The following sections provide details regarding failures on each type of trap.

4.3.2 Inverted Bucket Traps

Inverted bucket traps typically fail open, passing steam. The trap loses its prime, and steam and condensate steadily blow through together.

If this happens, a continuous rushing sound may be accompanied by the sound of the bucket banging against the enclosure. The temperature of the trap and of the condensate line will be at or very close to the steam temperature. The sound of the bucket linkage rattling inside the trap, without the rushing sound, may indicate that the trap components are beginning to wear and need closer inspection.

If an inverted bucket trap is passing steam, ensure a water seal exists inside. Remember that an inverted bucket trap must be primed with water and free of air or the trap will not work properly. Isolate the trap with a hand valve, wait for condensate to accumulate, and restart the steam flow. If this solves the problem, try to discover why all the water was forced out. It could be caused by sudden pressure fluctuations, or an improper installation that allows the water to drain out by gravity during the off cycle.

With the trap disassembled, examine the bucket. If the bucket or lever is distorted, the trap failure was probably caused by water hammer. Consider the prospect of water hammer and how it can be eliminated or reduced. Also check the bucket's air vent opening to ensure it is clear and hasn't been distorted. A clogged air vent causes air to be trapped in the bucket. This condition is called *air binding*.

If the trap will not pass condensate (won't open), air may be trapped inside. Since its ability to fall and open the discharge valve relies on the condensation of steam to reduce buoyancy, air keeps it afloat. Install an air vent to release the air if one is not currently installed.

Also ensure that the system pressure is not higher than the trap's rated maximum pressure. If it is, the valve cannot open. Install a trap with the proper pressure rating and capacity for the application.

4.3.3 Liquid-Expansion Thermostatic Traps

If a liquid-expansion thermostatic trap passes steam, check for fouling or wear of the valve and seat. Replace any worn internal parts. Remember that this type of trap can't respond to changes in pressure. If it has been set to close at a high pressure, it may not close off at a lower pressure. Try adjusting the trap to a lower setting, making sure that it does not waterlog (flood) excessively.

If it does not appear to react to this change, replace all the internal parts. If a liquid expansion trap will not pass condensate, make sure that it has not been adjusted to a setting that is too cold.

4.3.4 Balanced-Pressure Thermostatic Traps

If a balanced-pressure thermostatic trap blows steam, isolate the trap, and allow it to cool before inspecting it for dirt and sediment. If the seat is **wiredrawn**, replace all the internal parts, including the thermostatic element. The original parts have probably been stressed by continuous steam flow.

Wiredrawn: Wear that results in the formation of grooves in the seat of a valve, caused by fluid flow that erodes and abrades the seat as it repeatedly passes across the same spot at a high velocity.

If the valve and seat seem to be in good order, check the thermostatic element. You should not be able to compress it when it is cool. Weakness in the element indicates failure. Flattening of the convolutions indicates water hammer damage. If the water hammer cannot be eliminated at its source, a more durable type of trap must be used.

If the trap will not pass condensate, the element may be overextended due to excessive internal pressure, making it impossible for the valve to lift off its seat. An overextended element could be caused by excessive superheat or by someone opening the trap while it is still very hot and before the vapor inside has had time to condense.

4.3.5 Bimetal Thermostatic Traps

Bimetallic thermostatic traps respond to a temperature difference between condensate and steam. These traps open and close intermittently. When there is little condensate, they remain closed most of the time. When there is a lot of condensate, at startup for example, bimetallic traps may run continuously before settling down when live steam reaches them.

If a bimetal thermostatic trap freely passes steam, check for dirt and wear on the valve seat. A bimetal trap has limited power to close, and the valve may be held off its seat by an accumulation of sediment. If cleaning has no effect, replace all the internal parts.

This type of trap is usually factory preset. Check that any locking device on the manual adjusting stem is still secure. If it is not secure, see if the trap responds to adjustment. Replace it with a properly set trap if it does not respond as expected to a setpoint adjustment.

The valve on a bimetal trap is on the downstream side of the valve orifice, so these traps tend to fail in the open position. Failure to pass cold condensate indicates either gross maladjustment or blockage of the valve orifice or strainer.

4.3.6 Float and Thermostatic Traps

F&T traps discharge continuously when working properly. Most of the other types tend to open and close intermittently.

F&T traps tend to fail closed. A leaking float will not rise as the condensate level rises. Sometimes a float is crumpled by water hammer, also interfering with its ability to float. In either case, the F&T trap remains closed. The trap will remain cooler than usual as it fills with standing condensate.

An ultrasonic tester can also be used to diagnose an F&T trap. The thermostatic element in an F&T trap is normally quiet. A failed float will be silent in the absence of condensate flow. However, a rushing sound indicates that the thermostatic element has failed open, allowing live steam to pass. A rattling or metallic clanging noise might mean that the mechanical linkage has sustained damage.

If an F&T trap freely passes steam, check the trap for dirt at the main valve and at the thermostatic air vent. If the trap has a *steam lock release*, check to make sure that it is not open. A steam lock release forcibly opens the valve to release steam if all the condensate is pushed out and only steam is present. The float can't close the valve with the steam lock release open.

Make sure the float valve mechanism has not been knocked out of line by rough handling or water hammer, preventing the valve from seating properly. Ensure the float ball is free to rise and fall without contacting the casing, which could cause the mechanism to bind. Also ensure that there is enough condensate moving into the trap to make the float rise.

The thermostatic air vent closes in the presence of steam. If the trap is cool, and it is much cooler than steam if you are handling and inspecting the trap, the air vent should be open.

If the trap will not pass condensate as expected and the internal parts are in good shape, check to ensure the maximum operating pressure marked on the trap is not lower than the actual pressure on the system. If it is, the float valve can't open. Make sure that the trap has sufficient capacity to handle the maximum load.

4.3.7 Thermodynamic Disc Traps

Normal failure for a thermodynamic trap is open; that is, passing steam. The disc inside usually clicks shut audibly as it closes, once for each cycle. When the disc is cycling normally, it will close about 4 to 10 times per minute.

When the trap fails, the disc no longer clicks as it rises and falls, and steam consistently blows through. If the disc produces a continuous rattling sound, the disc is likely worn, and problems will likely increase.

If a continuous hissing sound is produced, the trap is not cycling closed. Again, the condensate line will heat up due to escaping steam and the absence of cooler condensate.

Check the trap and strainer for dirt and wipe the disc and seat clean. If this does not correct the problem, the disc may be worn and should be replaced. The extent of wear is evident by the amount of shiny surface that has replaced the normal machined surface.

If records show that thermodynamic traps are repeatedly suffering from rapid wear, the trap may be oversized, the associated piping may be undersized, or there may be excessive back pressure from the condensate drain line.

If the trap will not pass condensate, causing it to back up, the discharge orifices may be plugged with debris. It can also be caused by air binding, particularly if it occurs during startup. Check the air venting arrangements of the equipment. In extreme cases, it may be necessary to install an air vent parallel with the trap, or to use a float trap with a built-in air vent instead of a thermodynamic trap.

4.0.0 Section Review

1. Which of the following steps is part of the inverted bucket trap installation process?
   a. Prime it before using it.
   b. Mount it with the outlet up.
   c. Mount it with the inlet horizontal.
   d. Mount it with the inlet vertical.

2. True or False? Steam traps require a strainer downstream from the trap.
   a. True
   b. False

3. Which of the following describes the outlet temperature of a malfunctioning trap that is open, freely passing steam?
   a. It is the same as the inlet temperature.
   b. It is lower than the inlet temperature.
   c. It is 10°F higher than the inlet temperature.
   d. It is at least 25°F higher than the inlet temperature.

Module 03306 Review Questions

1. How much latent heat is required to vaporize 5 pounds of water?
   a. 970 Btus
   b. 1,114 Btus
   c. 1,294 Btus
   d. 4,850 Btus

2. The latent heat of condensation is defined as ______.
   a. the heat removed during the change of state from steam to water
   b. the heat gained during the change of state from water to steam
   c. any heat added after ice changes to water
   d. the heat lost in changing water to ice

3. Which of the following requires most of the maintenance attention in a steam system?
   a. Piping system
   b. Pressure regulating valves
   c. Steam traps
   d. Heat exchangers

4. Which of the following is *true* about steam boilers?
   a. Steam boilers are only partially filled with water.
   b. All steam boilers can operate at pressures up to 100 psig.
   c. Due to the nature of steam, all steam boilers are very clean inside.
   d. Steam boilers and hot water boilers use identical trim and safety devices.

5. Which of the following is generally installed between the boiler and an external pressure gauge to protect the gauge from contacting live steam?
   a. Trap
   b. Siphon
   c. Strainer
   d. Heat exchanger

6. Steam-to-water heat exchangers designed for potable hot water use are of ______.
   a. cast iron construction
   b. aluminum construction
   c. double-walled construction
   d. hermetically sealed construction

7. What is the correct term for a condensate return line that is located above the boiler water line?
   a. Steam line
   b. Dry return
   c. Wet return
   d. Hartford loop line

8. Which of the following causes excessive vacuum pump operation in a vacuum return system?
   a. Air leaks
   b. Heat exchangers
   c. Low steam pressure
   d. Too much condensate

9. To ensure a sufficient supply of water is available to the boiler, what can the condensate receiver be equipped with?
   a. Flash reservoir
   b. Vacuum pump
   c. Makeup water valve
   d. Thermodynamic trap

10. When more than one condensate line is connected to a flash tank, each line should be equipped with its own ______.
    a. sight glass
    b. check valve
    c. condensate pump
    d. pressure reducing valve

11. What happens when the steam system piping is too large?
    a. Increased trap failures
    b. Condensate pump overload
    c. Higher installation costs and greater heat loss
    d. Excessive pressure drop and increased noise

12. Mechanical steam traps take advantage of the difference in density between steam and ______.
    a. air
    b. gases
    c. debris
    d. condensate

13. Which of the following is an advantage of using fixed-orifice steam traps?
    a. They have no moving parts.
    b. They are practical for most conditions.
    c. They can quickly adjust to changing conditions.
    d. They respond to both temperature and condensate volume.

14. What are the four basic tools used for diagnosing problems with steam traps?
    a. Current, pressure, sound, and temperature
    b. Sight, sound, temperature, and conductivity
    c. Pressure, temperature, flow rate, and density
    d. Temperature, flow rate, sound, and conductivity

15. Bimetallic steam traps function based on the difference between condensate and steam _____.
    a. weight
    b. pressure
    c. air volume
    d. temperature

Answers to odd-numbered review questions are found at the Module Review Answer Key at the back of this book.

Answers to Section Review Questions

Answer	Section	Objective
Section One		
1. b	1.1.0	1a
2. a	1.2.0, Table 1	1b
Section Two		
1. c	2.1.0	2a
2. c	2.2.1	2b
3. a	2.3.2	2c
Section Three		
1. b	3.1.1	3a
2. a	3.2.3	3b
3. b	3.3.0	3c
4. b	3.4.1	3d
Section Four		
1. a	4.1.1	4a
2. b	4.2.0	4b
3. a	4.3.1	4c

MODULE 03214

Fiberglass and Fabric Duct Systems

Source: DuctSox Corporation

Objectives

Successful completion of this module prepares you to do the following:

1. Identify fiberglass duct fabrication standards and application considerations.
   a. Identify fiberglass duct fabrication and installation standards.
   b. Identify fiberglass duct application considerations.
2. Explain how to fabricate and repair fiberglass duct.
   a. Explain how to close and join fiberglass duct modules.
   b. Explain how to repair minor and major fiberglass ductboard damage.
3. Describe the methods used to suspend and support fiberglass duct systems.
   a. Explain how to suspend and support fiberglass duct modules.
   b. Explain how to suspend and support fiberglass duct fittings and risers.
4. Describe fabric-based air distribution products and installation methods.
   a. Identify various types of fabric-based air distribution products and systems.
   b. Explain how fabric-based air distribution products are installed.

Performance Tasks

This is a knowledge-based module. There are no Performance Tasks.

Overview

Fiberglass ductboard has been applied successfully in both residential and commercial installations for many years. It is lightweight, easy to handle, and creates a very quiet air distribution system. Fiberglass ductboard provides its own insulation, so it is not necessary to add liner or exterior insulation as it is with sheet metal duct.

Fabric duct systems continue to gain popularity in commercial, industrial, and institutional applications. Fabric duct systems are affordable, require limited labor to install, and excel in a variety of applications.

1.0.0 Fiberglass Duct Standards and Applications

Performance Tasks

There are no Performance Tasks in this section.

Objective

Identify fiberglass duct fabrication standards and application considerations.

a. Identify fiberglass duct fabrication and installation standards.
b. Identify fiberglass duct application considerations.

Fiberglass ductboard, like sheet metal, is used to create air distribution systems. Because fiberglass duct is inherently insulated, it is not necessary to wrap or line it with additional insulation like sheet metal duct. Fiberglass duct components are also lightweight, easy to transport and handle, and can often be fabricated in the field.

In this module, you will learn how fiberglass duct is fabricated into duct sections and fittings; how to properly hang and support fiberglass duct; and how to identify and repair material damage.

WARNING!

Take the appropriate precautions when working with fiberglass duct. Wear the appropriate personal protective equipment (PPE), such as gloves, dust masks, and goggles to protect your hands, lungs, and eyes.

1.1.0 Fiberglass Duct Standards

Fiberglass ductboard (*Figure 1*), sometimes referred to as *fibrous glass duct*, is made from flame-resistant glass fibers that are bonded with a heat-setting resin.

Rigid ductboard is typically available in thicknesses of 1", $1\frac{1}{2}$", and 2". The most common panel size is 4' × 10'. Fabricators use the rigid panels to fabricate rectangular duct sections and other shapes. Round fiberglass duct is no longer manufactured, but the material may be found in existing systems. However, round fiberglass duct can still be manufactured in the shop using ductboard grooving machines with that capability.

Figure 1 Fiberglass ductboard products.
Source: ©CertainTeed

The exterior of fiberglass ductboard is covered with a tough, abuse-resistant vapor retardant called **foil scrim kraft (FSK)**. FSK is a flame-retardant wrapper made from aluminum foil, fiberglass yarn (called scrim), and kraft paper.

Foil scrim kraft (FSK): A flame-retardant vapor retardant made from layers of aluminum foil, fiberglass yarn for reinforcement (scrim), and kraft paper.

Did You Know?

Fiberglass

Fiberglass has been commercially manufactured and marketed for more than 60 years. Fiberglass is used to make many products. The material absorbs sound, impedes heat transfer, removes impurities from liquids and gases, and reinforces other materials such as boat hulls. Uncontaminated fiberglass alone does not support the growth of mold or bacteria.

These qualities make fiberglass an important material in the HVACR industry.

Source: PragasitLalao/iStock/Getty Images

On Site

The Health and Safety Aspects of Fiberglass

In October 2001, the International Agency for Research on Cancer (IARC) removed fiberglass from its list of materials that may cause cancer. The US National Academy of Sciences reached a similar decision in 2000. This means that the fiberglass materials used in thermal and acoustical installation are not believed to pose a cancer risk to people who are exposed to them on the job.

However, this does not mean that lung damage can't occur from inhaling airborne fibers. For that reason, it is always best to wear a form of respiratory protection when working with or handling the material. Protect yourself.

Rigid ductboard used in HVAC work must be manufactured and tested in accordance with industry standards. UL Solutions (UL) tests the materials that are manufactured for use in air distribution systems. The North American Insulation Manufacturers Association (NAIMA) recommends standards for the fabrication and installation of fiberglass ductboard products. These organizations work to ensure that the materials used are safe, durable, and properly applied in the field.

The standards in this module primarily apply to air distribution systems designed for use in residential applications. These applications include constant-volume systems operating at maximum pressures of 0.5" water column (w.c.). For systems operating above this pressure, refer to the NAIMA Publication AH116, *Fibrous Glass Duct Construction Standards*, 5th Edition, for additional reinforcing requirements and details.

1.1.1 Strength, Deflection, and Fatigue

Rigid ductboard must be strong enough to withstand certain types of stress. It must also be able to span a specified distance without deflecting or deforming, and it must retain its shape and integrity when exposed to internal air pressure and system on-off cycles.

The strength of rigid ductboard is identified by its **EI rating**. This rating is the result of a calculation based on two tests: **Young's modulus of elasticity (E)**, and the **moment of inertia (I)**. Engineers and testing organizations use this calculation to determine the level of stress the ductboard can withstand. There are three different EI ratings for fiberglass duct: 475-EI, 800-EI, and 1400-EI. These ratings indicate the increasing rigidity of the product; higher numbers indicate greater rigidity. The EI rating for ductboard required on a project is written into the job specifications.

A **fatigue test** is also conducted on the ductboard under normal service conditions. A fatigue test gauges the ability of rigid duct to withstand expected or specified air pressures. These tests are important to establishing the expectations of performance and warranty provisions.

EI rating: A strength and rigidity rating assigned to rigid ductboard products that is based on a calculation called *Young's modulus of elasticity*.

Young's modulus of elasticity (E): A mathematical ratio that defines *tensile elasticity*, or the tendency of an object to deform along an axis when opposing forces are applied along the axis.

Moment of inertia (I): The sum of the products obtained by multiplying the mass of each particle in a material by the square of its distance from a rotational axis.

Fatigue test: A test in which a material is subjected to normal wear and tear to determine how long it will remain intact and effective before beginning to break down.

Did You Know?

EI Rating

Young's modulus of elasticity—the E in the EI rating—is a test. During this test, the ductboard is placed under stress and strain to test its rigidity. The result is a ratio of stress to strain. The I part of the rating is based on a mathematical calculation. The results of the test and the calculation are combined to assign the appropriate EI rating to rigid ductboard.

1.1.2 Thermal and Acoustical Performance

The American Society of Heating, Refrigerating and Air Conditioning Engineers (ASHRAE) developed standards for the thermal performance of fiberglass duct. These standards, which have been made part of many local building codes, recommend an **R value** for ductboard based on climate zones. The recommended R values also depend on where the ductboard is installed in a building, e.g., in an attic versus a crawl space.

R value: A measure of a material's ability to impede heat transfer, expressed as a number; the higher the number, the better the insulating properties. Also known as *thermal resistance value.*

Fiberglass insulation absorbs fan and air turbulence noise and minimizes the popping noises caused by expansion and contraction in metal ducts. ASTM International (formerly known as the American Society for Testing and Materials) has established typical acoustical performance standards for fiberglass ductboard. Because acoustical performance is a complex subject, it is best to contact the fiberglass manufacturer for detailed acoustical performance information.

1.1.3 Air Leakage

ASHRAE has established standards for the expected and acceptable amount of leakage along duct system seams and joints. Fiberglass duct systems fall into Leak Class 6, which means the expected rate of leakage is no more than 0.06 cubic feet per minute (cfm) of air for every square foot of duct surface area at a pressure of 1" w.c. The acceptable leak rate changes with the duct pressure. At 0.5" w.c., for example, the allowable leak rate is 0.038 cfm per square foot of duct surface area. Properly constructed and sealed fiberglass duct typically exhibits less leakage than sheet metal duct.

1.1.4 Fabrication and Installation Requirements

National Fire Protection Association (NFPA) Standard 90A and UL Solutions have set the following fabrication and installation requirements for fiberglass duct:

- Fiberglass duct shall be constructed of Class 1 duct materials as tested in accordance with UL Standard 181, *Factory-Made Air Ducts and Air Connectors. Class 1 duct material* refers to materials that have low flame spread and smoke development indexes.
- Fiberglass duct must be installed in accordance with the recommendations of the testing organization.
- Fiberglass duct cannot be used in air distribution systems that operate with a supply air temperature above 250°F (121°C).
- Vertical duct risers in buildings that are more than two stories high cannot be constructed of fiberglass ductboard.

Fiberglass duct systems must also be installed in accordance with any local codes and model codes that are incorporated by reference.

1.2.0 Fiberglass Ductboard Application Considerations

The following are some of the benefits related to the use of fiberglass duct, which play a role in application choices:

- Fiberglass ductboard combines the structural, thermal, and acoustic components needed in a duct system into a single material.
- Fiberglass ductboard provides consistent heating and cooling performance through controlled and highly consistent insulation thickness.
- The light weight of ductboard makes the duct and fittings easy to handle and imposes a reduced load on the building structure.
- Both fabrication and installation labor are significantly reduced due to the weight and fast, simple methods used to construct the components.

- Fiberglass does not support the growth of mold or provide any sort of nutrition to the organisms.
- Fiberglass duct absorbs the noise generated by air handling equipment and turbulence far better than metal duct.
- Fiberglass duct seams and joints exhibit a lower air leakage rate per square foot of surface area than sheet metal duct.

Fiberglass duct systems obviously have many benefits. However, fiberglass duct is not suitable for some applications, including but not limited to the following:

- In kitchen or fume exhaust ducts
- Transporting solids or corrosive gases
- Duct encased in concrete or buried below grade
- Outdoor applications
- Areas near electric heating elements (unless provided with heat radiation protection)
- Vertical risers in air distribution systems serving buildings with more than two stories
- With equipment of any type that does not include automatic, high-temperature safety controls
- With solid-fuel heating equipment
- For duct penetrations in wall, floor, and ceiling construction where fire dampers are required by codes

1.2.1 Working Safely with Fiberglass

According to NAIMA, fiberglass is safe to handle and install when the recommended work practices are followed. Recommended work practices and safety guidelines can vary from one manufacturer to another. Therefore, you must always read the safety data sheet (SDS) for the latest safety information and recommendations relative to the product in use.

Follow these guidelines to ensure your safety when working with fiberglass ductboard:

- If there are concerns about airborne fibers, use respiratory protection. Ensure that the dust mask or respirator is approved for use with fiberglass. Read and follow the manufacturer's recommendations for proper fit and use.
- Wear a cap, gloves, a long-sleeved shirt, and long pants. To help keep glass fibers from being trapped between your clothing and your skin, it is best if your work clothes fit loosely. Fiberglass can irritate your skin. This is not an allergic reaction but a temporary mechanical response from the skin when the glass fibers rub or become embedded in it.
- Wear safety glasses with side shields, safety goggles, or a face shield to protect your eyes.
- Do not rub or scratch your skin. Remove fiberglass particles from your skin by washing the area gently and thoroughly with warm water and mild soap. Applying a skin lotion may also help.
- Wash work clothing separately from other household laundry to prevent the transfer of glass fibers to other clothing. Rinse the washing machine thoroughly before using it again. If your work clothes have lots of fibers in them, it is best to soak and rinse the clothes separately before placing in the washing machine.
- Keep your work area clean. Avoid handling scrap fiberglass by keeping disposal bins near the work area. Don't let scrap material accumulate on the floor or other surfaces.
- Prevent airborne dust. Do not dry sweep or use compressed air to clean the work area. Use a properly filtered vacuum cleaner to clean up.

- Workers engaged in such operations as sawing and cutting fiberglass products are exposed to more airborne glass fibers than duct installers. Areas where consistent cutting and fabrication are done should be equipped with dust collection systems. Currently, NAIMA recommends that airborne exposures be kept below one fiber per cubic centimeter (1 f/cc).

More information regarding the safe handling of fiberglass insulation and ductboard products can be found on the NAIMA website at *https://insulationinstitute.org/*.

1.2.2 Moisture Concerns

Excessive moisture can quickly ruin fiberglass ductboard as well as other fiberglass insulation products. Avoid exposing ductboard to water, ice, or water vapor. If the material gets wet, it will quickly become ineffective. When working with fiberglass duct, follow these guidelines to avoid moisture-related problems:

- Use a drip pan and sheet metal sleeve to protect the immediate area near connections to evaporative coolers and humidifiers.
- Ensure that ducts passing through unconditioned spaces during the winter are tightly closed. This will prevent water vapor from condensing in the duct during the heating season.
- Do not install any wet fiberglass ductboard materials. Replace any sections of duct that get wet before or during installation.
- Find and correct the source of any water that appears in installed duct sections during normal service. Contact the manufacturer for help or additional information if needed.

1.2.3 Advantages of Modular Duct Construction

Fiberglass duct systems are typically assembled from sections and fittings commonly referred to as *modules*. Most modules are 4' long, matching the width of common fiberglass ductboard panels. This modular approach allows installers to take advantage of the standard-length modules for each straight and fitted section in a duct system.

To understand this concept, compare a run of sheet metal duct with a matching run of fiberglass duct (*Figure 2*). When working with sheet metal, fittings such as elbows and transitions of various lengths must be added to the straight,

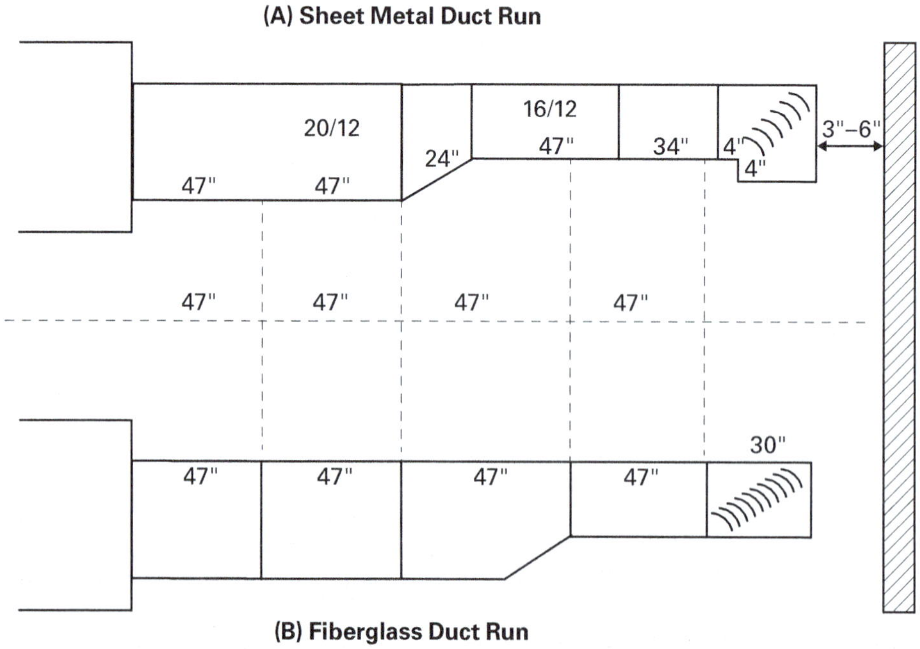

Figure 2 Modular fiberglass duct construction compared to sheet metal.

nominal length duct sections. Nominal lengths for straight sections of metal duct are 4' (fabricated from metal sheets) and 5' (fabricated from metal coils). However, these nominal lengths are not precise measurements. A nominal 4' section of metal duct typically has a finished length of 47".

In *Figure 2A*, a 2' long sheet-metal transition is installed after the two straight 47' sections. Farther down the air stream, the sections vary in length from 47", with different side dimensions than those used in the first two sections, to 34". The run terminates with an elbow that has a 4" throat.

When working with fiberglass duct (*Figure 2B*), you can fabricate all the modules using the same standard-length sections. Each piece can be cut to length on the job if needed, in seconds. Using a throat-less elbow in the fiberglass system also allows for easy adjustment to the prescribed layout. You can prepare a section of duct of any length (up to 47") to make up for any discrepancy in the plans versus field measurements.

Note that there are only five sections of fiberglass duct compared to six sections of metal duct. The fiberglass modules are also fabricated using fewer and simpler pieces of equipment and are fabricated faster.

When working with sheet metal, the throats of the elbows and the lengths of the fittings must be fabricated with precision. If there is a discrepancy in the field, either an additional section must be fabricated in the shop or a previously made fitting must be remade. With fiberglass, the pieces can be field modified to make up for minor changes in equipment location or obstacles. Learning how to plan for module fabrication from a set of plans and mastering the fabrication guidance in this module are critical to success when working with fiberglass ductboard.

NOTE

Although ductboard modules are made from 48" panels, the modules have an effective length of about 47⅛" when using 1" panels and an overlapping joint is created to join the modules.

1.0.0 Section Review

1. Which value refers to a material's thermal performance?
   a. I value
   b. R value
   c. IE rating
   d. EI rating

2. Which of the following applications is not recommended for fiberglass ductboard installation?
   a. indoor
   b. outdoor
   c. residential
   d. commercial

2.0.0 Fiberglass Duct Fabrication and Repair

Objective

Explain how to fabricate and repair fiberglass duct.

a. Explain how to close and join fiberglass duct modules.
b. Explain how to repair minor and major fiberglass ductboard damage.

Performance Tasks

There are no Performance Tasks in this section.

The extended-**plenum** supply air system is common in residential applications. Extended plenum refers to the fact that the main ducts leaving the plenum attached to the equipment have the same width. It is an ideal system for basement installations. The main duct size is not reduced after each branch line connection unless the length of the main ducts and number of branches demands it to ensure enough air volume remains at the end of the run. The extended-plenum approach will be the basis for discussion of fiberglass ductboard closing and joining methods.

Plenum: A box that acts as the point of connection for equipment to a duct system, or for multiple ducts.

2.1.0 Closing and Joining Fiberglass Duct Modules

Figure 3 is an example of an extended-plenum supply air duct layout. The plenum in the example is installed on an upflow air handler. Branch lines extend from the plenum to the registers or diffusers. The branch lines may be made from smaller rectangular ductboard modules, flexible round duct, or insulated metal duct. Flexible round duct is easily the most popular product to connect registers and diffusers unless the length of the run precludes its use.

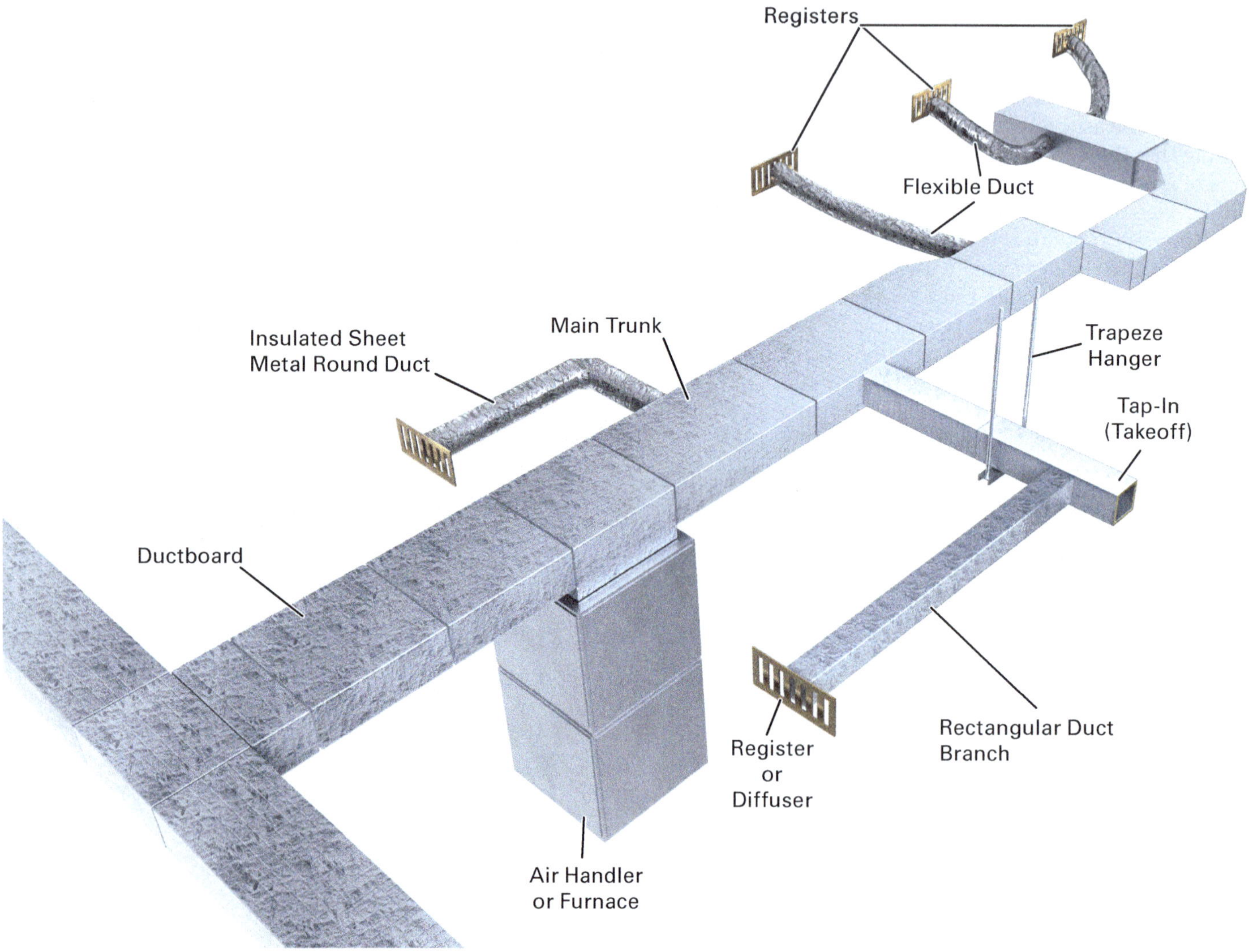

Figure 3 Typical extended-plenum system components.

The following items are identified in *Figure 3*:

1. Main trunk (extended plenums)
2. Connection of the plenum to the air handler or furnace
3. **Tap-ins** (takeoffs) for rectangular or round branch lines
4. Registers or diffusers
5. Flexible branch duct connections

Tap-ins: Rectangular or round takeoff fittings made of fiberglass ductboard or sheet metal that tap into a trunk duct to connect a branch duct.

Closure system: Any one of three types of sealing products used to seal fiberglass duct modules: pressure-sensitive tape, heat-activated tape, or mastic and glass fabric tape.

Note that *Figure 3* is an example only. A complete illustration of the system would include many more elements, such us duct hangers and supports. These elements have been left out for the sake of simplicity and will be discussed later in this module.

Whether fabricating and joining duct modules, all joints must be properly sealed. For maximum efficiency and performance, the correct **closure system** must be chosen and applied. Only closure systems that comply with UL

Standard 181A, *Closure Systems for Use with Rigid Air Ducts* can be used on fiberglass duct systems. Products such as pressure-sensitive foil tapes must also be UL listed for use in a fiberglass duct closure system.

The basic closure systems for fiberglass ductboard include the following:

- Pressure-sensitive tape
- Heat-activated tape
- Glass fabric tape and **mastic**

Mastic: A pasty substance used as a filler and sealant on both sheet metal and fiberglass duct components and joints.

Preparing the surface is an essential step in working with tape closure systems. The surfaces to be bonded must be clean and dry. Dust, dirt, oils, and moisture prevent the tape adhesive from sticking properly, compromising the quality and longevity of the joint. The surfaces around the joint or seam can generally be cleaned with an oil-free, lint-free cloth or a paper towel. An oily or greasy surface will require an appropriate cleaner to break down the oil so it can be effectively removed. When you encounter fiberglass duct joints with tape clearly separating from the surface and peeling after just a few years, you are usually seeing the results of poor surface preparation.

2.1.1 Tape Closure Products

Pressure-sensitive foil tape (*Figure 4*) is used to join and seal fiberglass duct. It is also used when connecting fiberglass modules to sheet metal.

The tape width varies with the thickness of the ductboard. A $2\frac{1}{2}$" wide tape is required for 1" ductboard. A 3" tape is recommended for $1\frac{1}{2}$" ductboard, while 2" ductboard requires 3" tape per the NAIMA standards.

Tapes and mastics have temperature and humidity storage requirements and a limited shelf life. Store these materials according to the manufacturer's guidance. Do not use expired products or those that have been stored in extreme temperature and humidity conditions. The adhesive material will dry out and fail to bond effectively.

Pressure-sensitive tape is normally pressed and rubbed into place to ensure adhesion. But in application temperatures below 50°F (10°C), it may need to be warmed before being applied. An iron set to roughly 400°F (~205°C) is used to preheat the surface. With the surface warmed, position the tape quickly before it cools. The iron is then moved rapidly over the tape to complete the application.

If the tape itself is cold, but the work area and ductboard are above 50°F, the tape can be warmed up and used without ironing. When an iron is not necessary, a plastic squeegee-like tool is used to rub the tape to improve adhesion (*Figure 5*).

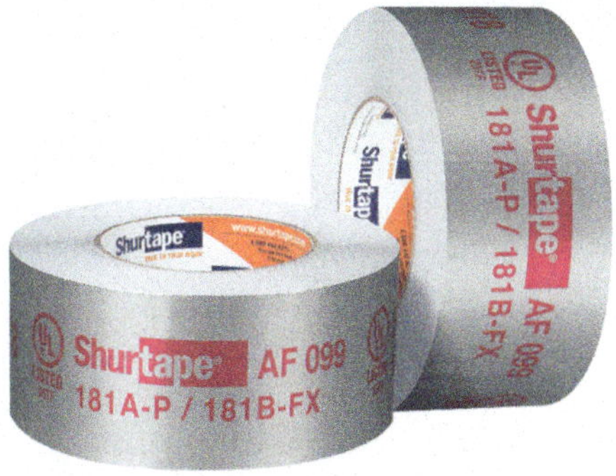

Figure 4 UL 181A-listed pressure-sensitive foil tape.
Source: Courtesy of Shurtape Technologies, LLC

CAUTION

When rubbing down pressure-sensitive tape, avoid placing too much pressure on the sealing tool. Too much pressure could tear the tape as it passes over staples that are often beneath the tape.

Figure 5 Using a plastic squeegee to seal tape.
Source: Courtesy of Shurtape Technologies, LLC

Figure 6 Heat-activated FSK tape.
Source: Courtesy of Ideal Tape

Foil tape for fiberglass ductboard has a paper backing. To keep the tape smooth as it is applied, first expose only a few inches of foil at the end of the tape and stick it in place. Then unroll the tape several feet, allowing you to see that it is straight. Then begin to remove the paper backing with one hand, keeping pace with the tape-sealing tool while using the other hand to keep the tape taut.

While heat may be needed to apply standard foil tape in cold temperatures, some tapes are designed specifically for heat activation (*Figure 6*). A heating iron is required for the adhesive to activate. Heat-activated tape is applied using a heating iron set around 500°F (260°C). Closure machines with heated rollers can be used in the shop to automate heat-activated tape application. After ironing, the joint or seam must be allowed to cool below 150°F (66°C) before significant handling.

WARNING!

Both the iron and the heated surfaces are very hot. To avoid serious burns, wear gloves that provide protection from burns. While positioning the tape, ensure that the iron is rested on a heat-resistant surface. Work so that the hot iron is always in front of the tool cord and ensure your feet do not get tangled in the cord. Never use a torch to apply heat-activated tape.

Most of the heat-activated products have a foil scrim kraft (FSK) construction, as shown in *Figure 6*. Threads serve as the scrim to make these products stronger and resist tearing. The tape shown also has heat-activated dots that change color when the required temperature is reached.

Staple flap: A section of ductboard facing from which the insulation has been removed that is then pulled over a joint or seam and stapled into place.

These tape products are used to reinforce **staple flap** joints (*Figure 7*) or are laid in a crossing pattern to seal butt joints without staple flaps (*Figure 8*).

CAUTION

Do not use heat-activated tape over standard pressure-sensitive tape or over mastic and glass fabric closure systems. The heat may cause the other closure systems to fail.

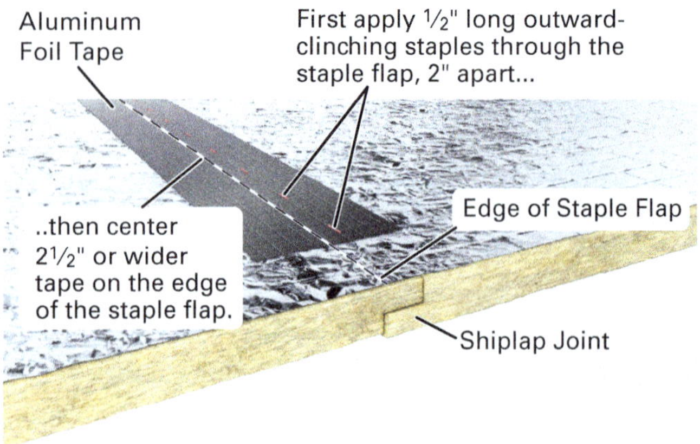

Figure 7 Tape closure with staple flap.

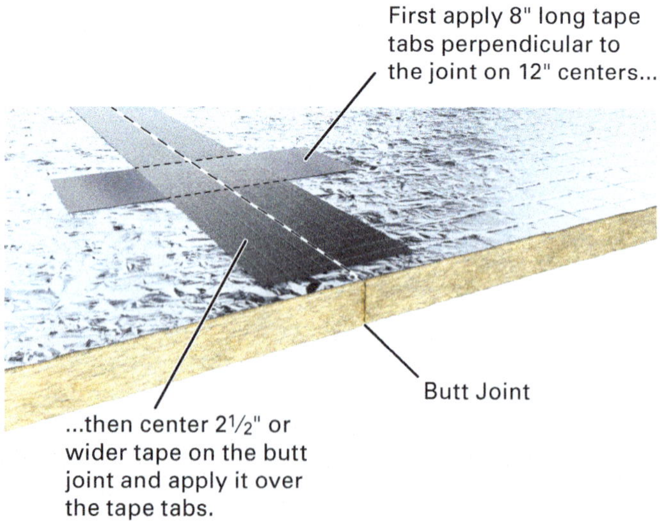

Figure 8 Tape closure without a staple flap or staples.

Glass Fabric Tape and Mastic

Mastic and glass fabric tape is a two-part closure system consisting of mastic and a non-adhesive fiberglass mesh tape (*Figure 9*). The standard tape width for ductboard joints is 3".

Although there are similarities among mastic products (*Figure 10*), the safety precautions, shelf-life limitations, and minimum setup time may vary from one manufacturer to another. Always read the manufacturer's instructions and recommendations to ensure the product is applied properly.

A layer of mastic wider than the tape (about 4" wide for 2" tape) is applied first as a bed for the tape (*Figure 11*). The mastic is applied with a disposable brush; cleaning the brush after use isn't practical, and most products are much too thick to apply with a roller. The glass fabric tape is then laid neatly in the mastic bed, and a second coat of mastic is applied over it. The product literature will specify the length of time necessary for curing before the module is transported or installed.

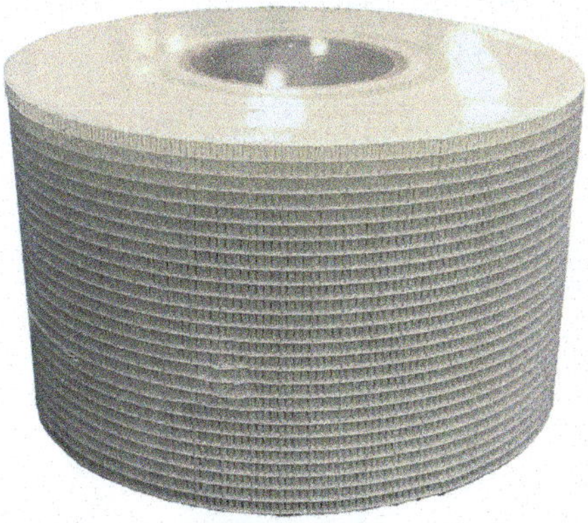

Figure 9 Fiberglass mesh tape (no adhesive).

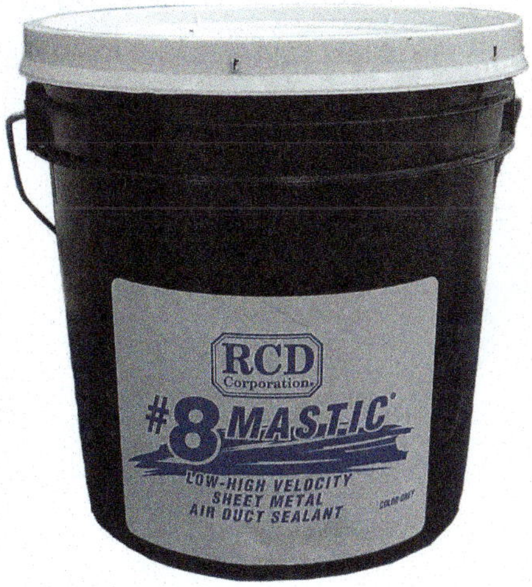

Figure 10 Duct mastic.

Figure 11 Embedding tape in a bed of mastic.

2.1.2 Closing and Joining a Duct Module

Closing a duct module refers to making the necessary cuts, folding the material into its final shape, and taping/sealing the seam.

To neatly fold and seal modules, some material must be removed from the corners. **Shiplap** corners (*Figure 12*) are more desirable than corners made using V-grooves (*Figure 13*). Shiplap corners do not compromise the thickness of the material at the corners. On the right side of each figure, a section of fiberglass must be removed from the facing to form a staple flap.

Shiplap: A groove cut into the end of a ductboard module that allows adjoining edges to overlap when the ductboard is folded or fit to another piece. A special cutting tool is used to create the profile.

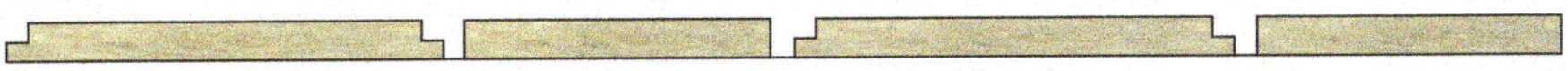

Figure 12 Ductboard panel profile with shiplap grooves prepared for folding.

Figure 13 Ductboard panel profile with V grooves prepared for folding.

Shiplap and V grooves are cut using either hand or machine tools. Ductboard grooving machines (*Figure 14*) are designed to accommodate the width of a 48" wide ductboard panel and cut a series of grooves of the desired shape across the full width. In *Figure 14B*, note the large letter A on the adjustable cutting tool for the machine. Each cutting tool is identified with a letter to identify its cutting profile, depth of cut, and orientation. For example, to cut the series of shiplap grooves shown in *Figure 12*, five cutting tools with three different profiles are required. Many of these cutting tools can be positioned across a machine, creating multiple grooves in a single pass.

Hand tools for cutting ductboard are best purchased in kits (*Figure 15*) that provide a proper cutter for each profile, depth of cut, and orientation. Hand tools are numbered rather than lettered. In addition, most are color-coded so that craftworkers can visually identify the proper tool for the needed cut.

(A) Ductboard Grooving Machine

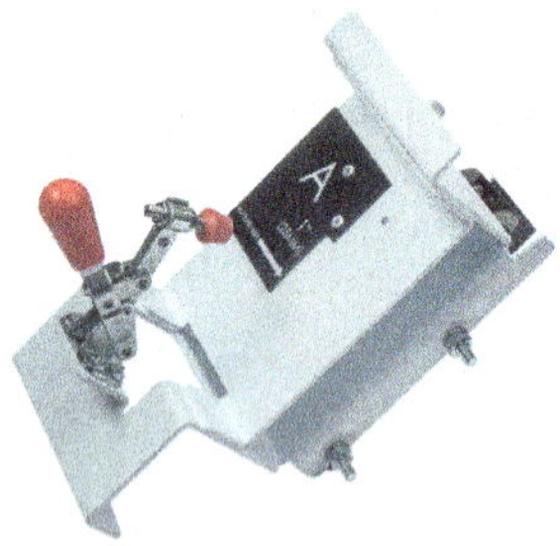

(B) Adjustable Cutting Blade

Figure 14 Ductboard grooving equipment.
Source: ©CertainTeed

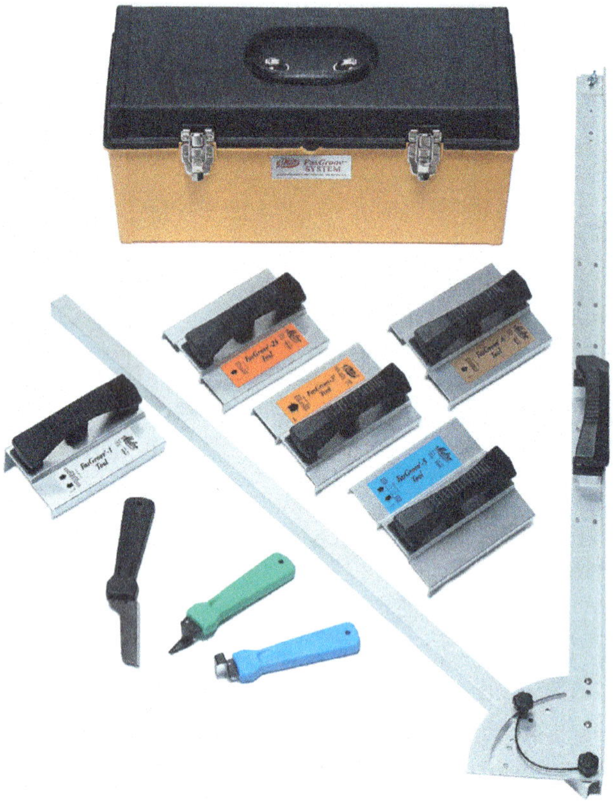

Figure 15 Ductboard hand cutting tools.
Source: Courtesy of Malco Tools

> **Did You Know?**
>
> **Shiplap Joints**
>
> Experienced woodworkers will recognize the shiplap joint. In woodworking, this groove profile is referred to as a *rabbet*.

The basic steps to fabricate and close a duct module are as follows:

Step 1 Cut shiplap or V grooves with hand or machine tools to accommodate the folds. Remove scrap from the grooves.

Step 2 Fold the panels to form the duct (*Figure 16*). Make sure that the ends are flush and properly seated in the shoulder of the shiplap grooves.

Step 3 Hold the duct at about a 30-degree angle (*Figure 17*). Staple the long flap with $\frac{1}{2}$" outward-clinching staples about 2" on center.

Step 4 Close the long seam using one of the closure methods discussed earlier in this module.

Figure 16 Folding the cut panels to form the duct module.
Source: ©CertainTeed

Figure 17 Securing the staple flap.

Once the individual duct modules are constructed (*Figure 18*), they must be joined to create a run of duct. This is a fairly simple task. However, it must be done carefully to ensure that the joints fit tightly and evenly. The joints must also be sealed properly with the chosen tape closure system. The quality of the joint has a significant effect on the duct's ability to maintain its shape.

Figure 18 Completed duct modules.
Source: ©CertainTeed

CAUTION

When rubbing down pressure-sensitive tape, avoid placing too much pressure on the sealing tool as you pass over the staples. Too much pressure could cause the tape to puncture as it passes over the staples beneath.

When stood on end, the large duct sections tend to wobble. To steady the sections, prop them against a wall while joining two modules. The basic steps for joining modules are as follows:

Step 1 Using an insulation knife or scissors, slit the stapling flap at each corner, down to the edge of the fiberglass.

Step 2 Place one section on top of the other, standing them up against a wall if possible. Make sure the male and female shiplap grooves mate properly.

Step 3 Using an outward-clinching staple gun, staple the flaps on all four sides. Use ½" outward clinching staples about 2" on center.

Step 4 Make sure the ductboard surface is clean and dry. Then tape the joint with pressure-sensitive tape. Using a plastic smoothing tool, rub the tape firmly until the facing scrim can be seen through the tape.

2.1.3 Connecting Ductboard to Sheet Metal

It is often necessary to connect fiberglass ductboard to a sheet metal component. *Figure 19* shows two common methods.

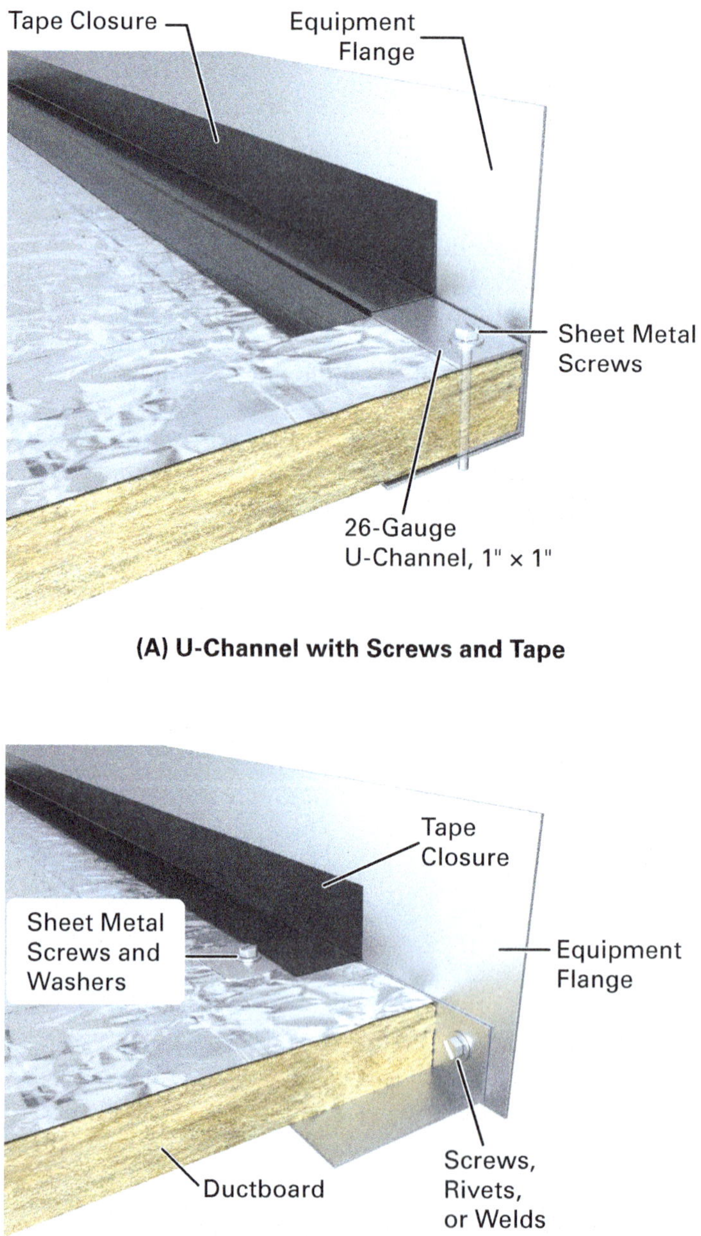

Figure 19 Connecting fiberglass ductboard to sheet metal.

Figure 19A shows a section of U-channel, or *nosing*, attached to the flange, providing a slot for the edge of the ductboard to slip into. Screws are then driven through the channel to capture the edge of the ductboard.

Figure 19B shows another method of connecting to a flange. In this case, the flange itself is first attached to the equipment. Screws with large, square washers are then driven through the board and into the flange beneath. The screws used to capture the ductboard edge are placed no more than 12" apart, on center. The size of the hardware and washers is determined from NAIMA tables. In all cases, pressure-sensitive tape is applied over the finished and properly cleaned joint.

2.2.0 Repairing Fiberglass Ductboard

Fiberglass duct can be damaged during handling or installation. This section will present repair procedures for different types of damage.

2.2.1 Repairing Minor Damage

Minor damage includes small, straight slits in the facing material that have not damaged or penetrated the fiberglass material. When making these repairs, use the same closure method used to assemble the module (pressure-sensitive tape, heat-activated tape, or mastic and glass fabric tape) to cover the damage. Clean and dry the surface before applying the closure material to ensure a good bond.

If the facing damage is more than just a straight slit but is no more than $\frac{1}{2}$" wide, use the procedure shown in *Figure 20*. The closure material must extend at least 1" beyond any point of the tear.

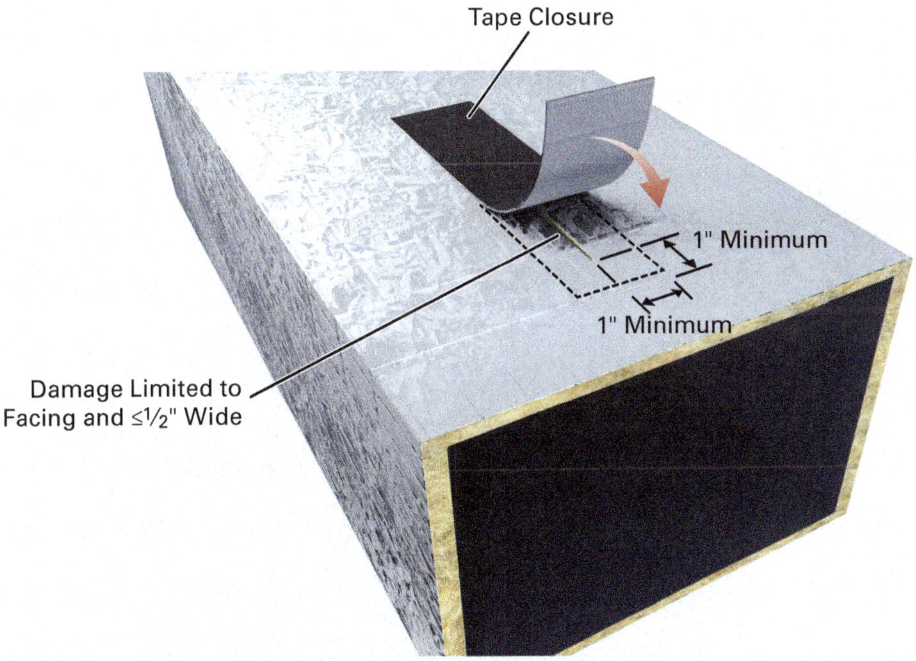

Figure 20 Repair of minor facing damage ≤½" wide.

If the facing damage is wider than $\frac{1}{2}$" but is less than the width of the tape in use, smooth the facing and make the repair as shown in *Figure 21*. Apply two layers of the tape. The first layer is a single piece of tape centered over the tear. The second layer is two pieces of closure material butted side by side so that the second layer covers the first layer with a 1" minimum overlap.

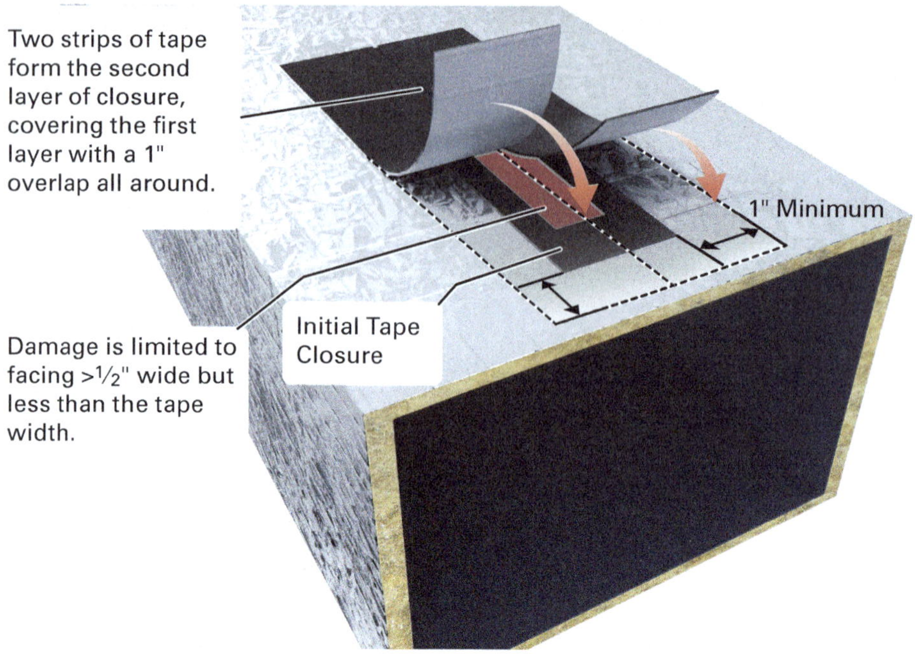

Figure 21 Repair of minor facing damage more than $\frac{1}{2}$" wide.

2.2.2 Repairing Major Damage

Major damage is identified by damage or displacement of the fiberglass portion of the ductboard. When this happens, the damaged section is removed, and a plug is fabricated to fit the space (*Figure 22*).

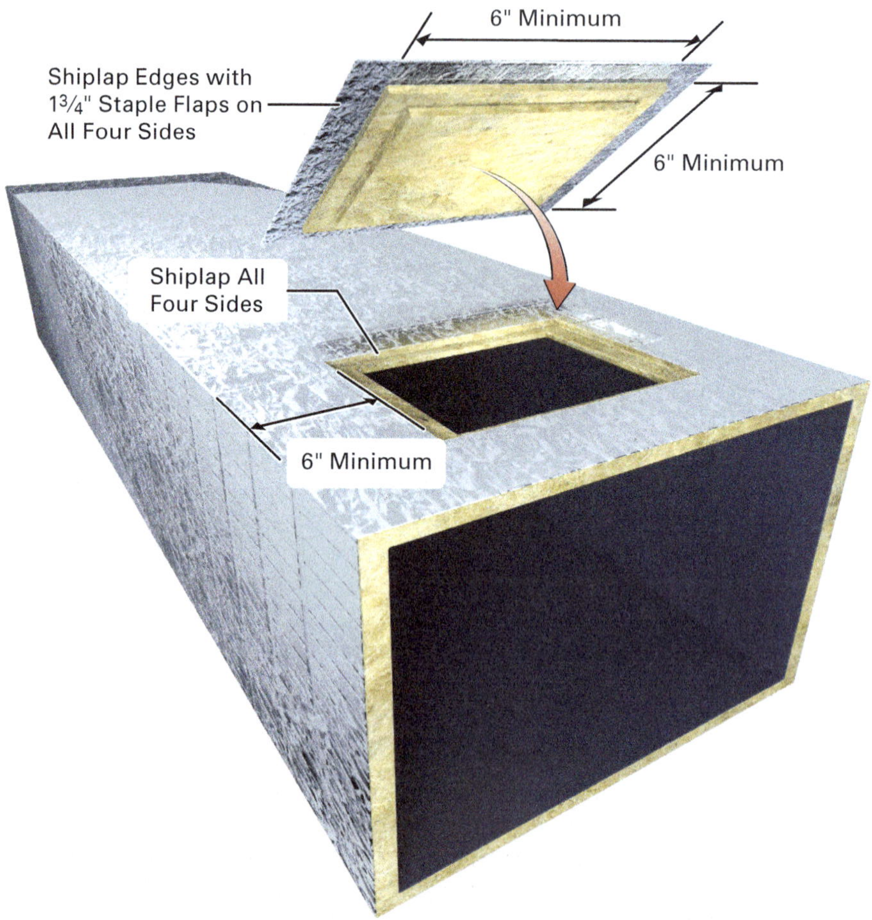

Figure 22 Repairing major damage to a duct module.

To replace a damaged panel, follow these steps:

Step 1 Using a shiplap tool, cut out the damaged area and discard it. Cut out a square or rectangle that is slightly larger than the damaged area, leaving shiplap grooves behind. The shiplap groves prevent the replacement plug from dropping through the hole and into the duct.

Step 2 Measure and cut a replacement plug from fiberglass ductboard. Be sure the replacement board thickness matches the thickness of the ductboard being repaired. Ensure the plug is large enough to accommodate the $1\frac{3}{4}$" staple flap on all four sides; the plug will need to be $3\frac{1}{2}$" larger, in both directions, than the opening. Shiplap the edges and leave a $1\frac{3}{4}$" staple flap on all four sides.

Step 3 Insert the plug and test the fit. The plug should fit smoothly and snugly.

Step 4 Staple the flaps and seal with the same closure system used to construct the duct module.

In *Figure 22*, note that the removed area must be at least 6" away from the edge of the module to use the preceding approach. If the damaged area extends to within 6" of the edge of the panel, cut out and repair the entire width of the module, using the previous steps. Make end cuts on the plug to match the ends of the damaged panel (*Figure 23*).

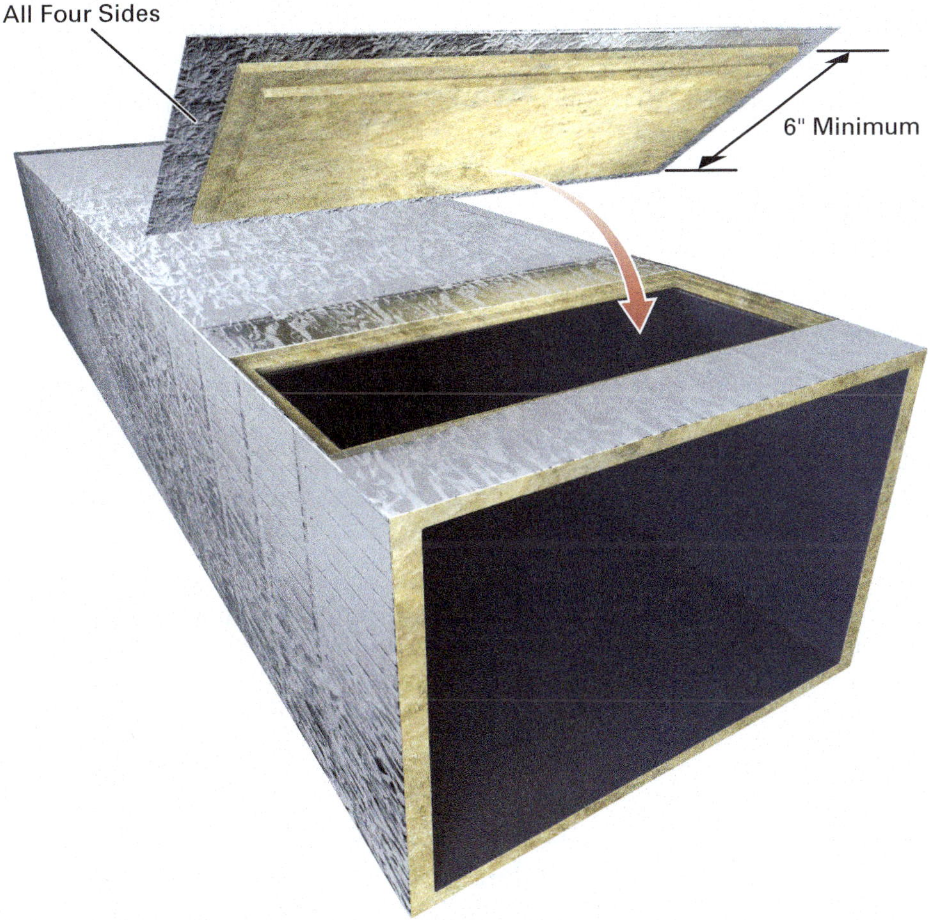

Figure 23 Repairing major damage, edge-to-edge.

2.2.3 Replacing an Entire Panel

Sometimes the damage to a duct panel is so great that the entire panel on a module must be replaced. This usually happens when something falls on the panel, causing it to buckle or crease.

Figure 24 Replacing an entire panel.

To fabricate the panel, you must observe the required dimensions of the shiplap and staple flap based on the ductboard thickness. Refer to *Table 1*. To make this repair, refer to *Figure 24* and follow these steps:

TABLE 1 Shiplap and Staple Flap Dimensions

Ductboard Thickness	Minimum Shiplap Width	Minimum Staple Flap Width
1" Ductboard	7/8"	1 3/8"
1 1/2" Ductboard	1 3/8"	1 11/16"
2" Ductboard	1 3/4"	2 1/8"

Step 1 Cut away and discard the damaged panel.

Step 2 Measure and fabricate a new panel. Refer to *Table 1* for dimensions. Remember that you will need staple flaps on the long edges, and likely a staple flap on one end of the new panel. Form shiplap grooves along the edges of the new panel.

Step 3 Remove insulation from the new panel to form the staple flap on the two long sides as shown in *Figure 24*.

Step 4 Remove any debris from the panel grooves and insert the replacement panel. It should fit snug and be smooth around the edges.

Step 5 Staple the flaps and seal with the same closure system used to fabricate the module.

You may also need to construct a panel with square edges instead of shiplap grooves. This depends on which part of the module is damaged. Always pay close attention to the dimensions and required allowances for staple flaps.

2.0.0 Section Review

1. Which of the following is *not* an acceptable closure system for fiberglass ductboard?
   a. Heat-activated tape
   b. Stapled duct joints
   c. Pressure-sensitive tape
   d. Mastic and glass fabric tape

2. When repairing minor facing damage with tape, how far must the closure material extend beyond any point of the tear?
   a. 1/2"
   b. 1"
   c. 2"
   d. 3"

3.0.0 Fiberglass Duct Suspension and Support

Performance Tasks

There are no Performance Tasks in this section.

Objective

Describe the methods used to suspend and support fiberglass duct systems.

a. Explain how to suspend and support fiberglass duct modules.
b. Explain how to suspend and support fiberglass duct fittings and risers.

Fiberglass duct is very lightweight and somewhat self-supporting. However, a minimum number of carefully placed hangers is still required to fully support the duct. Ensure that the hanging method used has adequate load-bearing capability for the installation, without placing too much stress on either the structure, hanger components, or the duct itself. Hangers must also be installed in the proper locations to support duct-mounted accessories, such as electric heat packages and motorized dampers.

Hanger treatment and spacing requirements depend on the duct dimensions. Channel gauge and profile vary with duct size. Always refer to your local code for specifications regarding the method and materials.

3.1.0 Hanging and Supporting Fiberglass Duct Modules

Fiberglass duct up to 48" wide can be suspended using either 12-gauge wire (minimum size) or 22-gauge sheet metal straps at least 1" wide (*Figure 25*). When using sheet-metal channel to form a trapeze hanger as shown, remember that the gauge and dimensions of the channel vary with the size of the duct and the amount of **hanger extension**.

Hanger extension: Refers to the sum of the distance between the wall of the duct and the hanger wires or straps on each side.

3.1.1 Hanger Channel Size and Spacing

To determine the proper channel dimensions for a trapeze hanger, one must first determine the hanger extension. The hanger extension (E) is found by summing the distance between the hanging wires and the duct walls on each side (*Figure 26*). Measure both sides and add the measurements together.

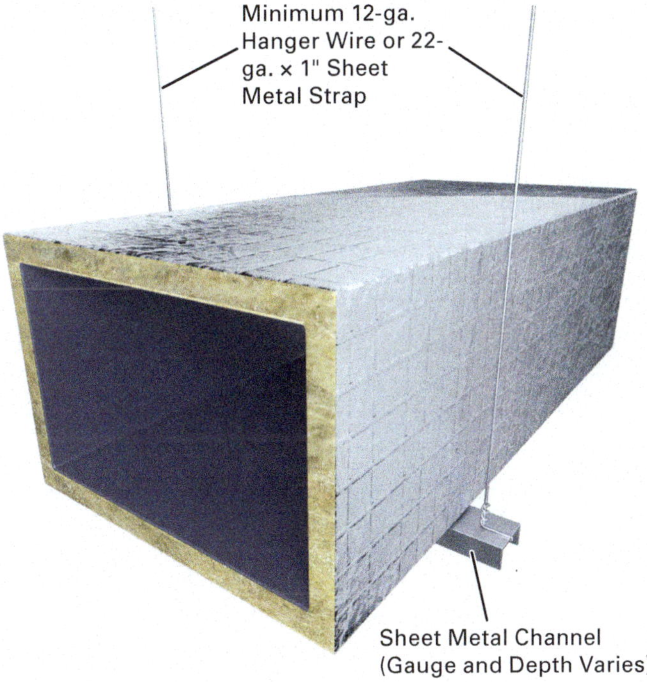

Figure 25 Suspending fiberglass duct on a trapeze hanger.

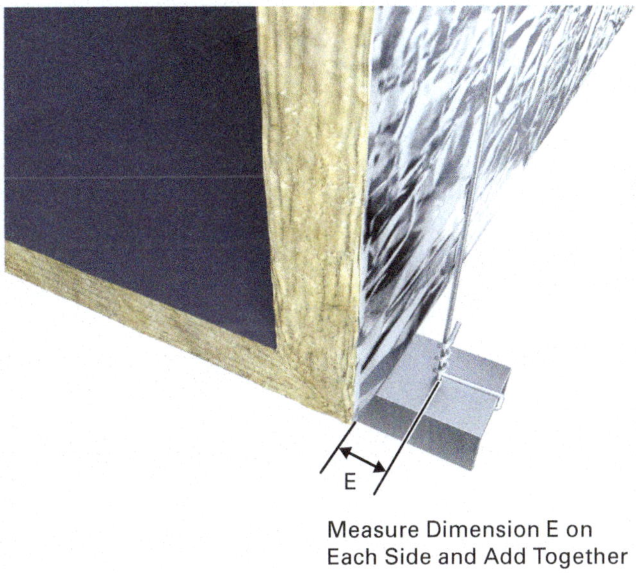

Figure 26 Hanger extension.

For rectangular fiberglass duct, the supporting channel should never be less than 2" wide × $1\frac{1}{2}$" deep. If the total hanger extension is 6" or less, 2" wide channel can be used on fiberglass ducts up to 48" × 24" high inside. Use 22-gauge wire to suspend the channel and place the hangers no more than 4' apart (*Figure 27*).

Applications that exceed any of the limitations shown in *Figure 27* require the use of 3" sheet-metal channel for the trapeze. *Table 2* summarizes the required channel dimensions and gauge for various duct sizes and hanger extensions. Remember that when using 2" channel, the maximum spacing between hangers is 4'.

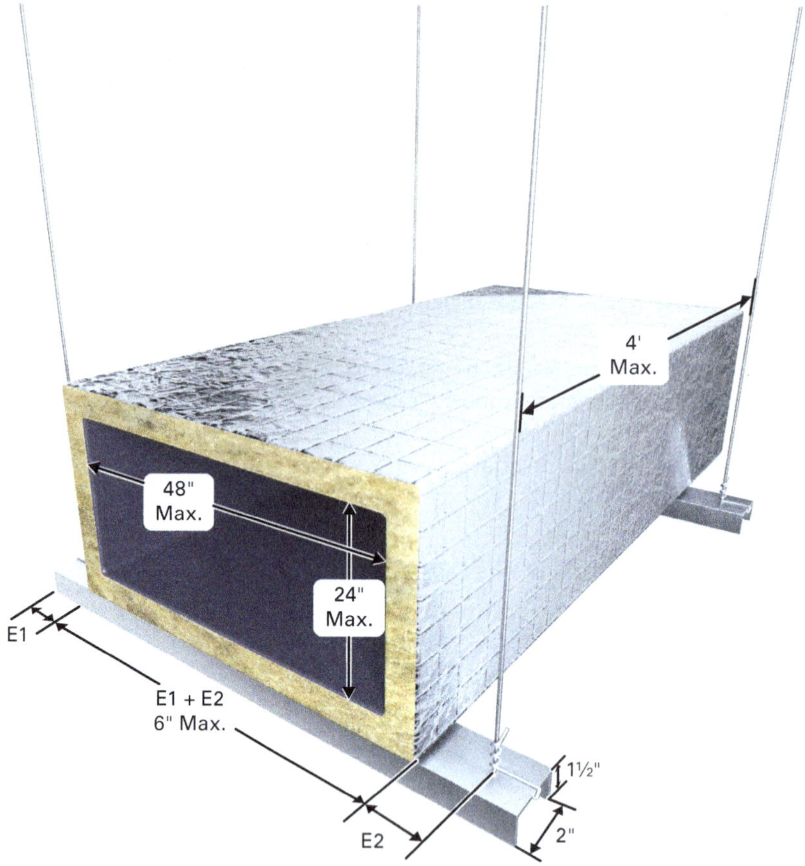

Figure 27 Restrictions to using 2" sheet-metal channel hangers.

TABLE 2 Hanger Channel Requirements for Fiberglass Duct

Duct Width (Inside)	Hanger Extension	Channel Width	Channel Depth	Minimum Channel Gauge
Up to 48" × 24"	≤6"	2"	1½"	22 ga.
Over 48" × 24"	≤6"	3"	1½"	24 ga.
	6" up to 18"	3"	2"	22 ga.
	18" up to 30"	3"	2"	18 ga.

With 3" channel, the hangers can often be spaced farther apart. Refer to *Table 3* to determine the maximum hanger spacing for straight duct using 3" channel for the trapeze.

TABLE 3 Hanger Spacing for Straight Fiberglass Duct (3" Channel Hangers)

Duct Width	Duct Height	Maximum Hanger Spacing
≥48"	N/A	4'
<48"	<12"	6'
≥24" but less than 48"	≥24"	6'
<48"	≥12" but less than 24"	8'
≤24"	≥12"	8'

3.2.0 Hanging and Supporting Fiberglass Duct Fittings and Risers

Fiberglass duct fittings often require additional support. Elbows, for example, usually require support in addition to the hangers holding the intersecting duct runs. The hangers supporting the intersecting ducts must also be within a certain distance of the supported fittings.

3.2.1 Duct Fittings

If the duct is wider than 18", support elbows with channel hangers located two-thirds of the diagonal distance from the throat to the heel (*Figure 28*). Note that the duct hangers must also be installed within 12" of the elbow throat.

If a trunk duct hanger should be placed at the same place a rectangular branch duct connects with the trunk (a tee), install hangers on either side of the branch duct instead (*Figure 29*). Hangers for each duct should be placed within 12" of the throat of the tee. Do not exceed the maximum hanger spacing when adding branch and tee supports, even if you must place two hangers relatively close together.

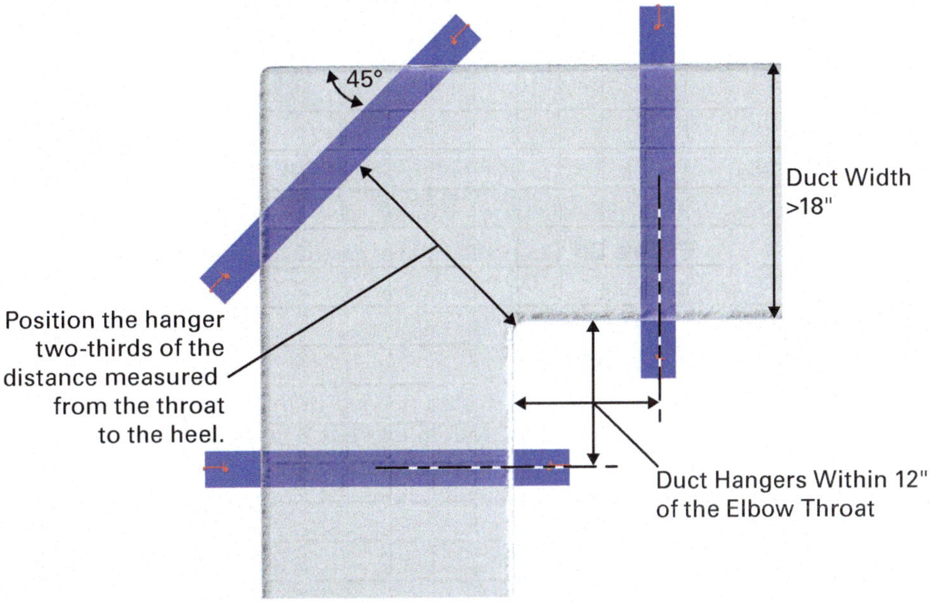

Figure 28 Support for fiberglass elbows with a width greater than 18".

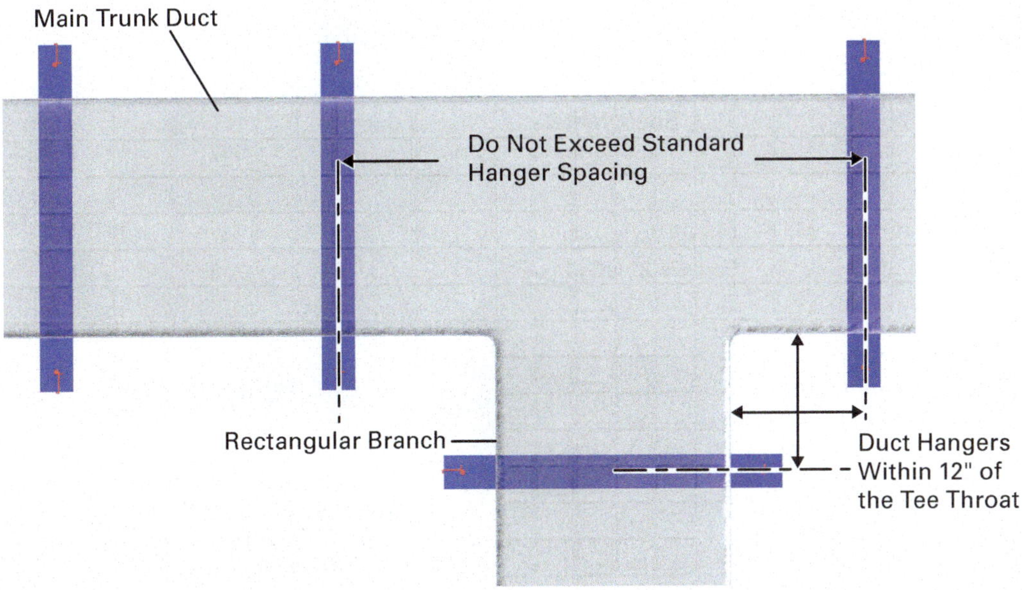

Figure 29 Fiberglass duct support for a tee.

Offset duct supports are needed when the angled offset module is longer than 48" (*Figure 30*). Hangers on the main duct on each side of the offset module must be installed 6" to 12" of the point of connection.

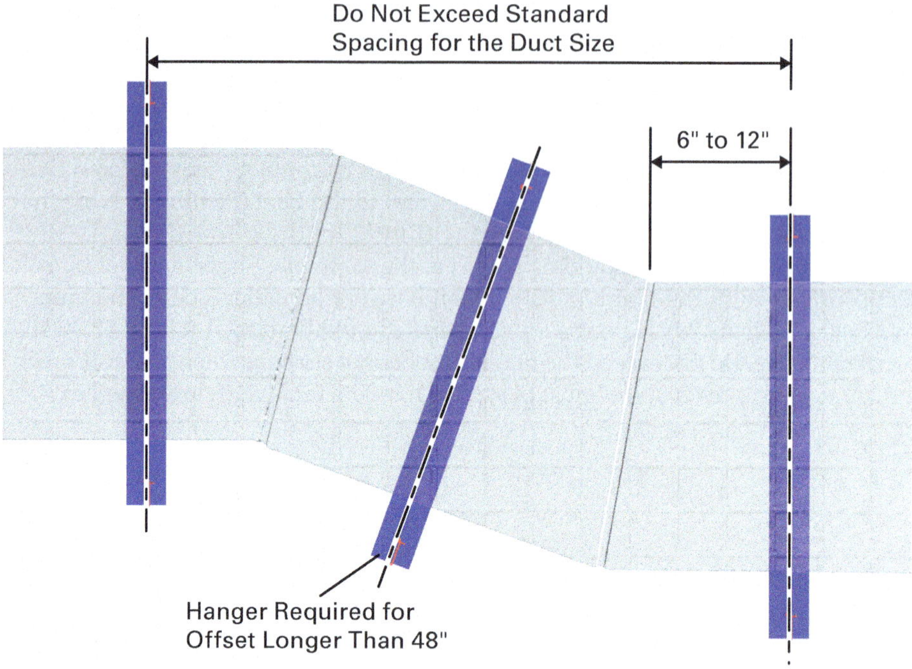

Figure 30 Supporting a fiberglass duct offset module.

NOTE

Remember that fiberglass ductboard can't be used for vertical risers in buildings that exceed two stories in height.

3.2.2 Risers

Vertical risers are often needed in multistory construction. If a vertical riser in a fiberglass duct system exceeds 8' in height, special support is needed. When one or more of these supports are required, the maximum distance between assemblies like the one shown in *Figure 31* is 12'.

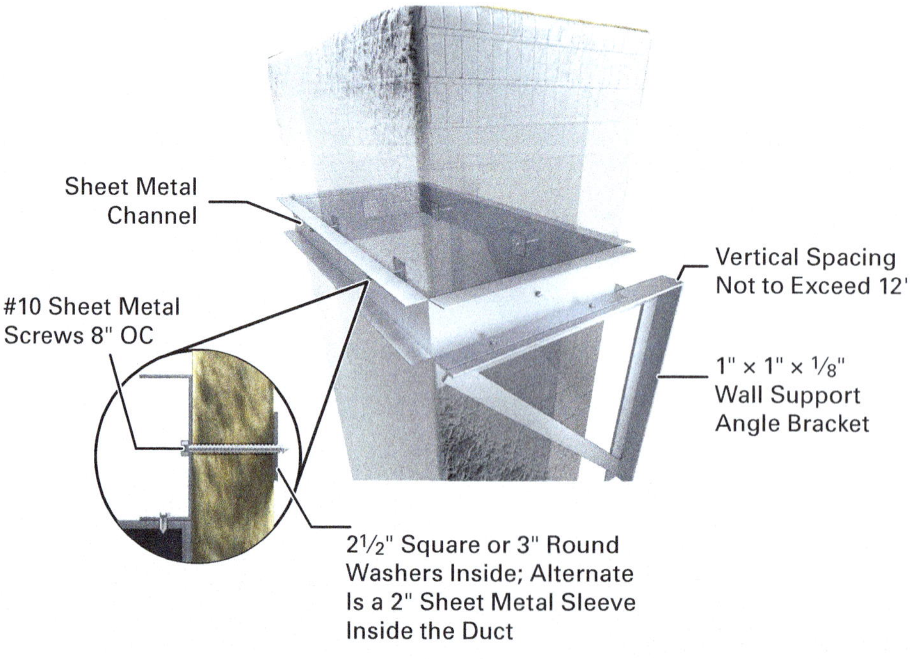

Figure 31 Vertical riser support assembly.

3.0.0 Section Review

1. To properly suspend rectangular fiberglass duct that is 60" × 30" with a hanger extension of 8", what size sheet metal channel is required?
   a. 2" × ½"
   b. 3" × ½"
   c. 3" × 2"
   d. 4" × 2"

2. When a rectangular branch duct connects to a main trunk, the first hanger on the branch must be located within how many inches of the tee throat?
   a. 6"
   b. 12"
   c. 18"
   d. 24"

4.0.0 Fabric-Based Air Distribution Products

Objective

Describe fabric-based air distribution products and installation methods.

a. Identify various types of fabric-based air distribution products and systems.
b. Explain how fabric-based air distribution products are installed.

Performance Tasks

There are no Performance Tasks in this section.

Fabric-based air distribution products (*Figure 32*) have been used for many years now, although these products may seem to be a recent innovation to some. In appropriate applications, the popularity continues to grow due to the advantages these products offer both contractors and end-users.

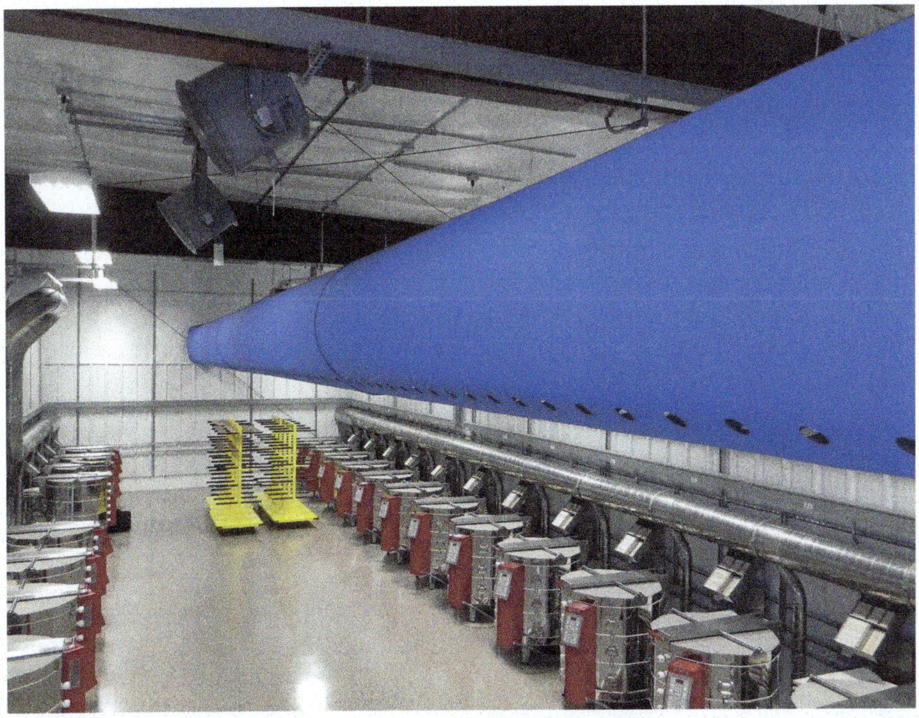

Figure 32 Fabric air-distribution duct in a food handling area.
Source: DuctSox Corporation

Fabric-based systems are also known as *air dispersion systems*. This is because many of the products are designed without obvious air openings. Instead, many fabric-based systems disperse the air along the entire surface through the tiny openings between the fabric threads. However, fabric systems can be constructed of fabrics that are not porous, with openings along the length of the duct that serve as supply air grilles.

Note that fabric duct can't be used on the return side of the system, as the duct collapses when subjected to a negative pressure. A positive pressure is needed to inflate the duct.

Although not perfect for every situation, fabric-based duct systems do offer some distinct advantages in many applications:

- *Gentle and quiet air distribution* — The characteristics of fabric duct allow for consistent and gentle air distribution. Little or no air balancing is generally required, but devices are available to balance the flow in sections of the duct. There is little noise and none of the resonance like that generated in sheet metal duct. Since most fabric systems are installed in spacious indoor areas, the acoustic performance is a significant benefit.
- *Clean and dry external surface* — Porous fabric duct keeps itself free of dust and dirt that would normally settle on fixed metal and fiberglass ducts. In addition, condensation on the surface of the duct is prevented. The fabric fibers can also be treated with antimicrobial products to prevent bacterial growth when the duct is used in food processing areas.
- *Lightweight* — A lightweight air distribution system reduces the load on building structures. Entire duct systems for a large facility, such as an arena, can easily be packed and shipped in cardboard boxes. The same amount of sheet metal duct would require multiple truckloads of heavy duct assemblies. Fabric duct is also not likely to be damaged during shipment like rigid products.
- *Flexible* — In environments where the duct is near workers and/or moving equipment, fabric duct is not subject to damage from hazards such as forklifts or material movement.
- *Ease of cleaning* — Fabric duct systems can be removed and cleaned with vacuum cleaners or compressed air. Sections can also be laundered. This avoids the extreme cost of special duct cleaning equipment used on other systems that are difficult to access and must be cleaned in place.
- *Low product cost and reduced installation labor* — Both are very important advantages to both contractors and end users. The material cost of a fabric duct system can be as much as 80% lower than an equal metal duct system. Its flexibility and light weight also make it far easier and faster to install. Metal duct systems require as much as ten times more installation labor, not to mention the cost of transporting and storing sheet metal assemblies prior to installation.
- *Eye-catching colors and marketing aids* — Fabric duct is available in a wide variety of colors to suit the environment. Advertising logos and messages can be printed directly on the fabric at the factory, providing income from sponsors for end users. Using the duct system as a form of sponsor or owner advertising also helps cover the cost of the material.

Going Green

Fabric Duct Systems

Many of the advantages of fabric duct are directly or indirectly related to positive environmental issues. For example:

- Lightweight fabric duct dramatically reduces fuel costs related to transportation, as opposed to shipping raw metal materials and finished assemblies.
- A duct system that can be laundered is far more likely to remain free of mold, mildew, and similar threats to indoor air quality.
- Some fabric products can be recycled when replacement is required. Indeed, many of the products are made from recycled materials.

4.1.0 Types of Fabric Duct

Fabric duct systems are ideal for some environments. Two important product characteristics to consider when selecting a fabric duct product are the air dispersion design and porosity of the material.

4.1.1 Air Dispersion Design

The application determines the best choice for air dispersion. For example, in a gymnasium or warehouse environment where activity levels are high and the duct is distant from occupants, a high-velocity airstream is unlikely to bother anyone. Thus, a nonporous product with strategically placed openings for the air to escape is an acceptable option.

In smaller areas where the duct is closer to occupants or processes, the airstream must be gentler to prevent discomfort. High-velocity airstreams may also produce slightly more noise if that is a consideration. A porous product and smaller openings are better options for this situation.

Several levels of air dispersion performance are available from which duct designers may choose. A *high-velocity / high-throw design* incorporates the largest and fewest openings in the fabric surface (*Figure 33*). High-velocity airstreams encourage more air mixing. The openings in the duct can be placed continuously along the duct or be strategically located to blow in specific directions. The high-velocity / high-throw design also offers the longest **throw** of the other styles. The airflow from the vents may be felt as much as 40' away.

Throw: Refers to the distance air travels from a register or diffuser based on its exit velocity.

Figure 33 A high-velocity, high-throw dispersion design.
Source: Courtesy of Air Distribution Concepts

A gentler flow of air can be achieved using smaller openings, but more openings are required to deliver the same air volume (*Figure 34*). This approach can be referred to as a *reduced-velocity / low-throw design*. Noticeable airflow is not likely to be felt more than 10' away. This style is best for areas where occupants may feel discomfort from a moving airstream. The openings can be placed in a continuous line down the duct surface or be placed in groups in specified locations.

Even smaller openings, again in greater numbers, can be used. This style results in airflow that is virtually imperceptible to occupants. Rather than relying on moving jets of air that encourage air mixing, the air from these products moves consistently as a mass, displacing air as it moves out from the duct.

Since the openings in the material are very small, air filtration is a more important consideration than with other dispersion designs. If the air entering a system is not well filtered, the duct collects too much dust and debris inside. Although the duct can be removed and laundered periodically if necessary, effective return air filtration is the correct solution. The airstream of all systems should be well filtered to protect coils as well as ductwork.

Air can also be dispersed through a continuous linear vent (*Figure 35*). The vent material is like a fabric screen. The vent can run the length of the duct or be installed only in specific sections and locations. An air throw distance up to about 20' can be achieved, depending on the air velocity as it exits the duct.

(A) Continuous Dispersion

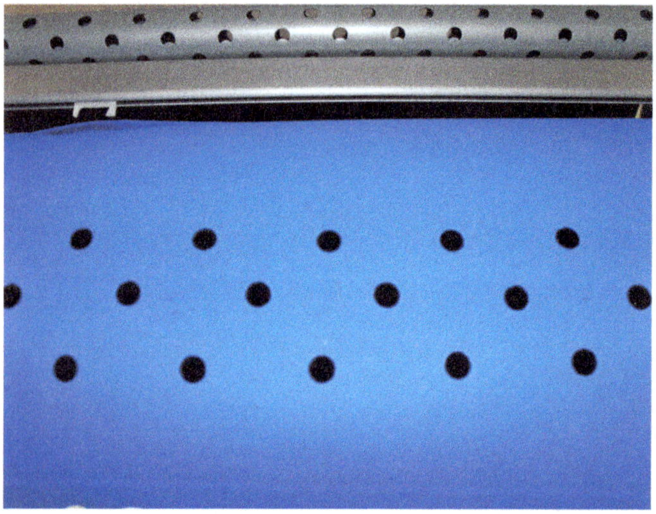

(B) Grouped Dispersion

Figure 34 A low-throw, reduced-velocity dispersion design.
Source: Courtesy of Air Distribution Concepts

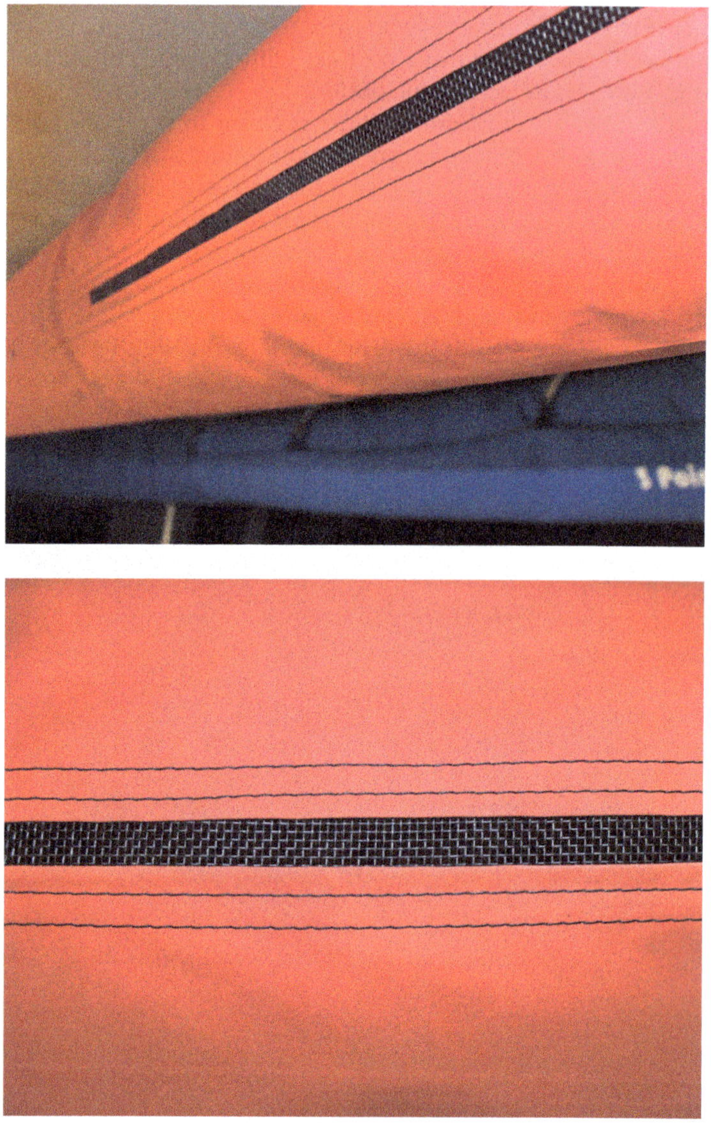

Figure 35 Linear air dispersion.
Source: Courtesy of Air Distribution Concepts

Manufacturers can use any of these approaches as well as mix some approaches in the same system. The duct designer must choose the best approach for the application and then determine other characteristics such as the number and position of openings. This would be determined based on the system airflow, and how much air is to be dispersed into each area served by the system.

4.1.2 Porosity

The duct material used for systems that rely on previously described methods of dispersion is typically nonporous. All air is forced out through the carefully sized and placed openings. However, nonporous systems are not able to keep the outer surface clear of dust. Dust is not likely to settle and remain on the outside of porous fabrics, since the air is being dispersed across the entire surface from the inside.

Fabrics that are porous must be rated in a sensible way to allow for comparison and support design efforts. Porosity for these products is rated in units of cubic feet per minute of airflow per square foot of surface area (cfm/ft^2). The porosity of the products ranges from 0 cfm/ft^2 (nonporous) to as high as 50 cfm/ft^2. These high-porosity products are used for fabric diffusers (*Figure 36*).

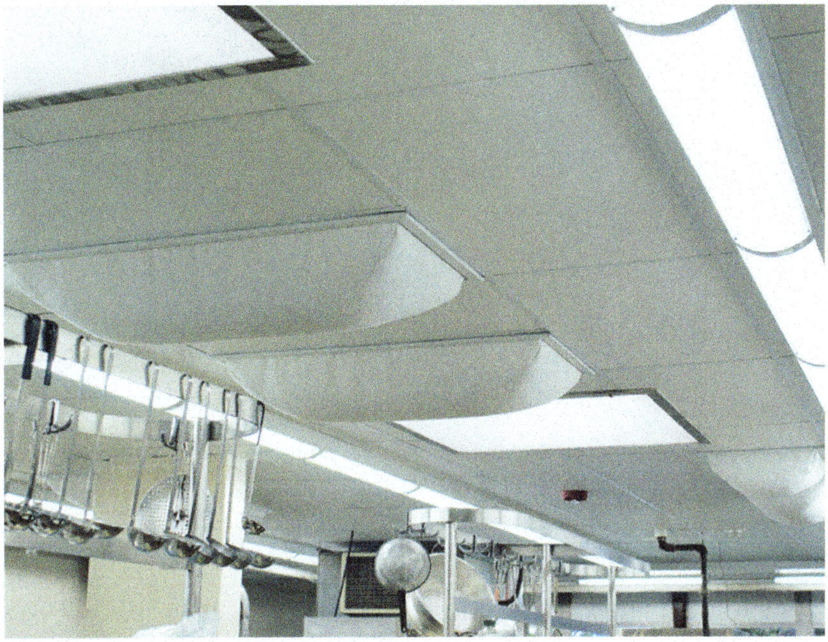

Figure 36 Fabric diffusers.
Source: DuctSox Corporation

Of course, the pressure applied to the duct affects the amount of airflow through the fabric. In most cases, a product's porosity is stated based on a static pressure of 0.5" w.c. If the pressure applied to the duct will be more or less than that, designers must take the difference in airflow into account. Manufacturers provide graphs that show the volume of air delivered per square foot of fabric across a range of pressures or provide correction factors for that purpose.

4.1.3 Other Characteristics

Fabric duct materials with special characteristics are also available. One option is the application of antimicrobial protection to the fabric. Products that have been treated for this purpose discourage the growth of bacteria. This is a valuable option for ducts serving food-processing areas (*Figure 37*). Fabric duct systems are very well suited for large food processing areas that must remain cool. The antimicrobial treatments are designed to last through many washings.

Fabrics that dissipate static electricity are also available. Facilities that house sensitive electronics are susceptible to damage from static electricity. Every precaution must be taken to prevent the accidental accumulation and discharge of high-voltage static around the equipment. Fabrics that resist chemical attack can be used in areas where acids and caustic materials are present.

On Site

Special Shapes

Although the typical fabric duct shape is round, D-shaped duct is also available. This type can be mounted flat against the ceiling, as shown here. The porous fabrics can also be used as diffusers, providing very gentle air distribution across the entire surface of the fabric.

Source: DuctSox Corporation

Figure 37 Fabric duct treated to provide microbial protection.
Source: DuctSox Corporation

4.2.0 Fabric Duct Assembly and Installation

Fabric duct systems are suspended in a variety of ways. The choice is based primarily on the structure above, as its characteristics may prevent some suspension options. Other factors such as cost and appearance are also considered.

Before exploring suspension options, note that standard fabric duct only inflates when the system blower is operating. Once the blower shuts down, the fabric simply hangs limp from its support (*Figure 38*). This can present an unsatisfactory appearance. In addition, when a blower starts and accelerates to its operating speed quickly, the duct will snap into shape with a loud popping sound. When used on a variable air volume (VAV) system, the duct is only fully inflated when the system is operating at maximum design airflow. At all other times, it will be partially inflated.

Figure 38 Fabric duct appearance without airflow.

The solution is an optional supporting structure placed inside the duct. The structure maintains its inflated appearance when there is little or no airflow. One approach uses independent hoops inside the duct, positioned about 5' apart (*Figure 39*). The hoops are attached to the duct at three points around the perimeter.

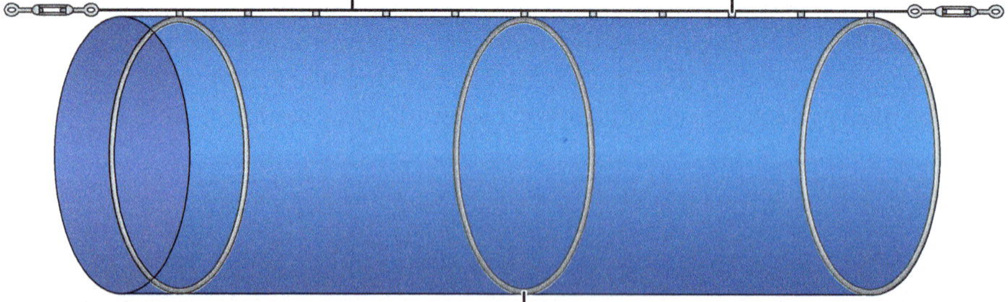

Figure 39 Hoops placed inside fabric duct.
Source: DuctSox Corporation

A more elaborate internal structure keeps the duct both taut and round along its full length, offering the best appearance (*Figure 40*). The hoops in this system are connected longitudinally with sections of tubing routed through the center of the duct. The suspension system is attached only to the internal structure, rather than to the duct directly.

Some hanging systems provide some external support for the duct. These devices are discussed with suspension systems.

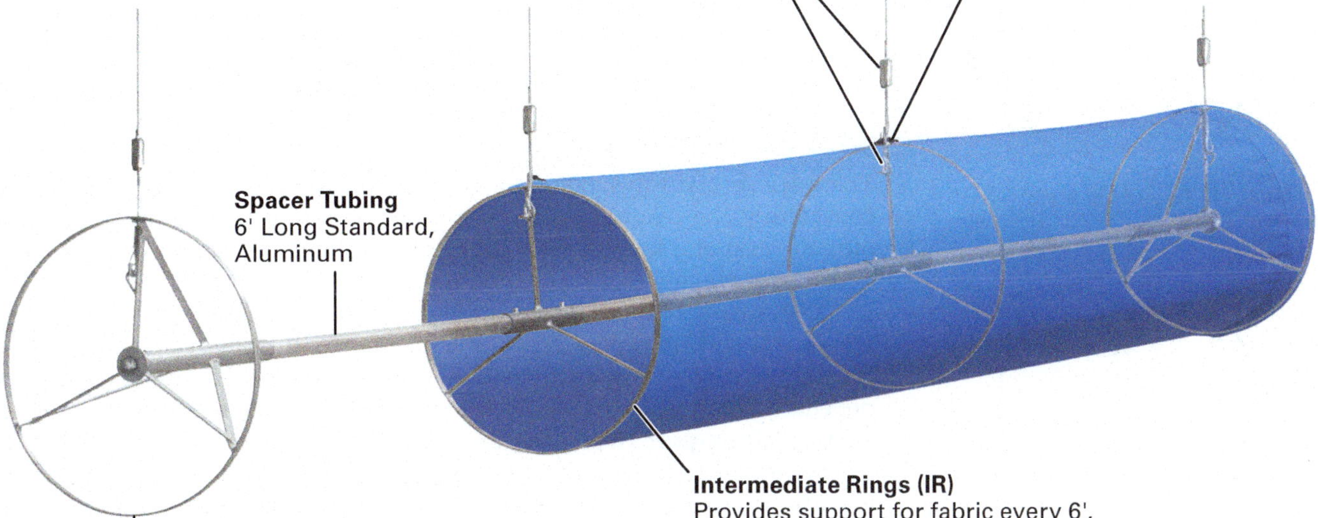

Figure 40 The Skelecore® fabric tensioning system.
Source: DuctSox Corporation

4.2.1 Duct and Fitting Assembly

Sections of fabric duct and fittings are attached with mating zippers. Two pieces are simply brought together, and the zipper engaged. A fabric flap covers the zipper once the connection is made. Common fittings such as elbows and end caps are readily available. Special fittings can be made upon request. Duct sections and fittings have hoops along the upper duct seam where hanging straps are attached.

The duct system usually begins with a connection to a round metal collar at the air handling unit or a metal duct adapter. The inlet collar for the fabric duct is made slightly larger than the metal so it slips over the opening. An integral belt is then cinched tight around the metal collar and anchored with screws to secure it (*Figure 41*).

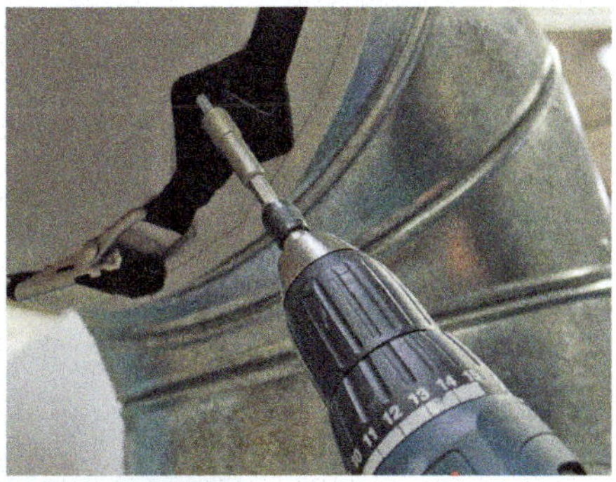

Figure 41 Anchoring the tightened belt on an inlet collar.
Source: DuctSox Corporation

One special, yet simple component is available to provide air balancing capability. A flow control device can be placed in the duct by simply adding it as a short section of duct, using the provided zippers (*Figure 42*). The opening size can be adjusted from outside the duct with a drawstring, allowing it to serve as a balancing damper. The added resistance can also help reduce the snapping sound from rapid inflation. The device is made from a mesh material, so air does pass through it. The resistance it adds, however, is sufficient for most balancing requirements.

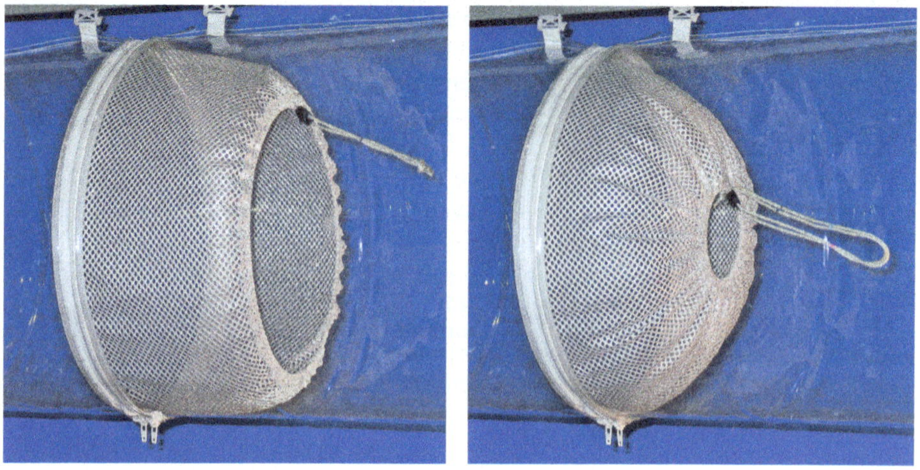

Figure 42 Flow control device.
Source: DuctSox Corporation

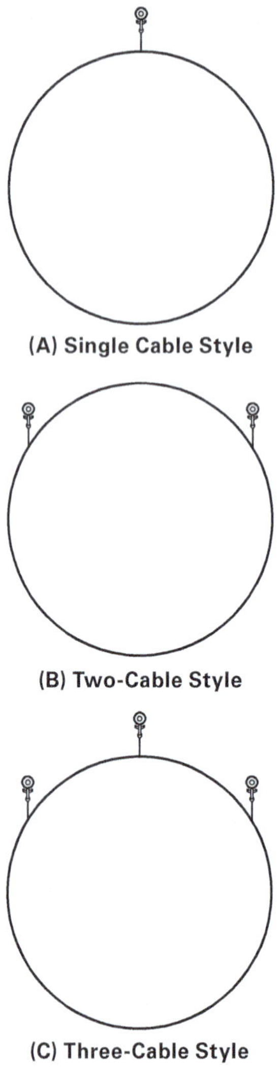

Figure 43 Single, double, and triple cable installations.
Source: DuctSox Corporation

4.2.2 Suspension Systems

There are several ways to suspend fabric duct. The selection process begins by considering the overhead structure type. Some installations require the use of more than one suspension system. Note that all suspension options are not necessarily available for each fabric type.

Cable suspension is typically the simplest and least expensive option. It works well with an exposed roof structure, where the cable can be suspended at the desired height from the floor. One, two, or even three cables may be needed, depending on the size of the duct (*Figure 43*). Duct diameters between 30" and 72" typically require two cables; diameters over 72" require three cables.

The cable(s) are routed along the length of the duct (*Figure 44*). Hanging straps are attached to the duct and then snapped over the cable. The straps are free to slide back and forth on the cable to accommodate expansion and contraction as the duct inflates and deflates. Cable supports are typically placed no more

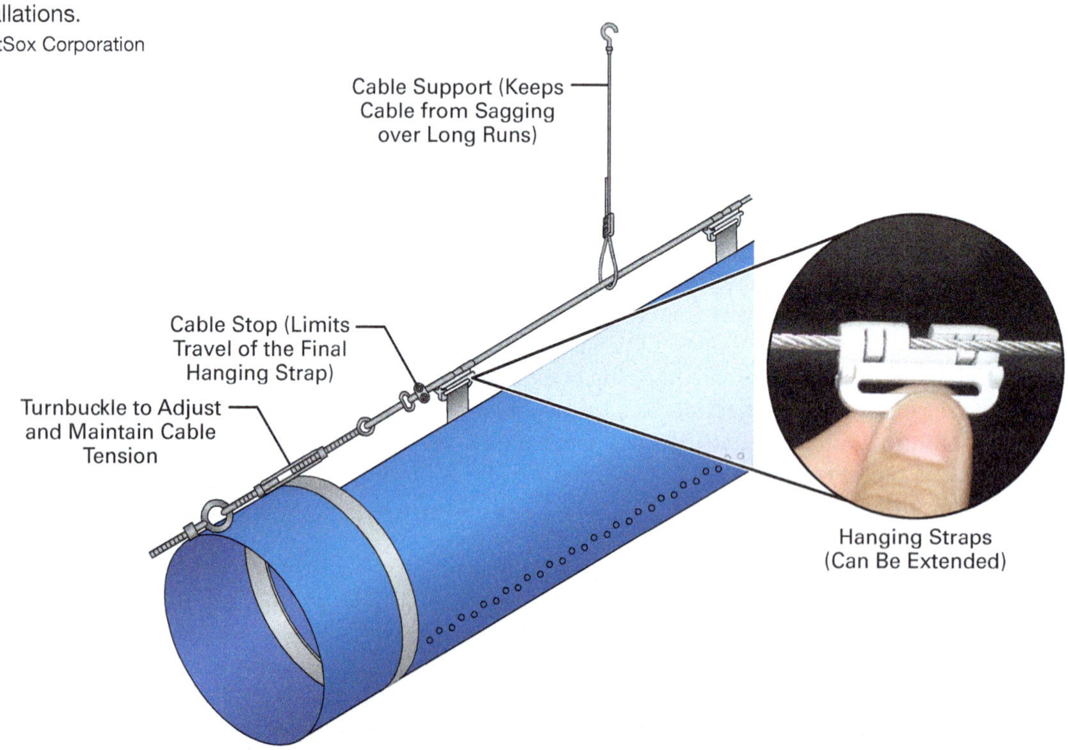

Figure 44 Single cable suspension.
Source: DuctSox Corporation

than 30' apart. Additional cables must be routed to follow the duct as it changes direction. Elbows are provided with extended hanging straps to accommodate vertical risers. (*Figure 45*).

Cable system hardware is available in galvanized steel, stainless steel, and plastic-coated stainless-steel cable. Corrosion-resistant hardware is necessary for indoor pools and similar applications where excessive humidity and air-borne chemical vapors corrode some metals.

For a more rigid suspension system with an improved appearance, a U-track system can be used (*Figure 46*). The track is made from aluminum for light weight. It can be suspended like a cable system, or surface-mounted to a flat ceiling. Clips are also available to mount the track to the T-bars of a suspended ceiling grid. The light weight of the duct and rail allow the suspended ceiling grid to handle the additional load.

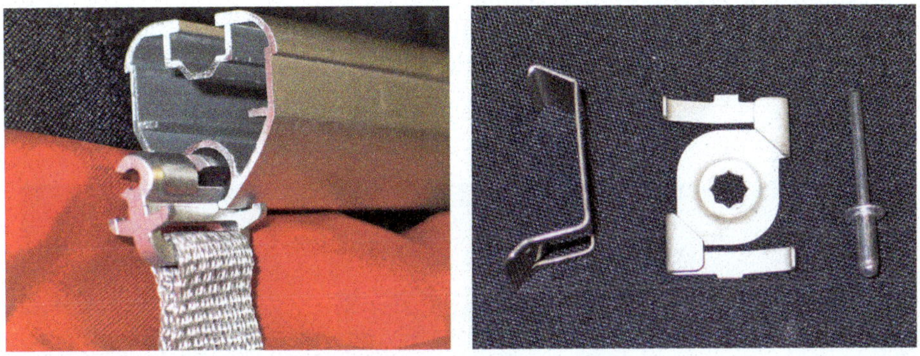

Figure 46 U-track and a T-bar mounting clip.
Source: DuctSox Corporation

Radiused hangers can be added to both the cable system and the U-track system (*Figure 47*). The hangers are attached to the cable or track and are equipped with strap connections to connect the duct. The connecting points on the duct are at or near the 10 o'clock and 2 o'clock positions. Hangers are available to fit single or double cables and tracks.

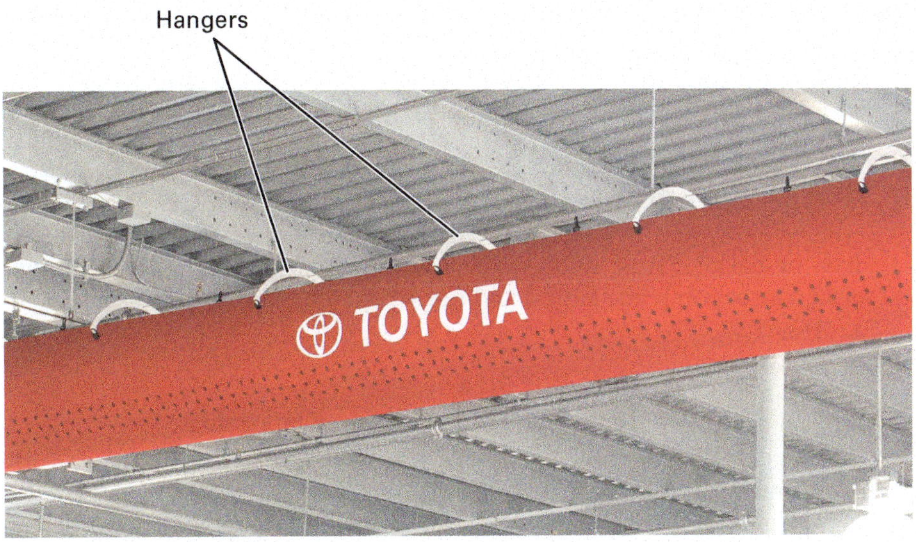

Figure 47 Added hangers to help maintain the duct shape when it is deflated.
Source: DuctSox Corporation

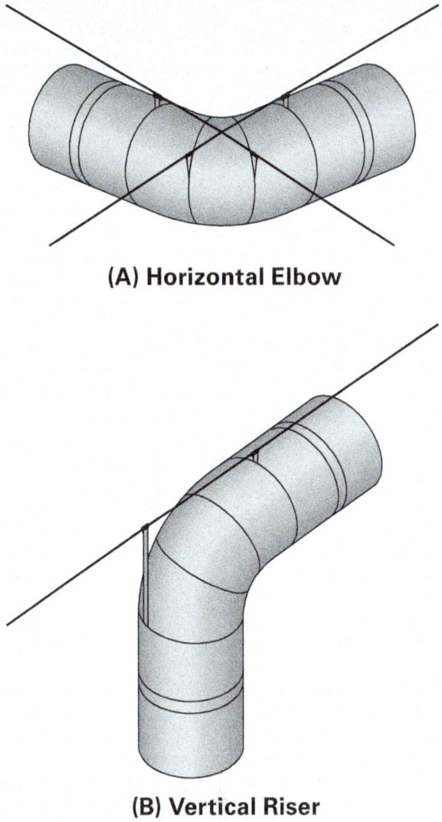

Figure 45 Directional changes—horizontal and vertical.
Source: DuctSox Corporation

4.2.3 Underfloor Installations

Underfloor air distribution systems have been used for many years in data centers and similar facilities. These areas are often placed on raised floors, with the space beneath used to route the many cables that are required out of sight. The area beneath the floor is also used to distribute cooling airflow.

Traditional raised-floor cooling systems rely on a pressurized floor plenum rather than duct. The air handling unit creates a positive pressure beneath the floor. Some of the floor panels, usually 24" × 24", are replaced with perforated panels to serve as diffusers. The panels are strategically placed near the heat-producing equipment. Conditioned air then flows out of the floor plenum and into the room through the perforated tiles.

One problem with these systems has always been the lack of control over air volume. It can be difficult to balance the air and encourage it to go where it is needed in sufficient volume. A fabric duct system can provide a reasonably priced solution to such a problem (*Figure 48*).

The duct can simply be placed under the floor. No elaborate suspension system is required; the duct is held in position by a few cables placed in strategic positions.

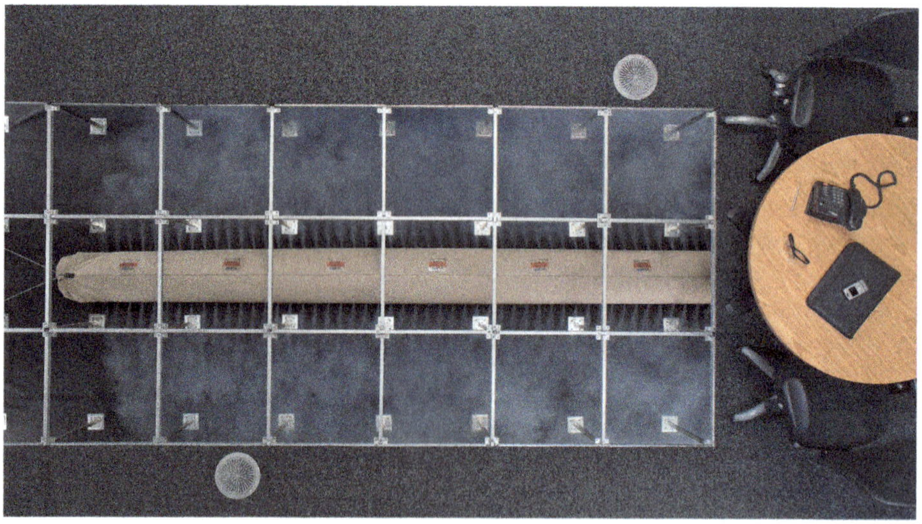

Figure 48 Fabric duct in an under-floor application.
Source: DuctSox Corporation

4.0.0 Section Review

1. True or False? Fabric duct systems with a high-velocity / high-throw design incorporate the largest and fewest openings in the fabric surface.
   a. True
   b. False

2. What is used to adjust flow balancing devices in fabric duct?
   a. Zippers
   b. A drawstring
   c. An opposed-blade damper
   d. A manually operated lever

Module 03214 Review Questions

1. What is the material called *scrim* in foil scrim kraft (FSK)?
   a. Wool
   b. Cotton
   c. Plastic wrap
   d. Fiberglass yarn

2. What is the expected leakage rate for a fiberglass duct system fabricated to the Leak Class 6 standard and operating at a pressure of 1" w.c.?
   a. 12 cfm per 100 ft^2 of duct surface
   b. 6 cfm per 10 ft^2 of duct surface
   c. 0.038 cfm per square foot of duct surface
   d. 0.06 cfm per square foot of duct surface

3. True or False? Fiberglass duct is the preferred material for use with heating equipment fueled by coal or wood.
   a. True
   b. False

4. Which of the following is an application suitable for fiberglass ductboard?
   a. Outdoor applications
   b. Kitchen or fume exhaust ducts
   c. Vertical risers in two-story buildings
   d. Penetrating walls where fire dampers are installed

5. Closure systems for fiberglass ductboard must conform to ______.
   a. UL Standard 90A
   b. UL Standard 181A
   c. ASTM Standard 90A
   d. ASHRAE Standard 15

6. For 1" thick fiberglass ductboard fabrication, how wide must the closure tape be?
   a. 1"
   b. $1\frac{1}{2}$"
   c. $2\frac{1}{2}$"
   d. 4"

7. Which of the following is the preferred joint style for closing ductboard corner seams?
   a. V grooves
   b. Butt joints
   c. Mitered joints
   d. Shiplap grooves

8. When connecting ductboard to sheet metal, how far apart must the screws be placed?
   a. 4" on center
   b. 6" center
   c. 8" on center
   d. 12" on center

9. Relevant to fiberglass ductboard, which of the following best defines minor damage?
   a. A hole penetrating the duct that is less than ½" in diameter.
   b. A hole penetrating the duct that is less than 1" in diameter.
   c. Any opening in the facing that has not damaged the fiberglass board beneath it.
   d. Any opening in the facing that goes less than halfway through the fiberglass board.

10. True or False? Fiberglass duct up to 48" wide can be suspended using 22-gauge sheet metal straps that are at least ½" wide.
    a. True
    b. False

11. What is the minimum width of the supporting channel used to suspend rectangular fiberglass duct?
    a. 1"
    b. 1½"
    c. 2"
    d. 3"

12. At what height do vertical risers in fiberglass duct systems require special support?
    a. 4' or less
    b. More than 8'
    c. More than 12'
    d. More than 14'

13. True or False? A high-velocity / high-throw fabric duct design is best for areas where occupants may feel discomfort from a moving airstream.
    a. True
    b. False

14. In most cases, the porosity of a fabric duct material is based on a duct static pressure of ______.
    a. 3" w.c.
    b. 2" w.c.
    c. 1" w.c.
    d. 0.5" w.c.

15. When using a cable suspension system for a fabric duct system, duct diameters between 30" and 72" require how many cables?
    a. One cable
    b. Two cables
    c. Three cables
    d. Four cables

Answers to odd-numbered review questions are found in the Module Review Answer Key at the back of this book.

Answers to Section Review Questions

Answer	Section	Objective
Section One		
1. b	1.1.2	1a
2. c	1.2.0	1b
Section Two		
1. b	2.1.0	2a
2. b	2.2.2	2b
Section Three		
1. c	3.1.1	3a
2. b	3.2.1	3b
Section Four		
1. a	4.1.1	4a
2. b	4.2.1	4b

MODULE 03304

Retail Refrigeration Systems

Source: Camij/iStock/Getty Images

Objectives

Successful completion of this module prepares you to do the following:

1. Describe retail refrigeration applications and the related refrigeration cycles.
   a. Describe medium-temperature systems and the refrigeration cycle distinctions.
   b. Describe low-temperature systems and the refrigeration cycle distinctions.
   c. Describe various approaches to defrosting refrigeration equipment.
2. Identify various types and functions of retail refrigeration equipment.
   a. Identify and describe reach-in coolers and freezers.
   b. Identify and describe various walk-ins and merchandisers.
   c. Identify various types of commercial ice machines.
3. Identify and describe common retail refrigeration system components.
   a. Identify and describe primary retail refrigeration system components.
   b. Identify and describe secondary retail refrigeration system components.
   c. Identify and describe common retail refrigeration system controls.
4. Describe the maintenance and operation of retail refrigeration equipment.
   a. Identify reach-in freezer maintenance activities and recount the typical sequence of operation.
   b. Identify cubed-ice machine maintenance activities and recount the typical sequence of operation.

Performance Tasks

Under supervision, you should be able to do the following:

1. Set up a freezer defrost timer.
2. Isolate a fault in a refrigeration unit or ice machine.
3. Clean an ice machine.

Overview

Retail refrigeration systems are found everywhere. Various refrigerated coolers, freezers, cases, and dispensers are used wherever food and beverages are served. Out of public view, much larger refrigeration systems protect stored products used to replenish the retail systems. The specialized field of retail refrigeration offers many career opportunities for HVACR service technicians.

Digital Resources for HVACR

SCAN ME

Scan this code using the camera on your phone or mobile device to view the digital resources related to this craft.

NCCER Industry-Recognized Credentials

If you are training through an NCCER-accredited sponsor, you may be eligible for credentials from NCCER. The ID number for this module is 03304. Note that this module may have been used in other NCCER curricula and may apply to other level completions. Contact NCCER at 1.888.622.3720 or go to **www.nccer.org** for more information.

You can also show off your industry-recognized credentials online with NCCER's digital credentials. Transform your knowledge, skills, and achievements into credentials that you can share across social media platforms, send to your network, and add to your resume. For more information, visit **www.nccer.org**.

1.0.0 Refrigerant Circuit Operation

Performance Task

1. Set up a freezer defrost timer.

Objective

Describe retail refrigeration applications and the related refrigeration cycles.

a. Describe medium-temperature systems and the refrigeration cycle distinctions.
b. Describe low-temperature systems and the refrigeration cycle distinctions.
c. Describe various approaches to defrosting refrigeration equipment.

Refrigeration has a significant impact on our daily lives. In the distant past, the preservation of food was limited to curing, dehydrating, or using natural resources such as snow and winter ice to avoid spoilage.

Since the mid-1800s, scientists have understood that bacteria are responsible for most food spoilage. After it was discovered that bacteria are unable to grow at low temperatures, the development of modern approaches to refrigeration began in earnest.

Throughout the 1800s, advances in the protection and preservation of foods continued. In 1882, the British sailing ship *Dunedin,* attempting to carry meat from New Zealand to England, was fitted with a coal-fired refrigeration apparatus to keep the meat frozen for the voyage. The successful delivery of frozen product by the *Dunedin* validated the practice of refrigerated food shipments.

Retail refrigeration systems and applications are now found around every turn. Although the priority is to preserve food, refrigerated fixtures must also offer style and convenience and be inviting to consumers. Quiet, visually appealing equipment attracts consumers, increases sales, and instills confidence that the food or beverage is safe and properly preserved.

Many types and styles of refrigerated equipment are used to cool or freeze foods and other perishables. Ideally, consumable products reach the retail level already cooled or frozen. The retail fixture is then used to maintain those conditions while providing consumer access to the products. They are not designed to freeze warm food.

NOTE

Review the details of refrigeration circuit operation as needed in NCCER Module 03107, *Introduction to Cooling*. Reviewing that material will help you better understand nuances in the cycle when it is applied at lower temperatures.

Figure 1 shows a basic air-cooled refrigerant circuit flow diagram. Primary refrigerant circuit components are the compressor, evaporator, condenser, and expansion (metering) device. However, the installed location of the primary components and interconnecting lines, evaporator temperatures, condensing environment, and types of refrigerants used are all significant variables in refrigeration.

Did You Know?

Ice the Hard Way

Before the development of mechanical refrigeration, people preserved food with blocks of ice. The ice would be harvested from nearby lakes in the winter and stored in icehouses. The ice blocks were covered with hay and sawdust to insulate them from the heat.

Throughout the year, ice was removed from storage and delivered to customers. The delivery was often done using horse-drawn ice wagons. In areas where the temperature seldom dropped below freezing, large masses of sawdust-insulated blocks of ice had to be delivered from colder locations in the north by rail or riverboat.

Sources: Cape Pond Ice Company

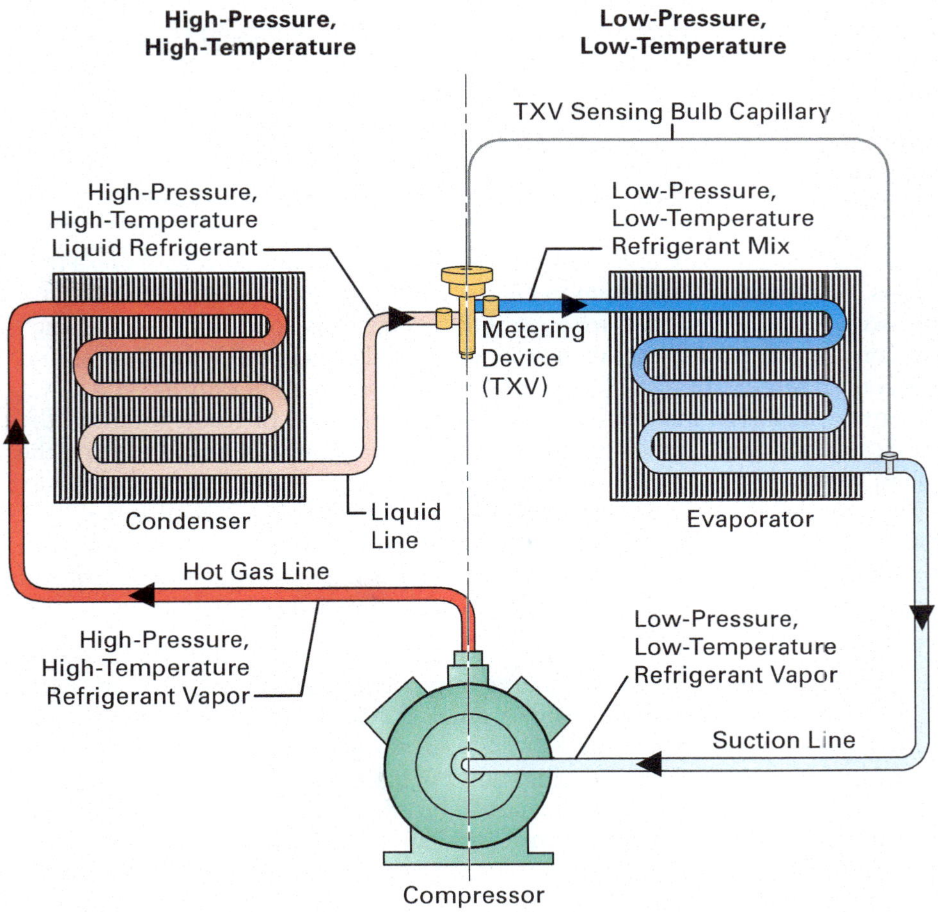

Figure 1 Basic refrigerant circuit.

Refrigeration applications are usually classified by the application's required space temperature. These classifications are important to remember because some components, such as compressors and metering devices, are classified for duty according to the following temperature ranges:

- *Air conditioning/high-temperature refrigeration applications* — Space temperatures above 60°F (16°C)
- *Medium-temperature refrigeration applications* — Space temperatures of 30°F to 60°F (–1°C to 16°C)
- *Low-temperature refrigeration applications* — Space temperatures of –40°F to 30°F (–40°C to –1°C)

These categories are broad and are not universally accepted for all purposes. For example, when selecting a thermostatic expansion valve (TXV) for a refrigeration application, the temperatures provided in the capacity tables reflect evaporator temperatures, not space temperatures. It is important to understand the difference in those temperatures and the perspective of various manufacturers regarding temperature.

The comfort cooling and basic refrigeration cycles are similar. However, some technicians are uneasy about servicing refrigeration systems because they have spent most of their time working with comfort cooling systems, where a 40°F (4°C) evaporator temperature (saturated suction temperature) is the established standard.

In refrigeration, lower evaporator temperatures must be reached to achieve the desired storage temperature. Different refrigerants are used as well, including some that are flammable. Competency in this regard is simply a matter of getting used to different pressure-temperature (PT) charts and knowing how far below the box temperature the evaporator temperature needs to be.

1.1.0 Medium-Temperature Refrigeration Cycle

An example of a typical medium-temperature refrigeration cycle, using HFC-134a as the refrigerant, is shown in *Figure 2*.

Assume that the system is for a retail fixture inside a conditioned space, such as a convenience store. The air temperature in the vicinity of the fixture and unit condenser is 75°F (24°C), and the desired box temperature is 38°F (3°C). Remember, however, that environmental conditions may fluctuate dramatically throughout the day. The box temperature is currently 40°F (4°C), prompting the system to run.

The numbers below correspond to the numbered points shown in *Figure 2*. Note that this example is theoretical and does not account for normal pressure drops through piping or components in an operating system. Also note that all temperature values have been rounded to the nearest whole number.

1. After the compression process, highly superheated vapor exits the compressor. In this example, the discharge vapor temperature has reached 160°F (77°C) at a pressure of 143 psig. A quick check of the PT chart reveals that the superheat is now at 52°F (29°C). All the superheat must be removed from the vapor before the refrigerant begins to condense. A small amount of superheat may be lost to the surrounding air as the vapor travels to the condenser through the discharge line. The hot gas line is rarely insulated, so a large temperature difference (ΔT) between the line and the surrounding air is common. In this example, the refrigerant arrives at point 2 slightly cooler than its discharge temperature.
2. As the refrigerant vapor enters the condenser coil, superheat is quickly transferred to the air. As the refrigerant temperature continues to fall toward the temperature of the entering outdoor air, the ΔT between the condenser air and refrigerant is reduced and the heat transfer process progressively slows. All the superheat must be removed, and the vapor temperature must fall to

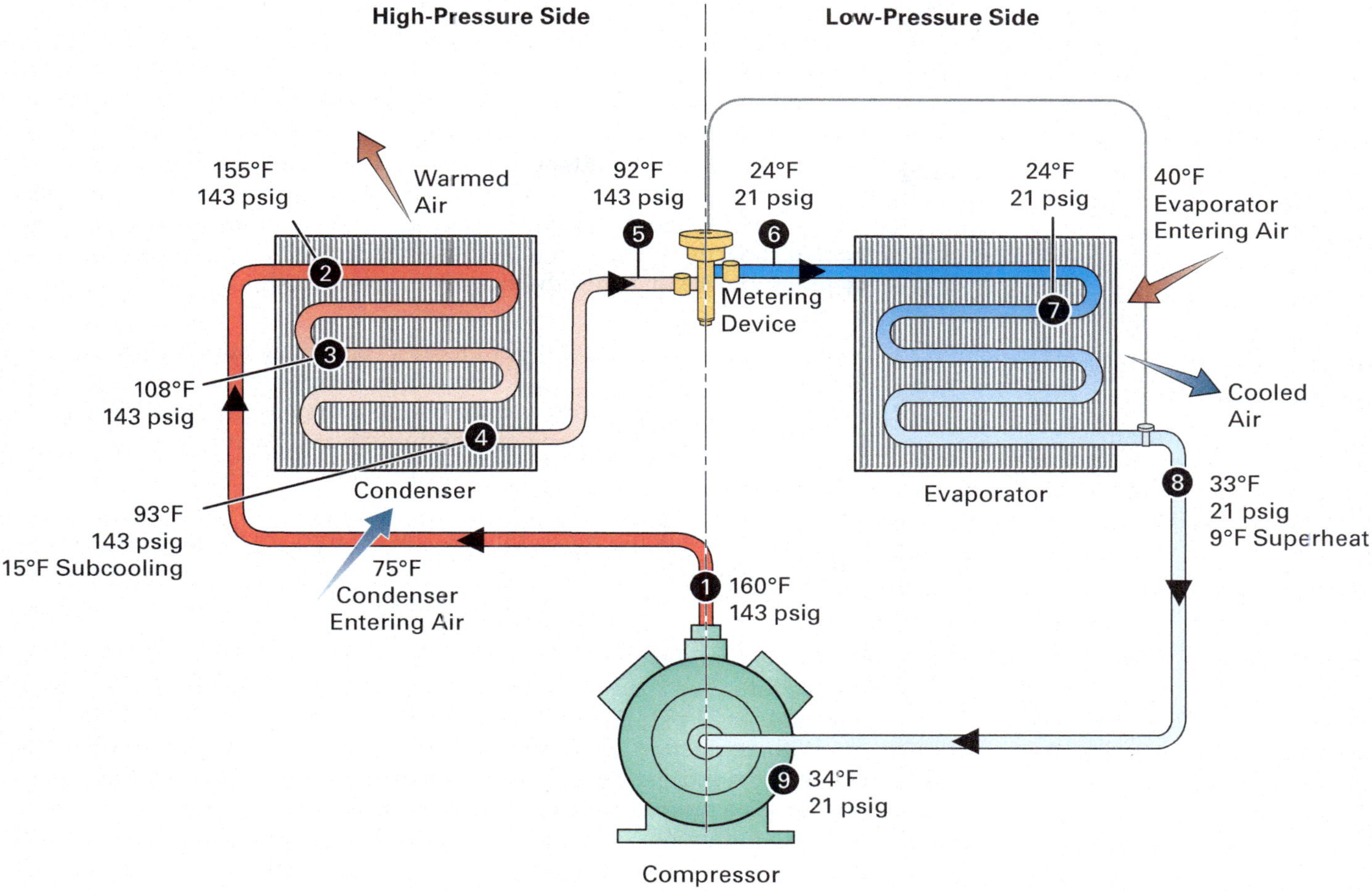

HFC-134a PT Chart

°F	°C	psig	°F	°C	psig
–5	–20.6	4.1	85	29.4	94.9
0	–17.8	6.5	90	32.2	103.9
5	–15.0	9.1	95	35.0	113.5
10	–12.2	12.0	100	37.8	123.6
15	–9.4	15.0	105	40.6	134.3
20	–6.7	18.4	110	43.3	145.6
25	–3.9	22.1	115	46.1	157.6
30	–1.1	26.1	120	48.9	170.3
35	1.7	30.4	125	51.7	183.6
40	4.4	35.0	130	54.4	197.6
45	7.2	40.0	135	57.2	212.4
50	10.0	45.3	140	60.0	227.9
55	12.8	51.1	145	62.8	244.3
60	15.6	57.3	150	65.6	261.4
65	18.3	63.9	155	68.3	279.5
70	21.1	70.9	160	71.1	298.4
75	23.9	78.4	165	73.9	318.3
80	26.7	86.4			

Figure 2 Example medium-temperature refrigeration cycle using HFC-134a.

the saturation temperature for HFC-134a at the given pressure before condensing begins. In this example, the condensing temperature is 108°F (42°C).

3. When the vapor has cooled to 108°F in the condenser, condensation begins, and the refrigerant temperature will not change throughout the process. As refrigerant flows through the condenser, the ratio of liquid to vapor increases until only liquid remains.

4. When the phase change is complete, the liquid refrigerant begins to measurably cool below its condensing temperature. In this example, the refrigerant temperature has fallen to 93°F (34°C) at the exit of the condenser. Because the saturation temperature for HFC-134a is 108°F (42°C) at a pressure of 143 psig, subtracting the 93°F actual temperature results in a subcooling value of 15°F (8°C). Remember that, without measurable subcooling as an indicator, the condensation process isn't complete, and some amount of vapor remains in the liquid. A lack of subcooling causes a serious loss of refrigerating capacity in the system because any vapor that travels through the metering device and into the evaporator is virtually useless as a cooling medium.
5. As is the case with other refrigerant lines connecting system components, the heat loss or gain in the liquid line can be large or small. In this example, the liquid line is very short, and the temperature has dropped 1°F (0.6°C). If the liquid line travels a long way or through an area at a significantly different temperature, a substantial increase or decrease in temperature can occur.
6. At this point, the liquid refrigerant has passed through the metering device. This resulted in a pressure drop of 122 psig. Although the refrigerant was subcooled and entirely in its liquid state as it entered the metering device, some of the refrigerant flashed to the vapor state to absorb heat from the remaining liquid. Both liquid and vapor are present here, so the refrigerant is in its saturated condition. Per the PT chart, the saturation temperature of HFC-134a at 21 psig is about 24°F (–4°C).

   With refrigerant temperature now at 24°F, a ΔT of 14°F (8°C) exists between the refrigerant and the desired box air temperature. In refrigeration, this is a more important value to be aware of. It can vary significantly in these applications, intentionally and for several possible reasons that will be presented later in this module.
7. As the refrigerant mixture flows through the evaporator, it absorbs heat from the warmer air in the box and the liquid refrigerant progressively vaporizes. While it vaporizes, the refrigerant temperature remains unchanged until the liquid is gone and only vapor remains.

   To maximize the refrigeration effect, most of the evaporator must be exposed to boiling liquid refrigerant. Ideally, the final passes through the coil are where the phase change from liquid to vapor completes. Once the phase change is complete, the vapor begins to increase in temperature and the superheat becomes measurable.
8. As the refrigerant exits the evaporator and enters the suction line, vapor alone is present, and the superheat is measurable. In this example, the refrigerant vapor has reached a temperature of 33°F (1°C). Subtracting the 24°F saturation temperature results in a superheat value of 9°F (5°C). As the environmental conditions and box loading change, the temperature here often falls below freezing. Thus, frost on the suction line may come and go. This does not indicate the presence of liquid floodback since there is plenty of superheat.
9. The vapor returning to the compressor gained a very small amount of heat as it traveled through the suction line. In actual practice, this value can vary considerably, depending on the length of the suction line, condition of the pipe insulation, and the ambient temperature around it. In a small retail fixture, the suction line may be as short as a few inches.

1.2.0 Low-Temperature Refrigeration Cycle

The cycle doesn't need to change much to produce freezing temperatures. Of course, the evaporator coil must be significantly colder. This section reviews the cycle for a small freezer using HFC-407C refrigerant.

Figure 3 represents a typical low-temperature HFC-407C refrigerant circuit, such as that used in a reach-in freezer. For this example, assume that the condenser entering air temperature is 75°F (24°C), since the condenser is located indoors. The desired box temperature is now 10°F (–12°C). The current box temperature is 12°F (–11°C), prompting the system to run.

The numbers below correspond to the numbered points shown in *Figure 3*. Note that this example is theoretical and does not account for normal pressure drops through piping or components in an operating system.

1. After compression, highly superheated vapor exits the compressor. In this example, the discharge vapor temperature has reached 160°F (71°C) at a pressure of 279 psig.

NOTE

Remember that HFC-407C is a zeotropic blend with a significant glide. The associated PT chart includes columns for all-vapor and all-liquid pressures. However, this characteristic does not significantly alter the refrigeration process.

NOTE

Freezers that protect ice cream products and protect food for a longer time typically operate at 0°F (–18°C) or lower.

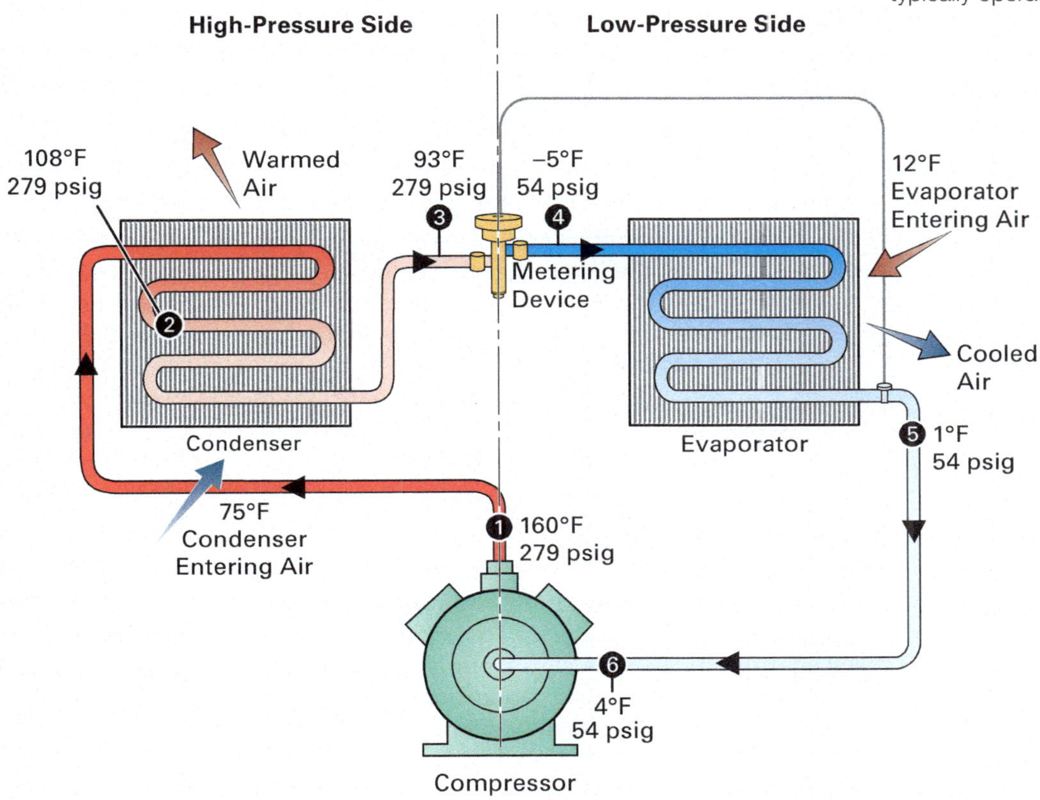

HFC-407C PT Chart

Temp.	Press. (psig)		Temp.	Press. (psig)		Temp.	Press. (psig)		Temp.	Press. (psig)	
°F	Liquid	Vapor	°F	Liquid	Vapor	°F	Liquid	Vapor	°F	Liquid	Vapor
–10	50.0	41.3	24	84.9	70.7	58	141.4	120.4	92	226.6	198.6
–8	51.5	42.6	26	87.5	73.0	60	145.5	124.2	94	232.7	204.3
–6	53.2	43.9	28	90.3	75.3	62	149.8	128.0	96	238.9	210.1
–4	54.8	45.3	30	93.1	77.8	64	154.1	131.9	98	245.2	216.1
–2	56.6	46.8	32	96.0	80.3	66	158.6	135.9	100	251.7	222.1
0	58.4	48.3	34	98.9	82.9	68	163.1	140.1	102	258.3	228.4
2	60.2	49.8	36	102.0	85.5	70	167.8	144.3	104	265.0	234.7
4	62.1	51.4	38	105.1	88.3	72	172.6	148.7	106	271.8	241.2
6	64.1	53.0	40	108.4	91.1	74	177.5	153.1	108	278.8	247.9
8	66.2	54.8	42	111.7	94.0	76	182.5	157.7	110	286.0	254.7
10	68.3	56.5	44	115.1	97.0	78	187.6	162.4	112	293.2	261.7
12	70.4	58.3	46	118.1	100.1	80	192.8	167.2	114	300.6	268.8
14	72.7	60.2	48	122.1	103.2	82	198.1	172.1	116	308.2	276.1
16	75.0	62.2	50	125.8	106.5	84	203.6	177.2	118	315.9	283.5
18	77.3	64.2	52	129.5	109.8	86	209.2	182.3	120	323.7	291.1
20	79.8	66.3	54	133.4	113.3	88	214.9	187.6	122	331.7	298.7
22	82.3	68.4	56	137.3	116.8	90	220.7	193.1	124	339.9	306.8

Figure 3 Example low-temperature refrigeration cycle for HFC-407C.

2. All refrigerant superheat must be transferred to the air and the vapor temperature must fall to the saturation temperature for HFC-407C at the given pressure before condensing can begin. Since this refrigerant is zeotropic, the more volatile component will begin condensing at a higher temperature than the other components. At 279 psig, the most volatile component of HFC-407C will begin condensing at 115°F while the least volatile will begin to condense as the refrigerant cools to 108°F (42°C). As refrigerant moves through the condenser, the ratio of liquid to vapor progressively increases until only liquid remains. Because the lowest condensing temperature in this example is 108°F, any temperature measured below this value at the condenser outlet represents subcooling.
3. In this example, the refrigerant temperature has fallen to 93°F (34°C) by the time it reaches the metering device. Subtracting 93°F from 108°F indicates the refrigerant is subcooled 15°F (8.3°C).
4. To achieve the desired box temperature of 10°F (–12.2°C), the evaporator coil must be at a substantially lower temperature. In this example, the evaporator temperature is –5°F (–21°C) and operating at a pressure of 54 psig. Thus, the subcooled refrigerant experienced a pressure drop of about 225 psig as it passed through the metering device. The resulting ΔT between the refrigerant in the evaporator and the desired box temperature is 15°F (8°C).
5. The mixture of liquid and vapor in the evaporator will not rise above –5°F (–21°C) until the last refrigerant component has vaporized. The refrigerant exiting the evaporator coil should be lightly superheated vapor. Because the temperature of the refrigerant leaving the evaporator is 1°F (–17°C), it has been superheated 6°F (3.3°C).
6. The vapor returning to the compressor gained some additional heat as it traveled through the suction line. As mentioned previously, this value can vary considerably and may not be a factor at all.

An important point must be made about superheat values for refrigeration applications. Although typical superheat values for comfort cooling are usually 10°F to 20°F (6°C to 11°C), refrigeration systems generally operate at lower superheat values. As the saturation pressure and temperature are reduced to create the needed refrigerating effect, compression ratios become higher and compressor energy efficiency is significantly reduced. Energy efficiency ratios (EERs) for medium-temperature applications are typically around 7.0 to 8.0, while low-temperature systems may operate with EERs as low as 4.0.

As a result, extracting the maximum amount of work per watt from the system is an essential goal for low-temperature applications. One way to achieve this is by reducing the superheat, thereby exposing the maximum amount of evaporator surface area to boiling liquid refrigerant.

Superheat values of 6°F to 12°F (3.3°C to 6.7°C), measured at the evaporator outlet, are generally acceptable for high- and medium-temperature applications. Superheat values for low-temperature applications typically range from 3°F to 8°F (1.7°C to 4.4°C) to maximize the refrigerating effect in the evaporator.

Low superheat values mean that greater care must be taken to protect the compressor from liquid floodback. However, the gain in energy efficiency makes the effort worthwhile.

1.3.0 Defrost Systems

As long as moisture is present in a refrigerated space, it will condense on the surface of the evaporator as it does in comfort cooling systems. However, when the coil surface is below freezing, the condensation becomes frost. This reduces heat transfer and eventually blocks airflow through the coil. The coil must be defrosted periodically for the system to perform properly.

It is important to understand several aspects of frost accumulation on evaporator coils. First, fixtures that were selected and designed to operate at large air-to-refrigerant ΔTs will generally build frost faster, since the coil temperature will likely be well below the dew point for the box.

Designing a system for a high air-to-refrigerant ΔT is done when the product is hermetically sealed and can't dry out, such as bottled drinks and sealed food packages. This design results in more condensation on the coil, but it is also more energy efficient. Fixtures with small air-to-refrigerant ΔTs will not accumulate as much condensate or frost. Foods that can be damaged by dehydration must be refrigerated with fixtures having small air-to-refrigerant ΔTs.

Another factor in frost accumulation is related to fixture usage and outside air infiltration. Many retail refrigeration fixtures experience heavy use. In hot, humid weather, a fixture that is opened many times each day is consistently exposed to warm, humid air. Not only does this significantly increase the volume of water vapor in the enclosure, but the infiltration of hot, moist air greatly increases the refrigeration load.

Such environmental conditions can temporarily overwhelm the refrigeration equipment and interfere with fixture performance. Consider the potential difference in performance of an identical reach-in refrigerated fixture installed in two different locations—one with very little consumer traffic in dry Arizona and the other in a busy, 24-hour convenience store in coastal Florida.

Also consider that any heat added to the refrigerated area during the defrost cycle, regardless of the heat source, must be removed by the system once the defrost cycle is complete. This has a substantial effect on the refrigeration load. In addition, the time a system spends in defrost mode reduces the time available for refrigeration. For example, if the total refrigeration load for a fixture is calculated as 192,000 Btus over a 24-hour period (8,000 Btuh), but the unit is expected to be in the defrost mode 2 hours each day, then the refrigeration equipment must be able to generate the same capacity in 22 hours instead of 24 hours. That would require 8,727 Btuh of refrigerating capacity instead of 8,000 Btuh.

There are four primary defrost methods used for retail refrigeration equipment, each of which is discussed in the sections that follow:

- Off-cycle defrost
- Timed defrost
- Electric defrost
- Hot-gas defrost

On Site

Take Care of That Fixture Too

Worn door gaskets and hinges can significantly affect the performance and energy efficiency of a refrigerated fixture. The constant infiltration of warm and moist outside air through poorly fitted doors and gaskets can prevent even the best fixture from reaching its setpoint. It also increases frost accumulation, reducing coil heat transfer and increasing refrigeration equipment runtime and operating costs.

Maintaining and replacing door gaskets and hinges is as important to a fixture's efficient operation as having the proper refrigerant charge.

1.3.1 Off-Cycle Defrost

Off-cycle defrost is a passive defrost method. In most fixtures that maintain temperatures at or above 36°F (2°C), the coil is simply allowed to defrost during the normal off cycle. Fixtures in this category usually operate at an evaporator coil/saturated suction temperature of 16°F to 31°F (–9 to –1°C), depending on the box design and desired humidity levels.

Because the fixture temperature is maintained above freezing, the coil might build some frost, but it should defrost naturally when the unit cycles off. The evaporator fan continues to run during the off cycle, helping the coil defrost faster.

Technicians and users alike can best support this natural defrost approach by ensuring the fixture temperature is not set lower than necessary. Maintaining the box just a few degrees colder than necessary can slow or defeat the natural defrost process, resulting in frozen coils and rising box temperatures. Remind users that thermostats are not accelerator pedals! Setting a refrigerator thermostat lower will not force the system to cool faster, with the notable exception of inverter-driven compressors that are not currently common in retail refrigeration applications.

1.3.2 Timed Defrost

Fixtures operating at lower temperatures than those relying on off-cycle defrost require a more active approach to defrost. Fixtures maintaining temperatures of 32°F to 36°F (0°C to 2°C) can usually rely on a timed defrost. A 24-hour time clock (*Figure 4*) can be placed in series with the compressor controls and set to shut the compressor down periodically, regardless of box temperature. The evaporator fan continues to run during the timed off-cycle to help melt the frost.

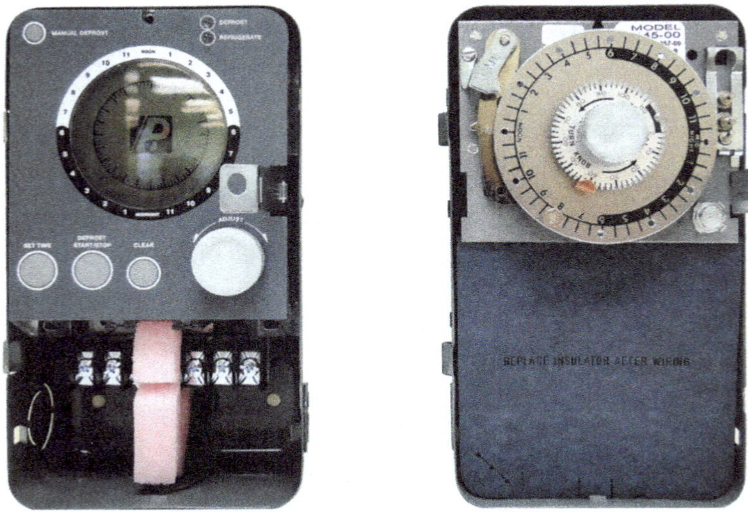

(A) Electronic Defrost Timer **(B) Electromechanical Defrost Timer**

Figure 4 24-hour defrost timers.

Electromechanical time clocks have a dial assembly (*Figure 5*) allowing the timer to be configured for the time, duration, and number of defrost periods during a 24-hour period. Trip pins are screwed into the outer dial at the time when each defrost cycle needs to begin. The inner defrost dial sets the duration of the defrost period; the defrost cycle terminates after the period has elapsed. The timer contacts then return to their normal position and the refrigeration cycle resumes. The contacts can generally handle the load of electric defrost heater(s) directly, when they are used; no additional relay or contactor is needed for most retail fixtures.

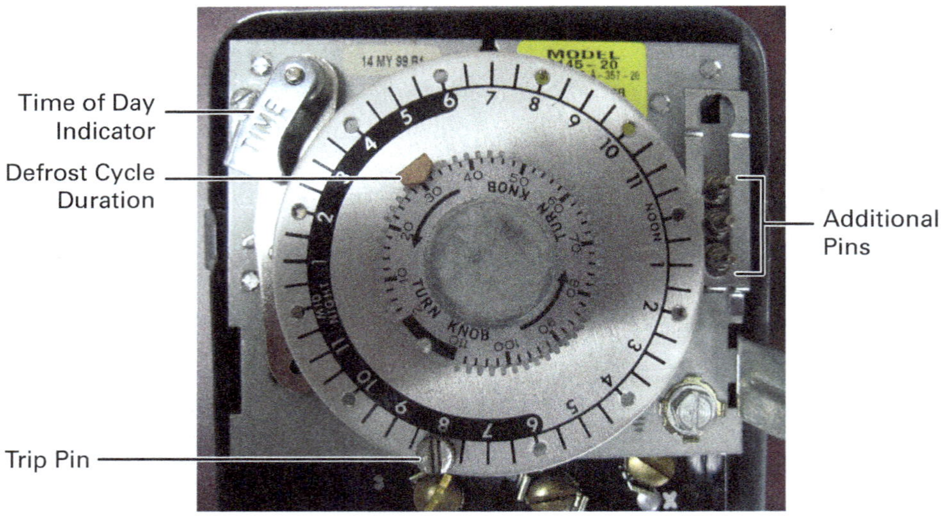

Figure 5 Electromechanical defrost timer dials.

The fixture temperature may rise a few degrees above normal during the defrost cycle, but it should recover quickly. Like all defrost scenarios, setting defrost periods that coincide with the fixture's lightest period of usage (late at night, for example) helps to prevent the fixture temperature from rising beyond an acceptable level. One or two defrost cycles per day are usually required.

1.3.3 Electric Defrost

Electric defrost systems are common for medium- and low-temperature fixtures that operate at temperatures too cold for off-cycle or timed defrosts to be effective. This includes freezers in the retail refrigeration class. Electric defrost components are inexpensive and reliable.

Figure 6 shows a typical arrangement for electric defrost heaters installed on both the face and back of an evaporator coil. For proper operation, the heater must be in good thermal contact with the coil fins to effectively distribute heat and minimize the time it takes to melt the frost.

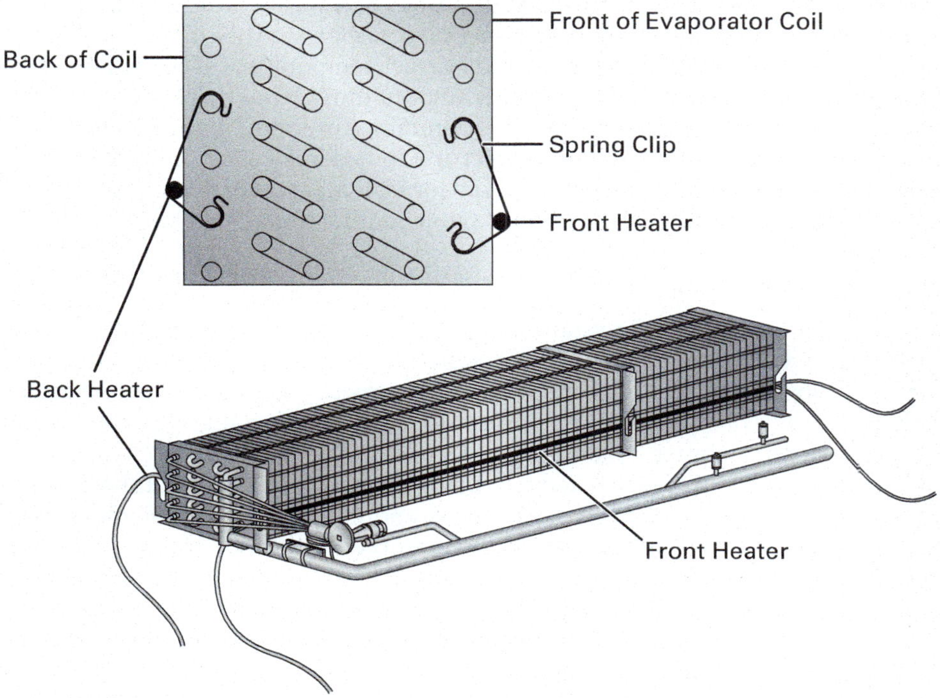

Figure 6 Electric heating elements on an evaporator coil.
Source: Hussmann Corporation

A defrost timer stops the compressor as well as the evaporator fans during the defrost cycle, then energizes the electric heaters. In most cases, the shape and dimensions of the defrost heater(s) are unique to a given fixture. Heating element replacement generally requires original parts.

Some coils have the heater placed deep inside the coil assembly rather than on the face. This results in a fast and effective defrost cycle, but the elements are more challenging to access and replace.

Some units are equipped with infrared electric heaters (*Figure 7*). The resistance heater wire is enclosed inside a glass tube, with the wires passing through the ends of the sealed tube cap. Using infrared heat, defrosting can be accomplished without the heater touching the coil. All surfaces with a line of sight to the heater are warmed. These heaters reach a high temperature very quickly and are corrosion resistant. However, the glass enclosure is somewhat fragile.

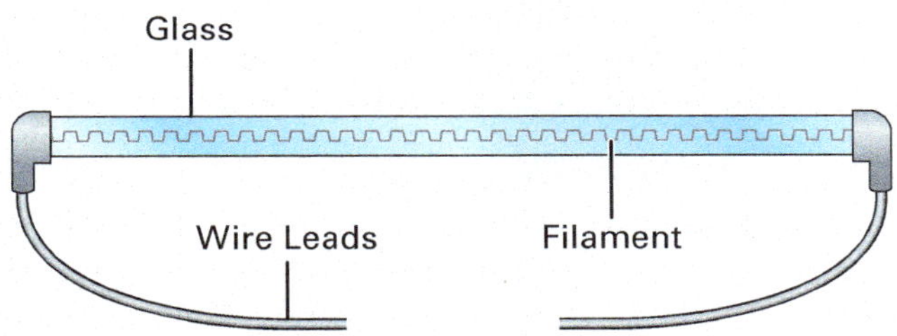

Figure 7 Infrared defrost heater.

Figure 8 Heated condensate pan for a freezer.
Source: Courtesy of Sealed United Parts Co., Inc.

When a defrost cycle is required for a freezer, condensate draining from the coil must also be kept above freezing until it exits the fixture. Many condensate collection pans, either as a component of the evaporator coil assembly or as a separate collection point inside the fixture, may also be equipped with electric heaters (*Figure 8*). The heater simply re-evaporates the water to the surrounding air.

Installations with the condensate drain line routed through a freezing area require the condensate drain line to be kept above freezing as well. For this, pipe heating cable, also referred to as *heat tape*, is attached to the exterior of the line.

The defrost cycle can be terminated by time or by temperature. By monitoring the temperature of the coil or the suction line close to the coil, the thermostat indicates that the coil has reached a target temperature. This allows the defrost cycle to terminate before the full amount of time has elapsed.

The temperature setpoint for termination may differ among products since the temperature is based on the manufacturer's chosen location and laboratory testing. Using the correct replacement sensor or temperature switch is important because line-mounted thermostats are generally non-adjustable.

1.3.4 Hot-Gas Defrost

Although not as common in retail refrigeration as the other defrost methods, hot-gas defrost is considered to be the fastest and most efficient method. Because of the additional refrigerant circuit components required, this method is more costly to install. The refrigerant circuit is also more complex. It is typically used only on large refrigeration systems.

Hot-gas defrost relies on a solenoid valve(s) to control the flow of discharge vapor directly from the compressor to the evaporator, bypassing the metering device (*Figure 9*). When the defrost cycle is initiated, the liquid-line solenoid valve closes, and the hot-gas discharge solenoid valve opens. The hot refrigerant gas condenses in the cold evaporator as the frost melts. The latent heat released as the refrigerant condenses raises the coil surface temperature above freezing.

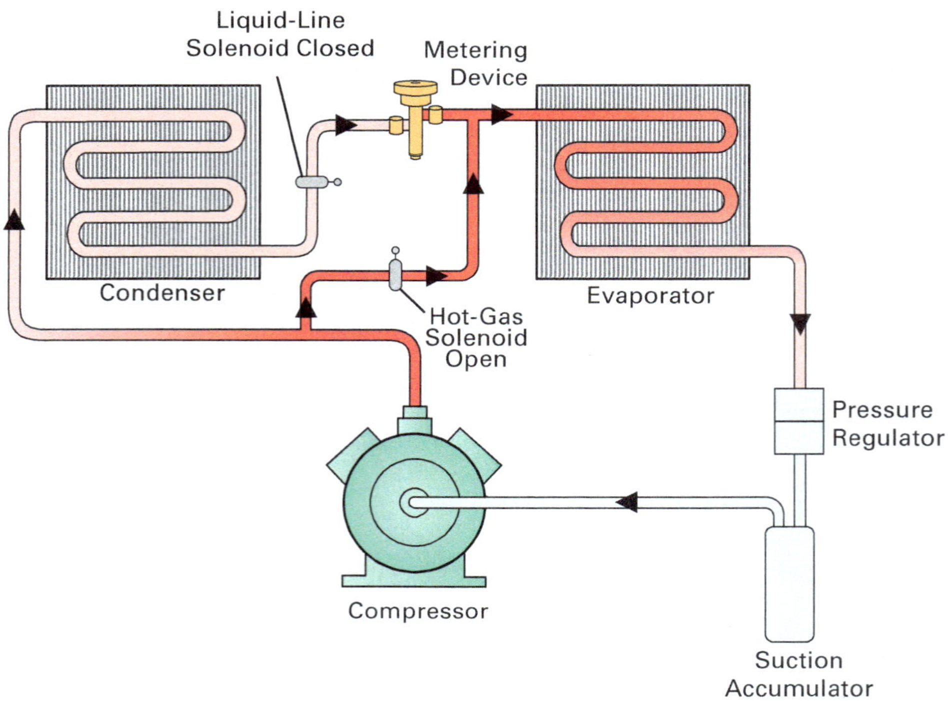

Figure 9 Basic hot-gas defrost schematic.

The suction pressure regulator shown in *Figure 9* is set to maintain an evaporator pressure (upstream of the valve) that is well above the saturated suction temperature for the refrigerant in use; it is operated using a solenoid coil, as it must open fully during the refrigeration cycle. A typical valve is shown in *Figure 10*. This valve also prevents the compressor from overloading while the defrost cycle is active.

Care must also be taken in these systems to avoid compressor overload after the defrost cycle is complete. A *crankcase pressure regulator* can be used to ensure the compressor is not overloaded following a defrost cycle or when the fixture has been off for an extended period.

A suction accumulator can protect the compressor from refrigerant floodback. *Figure 9* offers a very simple example of a hot-gas defrost system. Manufacturers differ in their approach, and you will likely encounter a number of variations. For example, some systems use vapor from the top of the receiver for defrost, rather than discharge vapor.

This approach to defrost is rarely found in small retail fixtures. Hot-gas defrost systems are more likely to be found in expansive grocery store systems and larger refrigeration applications. It is essential to understand the details of operation for any hot-gas defrost system you encounter to properly service and maintain it. This information should be acquired directly from the equipment and component manufacturers.

Figure 10 Suction pressure regulator.
Source: Courtesy of Sporlan Division - Parker Hannifin Corporation

NOTE

Refer to *Section 3.2.3* for a discussion of crankcase pressure regulators.

1.0.0 Section Review

1. In a medium-temperature refrigeration unit, the refrigerant phase change from liquid to vapor takes place in the ______.
   a. suction line
   b. evaporator
   c. liquid line
   d. condenser

2. What is the typical superheat range for low-temperature refrigeration applications?
   a. 3°F to 8°F (1.7°C to 4.4°C)
   b. 6°F to 12°F (3.3°C to 6.7°C)
   c. 10°F to 16°F (5.6°C to 8.9°C)
   d. 14°F to 20°F (7.8°C to 11.1°C)

3. Simple off-cycle defrost can usually be applied to refrigerated fixtures that maintain a minimum fixture temperature of ______.
   a. 28°F (–2°C)
   b. 32°F (0°C)
   c. 36°F (2°C)
   d. 40°F (4°C)

2.0.0 Retail Refrigeration Equipment

Performance Tasks

There are no Performance Tasks in this section.

Objective

Identify various types and functions of retail refrigeration equipment.

a. Identify and describe reach-in coolers and freezers.
b. Identify and describe various walk-ins and merchandisers.
c. Identify various types of commercial ice machines.

Many retail refrigeration fixtures are relatively mobile and can be installed in a variety of locations. For our purposes, most retail systems are considered packaged or self-contained in design. Although the refrigerated cases in supermarkets may be considered retail fixtures, supermarket systems represent some of the most complex systems in the industry and few are self-contained. As a result, supermarket systems are not reviewed in detail here. More training and experience will be necessary to master them. Advanced training for supermarket systems is typically offered by manufacturers.

One important consideration for packaged retail fixtures is the environment. Many manufacturers specify that indoor units are designed for indoor spaces maintained at 75°F (24°C) and 55% relative humidity. Although many units are placed in areas that conform to these requirements, many others are not. Technicians and end-users should not expect manufacturer-specified capacity and performance from units placed in areas that are excessively hot and/or humid.

This section, reviews the following types of equipment and their unique features:

- Reach-in coolers and freezers
- Merchandising walk-in systems
- Commercial ice machines

2.1.0 Reach-In Coolers and Freezers

Figure 11 Two-door upright reach-in merchandiser.
Source: Master-Bilt

Reach-in coolers and freezers are available in styles that suit almost any merchandising need. The refrigerated enclosure is typically insulated with urethane or polystyrene materials, either in board-form or sprayed.

Figure 11 shows a typical two-door reach-in unit. Some common features and options for these units include the following:

- Interior lighting
- Self-closing doors that can be locked open for stocking
- A variety of shelving options
- Shelf moldings to display price tags
- Digital thermometers
- A unitary condensing unit on top of or beneath the cabinet that can be removed for service or replacement
- Adjustable leveling legs

Figure 12 shows a typical undercounter unit used to store food and beverages. Because units of this type are not directly accessible by consumers, appearance is not as important as a durable and maintainable design.

Typical options and features for undercounter fixtures include the following:

- Stainless steel construction
- A unitary condensing unit that can be removed for service or replacement

Figure 12 Undercounter reach-in refrigerated cabinets.
Source: Suchada Tansirimas/iStock

- Casters for easy movement, or adjustable legs
- Epoxy-coated evaporator coils to protect the aluminum fins from corrosion when exposed to certain liquids and vapors, such as pickle juice
- Exterior thermometers

Condensing units for packaged fixtures (*Figure 13*) are mounted either above or below the refrigerated case, depending on manufacturer preference and fixture style. Medium-temperature units may incorporate either off-cycle or timed defrost, with electric defrost available as a factory-installed option. Low-temperature freezer units are almost exclusively equipped with electric defrost. The metering device is usually a thermostatic expansion valve (TXV).

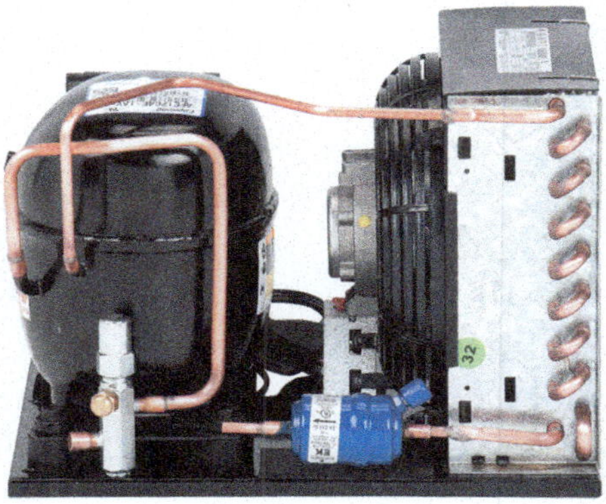

Figure 13 Packaged-fixture condensing unit.
Source: Courtesy of White-Rodgers by Copeland

2.1.1 Ice Merchandisers

Ice merchandisers for bagged retail ice have a simple design and function. Models are designed for either indoor or outdoor use in sizes ranging from roughly 25 ft^3 to over 100 ft^3 of internal volume. Note that these units are merely for ice storage and do not have the refrigeration capacity or means to create ice. Ideally, the ice is well subcooled before it is placed inside. This prevents the ice nuggets from easily fusing together while in storage.

Outdoor units (*Figure 14*) have metal doors that can be locked and are built to withstand the elements and some physical abuse. Both styles are controlled by a simple thermostat at box temperatures of 20°F to 25°F (–7°C to –4°C). Indoor models (*Figure 15*) usually have a glass door with internal lighting.

Ice merchandisers are generally offered with two major options: cold-wall construction (with no defrost cycle) and unitary evaporator assemblies with electric defrost. In cold-wall models, the cabinet walls are *plate evaporators*, with the serpentine evaporator tubing bonded to the inner cabinet wall. No evaporator fan is used and defrost is accomplished by emptying the contents and turning the unit off so it can warm up. An example of this type is shown in *Figure 14*.

Ice merchandisers with unitary evaporator coil assemblies have a defrost timer and electric heating elements on the evaporator coil. *Figure 15* is an example of this type; the evaporator assembly is visible at the top of the box and the condensing unit is under the cabinet.

Figure 14 Outdoor ice merchandiser.
Source: libre de droit/iStock/Getty Images

Figure 15 Indoor ice merchandiser.
Source: Master-Bilt

2.2.0 Merchandising Walk-In Systems

Many of today's convenience stores feature merchandising walk-in coolers and freezers (*Figure 16*). These units are simply walk-in coolers or freezers with one or more walls replaced by glass doors, allowing the consumer direct access to products. The consumer, however, does not walk inside. Stock is placed by walking into the cooler behind the display shelves.

Shelves placed directly behind the glass doors allow for rapid stocking from inside the cooler or freezer. Products are received and stored inside the same enclosure, never leaving the refrigerated area except in the consumer's hands. Well-stocked shelving at the door opening helps to block the consumer's view of stored product.

The merchandising shelf system may have product replacement feeding, allowing new product to fall or slide into place as one is removed. The doors of

Figure 16 Walk-in merchandiser.

the units are often equipped with a means to quickly defog the glass after the door has been opened and closed. Freezers have low-wattage electric heaters around the doorframe to prevent the door from freezing closed.

Supermarket open-face coolers (*Figure 17*) are medium-temperature coolers generally used for displaying perishable items such as vegetables, meats, and dairy items. These coolers provide the customer with more convenient access to examine product choices. Being open to the air, the warmth of the store environment places a significant demand on the refrigeration system. Reducing the load on these open cases is the primary reason that supermarkets are usually kept very cool.

Figure 17 A supermarket meat display case.
Source: Anna Krivitskaia/Dreamstime

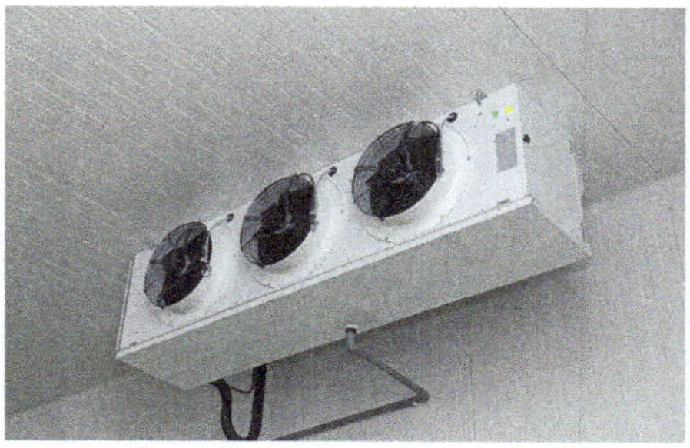

Figure 18 Unit cooler for a walk-in.
Source: Baloncici/iStock/Getty Images

Walk-in and open-faced coolers generally use a **unit cooler** (*Figure 18*) with a remote condensing unit. Remote condensing units are usually located outdoors. Because the refrigeration capacity and durability needed is greater than for small self-contained fixtures, semi-hermetic compressors are commonly used. TXVs are most often used as a metering device.

Unit cooler: A packaged unit containing the evaporator coil, metering device, and evaporator fan(s) that serves as the fan coil unit for a refrigeration system.

On Site

Refrigeration Rack Systems

Although many refrigerated merchandising systems are self-contained and operate independently, supermarkets generally use centralized refrigeration equipment to serve all the open and closed display cases as well as walk-ins used for mass storage.

Rack systems have multiple compressors connected to one or two common refrigerant circuits, staged as necessary to accommodate the changing loads. For efficiency, the medium- and high-temperature load circuit is separated from the low-temperature load circuit. These systems allow refrigeration to continue in some fixtures while other areas are defrosting using hot gas. Each refrigerated fixture that requires a higher temperature than the coldest fixture on its circuit is fitted with an evaporator pressure regulator. Advanced controllers stage compressors and continuously monitor system performance. Several racks are needed in large supermarkets.

Source: Courtesy of Zero Zone, Inc.

2.3.0 Commercial Ice Machines

Packaged ice machines are generally categorized by the form of ice they provide. Three common types used at the retail level are cubed-ice machines (cubers), tube-ice machines, and flaked-ice machines (flakers). All three are widely used in restaurants, convenience stores, hotels, and other point-of-use locations. Typically, ice is formed on a surface at roughly 10°F (–12°C). This is the target surface temperature for many ice machines.

The three types have similarities. All are designed to operate automatically, control the water supply, freeze water into the desired shape and size, collect the ice in a storage bin, and shut down when the bin is full. In most cases, the bin is simply an insulated enclosure with a tough plastic or vinyl liner, refrigerated only by the ice it contains. A drain removes any ice water. Ice machines are often combined with a mechanical ice dispensing system.

Most retail-class ice machines feature air-cooled condensers, but water-cooled models are also available. Self-contained, air-cooled units can usually be placed in ambient temperatures between 40°F and 115°F (4°C to 46°C). However, the ambient temperature can significantly affect the speed at which ice is made. Ice machine performance across the operating temperature range is not consistent. The temperature of the water supply also affects production speed because warmer water requires more sensible cooling before ice formation begins.

Care must be taken when locating air-cooled models to avoid the recirculation of hot condenser air. Larger ice makers produce a great deal of heat, so the condensing units are remote, outside of the conditioned space.

Most ice makers are not designed to operate in environments that approach or drop below freezing. If an ice machine is shut down during the winter but left in an area prone to freezing, the water supply must be turned off and the water drained from all internal components.

Did You Know?

Load Transference

Refrigeration equipment works hard to cool products in the refrigerated space. In the process, the equipment also creates additional heat as it consumes electrical energy. The heat of compression alone can add as much as 15% to the amount of heat that must be rejected through the refrigerant circuit.

But where does that heat go? All air-cooled condensing units located in the conditioned space reject the heat to the air around it. Consider that a small reach-in unit may have a capacity of 8,000 Btuh. Assuming it runs about 16 hours per day, the average amount of heat rejected is 5,333 Btuh. Add 15% for the heat of compression, and you arrive at 6,133 Btuh.

Now consider that there are seven units in an establishment working at the same pace. That brings the total to over 42,000 Btuh—about 3½ tons. The real problem is that the heat must be transferred out of the building by the air conditioning systems serving the establishment. A small store that might otherwise have a cooling load of 5 tons actually needs 8½ tons of capacity to maintain the conditions inside, and the owner pays twice for the energy required to refrigerate the products.

A little bit of math clearly demonstrates the benefit of locating the condensing equipment outdoors whenever possible.

Figure 19 Cubed ice machine with bin.
Source: Manitowoc Ice

2.3.1 Cubers

Commercial cubers (*Figure 19*) generally range in capacity from 65 to 2,000 lb of ice per day. Storage bins mated with cubers commonly store between 120 and 1,500 lb of ice. The selected combination of cuber and bin is based on the user's requirements. Applications where large quantities of ice are needed periodically versus those that experience steady consumption throughout the day lead a buyer to the right cuber and bin combination.

Cube sizes and shapes vary significantly. Although the machines themselves operate the same way, the refrigerated grid where ice is formed is changed to produce the desired shape and size. Although sizes vary by manufacturer, cube sizes generally fall into one of these three classes:

- Half-dice cube — $^3/_8$" × $^7/_8$" × $^7/_8$" deep
- Dice cube — $^7/_8$" × $^7/_8$" × $^7/_8$" deep
- Full cube — $1^1/_8$" × $1^1/_8$" × $^7/_8$" deep

Other sizes and shapes are also available, such as nuggets, tubular nuggets, crescents, and even octagons that are sometimes referred to by manufacturers as *gourmet ice*. Cubed ice should be very hard when harvested, which results from it being sufficiently subcooled during the freeze cycle. Ice that is harvested too early will be small and melt faster.

Cubers have four basic modes of operation:

- Fill
- Freeze
- Harvest
- Drain

In the fill mode, fresh water is supplied for the next batch of cubes. In the freeze mode, the water spilling across the evaporator surface freezes. When the ice has reached a predetermined size, harvesting begins by sending hot refrigerant vapor through the evaporator to quickly loosen the ice and allow it to fall into the bin. The drain, or *purge*, cycle rids the machine of any remaining water and the minerals it contains.

Some units harvest the ice in a single solid sheet, which falls from the evaporator onto a wire grid. The grid wires are heated, and the wires melt through the ice in knife-like fashion, separating the ice sheet into cubes of the desired size. *Figure 20* provides an example of this design. After the cubes are cut by the hot wire grid, they fall into the storage bin.

On Site

High-Quality Ice

Water quality is important to ice quality. Aerated water and water with a high mineral content results in ice that appears cloudy and unappealing to the consumer. The entering water supply should be filtered, free of excess air, and potentially conditioned to help avoid these problems.

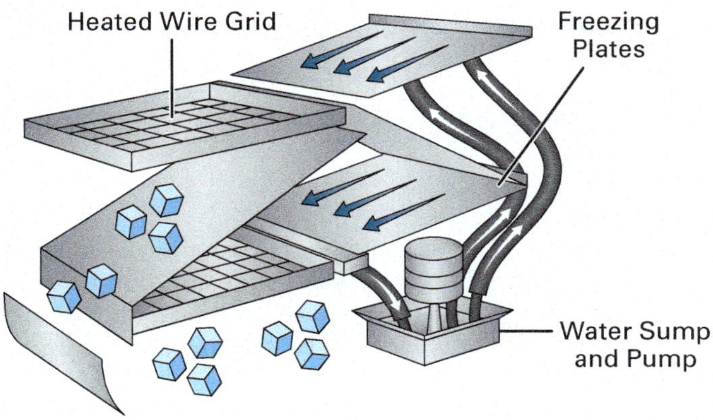

Figure 20 Cubed ice maker with an electrical cutting grid.

The other popular design either sprays water into cube-shaped chambers permanently attached to a horizontal plate evaporator (*Figure 21*), or water is pumped to the top of a vertically positioned evaporator of the same design and allowed to cascade down by gravity.

When the harvest mode begins, hot refrigerant vapor (or, rarely, electric heat) is directed to the plate to loosen the cubes, which then fall into the bin. With either style, ice cube size and shape are options when ordering the machine.

An ice thickness sensor is typically used on a cuber to initiate the harvest. It is set at a precise distance from the evaporator surface. As ice builds on the plate, the cascading water will eventually contact the sensor. When the sensor has been wetted continuously for a preset time (such as 10 seconds), the water pump shuts down, and the harvest mode begins.

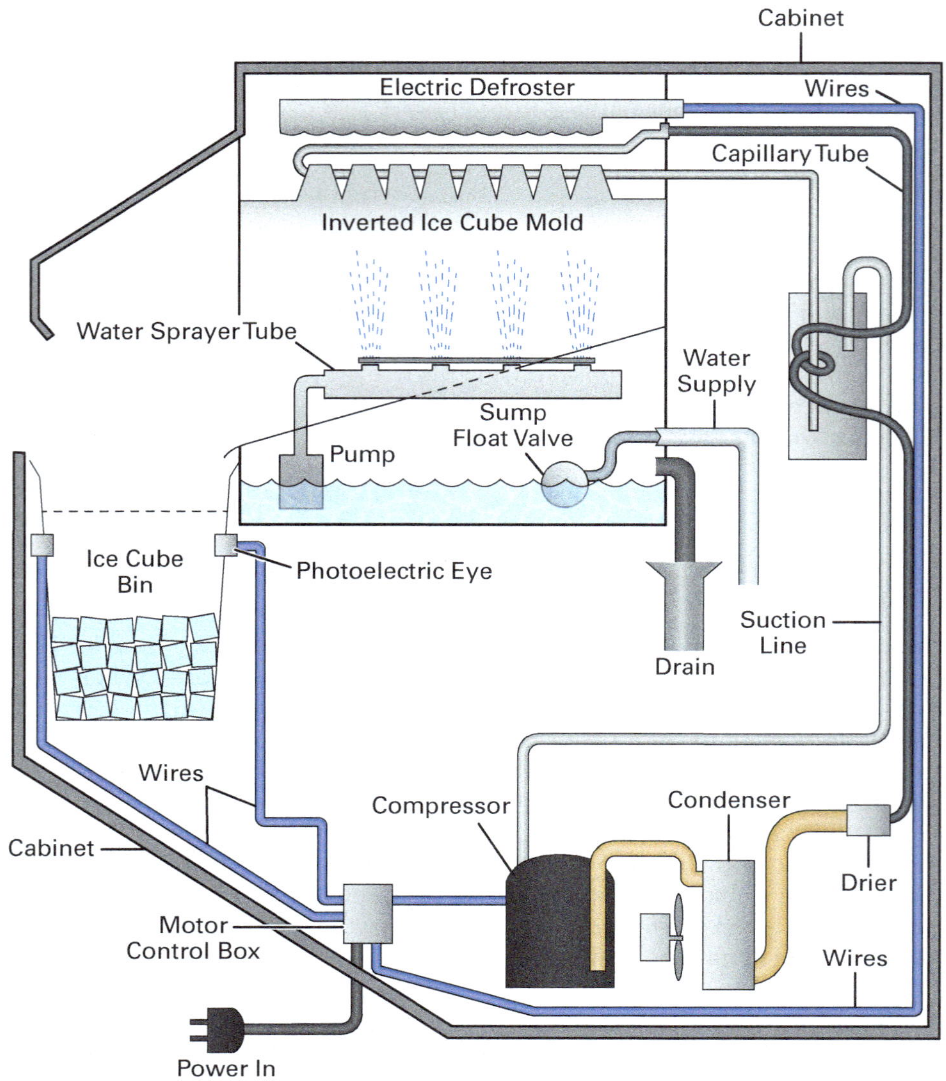

Figure 21 Cubed-ice maker with cube-shaped chambers on the evaporator plate.

Cubed-ice makers have a complex sequence of operation to ensure that a consistent ice cube is produced under a variety of operating conditions. Ice production is halted by either a temperature sensor located near the top of the bin or by using a photoelectric eye that is blocked by the presence of ice at a specific depth in the bin.

Figure 22 Tube ice.
Source: Shahizal Rizwan Ahmat Raslan/ Dreamstime

2.3.2 Tube Ice Machines

Tube ice is a popular form of ice all over the world. This form of ice (*Figure 22*) is delivered in short sections of clear, cylindrical ice. The process for creating tube ice is similar to that for cubes, except the ice freezes inside long cylinders instead of cube-shaped pockets. Water cascades down the inside walls of the cylinders.

When the ice reaches the desired thickness, the cylinders are warmed briefly by a burst of hot refrigerant vapor to harvest the ice. The ice in contact with the cylinder wall surface melts slightly, allowing it to fall into an ice cutter. The slight expansion of the tubes due to the hot vapor also helps to break the ice loose.

For commercial tube-ice machine applications, a series of rotary blades mounted on a common shaft cut the rods of ice into uniform lengths. The spacing of the blades determines the length of each section. One popular model can be user-adjusted to make ice in three sizes between 1" and 1½" long.

2.3.3 Flaked Ice Machines

Ice flakers (*Figure 23*) make ice continually, rather than alternate between freezing and harvesting. When the unit starts and the evaporator assembly reaches the proper temperature for freezing, flakes of ice continually fall into the bin.

The freezing assembly for flakers consists of an insulated vertical cylinder (*Figure 24*). The outer walls of the cylinder form the evaporator surface, which is filled with boiling, low-temperature refrigerant. Ice forms on the cylinder's inner walls, which are continuously bathed with fresh water. An auger, driven slowly by a motor through a gearbox that provides additional torque, scrapes the ice from the walls of the cylinder as it rotates. The ice flakes are pushed out the top of the cylinder by the auger and fall into the bin.

Figure 23 Flaked ice machine with bin.
Source: Manitowoc Ice

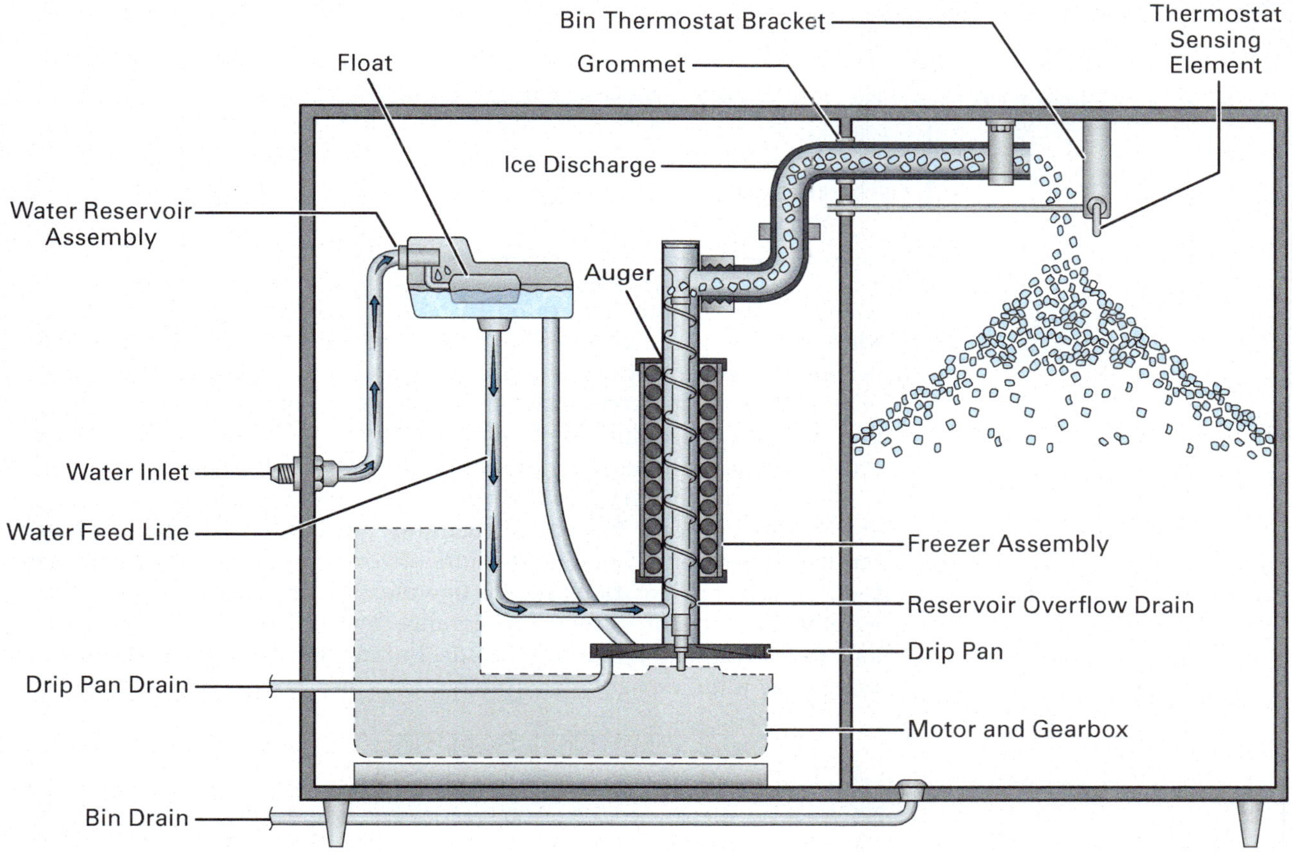

Figure 24 Flaked ice machine operation.

2.0.0 Section Review

1. What is the common box temperature range for ice merchandisers?
   a. –10°F to 5°F (–23°C to –15°C)
   b. 5°F to 15°F (–15°C to –9°C)
   c. 10°F to 20°F (–12°C to –7°C)
   d. 20°F to 25°F (–7°C to –4°C)

2. What type of metering device is typically used with walk-in coolers?
   a. Bypass piston
   b. Capillary tube
   c. Fixed-orifice piston
   d. Thermostatic expansion valve (TXV)

3. True or False? Ice is most often harvested in a cuber by applying electric heat to the evaporator to release the ice.
   a. True
   b. False

3.0.0 Retail Refrigeration System Components

Performance Tasks

There are no Performance Tasks in this section.

Objective

Identify and describe common retail refrigeration system components.

a. Identify and describe primary retail refrigeration system components.
b. Identify and describe secondary retail refrigeration system components.
c. Identify and describe common retail refrigeration system controls.

Many of the components found in retail refrigeration systems are similar or identical to components found in comfort cooling systems. Because retail refrigeration equipment operates at lower temperatures than comfort systems, the components must accommodate that difference. For example, compressors must be rated for refrigeration duty, which means they must be able to operate reliably at higher compression ratios and often around the clock.

Refrigerants that are ideal for comfort cooling are not often suited for refrigeration applications. System refrigerants are chosen by manufacturers, and today's refrigerant choices are more diverse than ever. Although many of today's retail refrigeration systems operate with HFC-134a or HFC-407C, as well as other hydrofluorocarbon refrigerants, hydrocarbon (HC) refrigerants such as HC-290 (propane) and HC-600a (isobutane) are now acceptable and used in many new retail refrigeration systems.

WARNING!

No refrigerant circuit operating on nonhydrocarbon refrigerants can be converted for use with a hydrocarbon refrigerant. Never attempt such a conversion.

3.1.0 Primary Refrigeration Circuit Components

Most retail refrigeration applications use air-cooled condensing units (*Figure 25*). Refrigeration condensing units are assembled on a single chassis containing the compressor and its start components, condenser coil and fan(s), refrigerant receiver and/or accumulator (if equipped with either), and various controls and valves.

Condensing units can be mounted on self-contained fixtures, or installed remotely and piped to the fixture. Used indoors, self-contained fixtures transfer

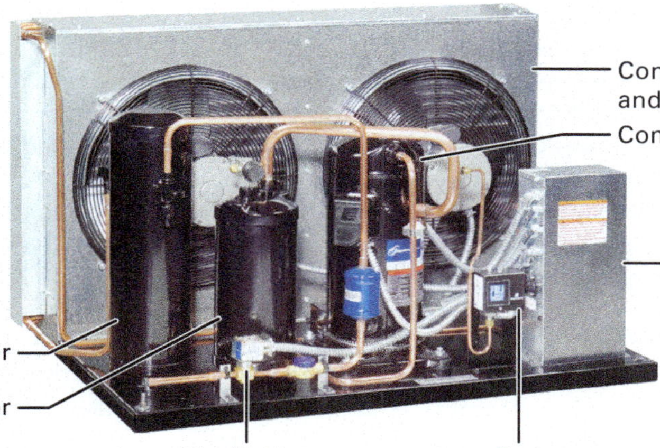

Figure 25 Refrigeration condensing unit.
Source: Courtesy of Emerson

heat from the box, as well as heat of compression, to the surrounding space. The added heat can be a significant comfort problem if not taken into consideration when loads for the comfort cooling system are calculated. Although more costly to install and certainly less mobile, remote condensing units have the advantage of rejecting heat outside of the conditioned space.

On the other hand, the heat from indoor units can benefit indoor comfort in the winter. Remote installations are generally reserved for larger applications such as walk-in coolers or freezers, as well as larger ice machines. Larger condensing units can serve multiple evaporators or fixtures.

3.1.1 Compressors

Hermetic and semi-hermetic reciprocating compressors (*Figure 26*) have long dominated the retail refrigeration market. Early attempts to use scroll compressors in refrigeration applications resulted in overheated and failed compressors because of the high compression ratios and the low density of the returning refrigerant vapor. However, scroll compressors have undergone design changes that improve their suitability for low-temperature applications. Several manufacturers now offer scroll compressors specifically designed for the refrigeration market.

For low-temperature use, many scroll compressors are fitted with a liquid- or vapor-injection system designed to improve internal cooling, reduce the refrigerant discharge temperature, and enhance reliability. The compressor shown in *Figure 27* can also modulate its capacity between 10% and 100%. This reduces energy consumption and provides very stable temperature control.

Figure 26 Copeland™ Discus semi-hermetic compressor.
Source: Courtesy of White-Rodgers by Copeland

Figure 27 Copeland™ scroll digital compressor.
Source: Courtesy of Emerson

Systems with hermetic reciprocating compressors can be fitted with capacity control arrangements. Semi-hermetic compressors with multiple cylinders can also provide capacity control. However, these approaches can't match the performance or energy efficiency of modulating scroll compressors.

Many scroll compressors used in refrigeration are not equipped with internal pressure relief valves. Therefore, providing reliable high-pressure safety controls on the condensing unit is necessary. Systems may also use a discharge-line thermostat to help prevent compressor damage or failure from overheating. The thermostats are generally set to open at 250°F to 260°F (121°C to 127°C).

3.1.2 Condensers

Most retail refrigeration systems use air-cooled condensers due to the initial cost, ease of installation, mobility, and maintenance simplicity. The condenser coil and fans can be mounted on the condensing unit, within inches of the compressor, or installed some distance from the compressor if necessary.

Proper condensing pressures and temperatures must be maintained for the refrigerant circuit to operate effectively. As condensing temperatures rise, capacity and efficiency are reduced. While some retail fixtures operate in a conditioned environment, many do not. This must be considered in the equipment selection.

Regardless of where the condenser is placed, do not assume the condenser coil will remain clean. In or near kitchens, for example, air-cooled condensers can foul rapidly with airborne grease. Regular condenser cleaning must be an integral part of retail refrigeration system service.

NOTE

Refer to *Section 3.2.5* of this module for more information about refrigerant-side head pressure controls.

Outdoor condenser installations can lead to performance problems. Low condensing pressures can result in a loss of performance due to insufficient pressure drop across the metering device. This condition often occurs in winter when condenser coils are exposed to outdoor conditions.

Either air or refrigerant controls can be used to maintain the minimum required condensing pressure. The condenser fan(s) can be cycled on and off by pressure switches to control the airflow. Alternatively, the condenser fan speed can be modulated to provide a more stable pressure. Cycling the fans on and off results in the head pressure constantly rising and falling between the setpoints.

On the refrigerant side, controls installed in the refrigeration circuit can be used to control the active surface of the condenser by flooding a portion of its internal volume with liquid refrigerant.

A few commercial refrigeration fixtures rely on water-cooled condensers. Manufacturers offer water-cooled condensers suitable for all applications. The condenser may be integral to the condensing unit (*Figure 28*) or be placed in another location that is not subject to freezing.

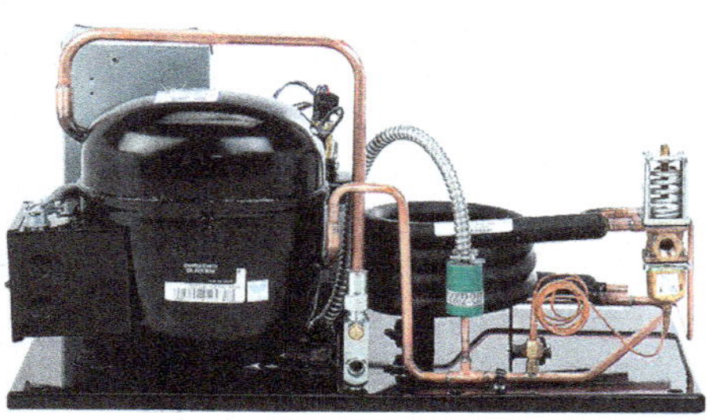

Figure 28 Copeland™ water-cooled refrigeration condensing unit.
Source: Courtesy of White-Rodgers by Copeland

3.1.3 Evaporators

Many types of evaporators are used in retail refrigeration systems. Although most are typical air-cooled fin-and-tube coils like those used for comfort cooling, evaporators used for liquid cooling and ice making have unique construction and design features.

For larger space-cooling applications, such as walk-in freezers or coolers, a unit cooler like the one shown in *Figure 29* is used. Unlike most comfort cooling

Figure 29 Walk-in refrigeration unit cooler.
Source: Hussmann Corporation

applications that require centrifugal blower wheels to overcome duct resistance and deliver air to specific locations, these units use propeller fans made of plastic or aluminum.

The bottom of the unit cooler serves as the condensate collection pan. In *Figure 29*, you can see the fitting on the bottom where the condensate drain line will be connected. For small, refrigerated fixtures, condensate is often collected in a pan installed in the fixture cabinet. An electric heating element inside the pan re-evaporates the condensate back to the air. An example of a heated condensate-collection pan is shown in *Figure 7*.

For unit coolers that operate below freezing, the pan under the coil has an electric heater to prevent freezing and ensure the condensate will flow away. The condensate line itself, where it passes through the refrigerated space, also requires freeze protection. Heat tape wrapped around the outside of the drain line ensures the condensate will leave the enclosure.

Maintaining the proper humidity in the refrigerated space is important, depending on the product stored. Products that are sealed in bags or buckets are not sensitive to the humidity. However, fresh produce, flowers, and similar products require moisture to prevent dehydration. Proper selection of the evaporator coil, with attention to the ΔT between the coil and box temperature, is essential,

To maintain high humidity levels, evaporator coils must operate at a narrow ΔT between the storage temperature and the saturated suction temperature of the coil. A ΔT between 4°F and 8°F (2°C – 4°C) is common for these applications. To achieve the cooling needed with such a low ΔT, the evaporator coil must have more surface area and produce greater air velocity.

Conversely, when low humidity is desirable, the evaporator coil can be smaller and operate at a reduced air velocity. When humidity isn't a concern, the air-to-refrigerant ΔT can be as high as 20°F–30°F (11°C–17°C). Evaporator coils that operate with a high ΔT are colder, and far more likely to develop frost. For example, if the design box temperature is 40°F (4°C), a ΔT of 25°F (14°C) results in a coil temperature of 15°F (–9°C). Frost will form quickly on the coil surface.

Most cubed-ice machines use plate evaporators (*Figure 30*). Refrigerant tubing (usually stainless steel for this application) is firmly attached to the plate on one side. The water flows over the vertical plate to form the individual cubes.

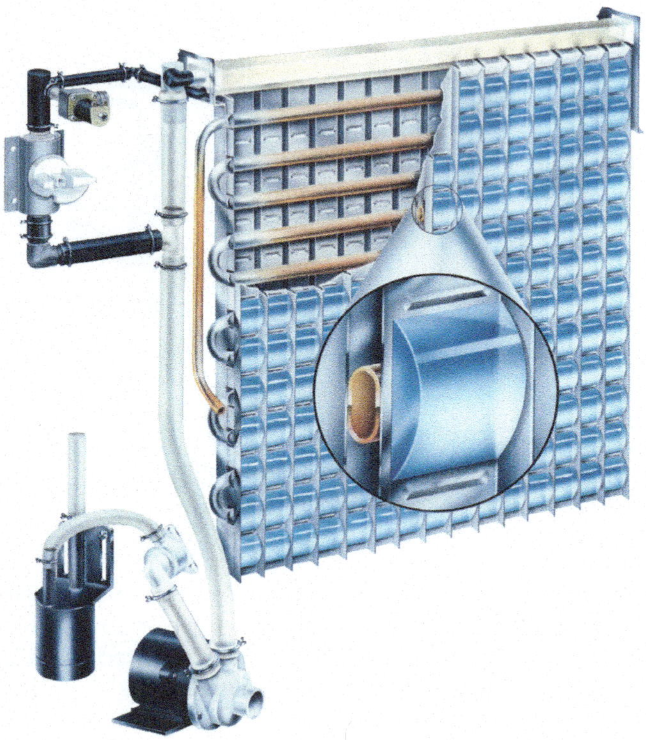

Figure 30 Cubed-ice machine evaporator.
Source: Hoshizaki America, Inc.

NOTE

All types of metering devices are covered in detail in NCCER Module 03303, *Metering Devices*.

Figure 31 Thermostatic expansion valve (TXV).

3.1.4 Refrigerant Metering Devices

Fixed metering devices such as capillary tubes are found on some small and domestic fixtures. However, they are not widely used in retail refrigeration because fixed devices cannot effectively modulate refrigerant flow as the operating conditions change. Fixed metering devices are also very sensitive to the volume of refrigerant charge in the system. For this reason, TXVs are widely used in retail refrigeration applications.

The TXV shown in *Figure 31* controls the amount of refrigerant flow using the superheat temperature sensed at the evaporator outlet. The device provides a flow rate to match the evaporator load. As the evaporator load changes, the TXV responds by opening or closing.

TXVs are not nearly as sensitive to the refrigerant charge volume as fixed metering devices. Using TXVs ensures that a specific superheat value can be maintained effectively under a wide range of conditions, even when the refrigerant charge is not precise.

Electronic expansion valves, or EEVs (*Figure 32*), are gaining popularity in the industry because they are not refrigerant-specific. Because the orifice size can be changed, each valve body can be used to cover a wide range of refrigerants and applications. EEVs offer precise operation and flexibility at extremely low superheat values, allowing maximum use of the evaporator's heat transfer surface. However, EEVs are currently used in more critical applications than most common retail units represent. Using an EEV increases the cost of the equipment, which represents a challenge in a very competitive market.

Figure 32 Electronically controlled expansion valves (EEVs).
Source: Courtesy of Sporlan Division - Parker Hannifin Corporation

EEVs can be very compact, offering thousands of possible positions, called *steps*. This term is derived from using a *stepper motor* to open the valve in tiny increments. While older EEVs offered a few hundred steps or less, over 6,000 steps are now possible. However, EEVs used in small retail fixtures only require 500 to 1,500 steps. The valve position is controlled by inputs from a microprocessor-based board and one or more temperature sensors.

On Site

EEV Superheat Control

There are two basic approaches for controlling superheat with EEVs. Superheat is all about pressure-temperature relationships, and these relationships are specific to each refrigerant. One common approach relies on a pressure input from a transducer at the coil outlet, a temperature input from a sensor at the same location, and the input of the proper refrigerant PT chart into the controller. This is considered the most accurate approach, since it is directly related to the thermodynamics involved.

A simpler approach relies on two temperature sensors. This method discounts the pressure-temperature relationship entirely. One sensor is located at the evaporator inlet, and the other at the evaporator outlet. Software compares the two and assumes any rise in temperature through the coil to be superheat. This method is not as precise and may require that a slightly higher superheat value be programmed in for safety's sake. However, this two-temperature method can be used with any refrigerant without reprogramming the valve for a specific compound or blend.

Systems with traditional TXVs are often equipped with a liquid-line solenoid valve upstream of the TXV. During the compressor off cycle, the solenoid valve closes to prevent refrigerant migration to the cold evaporator. Once in the evaporator and exposed to the cold temperatures of the refrigerated space, the refrigerant condenses. When the compressor starts again, the liquid can return to the compressor as floodback and cause damage.

Since EEVs can be driven closed during the off cycle, a solenoid valve is not needed to prevent migration when the system is off.

3.2.0 Secondary Refrigeration Circuit Components and Accessories

A variety of supporting components are needed in refrigeration systems. Although some of the components discussed here may also be found in comfort cooling applications, these components are more common in refrigeration applications.

3.2.1 Receivers

Receivers provide a place where liquid refrigerant can be stored on the high-pressure side of the system. Because variations in load conditions can demand rapid changes in refrigerant flow, it is important to have an adequate supply of refrigerant to satisfy the needs of the expansion valve and evaporator under all load conditions. Receivers serve no purpose in systems with a fixed metering device (capillary tube).

Many small receivers have a back-seating shutoff valve on the outlet for convenience. Receivers are also equipped with a dip tube, drawing liquid refrigerant from the bottom to ensure that liquid alone is sent to the metering device. The refrigerant level inside the receiver will fluctuate along with the system flow rate.

3.2.2 Accumulators

Because refrigeration systems tend to operate at lower superheat values, the chance of liquid refrigerant leaving the evaporator increases. The accumulator (*Figure 33*) is essentially a trap designed to capture liquid refrigerant on the low-pressure side of the system, protecting the compressor from floodback.

An accumulator has a U-shaped dip tube to separate liquid from vapor. The compressor draws refrigerant vapor from one end of the U-tube. The other end is contained within and terminates near the top of the vessel to ensure that only

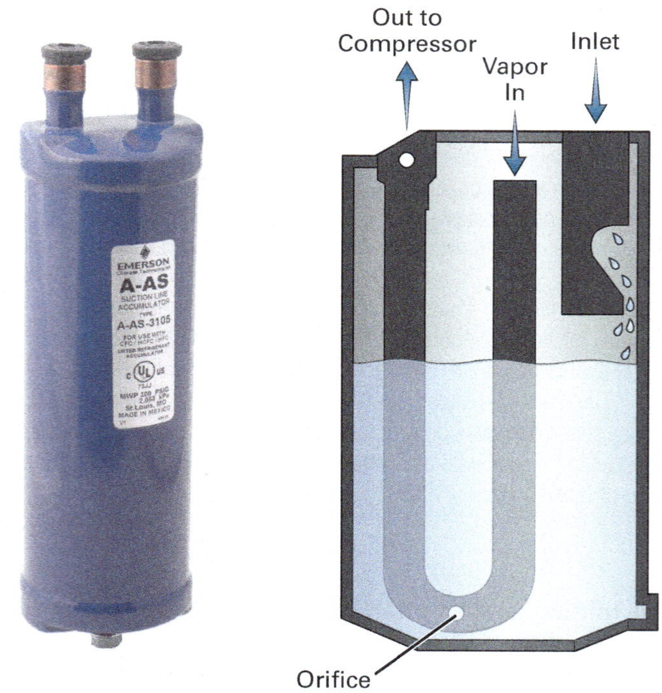

Figure 33 Suction accumulator.
Source: (Left) Courtesy of Emerson Climate Technologies

vapor is extracted by the compressor. A small orifice at the lowest point in the U-tube allows trapped compressor oil and liquid refrigerant to return to the compressor at a manageable rate.

3.2.3 Crankcase Pressure Regulators

Refrigeration compressors are selected for the load at or near the design temperature of the fixture. Consequently, when a freezer compressor starts with the box at ambient temperature, or following a defrost cycle, it is often overloaded. The current rises above its rated load amperage (RLA) and it may overheat. Although compressors can typically survive the event without damage, it's not something that should be repeated multiple times per day.

To prevent an overload of the compressor motor, a crankcase pressure regulating (CPR) valve can be installed in the suction line just upstream of the compressor (*Figure 34*). The CPR ensures that refrigerant vapor is fed to the compressor in a reasonable volume that prevents it from being overloaded. This adjustable valve controls the maximum pressure applied to the compressor suction inlet.

The valve setpoint is made by adjusting the internal spring pressure so that the valve moves towards its closed position on a pressure increase at the outlet. The adjusting stem is under the cap on top. CPR valves are very simple to set:

- Place a clamp-on ammeter on the compressor common lead or any one of the leads on a three-phase compressor to monitor the current. Note the RLA of the compressor.
- Remove the valve stem cap at the top of the valve and insert the correct size of Allen (hex) wrench into the valve stem.
- Connect a low-pressure refrigerant gauge to the hose fitting (if present).
- Start the system in a condition that should heavily load the compressor, while the fixture is warm.

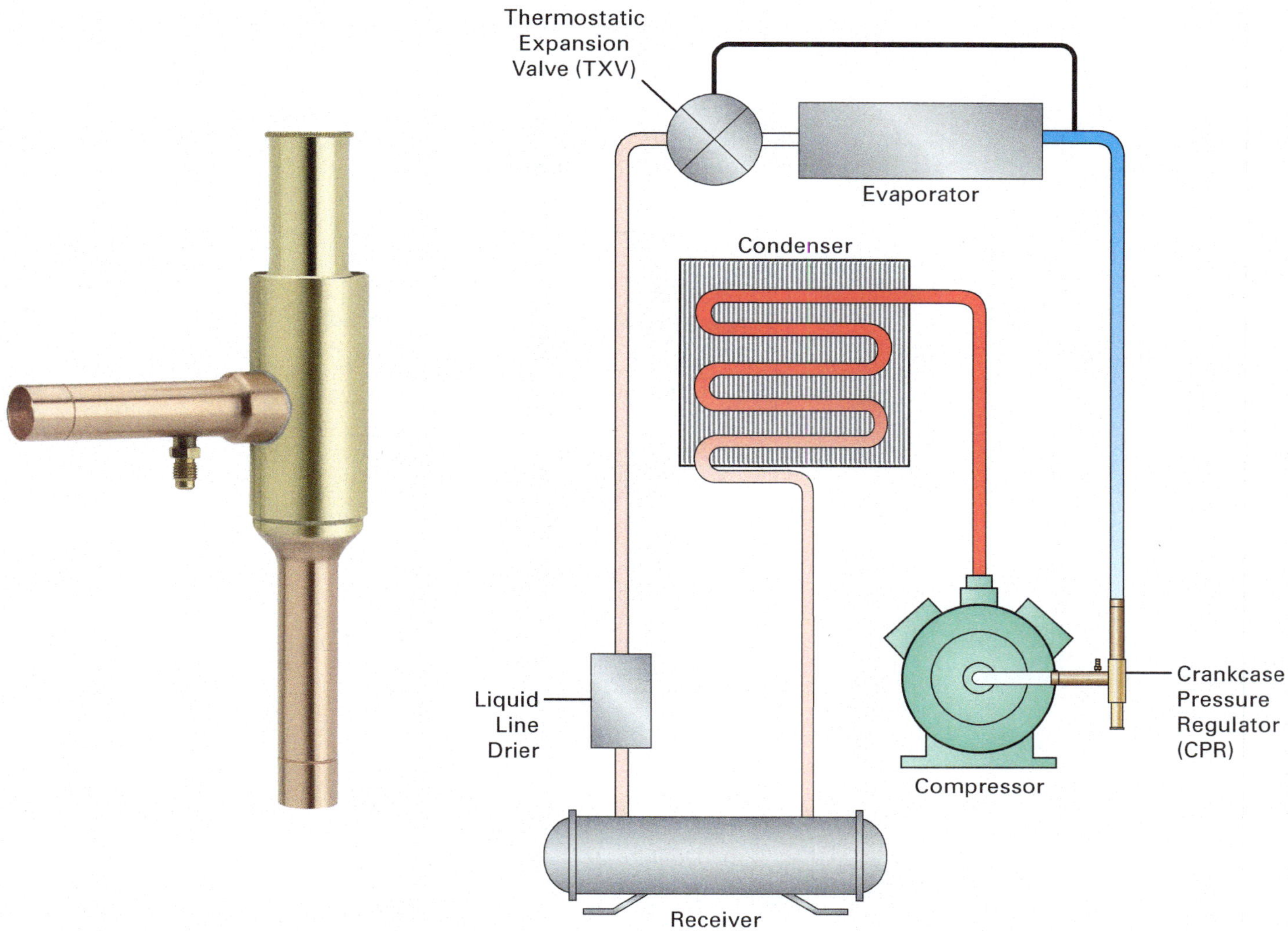

Figure 34 Crankcase pressure regulator (CPR) application.
Source: (Left) Courtesy of Sporlan Division - Parker Hannifin Corporation

- Watch the ammeter. Assuming the current is higher than the compressor RLA, adjust the valve to lower the outlet pressure until the current falls to or just below the RLA. When the suction pressure entering the valve is higher than this setting, the valve will regulate the outlet pressure to prevent an overload.
- If the compressor current is below the RLA, open the valve a turn to see if the current increases, as the valve may already be restricting the flow. Adjust the valve as necessary.

3.2.4 Evaporator Pressure Regulators

Larger refrigeration systems use a single compressor to maintain the temperature in multiple compartments or fixtures with different temperature requirements. The evaporator pressure regulator (EPR) allows an evaporator serving a higher temperature area to maintain the proper coil temperature by controlling the suction pressure. Fixtures operating in the lower temperature range do not need an EPR—their suction pressure will match that of the compressor and the main suction line. An example of such an installation is shown in *Figure 35*.

The pressure in the common suction line serving the fixtures must be low enough to properly refrigerate the coldest fixture. When the pressure downstream falls below the setpoint of an EPR, the valve throttles the refrigerant flow to hold the upstream pressure steady. When evaporator pressure falls, the valve moves toward its closed position. The evaporator will maintain the proper pressure that corresponds to the designated saturated suction temperature.

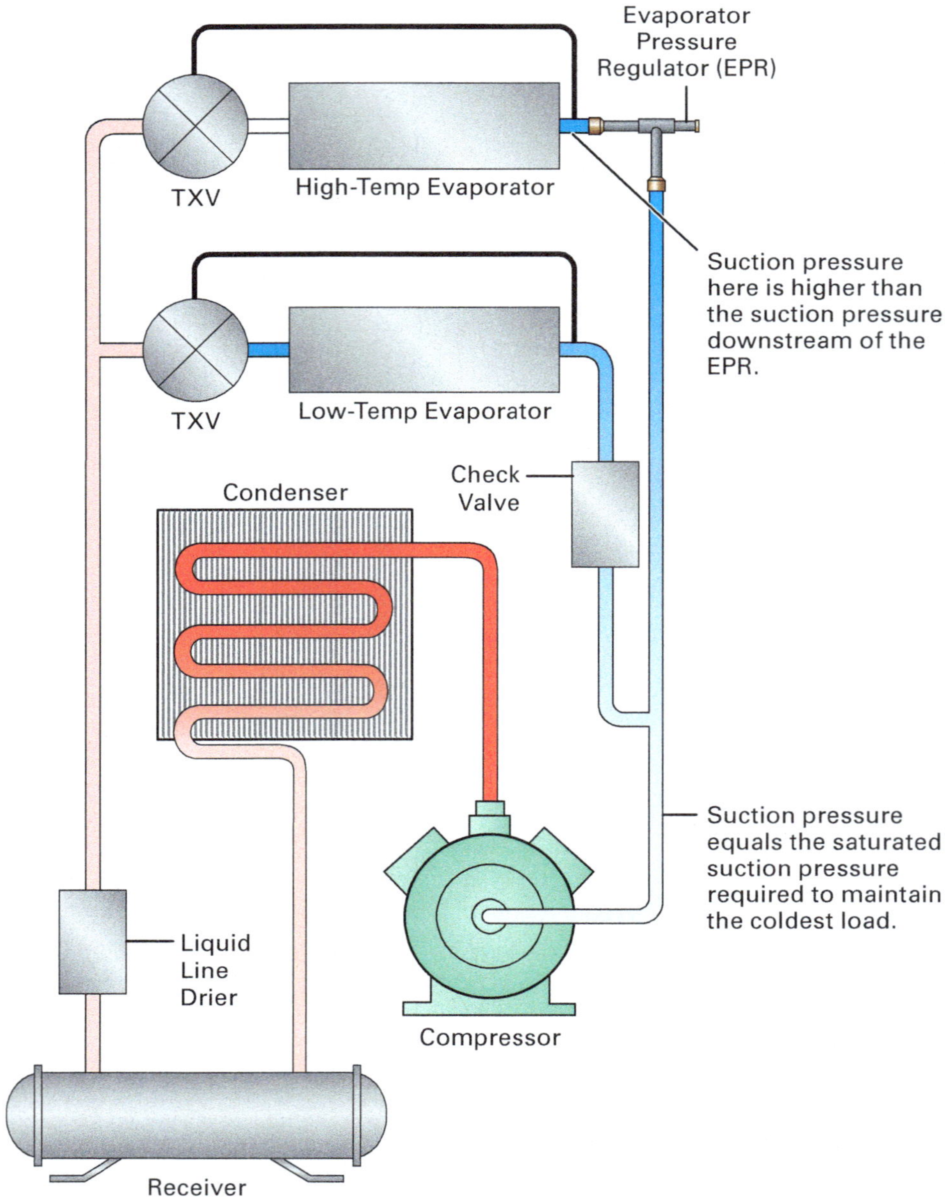

Figure 35 Evaporator pressure regulator (EPR) application.

For example, assume that two fixtures need to maintain a 0°F (–18°C) coil temperature to hold the fixture at 20°F (7°C). With HFC-134a, the saturated suction pressure needs to be about 6 psig. The saturated suction temperature for a third fixture needs to be 28°F (–2°C) to maintain a 40°F (4°C) fixture temperature while minimizing frost development on the coil. Thus, with HFC-134a, the saturated suction pressure needs to be about 24 psig.

In this scenario, most of the system operates at the lower required suction pressure of 6 psig. The high-temperature fixture would be equipped with an EPR set to maintain an inlet pressure no lower than 24 psig. If the suction pressure in the main suction line rises above that, the valve moves to its fully open position.

On Site

Pilot-Operated EPRs

Normally, EPRs are installed as close to the evaporator outlet as possible. If they must be installed some distance from the evaporator, or downstream from a suction riser, a pilot-operated EPR like the one shown here is typically used. The pilot line is connected to the evaporator to monitor the pressure at the coil and is routed to the valve location.

Source: Courtesy of Emerson Climate Technologies

3.2.5 Refrigerant-Side Head Pressure Control

Refrigeration units installed in locations subject to low ambient temperatures generally require a method of maintaining sufficient head pressure. Cold winter air entering the condenser reduces the head pressure significantly, often resulting in insufficient pressure to maintain the needed refrigerant flow through the metering device.

A head-pressure control valve (*Figure 36*) is the most accurate and stable means of maintaining adequate head pressure in cold weather. Although condenser fan cycling controls are often used for this function because they are less expensive, condensing pressures rise and fall through the fan-cycling control's range. The constantly changing condensing pressure causes the metering device to constantly modulate, or *hunt*, to keep up. A refrigerant-side head-pressure control maintains far more stable condensing pressures.

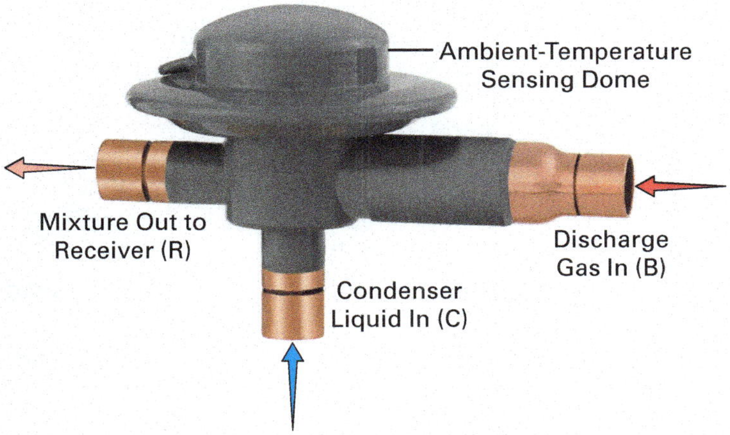

Figure 36 Refrigerant-side head-pressure control valve.
Source: Courtesy of Sporlan Division - Parker Hannifin Corporation

Figure 37 shows the typical installation of a three-way control valve. As the air temperature across the condenser falls, a corresponding decrease in the condensing pressure occurs. When the pressure falls to the valve's set point, a portion of the discharge gas is routed through port B of the valve and into the liquid line to the receiver through port R. Liquid from the condenser mixes with the vapor at the valve. This action causes a rise in pressure at the condenser outlet, which impedes the flow of refrigerant liquid out of the condenser. Condensed liquid begins to back up in the condenser, effectively reducing the available condensing surface and raising the head pressure.

Three-way head-pressure control valves are usually non-adjustable. This style is refrigerant-specific and is factory preset at an appropriate value for the application and refrigerant. It has a sensing dome, as shown in *Figure 36*, which must be exposed to the ambient temperature.

Adjustable valves are available. However, two separate valves that operate independently are needed to provide the same service. Since retail refrigeration systems are generally compact, the three-way arrangement shown in *Figure 37* is more popular unless the application is beyond the capacity of the three-way model.

Systems equipped with a head pressure control valve must include a receiver. Additional refrigerant, held in reserve in the receiver, is needed to partially flood the condenser coil as the valve modulates to maintain appropriate pressure.

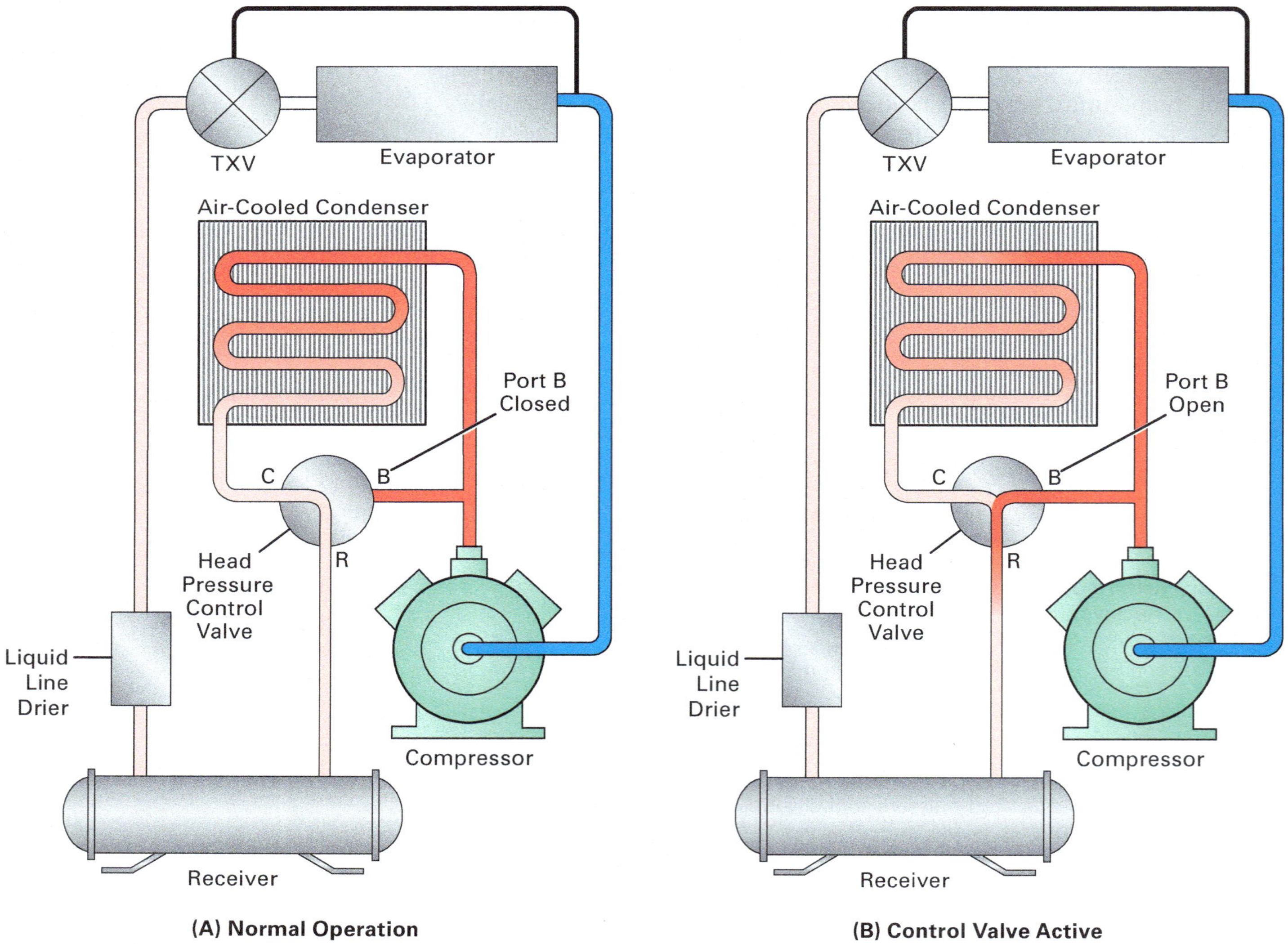

Figure 37 Head-pressure control valve application.

3.3.0 Common Refrigeration System Controls

Typical controls for retail refrigeration systems include thermostats, pressure controls, time delay relays, and solenoid valves. While these devices are similar to comfort cooling controls, they are often applied in different ways in retail refrigeration and may not have the same appearance.

3.3.1 Thermostats

Electronic temperature controls and thermostats are popular in the retail refrigeration industry today. Offering reliable operation and a wide temperature range, these controls allow identical controls to be installed on high-, medium-, and low-temperature systems. Installing a small temperature sensor and routing flexible wiring back to the control is much simpler than routing a long capillary tube through complex cabinet components, as is required for traditional remote-bulb thermostats.

Some applications for thermostats in refrigeration include the following:

- Controlling the refrigeration circuit and compressor to maintain the desired fixture temperature
- Providing a low-temperature safety device for a fixture
- Delaying the restart of evaporator fans following a defrost cycle — This allows time to freeze any remaining condensate on the evaporator coil before the fans start, preventing water from being blown off the coil and onto the product.
- Cycling of the condenser fan(s) based on ambient temperature
- Terminating the defrost cycle based on evaporator coil or suction line temperature
- Shutting down an ice-making process in response to the bin temperature

Thermostats for these applications are available with a wide variety of ranges and differentials. They can be adjustable or have a fixed setting chosen for a specific purpose.

3.3.2 Pressure Controls

Refrigeration systems require a variety of pressure controls. Traditional pressure controls rely on *Bourdon tubes*, which are curved, helical, or spiral-shaped flattened tubes to measure pressure. The Bourdon tube tends to straighten as the pressure increases. The end of the tube is mechanically connected to an indicator pointer or switch linkage.

Most often, pressure-sensing devices use bellows or diaphragm switches (*Figure 38*). These adjustable switches provide high- and low-pressure service for both control and safety purposes. Most accept a single pressure input, but differential pressure switches require two inputs. An oil-pressure safety switch is a good example of a differential pressure switch.

Like thermostats, pressure switches perform a variety of functions and are available in different pressure ranges to suit the application. Functions in retail refrigeration that pressure switches provide include the following:

- Controlling refrigeration equipment based on the suction pressure — Since suction pressure is related to saturated suction temperature, pressure switches can provide start-stop control of refrigerated fixtures. However, this method of system control is only applied to systems with a fixed metering device.
- Controlling one or more condenser fans by condensing pressure
- Low- or high-pressure safety protection
- Controlling pump-down cycles on systems equipped with a liquid-line solenoid valve — When a thermostat control opens and closes the solenoid valve, the compressor continues to run until most of the refrigerant has been

Figure 38 Typical refrigeration pressure switch.

pumped out of the evaporator and the suction pressure falls to the setting of the pressure switch. This minimizes the amount of refrigerant that can condense in the evaporator during the off cycle.

- Oil pressure safety protection

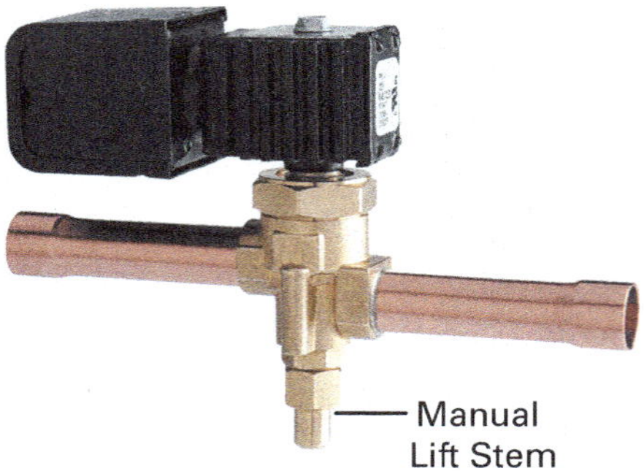

Figure 39 Solenoid valve.
Source: Courtesy of Sporlan Division - Parker Hannifin Corporation

3.3.3 Solenoid Valves

Solenoid valves (*Figure 39*) are used to stop or permit the flow of fluids in refrigeration systems. The valve shown has extended fittings that allow the valve to be brazed into the line with less chance of overheating and damaging the valve body.

Solenoid valves are selected for their rated refrigerant flow rate and resulting pressure drop. The chosen valve may also fit the system line size, but line size is a secondary selection factor.

Solenoid valves are available with or without a manual lift stem. The lift stem allows you to open the valve manually if an operating coil fails or the valve sticks closed. Although most solenoid valves are normally closed, normally open valves are also available. The manual lift stem of a normally open valve allows the user to close the valve.

Applications in refrigeration for solenoid valves include the following:

- Providing flow control of liquid refrigerant to the metering device, controlled by the fixture thermostat — This prevents refrigerant migration during the off cycle and enables a pump-down mode.
- Providing independent temperature control on systems with a single compressor serving multiple evaporators — The solenoid valves are placed in the liquid line to each evaporator.
- Controlling the flow of hot refrigerant vapor for defrost functions or capacity control

On Site

Solenoid Valve Service Tool

Several companies have developed a multifunction service tool for solenoid valves. The body of the tool is a powerful permanent magnet that simulates the electromagnetic coil of a solenoid valve. When placed over the valve plunger in place of the coil, the magnet can easily activate the valve. The body of the tool also acts as a spanner wrench for disassembling the valve body.

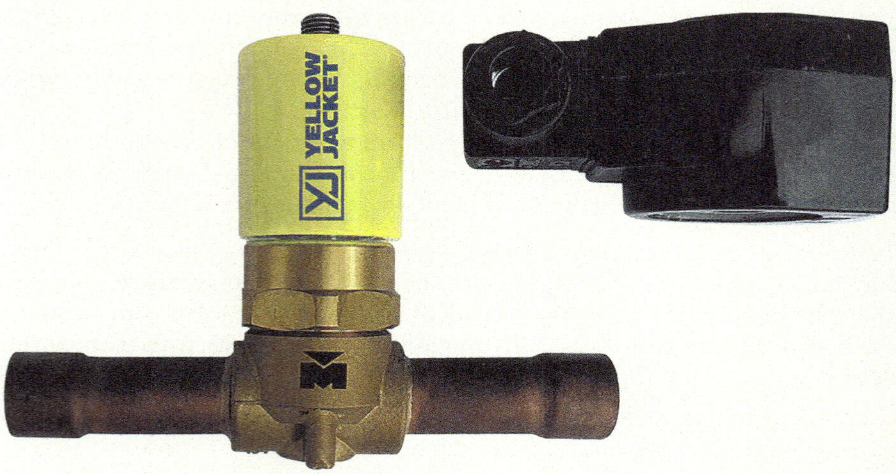

Source: Ritchie Engineering Company, Inc. / YELLOW JACKET

3.0.0 Section Review

1. What device is integrated into scroll compressors used in low-temperature applications to enhance compressor reliability?
   a. Suction unloader
   b. Cylinder unloader
   c. Hot-gas bypass valve
   d. Liquid- or vapor-injection device

2. Receivers are commonly equipped with which of the following?
   a. Dip tube
   b. Float valve
   c. Solenoid valve
   d. A bypass arrangement

3. Start-stop control of a refrigeration unit can only be done if the system is equipped with a(n) ______.
   a. receiver
   b. accumulator
   c. fixed metering device
   d. liquid-line solenoid valve

4.0.0 Retail Refrigeration Operation and Maintenance

Objective

Describe the maintenance and operation of retail refrigeration equipment.

a. Identify reach-in freezer maintenance activities and recount the typical sequence of operation.

b. Identify cubed-ice machine maintenance activities and recount the typical sequence of operation.

Performance Tasks

2. Isolate a fault in a refrigeration unit or ice machine.
3. Clean an ice machine.

All the basic troubleshooting skills previously presented for comfort systems apply to retail refrigeration. In this section, troubleshooting and maintenance procedures for two different retail refrigeration systems are examined: a reach-in freezer and a cubed-ice machine.

Note that systems can differ significantly, so the sequence of operation and service guidance for a system should always be consulted before beginning maintenance and troubleshooting processes.

4.1.0 Reach-In Freezer Operation and Maintenance

For a typical reach-in freezer (*Figure 40*) with electric defrost, a lack of simple maintenance is one of the primary causes of performance problems. Although the periodic maintenance required for unitary retail refrigeration equipment is not challenging, it is a critical task that is often overlooked. The most important and consistently needed of these tasks is cleaning the evaporator and condenser coils.

Depending on where the unit is located, the condenser coil can become fouled very quickly. The result is a continuing rise in condensing pressure. In the early stages, a fouled condenser coil may only increase the operating cost without affecting performance. But as fouling worsens, system performance suffers, the box temperature rises, and the system may run around the clock without reaching the setpoint.

Evaporator coils generally do not foul as quickly, but the product and amount of time the fixture is open are important factors. Remember that retail refrigeration systems are not fitted with air filters. It's best to have coils with wider fin

Figure 40 Reach-in freezer.
Source: Master-Bilt

spacing for this reason, but wider spacing also means the coil must be larger to provide the same refrigerating effect. These two factors are balanced by product engineers when a fixture is designed.

The aluminum fins on the evaporator may also corrode and deteriorate due to the off-gassing from certain foods, such as pickle juice. The fins then become coated with aluminum oxide or a similar residue, reducing heat transfer. Once the fins begin to corrode, performance steadily decreases.

Retail refrigeration systems of this size have a relatively small refrigerant charge. Refrigerant operating pressures should not be checked during every maintenance visit unless problems are suspected. Each time gauges are attached, a small loss of refrigerant occurs.

Maintenance requirements for a reach-in freezer include the following:

- Check the operation of condenser and evaporator fan motors. Ensure the blades are secure, intact, and turn freely. Most of the motors used in retail refrigeration applications are small and do not require lubrication. However, check each motor during a maintenance inspection and lubricate if necessary.
- Clean the condenser and evaporator coil as needed.
- Check the drain pan, drain line, and door heaters.
- Ensure the doors close tightly and the gaskets seal properly. Door hinges and gaskets are another often-overlooked maintenance item.
- Clean the inside and outside of the cabinet, as necessary. It is a best practice to leave serviced equipment cleaner than it was when you arrived, regardless of the type of equipment.

4.1.1 Reach-In Freezer Operation

Before troubleshooting a unit, review the manufacturer's data and normal operating characteristics. A wiring diagram for a reach-in freezer is shown in *Figure 41*. The typical features and operating characteristics for this freezer are as follows:

- The unit power switch is located on an electrical enclosure inside the lower louvered access panel. The condensing unit is also located in this area.
- The unit is equipped with an electronic temperature control that has no Off position. The electromechanical defrost timer and temperature control are located inside the same enclosure where the power switch is mounted. There is also an adjustable temperature differential setting (the difference between the cut-in temperature and the cut-out temperature) located on the back of the temperature control cover.
- The defrost timer has been factory preset for two defrost cycles per day—6:00 a.m. and 10:00 p.m., with a 40-minute fail-safe defrost termination period. The defrost cycle is normally terminated by a defrost thermostat located on the evaporator suction line. The three-wire thermostat will terminate defrost when a line temperature of 58°F (14°C) is reached. As refrigeration resumes, it will not allow the evaporator fans to operate until the line temperature has dropped below 32°F (0°C). This prevents condensate on the coil from being blown onto the product. Should the defrost thermostat fail to terminate defrost, the timer will terminate the cycle after 40 minutes.
- The electric defrost heaters are attached to the fins of the evaporator coil using spring clips. Refer to *Figure 6*. The unit is also equipped with low-wattage heaters in the condensate drain pan and drain line.
- Fluorescent lamps provide interior lighting and can be turned off with a switch located inside the cabinet above the door.
- A fan switch is located at the top of the doors. The evaporator fans stop when the door is opened to prevent cold air from being blown out while warm air is drawn in.

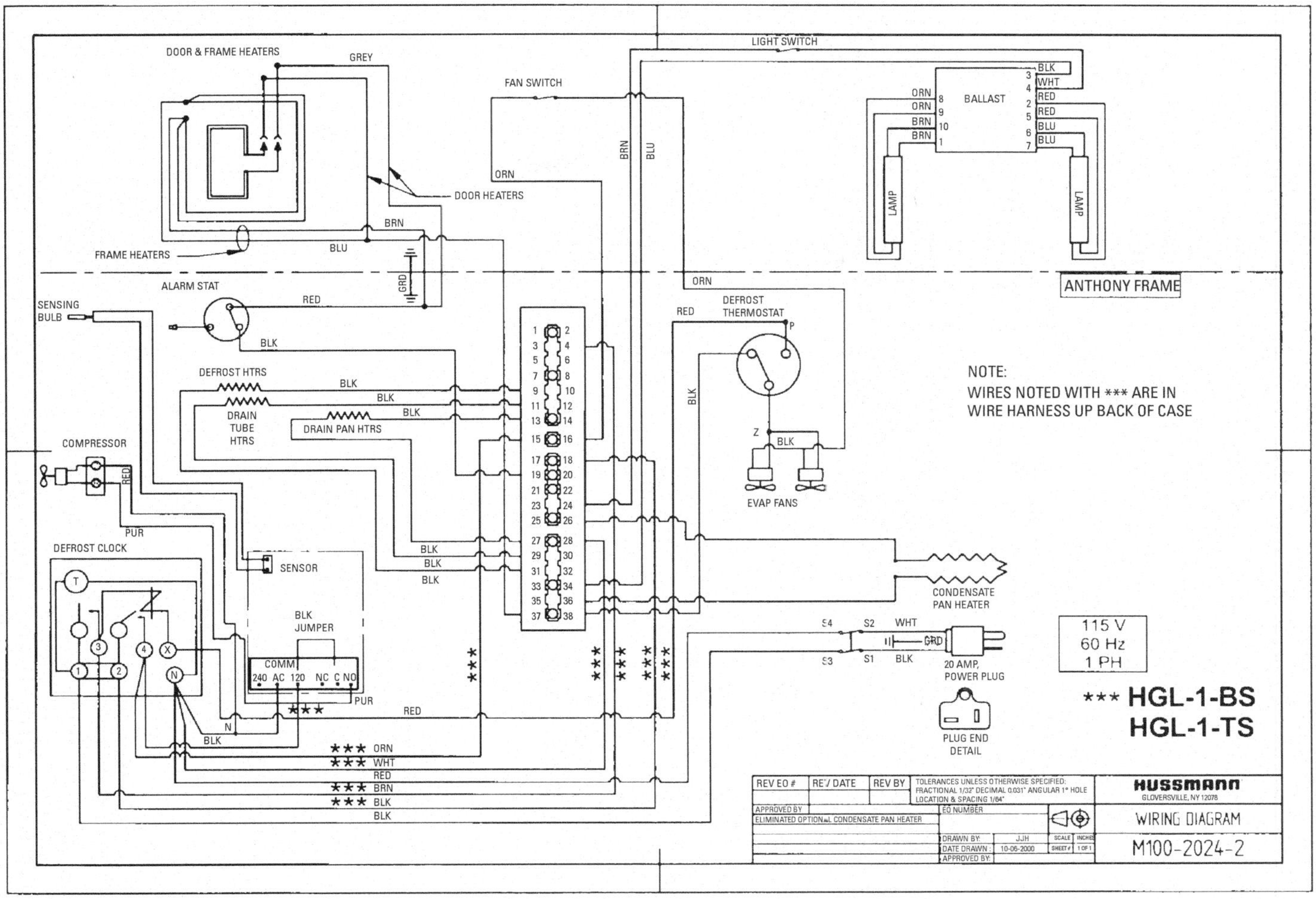

Figure 41 Freezer wiring diagram.
Source: Hussmann Corporation

- Low-wattage electric heaters are integrated into the doorframes and doors to prevent freezing. The heaters are not energized until the box has reached its normal operating temperature. The thermostat controlling heater operation is also referred to as an *alarm thermostat,* because it can be wired to also generate a visual or audible alarm when the temperature rises above 18°F (–8°C). The thermostat switch closes, energizing the door heaters, when the box temperature has reached 0°F (–18°C).
- The refrigeration circuit is equipped with a TXV metering device and a receiver.
- A small semi-hermetic compressor is used, and the system refrigerant is HFC-404A. To avoid compressor overloads, a crankcase pressure regulator is installed and set at 10 psig. This is the maximum suction pressure to which the compressor will be exposed during operation.
- The manufacturer indicates the standard ambient temperature should be 75°F (24°C) and lists the expected head pressure as 230 psig to 240 psig. This means the box was designed for use in a conditioned space. The expected suction pressure is 5 psig to 6 psig with the box temperature stable at –10°F (–23°C). The design air-to-refrigerant ΔT is 15°F (8.3°C), resulting in an expected evaporator air discharge temperature of –25°F (–32°C).

The wiring diagram provided is an essential troubleshooting tool. The example troubleshooting charts in *Figure 42* and *Figure 43* may also be helpful, but experienced technicians learn how to effectively match symptoms with possible causes. Although troubleshooting charts can't cover every potential situation, they can help guide your efforts while you gain experience.

Trouble	Probable Cause	Solution
Compressor Runs Continuously; Product Too Warm	1. Short of refrigerant 2. Inefficient compressor 3. Dirty condenser	1. Leak check, change drier, evacuate, and recharge 2. Replace 3. Clean
High Head Pressure	1. Cabinet location too warm 2. Restricted condenser air flow 3. Defective condenser fan motor 4. Air or non-condensable gases in system	1. Relocate cabinet 2. Clean condenser to remove air flow restriction 3. Replace 4. Leak check, change drier, evacuate, and recharge
Warm Storage Temperatures	1. Temperature control not set properly 2. Short of refrigerant 3. Cabinet location too warm 4. Too much refrigerant 5. Low voltage; compressor cycling on overload 6. Condenser dirty	1. Reset control 2. Leak check, change drier, evacuate, and recharge 3. Relocate 4. Change drier, evacuate, and recharge 5. Check power 6. Clean
Compressor Runs Continuously; Product Too Cold	1. Defective control 2. Control feeder tube not in positive contract 3. Short on refrigerant	1. Replace 2. Assure proper contract 3. Leak check, change drier, evacuate, and recharge
Compressor Will Not Start; No Noise	1. Blown fuse or breaker 2. Defective or broken wiring 3. Defective overload 4. Defective temperature control 5. Power disconnected	1. Replace fuse or reset breaker 2. Repair or replace 3. Replace 4. Replace 5. Check service cord or wiring connections
Compressor Will Not Start; Cuts Out on Overload	1. Low voltage 2. Defective compressor 3. Defective relay 4. Restriction or moisture 5. Inadequate air over condenser 6. Defective condenser fan motor 7. CPR not set properly	1. Contact electrician 2. Replace 3. Replace 4. Leak check, replace drier, evacuate, and recharge 5. Clean condenser 6. Replace 7. Reset to 10 psi
Icing Condition in Drain Pan	1. Low voltage 2. Cabinet not level 3. Defective drain tube heater 4. Defective drain pan heater	1. Check voltage at compressor 2. Check front to rear leveling; adjust legs accordingly 3. Replace 4. Replace

Figure 42 Example freezer troubleshooting chart.
Source: Hussmann Corporation

Troubleshooting Lighting System	
Lights Won't Start	1. Check light switch 2. Check continuity to ballast 3. Check to see if bulbs are inserted properly into sockets 4. Check voltage
Lights Flicker	1. New bulbs tend to flicker until used 2. Allow lamps to warm up 3. Check lamp sleeve for cracks 4. Check sockets for moisture and proper contact 5. Bulb replacement may be necessary 6. Check voltage
Ballast Hums	1. Check voltage 2. Replace ballast

Figure 43 Example refrigeration-fixture lighting troubleshooting chart.
Source: Hussmann Corporation

Did You Know?

Silver and Biological Protection

Silver is a well-documented antimicrobial. Studies have proven that silver can kill bacteria, fungi, and certain viruses. Silver is being used in products at an increasing rate to assist in the prevention of biological growth. Even clothing items can now be impregnated with silver for this purpose.

Many ice machine manufacturers are adding silver ions to plastic components such as ice bin walls to impede biological growth. One such product is known as AlphaSan®. These additives, compounded directly into a part during manufacturing, release silver ions slowly to the surface. Not only do the ions impede or prevent biological growth, but the additives also impede discoloration of the part over time. The additives do not impact the taste and flavor of the ice. However, the presence of these products must not be considered as an alternative to proper periodic cleaning.

4.2.0 Cuber Operation and Maintenance

Cubed-ice machines, or *cubers*, have a more complex sequence of operation than freezers. Troubleshooting and testing an ice machine can be time consuming since a full cycle can take 30 to 90 minutes. After any adjustments or changes are made, the unit should complete several complete cycles to ensure that it is operating properly. An air-cooled cuber is the basis for the maintenance procedures and operating sequence presented in this section.

WARNING!

It is important to remember that ice machines produce a consumable product. Ice machines must be properly maintained and kept sanitary. Failure to properly maintain and clean these units could lead to widespread illness.

NOTE

The following processes are generic in nature. Cleaning or maintenance should not be performed without access to unit-specific manufacturer's instructions.

4.2.1 Cuber Cleaning and Maintenance

Performing maintenance and cleaning of an ice machine can be a significant benefit to you as well as to the equipment. Cleaning provides an opportunity to become familiar with the location of components and to observe normal operation without the pressure of trying to correct a problem.

Interior cleaning and sanitizing should be conducted at least every 6 months. Environmental factors and water quality can increase the frequency to as often as every 30 days.

Commercial ice machine cleaners help remove lime and other mineral deposits left on surfaces by the water supply. Ice machine sanitizers disinfect and remove biological growth. Always ensure that cleaning and sanitizing agents are used according to the manufacturer's recommendations and are used in the proper concentrations. Also be sure you check the label and are certain about which product—cleaner or sanitizer—you are using.

WARNING!

Note that some ice machine cleaning and sanitizing products can be harmful. It is a violation of federal law to use these products in any manner not consistent with their stated purpose. Always wear rubber gloves and goggles or a face shield when handling and using these products.

Some general cleaning and maintenance guidelines follow:

- Turn off the ice machine immediately following a harvest so that the evaporator is clear of ice. Remove all the ice from the bin. Thorough cleaning requires significant disassembly of the unit and a significant amount of time to do it correctly. Schedule these activities to ensure that an adequate supply of ice will be on hand for users.
- Ice machine cleaners are added and circulated throughout the water distribution system using the machine's water pump. Most manufacturers provide a pump control switch specifically for this reason. The cleaning solution should be circulated for at least 30 minutes unless directed otherwise.
- Disassemble and remove all components possible that are in contact with water. This includes water curtains, water sumps, distribution tubes, water level probes, and pump assemblies. Clean these components manually using fresh cleaning solution, allowing them to soak 15 to 20 minutes and loosen mineral deposits. Use a soft-bristle brush or sponge for cleaning. While the components are soaking, use fresh cleaning solution to wipe down all the accessible surfaces of the ice machine interior and the bin. Then rinse or wipe the surfaces again with water.
- Prepare the sanitizer solution and use it to wipe down the removed components and all accessible interior surfaces of the unit and bin. Reassemble the unit and add sanitizing solution to the water sump. Circulate the sanitizing solution for approximately 30 minutes. Disassemble the unit a second time and rinse or wipe down the removed components and accessible areas of the bin and unit.
- While the sanitizing solution is circulating, the condenser coils can be cleaned. Clean the coils using a soft brush and a vacuum cleaner. A coil cleaning solution may be needed if grease or other deposits are evident. A few ice machines are equipped with washable filters to help protect the condenser coil. The air filters can be washed using a mild soap and water solution. Straighten any bent fins with a fin comb.
- Many ice makers are equipped with water filters to remove odor, taste, and particulates from the water supply. Filter elements should be replaced following the cleaning/sanitizing process, or when fouled.

- Reassemble the unit, restore the power and water supplies, and return the unit to service. Observe the unit through several operating cycles to ensure that all components have been properly reinstalled, and the unit is working as expected.
- While the unit is making ice, clean the external surfaces and cabinet. Users and owners alike appreciate a well-maintained exterior.

Specialized water treatment systems may be installed to help prevent biological growth by adding trace amounts of chlorine dioxide (ClO_2) to the water. Another system generates ozone gas (O_3) and disperses it into the ice maker's water supply line (*Figure 44*). Ozone is a very effective antibacterial chemical and is also effective as a demineralizing agent. Service these systems per the manufacturer's guidance.

Figure 44 Ozone disinfection unit.
Source: Courtesy of Sanzonate Global Inc.

4.2.2 Typical Cuber Operation

The electrical system of the ice machine includes circuits for the compressor, water pump, condenser fan(s), water and refrigerant solenoid valves, control transformer, control relays, and microprocessor (*Figure 45*). Basic electrical troubleshooting principles apply to all these items.

Knowing and understanding the precise timing and sequence of operation is essential. Ice machines can be challenging to troubleshoot, as there are many interactive components in a compact package. Although ice-making equipment has a similar appearance, the sequence of events often differs. Always ensure that proper literature is on hand for the equipment being serviced.

The following sequence of operation applies to the cubed-ice machine diagram shown in *Figure 45*. Also refer to the energized-parts chart shown in *Figure 46* for additional details about the operating sequence for the same machine. This is a resource that not every manufacturer provides. A watch or smartphone is helpful for timing each event while you're working on an ice machine.

1. The toggle switch must be in the Ice position. The water curtain covering the evaporator plate must be in position, closing the curtain switch, before the unit starts.
2. The water pump and water dump solenoid are energized for 45 seconds before the compressor starts to purge the unit of old water with elevated levels of minerals. The harvest valve is also energized during the water purge and remains energized for an additional 5 seconds after the purge process ends.
3. The compressor starts after the water purge process ends and remains running throughout the ice-making and harvest cycles. The water fill valve is also energized. The condenser fan motor circuit receives power, but the fan cycling control may cause it to cycle on and off, depending upon the ambient conditions.
4. The compressor runs to prechill the evaporator plate for 30 seconds before the water pump starts. The water fill valve remains energized until the probe indicates that a sufficient water level exists in the sump.
5. The water pump starts after the 30-second delay. Water is evenly distributed across the top of the plate and trickles down across and into the individual cube cells. Unfrozen water collects in the evaporator trough and is recirculated. Some water remains behind as ice in the cells, continuing to build in thickness as the cycle continues. The water fill valve may cycle open one more time to replenish the water level.
6. As ice builds, the cascading water is pushed out slightly away from the evaporator plate. Eventually, the flowing water contacts the ice thickness probe. After roughly 10 seconds of constant contact proves that ice has formed, the harvest cycle begins. The unit microprocessor allows at least six minutes to elapse in the ice-making cycle before initiating a harvest cycle. If the ice thickness probe indicates it is closed in less than six minutes of run time, the system assumes there is a problem, cycles off, and initiates an alarm LED.

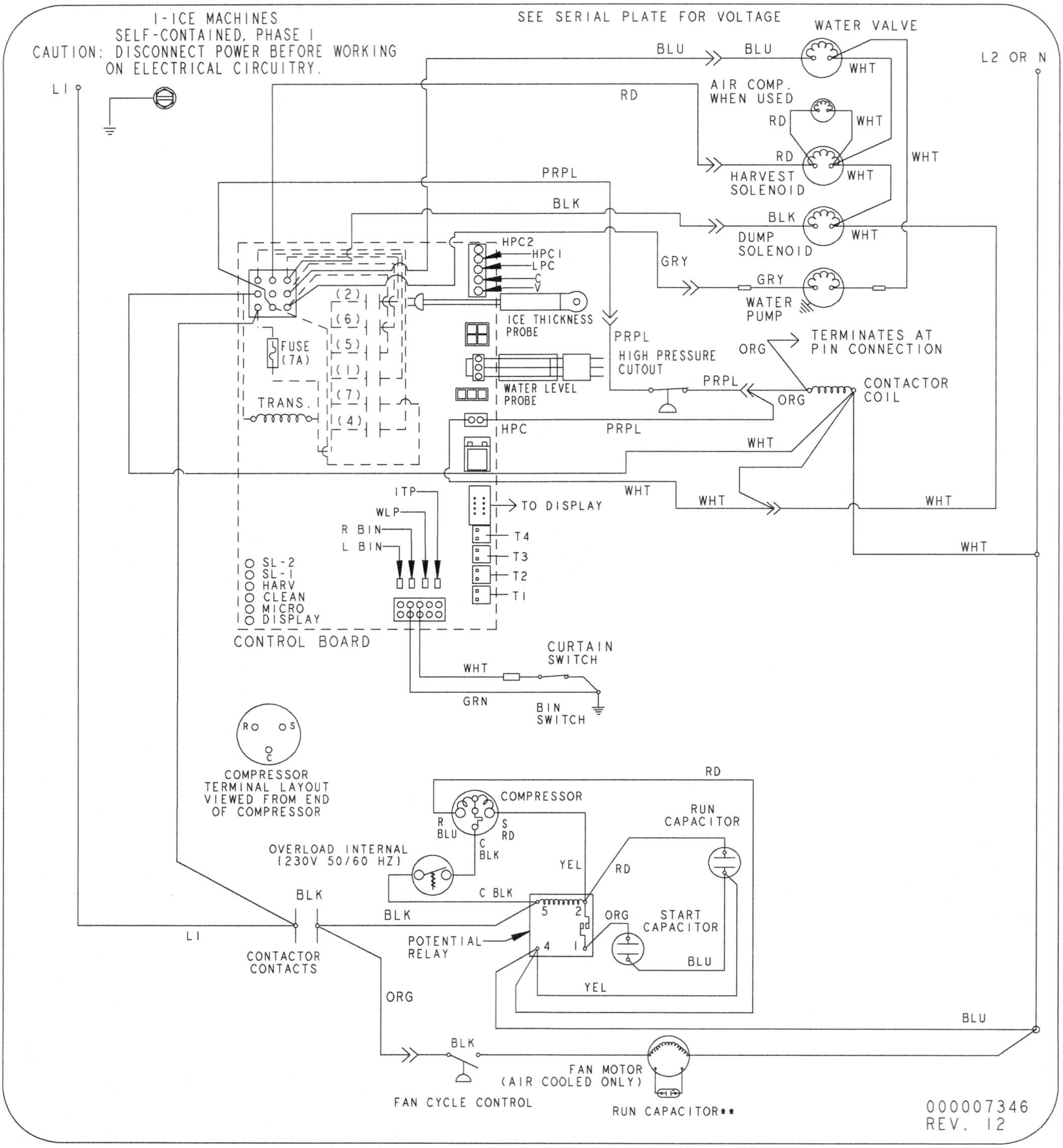

Figure 45 Cuber wiring diagram.

Source: Manitowoc Ice

Indigo Sequence of Operation

Air & Water-Cooled and Traditional Remotes
Energized Parts Chart

⚡ = Energized Part

Ice Making Sequence of Operation	Control Board Relays					Contactor		Length of "On" Time
	Water Pump	Harvest Valve(s) **HPR Solenoid** AirPump(s)*	Water Inlet Valve	Water Dump Valve	Contactor Coil **Liquid Line Solenoid**	Compressor	Condenser Fan Motor†	
Start-Up **1. Water Purge**	⚡			⚡				*45 Seconds*
2. Refrigeration System Start-up a. Equalization		⚡						*5 Seconds*
b. Compressor Start		⚡			⚡	⚡	⚡	*5 Seconds*
Freeze Sequence **3. Prechill**			May Cycle On / Off During Pre-chill		⚡	⚡	⚡	*Initial Start-Up Is 120 Seconds* *30 Seconds Thereafter*
4. Freeze Six-Minute Lock-In on Every Cycle (Max Freeze 35 Min)	⚡		Cycles Off Then On Two More Times		⚡	⚡	⚡	*Until Ice Contact with Ice Thickness Probe*
Harvest Sequence **5. Water Purge**	⚡	⚡		⚡	⚡	⚡	⚡	*Factory Set at 45 Seconds*
6. Harvest a. Normal (0–3.5 Minutes)		⚡			⚡	⚡	⚡	*Bin Switch Activation*
b. Water Assist (3.5–6.5 Minutes)	⚡	⚡	Energizes and Fills to the High WLP		⚡	⚡	⚡	
c. Water Dump (6.5–7 Minutes) *Max Harvest 7 Minutes*	⚡	⚡		⚡	⚡	⚡	⚡	
Bin Switch Open 30 Seconds **7. Automatic Shut-Off**								*Until Bin Switch Recloses & 3 Min. Delay*

*Not Used On All Models
** Remotes Only

† Condenser Fan Motor: Self-contained Air-Cooled - The fan motor is wired through a fan cycle pressure control; therefore, it may cycle on and off. Remotes do not have a fan cycle pressure control with the exception of the JC995, which does have condenser fan cycling control.
♦ 3 Minute Delay for Restart

Figure 46 Cuber energized-parts chart.
Source: Manitowoc Ice

7. The harvest valve opens, allowing hot discharge vapor to circulate through the evaporator. At the same time, the water pump continues to run, but the water dump valve is energized for 45 seconds, and the remaining mineral-laden water is purged from the sump. The water fill valve is energized during the last 15 seconds of this period to rinse the sump. After 45 seconds, the water pump, fill valve, and dump valve are de-energized.
8. The harvest valve remains open, continuing to warm the evaporator and break the bond between the ice and the plate. Cubes are formed as a sheet with a thin bridge connecting the individual cubes. The sheet slides out and breaks into individual cubes as it falls into the storage bin. The water curtain swings out as the sheet falls, actuating the bin switch in the process. If the curtain and the bin switch remain open for more than 30 seconds, this is a signal that the ice sheet has not completely fallen out and the ice sheet is resting atop a full bin of ice. When the bin is not full, the opening and closure of the bin switch takes only a few seconds. This is a signal for the unit to return to making ice.
9. If the bin switch is open for more than 30 seconds and the unit shuts down, it will remain off for at least 3 minutes before the compressor restarts. When enough ice is removed from the bin to allow the latest ice sheet to fall down, the curtain swings closed, and the machine starts a new cycle.
10. The maximum amount of time allowed to complete the ice-making cycle is 60 minutes, after which the unit will automatically initiate a harvest cycle.
11. The maximum amount of time allowed for ice to harvest is $3\frac{1}{2}$ minutes after which the unit will initiate a new ice-making cycle.

Over time, water vapor will collect and freeze on the back of the evaporator plate. After 200 complete cycles, the microprocessor will initiate a warm-water rinse as follows:

- Both the Clean and Harvest LEDs come on, indicating that the unit has entered a warm-water rinse cycle.
- The compressor and harvest valve are energized.
- The water pump is energized.
- The water inlet valve is energized until the water contacts the water level probe.
- The water is warmed by compressor discharge vapor for 5 minutes, then both the compressor and harvest valve are de-energized.
- The water pump remains on for an additional 5 minutes, then is de-energized.
- The process can be terminated by the user by moving the control switch to Off, then back to Ice.

The following tests and adjustments should be made during the initial startup, after cleaning and sanitizing, and after a prolonged shutdown:

- The water level sensor is designed to maintain the proper water level in the sump and is not adjustable on this machine. If the water level appears incorrect, check the probe for physical damage and clean it if necessary.
- The ice thickness probe on this machine is factory set to create an ice bridge thickness of $\frac{1}{8}$". This is sufficient to keep the ice cubes in a sheet until the mass falls from the evaporator, yet thin enough to break apart when it lands. If the ice bridge appears to be too thick or too thin, set the ice thickness probe distance using the adjusting screw.

Note that newer ice machines with advanced features allow the user to adjust the bridge thickness of the sheet electronically from the user interface or control panel on the front. These units use a different type of sensor to measure bridge thickness and report it continuously to the microprocessor.

This unit is controlled by a microprocessor, which handles the large number of timed events with precision. Microprocessor controls are very reliable, and technicians are not expected to repair the board itself. If a board component fails, the entire board must be replaced. However, this type of control can be costly, so be sure the control board has failed before replacing it.

Remember that troubleshooting a microprocessor-based control relies primarily on ensuring that the control is receiving the correct inputs and delivers the correct outputs. If it is doing both, but the equipment is not responding, the problem lies somewhere else.

One of the best ways to evaluate a problem is by using the sequence of operation and a timer to time each step in the cycle. Look for events that do not occur when expected. Use the LEDs on the microprocessor board (when equipped) to help determine what it is attempting to do, then see if the controlled devices respond.

To prevent unnecessary refrigerant loss, service gauges should only be connected when absolutely necessary. When refrigerant must be added, or when refrigeration cycle components must be replaced, use the proper service practices. Weigh in the refrigerant when possible to ensure accuracy. Superheat and subcooling should be checked even when the charge has been weighed in to ensure proper system operation.

Here are some other troubleshooting tips to consider when servicing ice machines and all other types of retail refrigeration equipment:

- Always listen carefully to the customer's complaints and experiences with the machine. While they may not understand the details of the operating sequence, valuable information can be gained from their observations. Remember that technicians who develop good listening skills are often the first to advance.
- Pay close attention to the unit when it is operating properly. If you are familiar with normal unit operation, you'll find it easier to recognize when things aren't functioning properly.
- Before troubleshooting an ice machine, ensure that the condenser coil and other heat transfer surfaces are clean. Fouled evaporator surfaces and clogged coils will cause abnormal operation and high refrigerant pressures.
- Develop a logical approach to troubleshooting all systems. Patience and logical thinking are important troubleshooting tools at your disposal, especially when working with ice machines and similar complex systems. Each step depends on the proper execution of previous steps. Once you know a particular machine, you'll find that you have ideas about what may be wrong as soon as the call comes in.

4.0.0 Section Review

1. Which of the following is the most important and consistently needed maintenance task for unitary retail refrigeration equipment?
   a. Cleaning the coils
   b. Greasing fan motors
   c. Changing compressor oil
   d. Checking the refrigerant charge

2. What substance is an effective antibacterial and demineralizing agent and is often dispersed into the water supply of an ice machine?
   a. Bleach
   b. Ozone
   c. Silver
   d. Fluoroquinolones

Module 03304 Review Questions

1. Which temperature classification does a system fall into if it maintains a space temperature of 25°F (–4°C)?
   a. High temperature
   b. Medium temperature
   c. Low temperature
   d. Ultralow temperature

2. After refrigerant vapor has cooled to the condensing temperature in the condenser, what happens while the vapor fully condenses to its liquid state?
   a. The liquid refrigerant boils as it forms.
   b. The condensing pressure consistently decreases.
   c. The refrigerant remains at its condensing temperature.
   d. The refrigeration oil vaporizes while the refrigerant condenses.

3. A technician can determine if all refrigerant in a condenser has changed to its liquid state by ______.
   a. the presence of superheat
   b. the presence of subcooling
   c. a sudden increase in pressure
   d. a sudden decrease in pressure

4. When a system operates at low superheat values, it is important to provide adequate protection against ______.
   a. contamination
   b. liquid floodback
   c. compressor overheating
   d. compressor current overload

5. The timer for a defrost period should be set ______.
   a. to at least six defrost cycles per day
   b. for longer defrost times in dry climates
   c. to coincide with the fixture's lightest period of usage
   d. to coincide with the fixture's heaviest period of usage

6. Ice merchandisers using cold-wall construction are defrosted ______.
   a. manually
   b. with hot gas
   c. twice each day
   d. with electric heaters

7. Which of the following components is *not* part of a unit cooler?
   a. Evaporator coil
   b. Condenser coil
   c. Metering device
   d. Evaporator fan(s)

8. In commercial ice makers, the target temperature for forming ice on a surface is roughly ______.
   a. 32°F (0°C)
   b. 10°F (–12°C)
   c. 0°F (–18°C)
   d. –10°F (–23°C)

9. How does a cubed-ice machine typically loosen ice from the grid in which it is formed?
   a. Vibration
   b. Hot water
   c. Hot refrigerant gas
   d. Electric heating elements

10. The primary function of the suction accumulator is to ______.
    a. store refrigerant for system operation at heavy load
    b. remove and store refrigerant oil from the suction line
    c. capture liquid refrigerant to prevent compressor damage
    d. prevent refrigerant vapor formation during a defrost cycle

11. Allowing the head pressure to get too low in low ambient conditions can lead to ______.
    a. a frozen metering device
    b. the compressor overheating
    c. damage to an air-cooled condenser
    d. insufficient flow through the metering device

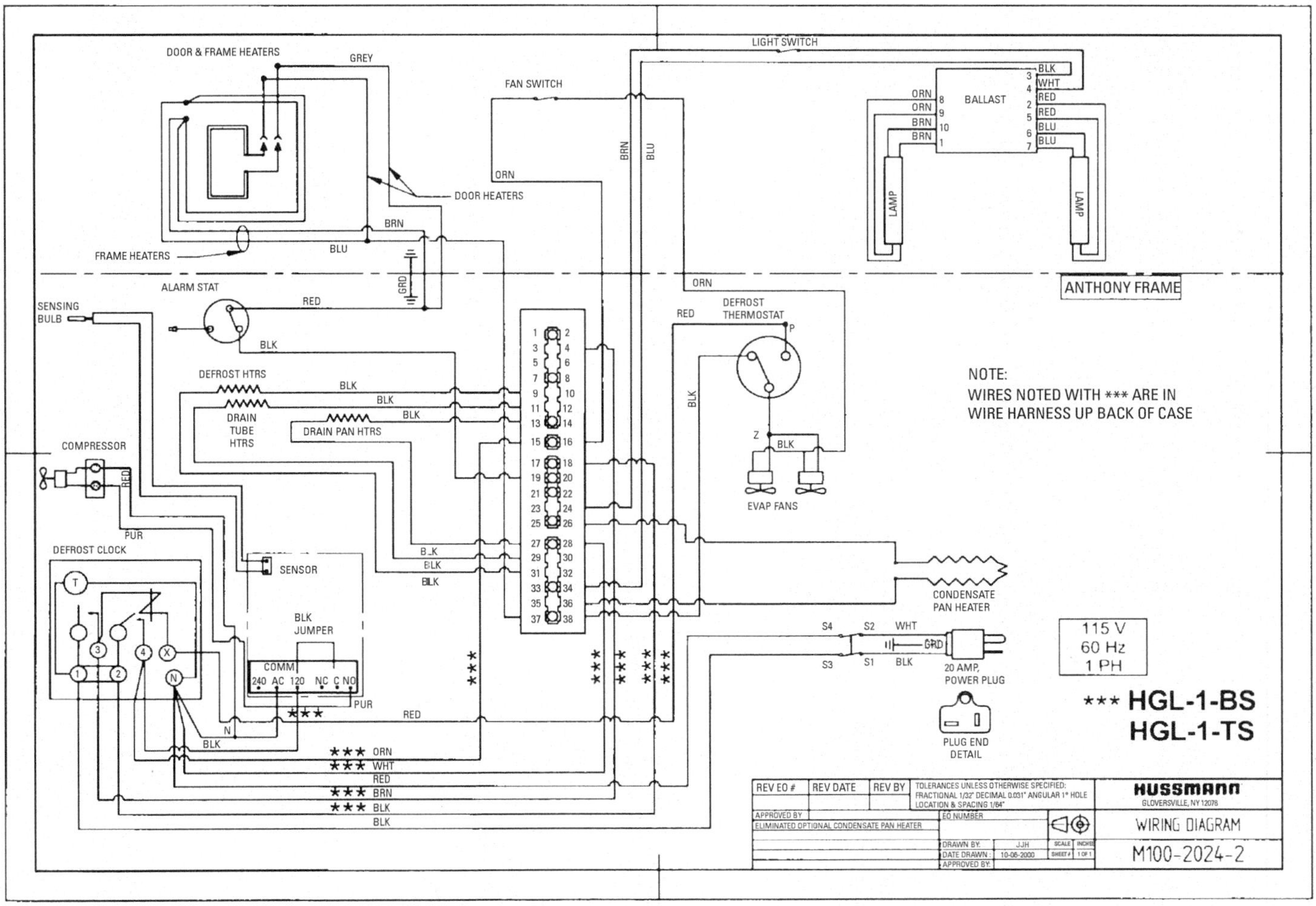

Figure RQ01
Source: Hussmann Corporation

12. Refer to *Figure RQ01*. When the Fan Switch is closed, what other device controls power to the evaporator fans?
    a. Light switch
    b. Defrost timer
    c. Alarm thermostat
    d. Defrost thermostat

13. Which of the following statements regarding maintenance of a cubed-ice machine is *not* correct?
    a. Ice machines are best cleaned annually.
    b. Proper cleaning requires significant disassembly of all components in direct contact with water.
    c. Components should be cleaned and soaked in fresh cleaning solution for 15 to 20 minutes.
    d. Sanitizers must be used according to the cleaning product and equipment manufacturer's instructions.

Indigo Sequence of Operation
Air & Water-Cooled and Traditional Remotes
Energized Parts Chart

⚡ = Energized Part

Ice Making Sequence of Operation	Control Board Relays					Contactor		Length of "On" Time
	Water Pump	Harvest Valve(s) **HPR Solenoid** AirPump(s)*	Water Inlet Valve	Water Dump Valve	Contactor Coil **Liquid Line Solenoid**	Compressor	Condenser Fan Motor†	
Start-Up 1. Water Purge	⚡			⚡				45 Seconds
2. Refrigeration System Start-up a. Equalization		⚡						5 Seconds
b. Compressor Start		⚡			⚡	⚡	⚡	5 Seconds
Freeze Sequence 3. Prechill			May Cycle On / Off During Pre-chill		⚡	⚡	⚡	Initial Start-Up Is 120 Seconds 30 Seconds Thereafter
4. Freeze Six-Minute Lock-In on Every Cycle (Max Freeze 35 Min)	⚡		Cycles Off Then On Two More Times		⚡	⚡	⚡	Until Ice Contact with Ice Thickness Probe
Harvest Sequence 5. Water Purge	⚡	⚡		⚡	⚡	⚡	⚡	Factory Set at 45 Seconds
6. Harvest a. Normal (0–3.5 Minutes)		⚡			⚡	⚡	⚡	Bin Switch Activation
b. Water Assist (3.5–6.5 Minutes)	⚡	⚡	Energizes and Fills to the High WLP		⚡	⚡	⚡	
c. Water Dump (6.5–7 Minutes) *Max Harvest 7 Minutes*	⚡	⚡		⚡	⚡	⚡	⚡	
Bin Switch Open 30 Seconds 7. Automatic Shut-Off								Until Bin Switch Recloses & 3 Min. Delay

*Not Used On All Models
** Remotes Only

† Condenser Fan Motor: Self-contained Air-Cooled - The fan motor is wired through a fan cycle pressure control; therefore, it may cycle on and off. Remotes do not have a fan cycle pressure control with the exception of the JC995, which does have condenser fan cycling control.
♦ 3 Minute Delay for Restart

Figure RQ02
Source: Manitowoc Ice

14. Refer to *Figure RQ02*. The ice machine compressor is de-energized during which operating event(s)?
    a. Freeze prechill cycle
    b. Harvest water purge cycle
    c. Initial water-purge cycle and the harvest water-purge cycle
    d. Initial water-purge, equalization, and automatic shut-off cycles

15. The cuber used as an example in the text periodically removes ice that forms on the back of the evaporator plate using ______.
    a. hot refrigerant gas
    b. a warm water rinse
    c. a lengthy off cycle
    d. electric heating elements

Answers to odd-numbered questions are found in the Review Question Answer Keys at the back of this book.

prepares you to install and service the related systems.

Answers to Section Review Questions

Answer Section Objective

1.0.0 HVAC Zoning Systems

Performance Task

1. Install and start up, or troubleshoot and repair, a residential zoning system.

Objective

Describe common zoning systems and the approach to troubleshooting.

a. Describe common zoning system components.
b. Recount the sequence of operation for common zoning systems.
c. Explain how to test and troubleshoot zoning systems.

A conventional HVAC system maintains the desired conditions in a home or business by answering to a single thermostat. The system has no idea of the conditions in other areas. The system can respond only to the conditions at the thermostat, which is typically placed in a central location. This often results in areas that are too hot or too cold, even with a properly balanced system. During a typical 24-hour cycle, the load in various areas can change dramatically as the sun moves across the sky, or as internal loads cycle on and off.

A zoning system allows a single heating and cooling system to respond to two or more thermostats, providing conditioned air only to areas (zones) that need heating or cooling. This enables a single system to precisely accommodate the needs of the structure while conserving energy.

For example, the bedrooms on one side of a home may form one zone, bedrooms on the opposite side form a second zone, and the living area in the central part of the home form a third zone. Based on their location, the rooms in each zone have similar load profiles. If the bedrooms are unoccupied all day, the temperature can be set up / set back in those two zones. The system can then focus primarily on providing conditioned air to the living area. At night, the temperature in the living area can be set up / set back while the system focuses on comfort in the bedrooms.

1.1.0 Zoning System Components

A typical zoned system (*Figure 1*) consists of an electronic control panel that is connected to several zone thermostats and multiple motorized dampers. Each zone thermostat reports its status to the control panel and calls for cooling or heating as needed by energizing the G, Y, and W terminals. The zone damper, which is installed in the supply duct serving each zone, controls the flow of supply air to that zone. The control panel is also connected to the building's heating/cooling equipment.

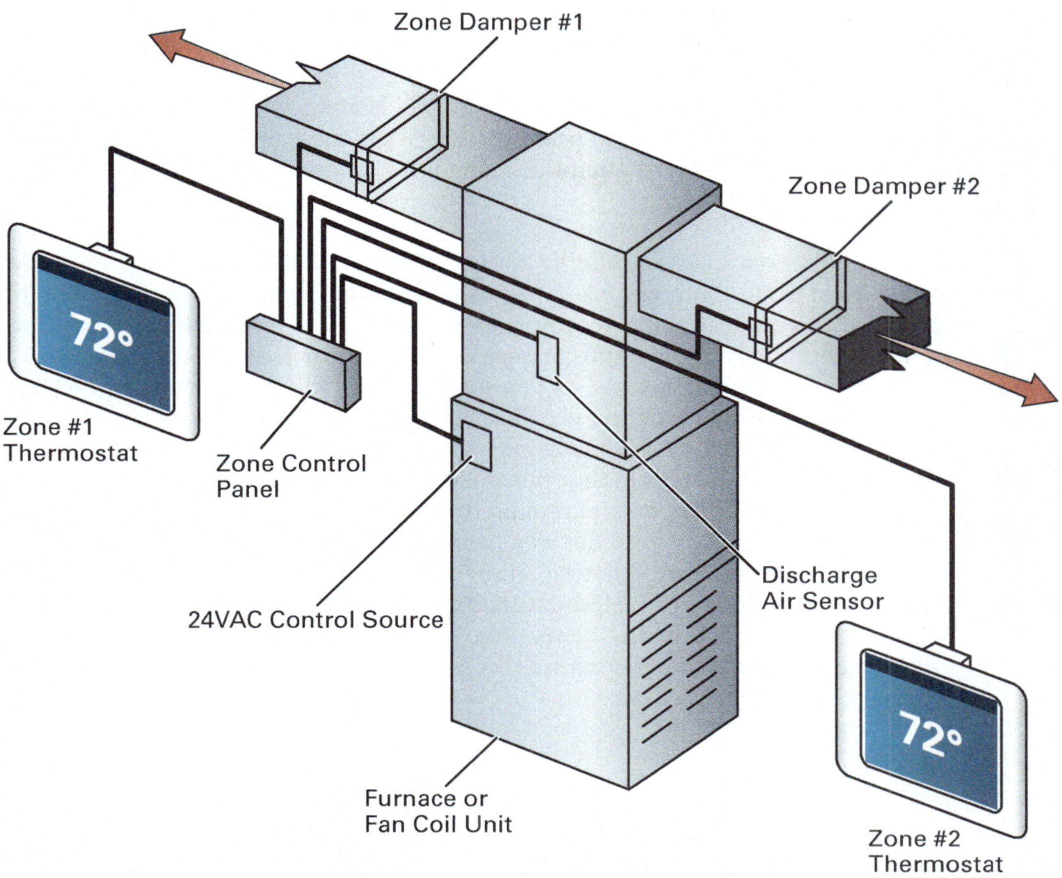

Figure 1 Basic zoned system schematic.

1.1.1 Zone Thermostats

Zone thermostats (*Figure 2*) look like any other thermostat. Indeed, most thermostats designed for conventional use can also serve as a zone thermostat. Programmable models allow the user to set schedules for setpoint changes in the zone. For some zoning systems, the thermostat selected must be directly compatible with the HVAC equipment. For example, if a heat pump is used, each zone thermostat must be compatible with a heat pumps. However, this is not true for every system.

Some systems do not offer the means to change the operating mode (heating versus cooling) from all zone thermostats. Instead, one thermostat is designated as a master thermostat (*Figure 3*), and only the master thermostat provides

Figure 2 Programmable zone thermostat.
Source: Image provided by Resideo

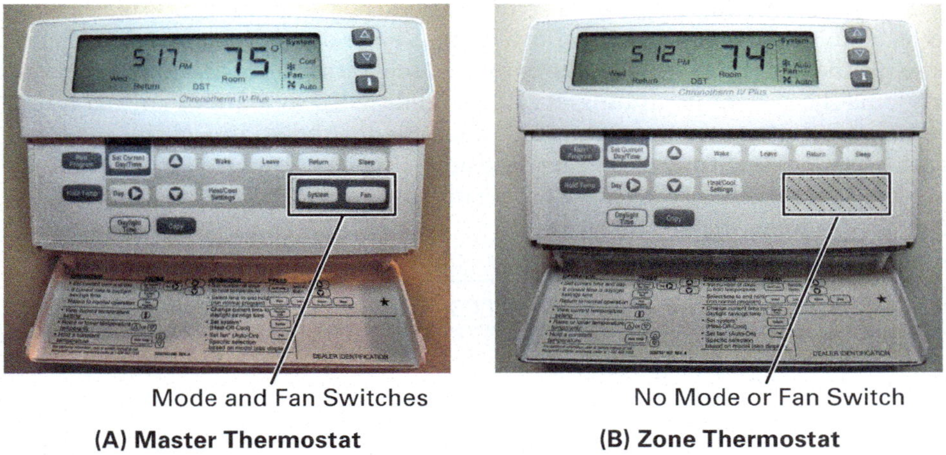

Figure 3 Master thermostat and matching zone thermostat.

mode-switching capability. The system will operate in the heating or cooling mode, or it can be set for auto-changeover if the master thermostat is equipped with this function.

When a zone calls for heating, for example, heating will only be provided if the master thermostat is set to heating. Otherwise, the system does not respond. However, many of today's systems offer mode switching at every zone thermostat.

Figure 4 Round zone damper.
Source: Image provided by Resideo

1.1.2 Zone Dampers

Zone dampers are rectangular or round (*Figure 4*) and are available in many sizes. Most zone dampers are two-position models. The damper actuator is powered to one position, typically closed, and are equipped with springs that return the damper to the opposite position when the actuator is de-energized.

This may seem backwards, as you might expect the damper to be powered open when heating or cooling is called for. However, by maintaining power on the damper when it is closed and using a spring to open it when it is de-energized, the damper naturally fails to the open position. If the actuator or its power source fails, the damper will open and remain in this position. This ensures that a zone can still receive air in cold or hot weather despite a damper or wiring failure.

Some dampers are modulating, allowing them to stop at any point between open and closed (*Figure 5*). Modulating dampers must be positioned by an energized actuator in both directions. Two-position dampers require two wires while modulating dampers require three wires. *Figure 6* shows the features of a typical zone damper actuator.

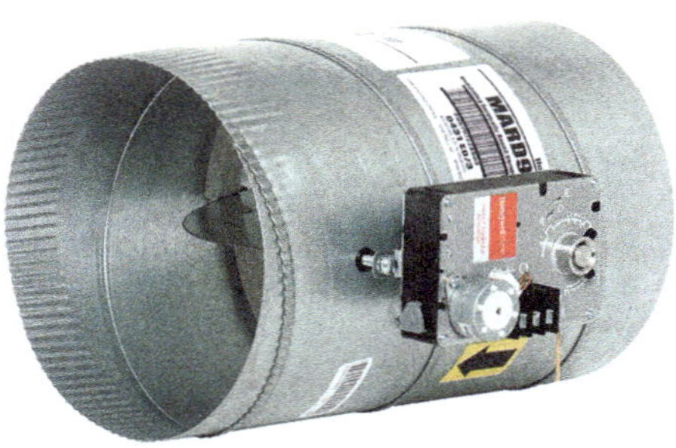

Figure 5 Modulating damper.
Source: Image provided by Resideo

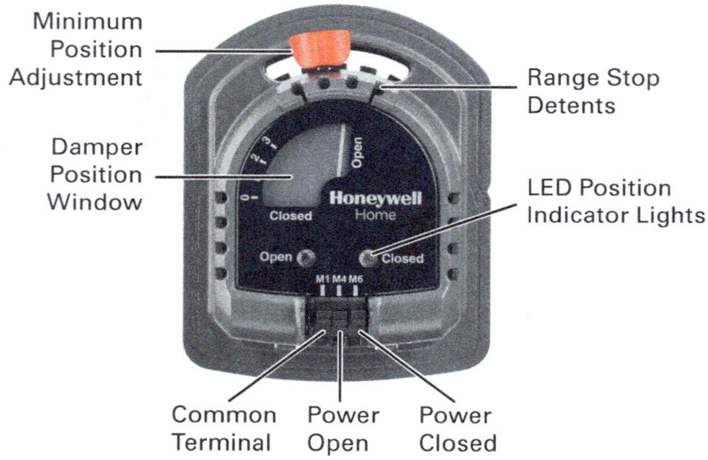

Figure 6 Typical zone damper actuator features.
Source: Image provided by Resideo

Many dampers can be adjusted so that they do not fully close. This is referred to as a *minimum position setting*. Preventing a zone damper from fully closing ensures that the zone will always receive a small amount of air for ventilation any time the system is running. It can also reduce the amount of air that must be bypassed when zones are satisfied. The actuator shown in *Figure 6* has four positions including fully closed. The red adjusting lever is moved to the desired minimum position. *Table 1* shows the relative air volume that will pass through the damper for each given position.

TABLE 1 Air Flow Settings

Range Stop Setting	Bleed Rate
0	0%
1	20%
2	30%
3	50%

Figure 6 also shows the wiring terminals. Although this is a two-wire damper, three terminals are provided for flexibility. The common terminal must always be connected to the actuator. Terminal M6 is used for the second wire when the damper will be powered closed, while terminal M4 is used if the damper will be powered open.

In addition to the window that shows the position of the damper blade relative to the duct, two LED lights are also provided to indicate when the damper is fully open or fully closed. The lights allow a quick check of position from a distance.

1.1.3 Bypass Dampers

Moving enough air volume through a heating and cooling system is very important to the equipment. A certain amount of airflow must be maintained across the heat exchanger of a furnace, for example, to prevent it from overheating.

A certain amount of airflow is also needed to prevent frost from developing on the evaporator coil in the cooling mode. A sufficient cooling load must exist to vaporize the liquid refrigerant entering the coil. This is especially true of systems using a fixed restrictor for a metering device. For heat pumps operating in the heating mode, insufficient indoor airflow can cause excessive head pressure. The issue of airflow across coils and/or heat exchangers is clearly a significant concern in zoning systems.

An increasing number of indoor blowers are equipped with electronically commutated motors (ECMs) that modulate their speed and therefore the air volume. As zone dampers close, the motor can simply slow down to maintain a consistent static pressure in the duct. However, the furnace heat exchanger or indoor coil still require a minimum air volume for the system to function properly.

As zones become satisfied and zone dampers close, the total amount of airflow is naturally reduced. However, the airflow through the furnace or air handling unit must remain reasonably consistent, without overheating or overcooling satisfied zones. Small fluctuations are acceptable, especially when a thermostatic expansion valve (TXV) is used as a metering device. Furnaces generally have at least a 30°F (17°C) range of acceptable temperatures rise, but reductions in airflow that increase the temperature rise outside of this range cannot be tolerated.

In typical zoning systems, a bypass damper (*Figure 7*) is used to maintain consistent airflow through the equipment. The bypass damper is arguably the

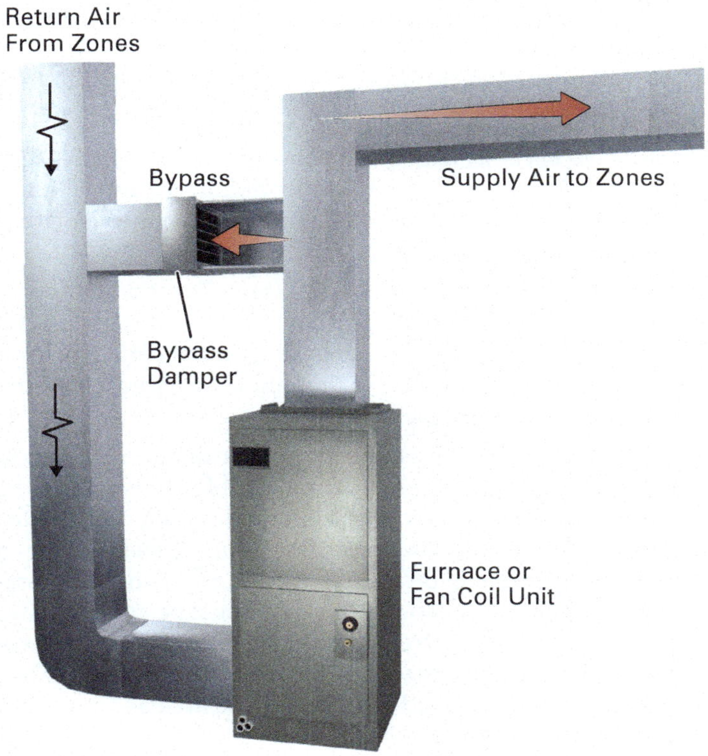

Figure 7 Bypassing air directly to the return duct.

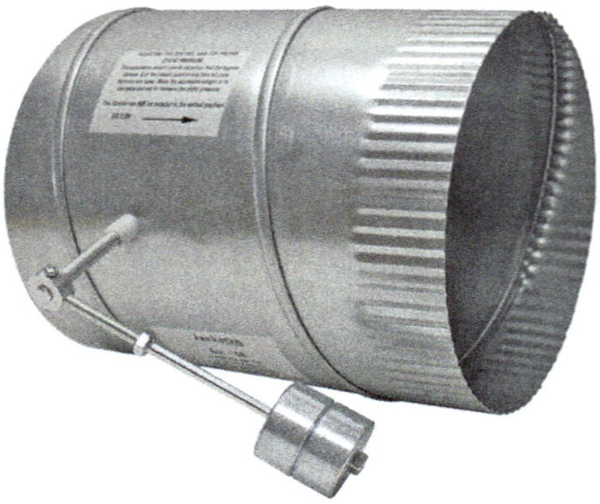

Figure 8 Barometric bypass damper.
Source: Jackson Systems, LLC.

Figure 9 Constant-pressure bypass damper.
Source: Image provided by Resideo

most important damper in a zoning system. A bypass damper is often installed in a short section of duct that connects the supply duct directly to the return duct. To maintain a consistent volume of airflow through the equipment as zone dampers close, the air must have a path around the closed dampers. Since the number of closed zone dampers is unpredictable moment to moment, the volume of air that needs to be bypassed also changes.

While zone dampers are usually two-position models (fully open or fully closed), bypass dampers must modulate their position to match current conditions. This means the bypass damper can stop at any point between fully open and fully closed to satisfy the needs of the system.

Three types of bypass dampers are commonly used: modulating motorized dampers (*Figure 5*), *barometric dampers* (*Figure 8*), and *constant pressure dampers* (*Figure 9*). The modulating damper must be positioned by a floating electrical input provided by a static pressure control that monitors pressure in the supply duct (*Figure 10*). As the pressure increases to the static pressure setpoint, the control signals the bypass damper to open as needed to maintain the pressure.

The barometric damper is often chosen as a cost-cutting measure because it does not require electrical connections or control wiring; it is self-contained. The barometric damper's balancing linkage is adjusted to open the damper as the system static pressure increases. Adjustments are made by sliding the weight along the length of the rod while monitoring the duct static pressure. Due to the influence of gravity on the weight and linkage, this damper cannot be installed in a vertical position. In addition, although the arrangement can be functional, barometric dampers deliver less than ideal performance and rarely accommodate variations in pressure effectively.

The constant-pressure damper shown in *Figure 9* is also self-contained, like a barometric damper; no power source is required. However, a carefully calibrated spring and mechanical linkage allows you to set the desired static pressure at the damper. This allows it to accommodate changing pressures better than a barometric damper but without the need for an external static pressure control.

The use of a conventional bypass damper is being reconsidered in the industry. When a great deal of air is consistently being bypassed, energy efficiency suffers. Ideally, a zoning system relies on other methods of managing airflow. However, the more zones a system generally has, the more likely a bypass damper will be needed.

Other airflow management techniques that can be used together to reduce or eliminate the need for a bypass damper include the following:

- *Oversized branch/zone ducts* — The size of zone ducts can be increased to allow zones to accommodate more air as other zone dampers close. While this may help, it has limited potential and can lead to noise when the velocity increases through the registers and diffusers.
- *Zone damper minimum position settings* — You can adjust the damper actuators to stop at a minimum open position, forcing satisfied zones to accept some airflow. Although this is helpful, it can lead to overheating or overcooling of zones. A minimum position setting that allows between 10% and 25% of design airflow for the zone is generally chosen.

Dump zone: An area where excess supply air from a zoning system is discharged to relieve excess pressure in a zoned air distribution system. The dump zone is typically located in the conditioned space, but in an area where some overheating or overcooling is not a problem.

- **Dump zone** — A dump zone refers to a room(s) or diffuser(s) without a zone damper, meaning that it receives air any time the blower is operating. A *managed dump zone* is a variation that does use a modulating damper to manage airflow through the bypass damper based on system static pressure (*Figure 11*). This is preferable to an unmanaged dump zone, allowing control over the system static pressure. The dump zone must be an area that does not require precise temperature control, such as a basement, large foyer, or hallway. In a commercial building that uses the area above a ceiling as a return air plenum, the air can be dumped into the ceiling area as far away from the unit's return air inlet as possible.

Figure 10 Bypass-damper static pressure control.
Source: Image provided by Resideo

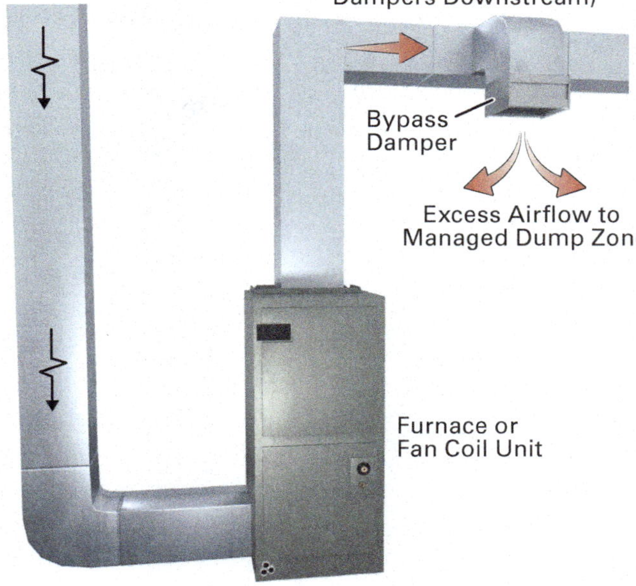

Figure 11 Managed dump zone application.

One significant advantage to using a dump zone is that the return air temperature to the furnace or coil does not change dramatically. When supply air is bypassed directly from the supply duct to the return duct, the return airstream is either heated or cooled. This can quickly overheat a furnace or cause an evaporator coil to develop frost. Air sent to the dump zone mixes with room air on its journey back to the blower. A return air grille is normally located in a dump zone area to avoid pressure buildup in the space.

The disadvantage of a dump zone is that the area where the air is discharged can become too hot or too cold. Not every installation scenario has an area that can be used as a dump zone.

Equipment Protection

When supply air is bypassed directly to the return duct of the furnace or fan coil unit, the heating or cooling equipment is immediately affected. For example, if a gas furnace is discharging 120°F (49°C) that is mixing with normal return air at 70°F (21°C), the mixed air temperature will likely be higher than 85°F (29°C). As a result, the furnace may cycle off and on due to the high-temperature limit switch. You should never allow a furnace to operate by allowing it to cycle on a primary safety control.

In some zoning installations, excessive temperature swings due to a bypass damper are unavoidable. To minimize the effect, a direct bypass is usually designed to handle only about 25% of the total system air volume. In addition, equipment protection is provided by using a duct temperature sensor (*Figure 12*), allowing the furnace's primary safety controls to provide final protection as designed.

The sensor is connected to the zoning panel and monitors the duct air temperature. It will cycle the equipment off when it reaches preset low and high setpoints. The equipment restarts once the duct temperature drops below the differential setting. As a result, the equipment may cycle on and off often when a single zone requires heating or cooling.

For cooling equipment, an additional temperature sensor can be used to cycle the condensing unit off when evaporator coil freezing is imminent (*Figure 13*). The thermostat shown here is mounted on the suction line as it leaves the evaporator coil. The device breaks the control circuit when the temperature falls below 38°F (3°C) and recloses its contacts when the temperature rises to 48°F (9°C).

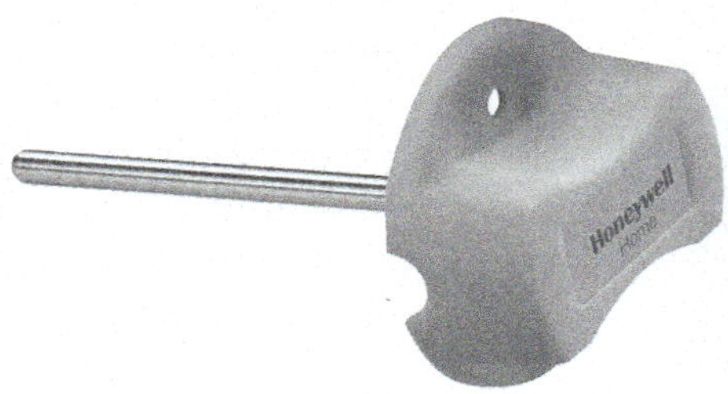

Figure 12 Discharge air temperature sensor.
Source: Image provided by Resideo

Figure 13 Freeze-protection temperature switch.
Source: Image provided by Resideo

On Site

HVAC Equipment for Zoned Systems

A single-speed compressor and a common multispeed indoor blower deliver a fixed amount of air and cooling capacity, regardless of the number of zones calling for conditioned air. If all the zones are calling for air, the system works fine. If only one or two zones call for conditioned air, the excess capacity must be handled in some way, normally with a bypass damper for the air and discharge air temperature protection for the cooling equipment.

Today's multispeed and inverter-driven compressors, two-stage and modulating gas furnaces, and variable-speed ECM blower motors can respond much better to a reduced load. When few zones call for conditioned air, the system can respond by reducing compressor speed, lower the gas input, and/or reduce the blower speed. The result is better indoor comfort, increased energy efficiency, and a quieter system.

1.1.4 Zoning Control Panels

The zoning control panel (*Figure 14*) provides overall system management. It serves as the point of connection for the thermostats, dampers, sensors, and HVAC equipment it controls. Although zoning control panels can vary significantly in their layout, most provide the same basic connections as shown for this panel in *Figure 15*.

Figure 14 TrueZONE® zoning system control panel with accessories.
Source: Image provided by Resideo

Zoning control panels are selected according to the application requirements. Manufacturers offer panels that can control a finite number of zones. For example, a two-zone panel may be appropriate for a simple two-story home. Another panel may handle up to six zones. Panels that can accommodate as many as 20 zones are available, but the need for so many zones is generally precluded by using multiple HVAC systems.

Note that panels capable of handling some number of zones do not need connections at every zone output. A six-zone panel, for example, can function just fine with fewer zones connected.

The type of HVAC equipment to be controlled is the second consideration. Standard, single-stage heating and cooling equipment can be handled by the simplest of panels. Multistage systems require a more robust panel with additional programming options. The panel shown in *Figure 14* can control up to

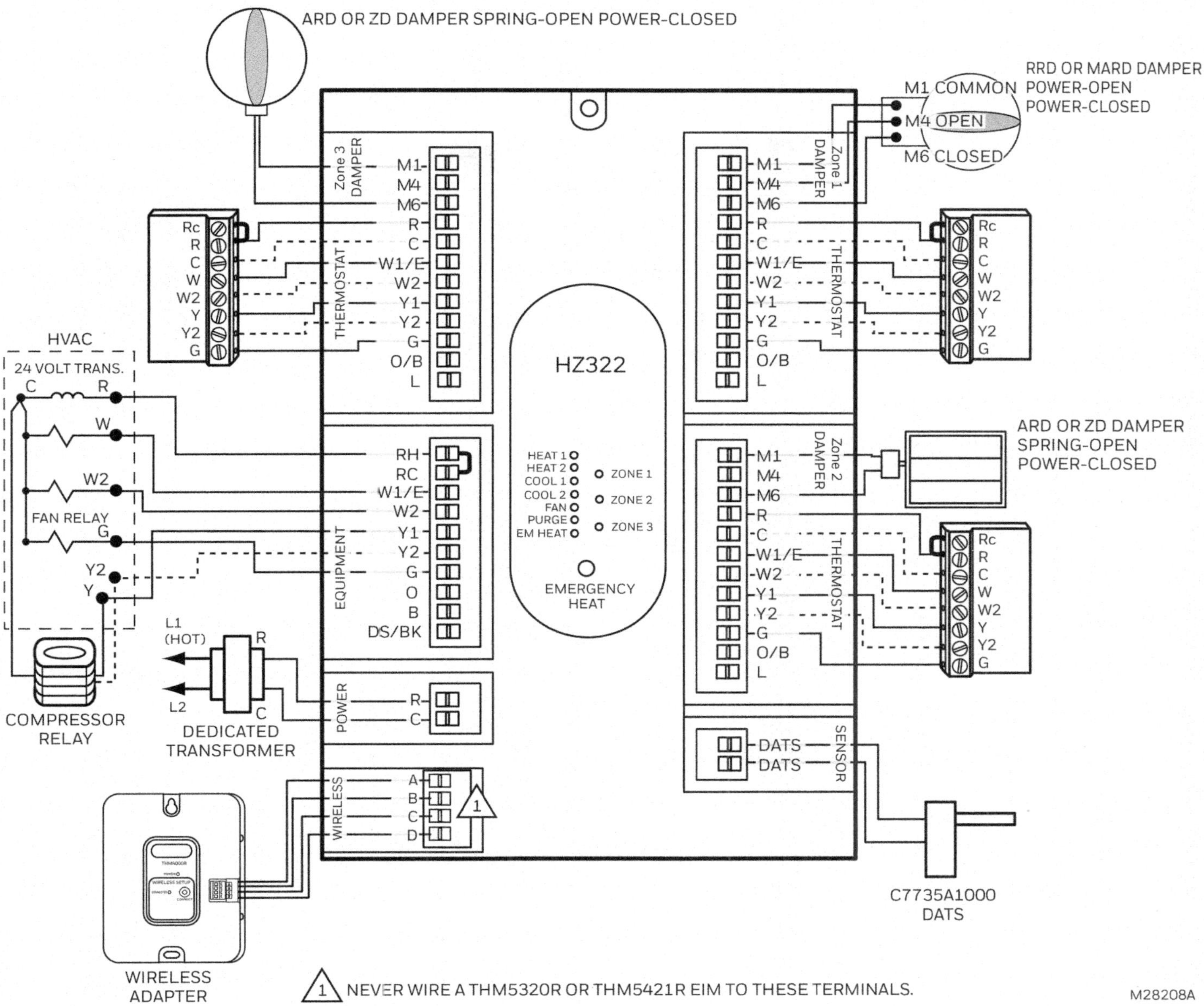

Figure 15 TrueZONE® zoning panel connections for conventional equipment.
Source: Image provided by Resideo

two stages of cooling and two stages of heating, as well as heat pump systems that require reversing valve and electric heat control. The panel is set up for the equipment configuration during the installation and commissioning process. Additional features potentially needed in a panel include blower-motor speed control based on demand, wireless capabilities, and dual-fuel system control.

The control panel must also be compatible with the dampers that will be used. Zone dampers that are powered closed and spring open when de-energized are often referred to as *two-wire dampers*; only two wires are required to operate them. Zone dampers that modulate their position require three wires. The panel shown in *Figure 14* and *Figure 15* can accommodate both three-wire dampers and two-wire dampers regardless of whether the dampers are powered closed or powered open.

One other option available in zoning systems is wireless communication. Wireless dampers, thermostats, and panels are available. However, dampers still require a low-voltage power source, meaning that some wiring is still required to power the damper and wireless receiver. Although using wireless components might seem like a better and more modern approach, the need for wiring to provide power for the dampers negates a great deal of the perceived advantage.

On Site

Zone Dampers for Existing Systems

Existing systems can often be retrofitted for zoning. However, installing common zone damper assemblies in existing systems can be labor-intensive. There are also times when a stand-alone installation of one or two dampers directly controlled by a thermostat can solve a comfort problem.

The damper shown here is specifically designed for quick installation in existing rigid round duct branches. The damper is compatible with zone panels but can function alone with a compatible thermostat and a control transformer for power.

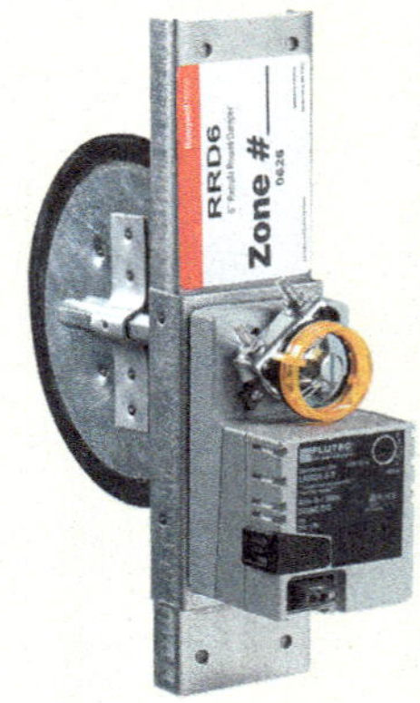

Source: Image provided by Resideo

1.1.5 Installation Considerations

Zoning system components are designed to operate reliably in a wide range of temperatures. Manufacturers understand that the components are often placed in attics or other unconditioned spaces that can reach very high or very low temperatures. Ensure that each component chosen for the system is rated for the expected temperature and humidity conditions.

Most panels require a 24VAC power supply to power both the panel and all connected dampers. Zone panels are usually installed near the indoor furnace or fan coil unit, and the line voltage power supply serving the unit is used to power the zoning system transformer. The transformer serving the equipment should not be used to power the zoning system.

The capacity of the transformer, in volt-amps (VA), must be considered. All the connected zone dampers as well as the panel can potentially operate at the same time. Thus, the rated load of the panel must be added to the sum of the rated loads for all connected dampers to determine the total VA load. Other accessory loads, such as a wireless adapter, must also be considered. Select a transformer that can provide the necessary current with a reasonable amount left over.

Being largely electronic, zone panels can generally be mounted in any position. Zone panels should always be mounted neatly and positioned straight and level for a professional appearance.

Zone dampers are typically rated for a load of 10VA at 24VAC, or about 0.4A. Thermostats represent an even smaller load as they are primarily switching devices. This allows 18 AWG solid thermostat wire to be used for most external wiring. Today's panels rarely have screw terminals, and instead rely on push-to-connect and similar wiring terminations. Stranded wire should not be used with these terminals.

Before installing any zoning panel or device, review the installation instructions carefully and follow the guidance relevant to the system being installed. Manufacturers make a significant effort to produce products and instructional materials that lead to a positive installation and user experience. Make use of these materials for the best possible results.

1.2.0 Zoning System Operation

Many variations are available in zoned systems, but the principles of operation are largely the same. Each zone damper is typically controlled by a dedicated thermostat that determines the position of the damper. The panel powers the damper actuator to the needed position and controls the HVAC equipment as needed to satisfy the demand. When all zones are satisfied and the HVAC equipment is idle, the damper actuators are de-energized, and springs return the dampers to their open position.

NOTE

Remember that dampers that are powered open and close when de-energized are also available and will occasionally be encountered.

When a single zone thermostat calls for heating or cooling, the control panel starts the equipment and powers the dampers serving all the satisfied zones to the closed (or minimum) position. The zone damper for the calling zone remains de-energized and open. When the zone thermostat is satisfied, all dampers are de-energized and return to the open position.

Indoor blower operation may be allowed to continue for a preset period to circulate some air in all zones. In addition, this strategy makes effective use of boiling refrigerant remaining in the evaporator coil or heat stored in the heat exchanger of a furnace. This is often referred to as a *purge cycle*.

It is very important to read the manufacturer's literature for each system thoroughly before installing or troubleshooting a zoned system. Never assume that one system works the same as another. Even minor differences in the sequence of operation can lead a technician to the wrong conclusion.

Before a zoning panel and system are placed in service, you must review the configuration carefully and ensure that each setting meets the needs of the environment and connected equipment. An example of a zoning panel configuration menu is shown in *Figure 16*.

Menu Name	Menu Title (LCD Top Line)	Menu Options (LCD Bottom Line; Defaults in Bold)	Menu Option Description	Notes:
Heat Fan	Heat Fan Control	**[HVAC]** Panel	Fan Control by HVAC	In heat mode, fan controlled by HVAC or turned on by panel on a call for heat.
		HVAC **[Panel]**	Fan Control by Panel	
Stage 2 Timer	Stage 2 Timer	**[5 Min.]** > — < [60 Min.] **[5 Min.]**	5 Minutes—60 Seconds	Number of minutes to delay before engaging second stage.
Purge Timer	Purge Time	**[2]** 3.5 5 Min.	2.0 Minutes	Number of minutes panel will purge following call for heat or cool.
		2 **[3.5]** 5 Min.	3.5 Minutes	
		2 3.5 **[5]** Min.	5.0 Minutes	
Purge Fan	Fan in Purge	**[HVAC]** Panel	Fan Control by HVAC	Fan controlled by HVAC or panel during purge.
		HVAC **[Panel]**	Fan Control by Panel	
Purge Dampers	Purge Dampers	**[Unchanged]** Open	Dampers Unchanged	Damper position unchanged or all dampers open during purge.
		Unchanged **[Open]**	Dampers All Open	
Auto-Changeover Delay	Changeover Delay	**[15]** 20 30 Min.	15 Minutes Auto-Changeover Timer	Number of minutes to delay auto-changeover when one zone is calling for heat and another is calling for cooling.
		15 **[20]** 30 Min.	20 Minutes Auto-Changeover Timer	
		15 20 **[30]** Min.	30 Minutes Auto-Changeover Timer	
DATS Enabled	Discharge Sensor	**[No]** Yes	Disabled	Enables or disables DATS. If disabled, the multistage DATS inhibit setting is disabled.
		No **[Yes]**	Enabled	
DATS High Limit	DAT High Limit	[110°F] >—< [180°F] < **[160°F]** >	110°F—180°F	For furnace systems, high limit should be set approximately 15° below the furnace high-limit setting.
DATS Low Limit	DAT Low Limit	[30°F] >—< [60°F] < **[40°F]** >	30°F—60°F	Low-temperature limit.
OT Temp Enabled*	OT Sensor	**[No]** Yes	Disabled	Enables or disables wireless outdoor temperature sensor. If disabled, the multistage OT temp lockout setting is disabled.
		No **[Yes]**	Enabled	
OT Trip Point for Multistage Lockout	OT Lockout (Temp)	[0°F] >—< [50°F] < **[50F]** >	0°F—50°F	Above this temperature, 2nd stage heat is locked out.
Multistage DATS Inhibit	DAT MSTG Inhibit	No **[Yes]**	Disabled	Allow panel to downstage multistage equipment when near DATS high or low limit.
		[No] Yes	Enabled	
LCD Contrast Adjust	LCD Contrast	[1] >—< [10] < **[5]** >	Contrast Value 1—10	Sets LCD display contrast for ease of viewing. Lowest contrast is 1, highest is 10.
Save Changes	Save Changes?	**[No]** Yes	Disabled	Saves or rejects the configuration settings.
		No **[Yes]**	Enabled	

Figure 16 Example zoning panel configuration menu.

Source: Image provided by Resideo

On Site

Additional Heating and Cooling

In many homes, there is one room or area that always seems to be a bit too cold or too hot when the rest of the home is comfortable. The hotpod™ is one way this problem can be solved. Equipped with an electric heating element, a small fan, and a dedicated wall thermostat, the device can boost cooling airflow and add additional heat to a cold space. Both wired and wireless control are options.

Two sizes are available—a 6" model with 1,000 watts of electric heat and an 8" model with 1,440 watts of electric heat. Both operate on a 120VAC power supply.

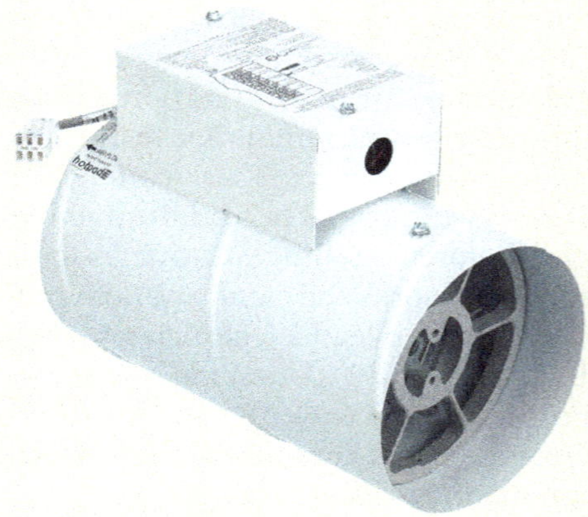

Source: TPI/HotPod

The hotpod™ is not connected to a zoning panel; it is a standalone device. In fact, using it with a zone system could create some problems. If the zone damper for the area the hotpod™ serves is closed, then its source of air is cut off. Although there would be little harm in the cooling mode, the electric heater would overheat without airflow across it, cycling off due to the high-limit switch.

1.2.1 Ventilation Modes

Each thermostat in most zoning systems is equipped with both fan and system mode switches. Any of the zone thermostats can start the indoor blower by turning the fan switch from Auto to On. The zone damper for any zone thermostat fan switch set to On will remain open and allow ventilation air to enter. Any zone thermostat with the fan switch still in the Auto position will power the zone damper closed during the ventilation mode.

However, this feature is often adjustable. Panel setup during the installation often offers the option of opening all the zone dampers in response to a call for ventilation.

If any of the zones call for cooling or heating while the ventilation mode is functioning, the system will respond as it normally would by closing the dampers for zones that do not need conditioned air. In other words, even with the fan switch set to On for ventilation, a zone's damper will close (or move to its minimum position setting) if unneeded conditioned air is being provided, despite the ventilation setting. Once the heating or cooling demand is satisfied, the damper serving the zone with the fan switch set to On will reopen to restore ventilation airflow.

1.2.2 Heating and Cooling Control

Different zones often have different needs at any given time, which is one of the primary reasons a zoning system makes sense. Sometimes one zone will request cooling while another requests heating. Generally, the zone calling first takes priority (first come, first served). If the first zone to call requests heating, the zone panel will start the appropriate equipment to satisfy the heating demand first. Once that zone is satisfied, the system will change modes and satisfy zones that are requesting cooling.

Some systems have programming that allows for mode control to be shared without conflict. For example, assume a call for both heating and cooling from different zones is made at nearly the same time, but the cooling request was made first. The cooling equipment will start, and the dampers for all zones that do not need cooling will close. Without the ability to share operating time, this status could continue indefinitely—the zone needing cooling will dominate until it is satisfied. This scenario is easily created by a user that has set a zone for an unachievable cooling temperature.

With programming that allows sharing, the dominant request will only be served for a fixed period; a 20-minute maximum is common. At that point, the system will stop cooling and start the heating equipment to serve the zone(s) requesting heat. If the call for heat is satisfied in less than 20 minutes, the system will switch back to cooling to continue serving the warm zone. Otherwise, it will switch back to cooling again once the maximum time limit is reached.

Some panels offer both minimum-on and minimum-off times for the cooling equipment. As you know, short-cycling HVAC equipment increases wear and tear, especially for the compressor. Cooling systems often need a 3- to 5-minute time delay before being restarted, allowing time for the refrigerant pressures to equalize. Technicians must pay attention to the setup of such systems. Thermostats often have built-in time delays for this purpose. Zone panels may also have short-cycle protection enabled, and a delay-on-break short-cycle timer may be installed in the condensing unit. The time delays are compounded as a result. With zoning systems, it is best to configure the zoning panel to provide the delay and disable the function on the individual thermostats.

1.2.3 Multistage Systems

Ideally, a zone's need for heating or cooling is taken care of by the first stage of heating or cooling, and in a reasonable amount of time. This is the energy-efficient point of having a multistage system. However, zone control panels need some sort of programming that determines when the second stage of heating or cooling should be called upon. This feature eliminates the need to use multistage thermostats in each zone.

One common approach is based on the amount of time that a call for heating or cooling remains active. The timing begins when the zone thermostat first calls for cooling or heating. After a preset period, if the zone thermostat has not been satisfied, the zone panel will start the second stage of heating or cooling. The timing is adjustable on most systems that use this approach.

Other systems allow for each zone to have different timing before the second stage of heating or cooling is called upon. For example, a living-area zone could be set up to allow a 5-minute run time before a second stage is energized. Another zone may then be set up to allow 11 minutes of run time before a second stage can start. Note that these are not suggestions, but simple examples; the environment and the habits of the users must be considered.

Sophisticated systems may also require a minimum number of zones to be calling for conditioned air before a second stage is called upon. This intelligent feature allows the installer to prevent the second stage of heating or cooling to start when only one or two zones in a large system are calling, regardless of demand.

Note that the operation of additional stages is also managed by a discharge air thermostat. For example, zone panels that manage multiple stages of capacity will de-energize a second stage of cooling if the low temperature limit is reached. The first stage would only be shut down if the temperature continues to fall or later reaches the low temperature limit.

1.2.4 High- and Low-Limit Operation

Many two-zone systems do not require bypass dampers or high- and low-limits for equipment protection. However, as the number of zones increases, both bypass dampers and temperature-limiting control are usually required.

High- and low-limit switches or sensors cycle the cooling and heating equipment to prevent overheating, freezing of the coil, and refrigerant floodback. One goal is to prevent the equipment safety controls, such as a furnace high-limit switch or a condensing unit low-pressure switch, from activating and shutting the equipment down. Some equipment safety devices require a manual reset. In addition, equipment safety devices are generally designed to activate near the performance limits of the equipment.

When applying a zoning system, there is no desire to push equipment performance to a risky level. By using its own temperature switches or sensors, the zoning panel can cycle the equipment off before the equipment safety devices are activated.

When a bypass damper is partially or fully open, very cool or warm air is mixed with return air and re-enters the equipment. Monitoring the discharge air temperature allows the panel to cycle the heating or cooling equipment off until the air temperatures return to a normal level. The rest of the zoning system continues to operate normally.

Once the air temperature changes sufficiently, the equipment is cycled back on. However, equipment startup may be subject to other limits, such as the minimum-off time programmed into the panel. In other words, both time and air temperature can play a role in how long the equipment remains off.

Low limits are occasionally fixed and cannot be changed by the installer. Panels offering adjustable setpoints provide knobs (*Figure 17*) or an opportunity to change the setpoint through an installer setup menu on a clear-language display.

Figure 17 High- and low-limit setpoint adjustment on a zoning panel.

1.3.0 Zoning System Testing and Troubleshooting

Troubleshooting a zoning system is not difficult since you already understand basic control troubleshooting. However, like thermostatic expansion valves, zoning controls are often suspected or blamed for problems caused by other components. This is often due to a lack of understanding of what the system is designed and programmed to do.

To successfully troubleshoot any zoning system, it is essential that the technician study the literature for that specific system. Without a clear understanding of the sequence of operation, troubleshooting cannot be accomplished effectively.

The troubleshooting process for a non-zoned system often starts as a no-heat or no-cooling service call. For a conventional single-zone system, technicians often begin the process by determining if the primary equipment is powered. If the equipment is powered, the process moves to determining if a control voltage is provided. Once the presence of line voltage and control power are both present, the next step is to determine if a call for cooling or heating is coming from one or more zone thermostats.

This initial troubleshooting process should remain the same when a zoning system is present. The existence of a zoning system should not dramatically change the process. When there is a lack of cooling or heating throughout the building, start with those simple tests to verify the presence of power and control voltage. If the heating or cooling equipment is operating, verify that it is operating correctly before proceeding to the zoning system controls. Check the refrigerant charge, air filter condition, and other common causes for comfort complaints.

It is best to test the equipment with all zone thermostats calling for either cooling or heating. This ensures that the maximum available load is applied to the equipment as you evaluate its performance.

1.3.1 Zone Panels

After reviewing the literature for the zoning system, use a voltmeter and verify the panel is being powered and the voltage is within the panel's operating range. Many panels also have LED lights to indicate the system status, also providing an indication that the panel is energized.

Next, it is good practice to review the current programming and setpoints. Many different panels are in use, each with different features and characteristics. If the programming or setup is incorrect for the application, problems will result.

This step also provides valuable information to the technician. During the rest of the process, it is helpful to know precisely what the settings are for various functions. Use the manufacturer's literature to review the panel programming carefully. Also check the position of any dual-inline package (DIP) switches located on the panel (*Figure 18*). Many panels are configured using switches like these.

Figure 18 Panel configuration DIP switches.

For example, if a zoning panel is set up to require at least two zones calling for cooling to start the equipment, then a single zone requesting cooling will not start the condensing unit or blower. A technician could mistakenly assume that the panel is defective when it doesn't respond to a verified call for cooling from a single zone. In reality, the panel is doing exactly what it was programmed to do. Scenarios like this are why it is important for you to review and consider the panel settings.

Some panels offer a built-in testing process. *Figure 19* provides an example of a step-by-step procedure to test various functions one at a time, conducted by the panel. For example, steps six through nine cycle the zone dampers with the fan running to ensure they are operating as expected. Built-in testing features save time and ensure the process is done in an orderly manner. These processes are also helpful when troubleshooting is required.

Checkout Step		Line 1 Display	Line 2 Display	Notes:
1.	Display Shows OT*	OT Sensor VAL	Current OT Temp (Dynamic)	All zone dampers open; all other relays off.
2.	Display Shows DATS**	DAT Sensor VAL	Current DATS (Dynamic)	
3.	Heat Stages Test	Test Heat	**[Off]** 1 2	Heat turns on (fan also turns on if configured For fan on in heat).
4.	EM Heat Stages Test	Test Emerg Heat	**[Off]** 1 2	Emergency heat turns on (fan also turns on).
5.	Cool Stages Test	Test Cool	**[Off]** 1 2	Cooling turns on (fan also turns on).
6.	Fan Test	Test Fan	**[Off]** On	Fan cycles on and off.
7.	Damper 1 Test	Test Z1 Damper	**[Open]** Closed	Cycles damper position with fan on.
8.	Damper 2 Test	Test Z2 Damper	**[Open]** Closed	Cycles damper position with fan on.
9.	Damper 3 Test	Test Z3 Damper	**[Open]** Closed	Cycles damper position with fan on.
10.	View Tstat1 Inputs	Zone1 Stat Inpts	Displays active Tstat1 terminals or displays wireless thermostat operation.	Tests thermostat wiring with HVAC off.
11.	View Tstat2 Inputs	Zone2 Stat Inpts	Displays active Tstat2 terminals or displays wireless thermostat operation.	Tests thermostat wiring with HVAC off.
12.	View Tstat3 Inputs	Zone3 Stat Inpts	Displays active Tstat3 terminals or displays wireless thermostat operation.	Tests thermostat wiring with HVAC off.
13.	Exit Checkout Mode?	Exit Checkout?	(Next = Exit)	

* Wireless outdoor Sensor C7089R and Wireless Adapter THM4000R required for OT Temperature.
** Discharge Air Temperature Sensor C7735A required to display Discharge Air Temperature.

Figure 19 Example zone panel functional-testing menu.
Source: Image provided by Resideo

The HVAC equipment outputs at the panel are checked in the same way as any thermostat. R is the leg of control power being broken through the panel relays; it is the source of power that is fed back to the equipment via the W, Y, and G wires (as well as other wires, if present) to control system operation.

For example, if an output to start the condensing unit is expected, check for voltage between R/R_C and Y; if Y is energized, 24VAC will not be shown on the meter and the compressor contactor should be energized. Note that most boards do not have a terminal for the common lead from the transformer, like many thermostats do. The panel has no need for 24VAC from the equipment since it operates on its own low-voltage power source.

Jumper wires can also be used at the panel outputs to check the integrity of controls downstream. For example, placing a jumper wire between R and Y at the panel should start the condensing unit. If it does, all the wiring and controls downstream are functional. If outputs to the HVAC equipment are not present at the panel when they should be, and you have confirmed that any other conditions have been met, the panel is likely defective.

Remember that temperature switches and sensors can also prevent heating and cooling equipment from operating. If a failed thermistor indicates to the panel that the supply air temperature is very high or low, equipment operation will be prevented. Before condemning a panel that does not provide outputs when all the inputs and factors are correct, check the accuracy of any temperature sensors. Remember that you will need a temperature/resistance chart or graph for a thermistor to determine if its resistance is correct.

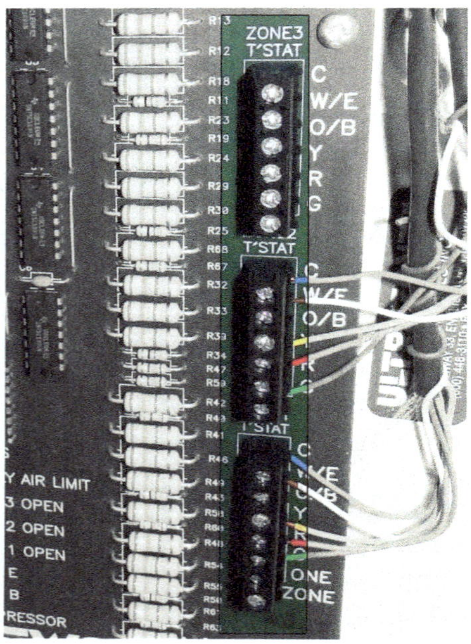

Figure 20 Zone thermostat terminal blocks.

1.3.2 Zone Thermostats and Dampers

Zone thermostats are typically wired back to the zoning control panel (*Figure 20*). The zone thermostats are tested like any other thermostat. From the panel, a technician can test to see if thermostat circuits are open or closed by measuring voltage at the appropriate terminals. If a thermostat is suspected of being faulty, it is always best to confirm this at the thermostat before replacing it. This eliminates the possibility of a problem in the interconnecting wiring.

A good way to start the process of checking thermostats and dampers is to begin with setting all zones to call for either heating or cooling. During this period, you may wish to de-energize the heating and cooling equipment. With all zones calling, evaluate system airflow and visually confirm that every zone damper is open and the bypass damper (if present) is closed. Remember that barometric bypass dampers are controlled only by weight and the applied static pressure. If barometric and constant-pressure bypass dampers are not fully closed, make the necessary adjustments.

Once proper operation has been verified with all zones calling, begin satisfying the setpoint for one zone thermostat at a time. Confirm that this damper closes fully (or to its minimum position setting). Dampers sometimes fail to open or close properly due to binding or stress placed on the housing during or after installation. The spring can only overcome so much resistance.

Continue the process for each of the connected zones. Visually confirm that each damper responds as expected.

The bypass damper may or may not move slightly as the first zone dampers close. This depends on the system layout and the effect on static pressure as each one closes. As zone dampers close, observe the bypass damper and the static pressure at the same location as the static pressure sensor, barometric damper, or constant-pressure damper. Make sure it begins to respond as the static pressure setpoint is reached. Remember that systems using a variable-speed blower motor may automatically reduce the system airflow. A bypass damper may not be necessary for these systems.

When a zone damper does not respond as expected, begin by determining whether the damper is powered open or closed. If it fails to operate in the powered direction, check for a 24VAC output to the damper from the panel. If power is present at the actuator, see if the damper is mechanically jammed or seized. If it is free to move, then the actuator has likely failed. Test it again to confirm a failure after it has been disconnected from the damper.

If the damper is not being powered by the panel, make sure that the inputs from the thermostat are correct. Before condemning a panel, confirm your findings through a second test.

If the spring is not moving the damper when the actuator is de-energized, binding is the likely problem. Ensure the damper is properly installed in the duct, no hardware can interfere with blade movement, and there is no physical damage to the duct walls. Resolve any interference.

Remember that three-wire dampers must be powered both open and closed (*Figure 21*). *PO* represents Power Open and *PC* represents Power Closed. These

Figure 21 Three-wire bypass damper terminal block.

dampers will not move unless powered in one direction or the other. Note that power should never be applied to the open and closed terminals at the same time.

When in doubt about the operation of a system or component, contact the manufacturer directly. Many companies offer technical assistance by phone for challenging problems. They prefer to do so while the technician is on site, so they can suggest specific troubleshooting steps in real time. However, you should not rely on the manufacturer to replace self-study of the system and a sincere effort to solve problems on your own.

1.0.0 Section Review

1. In a zoning system with a master thermostat, ______.
   a. an auto-changeover feature is not an option
   b. the master thermostat alone determines the mode
   c. each individual zone can determine the operating mode of the system
   d. the mode switches on the master and zone thermostats must be set to the same position

2. High- and low-temperature limit devices associated with zoning systems are needed to ______.
   a. avoid overheating or overcooling a zone(s)
   b. protect the zone actuators from thermal damage
   c. replace the safety devices installed in the equipment
   d. cycle the equipment off before the equipment safety devices activate

3. When a user with a zoning system complains they have no cooling at all, the best way to begin is to ______.
   a. check the individual zone dampers
   b. test the individual zone thermostats
   c. check the zoning panel programming
   d. ensure the HVAC equipment is powered

2.0.0 VVT Systems

Performance Tasks

There are no Performance Tasks in this section.

Objective

Describe VVT systems and their operating characteristics.

a. Describe common VVT system zoning components.
b. Describe VVT system operating characteristics.

Variable volume and temperature (VVT) systems have a lot in common with standard zoning systems, but VVTs are more suited to the commercial environment. Although there are many similarities, there are also significant differences. The components of a VVT system communicate in a more sophisticated way.

VVT systems have changed in several ways over the years. Each generation of VVTs offers different capabilities and may communicate differently. However, the task of programming most VVT systems is as simple as following a checklist and understanding the details of the design well enough to make good decisions. Even though a task may be somewhat consistent from one product or model to another, terminology and features can differ dramatically. Whenever working with a VVT system, it is crucial to obtain and review the technical data for that specific system and its components.

2.1.0 VVT System Components

Figure 22 shows the layout of a typical VVT system. VVT systems are best used in commercial settings and are generally applied to systems up to 25 tons of capacity. Since they are used on larger systems than common zoning systems, VVTs usually have more zones. The arrangement shown in *Figure 22* allows a single constant-volume packaged unit to meet individual zone cooling and heating needs. Each room is a controlled zone with its own communicating zone thermostat, zone damper controller, and room sensor.

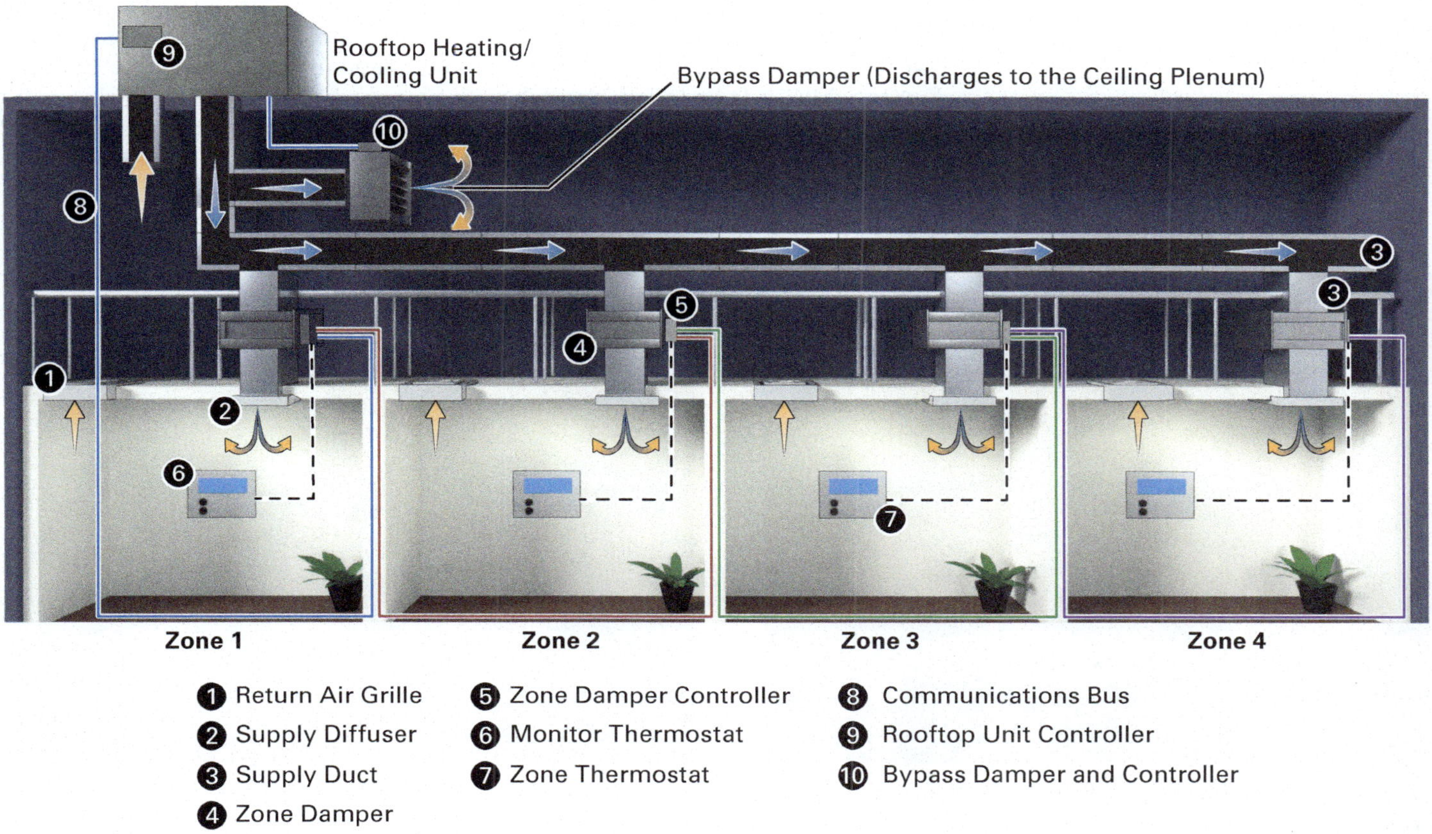

Figure 22 Basic variable volume/temperature (VVT) system.

Many of these components will look familiar because they are similar to those found in residential zoning systems. Zone dampers are in place along with a wall-mounted communicating thermostat in each zone. In a VVT system, two-way communication is required between the zone thermostats and the rest of the system.

Another notable difference is found in the interconnecting wiring. In a VVT system, the zone dampers are communicating with each other through a *bus*. The bus represents the communication chain connecting the HVAC unit and the bypass damper controller.

2.1.1 Bypass Damper

Because the HVAC system is a constant-volume system, a bypass damper must be provided. As zone dampers throttle, the bypass damper is opened to keep the air volume consistent through the unit. The bypass damper is controlled based on the static pressure of the supply air duct. The damper bypasses supply air directly to the return duct or uses the ceiling plenum as a dump zone.

In commercial buildings, the area above a suspended ceiling often serves as a return air plenum. This eliminates the need to install an elaborate return air

NOTE

In some cases, a 90-degree duct elbow is placed on top of a return grille to prevent occupants from looking into the grille and seeing an unattractive view of the area above the ceiling.

duct system. The unit simply draws its return air from the plenum through a single opening. One or more ceiling tiles in a room are replaced with grilles to allow room air to naturally migrate into the ceiling plenum as supply air enters the room. No duct is attached to the grilles.

Using a ceiling plenum for return air works in favor of a zoning or VVT system. The bypass air can be dumped into the ceiling plenum some distance away from the unit return air duct; the farther away, the better. As the air travels back toward the return duct inlet, there is plenty of time for the bypass air to mix with return air in the plenum before reaching the unit.

2.1.2 VVT Terminals

VVT zone dampers, also referred to as *terminals* (*Figure 23*), are available for VVT systems in both round and rectangular styles. The most economical installation cost is achieved when the terminal size and shape match those of the branch duct in which it is installed, and when the number of dampers is minimized. However, it is important to properly size a damper for the needed airflow, not to an existing duct size. Avoid choosing a zone damper based on duct size alone, although the size will match the system design most of the time.

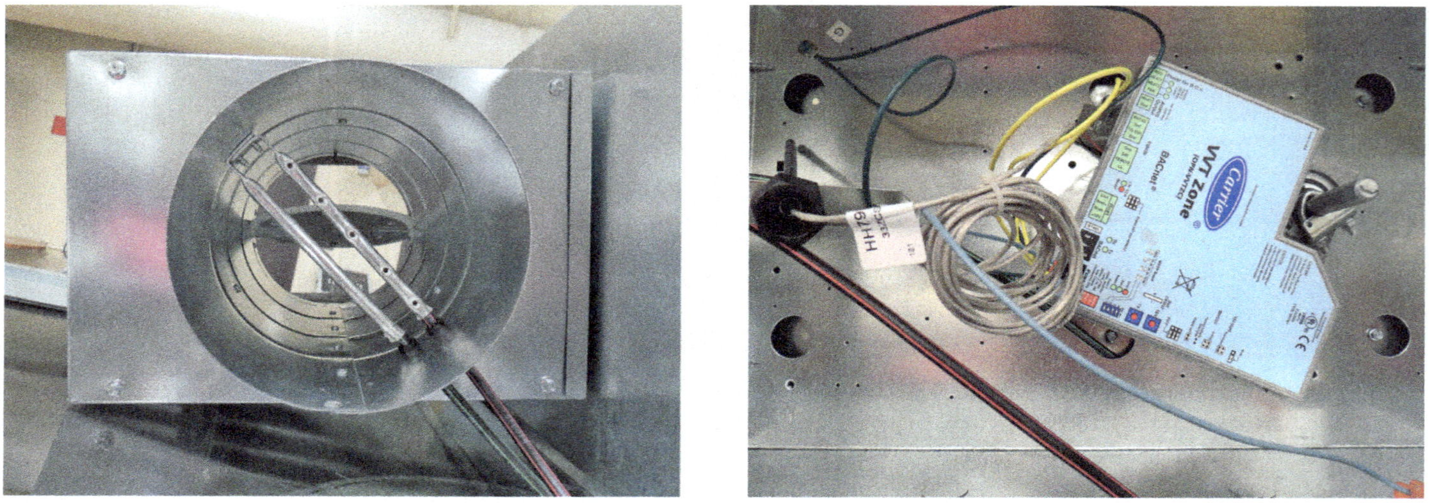

(A) VVT Terminal and Sensors **(B) VVT Damper Actuator**

Figure 23 A VVT terminal awaiting installation.

Round terminals are made in various sizes, typically ranging in capacity from 160 to 1,125 cfm. The dampers range in size from 6" to 16" in diameter, matching typical branch duct sizes. The damper assembly generally comes without insulation; insulation and a vapor barrier are applied externally after installation. These terminals are normally specified for systems operating at a pressure of 1" w.c. or less.

Rectangular terminals are readily available in sizes ranging from 8" × 10" to 8" × 24" with the 8" dimension representing the duct height (*Figure 24*). These terminals are available from the factory pre-insulated. The duct has internal insulation that complies with fire protection codes.

The damper has one or more opposed blades and is powered by a high-torque actuator. The actuator delivers sufficient torque to overcome high air velocity, static pressure, and turbulence. The actuators for VVT systems are significantly more powerful than those used in standard residential zoning systems.

Individual VVT terminals may be equipped with an auxiliary heat source. This is usually electric heat, but a hydronic heating coil can also be used. The hot water coil or electric duct heater is mounted directly to the discharge of a rectangular terminal. The auxiliary heat source is used to provide heat to a needy zone while other zones need cooling.

Figure 24 Rectangular VVT terminal (single blade).
Source: Carrier Corporation

2.1.3 Zone Controllers

Older VVT systems had a single *primary controller*. It may be referred to as a *linkage coordinator, master zone controller, master thermostat*, or *monitor thermostat*. All these terms are referring to a wall-mounted device that users perceive as the thermostat. Herein, we will refer to it as a monitor thermostat.

There are also zone, or secondary, controllers. We'll refer to these as *zone thermostats*. The monitor thermostat shares time of day, system current mode of operation, and outdoor temperature information with each of the zone thermostats. It also monitors the status of the zone thermostats on its communications bus and selects the mode of operation for the HVAC equipment based on the needs of its own zone and the other zones.

The monitor thermostat is also responsible for managing its own zone damper, like the rest of the zones. When enough zones call for heat (as required by the system configuration), the monitor thermostat switches the central unit into heat. When all the zones needing heat have been satisfied, it can switch the unit back to the cooling mode as needed.

Like a zoning system, a VVT system cannot provide simultaneous heating and cooling. On the other hand, recall that VVT terminals can be fitted with electric or hydronic auxiliary heat. These terminals can be used to reheat zones that are being overcooled by the system, but it isn't a good energy practice to use a heat source to offset a cooling source, pitting the two forms of energy against each other.

The monitor thermostat is where the system parameters are typically configured. It also has system mode switches, while the zone thermostats do not. Time-of-day scheduling can often be accomplished either at the zone thermostat or at the monitor thermostat, although only the monitor thermostat maintains the time and date for use by the zones. Each zone can operate on its own schedule.

Newer VVT control systems have done away with the concept of monitor and zone thermostats with independent programming. The monitor thermostat has been replaced by software that interfaces with zone sensors, zone dampers, and the HVAC equipment through a network. The system is configured through a PC or tablet. Once programmed, each zone control module stores its own information. The zone devices can then operate even if the controlling PC is disconnected.

On Site

Monitor-Only Operation

When a new VVT system is initially installed, comfortable conditions may need to be provided before the complete control system has been installed and configured. Monitor thermostats typically have a *monitor-only mode*. In this mode, the monitor thermostat merely controls the HVAC equipment like a regular thermostat; any dampers in place in the air distribution system remain de-energized and open.

2.1.4 Relay Packs and Unit Controllers

Common zoning systems are largely binary in that the devices communicate only through switch closures and the application or removal of power. VVT control systems must share more specific information, requiring an analog approach. Thus, communicating with the HVAC equipment requires a device that translates information into a binary form.

For most systems using a monitor thermostat, an add-on *relay pack* installed in the equipment provides this service. Signals from the monitor thermostat are accepted, and the relays open and close in response to start and stop the equipment in the correct mode.

Newer HVAC systems may use a digital controller inside the equipment to provide this function. Like newer zone damper controllers, the digital controllers also store their own programming information.

2.2.0 VVT System Operation

As mentioned previously, VVT systems must share some detailed information. Most common zoning systems operate by closing and opening switches to energize or de-energize components. Each zone damper opens or closes; they do not provide or consume any real information. However, VVT system components share important information and actual values with each other, always working as a team.

2.2.1 Communication

Figure 25 shows a connection diagram for a zone controller on a system equipped with a monitor thermostat. The zone controller electronics are housed in the front portion of the device, which is then connected to its wiring subbase by a ribbon cable.

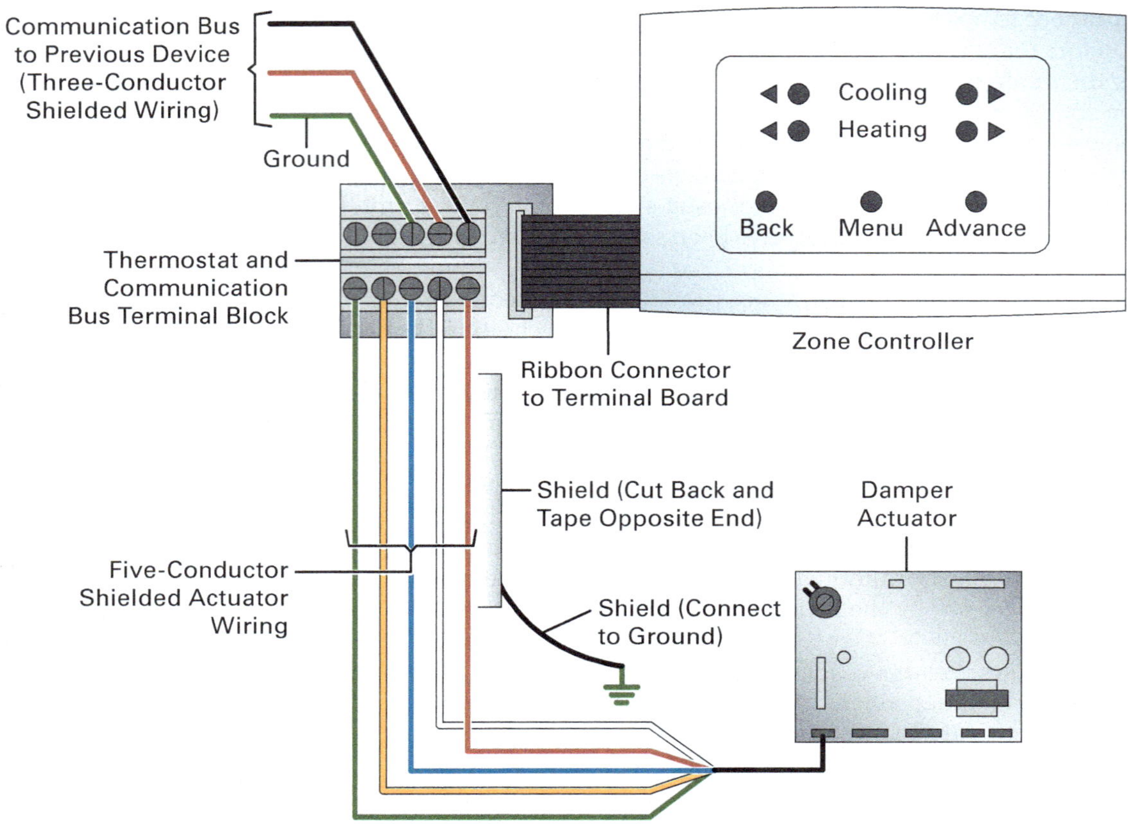

Figure 25 Zone controller and damper actuator wiring.

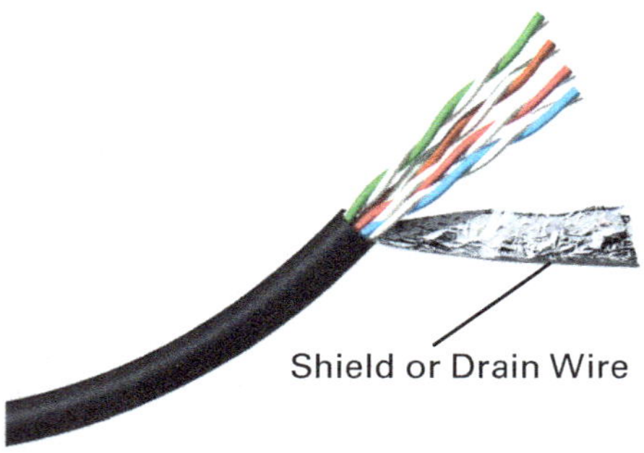

Figure 26 Shielded cable.

Polling: A network communications process where a VVT system seeks input from individual zones, identifies demand priorities, and then responds accordingly to satisfy heating or cooling loads. Electronic participants in the polling system are referred to as *voters*.

At the top of *Figure 25*, the red and black wires provide the path for communication between the devices. This is called the *communication bus*. Signals are transferred through the red and black wires from each device along the bus, and the devices share the data. Shielded cable (*Figure 26*) is used for communication to dissipate electrical noise generated by high-voltage and/or high-current load wiring nearby.

Figure 27 shows a basic VVT system communication bus diagram. The key to VVT system operation and control is the use of zone controllers that can communicate with each other through the bus. As shown, each zone has its own damper controller mounted on a VVT air terminal. A zone controller monitors temperature and allows the zone occupants to modify the cooling and heating setpoints. The monitor thermostat manages the data and shares the time with the other zones.

The VVT system uses a **polling** approach to determine the needs of the individual zones and to gather system information from other devices. In a *polling network*, the monitor thermostat requests, or polls, information from each of the other controllers, one at a time, and in a consistent order. Information from other devices, such as an outdoor temperature sensor, will also be polled. The polling process is executed in the same order repeatedly.

The response time for a heating/cooling request or an alarm may be a bit slow due to the time required to complete the polling sequence. This may be noticeable to a technician watching the system closely as it operates. For example, assume Zone 4 is being polled in a network consisting of 25 devices. If Zone

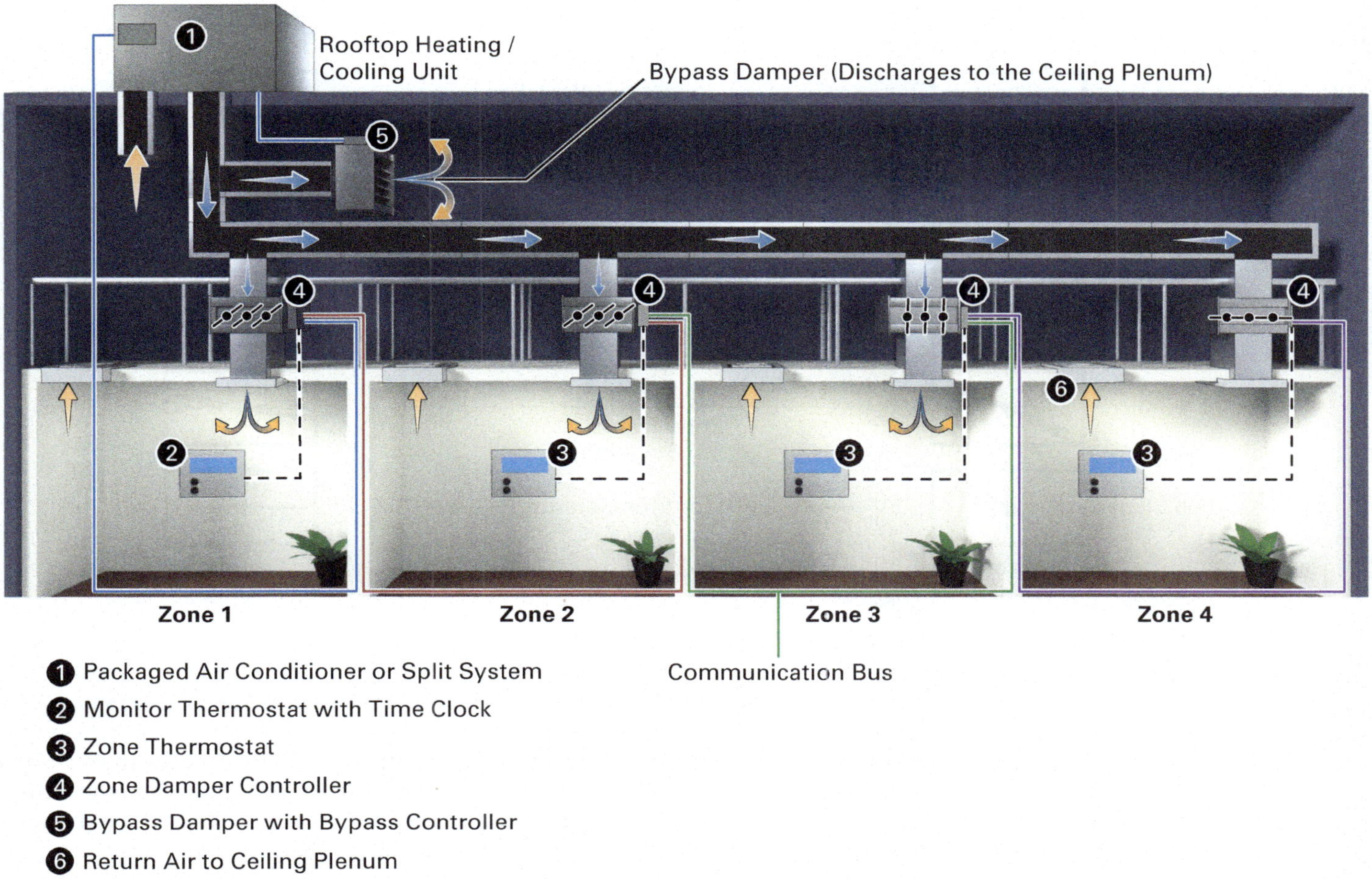

Figure 27 VVT communication bus diagram.

3 alarms at this moment (high or low zone temperature), the alarm cannot be communicated to the monitor thermostat until it has communicated with all the other devices and returns to Zone 3 again.

The results of the poll determine what the monitor thermostat does, and it communicates its decision to the HVAC unit.

2.2.2 Pressure-Dependent versus Pressure-Independent Operation

In a **pressure-dependent system**, the actual volume of airflow through a given zone damper is not monitored. The damper is positioned between the open and closed position based solely on the demand from the zone thermostat. A larger differential between the temperature setpoint and the room temperature causes the damper to open further. Most VVT systems work in a pressure-dependent operation mode. This is a simpler approach and requires fewer and less-sophisticated controls.

Pressure-dependent system: A VVT system design approach where dampers are positioned based on strength of demand only or the dampers operate in just two positions (open or closed). The volume of air flowing through each damper is dependent on the duct pressure, fluctuating as the pressure fluctuates, and the airflow is not measured.

Figure 28 shows an example of this. The damper position changes incrementally as the demand increases or decreases. Usually, when the setpoint and room temperature are 1.5°F (0.83°C) apart or more, the damper is fully opened. However, this is typically a programmable value. Remember though, that a pressure-dependent system has no idea how much air is flowing through a given damper and therefore cannot exercise any additional control. The actual volume flowing through a given damper will depend on the pressure present in the duct at any given moment.

A **pressure-independent system** is a little more sophisticated. In these systems, the airflow through each zone damper is monitored using a velocity sensor. Based on the measured velocity and the area of the duct in which the damper is installed, the system calculates the actual airflow. Programmed values for the minimum and maximum airflow in a zone are then used to determine damper position.

Pressure-independent system: A VVT-system design approach where dampers are positioned by demand and by the volume of air flowing through the damper. Maximum and minimum air volume setpoints are in effect, and the damper will modulate as necessary to remain within these boundaries.

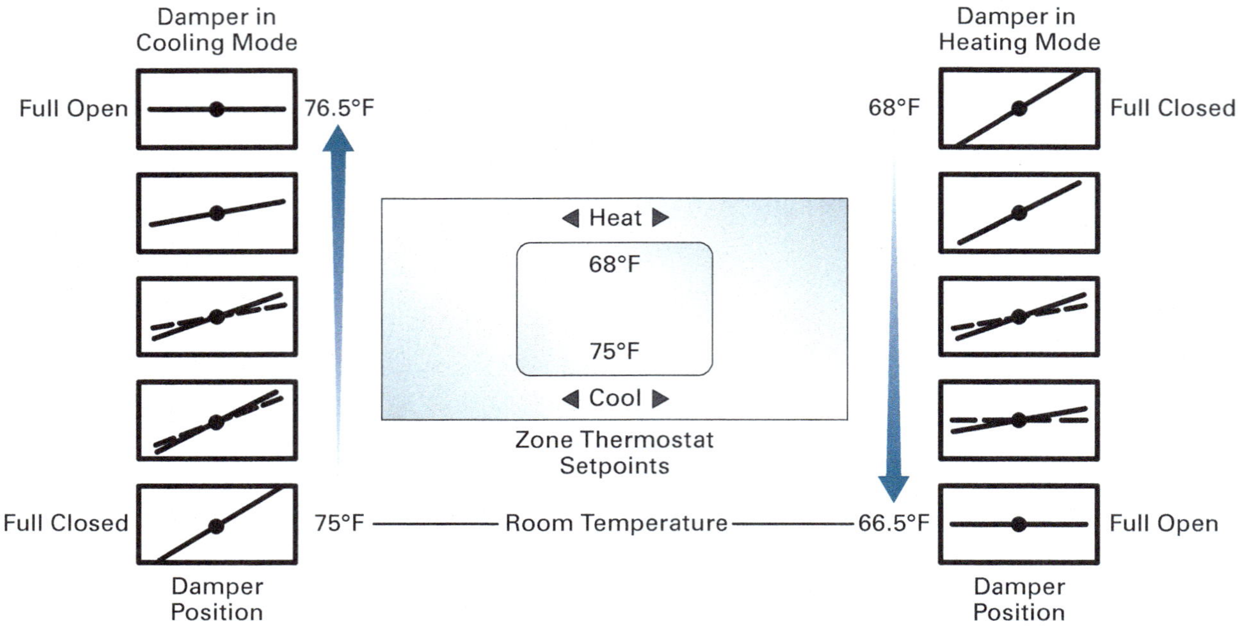

Figure 28 Example of pressure-dependent zone damper modulation.

The basic operation remains the same: more demand for a given mode of operation (heating or cooling) results in the damper opening further. However, with the air volume monitored, the damper will only open until the maximum airflow setpoint is reached. Regardless of how high the demand becomes, the damper will not open further unless the airflow is less than the maximum setting.

The air volume calculation is also used to set the damper at its minimum position for ventilation. In a pressure-dependent system, the only accurate way of determining the minimum position for the damper is to measure the airflow manually and set the minimum position of the damper accordingly. The minimum position of the damper will always be the same, regardless of pressure changes in the duct. In a pressure-independent system, the damper position will be positioned as needed to achieve the minimum air volume. In a manner of speaking, pressure-independent systems are self-balancing.

2.2.3 Occupied Cooling Mode

When one or more zones reach the time for occupancy, the monitor thermostat typically starts the indoor fan. In VVT systems, the indoor fan runs continuously when the building is occupied. At the same time, the HVAC unit outside air dampers open to the minimum position. The individual zone dampers also move to the minimum open position unless cooling is being called for or they have no preset minimum position. After being in setback overnight, many zone thermostats will likely be calling for cooling.

Running the fan continuously is the only way to ensure that each area gets the required amount of ventilation air, regardless of whether heating or cooling is needed. Some systems can also respond to the input of a CO_2 sensor and position the outside air dampers to meet the demand for outside air based on the CO_2.

Zone dampers are rarely programmed to move to the fully closed position during occupied hours. Even when a zone is not calling for heating or cooling, fresh air needs to be provided in a commercial setting. Therefore, each zone damper moves to its minimum position setpoint, leaving the damper open to some degree.

Each zone controller has its own heating and cooling setpoints. The zone controller works with the monitor thermostat to determine how far away the zone temperature is from the setpoint. This is referred to as the *demand*.

The monitor thermostat regularly polls the individual zone controllers, and some calculate the total average cooling demand and total average heating demand for the group. The demands from the zones may be displayed at the monitor thermostat or another user interface.

The demand from each zone is weighted based on the design airflow to be delivered to a zone. The demand from a zone designed to deliver a maximum of 400 cfm will count twice as much in the demand calculation as a zone designed for a maximum of 200 cfm.

Demand may also be based solely on individual zone votes. Each zone, regardless of design capacity, gets one vote. Zones vote for heating or cooling, and the majority wins. Cooling is usually the priority. If the average cooling demand exceeds the average heating demand and is greater than the minimum level of demand required for the cooling mode, the system enters the cooling mode and communicates that information to each zone controller.

More than one vote, and sometimes as many as three or four votes, is typically required to get the system to operate in either mode. The minimum level of required demand is a programmable feature of the system. This is true for systems that calculate weighted demand as well as for those that simply count individual votes. This prevents the system from operating a large HVAC unit and system to satisfy a very small demand, which often results in significant energy use. A great deal of air would need to be bypassed to serve the demand from a single small zone. However, it can also lead to some discomfort in small zones that consistently require heating or cooling while other zones do not.

A mode timer starts when the cooling mode begins. VVT systems can be programmed to operate in the cooling mode for a maximum specified period before switching to the heat mode to satisfy an increasing heating demand from other zones. This prevents a few zones from totally dominating the system. Many systems have minimum operating times as well. Once the cooling mode starts, for example, it will run for some preset minimum period before shutting down or switching to the heating mode.

Since this is the occupied mode, the HVAC unit outside air damper will be open to provide ventilation air. Additional stages of cooling, if available, are added by the monitor thermostat as it sees fit. If the system is equipped with an economizer, it may be activated as the first stage of cooling when the outdoor conditions are favorable.

As the zones cool and move closer to the setpoint, zone dampers begin to throttle the airflow. The supply duct pressure rises and airflow through the HVAC unit decreases. A bypass controller monitors the rise in duct pressure and opens the bypass damper accordingly to maintain the needed airflow through the unit. In this example system, the air is bypassed into the ceiling plenum where it mixes with ceiling air on its way back to the unit return duct.

Like common zoning systems, temperature sensors are used to monitor the discharge air temperature. Minimum and maximum values are programmed, and the heating and cooling equipment is cycled off as necessary to prevent problems.

The cooling mode ends when the average cooling demand has dropped below the field-configured minimum average cooling demand, when a greater average heating demand develops, or when a heating demand has existed for a specified period of time while the unit was in the cooling mode.

2.2.4 Occupied Heating

Should any zone require heating while the central unit is in cooling, the zone controller closes that air terminal's damper to the minimum position and sends the zone heating demand to the monitor thermostat or linkage coordinator. Since the damper must remain at a minimum position for ventilation and is not fully closed, some continued cooling can and does occur. This may drive that zone's temperature even further from the setpoint, increasing the demand.

Once the average demand for heating throughout the system becomes greater than the average demand for cooling, and the cooling-mode minimum time setting (if used) has elapsed, the monitor thermostat switches the HVAC unit from cooling to heating. It then informs the zone controllers that the heating mode is active. The mode timer then restarts.

Zones that need heat modulate their dampers open as needed to accommodate the zone setpoint. Zones that do not need heat move their dampers to the minimum open position. The monitor thermostat continues to poll the zones and accessory devices and responds when the conditions change.

If the average heating demand continues to rise, additional heating stages (if available) can be called upon. Multiple heating and cooling stages are common in commercial equipment over five tons of capacity. Heating stages are cycled on and off as needed to accommodate the demand. The heating mode ends when the average heating demand drops below the field-configured minimum average heating demand; when the average cooling demand exceeds the heating demand; or when any cooling demands have gone unanswered for a preset period of time.

During the heating mode, the bypass controller continues to position the bypass damper to maintain the duct static pressure, ensuring that a constant air volume flows across the heat exchanger and/or coil.

2.2.5 Unoccupied Period

Unoccupied periods offer an opportunity to save energy. Since the occupants are largely gone, fresh air is no longer a concern, so the outside air damper can be closed. This change alone provides energy savings. However, the temperature setpoints can also be changed for the unoccupied period to save additional energy.

When the monitor thermostat determines that all the zones are in the unoccupied mode and there are no calls for heating or cooling, it shuts down the HVAC unit and closes the outside air damper. The zone dampers move to the minimum position. The monitor thermostat operates just as it did during the occupied period to maintain the new zone setpoints, except that the outside air dampers remain closed as long as all zones are unoccupied.

2.2.6 Simultaneous Cooling and Heating

When the HVAC unit is in the cooling mode, reheating an overcooled zone is possible if the zone damper assemblies are equipped with a supplemental heat source. The heater can be activated by the zone thermostat. The zone damper remains at its minimum open position because ventilation must still be maintained. The zone heater, whether it be electric or hydronic, reheats the supply air to warm the zone. When the HVAC unit is in the heating mode, however, cooling cannot be provided to an overheated zone.

Normal operation of a VVT system does not require zone heaters. The system comes very close to satisfying a building's simultaneous need for heating and cooling by time-sharing the heating and cooling modes. In fact, few VVT systems are equipped with zone heaters as their installation increases the installation cost significantly due to the added costs of the equipment and electrical wiring or piping.

The need for simultaneous cooling and heating in commercial buildings is often seasonal, occurring primarily in the spring and fall. However, internal loads are also a significant factor. An FM radio station studio, for example, has a very different load profile than an administrative office in the same building. Thus, VVT systems are best suited for buildings with limited diversity in the heating or cooling loads. If one or two of many zones have extremely diverse requirements, unique solutions such as dedicated ductless systems for those zones are often required.

2.0.0 Section Review

1. The selection of a VVT damper size should be based on which factor?
   a. Existing duct size
   b. Air volume needed
   c. Dominant mode of operation (heating or cooling)
   d. Use of a monitor thermostat versus a zone thermostat

2. In which VVT system operating mode does the HVAC unit outside air damper remain closed?
   a. Ventilation
   b. Unoccupied
   c. Occupied cooling
   d. Occupied heating

3.0.0 Ductless HVAC Systems

Objective

Identify ductless systems and describe basic installation and troubleshooting tasks.

a. Identify and describe the operation of ductless HVAC systems.

b. Describe ductless system installation and troubleshooting procedures.

Performance Task

2. Install and start up, or troubleshoot and repair, a ductless cooling or heat pump system.

Ductless split systems have been extremely popular in Europe and Asia for many years, dominating the housing industry in many locations. These systems have even been used extensively in high-rise residential buildings overseas. The popularity of ductless systems has now increased in the United States significantly after a rather slow start.

3.1.0 Ductless Split Systems

It is difficult, impractical, or expensive to install a ducted air distribution system in some applications. Examples include homes heated with hydronic systems and have no existing air distribution systems, and historic buildings that do not offer space to install ductwork without undesirable structural modifications. Other unique and challenging applications for ducted systems include highway tollbooths, guard shacks, ATMs outside of banks, fast-food restaurant drive-through windows, and add-on sunrooms.

As the name implies, ductless systems use one or more compact, packaged fan coil units. The fan coil units can be mounted high on a wall, on the floor against a wall, and either on or in the ceiling. Wall-mounted indoor units (*Figure 29*) are by far the most popular choice.

Figure 29 Wall-mounted ductless indoor unit with front cover open.
Source: Kirilllutz/iStock/Getty Images

Although ductless systems, also referred to as *minisplits*, are normally installed without any duct, some manufacturers do offer indoor units that can accommodate a small run of duct routed to one or two grilles. However, the small blowers used cannot overcome substantial resistance to airflow, so the ductwork must be minimal.

Refrigerant tubing and wiring connect the indoor and outdoor units. Ductless systems are typically controlled by default with an infrared handheld remote. Most manufacturers also offer an optional wall-mounted thermostat that is either hardwired or wireless as well as control via a mobile app (*Figure 30*).

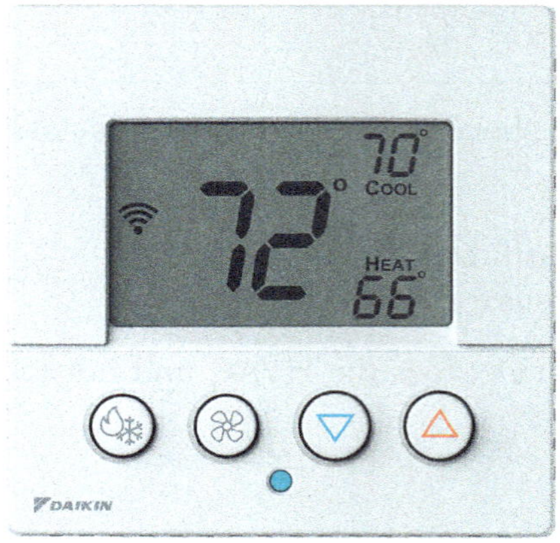

(A) Wired Wi-Fi Capable Thermostat

(B) Handheld Remote Control

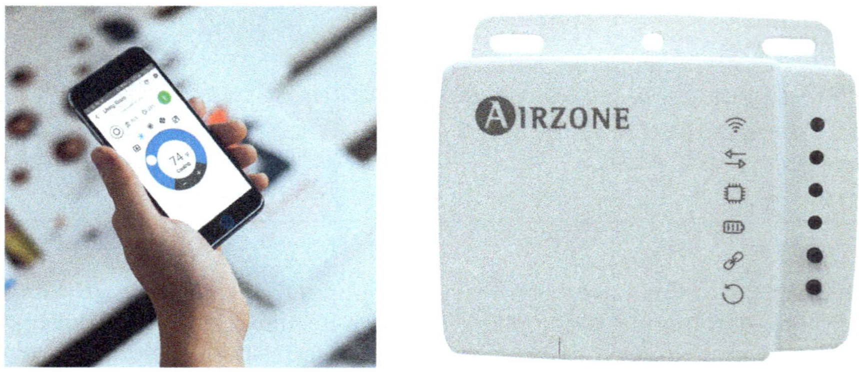

(C) Wi-Fi Adapter with Supported Mobile App

Figure 30 Common ductless-system control options.
Sources: (A-C) Daikin Comfort Technologies North America, Inc

On Site

Ductless System Efficiency

Inverter-driven ductless HVAC systems now represent some of the most efficient HVAC equipment available. At this writing, the ENERGY STAR® website shows the most efficient ductless system to be a Mitsubishi model with an astounding SEER2 rating of 35.0.

Figure 31 Typical ductless system condensing unit.
Source: Y-studio/iStock/Getty Images

3.1.1 Ductless System Condensing Units

The outdoor unit contains the compressor, condenser coil, condenser fan, and controls like other split-system condensing units (*Figure 31*). However, other construction aspects differ significantly. Condensing units for ductless systems are upright rectangular slabs that resemble a large suitcase. The condenser coil is mounted vertically in the cabinet and air is either blown or drawn through the coil and discharged horizontally. Higher-capacity units may look like two units stacked on top of each other, with two condenser fans.

Condensing units are typically powered by single-phase 115VAC or 230VAC power. Power for the indoor unit is typically supplied through the outdoor unit. An electronic expansion valve is the standard metering device, and it is usually located in the outdoor unit, not inside the fan coil or at the evaporator coil like traditional systems. The result is that the smaller refrigerant line that normally carries subcooled liquid refrigerant to the indoor unit in the cooling mode instead carries a saturated liquid / vapor refrigerant mix at low pressure. For this reason, both refrigerant lines must be insulated to avoid capacity loss and condensation on the lines.

Most ductless systems use a variable-speed, inverter-driven compressor. This provides the precise amount of capacity needed at any time. Since many ductless systems can support more than one indoor unit, the required capacity can vary widely as the indoor units respond to the individual temperature controls. Variable-speed compressors provide a major portion of the energy savings ductless systems offer, as well as precise capacity control.

Because the indoor unit(s) is typically powered from the outdoor unit, the interconnecting wiring between the two includes both line-voltage power as well as communication wiring.

NOTE

The *NEC*® and local codes may require a separate power disconnect switch for the indoor unit.

3.1.2 Ductless System Indoor Units

The indoor unit contains the evaporator coil, air filter, condensate collection pan, indoor blower, and controls. The centrifugal blower wheel is small in diameter but elongated. The refrigerant piping and wiring usually enters through the back of the unit, where it remains unseen on the inside. However, it can also enter from the side.

The variety of indoor units greatly enhances ductless system versatility. Indoor unit types include but are not limited to the following:

- High-sidewall
- Ceiling mount
- Ceiling cassette
- Floor mount

High-sidewall air handlers (*Figure 29*) are rectangular and shallow in depth. When mounted high on a wall, the fan in the air handler can more effectively move the air into and through the room. Moving motorized louvers on the air outlet help distribute the air more evenly throughout the room, if desired.

If the indoor unit is mounted on an outside wall, the condensate can be drained through the wall by gravity. If mounted on an inside wall, an accessory condensate pump can be used for condensate disposal. Some miniature condensate pumps fit inside the unit cabinet, while others can be mounted immediately beneath the indoor unit. Some pumps are also designed to be concealed inside the same duct that hides the refrigerant piping.

High-sidewall air handlers are typically attached to the wall on a mounting plate. The standard control furnished is a handheld remote. As noted earlier, most systems offer an optional hard-wired, wall-mounted controller as well as other control options. Regardless of which control type is used, room temperature is sensed at the indoor unit return-air inlet.

Surface-mounted overhead fan coils offer many of the same characteristics as high-sidewall air handlers. These units are often located adjacent to an outside wall so that refrigerant piping, wiring, and condensate piping can be routed to the outdoor unit without being exposed. If located away from an outside wall, a condensate pump is needed to dispose of condensate. In most cases, the piping can be routed above the ceiling, so it remains out of sight.

Ceiling cassettes (*Figure 32*) are usually mounted in drop ceilings with the face of the unit flush with the ceiling. The return air intake is in the

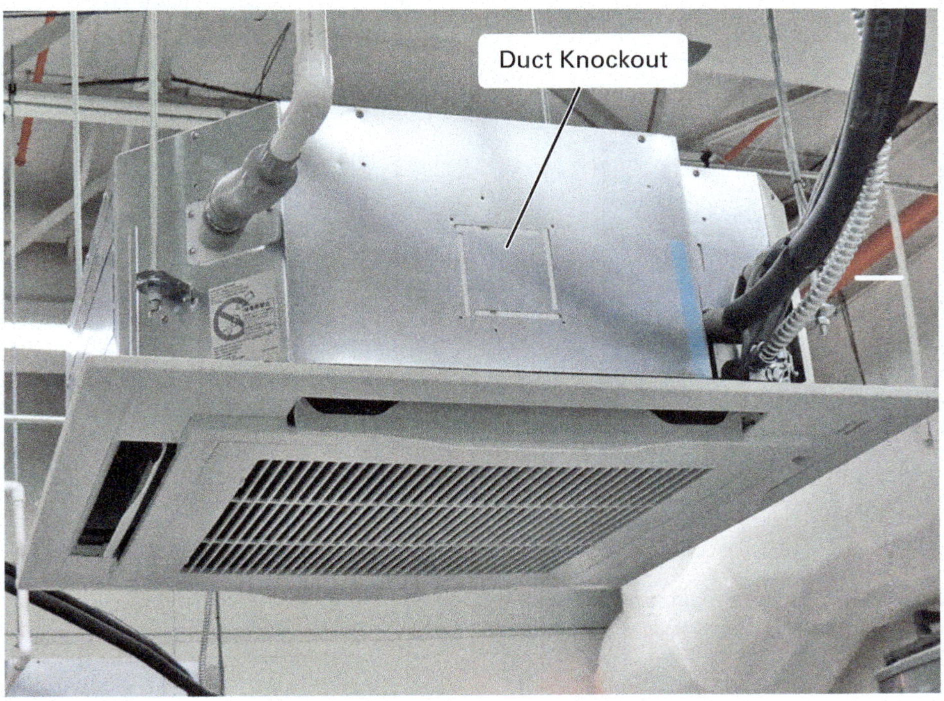

Figure 32 Ceiling-cassette ductless fan coil unit.

center and the four supply-air distribution louvers are located around the perimeter. When installed, the unit has the appearance of a concentric ceiling diffuser. Refrigerant lines, wiring, and condensate lines are all located above the ceiling tiles. A condensate pump is usually required for ceiling cassettes.

Note the duct knockout shown in *Figure 32*. Knockouts like this may also be found on other sides. Some ceiling cassettes allow the connection of a short branch duct to serve an adjoining small space. Most ceiling cassettes, since they are considered commercial units, also offer an outside air duct connection for ventilation. A powered booster fan may be added to overcome the resistance of the interconnecting duct (*Figure 33*). Otherwise, the blower is unlikely to draw enough air through the outside air duct.

Floor-mounted fan coils (*Figure 34*) direct the supply air into the room from the top. Return air enters through the bottom or front. These units are mounted against an outside wall, ideally adjacent to the outdoor unit. An outside air intake is also provided for this unit style.

Figure 33 Inline duct booster fan.
Source: Jackson Systems, LLC.

Figure 34 Floor-mounted ductless fan coil unit.
Source: Daikin Comfort Technologies North America, Inc

3.1.3 Multizone Systems

Ductless multizone systems have several air handlers attached to a single outdoor unit (*Figure 35*). Each zone has independent temperature control. The outdoor unit has multiple service valves and piping connections to accommodate a specified number of indoor units. The larger the capacity of the outdoor unit, the more fan coils it can support. Inverter-driven, variable-speed compressors can easily accommodate the changing load as the indoor units cycle on and off.

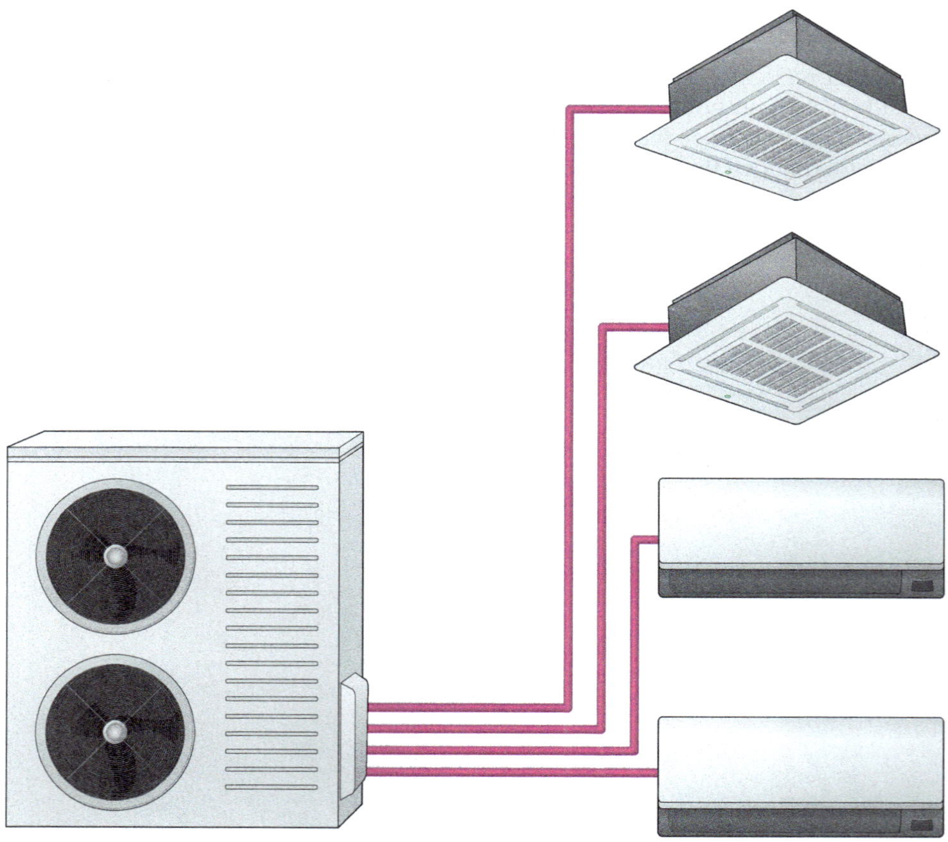

Figure 35 Ductless multizone arrangement.
Source: MeggiSt/iStock/Getty Images

On Site

Cold-Climate Heat Pumps

Cold-climate heat pumps can provide far more heat at colder temperatures than their predecessors. One of the secrets to their performance is referred to as *enhanced vapor injection*, or EVI.

Liquid refrigerant leaving the condenser is separated into two parts. A small portion of the liquid is routed through an additional expansion valve that feeds a counter-flow plate heat exchanger. The remaining liquid leaving the condenser passes through the heat exchanger and is further subcooled by the refrigerant from the added expansion valve.

The vapor resulting from the absorption of heat is routed back to the compressor at an intermediate pressure. Compressors designed for EVI have an additional inlet where the vapor is injected. Mitsubishi Electric refers to this innovative design as Hyper Heat (H2i®).

3.2.0 Ductless System Installation and Troubleshooting

Some generic installation and troubleshooting guidance are provided in this section. However, you must always check the manufacturer's instructions for the exact model being installed. The installation and service instructions vary from one manufacturer to another and from one model to another. The proper information must be acquired to successfully install and troubleshoot a system. Some manufacturers require factory technician training for their product line. Regardless of whether it is required, factory training is extremely helpful and a necessary part of a technician's career progression.

3.2.1 Installing Ductless Systems

The installation tasks for ductless systems are similar to conventional split systems. When locating the outdoor unit, installers must ensure that condenser airflow is not obstructed. The potential for snowfall and the prevailing wind direction must also be considered.

Ductless condensing units should be secured to their foundation because of their light weight and shape. Otherwise, the units are likely to be blown over by strong winds. Before securing the unit, ensure that all minimum clearances have been considered. Some ductless heat pumps require a short section of drain line to be attached to the pan of the outdoor unit for the discharge of defrost water.

In many countries, ductless condensing units can be seen mounted on the outside wall of residential high-rise buildings, from top to bottom (*Figure 36*). Servicing the equipment on these buildings is not without risk. However, thousands of residential buildings in Europe and Asia are served by these systems. Since the condenser air can be drawn in through the sides as well as the back or front, the unit can be placed close to a wall. Mounting the outdoor unit on a wall is also a very good way to avoid problems associated with heavy snowfall and occasional flooding (*Figure 37*).

The refrigerant piping is most often connected to the indoor and outdoor units by flare fittings, although different connections may be encountered. Larger systems may have brazed connections, due to the line size. Where flare fittings are used, the installation guidelines provide a specific torque value to be applied to the flare nuts. The flare nut threads should not be lubricated before torquing unless the instructions specifically direct it.

Figure 36 Many international high-rise buildings have ductless systems for each residence, making service challenging.

Source: Tomo Nogi/iStock/Getty Images

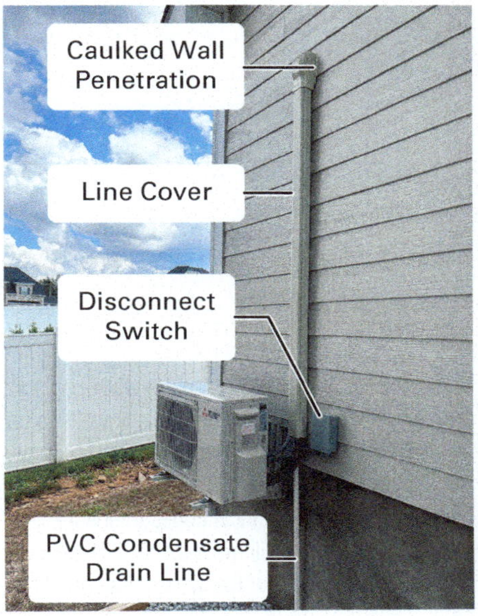

Figure 37 Wall-mounted ductless condensing unit.

The installation instructions will provide other refrigerant piping guidelines, such as the following:

- The proper refrigerant line size
- Maximum length of the refrigerant lines
- Maximum rise of the refrigerant piping
- The amount of refrigerant to be added for each foot of refrigerant piping beyond a certain distance

Both refrigerant lines are insulated, as both can produce condensation and transfer heat to or from the surrounding environment. The refrigerant lines must be properly leak-tested and evacuated before the refrigerant contained in the outdoor unit is released into the system.

Note that some manufacturers prohibit adding a filter-drier in the lines. Therefore, any refrigerant piping work must be done using best practices for cleanliness and joint quality. A nitrogen purge is required during brazing. Triple-evacuations are necessary to ensure that all the moisture and air has been removed.

The installation procedure for the indoor unit depends on the chosen style. The typical wall-mounted indoor unit hangs from a metal plate mounted securely to the wall (*Figure 38*). Note the torpedo level across the two tabs, used to level the mounting plate. The unit must be level to avoid condensate drainage problems and to present a neat appearance.

Figure 38 Ductless indoor unit mounting plate.
Source: Daikin Comfort Technologies North America, Inc

CAUTION

Mechanical codes typically require wall openings for piping through foundation or block walls to be sleeved with structurally sound material, such as Schedule 40 or Schedule 80 steel pipe. Before boring the hole, check local codes to ensure that the penetration will meet the requirements. The outside diameter of the sleeve material determines the size of the hole needed.

The typical installation requires a hole to be bored through a wall to accommodate the interconnecting piping and wiring. The hole through the wall is bored at a slight angle down towards the outside (*Figure 39*). This prevents rain from entering the opening and encourages condensate drainage. A plastic sleeve is used to line the hole, then insulating material and caulk is used to seal the inside and outside of the penetration.

Note the threads shown on the sleeve in *Figure 39*. This accessory sleeve consists of two threaded pieces that are adjusted to fit the thickness of the wall.

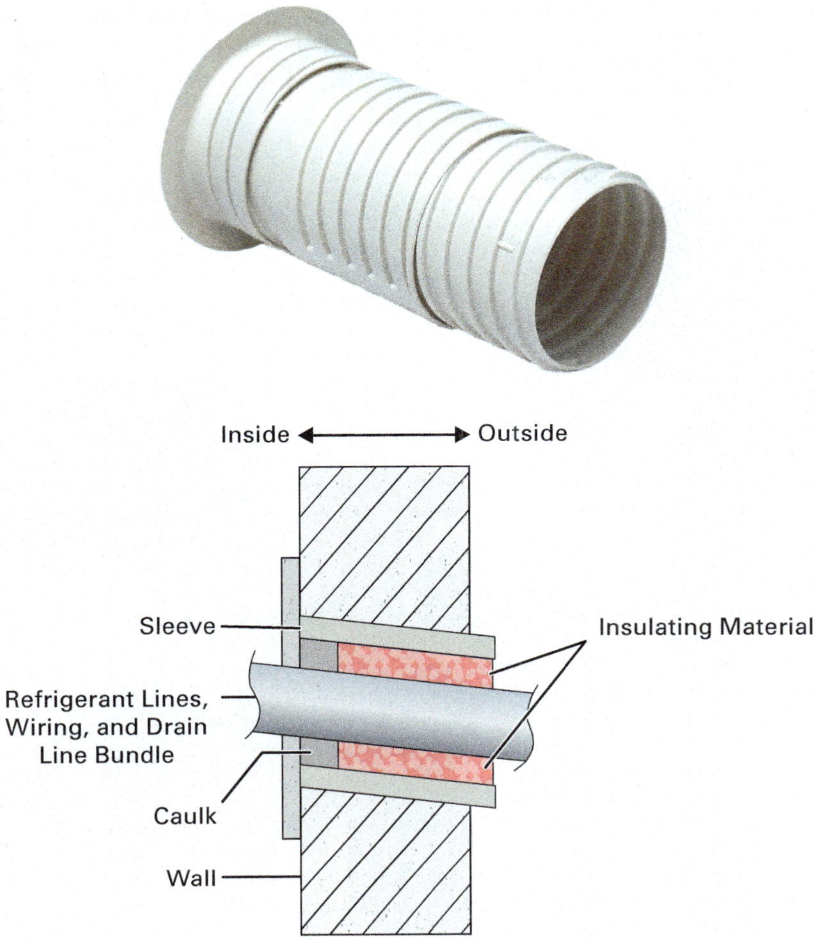

Figure 39 Wall sleeve and opening detail.
Source: (Top) RectoSeal, LLC

There are two approaches used to make refrigerant line connections at the indoor unit. One method uses flexible refrigerant lines about 18" long, factory-connected to the indoor unit. The lines are pushed through the wall penetration as you hang the indoor unit on its mounting bracket. The line set connections are then made outdoors where they are more easily accessed. The refrigerant lines can also be routed in other directions when the lines must be routed along the inside wall before passing through the outside wall. The installation instructions include options for connections as the unit is approached from different directions, such as right-side, right-back, or right-bottom.

Another approach is to route the refrigerant lines through the wall from the outside and make the connection at the indoor unit. These units are designed to hang from the top of the mounting plate with a spacer placed behind it to hold the bottom away from the wall while you work.

The condensate drain line is also flexible and is usually pushed through the outside wall where it must be connected to additional drain tubing or pipe. Condensate drains on ductless systems are not trapped because there is negative or positive pressure in the small unit to interfere with gravity drainage.

The wiring from the outdoor unit powers the indoor unit and provides communication between the two pieces. All applicable codes for electrical wiring must be followed. Marked terminal blocks are provided on both units to make the wiring connections straightforward. For protection and a neat appearance, the refrigerant piping and wiring can be concealed in special line set covers (*Figure 40*). A variety of fittings, colors, and sizes helps to accommodate almost any situation.

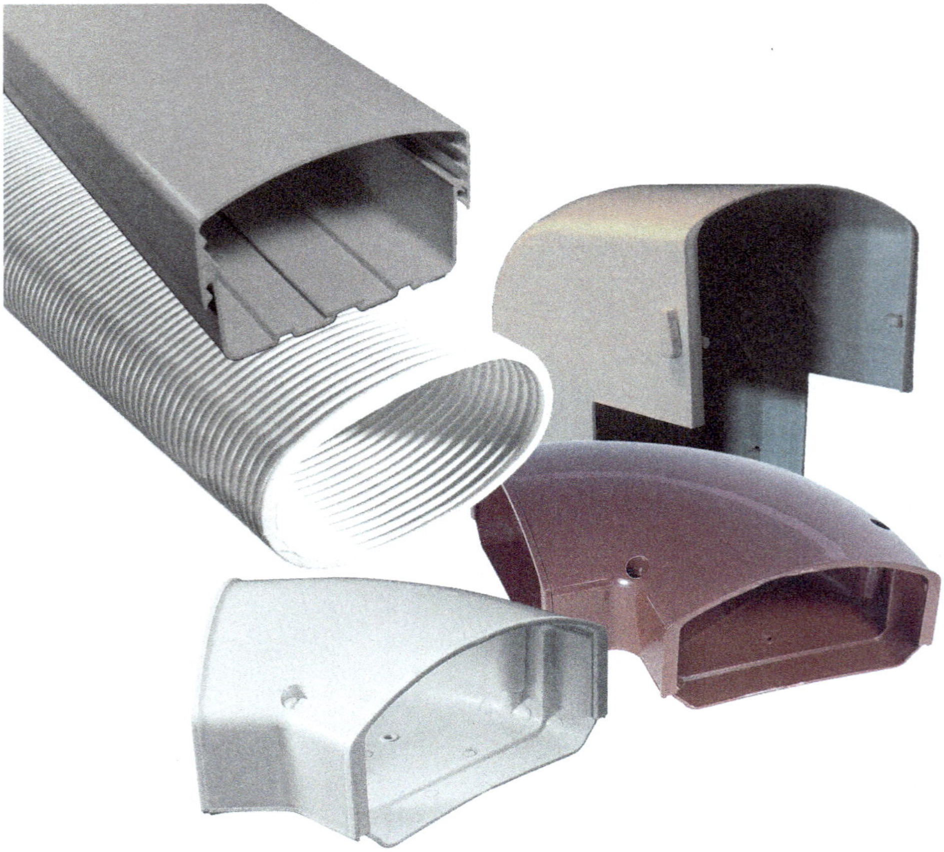

Figure 40 Line-set cover components.
Source: RectorSeal, LLC

Unlike conventional systems, most ductless split systems do not use a 24VAC control circuit. Instead, DC communication signals between the outdoor unit and indoor unit(s) control the system. The controls may have to be configured during installation to ensure proper communication between the outdoor unit and the indoor unit(s). Configuring the unit may include placing jumpers on the PC board across specific pins or setting a series of DIP switches. It is important to read the installation instructions carefully to ensure a proper setup.

The handheld remote control (when used) must also be prepared prior to startup. These remotes are battery-operated, usually requiring two or three AAA batteries. Once the outdoor unit is energized, the remote control must be programmed for the present day of the week and time of day. A typical handheld remote control is shown in *Figure 41*.

Terminology and features vary between companies and products. What one company refers to as the "powerful mode," another may refer to as "boost mode." Always follow the manufacturer's guidance for the specific unit. Take the time to assist the user of a new system with the controls and terminology, and make sure they know where to find information when they forget something.

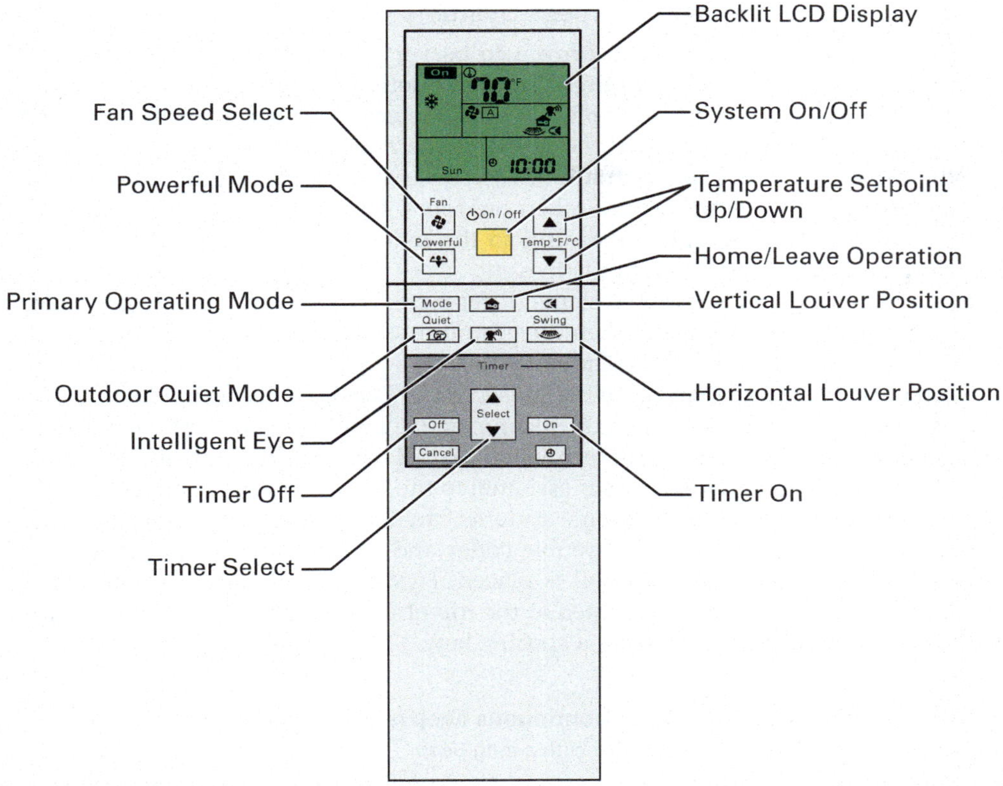

Figure 41 Handheld remote button layout.
Source: Daikin Comfort Technologies North America, Inc

Moving clockwise from the right side of *Figure 41*, some of the features of this remote are described as follows:

- The *Home Leave* feature allows the user to press a single button to revert to previously entered unoccupied temperature and airflow settings.
- The *Vertical/Horizontal Louver* and *Swing* buttons determine the position and action of the air discharge louvers. *Swing* moves the louvers left to right or up and down to consistently spread the air over a broad area.
- The two *Timer* buttons along with the *Timer Select* button allow the user to set a specific number of hours for the system to remain in operation.
- The *Intelligent Eye* or *Sensor* button controls an energy-saving feature. This feature activates a sensor that detects movement, acting as an occupancy sensor. After a predetermined period passes without sensed movement, the unit reverts to its unoccupied setpoints. The angle of the sensing eye on the face of the unit can often be adjusted to face a specific direction.
- At night or other times when the sound of the outdoor unit running may be a nuisance, the *Outdoor Quiet* button changes the speed and frequency of the condenser fan to reduce noise.
- The *Powerful* setting disregards any normal settings for airflow control and maximizes the cooling capacity for 20 minutes.

During the initial startup, all the functions and settings of the remote should be checked and proper operation confirmed. The remote control described here has a *Trial* mode that is available when the system is first started. In *Trial* mode, the system runs through a preset sequence on its own for roughly 30 minutes, performing various operational tests. Any problems are reported through fault codes displayed on the remote or on the outdoor unit control board.

3.2.2 Troubleshooting Ductless Systems

While ductless systems have proven to be very reliable, troubleshooting and repair will eventually be required. Troubleshooting the refrigerant circuit is not significantly different from common split systems. The service valves located at the outdoor unit provide the only refrigerant circuit access points. Since the refrigerant lines are virtually inaccessible at the indoor unit, all pressure and temperature readings are taken outdoors. Procedures such as refrigerant charging should be done according to the manufacturer's instructions using commonly accepted service practices.

The controls on ductless systems differ significantly from traditional HVAC systems. Ductless systems often have built-in diagnostics in the form of fault codes and/or flashing LEDs. The LEDs flash in a particular sequence or for varying periods of time to indicate the sensed fault. Labels on the equipment and in the service literature can help the technician decipher the codes. Manufacturers generally provide hotlines for technicians to call if further assistance is required. Understanding the fault codes is essential to the troubleshooting process.

The handheld remote on some systems can also query the system for faults. An astute user can view the trouble codes and report them to the HVAC service agency when a service call is placed. *Figure 42* shows an example list of fault codes for a system. As noted at the top of the chart, the fault codes can be viewed by repeatedly pressing a specific key.

2. Press the TIMER CANCEL button repeatedly until a continuous beep is produced.

- The code indication changes as shown below, and notifies with a long beep.

	Code	Meaning
System	00	Normal
	U0	Refrigerant Shortage
	U2	Drop Voltage or Main Circuit Overvoltage
	U4	Failure of Transmission (Between Indoor Unit and Outdoor Unit)
Indoor Unit	A1	Indoor PCB Defective
	A5	High-Pressure Control or Freeze Protector
	A6	Fan Motor Fault
	C4	Faulty Heat Exchanger Temperature Sensor
	C9	Faulty Suction Air Temperature Sensor
Outdoor Unit	EA	Cooling/Heating Switching Error
	E5	OL Started
	E6	Faulty Compressor Startup
	E7	DC Fan Motor Fault
	E8	Operation Halt Due to Input Overcurrent
	F3	High-Temperature Discharge Control
	F6	High-Pressure Control (Cooling)
	H6	Operation Halt Due to Faulty Position Detection Sensor
	H8	CT Abnormality
	H9	Faulty Suction Air Temperature Sensor
	J3	Faulty Discharge Temperature Sensor
	J6	Faulty Heat Exchanger Temperature Sensor
	L4	High Temperature at Inverter Circuit Heatsink
	L5	Output Overcurrent
	P4	Faulty Inverter Heatsink Temperature Sensor

Note:

1. A short beep and two consecutive beeps indicate non-corresponding codes.
2. To cancel the code display, hold the TIMER CANCEL button down for 5 seconds. The code display also cancels itself if the button is not pressed for 1 minute.

Figure 42 Example ductless system fault codes.

Source: Daikin Comfort Technologies North America, Inc

Ductless system controls can be very complex. Factory-supported, equipment-specific training provides the best possible foundation for effectively troubleshooting and repairing these systems. Always refer to the manufacturer's instruction manual for the specific unit being serviced. This literature usually provides troubleshooting charts, such as the one shown in *Figure 43*, that help to point technicians in the right direction.

Symptom	Check Item	Details of Measure	Reference Page
The unit does not operate.	Check the power supply.	Ensure the rated voltage is supplied.	–
	Check the type of the indoor unit.	Ensure the indoor unit type is compatible with the outdoor unit.	–
	Check the outdoor temperature.	Heating operation cannot be used when the outdoor temperature is 24°C (75.2°F) or higher, and cooling operation cannot be used when the outdoor temperature is below 10°C (50°F).	–
	Diagnose with remote controller indication.	–	94
	Check the remote controller addresses.	Ensure the address settings for the remote controller and indoor unit are correct.	388
Operation sometimes stops.	Check the power supply.	A power failure of 2 to 10 cycles stops air conditioner operation. (Operation lamp OFF)	–
	Check the outdoor temperature.	Heating operation cannot be used when the outdoor temperature is 24°C (75.2°F) or higher, and cooling operation cannot be used when the outdoor temperature is below 10°C (50°F).	–
	Diagnose with remote controller indication.	–	94
The unit operates but does not cool, or heat.	Check for wiring and piping errors between the indoor unit and outdoor unit.	–	–
	Check for thermistor detection errors.	Check if the thermistor is mounted securely.	–
	Check for faulty operation of the electronic expansion valve.	Set the unit to cooling operation, and check the temperature of the liquid pipe to see if the electronic expansion valve works.	–
	Diagnose with remote controller indication.	–	94
	Diagnose by service port pressure and operating current.	Check for refrigerant shortage.	104
Operating noise and vibrations.	Check the output voltage of the power module.	–	153
	Check the power module.	–	–
	Check the installation condition.	Check if the required space for installation (specified in the installation manual) are provided.	–

Figure 43 Example ductless system troubleshooting chart.
Source: Daikin Comfort Technologies North America, Inc

As shown in *Figure 43*, the chart references a particular page of the manual for additional information. A chart showing all the possible error codes (*Figure 44*) also guides technicians to a specific page that provides additional instructions to help resolve the problem.

	Error Codes	Description		Reference Page
System	00	Normal		—
	U0★	Refrigerant shortage		104
	U2	Low-voltage detection or over-voltage detection		107
	U4	Signal transmission error (between indoor unit and outdoor unit)		109
	UA	Unspecified voltage (between indoor unit and outdoor unit)		112
Indoor Unit	A1	Indoor unit PCB abnormality		95
	A5	Freeze-up protection control or heating peak-cut control		97
	A6	Fan motor or related abnormality	DC motor (FTXS series)	99
			AC motor (FDXS series)	101
	C4	Indoor heat exchanger thermistor or related abnormality		103
	C9	Room temperature thermistor or related abnormality		103
Outdoor Unit	E1	Outdoor unit PCB abnormality		113
	E5★	OL activation (compressor overload)		115
	E6★	Compressor lock		116
	E7★	DC fan lock		117
	E8	Input overcurrent detection		118
	EA	Four-way valve abnormality		119
	F3	Discharge temperature control		121
	F6	High-pressure control in cooling		123
	H0	Compressor system sensor abnormality		125
	H6	Position sensor abnormality		128
	H8	DC voltage / current sensor abnormality (09/12 class only)		131
		CT or related abnormality (24/30/36 class only)		132
	H9	Outdoor temperature thermistor or related abnormality		134
	J3★	Discharge thermistor or related abnormality		134
	J6	Outdoor heat exchanger thermistor or related abnormality		134
	L3	Electrical box temperature rise		136
	L4	Radiation fin temperature rise		138
	L5★	Output overcurrent detection		140
	P4	Radiation fin thermistor or related abnormality		134
	U7	Signal transmission error on outdoor unit PCB (24/30/36 class only)		111

★: Displayed only when system-down occurs.

Figure 44 Example fault-code guide.
Source: Daikin Comfort Technologies North America, Inc

Once an error code is defined and a referenced page number accessed, a flow chart is often found to guide the technician through the appropriate actions to take. For example, note the error code EA in the second column of *Figure 44*. The code refers to a four-way (reversing) valve abnormality. The code description references page 119 of the service manual. A more detailed description of the problem is found on this page, as shown in *Figure 45*. From there, a flow chart is provided to help resolve the problem (*Figure 46*). Due to the complexity of these systems, troubleshooting aids like these are indispensable.

Did You Know?

Ductless System Sophistication

Here's a clue to the level of sophistication that can be found in ductless systems. The service manual for a relatively standard indoor and outdoor matched combination, portions of which were referenced in this module, is 414 pages in length!

4.16 Four-Way Valve Abnormality

Remote Controller Display	EA
Method of Malfunction Detection	The room temperature thermistor and indoor heat exchanger thermistor are checked to ensure they function within their normal ranges in each operating mode.
Malfunction Decision Conditions	One of these conditions continues over 1 to 10 minutes after operating for 5 to 10 minutes: ■ Cooling / Dry (room thermistor temp — indoor heat exchanger temp) < –5°C (–9°F) ■ Heating (indoor heat exchanger temp — room thermistor temp) < –5°C (–9°F) If the error repeats, the system is shut down. Reset condition: Continuous run for about 60 minutes without any other error
Supposed Causes	■ Disconnection of four-way valve coil ■ Defective four-way valve, coil, or harness ■ Defective outdoor unit PCB ■ Defective thermistor ■ Refrigerant shortage ■ Water mixed in refrigerant ■ Defective stop valve

Figure 45 Example problem description.

Source: Daikin Comfort Technologies North America, Inc

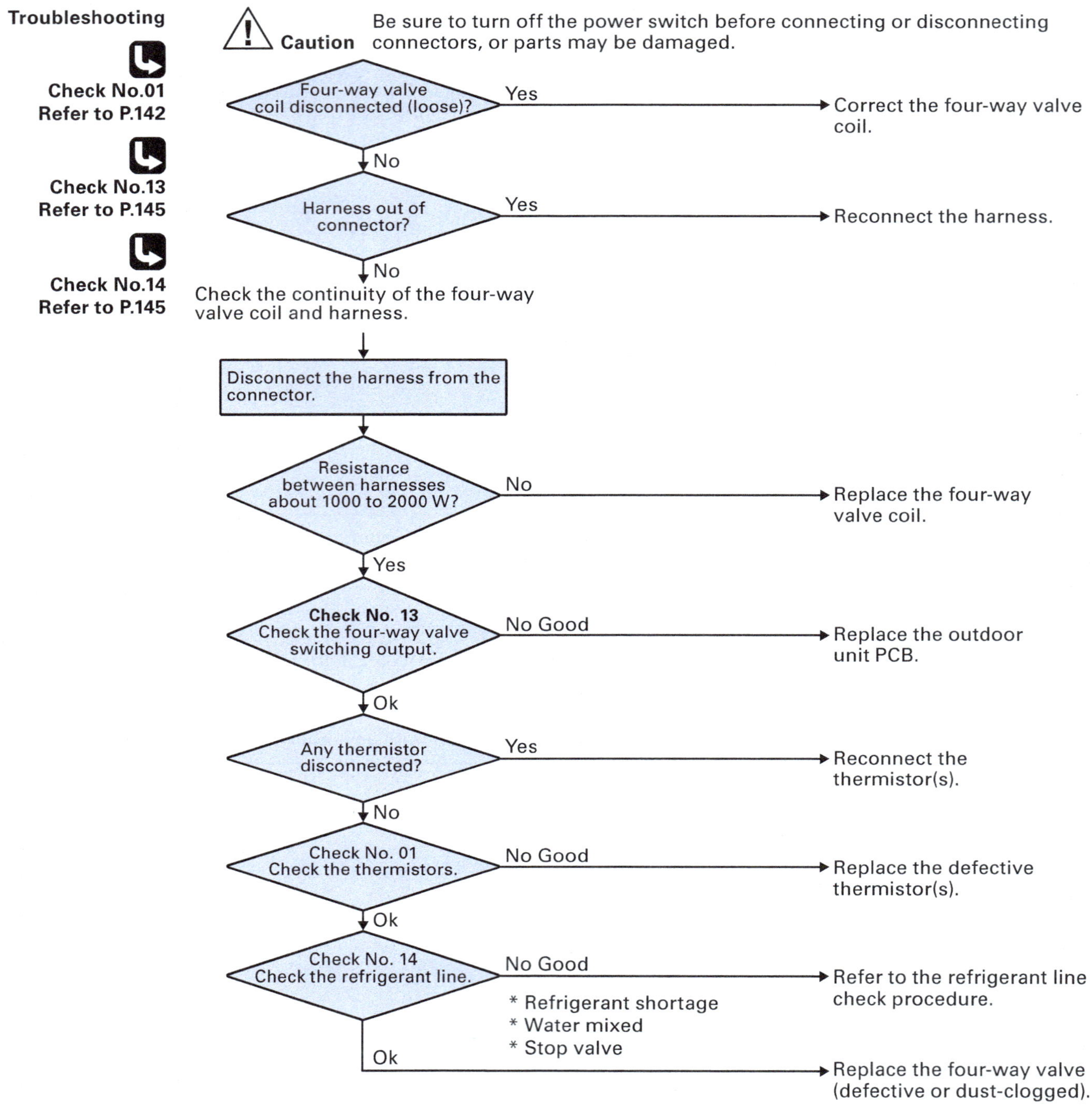

Figure 46 Example troubleshooting flow chart.

Source: Daikin Comfort Technologies North America, Inc

3.0.0 Section Review

1. What typically flows through the smaller refrigerant line in a ductless split system?
   a. Hot refrigerant vapor
   b. Cold refrigerant vapor
   c. Saturated liquid/vapor mixture
   d. Fully condensed liquid refrigerant

2. Which of the following statements about ductless systems is correct?
   a. The condensate drain should not be trapped to ensure proper water flow.
   b. The power supply connects to the indoor unit and is then extended to the outdoor unit.
   c. Refrigerant-line flare nuts should always be lubricated before assembly and torquing.
   d. All system configuration settings for a system are made through the handheld remote.

4.0.0 Variable Refrigerant Flow Systems

Objective

Identify variable refrigerant flow systems and describe basic installation and troubleshooting tasks.

a. Identify and describe the operation of variable refrigerant flow systems.

b. Describe variable refrigerant flow system installation and troubleshooting procedures.

Performance Tasks

There are no Performance Tasks in this section.

Variable refrigerant flow (VRF) systems represent a completely different approach to zone control. Although VRF systems and ductless systems share some characteristics, a larger assortment of ducted indoor units is available. In fact, ducted indoor systems represent roughly half of the equipment connected to VRF systems worldwide.

4.1.0 VRF Equipment and Operation

Like some ductless systems, a single VRF outdoor unit can support a number of indoor units. With ductless systems, each indoor unit has an independent set of refrigerant lines that must be connected directly to the outdoor unit. However, a VRF system uses only one set of refrigerant lines and distributes the refrigerant to each indoor unit from the same circuit (*Figure 47*). VRF systems can also support more indoor units than ductless systems.

The primary task of the outdoor unit is to produce a varying amount of refrigerant flow at a preset subcooling value. Daikin, the original developer of this technology, uses a subcooling value of 9°F (5°C). Rather than relying solely on data gathered from the indoor units to determine the refrigerant flow rate, the condensing unit modulates the refrigerant flow to maintain the 9°F (5°C) subcooling value.

The indoor units are equipped with electronic expansion valves (EEVs) that are set to maintain a 9°F (5°C) superheat value. The subcooling and superheat setpoint for other brands can vary.

As individual indoor units start and stop, the refrigerant flow rate in the primary refrigerant circuit changes. This change in the flow rate causes a change in the subcooling value of refrigerant leaving the condenser. A faster flow rate results in a reduction of subcooling.

NOTE

Instead of using the generic term *variable refrigerant flow*, Daikin has trademarked the term *variable refrigerant volume* (VRV™).

On Site

Metering Devices

Ductless systems place the system metering device(s) in the outdoor unit. In VRF systems, the metering device responsible for metering refrigerant to the indoor coil is located there, not in the outdoor unit. However, EEVs may also be found in the outdoor unit to control the flow of refrigerant within the unit itself for various purposes.

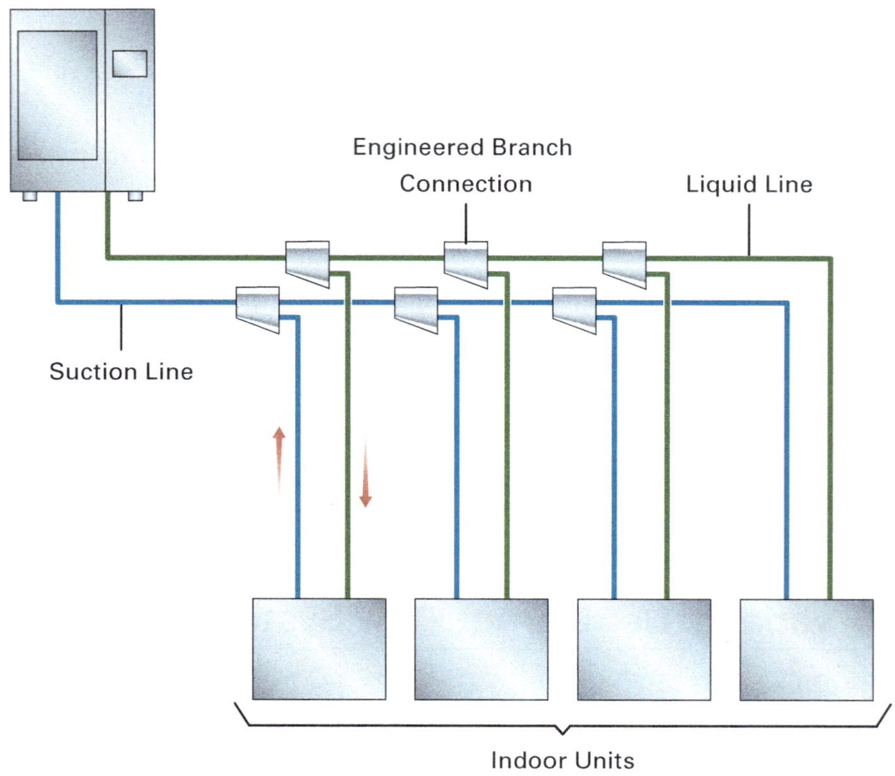

Figure 47 VRF system refrigerant flow schematic.

For example, assume two indoor units rated at 9,000 Btuh each are in steady operation. A third unit rated at 12,000 Btuh then starts to cool a third area. As the third unit's EEV opens to admit refrigerant, the subcooling monitored at the condensing unit drops. This is due to the refrigerant now moving faster through the condenser, with less time allowed for the refrigerant to cool. The condensing unit responds by increasing the speed of the compressor(s) to move more refrigerant, settling at a speed that again results in a 9°F subcooling value.

The outdoor unit contains several temperature and pressure sensors and adjusts the compressor speed to maintain the target subcooling for the combined load of operating indoor units. The control programming for the outdoor unit determines the ideal compressor speed. Subcooling can be enhanced by a secondary heat exchanger positioned downstream of the condenser coil. When the subcooling is lower than desired and cannot be attained using compressor speed alone, the heat exchanger is activated to provide additional cooling.

VRF systems are available in straight cooling, heat pump, and *heat-recovery* models. Like conventional heat pumps, VRF heat pumps can provide both heating and cooling, but not at the same time. Heat recovery systems, however, can provide both heating and cooling simultaneously.

Heat recovery systems do this by also routing a hot-gas refrigerant line to each indoor unit, along with the liquid and suction lines. Heat that is absorbed by the system from units in the cooling mode is then transferred to areas that require heating. This feature is unique among heating and cooling systems, other than in specialized swimming pool systems that transfer recovered heat to pool water.

Figure 48 VRF condensing unit.
Source: Daikin Comfort Technologies North America, Inc

4.1.1 VRF Equipment

Single VRF condensing units (*Figure 48*) are typically available up to about 20 tons in size. VRF outdoor units can be connected in a modular arrangement by a piping manifold to achieve significantly more capacity.

Individual units exceeding 8 tons of capacity usually have multiple compressors. When two compressors are used, one may be a fixed-speed model, while the other is an inverter-driven variable-speed model. Both compressors may be variable-speed in other models or brands.

Other benefits associated with inverter-driven variable-speed compressors include the following:

- Very low starting current
- Soft, quiet startup that does not cause a significant voltage drop on a circuit
- System pressures that build slowly, reducing stress
- Fewer start and stop cycles that cause stress in traditional compressors
- Enhanced ability to dehumidify without overcooling

In a system where the refrigerant flow rate is constantly changing, the return of refrigeration oil to the compressor is a concern. If the refrigerant lines were sized to maintain a proper velocity for oil flow at the lowest possible flow rate, the lines would be too small to support the flow needed at maximum capacity. Thus, VRF systems incorporate oil separators at the outdoor unit to minimize the amount of oil in circulation.

Some manufacturers may also apply what is referred to as an *oil recovery mode*. After a predetermined number of operating hours, the system opens all the indoor EEVs to maximize the flow rate, and then increases the speed of the compressor(s) to 100%. This maximizes the refrigerant velocity through the refrigerant circuit, giving any oil trapped in the circuit an opportunity to return. Oil recovery cycles typically last less than 10 minutes.

Most VRF condensing units are air-cooled, but water-cooled units are also available. Like traditional water-cooled systems, heat rejection equipment such as a cooling tower or dry cooler is required. In addition, a boiler or similar source of heat may be needed to heat the water loop in the winter. The two parts of the system work to maintain the loop water temperature in the appropriate range—such as 67°F to 95°F (19°C to 35°C), for example.

VRF indoor units are available in a variety of sizes and configurations. Models that are ductless and similar in appearance to standard ductless units are popular. Indoor units that are virtually the same as traditional fan coil units are also available, as are units designed to be concealed in small spaces (*Figure 49*). Some units can be fitted with supplemental electric heat. Most ducted models can accommodate outside air connections to satisfy indoor air quality requirements.

Figure 49 VRF fan coil unit designed for concealment.
Source: Daikin Comfort Technologies North America, Inc

VRF indoor and outdoor units must consistently communicate with each other. Most use a DC control protocol that facilitates communication between the units. Shared data includes current mode of operation, calls for specific functions, and error or fault codes. Due to the collective nature of the programming, data is available for monitoring system operation and identifying malfunctions from all connected components.

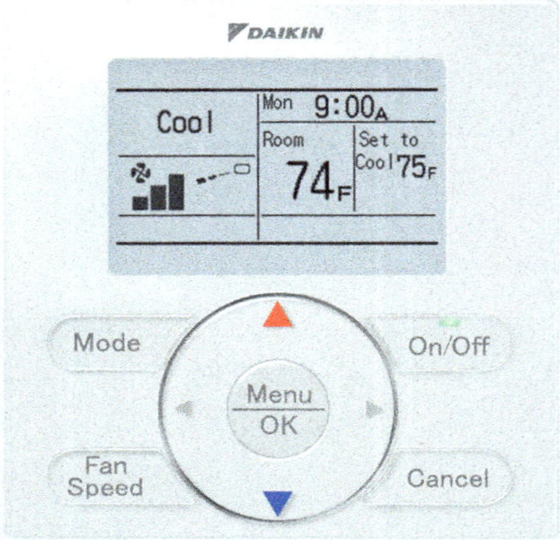

Figure 50 Navigation Remote Controller™.
Source: Daikin Comfort Technologies North America, Inc

Most VRF manufacturers offer a data-interface device that allows the technician to connect and analyze system operation and performance and troubleshoot problems through an app. Tablets and laptops are common ways to access the data.

If a specific indoor unit is having problems, you can use the app to see the readings from each thermistor, the current superheat, the precise position of the EEV, and information related to the blower speed.

VRF indoor units can be controlled in several different ways. One way is through a wireless remote control, much like those used with ductless systems. A simple wall-mounted, wired controller is also available. Depending on the manufacturer, even the simple control devices for VRF systems can often control up to 16 indoor units from a single device. The wired control shown in *Figure 50* is the Navigation Remote Controller™, a trademarked device from Daikin. This controller provides advanced scheduling in addition to controlling up to 16 indoor units.

4.2.0 VRF System Installation and Troubleshooting

NOTE

Manufacturers of VRF and similar systems often require that the installers and service technicians be certified through a corporate training program. This is an important warranty consideration, as a manufacturer may deny warranty claims on systems installed and serviced by technicians who are not certified on a manufacturer's specific product line. If you are not intimately familiar with these systems and do not possess the appropriate training, do not attempt to troubleshoot and repair VRF systems without assistance.

VRF systems have unique refrigerant piping requirements. Failing to follow the specific installation guidelines for the piping will undoubtedly result in functional problems. Sophisticated piping systems present many opportunities for errors. The quality of the overall installation is of the utmost importance to the operation and longevity of a VRF system.

The interconnecting communications wiring must also be done with great care and attention to detail. Thanks to the sophisticated nature of these systems, troubleshooting is often done with very few tools in hand; the system itself can provide much of the information needed to analyze a problem and make an informed decision.

4.2.1 VRF System Installation

System designers often work with equipment manufacturers in the design phase by providing the building plans along with a preferred equipment layout. The manufacturer then provides an engineered piping and control layout, identifying the proper joints and other components that ensure optimal performance. VRF system installations must be planned with great care and the proper knowledge. As one industry professional puts it, "Unlike traditional systems, you cannot properly design a VRF system sitting on the tailgate of a pickup truck."

Engaging the manufacturer in the process also helps to shift some of the contractor's burden for performance to the manufacturer. If the manufacturer's guidance and plans are carefully followed, fewer problems are likely to occur. Problems that do arise are resolved in partnership with the manufacturer.

The most unique installation requirements for VRF systems are associated with the refrigerant piping and control wiring. *Figure 51* provides an example of the refrigerant piping and control wiring for a typical VRF system. Note that the refrigerant lines are represented by a single line in this simplified schematic.

The same refrigerant circuit serves all the indoor units. This requires careful management of the refrigerant flow and great attention to detail during the installation process. Special fittings and piping assemblies are engineered for VRF systems and must be applied as designed. These special piping accessories are provided by the product manufacturer; they are not typically sold generically.

All refrigerant lines, including the hot-gas line used for heat recovery systems, must be insulated. The insulated lines are sometimes wrapped together as a group with a special tape, with communication wiring also included in the bundle.

Using a nitrogen purge during brazing is essential with VRF systems. Because these systems have many strainers, check valves, capillary tubes, and orifices, solid contaminants of any kind must be avoided. It is also important to avoid traps in the refrigerant lines. Remember that the refrigerant flow rate is less than ideal under light loads, and this hinders the return of oil. Traps

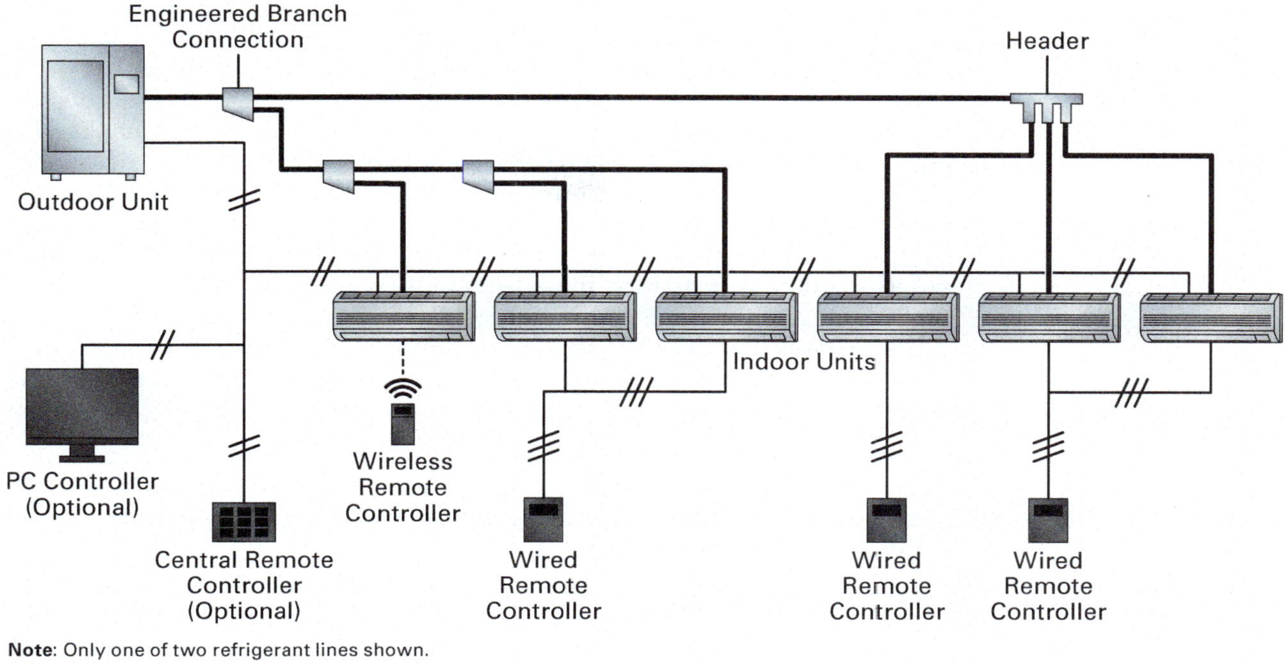

Figure 51 VRF system piping and wiring overview.

further hinder the movement of oil in this case. The installation of refrigerant filter-driers in the lines is not usually recommended.

Engineered branch-line connections and headers (*Figure 52*) are either fabricated or approved by the manufacturer and must be used to create branch lines

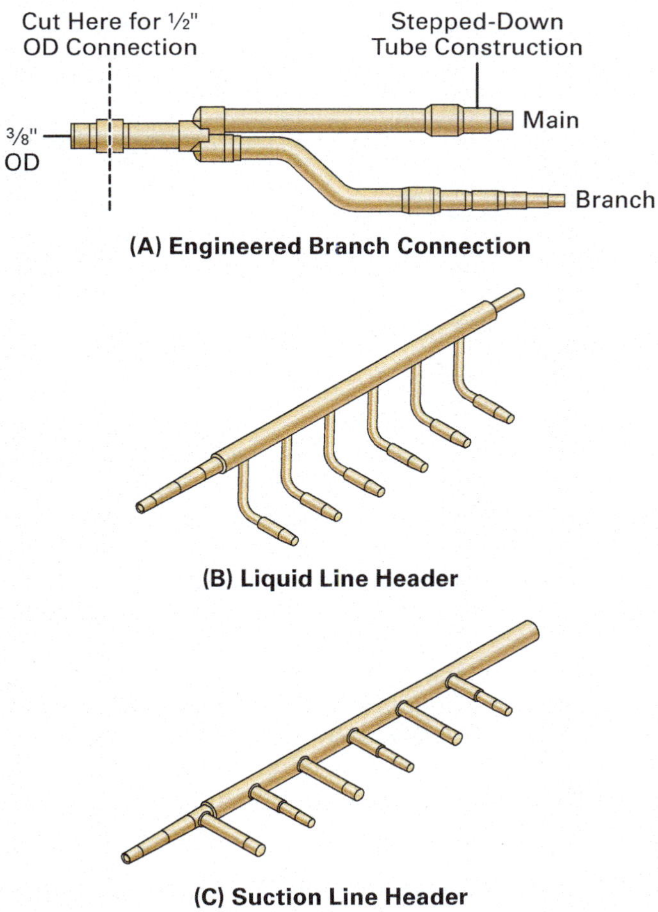

Figure 52 Refnet joints and headers.

Refnet joints: Engineered joints used as connection points in a VRF-system refrigerant network. Refnet joints connect the main refrigerant lines to branch lines or indoor units.

where directed. These engineered connections are often referred to as **refnet joints**. The refnet (refrigerant network) joints include a special distributor to help ensure the balanced distribution of refrigerant in both directions.

Note the stepped-down construction of the refnet joints at the points of connection. The components are designed so that a single style of refnet joint can be used with multiple tubing sizes. The installer simply uses the tube size required for the installation, cutting off any portion of the connection that is smaller than the desired size. This reduces the required inventory of installation components. Manufacturers also provide preformed insulation that fits these joints properly; properly insulating these parts in the field would otherwise be very challenging and time consuming.

The refnet joints must be installed according to the manufacturer's instructions. The system designer must size the piping and choose the location of the joints wisely. When installing refnet joints and headers, consider the following:

- The refnet joint can be oriented vertically in either direction or horizontally with the main and branch lines side-by-side and level with each other. It should never be installed in a manner that places the main line and branch line directly above or below each other.
- Liquid and suction headers must always be installed so that the branch line connections are horizontal, never vertical. Remember that liquid and suction headers are constructed differently and are not interchangeable. Both headers are shown properly oriented for installation in *Figure 52*. Unused ports on a header must be closed by brazing or soldering a cap over the connection.
- A straight length of tubing at least 20" long is generally required ahead of a refnet joint to reduce turbulence. The next refnet joint encountered must then be at least 40" downstream. Refer to *Figure 53*.

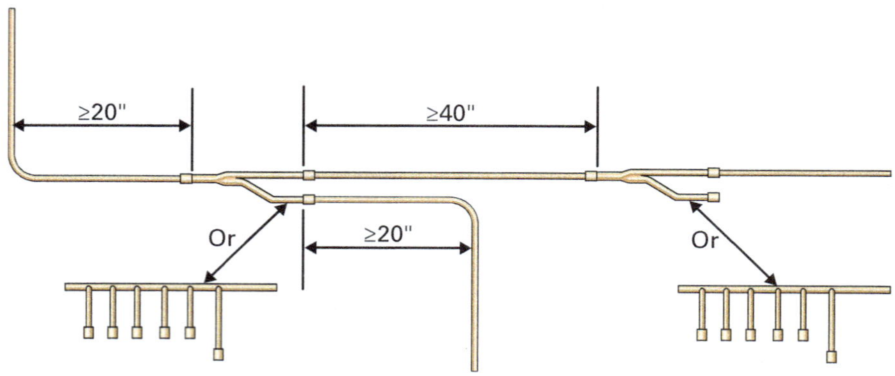

Figure 53 Refnet joint positioning.

Manufacturers also have requirements for the size and length of each segment of the refrigerant piping network. The requirements will differ from one brand to another, and possibly from one model to another. It is essential to follow the manufacturer's specifications for the specific system being installed. Installation technicians are not responsible for sizing or designing the layout of the refrigerant piping, but they should understand the requirements behind the design.

The piping layout may include expansion loops or flexible refrigerant line segments in key locations that could be subjected to expansion and contraction. Otherwise, the piping may be damaged by these powerful forces.

For most VRF systems, maximum and/or minimum values will be provided for the following portions of the refrigerant network:

- The maximum distance from the first refnet joint to the farthest indoor unit
- The maximum distance from the outdoor unit to the nearest refnet joint
- The maximum distance from a refnet joint to the indoor unit it serves

- The minimum distance between refnet joints
- The maximum difference in length between the longest and shortest run connecting refnet joints to an indoor unit
- The maximum difference in height from top to bottom

The refrigerant piping connecting modular condensing units to a single refrigerant circuit also requires attention to detail. When multiple units are improperly connected, oil return problems result and balance in the diverging refrigerant flow paths suffers.

For systems that provide heat recovery with simultaneous heating and cooling, an additional device is required. A *branch selector box* (*Figure 54*) or similar apparatus is required to manage the flow of refrigerant at the indoor unit, facilitating the changeover from one mode to the other.

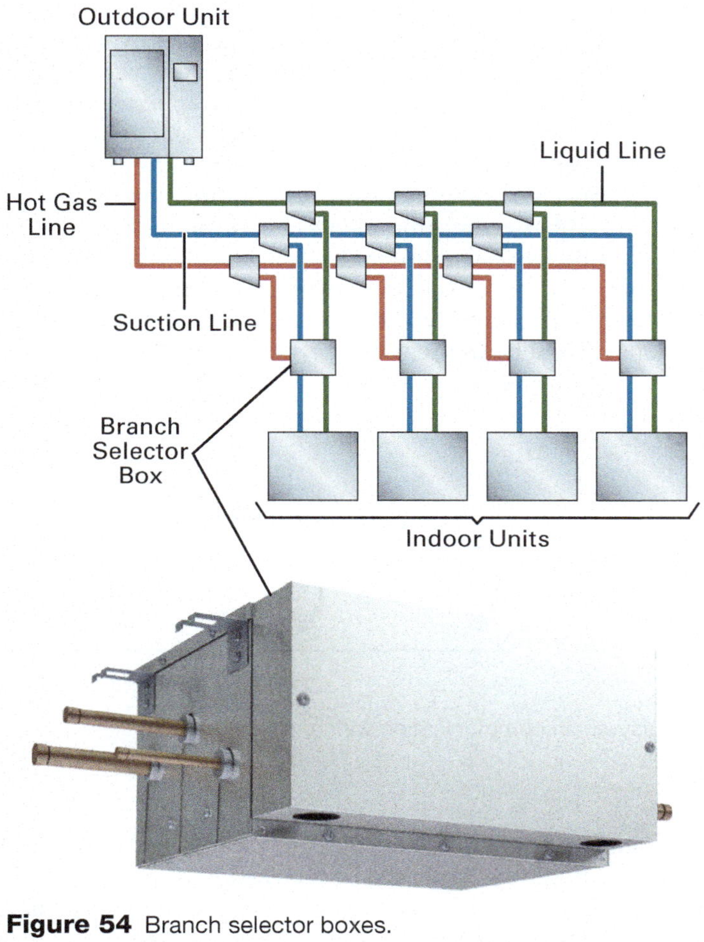

Figure 54 Branch selector boxes.
Source: (Bottom) Daikin Comfort Technologies North America, Inc

(A) Indoor Unit Terminals

(B) Branch Selector Box Terminals

(C) Outdoor Unit Terminals

Figure 55 Example of VRF system communication terminals.
Sources: (A-C) Daikin Comfort Technologies North America, Inc

A heat-recovery system has three refrigerant lines, including the hot-gas line. However, indoor units have only two piping connections—one for liquid and one for vapor. The branch selector box manages what refrigerant sources are actively connected to the indoor units. To know what mode to choose, it must be integrated into the communications network. While branch selector boxes often support a single indoor unit, some models can support up to twelve indoor units.

Terminals are provided on each indoor, outdoor, and branch selector box for the communication wiring. *Figure 55* shows an example of the terminal designations. Note that the communication wiring terminals are designated as *transmission terminals* by some manufacturers.

The terminals designated FORCED OFF are used to interlock the equipment with safety devices, such as the float switch of a condensate pump or a fire alarm. On the outdoor unit, the terminals marked Q1 and Q2 are used when multiple condensing units are connected on the same system. The outdoor units communicate with each other through this circuit. The P1 and P2 terminals are used to connect to a wired controller.

As the wiring travels between one unit and another, the polarity must remain the same; the same wire must be connected to all F1 terminals throughout the system, for example. Remember that these terminal designations will differ from manufacturer to manufacturer.

The wire typically used for communication is two-conductor 18 AWG stranded, non-shielded wire. If the length is excessive, 16 AWG may be needed to avoid voltage loss. Always follow the manufacturer's requirements carefully, as the requirements can differ. Three conductors may be required for some brands or models.

Communication wire rarely needs to be in conduit other than for its own protection. Indoors, it is often routed through ceiling cavities by securing it to other runs of piping or building structural components. However, to avoid interference, the communications wiring should never be strapped along the outside of conduit carrying AC load wiring that generates electrical noise.

Shielded wire is not usually required. If it is used, the shield must be connected to every connected component's ground point along the way. The wire should not be spliced between components; a continuous piece of wire should be routed from component to component.

Communication wiring must be routed in a daisy-chain configuration (*Figure 56*). All terminations and connections occur at the equipment. No connections or junctions should be in the wiring elsewhere. Although other strategies may be fine for distributing power, they do not allow for proper communication signals in this type of system.

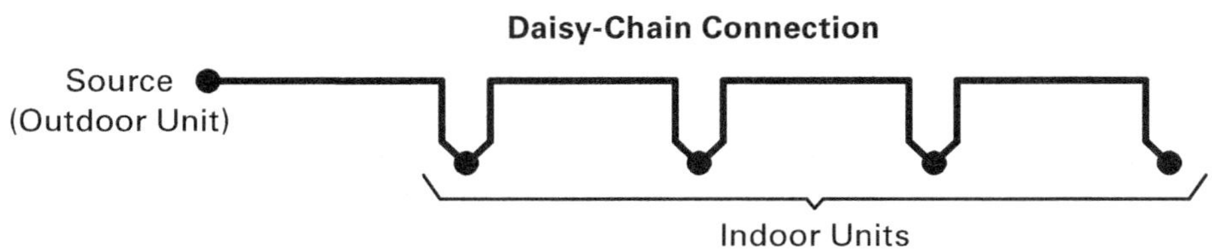

Figure 56 Daisy-chain communication wiring.

Once the installation is complete and the piping has been carefully leak tested and evacuated, the system must be charged with refrigerant. Traditional HVAC and ductless units of five tons capacity or less come with enough refrigerant to handle a specific length of piping, such as 15'. This is not the case with VRF systems.

When charging VRF systems, the entire piping system all the way back to the condensing unit must be considered. The required weight of the refrigerant charge is the sum of the following:

- Weight of the refrigerant liquid that the entire liquid line network can hold
- Weight of the refrigerant vapor that the entire suction line network can hold
- Weight of the refrigerant vapor that the entire hot-gas line network can hold (heat recovery systems only)
- Weight of the refrigerant that each component in the system, such as the branch selector boxes and indoor coils, is rated to hold

Most manufacturers require charging to be done exclusively by weight. A proper refrigerant charge in systems with this level of complexity is critical. When the charge is incorrect due to leaks or other issues, the remaining charge must be recovered, and the system fully recharged by weight. Attempting to charge a system using traditional superheat and subcooling methods is not effective. This significant and time-consuming task is yet another reason the quality of piping and joints must be at its highest for VRF systems.

Did You Know?

HFC-410A Flare Joints

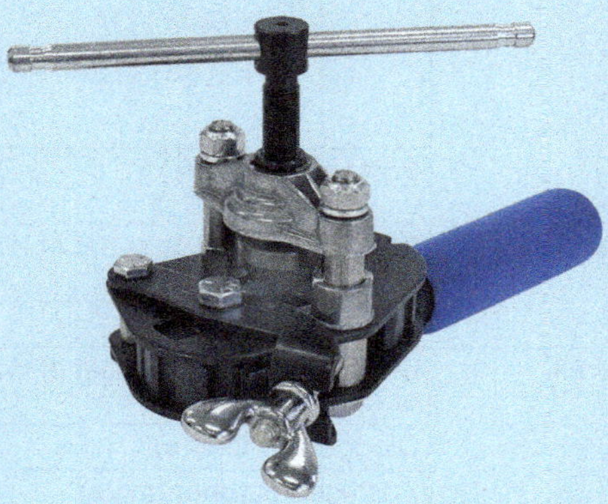

Source: Ritchie Engineering Company, Inc. / YELLOW JACKET

Due to the higher pressures associated with HFC-410A, the requirements for forming reliable flares are slightly different. The outer diameter of the flare should generally be larger, and the flare deeper. With most flaring tools, the user must take care to set the tube in the block at the proper depth. The 45-degree flaring tool shown here takes the guesswork out of the task, automatically setting the tube at the appropriate depth for HFC-410A flares.

For example, the SAE International specification for the diameter of a 3⁄8" flare cone is 0.488" (12.40 mm). The diameter specified by most ductless system manufacturers for their HFC-410A equipment is 0.520" (13.21 mm). This tool from Yellow Jacket® creates flares to that specification. Ductless system manufacturers may also offer flaring tools that are specifically designed to create flares meeting their specifications. Be sure that your tool can form the correct flare to avoid consistent problems due to joint integrity.

4.2.2 VRF System Troubleshooting

Proper training is essential to successfully troubleshoot VRF systems. Although VRF systems are mechanically and electrically complicated, the sophisticated controls and communications can do a great deal of the troubleshooting for you. When troubleshooting traditional systems, technicians often reach for the gauge manifold first. With VRF systems, troubleshooting should never begin with connecting gauges.

Troubleshooting common problems in VRF systems should begin with the symptoms at hand. Charts and tables that provide guidance based on symptoms are offered by many manufacturers (*Figure 57*). In some cases, the suggested solution will seem overly simplified, but many problems are caused by simple errors or a misunderstanding of how the system works.

For more complex problems or those that do not have a simple solution (such as clearing a blocked return air grille), troubleshooting often begins at the zone thermostat or controller. This is where the real value of the communications network for technicians is found. In many VRF systems, the operating parameters of each piece of equipment can be viewed from a single point, regardless of how large or expansive the system may be.

Each manufacturer has a different approach to accessing the needed information from a particular device. It is done through menus to access test functions where error codes related to the problem will be displayed, or by using a unique sequence of keypresses to access the desired information. In some cases, error codes are displayed on a user's main display. However, unique methods of accessing service and maintenance menus provide a layer of protection from user error. Many settings are not understood by the users and their changes can lead to serious problems.

NOTE

Remember that many ductless and VRF systems are manufactured outside of the US. Thus, many troubleshooting charts and tables have been translated to English with mixed results. Expect to see unusual terminology and questionable grammar in some ductless and VRF system manuals.

	Symptom		Supposed Cause	Countermeasure
1	The system does not start operation at all.		Blowout of fuse(s)	Turn Off the power supply and then replace the fuse(s).
			Cutout of breaker(s)	• If the knob of any breaker is in its OFF position, turn ON the power supply. • If the knob of any circuit breaker is in its tripped position, do not turn ON the power supply. 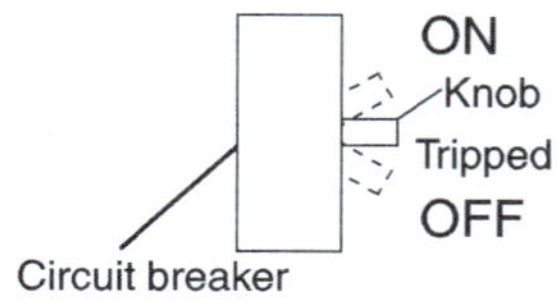
			Power failure	After the power failure is reset, restart the system.
2	The system starts operation but makes an immediate stop.		Blocked air inlet or outlet of indoor or outdoor unit	Remove obstacle(s).
			Clogged air filter(s)	Clean the air filter(s).
3	The system does not cool or heat air well.		Blocked air inlet or outlet of indoor or outdoor unit	Remove obstacle(s).
			Clogged air filter(s)	Clean the air filter(s).
			Enclosed outdoor unit(s)	Remove the enclosure.
			Improper set temperature	Set the temperature to a proper degree.
			Airflow rate set to **LOW**	Set it to a proper airflow rate.
			Improper direction of air diffusion	Set it to a proper direction.
			Open window(s) or door(s)	Shut it tightly.
	[In cooling]		Direct sunlight received	Hang curtains or shades on windows.
	[In cooling]		Too many persons staying in a room	
	[In cooling]		Too many heat sources (e.g. OA equipment) located in a room	
4	The system does not operate.	The system stops and immediately restarts operation.	If the **OPERATION** lamp on the remote controller turns ON, the system will be normal. These symptoms indicate that the system is controlled so as not to put unreasonable loads on the system.	Normal operation. The system will automatically start operation after a lapse of five minutes.
		Pressing the **TEMP ADJUST** button immediately resets the system.		
		The remote controller displays **UNDER CENTRALIZED CONTROL**, which blinks for a period of several seconds when the **OPERATION** button is depressed.	The system is controlled with centralized controller. Blinking display indicates that the system cannot be operated using the remote controller.	Operate the system using the **COOL/HEAT** centralized remote controller.
		The system stops immediately after turning **ON** the power supply.	The system is in preparation mode of micro computer operation.	Wait for a period of approximately one minute.
5	The system makes	The remote controller displays malfunction codes **U4** and **U5**,	The system stops due to an interruption in communication	Remove causes of electrical noises.

Figure 57 Partial symptom-based troubleshooting guide.
Source: Daikin Comfort Technologies North America, Inc

The first menus encountered can provide operating and maintenance information for one or more units on the network (*Figure 58*). All communicating components in the network have a unique address, allowing the technician to access the information for a specific unit.

Problems that affect the entire system are likely related to the outdoor unit(s) or the refrigeration cycle. When only one zone experiences a performance issue, the problem likely lies with the indoor unit serving that area. When a unit number or address is selected from an interface device, the information for that unit is displayed. As shown in *Figure 58*, even the operating hours for each piece of equipment can be viewed.

Maintenanace Menu	Item 2	Remarks	
1. Model Name	1. Unit No.	Select the Unit No. you want to check.	
	2. Indoor unit		
	3. Outdoor unit		
2. Operating Hours	1. Unit No.	Select the Unit No. you want to check.	
	2. Indoor unit operating time	All of these are displayed in hours.	
	3. Indoor fan operation		
	4. Indoor unit energized time		
	5. Outdoor operating time		
	6. Outdoor fan 1 operation		
	7. Outdoor fan 2 operation		
	8. Outdoor comp. 1 operation		
	9. Outdoor comp. 2 operation		
3. Indoor Unit Status	1. Unit No.	Select the Unit No. you want to check.	
	2. FAN	Fan tap	
	3. FLAP	Swing, fixed	
	4. Speed	Fan speed (rpm)	
	5. EV	Degree that electronic expansion valve is open (pls)	
	6. MP	Drain pump ON/OFF	
	7. EH	Electric heater ON/OFF	
	8. Hu	Humidifier ON/OFF	
	9. TBF	Anti-freezing control ON/OFF	
	10. FLOAT		
	11. T1 / T2		
	12. Unit No.	Select the Unit No. you want to check.	
		SkyAir	VRV
	13. Th1	Suction air thermistor	Suction air thermistor
	14. Th2	Heat exchanger thermistor	Heat exchanger liquid pipe thermistor
	15. Th3	—	Heat exchanger gas pipe thermistor
	16. Th4	Discharge air thermistor	Discharge air thermistor
	17. Th5	—	—
	18. Th6	—	—
4. Outdoor Unit Status	1. Unit No.	Select the Unit No. you want to check.	
	2. FAN step	Fan tap	
	3. COMP	Compressor power supply frequency (Hz)	
	4. EV1	Degree that electronic expansion valve is open (pls)	
	5. SV1	Solenoid valve ON/OFF	
		SkyAir	VRV
	6. Th1	Outdoor air thermistor	—
	7. Th2	Heat exchanger thermistor	—
	8. Th3	Discharge pipe thermistor	—
	9. Th4	Heat exchanger deicer thermistor	—
	10. Th5	Heat exchanger gas pipe thermistor	—
	11. Th5	Liquid pipe thermistor	—
5. Forced Defrost (SkyAir only)	1. Forced defrost ON	Enables the forced defrost operation.	
	2. Forced defrost OFF	Disables the forced defrost operation.	

Figure 58 Partial remote-control maintenance menu.
Source: Daikin Comfort Technologies North America, Inc

A distinction is usually made between problems that cause system shutdown and those that merely represent cause for concern. The latter are usually referred to as *warnings*. In this case, a light on the remote and/or indoor unit may remain on, but a warning message is displayed on the thermostat or remote control. For problems that cause system or unit shutdown, this same light may blink rapidly, and an error code will be displayed. Problem indicators will differ from system to system.

Once the correct menu is accessed, one or more error codes are displayed. Each error code should be carefully considered. In some cases, one problem causes other problems that result in additional error codes. For example, on the error code chart shown in *Figure 59*, several error codes are related to

	Malfunction Code	Operation Lamp	Malfunction Contents	Page Referred
Indoor Unit	A0	l	Error of External Protection Device	109
	A1	l	PCB Defect	110
	A3	l	Malfunction of Drain Level Control System (S1L)	111
	A6	l	Fan Motor (M1F) Lock, Overload	113 115 116
	A7	k	Malfunction of Swing Flap Motor (M1S)	119
	A9	l	Malfunction of Moving Part of Electronic Expansion Valve (Y1E)	121
	AF	k	Drain Level Above Limit	123
	AJ	l	Malfunction of Capacity Determination Device	124
	C4	l	Malfunction of Thermistor (R2T) for Heat Exchanger	125
	C5	l	Malfunction of Thermistor (R3T) for Gas Pipes	126
	C9	l	Malfunction of Thermistor (R1T) for Suction Air	127
	CJ	k	Malfunction of Thermostat Sensor in Remote Controller	128
Outdoor Unit	E1	l	PCB Defect	129
	E3	l	Actuation of High-Pressure Switch	130
	E4	l	Actuation of Low-Pressure Sensor	133
	E5	l	Inverter Compressor Motor Lock	135
	E6	l	STD Compressor Motor Overcurrent/Lock	—
	E7	l	Malfunction of Outdoor Unit Fan Motor	138
	E9	l	Malfunction of Moving Part of Electronic Expansion Valve (Y1E, Y2E)	141
	F3	l	Abnormal Discharge Pipe Temperature	143
	F6	l	Refrigerant Overcharged	145
	H3	k	Failure of High-Pressure Switch	—
	H4	l	Actuation of Low-Pressure Switch	—
	H7	l	Abnormal Outdoor Fan Motor Signal	—
	H9	l	Malfunction of Thermistor (R1T) for Outdoor Air	146
	J2	l	Current Sensor Malfunction	—
	J3	l	Malfunction of Discharge Pipe Thermistor (R2T)	147
	J5	l	Malfunction of Thermistor (R3T, R5T) for Suction Pipe	148
	J6	l	Malfunction of Thermistor (R4T) for Outdoor Unit Heat Exchanger	149
	J7	l	Malfunction of Liquid Pipe Thermistor (R7T)	150
	J9	l	Malfunction of Subcooling Heat Exchanger Gas Pipe Thermistor (R6T)	151
	JA	l	Malfunction of High-Pressure Sensor	152
	JC	l	Malfunction of Low-Pressure Sensor	154
	L0	l	Inverter System Error	—
	L1	l	Malfunction of PCB	157
	L4	l	Malfunction of Inverter Radiating Fin Temperature Rise	159
	L5	l	Inverter Compressor Abnormal	162
	L8	l	Inverter Current Abnormal	164

Figure 59 Example error code chart.
Source: Daikin Comfort Technologies North America, Inc

the refrigerant condensing section. Error code E3 indicates the activation of a high-pressure switch; code E7 is related to a malfunction of the outdoor unit fan motor; and code H3 is related to the failure of the high-pressure switch. Only one problem may truly exist, but it can cause more than one error code to be displayed.

Once the error code is determined, the technician is guided to specific pages in the service manual that provide more detailed information. An explanation of the circumstances that caused the error code to be generated is usually found there. A flow chart to guide the technician's troubleshooting activities is often included as well.

The earlier example of error code E7 will be used here to demonstrate the use of the troubleshooting guide. Code E7 refers to a malfunction of the outdoor unit fan motor. The technician is directed to page 138 of the service manual where an explanation of the error code is provided (*Figure 60*). The text explains that this error code is displayed when there are abnormal currents from the main inverter or the fan motor inverter, or when there is a lower-than-expected fan rotational speed.

3.17 "E7" Malfunction of Outdoor Unit Fan Motor

Remote Controller Display	E7
Applicable Models	RXYMQ36 · 48PVJU
Method of Malfunction Detection	Malfunctions detected based on the current value in the inverter PCB or fan PCB, or based on an incorrect rotational speed as detected during fan operations. Five minutes of normal operation following a detection clears the incident.
Malfunction Decision Conditions	■ Overcurrent is detected for inverter PCB (A2P) or fan inverter PCB (A3P). Systems shut down if the condition is detected four times. ■ Rotational speed is below specifications for more than six seconds. System shuts down if the condition is detected four times.
Supposed Causes	■ Fan motor failure ■ Flaw or connection error in the wiring harness connecting the fan motors to the PCB ■ Fan physically impeded by debris or mount failure

Figure 60 Error code E7 explained.
Source: Daikin Comfort Technologies North America, Inc

As noted under Malfunction Decision Conditions in *Figure 60*, if this malfunction is detected four times, the outdoor unit is shut down to avoid further damage. Otherwise, only a warning message will be displayed on the remote and the unit will continue to operate.

Possible causes of the problem listed include a fan motor failure; a defect in the wiring or connections to the motor; or a foreign object interfering with motor rotation. The final note indicates that the error will clear itself if normal motor operation is detected for a full five minutes.

A flow chart is provided on the next page of the service manual to further guide the technician (*Figure 61A* and *Figure 61B*). Specific steps to follow are shown, with a resolution for each step based on the result. In some cases, the technician may be guided to another page for a procedure.

For example, the final step in the flow chart shown in *Figure 61B* and entitled *Check 6* directs the technician to page 215 for guidance on checking the

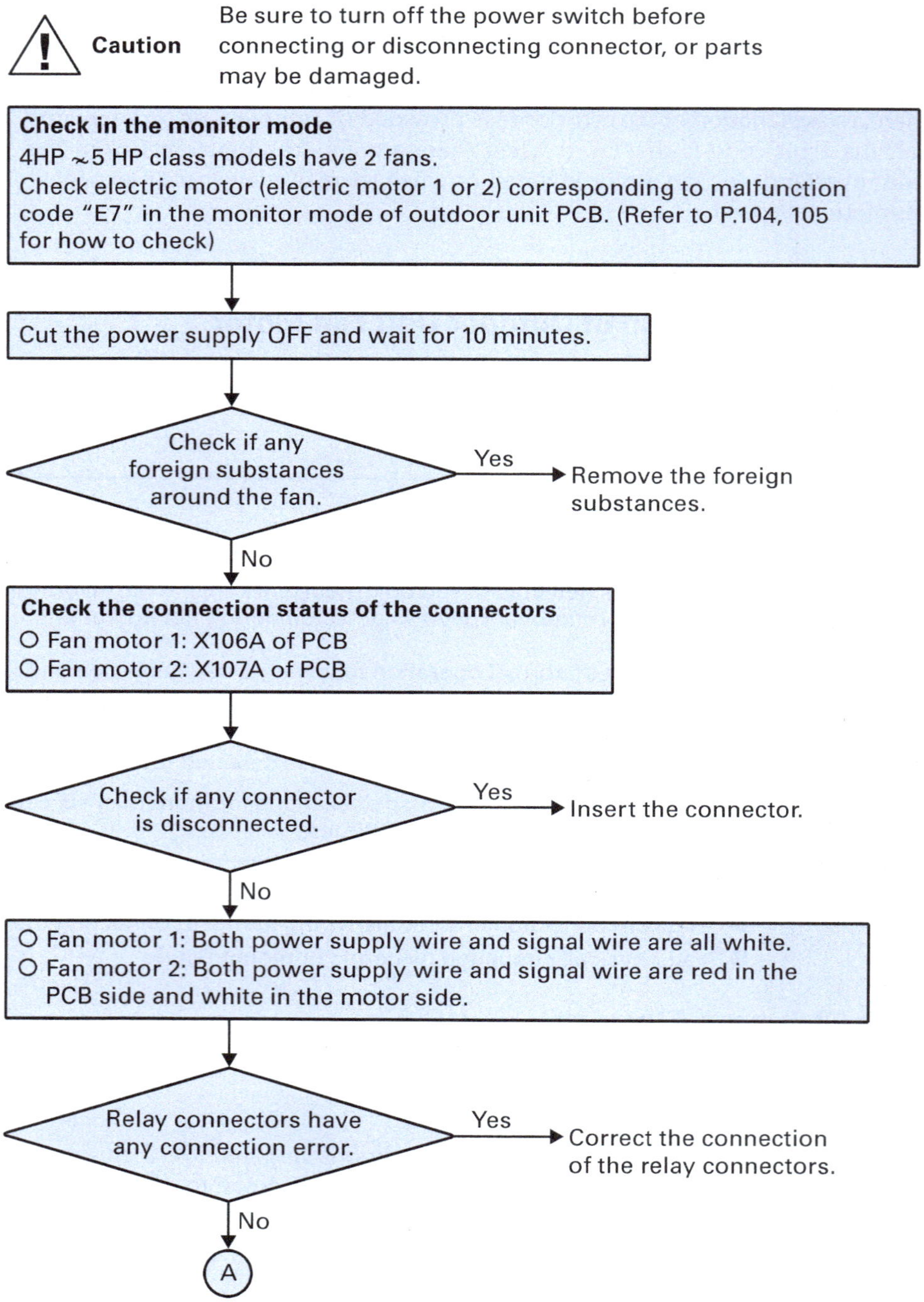

Figure 61A Flow chart to troubleshoot the condenser fan motor (1 of 2).
Source: Daikin Comfort Technologies North America, Inc

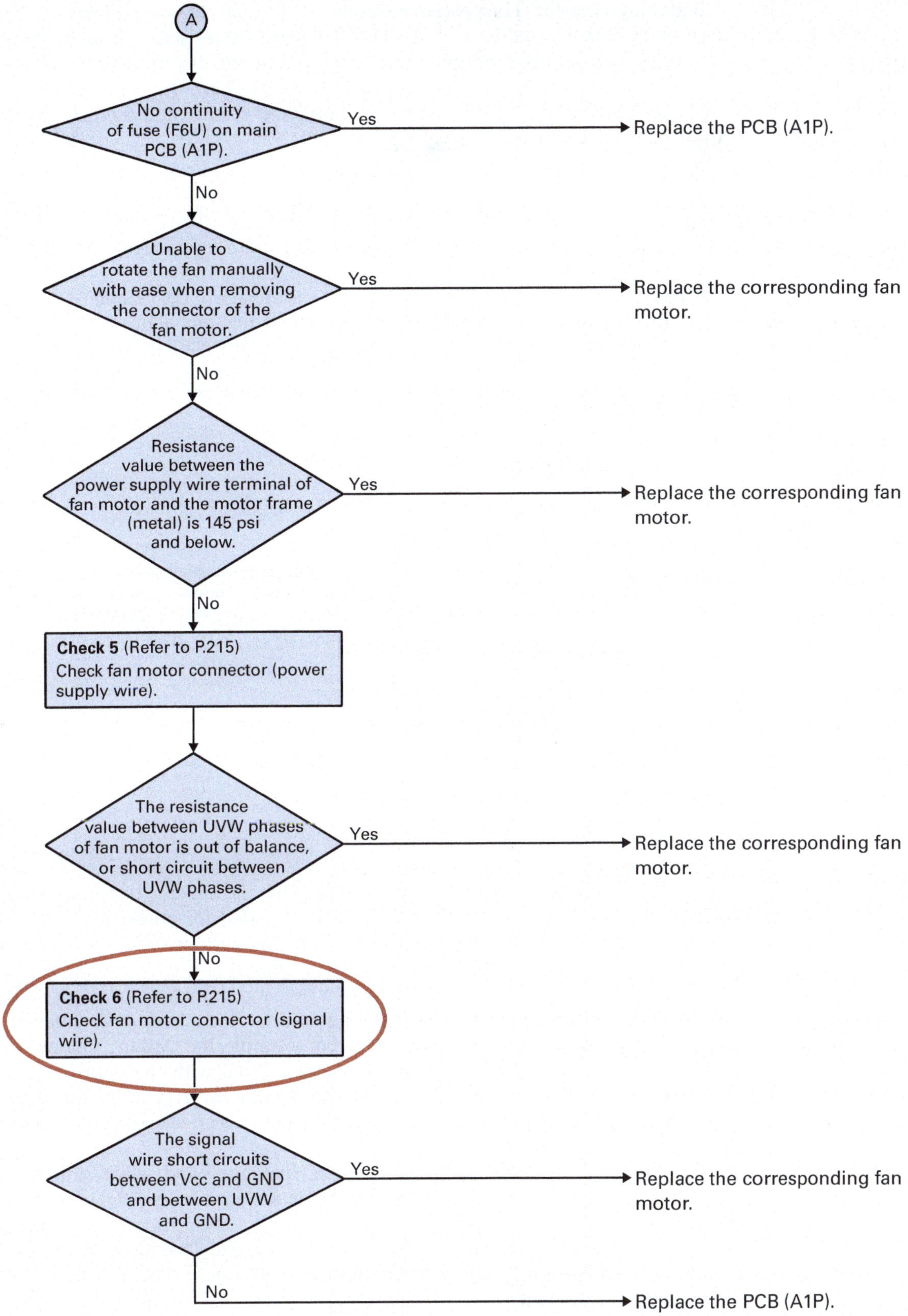

Figure 61B Flow chart to troubleshoot the condenser fan motor (2 of 2).
Source: Daikin Comfort Technologies North America, Inc

NOTE

The letters U, V, and W are routinely used to identify the three windings of an inverter-driven motor.

fan motor connector. The procedure from that page is shown in *Figure 62*. The technician is instructed to use a meter to measure the resistance from terminal Vcc to the U, V, and W terminals, and then measure the resistance from ground to the same three terminals. Returning to the flow chart, the technician reads the statement, answers yes or no based on the test results, and then completes the repair as prescribed.

Check 6

(1) Turn off the power supply.

(2) Measure the resistance between Vcc and each phase of U,V,W, and GND and each phase at the motor side connectors to ensure that the values are balanced within the range of ±20% while the connector is disconnected.

To use a multimeter for measurement, connect the negative probe to Vcc and the positive probe to GND.

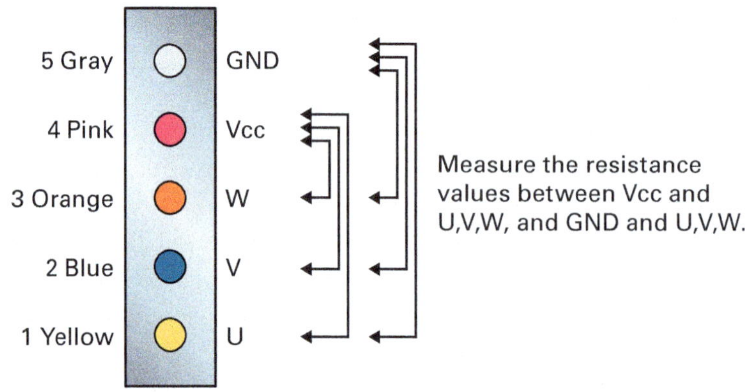

Figure 62 The "Check 6" test procedure.
Source: Daikin Comfort Technologies North America, Inc

For testing thermistors, charts are provided that indicate the resistance reading for a given temperature. Charts are also provided for pressure sensors to check the accuracy (*Figure 63*). In this case, the high-pressure and low-pressure sensors differ in their voltage output at a given pressure.

More powerful ways are available to view and troubleshoot many VRF systems. Some manufacturers have test equipment that serves as an interface between the system and external devices. For example, the Daikin^{TM} Bluetooth Service Checker allows a technician to communicate with the entire system using a laptop, tablet, or smartphone instead of using the very limited displays and keys available from a system device. The device is pocket-sized and connects to a system in seconds.

Troubleshooting VRF systems requires patience and clear thinking. Manufacturer training in the product line being serviced is essential to success. However, properly installed VRF systems also have a high degree of reliability, and many problems are simpler than initially thought. That said, ensure you have the needed training for servicing these systems to avoid liability for errors and accidental damage.

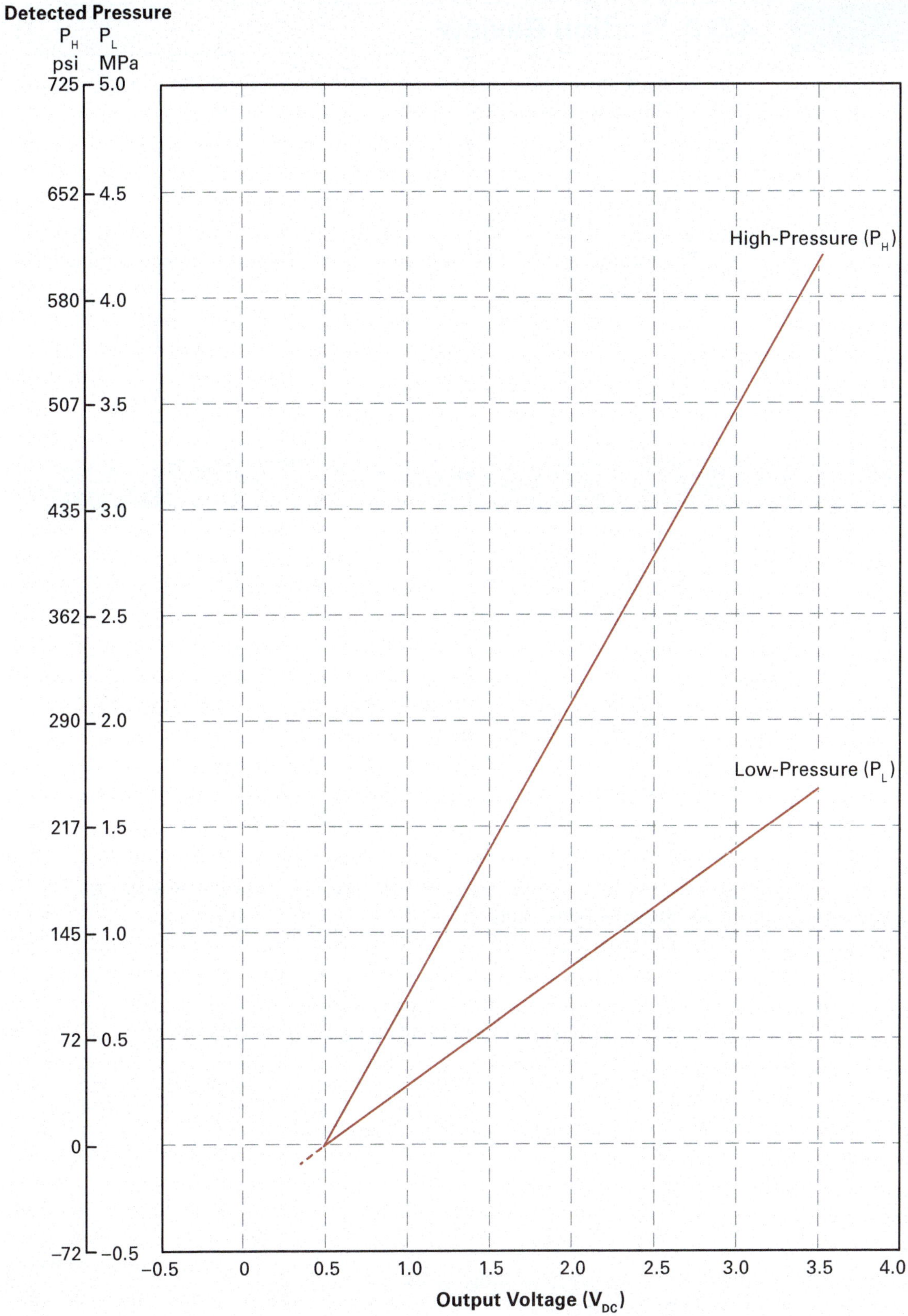

Figure 63 Pressure sensor test chart.

Source: Daikin Comfort Technologies North America, Inc

4.0.0 Section Review

1. At least one manufacturer designs VRF outdoor units to control refrigerant flow based on which factor?
   a. Superheat
   b. Subcooling
   c. Number of connected indoor units
   d. Capacity of the largest indoor unit

2. The wiring used for communications between VRF system components is most often 18 AWG _____.
   a. five-conductor shielded wire
   b. five-conductor solid wire
   c. two-conductor solid wire
   d. two-conductor stranded wire

Module 03315 Review Questions

1. True or False? In a standard residential zoning system, dampers that modulate are powered in one direction and move the other direction with the help of a spring.
   a. True
   b. False

2. To avoid air pressure buildup in a zoning-system dump zone, _____.
   a. another bypass damper is used to control pressure
   b. a return air grille is normally installed in the same area
   c. at least two-thirds of the zone dampers must remain open
   d. a pressure transmitter in the dump zone controls bypass damper position

3. Which type of bypass damper is the least expensive and likely to deliver inconsistent performance?
   a. Barometric damper
   b. Constant pressure damper
   c. Non-modulating motorized bypass damper
   d. Motorized damper with external static pressure control

4. True or False? In a standard zoning system that is operating in the ventilation mode due to the Zone 2 thermostat fan switch set to On, a call for cooling from Zone 3 results in the Zone 2 damper closing.
   a. True
   b. False

5. The control wiring between a standard zone control panel to the condensing unit can be checked by ensuring the compressor contactor is energized when a jumper wire is placed across terminals _____.
   a. R/R_C and G
   b. R/R_C and C
   c. R/R_C and Y
   d. R/R_C and O

6. The individual zone thermostats VVT systems receive time of day and system operating information from _____.
   a. the HVAC unit
   b. a monitor thermostat
   c. their own internal programming
   d. no source; the information is not necessary

7. Pressure-dependent VVT systems _____.
   a. do not require a monitor thermostat
   b. use only two-position zone dampers
   c. do not monitor the volume of air flowing through a zone damper
   d. will cease to function if the duct pressure is higher than 0.5" w.c.

8. In VVT systems, in what position is the outside air damper during the occupied cooling mode?
   a. Fully open
   b. Fully closed
   c. Minimum open position
   d. Modulating based on zone damper activity

9. True or False? When a typical ductless system is operating in the cooling mode, the smaller refrigerant line contains subcooled liquid.
   a. True
   b. False

10. The metering device used in ductless split systems is typically a(n) _____.
    a. capillary tube
    b. automatic expansion valve
    c. electronic expansion valve
    d. thermostatic expansion valve

11. What type of connection is used to connect the refrigerant lines to most ductless systems?
    a. Push-to-connect fittings
    b. Soldered joints
    c. Flare fittings
    d. Pressed fittings

12. Which of the following statements about ductless HVAC systems is correct?
    a. Ductless condensing units should be secured to their foundation.
    b. Wall penetrations should be angled down toward the inside wall.
    c. Ductless indoor units typically require a power source independent of the outdoor unit.
    d. Condensate drain traps should be installed where the drain line passes through the outside wall.

13. In a VRF system, what happens when the system initiates an oil recovery mode?
    a. All the indoor unit EEVs close.
    b. All the valves inside the branch selector boxes close.
    c. The compressor operates at 100% of its rated speed.
    d. The compressor crankcase heater is energized for a specified time.

14. What communication wiring configuration is required for VRF systems?
    a. Star configuration
    b. Homerun configuration
    c. Web configuration
    d. Daisy-chain configuration

15. Which of the following statements about VRF system installation is correct?
    a. Refnet joints can only be installed in specified orientations.
    b. Refrigerant charging is best done using the superheat method.
    c. Shielded wiring is required for communication between components.
    d. A branch selector box is required for every indoor unit in all VRF systems.

Answers to odd-numbered questions are found in the Review Question Answer Key at the back of this book.

Answers to Section Review Questions

Answer	Section	Objective
Section One		
1. b	1.1.1	1a
2. d	1.2.4	1b
3. d	1.3.0	1c
Section Two		
1. b	2.1.2	2a
2. b	2.2.5	2b
Section Three		
1. c	3.1.1	3a
2. a	3.2.1	3b
Section Four		
1. b	4.1.0	4a
2. d	4.2.1	4b

APPENDIX 03317A

Common Schematic Symbols

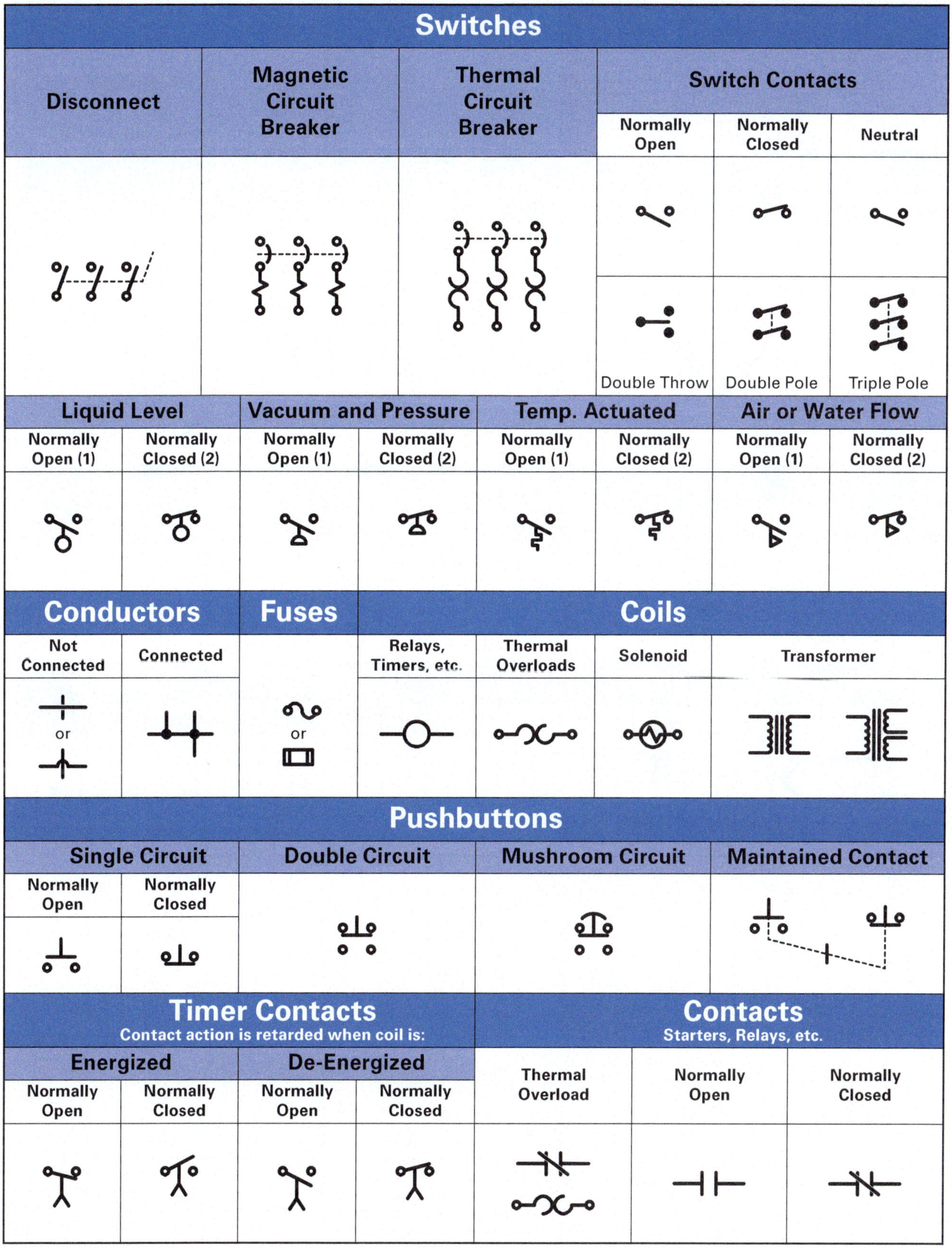

(1) Closes on Rise
(2) Opens on Rise

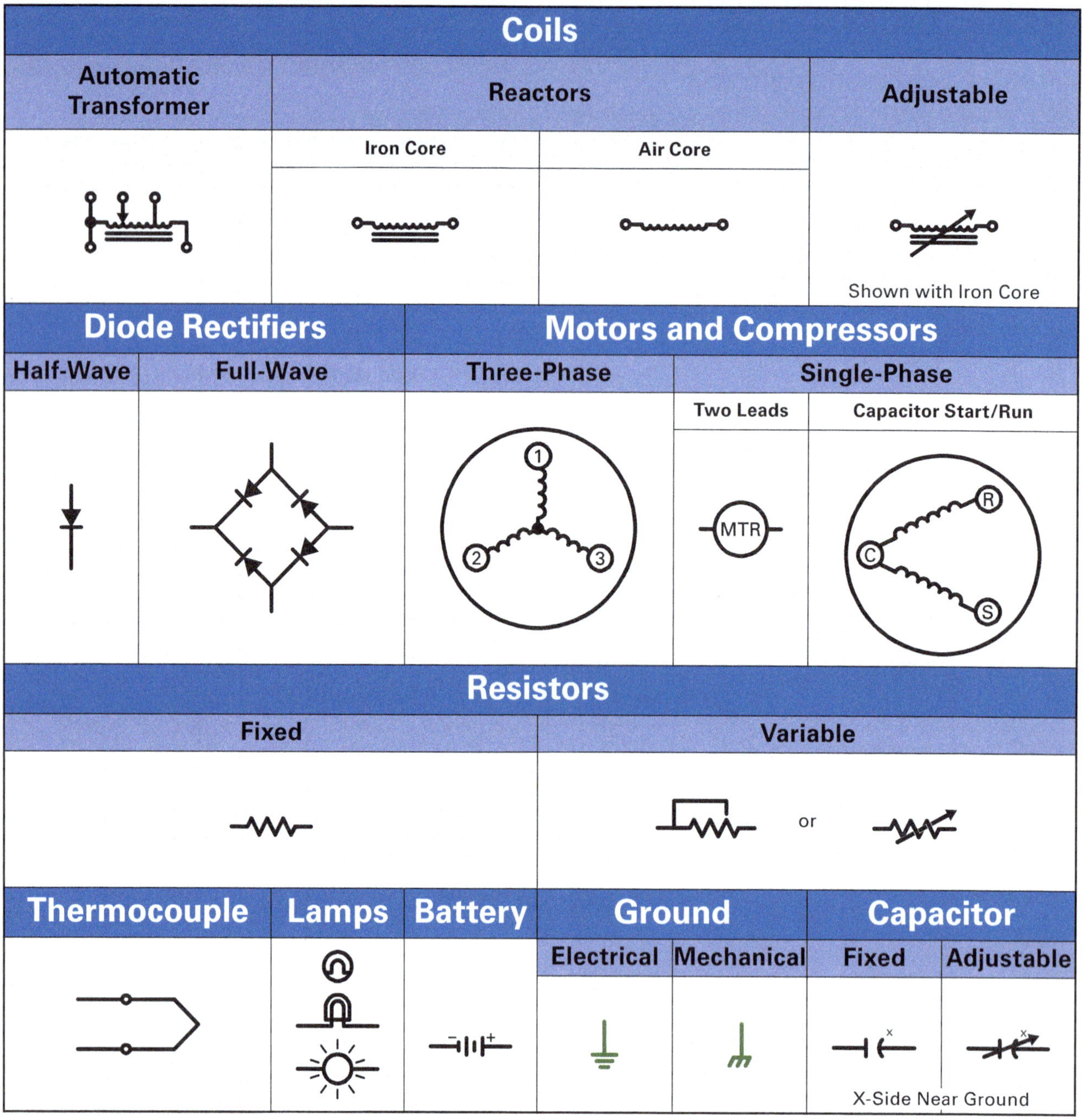

Coils
Automatic Transformer
Reactors
Iron Core
Air Core
Adjustable
Shown with Iron Core
Diode Rectifiers
Half-Wave
Full-Wave
Motors and Compressors
Three-Phase
1
2
3
Single-Phase
Two Leads
MTR
Capacitor Start/Run
R
C
S
Resistors
Fixed
Variable
or
Thermocouple
Lamps
Battery
Ground
Electrical
Mechanical
Capacitor
Fixed
Adjustable
X-Side Near Ground

APPENDIX 03318A

Sample VFD Application Checklist

VFD Application Checklist

Motor

New:____________ Existing:____________ Horsepower: _______ Base Speed: _______ Voltage: _______

FLA: ____________ LRA: ____________ NEMA Design: _____ Gearbox/Pulley Ratio:____________

Service Factor:____________

Load

Application:____

Load Type: Constant Torque________ Variable Torque ________ Constant Horsepower ____________

Load inertia reflected to motor:____

Required breakaway torque from motor:____

Running load on motor:____

Peak torques (above 100% running):____

Shortest/longest required accel. time:________ /______ secs up to _________ Hz from zero speed

Shortest/longest required decel. time:________ /______ secs down to _______ Hz from max. speed

Operating speed range:______ Hz to ______ Hz

Time for motor/load to coast to stop: _________ secs

VFD

Source of start/stop commands: ____

Source of speed adjustment: ____

Other operating requirements: ____

Will the motor ever be spinning when the VFD is started? ____________________________

Is the load considered to be high inertia? ____________________________________

Is the load considered to be hard to start? ____________________________________

Distance from VFD to the motor: ______________ feet

Type of VFD (V/Hz, Flux Vector, Closed Loop Vector): ____________________________

Options desired: ____

____

____

Other special requirements/conditions: ____________________________________

VFD Application Checklist (Continued)

Power Supply

Supply Transformer: ______ kVA and ______ % Z or short circuit current at drive input: ______ amps

(If the drive does not include a built-in line reactor and the available feeder short circuit current is more than 100 times the drive FLA rating, a 1% line reactor or a drive isolation transformer is required.)

Total horsepower of all drives connected to supply transformer or feeder: ______ hp

Is a drive transformer or line reactor desired? ______

Any harmonic requirements? ______ % Voltage THD: ______ % Current THD: ______ IEEE 519: ______

Total non-drive load connected to the same feeder as drive(s): ______ amps

Service

Start-up Assistance: ______ Customer Training: ______

Preventive Maintenance: ______ Spare Parts: ______

Additional Issues

Will the VFD operate more than one motor? ______

Will the power supply source ever be switched with the VFD running? ______

Is starting or stopping time critical? ______

Are there any peak torques or impact loads? ______

Will user-supplied contactors be used on the input or output of the VFD? ______

Does the user or utility system have PF capacitors that are being switched? ______

Will the VFD be in a harsh environment or high altitude? ______

Does the utility system experience surges, spikes, or other fluctuations? ______

APPENDIX 03209A

Gas Furnace Troubleshooting Flowchart

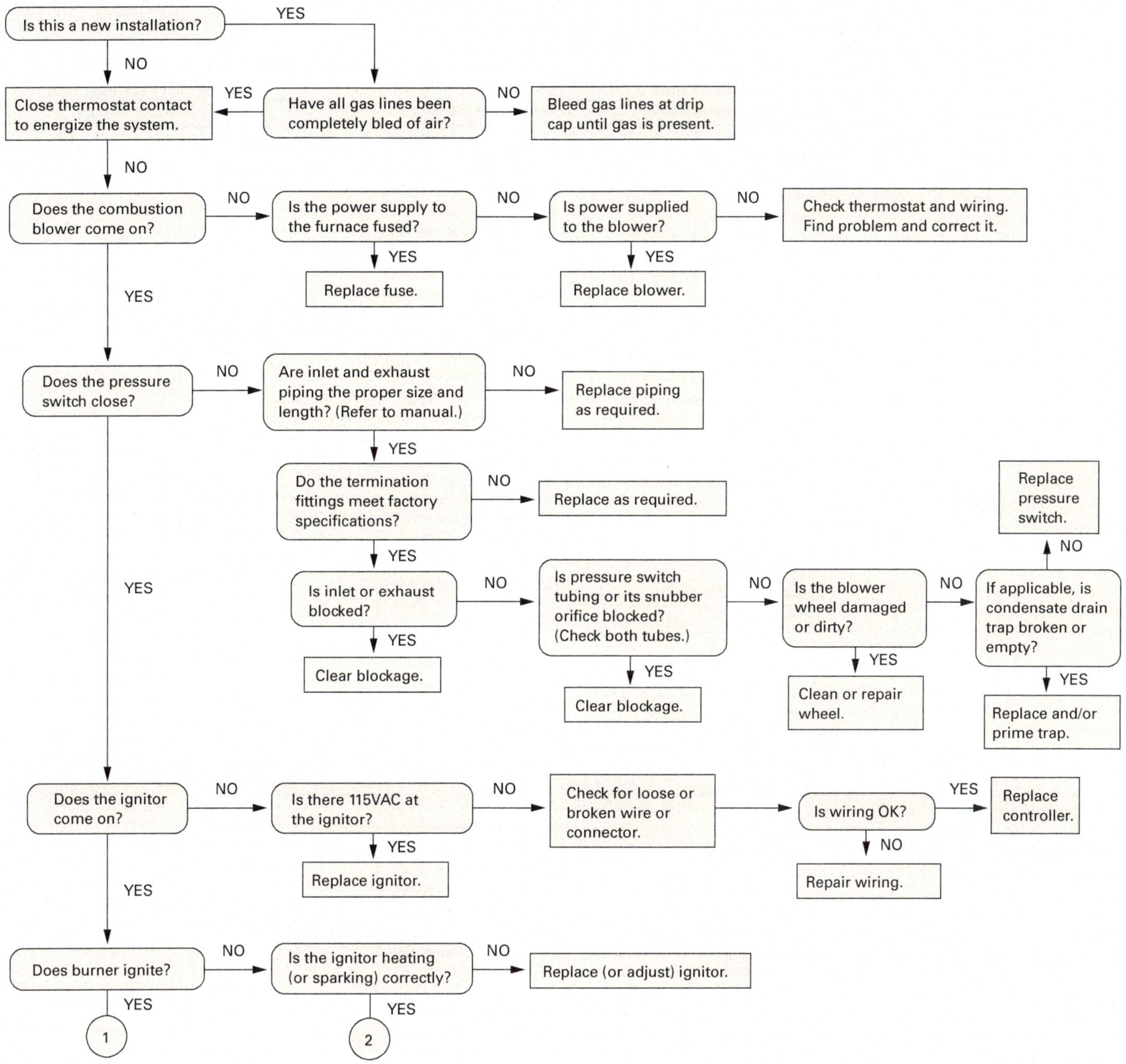

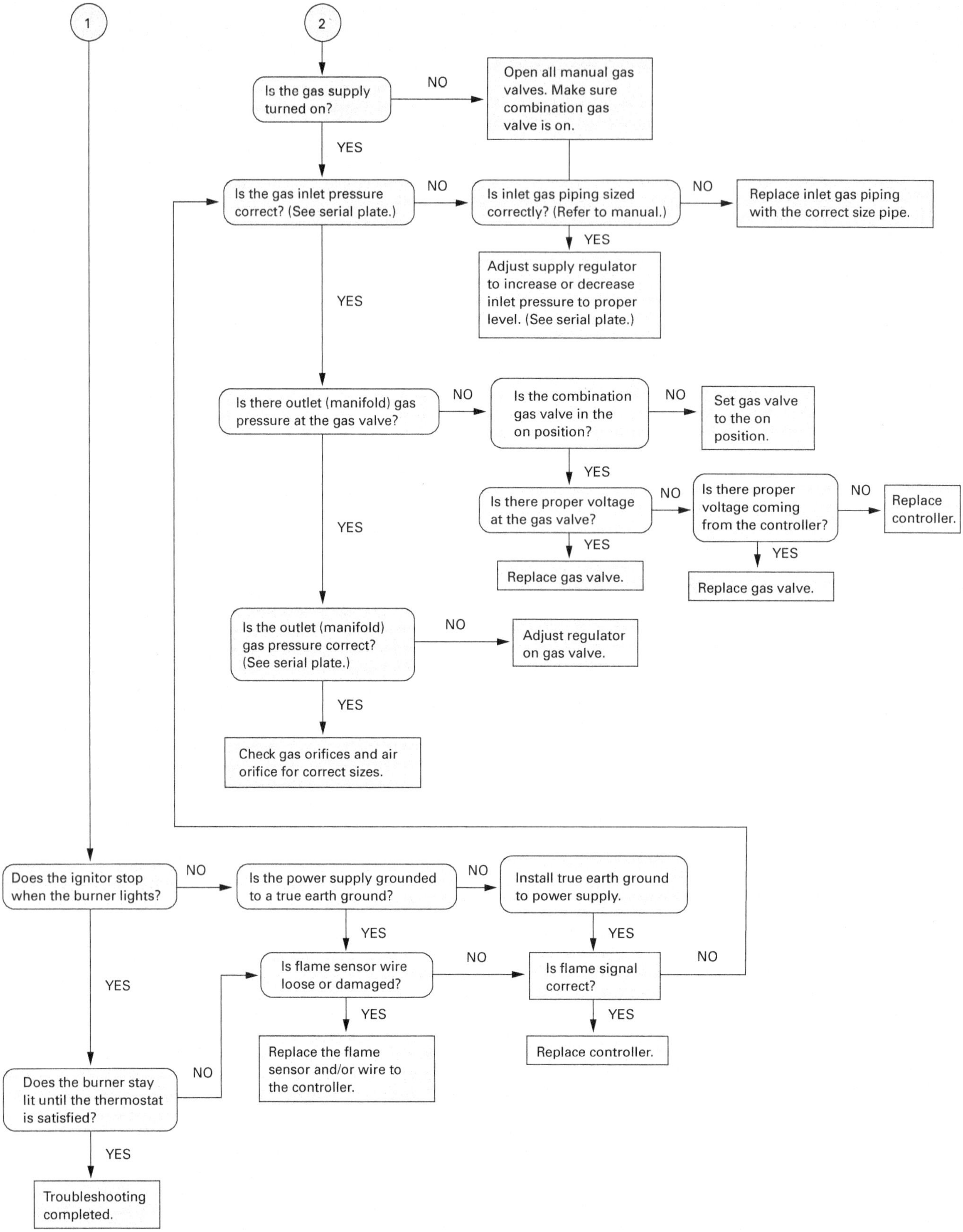
1
2
Is the gas supply turned on?
NO
Open all manual gas valves. Make sure combination gas valve is on.
YES
Is the gas inlet pressure correct? (See serial plate.)
NO
Is inlet gas piping sized correctly? (Refer to manual.)
NO
Replace inlet gas piping with the correct size pipe.
YES
Adjust supply regulator to increase or decrease inlet pressure to proper level. (See serial plate.)
YES
Is there outlet (manifold) gas pressure at the gas valve?
NO
Is the combination gas valve in the on position?
NO
Set gas valve to the on position.
YES
Is there proper voltage at the gas valve?
NO
Is there proper voltage coming from the controller?
NO
Replace controller.
YES
Replace gas valve.
YES
Replace gas valve.
YES
Is the outlet (manifold) gas pressure correct? (See serial plate.)
NO
Adjust regulator on gas valve.
YES
Check gas orifices and air orifice for correct sizes.
Does the ignitor stop when the burner lights?
NO
Is the power supply grounded to a true earth ground?
NO
Install true earth ground to power supply.
YES
YES
Is flame sensor wire loose or damaged?
NO
Is flame signal correct?
NO
YES
Replace the flame sensor and/or wire to the controller.
YES
Replace controller.
YES
Does the burner stay lit until the thermostat is satisfied?
NO
YES
Troubleshooting completed.

APPENDIX 03310A

Oil Furnace Troubleshooting Summary

Problem	Possible Cause	Corrective Action
Burner Motor Does Not Start		
Trouble in primary control	Check cad cell	Replace if resistance is incorrect
Trouble in thermostat	Broken wires or loose connections Defective thermostat	Replace wires or tighten connections Replace thermostat
Faulty burner components	Broken wires or loose connections Motor start switch or thermal overload switch open Defective motor Defective pump	See manufacturer's instructions
Burner Motor Starts – No Flame Established		
Trouble in primary control	Check cad cell or stack relay Loose connections or broken wires between ignition transformer and primary control	Replace wires or tighten connections
Trouble in ignition system	Defective transformer Ignition electrodes: • Improperly positioned • Spaced too far apart • Loose • Dirty Dirty or damaged ceramic insulators	Replace transformer Check manufacturer's instructions Replace or clean electrode assembly
Faulty burner components	Dirty nozzle Loose, misaligned, or worn nozzle Clogged oil pump strainer Clogged oil line Air leak in suction line Defective pressure regulator valve Defective pump Improper draft Water in oil Oil too heavy	Correct per manufacturer's instructions
Burner Motor Starts – Flame Goes On and Off After Start-Up		
Trouble in primary control	Cad cell problems	Check and/or clean cad cell
Trouble in ignition system	Limit settings too low Dirty oil filter	Readjust setting or differential Replace filter
Trouble in thermostat	Broken wires or loose connections Defective anticipator Defective thermostat	Replace broken wires or tighten connections Adjust anticipator Replace thermostat
Faulty burner components	Low oil pressure Defective pump, clogged or dirty oil lines Fluctuating water level (hydronic systems) Weak transformer Cracked electrodes	See manufacturer's instructions Replace transformer Replace porcelain electrodes

Problem	Possible Cause	Corrective Action
System Overheats House		
Trouble in primary control	Defective primary	With burner running, disconnect one low-voltage thermostat lead from primary control terminal; if relay does not drop out, replace primary
Trouble at thermostat		
	Wiring shorted	Repair or replace wiring
	Thermostat out of calibration	Recalibrate thermostat with an accurate thermometer
	Defective anticipator or anticipator improperly set	Replace or readjust anticipator; replace thermostat, if necessary
	Thermostat stuck in call-for-heat	Repair or replace thermostat
	Thermostat improperly located in area under control, or draft on stat through wall	Relocate thermostat: must be out of drafts, away from radiating surfaces, ducts, steam pipes, and sunny locations
	Thermostat not level	Level thermostat
Trouble in distribution system	Ductwork close to thermostat	Relocate thermostat
	Flow control valve (hot water) stuck in ON position	Repair or replace valve
	Circulator does not stop running	Check circulator circuits, replace switching device
Faulty burner components	Extremely oversized system	Replace system
	System control valves stuck in ON position	Repair or replace control valve
System Underheats House		
Trouble at thermostat	Open or loose wiring	Repair or replace wiring
	Thermostat out of calibration	Recalibrate thermostat with an accurate thermometer
	Defective anticipator or anticipator improperly set	Replace or readjust anticipator; replace thermostat, if necessary
	Thermostat not calling for heat	Repair or replace thermostat
	Thermostat improperly located	Relocate to sense more accurately
	Thermostat not level	Level thermostat
Trouble at high- or low-limit controller	Controller set too low	Raise setpoint
	Slow to return to ON position	Adjust differential
	Controller defective	Replace controller
Trouble at blower (for forced-air system)	Cutting out or cycling on overload	Check air filters
	Burned out blower motor	Replace motor
	Running too slow or inadequate capacity	Check/replace motor
	Fan belt broken or loose	Replace or tighten belt
System defects	Undersized furnace or boiler system (possible if addition has been made to house)	Replace furnace or boiler system
	Distribution system inadequate	Correct problem
	Sooted heat exchanger	Clean heat exchanger
	Dirty warm-air filters	Replace filters
	Dirty boiler water (steam system)	Drain/refill boiler

Problem	Possible Cause	Corrective Action
System Underheats House (*Continued*)		
Faulty burner components	Fuel line clogged or strainer plugged	Clear clog
	Air in fuel line or excess air in system	Check fuel lines for leaks
	Dirty or improperly sized nozzle	Replace nozzle
	Low oil pressure or defective pump	Adjust or replace pump
	Motor does not come up to proper speed	Replace motor
	Inoperative or inadequate circulator (hydronic system)	Replace circulator
System Cycles Too Frequently (This Problem Is Outside the Primary Control; The Most Common Problems Are Listed Below)		
Trouble at thermostat	Poor location	Relocate thermostat
	Poorly adjusted anticipator	Readjust anticipator; replace thermostat, if necessary
	Loose connections or defective wiring	Tighten connections or replace broken wires
Trouble at high limit	Limit set too low	Reset limit
	Differential too narrow	Reset differential; replace control, if necessary
Faulty burner system	Dirty or clogged air filter	Replace filter
	Loose connections	Repair connections
	Fluctuating voltage	Contact utility
	Intermittent shorts	Find short circuit
Relay Chatters After Pulling In		
Trouble at thermostat	Open contacts	Replace thermostat
Trouble at primary control	Supply voltage too low	Check for loose connections
	Defective load relay	Replace primary

Oil Furnace Troubleshooting Chart

Generally the Cause Make these checks first.	**Occasionally the Cause** Make these checks only if the first checks failed to locate the trouble.	**Rarely the Cause** Make these checks only if other checks failed to locate the trouble.
Problem: No Heat – burner fails to start		
Power failure Blown fuses or tripped circuit breaker Open disconnect switch or blown fuse Thermostat set too low Thermostat switch in improper position Primary or safety controls Flame detector	Control relay or contactor Limit controls Thermostat not level Fuel pump Burner motor	Faulty wiring Loose terminals Low voltage Thermostat faulty
Problem: No Heat – burner starts but fails to ignite		
Electrodes Ignition transformer Improper burner adjustment Oil supply and oil pressure Manual fuel valve	Fuel pump Nozzle Burner motor Oil line filter Fuel lines	Low voltage Water in oil
Problem: No Heat – burner starts and fires, then loses flame before locking out on safety		
Fuel pump Nozzle Oil supply and oil pressure Oil line filters Water in oil Primary or safety controls Flame detector Improper burner adjustment	Fuel lines Nozzle Displaced or damaged baffles	Faulty wiring Loose terminals Vent or flue problem
Problem: Not Enough Heat – burner cycle too short		
Control relay or contactor Thermostat heat anticipator setting Thermostat not level Thermostat location Dirty filters	Limit controls Fan controls Blower bearings Blower motor Defective blower wheel Dirty blower wheel	Faulty wiring Loose terminals Low voltage Low air volume Ductwork small or restricted
Problem: Not Enough Heat – burner cycle too long		
Blower belt broken or slipping Dirty filters Dirty or plugged heat exchanger	Fuel pump Low air volume Dirty blower wheel Oil line filter	Ductwork small or restricted Oil supply and oil pressure

Generally the Cause Make these checks first.	Occasionally the Cause Make these checks only if the first checks failed to locate the trouble.	Rarely the Cause Make these checks only if other checks failed to locate the trouble.
Problem: Too Much Heat – Burner Cycles Too Long		
Thermostat heat anticipator setting too high Thermostat out of calibration	Thermostat not level Thermostat location	
Problem: Too Much Heat – Burner Runs Continuously		
Faulty wiring Control relay or contactor	Thermostat faulty	Thermostat not level Thermostat location
Problem: Combustion Noise		
Nozzle Electrode Dirty or plugged heat exchanger Fuel pump Burner air adjustment	Burner motor Ignition transformer	Faulty wiring Loose terminals Vent or flue problem Oil supply
Problem: Air Noise		
Blower Ductwork small or restricted	Cabinet	Blower wheel Dirty filters
Problem: Cost of Operation		
Burner motor Barometric control Blower motor Dirty filters Improper burner adjustment Inefficient burner	Low air volume Insufficient insulation Excessive infiltration	Low voltage Fan control Nozzle Ignition transformer Blower belt broken or slipping Dirty blower wheel Ductwork small or restricted Vent or flue problem Input too low
Problem: Mechanical Noise		
Blower bearings Blower motor Blower belt broken or slipping Blower wheel out of balance	Cabinet	Low voltage Control relay or contactor Fuel pump Burner motor Oil valve Ruptured heat exchanger Displaced or damaged baffles Fuel lines
Problem: Odor		
Barometric control Vent or flue problems Flue gas spillage Fuel lines Oil supply Oil pump shutoff valve	Low air volume Improper burner adjustment Ruptured heat exchanger Displaced or damaged baffles Blocked heat exchanger Stagnant water in humidifier Water or moisture in air system	Faulty wiring or loose terminals Control transformer Fuel pump Dirty filters Dirty heat exchanger Input too high Input too low Outdoor odors

APPENDIX 03213

A. Specifications for Minimum Fasteners for Lap-Joined Straps

When hanging straps are lap-joined to increase their length for hanging rectangular metal duct, you must place the fasteners along the length of the lap in a single line, not side-by-side across the width of the strap. The minimum recommended number and size of fasteners used for lap-joining the straps are as follows:

Strap Width and Gauge	Connecting Hardware
1" × 22 ga, 20 ga, or 18 ga metal	(2) #10 screws or (1) ½" bolt
1" × 16 ga metal	(2) ½" bolts
1½" 16 ga metal	(2) ⅜" bolts

B. Maximum Loads for Single Metal Hangers

Hanger Size	Hanger Form	Maximum Load
1"	22 ga metal strap	260 lb
1"	20 ga metal strap	320 lb
1"	18 ga metal strap	420 lb
1"	16 ga metal strap	700 lb
1½"	16 ga metal strap	1,100 lb
0.106"	Rod	80 lb
0.135"	Rod	120 lb
0.162"	Rod	160 lb
¼"	Rod	270 lb
⅜"	Rod	680 lb
½"	Rod	1,250 lb
⅝"	Rod	2,000 lb
¾"	Rod	3,000 lb

C. Strap Sizes and Spacing

The minimum metal hanger sizes for round sheet-metal duct, based on conventional wall thickness and an allowance for insulation, with a maximum spacing of 12', are as follows:

Duct Diameter	Strap Hanger(s)	Rod Size(s)	Wire Hanger(s)
Sizes up to 24"	(1) × 22 ga	(1) ¼"	(2) 10 ga
25" through 36"	(1) × 20 ga	(1) ⅜"	(2) 8 ga
37" through 50"	(2) × 20 ga	(2) ⅜"	Not Allowed
51" through 60"	(2) × 18 ga	(2) ⅜"	Not Allowed
61" through 84"	(2) × 16 ga	(2) ⅜"	Not Allowed

Module Review Answer Key

MODULE 01 (03317)

Answer	Section
1. a	1.1.1
3. d	1.1.2
5. a	2.1.3
7. c	2.2.1
9. c	3.2.1
11. b	3.2.1, *Figure 48*
13. c	3.2.1, *Figure 48*
15. a	3.6.0

MODULE 02 (03318)

Answer	Section
1. a	1.1.0
3. c	1.2.2
5. a	1.3.2
7. a	2.0.0
9. d	2.1.0
11. c	2.3.1
13. b	3.1.0
15. d	3.4.3

MODULE 03 (03210)

Answer	Section
1.d	1.1.0
3.d	1.4.0; *Figure 3*
5.b	1.5.1
7.d	1.7.2
9.d	2.1.0
11.b	2.1.3
13.b	2.2.3
15.c	2.4.2

MODULE 04 (03311)

Answer	Section
1. b	1.1.1
3. b	1.2.0
5. d	1.2.2
7. c	2.1.0
9. b	2.3.0
11. b	3.1.1
13. c	3.2.0
15. c	3.3.0

MODULE 05 (03209)

Answer	Section
1. d	1.2.0
3. d	1.2.1
5. c	1.2.3
7. c	2.0.0
9. b	3.0.0

MODULE 06 (03310)

Answer	Section
1. b	1.0.0
3. b	1.2.2
5. c	1.3.0
7. b	1.4.2
9. c	2.1.5
11. d	2.2.1
13. c	3.1.2
15. c	3.2.2

MODULE 07 (03312)

Answer	Section
1. d	1.2.0
3. a	1.2.0
5. c	1.3.0
7. c	1.4.0
9. b	2.1.2

MODULE 08 (03213)

Answer	Section
1. c	1.1.0
3. a	1.2.0; *Table 1*
5. a	2.0.0
7. d	2.1.3
9. b	3.1.1
11. c	4.0.0
13. b	4.1.3
15. a	4.2.0
17. c	4.2.2
19. c	5.1.2
21. b	5.2.2
23. d	6.1.0
25. b	6.2.2

MODULE 09 (03203)

Answer	Section
1. c	1.2.0
3. c	2.1.2
5. b	2.2.1
7. b	2.4.4
9. b	3.1.2
11. c	3.2.0
13. d	4.1.2
15. a	4.2.1

MODULE 10 (03306)

Answer	Section
1. d	1.1.0
3. c	2.1.0
5. b	2.2.2
7. b	3.1.1
9. c	3.2.1
11. c	3.3.0
13. a	4.1.4
15. d	4.3.5

MODULE 11 (03214)

Answer	Section
1. d	1.1.0
3. b	1.2.0
5. b	2.1.0
7. d	2.1.2
9. c	2.2.1
11. c	3.1.1
13. b	4.1.1
15. b	4.2.2

MODULE 12 (03304)

Answer	Section
1. c	1.0.0
3. b	1.1.0
5. c	1.3.2
7. b	2.2.0
9. c	2.3.1
11. d	3.2.5
13. a	4.2.1
15. b	4.2.2

MODULE 13 (03315)

Answer	Section
1. b	1.1.2
3. a	1.1.3
5. c	1.3.1
7. c	2.2.2
9. b	3.1.1
11. c	3.2.1
13. c	4.1.1
15. a	4.2.1

GLOSSARY

Adsorb: To attract and hold gas or liquid molecules on the surface of a solid.

Air-free CO: A measure of CO volume in flue gases that accounts for the dilution caused by air and other gases. The air-free CO measurement from a combustion analyzer will always be higher than a standard CO reading, and air-free CO is the value of interest during a combustion analysis.

Alloy: A metal mixed with other metals or elements to create strength and durability.

Analog: Refers to continuously variable signals that represent measured quantities such as voltage, temperature, and pressure.

Angle bracket: An L-shaped metal supporting member used to support vertical risers. Also called angle iron.

Annual Fuel Utilization Efficiency (AFUE): An industry standard used to define furnace combustion efficiency.

Aquastat: A thermostatic control for water.

Atomizes: Reduces a liquid to a very fine spray.

Balance point: The outdoor temperature at which the heating capacity of a heat pump is equal to the heat loss of the structure it serves.

Base speed: The rotational speed listed on a motor's nameplate at which the motor will develop its rated horsepower at its rated load and voltage.

Binary: Refers to only two possible conditions, such as open/closed or off/on.

Blowdown: To intentionally drain water from a boiler under pressure to reduce the concentration of sediment and impurities.

Brake: A sheet metal-working machine, either manually operated or powered, used to bend metal.

Burnout: A compressor failure resulting from the breakdown of the motor winding insulation, grounding the motor windings, and releasing contaminants into the refrigerant circuit. The breakdown of the insulation typically results in the formation of strong acids.

Cad cell: A light-sensitive device (photocell) in which the resistance reacts to changes in the amount of light. In an oil furnace, it acts as a flame detector.

Cavitation: A condition caused by liquids reaching their vapor pressure at the pump inlet, causing bubbles to form that then collapse violently as they pass through the pump impeller.

Channel: A U-shaped piece of structural steel used as a supporting device.

Closure system: Any one of three types of sealing products used to seal fiberglass duct modules: pressure-sensitive tape, heat-activated tape, or mastic and glass fabric tape.

Cold-rolled steel: Metal that has been formed by rolling at room temperature, usually to obtain an improved surface or higher tensile strength.

Conductivity: The degree to which a substance conducts electricity. The conductivity of steam and water differs significantly.

Cooling compensator: A fixed resistor installed in a thermostat to act as a cooling anticipator.

Coping: The upper masonry course of a chimney designed to keep water out.

Corrosion: The breakdown or destruction of a material, especially metal, through chemical reactions. A common form of corrosion is *rust* (iron oxide), which forms when iron combines with oxygen.

Deadband: A temperature range in wall thermostats that separates heating and cooling modes in an automatic changeover thermostat. The deadband ensures that the system does not consistently alternate between modes due to competing setpoints.

Demand defrost: A defrost control strategy that does not depend on time to determine when defrosting is needed. Temperatures and temperature differentials determine when frost is present, and defrosting is required.

Dielectric: A material that strongly resists the passage of current.

Differential: The difference between the cut-in and cut-out temperature of a temperature control. Differentials also apply to pressure controls.

Downdraft: A situation in which outdoor air is forced down a chimney, preventing proper draft. A downdraft can occur if the top of the chimney does not extend above the roof line.

Drip leg: A drain line for condensate that forms naturally in a steam line that is routed to a steam trap for elimination.

Droop: A condition that causes a thermostat's control point to drift away from its setpoint, caused by heat developed from current flow through the thermostat.

Dry steam: The steam that exists after all the water has been vaporized and is moving beyond its saturation temperature (exhibits superheat).

Dump zone: An area where excess supply air from a zoning system is discharged to relieve excess pressure in a zoned air distribution system. The dump zone is typically located in the conditioned space, but in an area where some overheating or overcooling is not a problem.

EI rating: A strength and rigidity rating assigned to rigid ductboard products that is based on a calculation called *Young's modulus of elasticity*.

Enthalpy: A measure of the total heat contained in an air source, accounting for both the dry-bulb temperature and the volume of water content in the sample. Moist air, for example, contains more heat than dry air at the same temperature. Enthalpy is measured in units of Btus per pound of air (Btu/lb air).

Equivalent feet: The sum of the measured length of pipe in a system plus the length assigned to fittings and other components that relate the fitting loss to feet of straight pipe. For example, a threaded 1" 90-degree elbow has an equivalent length of 5.2'.

Fatigue test: A test in which a material is subjected to normal wear and tear to determine how long it will remain intact and effective before beginning to break down.

Fault isolation diagrams: Troubleshooting aids often contained in the manufacturer's installation, operation, and maintenance (IOM) instructions for equipment. Fault isolation diagrams typically outline failure symptoms and guide the technician through a logical decision and action process to isolate the cause of failure.

Flame rectification: A process in which a small alternating current is rectified to a direct current through the exposure of a sensing rod to a flame, resulting in a signal that proves the existence of a flame.

Flame-retention burner: A high-efficiency oil burner with a flame retention ring surrounding the oil nozzle.

Flash steam: Steam formed instantaneously when hot water is released under pressure to an area of lower pressure.

Floodback: A condition in which liquid refrigerant is returned to an operating compressor.

Flooded starts: Compressor starts during which slugging, oil foaming, and/or inadequate lubrication occur due to refrigerant migrating to the crankcase and condensing during the off cycle. When the compressor starts and the pressure in the crankcase drops rapidly, the condensed refrigerant flashes to the vapor state and causes the oil to foam, forcing oil into the compression area and reducing lubricity.

Foil scrim kraft (FSK): A flame-retardant vapor retardant made from layers of aluminum foil, fiberglass yarn for reinforcement (scrim), and kraft paper.

Free cooling: A mode of economizer operation when the system compressor is turned off and the indoor fan alone is used to bring outside air into the building as the sole source of cooling.

Fusible link: A connector designed to melt at a specific temperature to release the spring and allow a fire damper to close.

Hanger extension: Refers to the sum of the distance between the wall of the duct and the hanger wires or straps on each side.

Head: The height to which a pump can lift water; an indication of the energy that a pump can impart to the fluid.

Heat sink: A substance or structure that absorbs or disperses heat.

Hunting: Erratic changes in refrigerant flow rate through a TXV as it attempts to find the proper valve position to maintain the superheat setting. Hunting is characterized by consistent swings in superheat of several degrees, up and down.

Hydronic system: A system that uses water or water-based solutions as a medium to transport heat.

Insulated gate bipolar transistor (IGBT): A three-terminal power semiconductor that is highly efficient (low power losses) and designed to operate at high switching frequencies.

Invar®: A steel alloy containing 36% nickel. Invar® is one of the two metals in a bimetal thermostatic element.

Label diagrams: Troubleshooting aids placed in a convenient location inside the equipment. The diagrams typically consist of a wiring diagram, a component arrangement diagram, a legend, and notes pertaining to the equipment and optional components.

Ladder diagram: An electrical diagram that provides a schematic of power and control circuits with the loads arranged like the rungs of a ladder between vertical lines representing the power source.

Lap-joined: A condition in which one piece is joined to another by overlapping one piece with the other before joining.

Latent heat: The heat energy absorbed or rejected when a substance is in the process of changing state (e.g., solid to liquid, liquid to gas, or the reverse of either change).

Mastic: A pasty substance used as a filler and sealant on both sheet metal and fiberglass duct components and joints.

Megohmmeter (megger): A test instrument used to test high-resistance circuits using a high-voltage test signal.

Moment of inertia (I): The sum of the products obtained by multiplying the mass of each particle in a material by the square of its distance from a rotational axis.

National Electrical Manufacturers Association (NEMA): An electrical standards-developing organization made up of business leaders, electrical experts, engineers, scientists, and technicians from the electrical industry.

Net stack temperature: The measured stack (vent) temperature minus the entering air temperature of the combustion air.

Non-ferrous: Identifies a metal that contains little or no iron.

Ozone: A colorless and unstable gas with a pungent odor, formed from oxygen in response to electrical discharges in the atmosphere and UV light. It contains three atoms on its molecule (O_3) as opposed to oxygen with two atoms (O_2). Ozone in the upper atmosphere provides protection from UV radiation, but ground-level ozone is a significant and concerning pollutant.

Part-winding start: An approach to starting a three-phase motor that uses only a portion of the motor windings to initiate rotation, reducing starting current and heat production. Motors must be wound for part-winding start and have more leads than a standard three-phase motor.

Pickup voltage: The specific voltage at which a potential relay coil is sufficiently energized to reposition the relay contacts.

Pilot-operated valve: A valve that is positioned by a second, smaller valve that ports fluid pressure to the main valve to cause movement. Larger valves and those operating against high differential pressures use this strategy due to the limitations in the magnetic field strength of solenoid coils.

Plenum: A box that acts as the point of connection for equipment to a duct system, or for multiple ducts.

Plugging: An electric motor braking method in which reverse torque is applied by reversing the voltage or the phase sequence serving the motor. Although it provides rapid braking, the process places a great deal of stress on both the motor and the controlling device.

Polling: A network communications process where a VVT system seeks input from individual zones, identifies demand priorities, and then responds accordingly to satisfy heating or cooling loads. Electronic participants in the polling system are referred to as *voters*.

Potential relay: A relay energized by back EMF voltage generated during startup by the start winding in a capacitor-start motor.

Potentiometers: Three-terminal variable resistors used to change the resistance in a circuit. The recovery ventilator interprets the resistance and determines the speed output signal sent to the motors.

Pressure drop: The reduction in pressure between two points in a moving water stream, resulting from friction and other factors.

Pressure transducers: Devices that convert a monitored pressure to a variable analog signal.

Pressure-dependent system: A VVT system design approach where dampers are positioned based on strength of demand only or the dampers operate in just two positions (open or closed). The volume of air flowing through each damper is dependent on the duct pressure, fluctuating as the pressure fluctuates, and the airflow is not measured.

Pressure-independent operation: A VVT-system design approach where dampers are positioned by demand and by the volume of air flowing through the damper. Maximum and minimum air volume setpoints are in effect, and the damper will modulate as necessary to remain within these boundaries.

Primary control: A combination switching and safety control that responds to signals from the thermostat and other safety controls.

Pulse-width modulated (PWM): A method of controlling the average power or amplitude of an electrical signal, providing control of an electrical load. The speed at which the modulated signal changes (frequency of change) depends on the type of load, but for motor drives, the frequency of change is typically 4 kilohertz (kHz) to 16 kHz.

R value: A measure of a material's ability to impede heat transfer, expressed as a number; the higher the number, the better the insulating properties. Also known as *thermal resistance value*.

Redundancy: In HVACR systems, the provision of equipment to act as a backup to primary equipment, allowing system operation to continue despite a failed unit. An example of 100% redundancy is the installation of two boilers, each sized to handle the complete heating needs of the structure.

Refnet joints: Engineered joints used as connection points in a VRF-system refrigerant network. Refnet joints connect the main refrigerant lines to branch lines or indoor units.

Refractory: Materials designed to withstand extremely high temperatures and used to line or fabricate combustion chambers.

Reverberator: A wire or corrugated metal screen used to cover the ceramic grids in a gas-fired radiant heater to boost the radiant heat output by converting convective heat to radiant heat.

Saturated steam: The steam produced and existing at the temperature that corresponds to the boiling temperature of water at the existing pressure (saturation).

Scale: Chalk-like mineral deposits, typically a combination of calcium and magnesium compounds, that form and collect on surfaces inside hydronic systems of all types.

Sensible heat: Heat energy that can be measured by a thermometer or sensed by touch.

Shear load: The amount of weight or pressure that causes an exposed piece of metal to break or shear off.

Sheet steel: An uncoated sheet of steel with a bluish-black surface; commonly referred to as *black iron* in the trade.

Shiplap: A groove cut into the end of a ductboard module that allows adjoining edges to overlap when the ductboard is folded or fit to another piece. A special cutting tool is used to create the profile.

Skimming: Removing lighter-than-water impurities from the surface of the water in a steam boiler.

Slugging: The entrance of excess liquid refrigerant and/or oil into the compression area of a compressor Slugging can occur at startup due to condensed refrigerant and/or oil collected near the compressor inlet.

Smoke spot test: A test used to determine the density of smoke in the flue gases.

Snubber: A simple type of shock absorber used with fluids to suppress, or snub, rapid pressure changes that could damage sensitive instruments and gauges.

Soot: Fine carbon particles resulting from incomplete combustion. An accumulation of soot on heat exchangers will reduce combustion efficiency and result in increased carbon monoxide production.

Sound attenuation: The reduction in the level of sound that is transmitted from one point to another.

Specific heat: The amount of heat required to raise the temperature of one pound of substance 1°F. Expressed in units of Btu/lb/°F. At sea level, water has a specific heat of 1 Btu/lb/°F.

Staple flap: A section of ductboard facing from which the insulation has been removed that is then pulled over a joint or seam and stapled into place.

Static pressure: Reflects the pressure exerted by the weight of the water that is vertically above the gauge in the system. The static pressure of water is 0.43 pounds per square inch, per foot of water height at standard atmospheric conditions.

Subcooling: The reduction in temperature of a liquid after it has completed the change in state from a vapor. Only after the phase change is complete can the liquid begin to decrease below the saturation temperature as additional heat is removed.

Superheat: The increase in temperature of a vapor after it has completed the change of state from a liquid to a vapor. Only after the phase change is complete can the vapor begin to increase beyond the saturation temperature as additional heat is added.

Tap-ins: Rectangular or round takeoff fittings made of fiberglass ductboard or sheet metal that tap into a trunk duct to connect a branch duct.

Temperature rise: The difference between the return air temperature and the supply air temperature across the heat exchanger. The acceptable temperature rise for each furnace is listed on the furnace nameplate.

Tensile load: The weight required to cause metal to stretch or compress.

Tensile strength: The resistance of a material against rupture when placed under tension.

Thermal circulation: The circulation of water in a hydronic system resulting from differences in water temperature and the resulting change in weight.

Throw: Refers to the distance air travels from a register or diffuser based on its exit velocity.

Torsion: The twisting of an object along its length by two equal and opposite forces at each end.

Triacs: Three-terminal electronic semiconductors that support current flow in either direction when enabled by an input to the gate electrode.

Trim: External controls and accessories attached to the boiler, such as sight glasses and water feeder controls.

Turndown ratio: In steam systems, the ratio of downstream pressure to upstream pressure as it applies to steam pressure-reducing valves. More than a 50% turndown ratio is generally considered a large pressure reduction.

Unit cooler: A packaged unit containing the evaporator coil, metering device, and evaporator fan(s) that serves as the fan coil unit for a refrigeration system.

Vacuum breaker: A device used on a steam system to allow air to enter the system in the event a component, vessel, or section of piping falls below atmospheric pressure, thus breaking the vacuum.

Vapor retarder: A barrier, such as foil, plastic, or vinyl, placed between insulation and the surrounding air to prevent moisture in the surrounding air from contacting the insulation or the cold surfaces beneath.

Water hammer: A condition that occurs when hot steam comes into contact with cooled condensate, builds pressure, and pushes the water through the line at high speeds, slamming into valves and other devices. Water hammer can also occur in residential water systems. If the cause is not corrected, water hammer can damage the system.

Weatherized furnace: A furnace that has been well insulated and has a weather-resistant case as specified by government standards.

Wiredrawn: Wear that results in the formation of grooves in the seat of a valve, caused by fluid flow that erodes and abrades the seat as it repeatedly passes across the same spot at a high velocity.

Wiring diagram: A troubleshooting aid that provides pictorial representation of the physical wiring and connections in a unit.

Young's modulus of elasticity (E): A mathematical ratio that defines *tensile elasticity*, or the tendency of an object to deform along an axis when opposing forces are applied along the axis.

REFERENCES

Module 1 03317 Control Circuit Troubleshooting

Refrigeration and Air Conditioning: An Introduction to HVAC/R. 4th Edition. Larry F. Jeffus. New York, NY: Pearson.

HVAC Control Systems. 4th Edition. Ronnie J. Auvil. Orland Park, Illinois: American Technical Publishers.

Module 2 03318 Motor Testing and Troubleshooting

Application Guide for AC Adjustable Speed Drive Systems. 2021. Arlington, VA: National Electrical Manufacturers Association (NEMA).

Electrical Motor Controls for Integrated Systems. Gary Rockis and Glen A. Mazur. Latest Edition. Homewood, IL: American Technical Publishers, Inc.

Electric Motors and Drives: Fundamentals, Types, and Applications. Austin Hughes and Bill Drury. Latest Edition. Atlanta, GA: Newnes.

The ECM Textbook. 2007. Fort Wayne, IN: GE-ECM by Regal-Beloit.

Understanding Electrically Commutated Motors. Christopher Mohalley. 2014. Rolling Meadows, IL: Refrigeration Service Engineers Society (RSES).

Module 3 03210 Troubleshooting Cooling

Refrigeration and Air Conditioning, An Introduction to HVAC/R. 4th Edition. Larry F. Jeffus. Air Conditioning and Refrigeration Institute. Upper Saddle River, NJ: Pearson College Div.

System Diagnostics and Troubleshooting Procedures. 2002. John Tomczyk. Mt. Prospect, IL: Esco Press.

Module 4 03311 Troubleshooting Heat Pumps

Building America Solution Center. Access expert information on hundreds of high-performance construction topics, including air sealing and insulation, HVAC components, windows, indoor air quality, and much more. A service of the US Department of Energy, Office of Energy Efficiency and Renewable Energy. *https://basc.pnnl.gov/*

Building Science Education Solution Center. Relevant training modules from the US Department of Energy, Office of Energy Efficiency and Renewable Energy. *https://bsesc.energy.gov/training-modules*

Heat Pumps—Fundamentals and Applications. Walter Grassi. 2018. New York, NY: Springer Publishing.

NCCER Module 03211, *Heat Pumps*.

US Department of Energy, Office of Energy Efficiency and Renewable Energy. https://www.energy.gov/eere/office-energy-efficiency-renewable-energy

Module 5 03209 Troubleshooting Gas Heating

Refrigeration and Air Conditioning, An Introduction to HVAC/R. 4th Edition. Larry F. Jeffus. Air Conditioning and Refrigeration Institute. Upper Saddle River, NJ: Pearson College Div.

Module 6 03310 Troubleshooting Oil Heating

R. W. Beckett. *www.beckettcorp.com/*

Fundamentals of HVAC/R. Carter Stanfield and David Skaves. Upper Saddle River, NJ: Pearson.

Refrigeration and Air Conditioning, an Introduction to HVAC/R. 4th Edition. Larry F. Jeffus. Air Conditioning and Refrigeration Institute. Upper Saddle River, NJ: Pearson College Div.

The Complete Guide to Residential Oil Burners. 4th Edition. James Ries. Ronkonkoma, NY: On-Line Graphics Inc.

Module 7 03312 Troubleshooting Accessories

Refrigeration and Air Conditioning, An Introduction to HVAC/R. 4th Edition. Larry F. Jeffus. Upper Saddle River, NJ: Pearson College Div.